Introductory Nanoelectronics

Introductory Nanoelectronics
Physical Theory and Device Analysis

Vinod Kumar Khanna

CRC Press
Taylor & Francis Group
Boca Raton London New York

CRC Press is an imprint of the
Taylor & Francis Group, an **informa** business

First edition published 2021
by CRC Press
Taylor & Francis Group
6000 Broken Sound Parkway NW, Suite 300
Boca Raton, FL 33487-2742

First issued in paperback 2022

ISBN 13: 978-0-367-50403-8 (pbk)
ISBN 13: 978-0-8153-8426-7 (hbk)
ISBN 13: 978-1-351-20467-5 (ebk)

DOI: 10.1201/9781351204675

Visit the Taylor & Francis Web site at
http://www.taylorandfrancis.com

and the CRC Press Web site at
http//www.crcpress.com

To my grandson Hansh, daughter Aloka, and wife Amita,

for all the cherishing moments in my life they brought for me

Contents

Part II   Condensed Matter Physics for Nanoelectronics

Part V   Fabrication and Characterization of Nanostructures

Part VI Exemplar Nanoelectronic Devices

Preface

Have you ever given up reading a nanoelectronics book because you could not work out the monstrous mathematical steps in lengthy derivations of important formulae. Did you feel the mathematical hurdles were too formidable to overcome, and so you abandoned the study? If yes, this is the right book for you. It moves past the milestones of nanoelectronics by striking a balance between physical concepts and mathematics so that the reader's efforts are not frustrated by the inability to cope up with comprehending the unwieldy mathematical steps, which are often left over and relegated as exercises for the reader. Inability to derive and assimilate these steps often deprives the reader of the full satisfaction of digesting the subject matter in its entirety and depth leading to the feeling of some lacunae in knowledge which need to be filled. The readers' thirst for knowledge is not fully quenched. The reader has grasped the physics but will feel more confident after acquisition of expertise in the underlying mathematics.

With these forthright remarks, the present book is a humble attempt to systematically build up the physical cum mathematical superstructure of nanoelectronics following a straightforward and candid approach. The approach adopted is stepwise infrastructure development through mathematical organization of physical ideas in a lucid way without skipping the simplest, even the seemingly obvious steps. The main idea here is that being obvious or unclear is a relative perception; whatever is easy for someone may be difficult for some other possessing a different background knowledge.

Not only mathematics, another prime difficulty faced concerns building up the requisite conceptual background for students from diverse disciplines aspiring to understand nanoelectronics. Classical Newtonian mechanics had been successful in describing the behavior of microelectronic devices but felt short of expectations in the nanotechnology era. But the end of something marks the beginning of another good thing. A good bye in a relationship may be accompanied by a greeting of hello in another. And the hello came from the genesis of linkage between nanoelectronics and quantum mechanics. The new mechanics of Heisenberg, Schrodinger, and Dirac brought in its wake many interesting developments. It had far-reaching impacts on the physics of solids, which forms the body of present-day devices. Rudimentary knowledge of quantum mechanics as well as condensed matter physics is therefore indispensable for a thorough grasping of nanoelectronics.

The book seeks to remedy this lacking in knowledge of fundamentals by laying down the foundations of nanoelectronics in two preparatory parts, first on the essentials of quantum mechanics and second on those of condensed matter physics. The starting two parts strengthen the theoretical framework and pave the way to the third and fourth parts on electron behavior in nanostructures and the non-equilibrium Green's function formalism for nanoelectronic device design and simulations.

After reinforcing and consolidating the theoretical infrastructure for nanoelectronics, we turn around to explore how experiments are practically performed in the laboratory. So the fifth part on nanofabrication and characterization tools surveys these topics for experimental work. The concluding sixth part is devoted to nanoelectronic devices which are the precious fruits harvested from all theoretical and practical accomplishments. Here we select a few examples of nanoelectronic devices and try to delve deeply into their operating mechanisms, thereby providing physical insights into their *modus operandi*, and presenting mathematical analysis of their working.

All throughout the pages of the book, particular care is taken to ensure that no aspect of the development, neither theoretical nor mathematical, is overlooked, and the reader comfortably progresses through the book. Illustrative exercises appended at the end of each chapter will greatly help in proper application of formulae and correct selection of units. It is earnestly hoped that the book will be able to meet the objectives laid down, offering a refreshing change from traditional treatments.

Wishing the readers an enjoyable journey from cover to cover! Remember the inspirational lines:

Persevere, burgeon and prosper
Learn from failure, work hard and conquer
The goal is elusive and tough
The terrain is thorny and rough
But the spirits are invincible and high
Never accept defeat, never say die
Leap with full muscle power and touch the sky!

Vinod Kumar Khanna
Chandigarh, India

Acknowledgements

Above all, I am indebted to God for giving me energy and vigor to complete this work.

My deep thanks are tendered to the authors of all articles in printed or electronic media, whose work constitutes the body of this book. These authors are cited in the 'References' at the end of respective chapters. Their diligence and sacrifices brought many glories to nanoelectronics and made it the coveted field of contemporary science and technology. Three cheers for the pioneering workers in this field!

I am thankful to my editor and staff at the CRC press for giving me the opportunity to undertake this project.

With all warmth and love, I thank my family for their moral support.

Vinod Kumar Khanna
Chandigarh, India

Author

Vinod Kumar Khanna (born 25th November, 1952, Lucknow, India) is a former emeritus scientist, CSIR (Council of Scientific & Industrial Research) and emeritus professor, AcSIR (Academy of Scientific & Innovative Research), India. He is a retired Chief Scientist and Head, MEMS & Microsensors Group, CSIR-CEERI (CSIR-Central Electronics Engineering Research Institute), Pilani (Rajasthan) and Professor, AcSIR, India.
During his tenure of work at CSIR-CEERI, Pilani spanning over 37 years 7 months, he worked on a large number of CSIR and sponsored projects on power semiconductor devices, notably high-current and high-voltage rectifier, high-voltage transistor, fast-switching inverter grade thyristor, power DMOSFET, and IGBT. Another key research area included the device and process design, and fabrication of micro- and nanoelectronics and MEMS technology-based sensors and dosimeters. His contributions focused primarily on the development of technology and characterization of moisture and microheater-embedded gas sensor, ion-sensitive field-effect transistor pH sensor, MEMS acoustic sensor, capacitive MEMS ultrasonic transducer, PIN diode neutron dosimeter, and PMOSFET gamma ray dosimeter.

He is widely travelled and has worked at Technische Universität Darmstadt, Germany, 1999 and Kurt-Schwabe-Institut für Mess- und Sensortechnik e.V., Meinsberg, Germany, 2008. He also visited the Institute of Chemical Physics, Novosibirsk, Russia, 2009 as a member of Indian delegation and Fondazione Bruno Kessler (FBK), Trento, Italy, 2011, under India-Trento Programme of Advanced Research (ITPAR). He participated and presented research papers in IEEE-IAS Annual meeting at Denver, Colorado, USA, 1986.

Awarded National scholarship by the Ministry of Education and Youth Services, Govt. of India in 1970, he received his M.Sc. degree in Physics in 1975 from University of Lucknow, India and Ph.D (Physics) degree from Kurukshetra University, India in 1988.

Prior to the present book, he has authored 15 books. Author/ co-author of 192 research papers in prestigious refereed journals and national/international conference proceedings, he has 5 patents to his credit, including 2 US patents. He is a life member of several leading professional societies, and fellow of the Institution of Electronics & Telecommunication Engineers (IETE), India.

After superannuating as Head, MEMS and Microsensors Group, CSIR-CEERI in November 2014, and completing his tenure as Emeritus Scientist, CSIR, in November 2017, he presently resides at Chandigarh, India. He is a passionate author, avidly reading and writing.

Abbreviations, Acronyms, and Chemical Symbols

A	Ampere
A°	Angstrom unit
ABMO	Antibonding molecular orbital
AC	Alternating current
aF	AttoFarad (10^{-18} F)
AFM	Atomic force microscope
Al	Aluminum
AlAs	Aluminum arsenide
ALD	Atomic layer deposition
AlGaAs	Aluminum gallium arsenide
AlN	Aluminum nitride
Al-Si	Aluminum-Silicon
AMR	Anisotropic magnetoresistance
APCVD	Atmospheric pressure chemical vapor deposition
ArF	Argon fluoride
As	Arsenic
AsH$_3$	Arsine
As$_2$O$_3$	Arsenic trioxide
Au	Gold
B	Boron
BBr$_3$	Boron tribromide
BCPs	Block copolymers
Be	Beryllium
BF$_3$	Boron trifluoride
B$_2$H$_6$	Diborane
BMO	Bonding molecular orbital
B$_2$O$_3$	Boric anhydride
BOE	Buffered oxide etch
Br$_2$	Bromine
BSEs	Backscattered electrons
C	Coulomb, Carbon (diamond)
CBE	Chemical beam epitaxy
CCD	Charge-coupled device
C$_2$ClF$_5$	Chloropentafluoroethane
CdSe	Cadmium selenide
CF$_4$	Tetrafluoromethane
CH$_4$	Methane
CHF$_3$	Fluoroform
C$_{24}$H$_{39}$NO$_4$	4-nitrophenyl octadecanoate
(CH$_3$O)$_3$B	Trimethyl borate
CIP	Current-in-plane
Cl$_2$	Chlorine
CNT	Carbon nanotube
CO	Carbon monoxide
CO$_2$	Carbon dioxide
CPP	Current perpendicular to plane
Cr	Chromium
Cu	Copper
CVD	Chemical vapor deposition
C-Z Si	Czochralski silicon
0D	Zero-dimensional
1D	One-dimensional

2D	Two-dimensional
3D	Three-dimensional
DC	Direct current
dHEMT	Depletion HEMT
DIBL	Drain-induced barrier lowering
DOF	Depth of focus
DOS or	
DoS	Density of states
DPN	Dip Pen Nanolithography
DSA	Directed self-assembly
DUV	Deep ultraviolet
DXRL	Deep X-ray lithography
E-beam	Electron beam
EDX	Energy Dispersive X-ray (Analysis)
EGS or	
EG-Si	Electronic grade silicon
eHEMT	Enhancement HEMT
erf	Error function
erfc	Complementary error function
E-T	Everhart-Thornley (detector)
EUV	Extreme ultraviolet
eV	Electron volt
F	Farad
FDM	Finite difference method
FD-SOI-	
MOSFET	Fully depleted SOI-MOSFET
Fe	Iron
FeCl$_3$	Ferric chloride
Fe$_3$O$_4$	Ferrosoferric oxide
FET	Field-effect transistor
FIB	Focused ion beam
Ga	Gallium
GAA	Gate-all-around
GaAs	Gallium arsenide
GaN	Gallium nitride
GaP	Gallium phosphide
Ge	Germanium
GMR	Giant magnetoresistance
GΩ	Giga ohm
GRIN-SCH	Graded index separate confinement heterostructure
H$_2$	Hydrogen
h	Hour
HCl	Hydrochloric acid
HCl$_3$Si or	
SiHCl$_3$	Trichlorosilane
HEMT	High-electron-mobility transistor
HFET	Heterostructure field-effect transistor
HfO$_2$	Hafnium oxide
H$_2$O	Water
HOMO	Highest occupied molecular orbital
H$_2$SiCl$_2$	Dichlorosilane

InAs	Indium arsenide
InGaAs	Indium gallium arsenide
InN	Indium nitride
InP	Indium phosphide
InSb	Indium antimonide
IPL	Ion projection lithography
J	Joule
JFET	Junction field-effect transistor
K	Kelvin
keV	Kilo electron volt = 10^3 eV
kg	Kilogram
K, L, M, N, ...	Electron orbits in an atom
KrF	Krypton fluoride
LAD	Laser ablation deposition
LCAO	Linear combination of atomic orbitals
LED	Light emitting diode
LDD	Lightly doped drain (structure)
LEdit	Layout Editor
LHS	Left-hand side
LPCVD	Low-pressure chemical vapor deposition
LUMO	Lowest unoccupied molecular orbital
m	Meter
mA	Milliampere
MBE	Molecular Beam Epitaxy
MESFET	Metal-semiconductor field-effect transistor
MeV	Megaelectron volt = 10^6 eV
MGS or MG-Si	Metallurgical grade silicon
MHz	Megahertz
MOCVD	Metal-organic chemical vapor deposition
MΩ	Mega ohm
Mo	Molybdenum
MODFET	Modulation-doped FET
MO-line	Metal-organic line
MOMBE	Metal-organic molecular beam epitaxy
MOSFET	Metal-oxide semiconductor field-effect transistor
MOVPE	Metal-organic vapor phase epitaxy
MQW	Multiple quantum wells
MRAM	Magnetoresistive random access memory
mV	Millivolt
MWCNT	Multi-walled carbon nanotube
N	Newton
N$_2$	Nitrogen
nA	Nanoampere
NDR	Negative differential resistance
NEGF	Non-Equilibrium Green's Function
NH$_3$	Ammonia
Ni	Nickel
NIL	Nanoimprint Lithography
nm	Nanometer (10^{-9} m)
N-n heterojunction	Heterojunction in which both the wide bandgap and low bandgap materials are N-doped
N$_2$O	Nitrous oxide
N-p heterojunction	Heterojunction in which wide bandgap material is N-doped and narrow bandgap material is P-doped
O$_2$	Oxygen
OH	Hydroxyl group
P	Phosphorus
4PP	Four-point probe
Pa	Pascal
PC	Personal computer
Pd	Palladium
PDMS	Polydimethylsiloxane
PD-SOI-MOSFET	Partially depleted SOI-MOSFET
PECVD	Plasma-enhanced chemical vapor deposition
PH$_3$	Phosphine
pHEMT	pseudomorphic HEMT
pm	Picometer (10^{-12} m)
PMMA	Poly (methyl methacrylate)
P-n heterojunction	Heterojunction in which wide bandgap material is P-doped and narrow bandgap material is N-doped
P$_2$O$_5$	Phosphorus pentoxide
POCl$_3$	Phosphorus oxychloride
PolySi	Polysilicon
PS-b-PMMA	Polystyrene-block-poly(methyl methacrylate)
PSPD	Position-sensitive photodetector
P(S-r-MMA)	Poly(styrene-random-methyl methacrylate) copolymer
Pt	Platinum
P. V.	Principal value
PVD	Physical Vapor Deposition
PVDF	Polyvinylidene difluoride or Polyvinylidene fluoride
PVR	Peak-to-valley ratio of the diode
PZT	Lead zirconate titanate
QB	Quantum barrier
QCSE	Quantum-confined Stark effect
QD	Quantum dot
QW	Quantum well
RCP	Retrograded channel profile
Res	Residue
RF heating	Radio-frequency heating
RHS	Right-hand side
RIE	Reactive ion etching
RTA	Rapid thermal annealing
RTD	Resonant tunneling diode
Ry	Rydberg constant
S	Siemen
1S, 2S, 2P, 3S, 3P, 3D, ...	Internal states of the exciton
s, p, d, f	Atomic orbitals
1s, 1p, 1d, ... 2s, 2p, 2d, ...	External states of exciton
S(s), P(p), D(d), F(f)...	Combined states of the exciton with l = 0, 1, 2, 3, ...
s	second
Sb	Antimony
SB-CNT-FET	Schottky-barrier CNT-FET
SbH$_3$	Stibine
Sb$_2$O$_3$	Antimony (III) oxide
SEM	Scanning electron microscope
SET	Single-electron transistor

SEs	Secondary electrons
SF$_6$	Sulfur hexafluoride
SGS	Semiconductor grade silicon
Si	Silicon
SiC	Silicon carbide
Si$_2$Cl$_6$	Hexachlorodisilane
SiCl$_4$	Silicon tetrachloride
SI-GaAs	Semi-insulating gallium arsenide
SiH$_4$	Silane
SIMOX	Separation-by-implantation of oxygen
Si$_3$N$_4$	Silicon nitride
SiO$_2$	Silicon dioxide
SOI	Silicon-on-insulator
SSR	Super-steep retrograde (channel)
STM	Scanning tunneling microscope
SWCNT	Single-walled carbon nanotube
Ta$_2$O$_5$	Tantalum pentoxide
tDPN	Thermal DPN
TDSE	Time-dependent Schrodinger equation
TEAl	Triethylaluminum, Al(C$_2$H$_5$)$_3$
TEGa	Triethylgallium, Ga(C$_2$H$_5$)$_3$
TEGFET	Two-dimensional electron gas field-effect transistor
TEIn	Triethyl indium, In(C$_2$H$_5$)$_3$
TEM	Transmission electron microscope
TEOS	Tetraethyl orthosilicate, Si(OC$_2$H$_5$)$_4$

THz	Terahertz
Ti	Titanium
TiO$_2$	Titanium dioxide
TISE	Time-independent Schrodinger equation
TiW	Titanium-tungsten
TMAH	Tetramethylammonium hydroxide
TMAl	Trimethylaluminum, Al(CH$_3$)$_3$
TMGa	Trimethyl gallium, Ga(CH$_3$)$_3$
TMIn	Trimethyl indium, In(CH$_3$)$_3$
T-NIL	Thermal nanoimprint lithography
UHVCVD	Ultrahigh vacuum chemical vapor deposition
UV	Ultraviolet
UV-NIL	Ultraviolet nanoimprint lithography
V	Volt
W	Watt
XeCl	Xenon monochloride
XRD	X-Ray Diffraction (Crystallography)
XRL	X-ray lithography
Y$_2$O$_3$	Yttrium (III) oxide
ZnO	Zinc oxide
ZnS	Zinc sulfide
ZrCl$_4$	Zirconium(IV) chloride
ZrO$_2$	Zirconium oxide
μS	Microsiemens

Mathematical Notation

A	Area, Cross section, spectral function $= G^n + G^p$
A^+, A^-	Amplitudes of the waves
$|A|^2$	Incident intensity
$A_i(\xi)$	Airy function
$A_{n,l}$	Normalization constant
$\mathbf{A}(\mathbf{r}, t)$	Vector potential
a	Acceleration, Lattice spacing, radius of the sphere, spacing between points
a_0	Bohr radius
$\mathbf{a}_e$	Acceleration of the electron
a_{Electron}	Bohr radius of electron
$\mathbf{a}_h$	Acceleration of the hole
a_{Hole}	Bohr radius of hole
a_X	Exciton Bohr radius
C	Capacitance, number of points inside the conductor
C_0	Stray capacitance in the circuit referred to ground
C_α	Coefficients of the wave functions
C_{G1}	Capacitance of the controlling gate in SET
C_{ox}	Capacitance of the silicon dioxide film
C_Σ	Total capacitance
c	Velocity of light, phase velocity
$\hat{D}$ or D	Differential operator, diffusion coefficient of impurity in silicon at the diffusion temperature, diffusion coefficient of the carrier
D_1	Diffusion coefficient of the impurity in silicon at the pre-deposition temperature
D_2	Diffusion coefficient of the impurity in silicon at the temperature of drive-in
$D(E)_{3D}$	Density of states in bulk material (3-dimensional)
2DEG	Two-dimensional electron gas
$D(\mathbf{r}, \mathbf{r}'; \hbar\omega)$	A function representing the spatial correlation and energy spectrum of the phase-breaking scatterers
d	Spacing between atomic planes in a crystal, the distance of separation between the mirrors, Undoped thickness of large bandgap semiconductor
d_2	Thickness of the large bandgap semiconductor
$d^2 k_{t,L} dk_{z,L}$	An infinitesimally small volume of momentum space
E	Energy, Electric field

E_0	Rest energy of a particle, Peak value of electric field, Vacuum level of energy, Energy of non-degenerate ground state
E_1	Energy level of electrons
E_C	Energy of minimum conduction band called conduction band edge, Critical electric field at which the velocity becomes constant with variation of field.
$E - E_C$	Energy relative to the bottom of conduction band.
E_{CA} and E_{CB}	Bottoms of the conduction bands of materials A, B
$E_{C, \text{Direct}}$	Conduction band minimum at $k = 0$
$E_{C, \text{Indirect}}$	Conduction band minimum which is at $k \neq 0$
$E_{C,L}$	Conduction band minimum on the left
E_n^C	$= E_n$ for $m = m_C$
$E_{C,R}$	Conduction band minimum on the right
$E_{\text{Depletion}}$	Electrical potential energy of the field across the depletion region
E_{Direct}	Energy bandgap of a direct bandgap semiconductor
E_F	Fermi energy level or Fermi level
E_{Fp}	Fermi level at terminal p
E_{Fq}	Fermi level at terminal q
E_F^0 or E_{F0}	Fermi level at $T = 0$ K
E_{FL}	Quasi-Fermi level on left side
E_{FR}	Quasi-Fermi level on right side
E_g or E_G	Energy gap
$E_{G, A}$ and $E_{G, B}$	Bandgaps of the materials A, B
E_{Indirect}	Energy bandgap of indirect bandgap semiconductor
E_{k_x}	Longitudinal energy
$E_{k_x, n, m}$	Energy for the state having quantum numbers k_x, n, m
E_n	Eigenvalues
$E_{n,l}$	Energy level of electrons with quantum numbers n, l
$E_{n-1,l}$	Energy level of electrons with quantum numbers $(n-1), l$
$E_{n,m}$	Transverse energy

E_{1S1s}	Energy for the 1S1s state
E_T	Thermal energy
$\mathbf{E}(t)$	Instantaneous electric field
E_t	Energy of electron in the transverse direction
E_v	Total electron energy
E_n^V	$=E_n$ for $m = m_v$
E_x, E_y, E_z	X-, Y-, Z-components of energy E
E_z	Kinetic energy of perpendicular motion, i.e. in the Z-direction
$E_{z,L}$	Component of the electron energy along the Z-direction on the left side
$E_{z,R}$	Component of the electron energy along the Z-direction on the right side
e	Electronic charge (1.602×10^{-19} C)
F	Force, Electric field, free energy
$|F|^2$	Transmitted intensity
$F(\alpha a)$	A function defined in the derivation of Kronig-Penney model
$F_\alpha(x)$	Gaussian function
F_{eh}	Force exerted by the hole on the electron
F_{he}	Force exerted by the electron on the hole
f	Resonant frequency of a cantilever
$f_0(E)$	Fermi-Dirac distribution function at equilibrium
$f(E_v)$	Energy-dependent electron distribution function
$(f * g)(t)$	Convolution of function $f(t)$ with another function $g(t)$
$f(\mathbf{k})$	Distribution function used to specify the number of electrons occupying a particular state $\mathbf{k}$
$f_L(E)$	Fermi-Dirac distribution function for electron energies on the left side contact
$(f_L - f_R)_{Eq.}$	Difference between Fermi-Dirac distribution function for electron energies on the left side and right side contacts at equilibrium condition
$f_L(E_z, E_t)$	Fermi-Dirac distribution function of electrons on the left side having energies E_z, E_t in the Z-and transverse directions, respectively
$f_L(k_{z,L}, k_{t,L})$	Electron distribution function on the left side of the barriers
$f_R(E_z, E_t)$	Fermi-Dirac distribution function of electrons on the right side having energies E_z, E_t in the Z-and transverse directions, respectively
$f_p(E)$	Fermi function at terminal p
$f_q(E)$	Fermi function at terminal q
$f_R(E)$	Fermi-Dirac distribution function for electron energies on the right side contact
$f(\tau_n)$	Series of discontinuous strips of impulse
G	Gravitational constant, conductance, Green's function
G_0	Quantum of conductance
$G^A(x,x')$ or G^A	Advanced Green's function
G_C	Isolated conductor matrix
G^n	Electron correlation function
$G^n(k,k';t,t')$	A two-time correlation function G^n which correlates the amplitude in state $\mathbf{k}$ at time t with that in state $\mathbf{k}'$ at time t'
$G^n(k,k';\omega)$	Forward Fourier transform of the correlation function
$\left. G^n(k,k';t,t') \right|_{t=t'}$	Inverse Fourier transform of the correlation function
$G^n(\mathbf{r},\mathbf{r}';E)$	Electron correlation function
G_p	Isolated lead matrix
G^p	Hole correlation function
G_{pC} or G_{Cp}	Coupling matrix
G_{pq}	Conductance between terminals p, q
$G^R(i,j)$	Matrix representation of Green's function $G^R(\mathbf{r},\mathbf{r}')$
$G^R(\mathbf{r},\mathbf{r}')$	Green's function is the wave function at the point $\mathbf{r}$ produced by a unit excitation at the point $\mathbf{r}'$
$G(t, \tau)$	Green's function $g(t - \tau)$
$G^R(x,x')$ or G^R	Retarded Green's function
$G^R(x, y; x, y')$	Green's function between two points having the same x-coordinate
$G(x, x')$	Green's function meaning the wave function at x obtained from application of unit excitation at x'
$\{g(E)\}_{2D}$	Two-dimensional density of states
g_p^R	Green's function for the isolated semi-infinite lead
$g_p^R(p_i, p_j)$	Green's function for a discrete lattice
$g(t)$	Response function to $\delta(t)$ input
$g(z)$	An analytic function at z_0 called the pole of order m
H	Thermal energy
$[H]$	Matrix representation of Hamiltonian operator H_{op}
H_C	Hamiltonian describing a conductor
$[H_C]$	Matrix representation of the Hamilton operator H_{op} for the conductor
H_{op}	Hamiltonian operator
H_p	Hamiltonian for lead p connected to a conductor
$H(x)$	Heaviside step function
$\hat{H}_x, \hat{H}_y, \hat{H}_z$	Total energy or Hamiltonian operators in X-, Y-, Z-directions
h	Planck's constant
$\hbar$	Reduced Planck's constant $= h/2\pi$

I	Current	$\mathbf{K}$	Wave vector of the exciton
$[I]$	Identity matrix	k	Wave number, $\dfrac{\sqrt{2mE}}{\hbar}$ or $\dfrac{\sqrt{2m\Phi}}{\hbar}$; spring constant or stiffness of the cantilever
I_0	Function (Applied voltage×density of states in the sample and tip of STM)		
I_{DS}	Drain-source current of a MOSFET	k_0	$\dfrac{\sqrt{2mV_0(x)}}{\hbar}$
$(I_{DS})_{\text{Long channel}}$	Drain-source current of a long-channel MOSFET		
$(I_{DS})_{\text{Short channel}}$	Drain-source current of a short-channel MOSFET	k_1	A constant, $k_1 = C_{ox}W\mu_n\left(\dfrac{v_{Sat}}{\mu_n}\right)$
$I_{DS(\text{Velocity−saturated})}$	Drain-source current of a velocity-saturated MOSFET	k_2	A constant, $k_2 = \dfrac{1}{2}V_{DS(Sat)}$
I_L	Electric current due to electrons entering the quantum wire from left contact	k_B	Boltzmann constant
		k_F	Fermi wave vector
I_L^+	Electron influx from lead L into the wire	k_L	Wave vector of electron wave moving from left to right
I_L^-	Electron flux reflected back to contact L		
$\mathbf{Im}$	Imaginary	k_R	Wave vector of electron wave moving from right to left
I_{op}	A current operator		
I_p	Electric current at terminal p, Peak current	k-space	Momentum space
I_R	Electric current due to electrons entering the quantum wire from right contact	$k_{t,L}$	Component of the wave vector along the transverse direction on the left side
I_R^+	Electron outflux from lead R into contact R	$k_{t,R}$	Component of the wave vector along the transverse direction on the right side
$I_t(z)$	Tunneling current	k_x, k_y, k_z	X-, Y-, Z- components of wave number/wave vector k
I_v	Valley current		
i	$\sqrt{-1}$	$k_{z,L}$	Component of the wave vector along the Z-direction on the left side
i, j	Integers		
$i\eta$	A positive imaginary component added to the wave number k for inclusion of boundary conditions in the retarded/advanced Green's function	$k_{z,R}$	Component of the wave vector along the Z-direction on the right side
		$\mathcal{L}$	Linear differential operator
$i_p(E)$	Current at terminal p per unit energy E	L	Length, angular momentum, Lorentz number, width of the quantum well, channel length of a MOSFET, channel length measured from edge of the source to drain edge
$i_\phi(E)$	Current component due to interactions of electrons with their surroundings		
$\mathbf{i}, \mathbf{j}, \mathbf{k}$	Unit vectors along X-, Y-, Z-axes		
$\mathbf{i'}, \mathbf{j'}, \mathbf{k'}$	Unit vectors along x-, y-, z-axes	l	Length of the smaller parallel side of the trapezium away from the oxide-silicon interface
$\mathbf{J}$ or $\mathbf{J}(\mathbf{r},E)$	Current density		
J_0	Peak current density	L_D	Diffusion length of the carrier
$j_l(kr)$	Bessel function of the first kind	$(L_D)_{\text{Intrinsic}}$	Intrinsic Debye length
$\mathbf{J_P}$	Probability current density	$(L_D)_{\text{Extrinsic}}$	Extrinsic Debye length
$\mathbf{J}(\mathbf{r},t)$	Current density source at space-time point $(\mathbf{r}, t)$	L_ϕ	Phase coherence length
		L_m	Mean free path
$j_{\text{Incident, Left}}$	Incident current density perpendicular to the barrier from the left	L_{Line}	Length of line in k-space
$j_{\text{Transmitted, Left}}$	Transmitted current density perpendicular to the barrier from the left	$L(\rho)$	A function of $\rho = 2r\kappa$ chosen to obtain the radial solution of the wave function in hydrogen atom problem
$j_{\text{Transmitted, Right}}$	Transmitted current density perpendicular to the barrier from the right	$L_\eta^v(\rho)$	Associated Laguerre polynomials
		$L_{\text{Single state}}$	Length of single state in k-space
j_{Total}	Total current density	L_x, L_y, L_z	Dimensions of the box in X-, Y-, Z-directions, dimensions of quantum wire in these directions
K	Kinetic energy, electrical component of thermal conductivity		
		$\hat{L}_x, \hat{L}_y, \hat{L}_z$	Angular momentum operators in X-, Y-, Z-directions

l	Azimuthal quantum number, Mean free path	N_s	Impurity concentration at the surface of the wafer as determined by its solid solubility limit at the pre-deposition temperature, Sheet density of electrons trapped in the triangular potential well
M	Total mass of electron and proton $= m_e + m_p$, the sum of electron and hole effective masses, number of modes in a multimode wire		
M_A	Atomic mass	N_V	Number of valence electrons per atom
M_M	Molar mass	$N(x, t)$	Impurity concentration at a depth x at time t
m	Mass, quantum number		
m_0	Rest mass of electron, mass of free electron	n	principal quantum number, free electron concentration per unit volume, refractive index of an optical medium, the number of electrons left behind on the island
m^*	Effective mass of the carrier		
m_C	Conduction band mass		
m_e, m_{e*}	Mass of electron, effective mass of electron		
$m_{Electron}$	Mass of electron	$n(0)$	Number of electrons at the source of the pulse
m_{el}^*	Longitudinal electron effective mass	n_1	Number of electrons entering the island through junction J_1
$m_{el, \, Conductivity}^*$	Conductivity effective mass of electron		
$m_{el, \, DOS}^*$	Density-of-states effective mass of electron	n_2	Number of electrons leaving the island through junction J_2
m_{et}^*	Transverse electron effective mass	n_{2D}	Density of electrons per unit area
m_h	Mass of hole	n_F	Value of n for the topmost energy state
m_{hh}^*	Effective mass of heavy holes	n_i	Intrinsic carrier concentration of a semiconductor
m_l	Magnetic quantum number		
m_{lh}^*	Effective mass of light holes	n, m	Transverse quantum numbers representing the wave vectors k_y and k_z
m_n^*, m^*	Effective mass of electrons		
$m_{Nucleus}$	Mass of nucleus	$n(r)$	Electron density
m_p	Mass of proton	$n(r; E)$	Electron density per unit energy
m_p^*	Effective mass of holes	$n_{S,j}(E_F)$	Sheet density of electrons for the level j as a function of Fermi level E_F
m_s	Spin quantum number		
m_V	Valence band mass	$n_{S,j}(E_F^0)$	Sheet density of electrons for the level j as a function of Fermi level at zero temperature E_F^0
$m_{V,so}^*$	Split-off hole effective mass		
N	Number of valence electrons, 3-D electron density of the system, number of atoms in one-dimensional ring, number of uncompensated electrons, number of filled states in the circular disk	$n(x)$	Number of electrons at a distance x from the source of the pulse
		n_x, n_y, n_z	Quantum numbers for motion quantization in X-, Y- and Z-directions
N_0	Surface concentration of impurity after drive-in	$\hat{O}$	An operator
N_2	Doping concentration of semiconductor 2 (large bandgap semiconductor)	P	Probability, a variable in Kronig-Penney model derivation, $\dfrac{\beta^2 ba}{2} = P$
NA	Numerical aperture of a lens		
N_A	Avogadro's number, acceptor dopant density	$P_{l, m}(x), P_l^m(\cos\theta)$	Associated Legendre polynomials
$N_A(z)$	Concentration of acceptor impurities in the Z-direction	$P(r, t)$	Probability density in quantum mechanics
N_{Cell}	Number of cells accommodated in a space of unit volume	p	Momentum
		p_0	Peak value of momentum
N_D	Donor dopant density	p_i	Point in the lead p adjacent to the point i inside the conductor
$(N)_{3D}$	3-D density of states		
$|(N)_{3D}|_{Fermi}$	3-D density of states at Fermi energy	p_i, p_j	Points in the lead nearby the points i, j inside the conductor
$N_D(z)$	Concentration of donor impurities in the Z-direction	$p(t)$	Instantaneous momentum at time t
N_q	Number of phonons having wave vector q and frequency ω_q	$\hat{p}_x, \hat{p}_y, \hat{p}_z$	Linear momentum operators in X-, Y-, Z-directions

Q, q	Charge, impurity dose introduced into silicon during pre-deposition	S_p	Cross section of the contact separating the conductor from the lead p
Q_s	Sheet electron charge, Net charge enclosed by the surface	$S(\mathbf{r}')$	A source term at $\mathbf{r}'$
q	Electronic charge, Wave vector of phonon	s	Speed, distance between probes
q_0	Background charge on the island	T	Absolute temperature, Kinetic energy, transmission probability per mode at the Fermi energy, transmission probability of electron influx from lead L to go into contact R
q_1	Charge on junction J_1		
q_2	Charge on junction J_2		
q_{island}	Charge on the island	$\hat{T}$	Translation operator
q_P	Polarization charge	T'	Transmission probability of electron influx from lead R to go into contact L
q_R	Replacement charge		
R	Radius, resistance	$T_{Coherent}(E_z)$	Probability of coherent tunneling of the electron having energy E_z
R–C	Resistance-capacitance circuit		
R_D	Drain access resistance	$T(E_x)$	Probability of tunneling of the electron having energy E_z, Transmission coefficient as a function of energy in the X-direction
R_d	Differential resistance		
R_{DS}	Drain-source resistance		
$\mathbf{Re}$	Real	$T(E_F, n, m)$	Transmission coefficient as a function of Fermi level, and quantum numbers n, m
R_H	Rydberg constant		
$R_{L\rightarrow R}$	Reflection coefficient of electrons moving from left to right	$\{T(E_x)\}_{Eq.}$	Transmission coefficient as a function of energy in the X-direction at equilibrium condition
$R_{Parasitic}$	Parasitic resistance associated with source/drain junctions	$\Gamma_\phi^H(\mathbf{r},\mathbf{r}';E)$	Hilbert transform of $\Gamma_\phi(\mathbf{r},\mathbf{r}';E)$
$R(r), R_{n,l}$	Radial part of the wave function $\Psi(r, \theta, \phi)$; n, l are quantum numbers	$T_{Incoherent}(E_z)$	Probability of incoherent tunneling of the electron having energy E_z
		$T(k_{z,L})$	Transmission coefficient perpendicular to the barrier from the left
R_S	Source access resistance		
R_{SX}	Resistance connected in series with the source terminal of ideal long-channel MOSFET	$T(L, E)$	Tunneling probability or transmission coefficient for a particle of energy E incident on a barrier of width L
R_{T1}	Tunnel resistance of junction J_1		
R_{T2}	Tunnel resistance of junction J_2	$T_{L\rightarrow R}$	Transmission coefficient of electrons moving from left to right
R_y^*	Exciton Rydberg energy		
r	Distance between electron charge and proton in a hydrogen atom, radius, length of the radius vector from the origin to the point (x, y, z) in spherical polar coordinates, Rayleigh resolution, the distance between the centers of the molecules	$\bar{T} = M(E)T(E)$	Number of modes of energy $E \times$ transmission probability for energy E
		$T_{R\rightarrow L}$	Transmission coefficient of electrons moving from right to left
		$\bar{T}_{p\rightarrow q}$	Transmission for all modes from terminal p to terminal q
r_e	Radius of the electron	$\bar{T}_{pq}(E)$	Total transmission from q to p at energy E
(r, θ, ϕ)	Spherical polar coordinates		
r_{min}	Distance at which the potential attains the minimum value corresponding to equilibrium separation of the molecules	$\bar{T}_{q\rightarrow p}$	Transmission for all modes from terminal q to terminal p
		$\bar{T}_{qp}(E)$	Total transmission from p to q at energy E
$\mathbf{r}_i, \mathbf{r}_j$	Position vectors for lattice sites i, j	$Tr[I_{op}(E)]$	Trace of the current operator $I_{op}(E)$
r_j	Junction depth of the N⁺source or N⁺drain	$T(t)$	Function of t
r_L	Amplitude of reflected wave	$\hat{T}_x, \hat{T}_y, \hat{T}_z$	Kinetic energy operators in X-, Y-, Z-directions
S	Surface area, area of the heterojunction, the excitation due to a wave incident from one of the leads of the conductor, source or supply term representing the inflow of electrons from the contacts		
		t	Time, a symbol defined as $t = \dfrac{\hbar^2}{2ma^2}$, time for which the diffusion is performed, thickness
$S^{in}(\mathbf{k},t)$	Inscattering function in classical theory		
$S^{out}(\mathbf{k},t)$	Scattering function in classical theory expressing the rate of scattering of electrons out of an initially full state $\mathbf{k}$	$-t$	A coupling or hopping parameter between adjacent lattice points

t_1	Time for which pre-deposition of impurity is carried out
t_2	Time for which drive-in of impurity is performed
t_L	Amplitude of transmitted wave for electrons moving from right to left
t_R	Amplitude of transmitted wave for electrons moving from left to right
U	Potential energy
$U(r)$	Lennard-Jones potential
$U(y)$	Confining potential acting in the Y-direction
$U(z)$	Potential energy of the electron
u	A variable, $\frac{1}{2}(kL)$
u_0	$\frac{1}{2}(k_0 L)$
$u_j(z)$	Wave function factor along the Z-direction
$u(\rho)$	A substitution $= \rho R(\rho)$ defined in hydrogen atom problem
V	Volume, Potential, Potential energy
V_0	Potential barrier height
V_1, V_2	Voltage sources
V_{bi}	Built-in potential at a junction between two materials
$V_{Channel}(x)$	Channel voltage at the point x under the gate
$V_{Channel}(0)$	Channel voltage at the point $x = 0$ under the gate
V_{DS}	Drain-source voltage applied to a MOSFET
$V_{DS(Sat)}$	Drain-source saturation voltage
$V_{Effective}(L)$	Effective gate voltage controlling the charge at location $x = L$
V_{FB}	Flat-band voltage
V_{G1}	Controlling gate voltage in a single-electron transistor
V_{G2}	Gate voltage for choosing background charge in a single-electron transistor
V_{GS} or V_G	Gate-source voltage applied to a MOSFET
V'_{GS}	Gate-source voltage of the ideal long-channel MOSFET
V_L and V_R	Electrical potentials of the left and right contacts
$V(n)$	Voltage of the island
V_{off}	Voltage at which the 2DEG is annihilated
V_{ox}	Voltage drop across the silicon dioxide film
V_p	Peak voltage
V_p, V_q	Voltages at terminals p, q
V_{PT}	Punchthrough voltage
V_q	Potential felt by an electron due to a phonon with wave vector q
$V(r)$	Spherically symmetric potential
V_s	Supply voltage
V_{State}	Volume occupied by a cell of dimensions L_x, L_y, L_z in k-space
V_{Th}	Threshold voltage of MOSFET
$V_{Th}^{High\,DD}$	Threshold voltage of a MOSFET measured at a high drain bias $V_{High\,DD}$
$V_{Th}^{Low\,DD}$	Threshold voltage of a MOSFET measured at a very low drain bias $V_{Low\,DD}$
V_v	Valley voltage
$\hat{V}_x, \hat{V}_y, \hat{V}_z$	Potential energy operators in X-, Y-, Z-directions
v	Velocity
v_d	Drift velocity of electrons
v_F	Fermi velocity
v_g	Group velocity
v_{sat}	Saturation velocity of electrons
$v_{Thermal}$	Thermal velocity of electrons
$v_z(k_{z,L})$	Electron velocity in the Z-direction from left side
W	Width, width of the depletion region, channel width in a MOSFET, work function of a material, Work done
W_1	Work done by the voltage source for n_1 electrons tunneling across junction J_1
W_2	Work done by the voltage source for n_2 electrons tunneling across junction J_2
W_{-e}	Electrostatic energy required to transfer an electron of charge $-e$ from ground potential to the island
W_{+e}	Electrostatic energy required to remove an electron of charge $-e$ from the island to ground potential
W_C	Charging energy
W_{dD}	Depth of the N+drain-P substrate depletion region
W_{dS}	Depth of the N+source-P substrate depletion region
W_{dm}	Vertical depth of the depletion region from the oxide-silicon surface into the P-substrate
W_N	Quantum confinement energy
W_Σ	Total energy
W_V	Work done by voltage sources in the circuit
$W(y, z)$	Confinement potential for motion along the Y- and Z-directions, i.e. potential confining the carriers in the YZ-plane
w	Width
$X(x)$	Function of x
X, Y, Z	Center-of-mass coordinates
x	Position of a particle along the X-axis, aluminum content in AlGaAs
x	Expectation value of position of a particle
x, y, z	Relative position coordinates
$\hat{x}, \hat{y}, \hat{z}$	Position operators in X-, Y-, Z-directions
x_e, y_e, z_e	Coordinates of electron in hydrogen atom
x_j	Junction depth
$x_{m,l}$	Roots of the Bessel function; quantum number m is concerned with center-of-mass motion
$x_{n,l}$	nth zero of $j_l(x)$; quantum number n is concerned with reduced mass motion

x_p, y_p, z_p	Coordinates of proton in hydrogen atom
Y	Young's modulus of the cantilever material
$Y(\theta, \phi)$, $Y_{l,m}(\theta, \phi)$	Angular part of the wave function $\Psi(r, \theta, \phi)$; l, m are quantum numbers
$y_l(kr)$	Bessel function of the second kind
Z	Atomic number of an element, width of the gate of a field-effect transistor
z	Number of closest neighbors, the distance between the tip and the sample in STM
α or β	A number, $\sqrt{\dfrac{2m}{\hbar^2}\{V_0(x) - E\}}$
α/k	$\sqrt{\dfrac{V_0(x)}{E} - 1}$
Γ	A function representing the rate of loss of carriers by scattering
Γ_n	Full width of the energy of quasi-stationary state at half maximum when the transmission is purely coherent
Γ_s	Full width of the energy at half maximum for inelastic scattering
Γ_T	Total full width of the energy at half maximum
γ	Eigenvalue of the translation operator
$\Delta E, \Delta t, \Delta p, \Delta x$	Uncertainties in energy, time, momentum, position
ΔE_C	Energy difference between the edges of the conduction bands of the two semiconductors
$\Delta E_{C(A,B)}$	Discontinuity of conduction bands at an abrupt heterojunction of materials A, B
$\Delta E_{C \to F}$	Energy difference between conduction band edge and Fermi level:
$\Delta E_{G(A,B)}$	Bandgap discontinuity of materials A, B
ΔE_{N_s}	Energy Difference $(= E_F - E_0)$ caused by sheet density of electron charge
ΔE_{Shift}	Shift in energy
$\Delta E_{V(A,B)}$	Discontinuity of valence bands of materials A, B
ΔF	Free energy change
$\Delta F_1^{\pm}$	Free energy changes for an electron tunneling across junction J_1 (inward or outward)
$\Delta F_2^{\pm}$	Free energy changes for an electron tunneling across junction J_2 (inward or outward)
$\Delta W_{1L}(n)$	Energy change during an electron from right lead tunneling leftward from J_1
$\Delta W_{2L}(n)$	Energy change during an electron from right lead tunneling leftward from J_2
$\Delta W_{1R}(n)$	Energy change during an electron from left lead tunneling rightward from J_1
$\Delta W_{2R}(n)$	Energy change during an electron from left lead tunneling rightward from J_2
δ	Dirac delta function
$\delta_{k,l}$ or $\delta_{\beta\alpha}$	Kronecker delta
$\delta[m]$	A sequence of numbers
$\delta(t - \tau)$	A delta function displaced from the origin by a time $t = \tau$
$\delta(t)$	Delta function located at the origin $t = 0$
$\delta(x)$	Dirac delta function or impulse function, located at the origin
$\delta(x - s)$	Dirac delta function or impulse function, located at a distance $x = s$ from the origin
ε_0	Permittivity of free space
ε_1	Energy of the lowest quasi-bound state
ε_2	Dielectric constant of the large bandgap semiconductor
ε or ε_r	Relative permittivity or dielectric constant, Energy, depth of the potential well
ε_F	Fermi energy
ε_j	Electron energy minus sum of free electron energies in X- and Y-directions
$\varepsilon_n, \varepsilon_{n+1}$	Energies of the nth and $(n + 1)$th quasi-energy levels
ε_{n_x,n_y}	Discretized electron energies in X- and Y-directions
ε_s	Relative permittivity of silicon
η	Electron affinity, internal quantum efficiency of the LED
Θ	Heaviside step function
Θ-equation	Polar angle equation (in hydrogen atom problem)
θ	Angle, angle between radius vector and $+Z$-axis in spherical polar coordinates
κ, κ_n	A substitution $= \sqrt{-\dfrac{2\mu E}{\hbar^2}}$ in hydrogen atom problem
λ	Wavelength
μ	Reduced mass of two particle system $= \dfrac{m_e m_p}{m_e + m_p}$, Mobility of charge carrier
μ_n	Electron mobility
ν	Linear frequency
v_m	Electron velocity $= \dfrac{\hbar k_m}{m}$
ξ	A constant defined in Kronig-Penney model, $\xi = \dfrac{\hbar^2}{2ma^2}$
ρ	Resistivity, a substitution $= 2r\kappa$ made in hydrogen atom problem, density
ρ_0	A substitution $= \dfrac{\mu e^2}{2\pi\varepsilon_0 \kappa \hbar^2}$ made in hydrogen atom problem
$\rho_{Acceptors}(z)$	Electrical charge density of acceptor ions in the Z-direction
$\rho_{Donors}(z)$	Electrical charge density of donor ions in the Z-direction

$\rho_{\text{Electrons}}(z)$	Electrical charge density of electrons in the Z-direction
$\rho_{\text{Impurities}}(z)$	Electrical charge density due to impurity ions in the Z-direction
$\rho(\mathbf{k}, \mathbf{k}')$	A density matrix
$\rho(\mathbf{k}, \mathbf{k}'; t)$	A time-varying density matrix
$\rho\left(k_{z,\text{L}}, k_{t,\text{L}}\right)$	Density of states in k-space
$\rho_{\text{Total}}(z)$	Total electrical charge density in the Z-direction
$\rho(\mathbf{x})$	Charge density describing a continuous distribution of charges in 3-D space
Σ_0	Self energy for interaction of electrons with their surroundings
Σ^A	Advanced self-energy
Σ^{in}	Total inscattering function, total current inflow
Σ_0^{in}	Current inflow from surroundings
Σ_1	Self energy for contact 1
Σ_1^{in}	Current inflow from contact 1
Σ_2	Self energy for contact 2
Σ_2^{in}	Current inflow from contact 1
$\Sigma^{\text{in}}(\mathbf{k}, \mathbf{k}'; t, t')$	Inscattering function dependent on time coordinates
$\Sigma_\phi^{\text{In}}(\mathbf{r}, \mathbf{r}'; E)$	Inscattering function due to phase-breaking interactions in the conductor
Σ^{Out}	Total outscattering function
$\Sigma^{\text{Out}}(\mathbf{k}, \mathbf{k}'; t, t')$	Outscattering function dependent on time coordinates
$\displaystyle\sum_p \sum_p^{\text{Out}}$	Outscattering function due to interaction of electrons with lead p
$\Sigma_\phi^{Out}(\mathbf{r}, \mathbf{r}'; E)$	Outscattering function due to phase-breaking interactions in the conductor
$\Sigma^R = \displaystyle\sum_p \Sigma_p^R(i,j)$	Self-energy arising from the leads
$\Sigma_p^R(i,j)$	Symbol for $t^2 g_p^R(p_i, p_j)$
$\Sigma\Psi$	Term representing the outflow of electrons
σ_0	DC conductivity
σ	Conductivity, A constant, distance at which the intermolecular potential becomes zero
σ_{n}	Conductivity due to electrons
σ_{p}	Conductivity due to holes
σ_{Th}	Thermal conductivity
τ	Relaxation time, mean free time, transit time of electrons from left to right contact
τ_{n}	Lifetime characterizing each quasi-stationary state
τ_{p}	Coupling matrix G_{pC}
τ_{p}^+	Coupling matrix G_{Cp}
τ_{s}	Scattering time of the electron in the 2-D subband related to the nth quasi-energy level
Φ	Barrier height, electric flux across the Gaussian surface
Φ_0	Voltage
Φ-equation	Azimuthal angle equation (in hydrogen atom problem)
$\Phi(\mathbf{r}, t)$	Scalar potential
$\Phi(z)$	Electrostatic potential in the Z-direction
ϕ	Phase angle, Work function of a material, Angle between projection of radius vector on XY-plane and $+X$-axis in spherical polar coordinates
ϕ_A, ϕ_B	Work functions of the materials A, B
ϕ_B	Bulk potential, Schottky barrier height
ϕ_{bi}	Built-in potential
$\phi_{\text{Envelope}}(x)$	Envelope function
$\phi_{\text{m}}, \phi_{\text{M}}$	Work function of the metal
ϕ_{S}	Work function of the semiconductor
χ	Electron affinity of a material
χ_A and χ_B	Electron affinities of two materials A, B
χ_2	Electron affinity of large bandgap semiconductor
Ψ_{ABMO}	Wave function of the antibonding molecular orbital
$\Psi(\mathbf{r}, \mathbf{r}_1)$	Wave function at $\mathbf{r}$ due to a source term $S(\mathbf{r}_1)$ at $\mathbf{r}_1$
$\Psi(\mathbf{r}', \mathbf{r}_1')^*$	Complex conjugate wave function at $\mathbf{r}'$ due to a conjugate source term $S(\mathbf{r}_1')^*$ at $\mathbf{r}_1'$
$\Psi_\alpha(\mathbf{r}), \Psi_\beta(\mathbf{r})$	Two members of a set of wave functions
$\Psi_{\text{Bloch}}(x)$	Bloch wave function
Ψ_{BMO}	Wave function of the bonding molecular orbital
$\Psi_{\text{Corrected}}(x)$	Corrected wave function
$\Psi_{\text{L}}(x)$	Wave function for the left region
$\Psi_{\text{R}}(x)$	Wave function for the right region
$\Psi(x, t)$	Wave function defined over space and time
$\Psi^*(x, t)$	Complex conjugate wave function defined over space and time
$\Psi_{\text{I}}(x, t),$ $\Psi_{\text{II}}(x, t),$ $\Psi_{\text{III}}(x, t)$	Wave functions in regions I, II, III
$\Psi_\nu(\mathbf{r})$	Wave function of the electron where ν denotes the set of quantum numbers of the electron
Ω	Ohm, Volume of the box
ω	Angular frequency
ω_{q}	Frequency of phonon
∇	Del or nabla operator
∂	Partial differential operator
∇^2	Del squared or Laplacian operator $\nabla_{\text{R}}^2 \Psi : \dfrac{\partial^2 \Psi}{\partial X^2} + \dfrac{\partial^2 \Psi}{\partial Y^2} + \dfrac{\partial^2 \Psi}{\partial Z^2}$

About the Book

Introductory Nanoelectronics: Physical Theory and Device Analysis is an ardent endeavor to explain the mathematical steps in the derivation of key formulae in nanoelectronics in an easy-to-understand format. Beginning with recapitulation of quantum mechanics and condensed matter physics necessary for understanding nanoelectronics, the book deals with electron transport in nanostructures and the Green's function method extensively. Changing gears to experimental aspects, the nanofabrication and characterization techniques are addressed. The treatment culminates with in-depth description and analysis of state-of-the-art nanoelectronic devices.

Catering to the needs of a diverse audience, the book will be of immense value to students as well as professional categories. The systematic organization of subject matter in six parts containing a total of 22 chapters together with the broad, up-to-date, and comprehensive topical coverage illustrated with 172 meticulous line diagrams will make the book tremendously useful for graduate students as well as practicing engineers and scientists engaged in this fast-moving field. End-of-chapter worked out illustrative exercises and bibliographies for further reading add to the utility of the book.

1

Nanoelectronics and Mesoscopic Physics

Beginning from the comparison of size of atomic/subatomic particles, the nano size regime will be defined and nanoelectronics will be introduced.

1.1 Ultra-Small Objects

To get an idea of relative magnitudes of atomic/subatomic particles, let us take a quick look at the dimensions of some ultra-small objects.

1.1.1 Electron Radius

The radius of the electron r_e is $< 10^{-15}$ m (Pauling 1964). The classical electron radius $= 2.82 \times 10^{-15}$ m (see reference: Classical electron radius) is obtained by equating the relativistic rest mass energy of an electron (mass × square of velocity of light) with the electrostatic potential energy of a sphere carrying one electron charge and having radius = electron radius. Another approach (see reference: Determining the electron structure), which appears more realistic, takes the measured radius of proton $= 1.15 \times 10^{-15}$ m, uses the ratio of proton mass/electron mass = 1836, and divides the proton radius by $\sqrt[3]{1836} = 12.24$ to get the electron radius $= 9.1 \times 10^{-17}$ m because mass increases with cube of radius as

$$\text{Mass} = \text{Density} \times \text{Volume} = \text{Density} \times \frac{4}{3}\pi r_e^3 \qquad (1.1)$$

1.1.2 Atomic Nucleus Diameter

The nucleus of the hydrogen atom has a diameter $= 1.6 \times 10^{-15}$ m which is 1.45×10^5 times smaller than the hydrogen atom itself (see reference: Atomic nucleus). The diameter of the nucleus of a uranium atom is 1.5×10^{-14} m, which is smaller by a factor of 2.3×10^4 than the uranium atom.

1.1.3 Atomic and Molecular Sizes

Looking at the list of atomic radii of the elements (see reference: Atomic radius of the elements), we come across values such as helium (31 pm), hydrogen (53 pm), silicon (111 pm), germanium (125 pm), copper (145 pm), iron (156 pm), cesium (298 pm), …. From the list of kinetic diameters of molecules (kinetic diameter), we note the diameters of molecules of helium (260 pm), water (265 pm), hydrogen (289 pm), oxygen (346 pm), …. These values representing the subatomic, atomic, and molecular regimes are typically in the sub-nm range. The fields of subatomic particles and the atomic and molecular regimes constitute well-established disciplines.

1.2 The Nanoworld

Moving upward on the dimensional scale, there is a range of sizes ~ 1–100 nm, which spans 10–100 times the molecular diameter. It is an interesting size range with numerous possibilities, some utilized and many under exploration. This size regime is labeled as the nanoworld and when we delve into the behavior of electrons in materials of nanoscale dimensions, we come to nanoelectronics, the branch of nanoscience and nanotechnology dealing with the special properties displayed by electrons in matter at the nanoscale only, and the utilization of these properties to control electron flow in building devices and circuits exhibiting functionalities, which are not observed by those with bulk matter-built structures.

Nanoelectronics evolved from its predecessor, microelectronics. The micrometer is a unit of length equaling 1×10^{-6} m, i.e. one millionth of a meter and microelectronics is concerned with electronic components having minimum feature sizes of a few or several microns. As the electronic components shrunk further in size came the nanoscale. The nanometer is a unit of length = 1×10^{-9} m, which is a billionth of a meter. The smallest features in electronic devices and circuits became extremely small ~ a few nm to several nm, moving from millionth to billionth of a meter. Nanoelectronics describes the properties of electronic devices/circuits with smallest feature sizes in the 1–100 nm range. The advancement from microelectronics to nanoelectronics does not merely represent a thousand-fold shrinkage in feature sizes of devices as a leap toward miniaturization of electronic components but is intended to leverage many additional phenomena and unique properties of materials to build new capabilities, hitherto not possible.

1.3 Mesoscopic Physics

The heart and soul of nanoelectronics is mesoscopic physics. Meso means 'intermediate or middle'. Mesoscopic physics is the physics of structures of intermediate dimensions falling between the microscopic (individual atoms and molecules) and macroscopic (bulk matter) kingdoms, i.e. covering the submicron and nanoscale sizes. Mesoscopic physics bridges the microscopic and macroscopic regions (Das 2010). It is a branch of condensed matter physics. In contrast to the microscopic dominion, objects of both the mesoscopic and macroscopic realms contain a large number of atoms, mesoscopic less and macroscopic more. Below we mention some distinguishing features of mesoscopic objects that distinctly tell them apart from macroscopic ones.

1.3.1 Averaging of Behavior

The properties of macroscopic objects are usually measured on large congregations $\sim 10^{23}$ of atoms and molecules. Such objects are describable in terms of averaged properties of constituent atoms. Due to the presence of fewer number of atoms, mesoscopic objects do not permit description in terms of averaged behavior and are susceptible to fluctuations around the mean value.

1.3.2 Surface Area-to-Volume Ratio

The properties of a material undergo drastic changes when crossing from the macroscopic to the mesoscopic level. At the macroscopic level, the number of atoms at the surface of a solid is a negligible fraction of the total number of atoms in the solid. The surface area-to-volume ratio is small. But the same is not true at the mesoscopic level where the number of atoms on the surface of the solid becomes a significant percentage of the total number of atoms in it. The surface area-to-volume ratio attains a high value. For a cube of side L, the surface area-to-volume ratio is

$$\left(\frac{S}{V}\right)_{\text{Cube}} = \frac{6L^2}{L^3} = \frac{6}{L} \tag{1.2}$$

As the side length is decreased from 1 m to 1 nm, the S/V ratio increases from 6 to $6 \times 10^9 \, \text{m}^{-1}$.

The surface area-to-volume ratio for a sphere of radius R is

$$\left(\frac{S}{V}\right)_{\text{Sphere}} = \frac{4\pi R^2}{\frac{4}{3}\pi R^3} = 4\pi R^2 \times \frac{3}{4\pi R^3} = \frac{3}{R} \tag{1.3}$$

The calculation of surface area-to-volume ratios for spheres of radii 5 cm and 1 cm is shown in Figure 1.1. A sphere of radius 1 nm has 10^9 times the surface area-to-volume ratio of sphere of radius 1 m.

Due to the large difference in surface area-to-volume ratios between macroscopic and mesoscopic objects, the latter reveal novel properties which are not observed in the bulk matter. Due to their higher chemical reactivity, nanoparticles are efficient catalysts because a greater area of material per unit volume can participate in the reaction. Nanomaterials find wide applications in catalyzing chemical reactions.

1.3.3 Dominance of Electromagnetic Force over Gravitational Force

Owing to the exceedingly small mass of nanoparticles, the gravitational force dependent on masses of particles and distance between them is very weak. But the electrostatic attractive or repulsive force dependent on the charges and separation distance between them is very strong.

The electric force between two electrons is 10^{42} times the gravitational force irrespective of the distance between them. This result is deduced by considering two electrons (charge $-e$, mass m) at a certain distance r apart in air. They are mutually repelled by electric force but attracted toward each other by the gravitational force.

The electric force is given by Coulomb's law of electrostatics as

$$F_{\text{Electric}} = \frac{-e \times -e}{4\pi\varepsilon_0 r^2} = \frac{e^2}{4\pi\varepsilon_0 r^2} \tag{1.4}$$

where ε_0 is the permittivity of free space $= 8.854 \times 10^{-12} \, \text{m}^{-3} \text{kg}^{-1} \text{s}^4 \, \text{A}^2$.

The gravitational force obeys Newton's law of universal gravitation

$$F_{\text{Gravitational}} = G\frac{m \times m}{r^2} = G\frac{m^2}{r^2} \tag{1.5}$$

where G is the gravitational constant $= 6.6741 \times 10^{-11} \, \text{m}^3 \text{kg}^{-1} \text{s}^{-2}$.

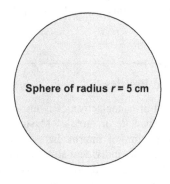

Sphere of radius r = 5 cm

Surface area = $4\pi r^2 = 4 \times 3.14 \times 5^2 \, \text{cm}^2$
Volume = $(4/3)\pi r^3 = (4/3) \times 3.14 \times 5^3 \, \text{cm}^3$
Surface area/Volume = $4 \times 3.14 \times 5^2 \times 3/(4 \times 3.14 \times 5^3) = (3/5)\text{cm}^{-1} = 0.6 \, \text{cm}^{-1}$

(a)

Sphere of radius r = 1cm

Surface area = $4\pi r^2 = 4 \times 3.14 \times 1^2 \, \text{cm}^2$
Volume = $(4/3)\pi r^3 = (4/3) \times 3.14 \times 1^3 \, \text{cm}^3$
Surface area/Volume = $4 \times 3.14 \times 1^2 \times 3/(4 \times 3.14 \times 1^3) = (3/1)\text{cm}^{-1} = 3 \, \text{cm}^{-1}$

(b)

FIGURE 1.1 Comparing the surface area-to-volume ratio of a spherical particle of radius 5 cm (a) with that of a sphere of radius 1 cm; the ratio of the surface area-to-volume ratio of smaller to larger sphere is = 3/0.6 = 5, which is the same as the ratio of their radii.

From eqs. (1.4) and (1.5), it is easy to see that

$$\frac{F_{\text{Electric}}}{F_{\text{Gravitational}}} = \frac{\dfrac{e^2}{4\pi\varepsilon_0 r^2}}{G\dfrac{m^2}{r^2}} = \frac{e^2}{4\pi\varepsilon_0 r^2} \times \frac{r^2}{Gm^2} = \frac{e^2}{4\pi\varepsilon_0 Gm^2}$$

$$= \frac{\left(1.602\times10^{-19}\,\text{C}\right)^2}{4\times3.14159\times8.854\times10^{-12}\ \text{m}^{-3}\ \text{kg}^{-1}\ \text{s}^4\ \text{A}^2 \times 6.6741\times10^{-11}\ \text{m}^3\ \text{kg}^{-1}\ \text{s}^{-2} \times \left(9.109\times10^{-31}\ \text{kg}\right)^2}$$

$$= \frac{2.566\times10^{-38}}{61614.528\times10^{-85}} = 0.000041646\times10^{47} = 4.17\times10^{42} \tag{1.6}$$

1.3.4 Size Dependence of Properties

Mechanical properties such as strength/flexibility, optical properties like color, transparency, fluorescence, and electrical/magnetic properties like electrical conductivity and magnetic permeability of a material are altered when the particle size reaches the nanoscale.

Thin, long cylinders of carbon known as carbon nanotubes (CNTs) are of two types: single-walled carbon nanotubes (SWCNTs) and multi-walled carbon nanotubes (MWCNTs). SWCNTs have average diameters in the range from 0.4 to 2 nm; outer diameters of MWCNTs are in the range from 2 to 30 nm (Eatemadi et al 2014). The extremely thin CNTs exhibit 100 times tensile strength of steel. CNTs are several micrometers long with aspect ratio of 1000:1 and much higher. Their electrical properties change with the diameter, the number of walls, and chirality or degree of twist depending on which they are either metallic, quasi-metallic, or semiconductors.

Gold nanoparticles are no longer yellow in color, rather purple or red. Large-size zinc oxide particles scatter visible light and appear white but ZnO nanoparticles do not scatter visible light and look clear. The variation of properties with size offers a valuable tool to tailor or tune the properties as desired, simply by taking particles of the required dimensions. Selective accumulation of gold nanoparticles in tumors is exploited for precise imaging and destruction of malignant cells by laser.

1.3.5 Classical and Quantum-Mechanical Laws

Macroscopic objects are subject to the laws of classical mechanics. Classical mechanics is not applicable to mesoscopic objects. Their theoretical treatment comes under the purview of quantum mechanics, also called wave mechanics, which is a fundamental theory in physics. Quantum mechanics is a branch of mechanics for mathematical formulation of motion and interaction of subatomic particles to describe nature at the smallest scales of energy levels of atoms and constituent particles. It incorporates:

i. Quantization of energy, momentum, angular momentum, and other parameters of a bound system with constraint to allow discrete values only.

ii. Wave-particle duality of matter: All matter including electrons and electromagnetic fields behave as waves and particles.

iii. Uncertainty principle: Limit to the precision with which certain pairs of physical properties of a particle called complementary variables, e.g. position x and momentum p, energy E and time t, can be simultaneously measured.

As we live in a three-dimensional (3D) world, we are accustomed to think about electrons moving in three dimensions. Nanostructures can be built to confine electron motion to two dimensions, one dimension, and even zero dimensions. These nanostructures are known as low-dimensional structures. They act as the basic ingredients of nanoelectronics.

The 3D structure is known as the bulk structure. In this structure, the electrons are free to move in all directions. Electron motion is not quantized. The two-dimensional (2D) structure is called quantum well (Figure 1.2). In the quantum well, electrons have freedom of motion in two directions while the electron motion in third direction is restricted and quantized. The one-dimensional (1D) structure is termed a quantum wire. In the quantum wire, the electron motion is free in one direction only while in the remaining two directions, the electron movement is delimited and quantized. The zero-dimensional (0D) structure is known as a quantum dot in which electrons do not have freedom to move in any direction leading to quantization of electron motion in all directions.

1.3.6 Quantization of Conductance

Familiar with the macroscopic level, we think that the conductance of a wire continuously decreases as its diameter is reduced. But when dealing with the mesoscopic level, this line of thinking needs to be changed because the conductance change is no longer continuous. Instead, conductance changes as a series of discrete steps or quanta. Let us see how this happens.

The resistance R of a metal wire of length L and cross-sectional area A made from a material of resistivity ρ, as measured with two contact electrodes, is given by the well-known formula

$$R = \rho\frac{L}{A} \tag{1.7}$$

On applying this formula we find that the resistance of the wire should go on decreasing as the wire is shortened in length, keeping area A fixed. However, in practice, the resistance decreases

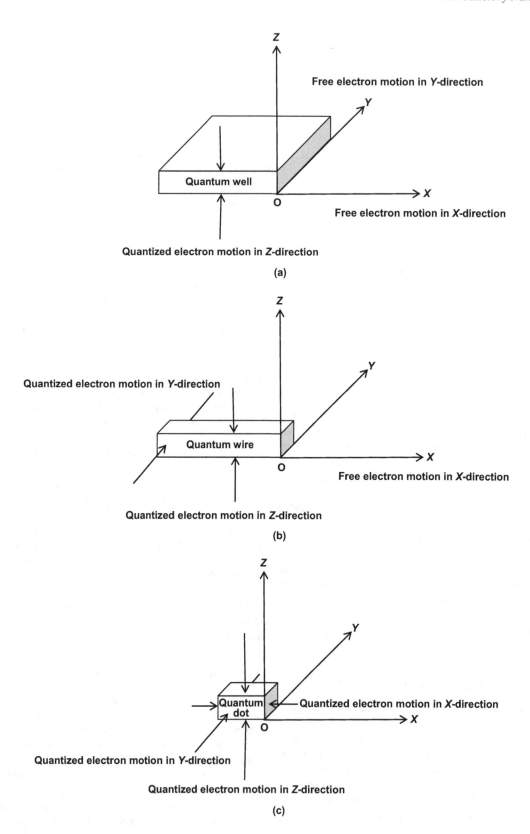

FIGURE 1.2 Low-dimensional structures: (a) two-dimensional, (b) one-dimensional, and (c) zero-dimensional.

up to a limiting value but flattens out beyond this value. This value of resistance is reached when the length of the wire becomes smaller than the mean free path of electrons. We have now come to the stage of ballistic transport. As the electrons do not suffer any scattering, it is expected that the resistance of the wire should be zero, which actually does not happen. The small resistance now obtained is the resistance of the contact electrodes, i.e. contact resistance.

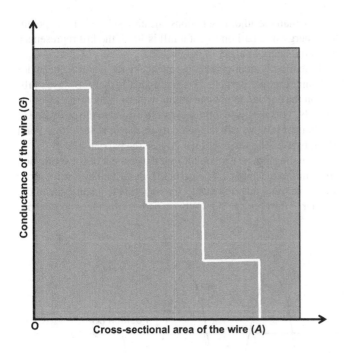

FIGURE 1.3 Fall of conductance of the wire as a series of discrete steps.

If now the cross-sectional area of the wire is decreased by thinning the wire, we anticipate that the resistance should increase, i.e. conductance should decrease continuously as A decreases. But again, the expected outcome is found to be untrue. Instead of a continuous fall, the conductance decreases in steps or quanta (Figure 1.3). This discretization of values is called quantization of conductance. The quantum of conductance denoted by the symbol G_0 is expressed in terms of the elementary charge e and the Planck's constant h as

$$G_0 = \frac{2e^2}{h} = \frac{2\left(1.602 \times 10^{-19}\,\mathrm{C}\right)^2}{6.626 \times 10^{-34}\,\mathrm{J\,s}} = \frac{5.133 \times 10^{-38}}{6.626 \times 10^{-34}}$$

$$= 0.775 \times 10^{-4}\,\mathrm{S} = 7.75 \times 10^{-5}\,\mathrm{S} \tag{1.8}$$

It is derived by analyzing a one-dimensional wire connected with two electron reservoirs (Chapter 11, eq. (11.187)).

1.3.7 Quantum Confinement

The electrical and optical properties of a material change when the size of the sample of material comes close to the de Broglie wavelength of electrons, typically 10 nm or less, owing to a restriction in the random motion of electrons in a continuum of energy levels called the energy band to a series of discrete energy levels, as found in isolated atoms (Figure 1.4). The range of forbidden energies increases as the material sampled becomes smaller. In semiconductors, the characteristic length scale at which quantum confinement is observed is the exciton Bohr radius (the distance between an electron and hole in the electron-hole pair). In ferromagnetic materials, the typical domain size (small regions in a magnetic material in which the

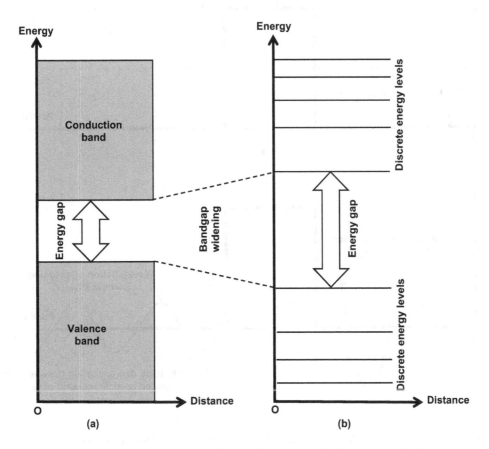

FIGURE 1.4 Energy-distance models of a material: (a) energy bands in the bulk material and (b) discrete energy levels in a nanomaterial.

magnetic moments of atoms are aligned in the same direction) is the decisive parameter. In superconductors, the coherence length of Cooper pairs (the average distance between two electrons in a Cooper pair) plays the same role.

1.3.8 Quantum-Mechanical Tunneling

It is a quantum-mechanical phenomenon in which a particle possessing less total energy than the height of a potential energy barrier has a finite probability to penetrate through the barrier and be found on the opposite side of the barrier, in violation of the laws of classical mechanics (Figure 1.5). A potential energy barrier is a region in a force field possessing higher potential

energy than the adjoining regions on either side, e.g. if the potential energy of a ball on top of a hill is 100 J, the hill represents a potential energy barrier of height 100 J.

Tunneling is an obvious consequence of the wavelike behavior of a microscopic particle. The main idea is that the waves falling on the barrier do not end abruptly there; rather they taper off quickly. This means that the waves can extend to the other side of the barrier but this can happen only in case it is very thin. For a thick barrier, the wave will not go past the barrier.

Tunneling can be explained from the concept of wave function in quantum mechanics. The particle is described by a wave function. The wave function must vary smoothly from one side of the barrier to the opposite side. When a particle impinges on a barrier

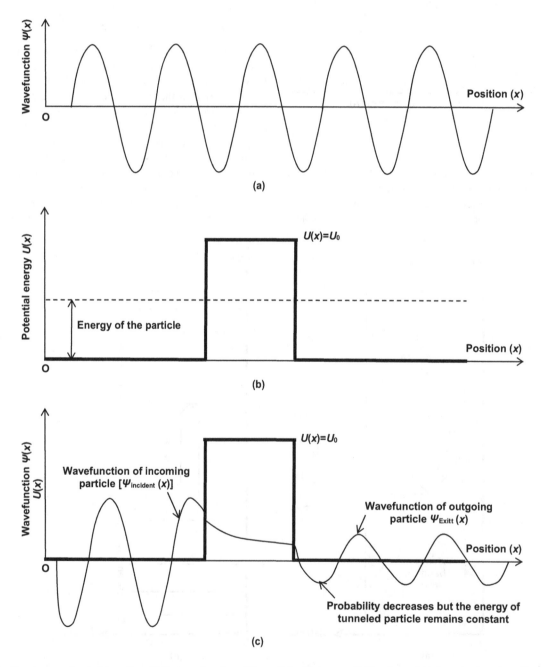

FIGURE 1.5 Quantum-mechanical tunneling: (a) the wave function of the particle, (b) the potential barrier, and (c) the reduction of amplitude of the wave function after passing through the barrier.

it cannot surmount, the wave function changes from sinusoidal shape to exponentially decaying form. Inside the barrier the wave function does not suddenly become zero but decays exponentially by rapid attenuation. Exiting the barrier on the other side, the wave function is again a sinusoidal wave with diminished amplitude. The amplitude of the wave function, however, decreases due to reflection of the incident wave. Since the wave function can extend to the other side of the barrier, and the square of the wave function represents the probability of a particle being found in a certain location of space, a non-zero probability is predicted for the particle being found on the opposite side of the barrier. The non-zero probability value is construed as if the particle has tunneled through the barrier. However, the value of this probability depends on certain properties of the particle and the barrier so that tunneling is appreciable only under specific conditions. The probability of a particle of mass m possessing a kinetic energy E to tunnel through a potential barrier of height $V > E$ and thickness r is approximately given by (Tunneling 2019)

$$P = \exp\left\{\frac{-4r\pi\sqrt{2m(V - E)}}{h}\right\} \qquad (1.9)$$

where h is the Planck's constant. Equation (1.9) shows that the tunneling probability exponentially decreases: (i) as the thickness of the barrier increases (hence possible only at nanoscale), (ii) the square root of mass of the particle increases (hence restricted to small subatomic particles), (iii) the square root of difference between barrier height and particle energy increases. More exact equations for tunneling probability will be derived in Chapter 5, Section 5.5 (eqs. (5.385)–(5.388)).

1.3.9 Giant Magnetoresistance Effect

Anisotropic magnetoresistance (AMR) is the increase in electrical resistance of a current-carrying magnetic material when the magnetic field acts in the same direction as the current flow and the decrease in its resistance of the material when the magnetic field acts at right angles to the direction of current. The resistance change is generally a few percent.

Giant magnetoresistance (GMR) effect is a quantum-mechanical effect observed as an appreciable decrease of ~ 10%–80% in electrical resistance, of a multilayer structure with typical layer thicknesses of a few nm, comprising a thin nonmagnetic spacer layer (e.g. Cr) separating two ferromagnetic layers (e.g.Fe), when subjected to a magnetic field, either parallel to (current-in-plane (CIP) configuration) or perpendicular to the plane of the layers (current perpendicular to plane (CPP) configuration) (Tsymbal and Pettifor 2001, Ennen et al 2016). The GMR effect is illustrated in Figure 1.6.

The GMR effect is qualitatively explained using the Mott model based on the spin state of electrons (Figure 1.7). As we know, the electrons possess an intrinsic angular momentum. The spin quantum number can acquire values $m_s = \pm\hbar/2$. In a domain

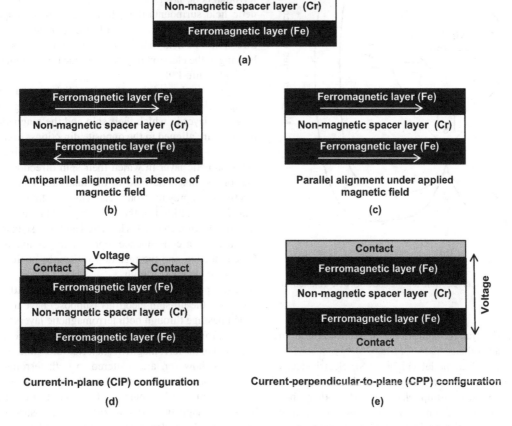

FIGURE 1.6 Giant magnetoresistance effect: (a) the double layer structure, (b) without magnetic field, (c) with magnetic field, (d) current-in-plane configuration, and (e) current perpendicular to plane configuration.

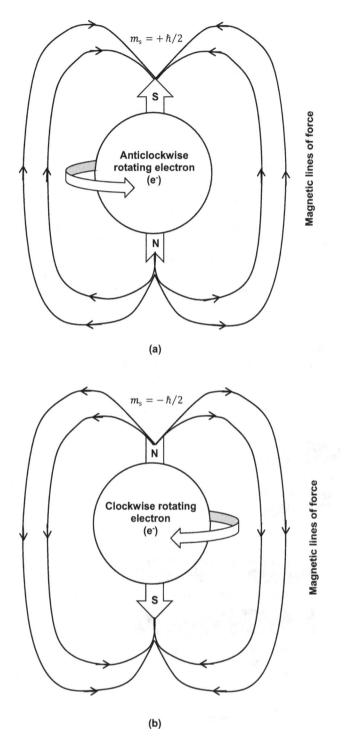

(a)

(b)

FIGURE 1.7 Spin states of electron for (a) anticlockwise rotation and (b) clockwise rotation.

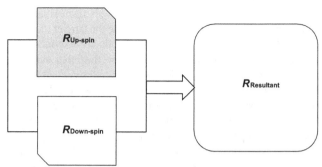

FIGURE 1.8 Mott model of giant magnetoresistance effect.

Therefore their contribution to the overall conductance is independent. Furthermore, they add parallely to it. The densities of states for the two types of electrons are different at the Fermi level. The probability of scattering for any type of electrons is proportional to the density of states of that type. So, the scattering rates for the two types differ. To reflect the asymmetry in the density of states at the Fermi level, the scattering is assumed to be feeble for electrons with spin parallel to the direction of magnetization and intense for electrons having spins antiparallel to it.

Consider a GMR structure consisting of a thin spacer film of non-magnetic material sandwiched between two layers of a ferromagnetic material, in absence of an externally applied magnetic field. In this condition, the internal magnetic fields in the two ferromagnetic layers are aligned in opposite directions through the action of interlayer exchange coupling, viz. the interaction of ferromagnetic layers across the spacer film. The external magnetic field surmounts the effect of exchange coupling to align the magnetic fields in the two ferromagnetic layers in the same direction. The field-induced resistance lowering is understood by looking at the electron flow through the two channels of the Mott model (Figure 1.9).

In the initial condition of antiparallel-aligned multilayer, the up-spin electrons experience intense scattering in one ferromagnetic layer in which their spin and the magnetic field in the material are aligned in the opposite directions. In the same manner, down-spin electrons undergo strong scattering in the other ferromagnetic layer in which their spin direction is opposite to that of the magnetic field in the material. Thus both kinds of electrons fall prey to strong scattering, the first kind in one layer and the second kind in the other layer. The current flow in the spin-up electron channel is hindered and so also in the spin-down channel. As a consequence, none of the channels is favorable for conduction and the resultant resistance of the two parallel-connected channels is high.

In the parallel-aligned multilayer (Figure 1.10), one kind of electrons, the up-spin electrons (suppose) have a smooth unimpeded passage through both ferromagnetic layers. The easy passage is provided because the electron spin and the magnetic field direction in the ferromagnetic materials are same. The down-spin electrons, however, are scattered in both ferromagnetic layers because their spin is opposed to the magnetic field direction. Due to the unruffled journey of one kind of electrons through both ferromagnetic layers, as opposed to obstructed movement of both kinds of electrons in the antiparallel arrangement, the electron motion in the material is comparatively easier now than

of ferromagnetic material, the spins of the electrons in the atoms are aligned with each other.

According to the Mott model (Figure 1.8), electrical conduction in metals involves two independent channels. One of these channels contains spin-up electrons. The other channel contains spin-down electrons. The two kinds of electrons are discriminated from the projections of their spins along the quantization axis. They do not blend over large distances.

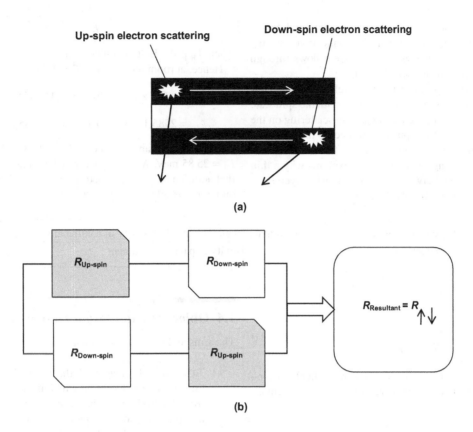

FIGURE 1.9   Electron scattering in an antiparallel aligned structure (a) and its representation in the Mott model (b).

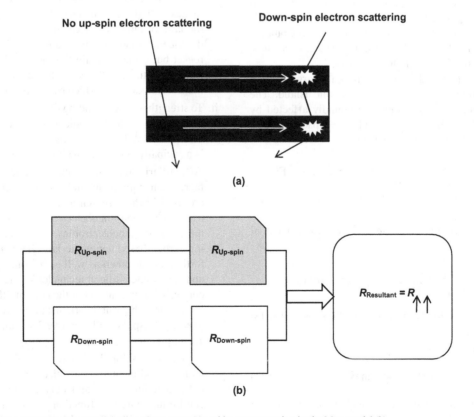

FIGURE 1.10   Electron scattering in a parallel-aligned structure (a) and its representation in the Mott model (b).

in the previous case. Hence, a lower resistivity is obtained. Since, the up-spin electron channel is in parallel connection with the down-spin electron channel, most of the current flows through the low-resistance up-spin electron channel.

Thus the GMR effect arises from the correlation of the conduction properties of a ferromagnetic material with the spin state of the electrons, i.e. the dependence of electron scattering on the relative orientation of electron spin with respect to the magnetic field in the material. Another prerequisite for GMR is the existence of interlayer exchange coupling. This explains why a thin spacer layer must be used between the ferromagnetic layers.

The GMR effect is quantified by the ratio

$$\frac{\text{Change in resistance}}{\text{Original resistance}} = \frac{\text{Resistivity in antiparallel alignment} - \text{Resistivity in parallel alignment}}{\text{Resistivity in parallel alignment}} \qquad (1.10)$$

or

$$\frac{\Delta R}{R} = \frac{\rho_{\uparrow\downarrow} - \rho_{\uparrow\uparrow}}{\rho_{\uparrow\uparrow}} \qquad (1.11)$$

The GMR effect finds applications in magnetic field sensors, hard disk drives, and magnetoresistive random access memory (MRAM).

1.3.10 Single-Electron Effects

The effect of device downscaling is not merely appearance of quantum-mechanical effects. The number of electrons participating in current conduction also decreases in ultra-small devices and can be reduced down to a single electron. At this stage, discreteness of charge plays a vital role. Due to Coulomb interaction between individual electrons, the transfer of a single electron becomes correlated with that of others.

Single-electron effects take place through charging of a miniscule conducting sample called the metallic island. The electrostatic potential of this island is profoundly affected by the entry of a single electron on the island. The capacitance of a spherical metallic island of radius $r = 10\,\text{nm}$ is

$$C = 4\pi\varepsilon_0 r = 4 \times 3.14159 \times 8.854 \times 10^{-12} \times 10 \times 10^{-9}\,\text{F}$$

$$= 1112.63 \times 10^{-21}\,\text{F} = 1.11 \times 10^{-18}\,\text{F} \qquad (1.12)$$

Addition of a single electron charge $\Delta q = -e$ changes its potential by

$$\Delta V = \frac{\Delta q}{C} = -\frac{e}{C} = -\frac{1.602 \times 10^{-19}}{1.11 \times 10^{-18}} = -1.4432 \times 10^{-1}\,\text{V}$$

$$= -0.1443\,\text{V} = -0.1443 \times 10^3\,\text{mV} = -144.3\,\text{mV} \qquad (1.13)$$

The change in energy is

$$\Delta E = -e\Delta V = 144.3\,\text{meV} \qquad (1.14)$$

This energy is compared with the thermal noise energy at a temperature T given by

$$E_{\text{T}} = k_{\text{B}}T \qquad (1.15)$$

where k_{B} is the Boltzmann constant.

Hence, at room temperature $T = 300\,\text{K}$, we have

$$E_{\text{T}} = 8.617 \times 10^{-5}\,\text{eVK}^{-1} \times 300\,\text{K} = 2585.1 \times 10^{-5}\,\text{eV}$$

$$= 2585.1 \times 10^{-5} \times 10^3\,\text{meV} = 25.85\,\text{meV} \qquad (1.16)$$

Clearly, for the small island, $\Delta E = 144.3\,\text{meV}$ is greater than $E_{\text{T}} = 25.85\,\text{meV}$. A nanostructure whose capacitance is so small that addition of a single electron to it yields a measured voltage output is a single-electron device.

1.4 Objectives and Organization of the Book

The aims and goals of the book are as follows:

i. To apprise the reader of the distinctive cardinal features of a new discipline which is profoundly different from day-to-day experiences on a bulk scale. The reader is introduced to the fundamental theory of quantum mechanics for describing nature at the nanoscale, the body of scientific laws serving as the root of nanoelectronics tree. As spatial dimensions shrink to the scale of atomic and subatomic particles, the laws of classical physics crumble down and become inadequate to describe the phenomena. They are superseded by the laws of quantum mechanics. Classical physics can describe the motion of an automobile, train, airplane, or rocket but cannot explain the motion of electrons in a solid. Quantum mechanics is the language and mathematical framework in which nanoscale world is sculpted.

ii. To strengthen understanding of condensed matter physics because nanoelectronic devices and circuits are essentially solid-state components. So, nanoelectronics is principally concerned with matter in the solid state.

iii. To familiarize the reader with the semiconductor device fabrication equipment and process technologies that are used in the research and development and industrial production of nanoelectronic structures. Although the scope of nanoelectronics is very vast encompassing carbon-based nanomaterials, 2D materials, and so forth, and top-down as well as bottom-up approaches, and various non-silicon technologies, nevertheless silicon technology remains at the core of all development and a thorough groundwork in it provides a springboard to easily grasp the collaborative disciplines.

iv. To build the theoretical background and analytical techniques for the behavior of electrons in three basic embodiments of nanoelectronics, viz. quantum dots, quantum wells, and quantum wires. The non-equilibrium Green's function formalism provides a rigorous theoretical foothold for modeling carrier transport in low-dimensional materials and nanoscale devices.

v. To provide the reader with application of nanoscale effects and principles for realization of various electronic and optoelectronic devices. Glimpses of the utilization of these effects can be caught as a prelude to advanced literature.

In pursuit of the objectives and goals laid down, the book is subdivided into six parts:

i. Part I: Quantum mechanics for nanoelectronics, welcoming the reader to the nanoworld. Quantum mechanics for nanoelectronics reviews and recapitulates elementary quantum-mechanical principles needed to describe the motion and interaction of atomic and subatomic particles.

This part is divided into five chapters. Chapter 2 in this part describes the various experiments which could not be interpreted in terms of classical theories and necessitated the introduction of quantum-mechanical concepts. Chapter 3 in this part explains the Schrodinger equation which will be our favorite tool throughout the book, providing the mathematical framework for analyzing the physical situations faced in different nanoelectronic devices. The solutions to these situations are obtained by determining the eigenfunctions or wave functions and the eigenvalues. Chapter 4 is concerned with operator methods in quantum mechanics. Chapter 5 is devoted to the treatment of free particles in quantum mechanics and the particles enclosed in boxes, subjected to predefined conditions, viz. the infinite potential well and finite potential well type circumstances. Many problems encountered in nanoelectronics fall into either one of these categories. So they form a mathematical basis to solve such problems. In addition, the non-termination of the wave function abruptly at the walls of the finite potential well and its continuation beyond the walls lead to the quantum-mechanical tunneling idea. Chapter 6 addresses the solution of Schrodinger equation for hydrogen atom and explains how the quantum numbers appear naturally during the solution of Schrodinger's equation. The solution is highly instructive and has a great value for understanding these numbers.

At this stage the reader habituated to the use of Newtonian mechanics becomes well-equipped with the contrasting quantum mechanics of subatomic particles, and so can easily describe electron motion. But most nanoelectronics devices are solid-state devices and so the properties of matter in solid-state need to be explored to progress further.

ii. Part II: Condensed matter physics for nanoelectronics provides the physics background for solid-state devices covering the main theoretical models, namely the Drude model, the Sommerfeld model and the Kronig-Penney model necessary to understand the operation of nanoelectronic devices. The three chapters comprising this part, viz. Chapters 7–9, deal respectively with the three models mentioned earlier.

Now the reader is geared up with the knowledge of mechanics of electrons and the behavior of electrons in solids.

iii. Part III: Electron behavior in nanostructures begins the application of knowledge acquired in previous two parts to study transport of electrons in nanostructures. This part contains Chapters 10–12 dealing elaborately with electrons in quantum dots (3D quantum confinement), quantum wires (2D quantum confinement), and quantum wells (1D quantum confinement), i.e. in the descending order of dimensions in quantum confinement.

iv. Part IV: The Green's function method paves the foundation for conceptual development of the quantum transport model for nanoelectronic device modeling. Chapters 13–15 in this part are dedicated to the Dirac delta function and Green's function, the finite differences method and derivation of self-energy equations, and the non-equilibrium Green's function approach.

v. Part V: Fabrication and characterization of nanostructures: This part contains Chapters 16 and 17, providing an opportunity to the reader to peep into how a silicon nanoelectronics laboratory or industry looks like.

vi. Part VI: Exemplar nanolectronic devices: Here examples of analysis of selected nanoelectronic devices are presented in Chapters 18–22, exposing the reader to the exciting outcomes of the applications of core nano concepts to produce devices capable of enhanced and new functionalities.

Parts (i) and (ii) build the infrastructural support upon which the superstructure comprising parts (iii)–(vi) resides. Stated in another way, the first two parts help in getting conversant with the physics and mathematical language to easily follow the developments in the last four parts. Thus the book provides a comprehensive coverage encompassing topics from the roots of nanoelectronics to its fruits.

1.5 Discussion and Conclusions

The salient characteristics of mesoscopic objects were highlighted. Nanoelectronics should not be misconnotated in a narrow perspective as simply a dimensional cutback to reach nanometer scale. Undoubtedly, the dimensional reduction uplifts operational speed and brings down the power consumption of devices and circuits. It also boosts the packing density of circuits. But the dimensional reduction does much more than this and has a far-reaching impact. Let us not overlook that it is the harbinger of many exclusive properties and phenomena. These phenomena comply with quantum-mechanical laws and fall under the solo province of nanoworld. They will not be exhibited nor will they occur without miniaturization. Utilization of these properties and phenomena to build novel devices is what nanoelectronics is concerned about. Thus a nanoelectronic device is not merely one with smaller geometrical features but a contrivance where the peculiarity of being small gives it the ability to perform certain tasks which are possible only because of this trait, and will not happen with a larger feature.

Illustrative Exercises

1.1 How is the electron radius calculated from the measured mass of proton? What is the value obtained? Please see Section 1.1.1

1.2 Compare the diameters of: nucleus of hydrogen atom, hydrogen atom, and hydrogen molecule. Please see Sections 1.1.2 and 1.1.3

1.3 Which of the following devices qualifies as a nanoelectronic device: (i) feature size = 150 nm, (ii) feature size= 75 nm, (iii) feature size = 300 nm? Device with feature size (ii)

1.4 According to the theory of relativity, an equivalence relation exists between the mass and energy of a particle. The rest energy E_0 of a particle is the product of its mass m_0 and the square of velocity of light (c^2). Find the energy equivalents of rest mass m_0 of electron and proton. Also express the Planck's constant in eV nm/c units.

Rest energy of electron

= Rest mass of electron $\times$ Square of velocity of light

$$(E_0)_{\text{Electron}} = (m_0)_{\text{Electron}} \times c^2$$

$$= 9.109 \times 10^{-31} \text{ kg} \times (2.99792 \times 10^8 \text{ m s}^{-1})^2$$

$$= \frac{9.109 \times 10^{-31} \text{ kg} \times (2.99792 \times 10^8 \text{ m s}^{-1})^2}{1.602 \times 10^{-19}}$$

$$= \frac{9.109 \times 8.9875 \times 10^4}{1.602} \text{ eV}$$

$$= 51.10 \times 10^4 \text{ eV} = 511 \times 10^3 \text{ eV} = 511 \text{ keV} \quad (1.17)$$

Hence,

$$(m_0)_{\text{Electron}} = (511/c^2) \text{ keV} \quad (1.18)$$

Similarly,

$$(E_0)_{\text{Protron}} = (m_0)_{\text{Protron}} \times c^2$$

$$= 1.6726 \times 10^{-27} \text{ kg} \times (2.99792 \times 10^8 \text{ m s}^{-1})^2$$

$$= \frac{1.6726 \times 10^{-27} \text{ kg} \times (2.99792 \times 10^8 \text{ m s}^{-1})^2}{1.602 \times 10^{-19}}$$

$$= \frac{1.6726 \times 8.9875 \times 10^8}{1.602} \text{ eV}$$

$$= 9.3836 \times 10^8 \text{ eV} = 938.36 \times 10^6 \text{ eV} = 938 \text{ MeV}$$

$$(1.19)$$

$$\therefore (m_0)_{\text{Protron}} = (938/c^2) \text{ MeV} \quad (1.20)$$

hc = Planck's constant $\times$ Velocity of light

$$= 6.62607 \times 10^{-34} \text{ J s} \times 2.99792 \times 10^8 \text{ m s}^{-1}$$

$$= \frac{6.62607 \times 10^{-34} \text{ J s} \times 2.99792 \times 10^8 \times 10^9 \text{ nm s}^{-1}}{1.602 \times 10^{-19} \text{ C}}$$

$$= \frac{6.62607 \times 2.99792 \times 10^2}{1.602} \text{ eV nm}$$

$$= 12.39977 \times 10^2 \text{ eV nm} = 1239.977 \text{ eV nm}$$

$$= 1240 \text{ eV nm} \quad (1.21)$$

$$\therefore h = (1240/c) \text{ eV nm} \quad (1.22)$$

1.5 (a) An electron with energy 8 eV encounters a potential barrier of height 10 eV and width 2 nm. Estimate its probability of tunneling using the energy equivalent of electron rest mass. What is the tunneling probability when the energy of the electron is increased to 9.5 eV and the barrier width is reduced to 0.5 nm?

(b) Solve the earlier problem using regular units.
(a) Here $r = 2$ nm, $m = (511/c^2)$ keV, $V = 10$ eV and $E = 8$ eV. Also, $h = (1240/c)$ eV nm

$$(P)_{r=2 \text{ nm}, E=8 \text{ eV}} = \exp\left\{ \frac{-4r\pi\sqrt{2m(V-E)}}{h} \right\}$$

$$= \exp\left\{ \frac{-4 \times 2 \text{ nm} \times 3.14159\sqrt{2 \times (511/c^2) \times 10^3 \text{ eV}(10-8) \text{ eV}}}{(1240/c) \text{ eV nm}} \right\}$$

$$= \exp\left\{ \frac{-25.1327\sqrt{2044000}}{1240} \right\}$$

$$= \exp(-28.977) = 2.6 \times 10^{-13} \quad (1.23)$$

For $r = 0.5$ nm, $m = (511/c^2)$ keV, $V = 10$ eV and $E = 9.5$ eV. Also, $h = (1240/c)$ eV nm

$$(P)_{r=0.5 \text{ nm}, E=9.5 \text{ eV}}$$

$$= \exp\left\{ \frac{-4 \times 0.5 \text{ nm} \times 3.14159\sqrt{2 \times (511/c^2) \times 10^3 \text{ eV}(10-9.5) \text{ eV}}}{(1240/c) \text{ eV nm}} \right\}$$

$$= \exp\left\{ \frac{-6.28318\sqrt{511000}}{1240} \right\}$$

$$= \exp(-3.62216)$$

$$= 0.0267 = 2.67 \times 10^{-2} \quad (1.24)$$

(b)

$$(P)_{r=2\,\text{nm},\,E=8\,\text{eV}} = \exp\left\{\frac{-4r\pi\sqrt{2m(V-E)}}{h}\right\}$$

$$= \exp\left\{\frac{-4\times2\times10^{-9}\times3.14159\sqrt{2\times9.109\times10^{-31}\,\text{kg}\times(10-8)\times1.6\times10^{-19}\,\text{J}}}{6.626\times10^{-34}\,\text{J}}\right\}$$

$$= \exp\left\{\frac{-25.1327\times10^{-9}\sqrt{2\times9.109\times10^{-31}\times2\times1.6\times10^{-19}}}{6.626\times10^{-34}}\right\}$$

$$= \exp\left\{\frac{-25.1327\times10^{-9}\sqrt{2\times9.109\times2\times1.6\times10^{-50}}}{6.626\times10^{-34}}\right\}$$

$$= \exp\left\{\frac{-25.1327\times10^{-9}\times7.6353\times10^{-25}}{6.626\times10^{-34}}\right\} = \exp\left\{\frac{-25.1327\times7.6353}{6.626}\right\}$$

$$= \exp(-28.961) = 2.64\times10^{-13} \tag{1.25}$$

Similarly, the probability can be found for $r = 0.5\,\text{nm}$ and $E = 9.5\,\text{eV}$.

REFERENCES

Atomic nucleus. https://simple.wikipedia.org/wiki/Atomic_nucleus, last changed on 30 June 2018.

Atomic radius of the elements. http://periodictable.com/Properties/A/AtomicRadius.v.html, © 2019 Wolfram.

Classical electron radius. https://en.wikipedia.org/wiki/Classical_electron_radius, last edited on 14 March 2019.

Das, M. P. 2010. Mesoscopic systems in the quantum realm: fundamental science and applications. *Advances in Natural Sciences: Nanoscience and Nanotechnology* 1: 043001.

Determining the electron structure. http://www.alternativephysics.org/book/ElectronStructure.htm, © 2010 Bernard Burchell

Eatemadi, A., H. Daraee, H. Karimkhanloo, M. Kouhi, N. Zarghami, A. Akbarzadeh, M. Abasi, Y. Hanifehpour and S. W. Joo. 2014 Carbon nanotubes: Properties, synthesis, purification, and medical applications. *Nanoscale Research Letters* 9(1): 393.

Ennen, I., D. Kappe, T. Rempel, C. Glenske and A. Hütten. 2016. Giant magnetoresistance: Basic concepts, microstructure, magnetic interactions and application. *Sensors* 16(904): 1–24.

Kinetic diameter. https://en.wikipedia.org/wiki/Kinetic_diameter, last edited on 28 September 2018.

Pauling, L. 1964. *College Chemistry*. W.H. Freeman and Company: San Francisco, CA, vol. 57, pp. 4–5.

Tsymbal, E. Y. and D. G. Pettifor. 2001. Perspectives of giant magnetoresistance. In: Ehrenreich H. and F. Spaepen (Eds.), Academic Press, vol. 56, pp. 113–237.

Tunneling. 2019. https://chem.libretexts.org/Bookshelves/Physical_and_Theoretical_Chemistry_Textbook_Maps/Supplemental_Modules_(Physical_and_Theoretical_Chemistry)/Quantum_Mechanics/02._Fundamental_Concepts_of_Quantum_Mechanics/Tunneling#mjx-eqn-prob.

Part I

Quantum Mechanics for Nanoelectronics

2

Origins of Quantum Theory

Epoch-making experiments, discoveries and hypotheses that culminated in the propounding of the quantum theory will be described.

2.1 Young's Double-Slit Experiment on Light Diffraction

Thomas Young was an English physicist and physician. Between 1801 and 1803, he delivered several lectures to the Royal Society approving of the wave theory of light (Ananthaswamy 2018). In Young's experiment (Figure 2.1), a point source of light illuminates two adjacent slits on a screen. The image of light emerging from the slits is observed on a second screen placed behind the first screen. The image is found to consist of light and dark bands called interference fringes. The explanation for appearance of fringes is that they are produced by constructive and destructive interference of light waves. The experiment cogently demonstrates that light has wave-like nature. The convincing argument showing that light is a wave is that if light had consisted of particles, it will not be diffracted, and there will be no interference phenomenon. Then we will observe only two stripes of light on the second screen for the two slits, not an interference pattern.

2.2 Blackbody Radiation and Planck's Quantum Hypothesis

A black body is an idealized concept of a body, which absorbs all the electromagnetic radiation falling on it, allowing none to be transmitted through or reflected from it (Planck 2013). Because such a body has a black appearance at room temperature, it is referred to as a black body. An example of a blackbody is a cavity with a tiny hole because any radiation entering the cavity through the hole suffers repeated reflections from the walls of the cavity until it is absorbed. Good absorbers being good emitters, it is evident that a blackbody emits the maximum radiation of all objects. The radiation emitted by a blackbody is known as blackbody radiation.

The total energy of radiation emitted from the surface of a blackbody does not depend on the material of which the body is made. The total energy emitted at all wavelengths is proportional to the fourth power of absolute temperature of the body (Stefan-Boltzmann law). As the temperature increases, the peak of the wavelength distribution of radiation shifts toward shorter wavelengths or higher frequencies (Wien's displacement law).

The distribution of energy from a blackbody is described in terms of the intensity $I(\lambda, T)$ defined as the power per unit area emitted in the wavelength interval $d\lambda$. In classical theory, it is given by the Rayleigh-Jeans law according to which the intensity of radiation is inversely proportional to fourth power of wavelength. For long wavelengths, the Rayleigh-Jeans law agrees well with experimental results. But this law implies that the intensity will be infinite as $\lambda \to 0$. In reality, the intensity $\to 0$ as $\lambda \to 0$. This gross disagreement between theory and experiment at small wavelengths is known as ultraviolet catastrophe. The word 'ultraviolet' is used because in the ultraviolet region, the wavelengths are short.

In 1900, the German theoretical physicist Max Karl Ernst Ludwig Planck developed a theory of blackbody radiation, which was in conformity with experiment at all wavelengths, low or high. The main idea of Planck's theory is that cavity energy arises from atomic resonators in the walls of the cavity. The energy of an oscillator cannot acquire continuous values but varies in a discrete fashion in steps of energy packets called quanta having energy

$$E = h\nu \tag{2.1}$$

where E is the energy, h is a constant, now called the Planck's constant, and ν is the frequency of radiation. The oscillators emit or absorb energy E when they undergo transition from one quantum state to another. Planck's idea marked the birth of the quantum theory. Planck received the Nobel Prize in physics in 1918 for his discovery of the energy quanta.

2.3 Photoelectric Effect

It is the emission of electrons from the surface of a material usually a metal when light falls on it (Figure 2.2). It is also called photoemission and the ejected electrons are known as photoelectrons. The photoelectrons are not different from ordinary electrons. The prefix 'photo' only means that they were emitted by the effect of light.

According to classical description of light as a wave, upon increasing the intensity of light, its energy or brightness rises, i.e. the amplitude of the light wave becomes large. Hence, greater number of photoelectrons should be emitted. This is found to be untrue because howsoever strong in intensity the light beam may be, there is no photoemission unless its frequency exceeds a threshold frequency or energy for the material. Feeble intensity light of high frequency may cause photoemission while high intensity light of low frequency is unable to do so. Thus classical theory was faced with contradictions and was unable to explain the observed photoelectric phenomena.

Albert Einstein was a German-born American theoretical physicist. In 1905, he proposed an explanation of the photoelectric effect. The explanation is based on Planck's quantum hypothesis about the constitution of light as being made of wave packets. These packets are termed photons or quanta (Willet 2005). To find the maximum kinetic energy KE_{max} of an electron ejected from the surface of a metal, we need two parameters,

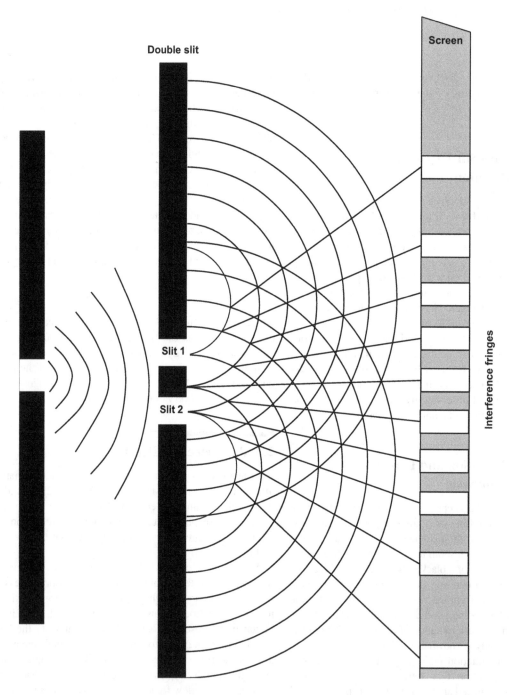

FIGURE 2.1 The double-slit experiment.

namely, the frequency ν_0 of the incident photon and the work function W of the metal defined as the minimum energy

$$W = h\nu_0 \qquad (2.2)$$

that must be expended for removal of an electron from the surface of the metal. Thus

$$KE_{\max} = h\nu - W = h\nu - h\nu_0 = h(\nu - \nu_0) \qquad (2.3)$$

Photoelectric effect occurs only when $\nu > \nu_0$. The 1921 Nobel Prize in physics was awarded to Einstein for his contributions to theoretical physics and particularly for discovery of the law of photoelectric effect.

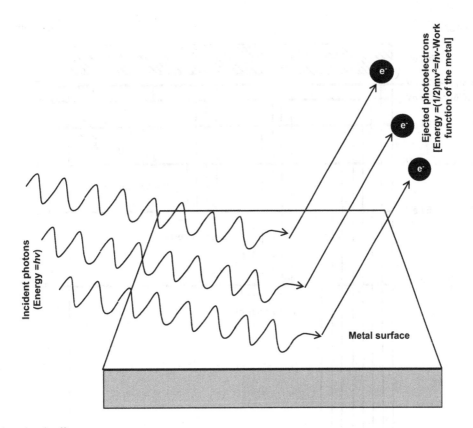

FIGURE 2.2 The photoelectric effect.

2.4 Emission Spectrum of Atomic Hydrogen and Bohr's Atomic Model

The spectrum of an element is obtained either by heating the element or exciting it by an electric current. When a high voltage ~5000 V is applied across a discharge tube containing hydrogen gas, the tube glows with a bright pink color. On passing this light through a prism/diffraction grating, it splits up into several lines at different wavelengths, which constitute the hydrogen spectrum (Figure 2.3). Each series of lines is named after its inventor, e.g. the Lyman series, Balmer series, Paschen series, Brackett series, Pfund series, Humphreys series. The wavelength λ of any line in the hydrogen spectrum is given by the Rydberg formula

$$\frac{1}{\lambda} = R_H \left(\frac{1}{n_1^2} - \frac{1}{n_2^2} \right) \tag{2.4}$$

where R_H is the Rydberg constant; n_1, n_2 are integers $n_2 > n_1$.

The line spectrum of hydrogen is explained from its atomic structure. In 1913, the Danish physicist Niels Henrik David Bohr proposed the atomic structure model (Figure 2.4) in which the atom consists of a central positively charged nucleus containing protons and neutrons around which negatively charged electrons revolve in circular orbits like the planets of the solar system revolve around the Sun (Kragh 2012). The gravitational force between the Sun and the planets holding the solar system together is replaced in the atomic model by the attractive Coulomb electrostatic force between the nucleus and electrons.

Bohr's model is instituted on several basic postulates or assumptions. Electrons can revolve around the nucleus in certain stipulated orbits only. So any arbitrarily chosen orbit is inadmissible. In revolving around the nucleus, the electrons do not radiate energy by acceleration. The permitted radiationless orbits are called stationary orbits. Each prescribed orbit is associated with a definite energy and is known as an energy shell or energy level. Energy shell refers to the surface formed by an allowed orbit in 3D space. The energy of an orbit is correlated with its size. The smallest orbit has the lowest energy. The energy linked with an orbit increases with its size.

Absorption or radiation of energy occurs when an electron makes a transition from one stationary orbit to another. By absorbing electromagnetic energy, an electron is promoted from a low energy orbit E_1 to a high energy orbit E_2. Radiation is emitted when the electron jumps from a high energy orbit E_2 to a low energy orbit E_1. The energy gained or lost ΔE in a transition equals the energy of a quantum of light:

$$\Delta E = E_2 - E_1 = h\nu \tag{2.5}$$

From inside and outward, the energy levels are designated by the integers 1, 2, 3, 4, … or alphabets K, L, M, N, …. The permissible orbits obey the condition that the angular momentum L of the electron is an integral multiple of $h/2\pi$, i.e.,

$$L = nh/(2\pi) \tag{2.6}$$

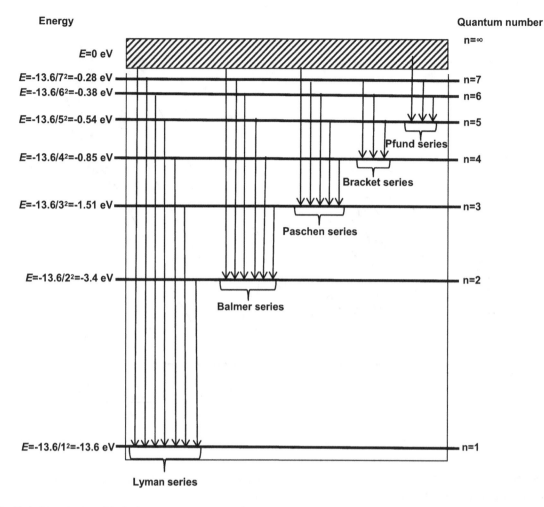

FIGURE 2.3   Emission spectrum of the hydrogen atom.

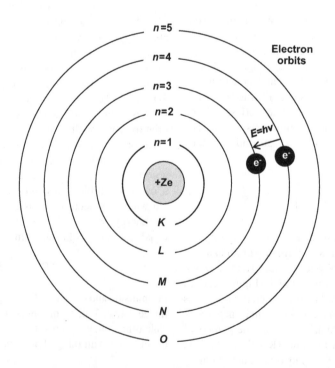

FIGURE 2.4   Bohr's model of the atom.

where n is an integer. It is known as the principal quantum number. Bohr was bestowed the Nobel Prize in physics in 1922. The prize was given for his investigation of atomic structure and the radiations emitted from atoms.

2.5 Compton Scattering

When a high-energy photon such as an X-ray or γ-ray photon strikes a target, e.g. carbon, containing free or loosely bound electrons on its outer shell, the incident photon is scattered (Wissmann 2004). The scattered photon has a longer wavelength or lower frequency, and hence lower energy than the incident photon. It is a case of inelastic scattering in which the kinetic energy of the incident particle is not preserved. A part of the energy/momentum of the photon is transferred to the recoiling electron. This effect was demonstrated by Arthur Holly Compton, an American physicist, in 1923 and is called the Compton effect or Compton scattering (Figure 2.5).

Compton's work persuasively validated that light consists of a stream of particles or photons. In Compton's experiment, some fraction of energy of the incident photon is imparted to the electron. Concurrently, a photon is released. Its direction of motion is different from the original photon while its energy equals the left over energy of the incident photon after giving energy to electron. Photon liberation maintains compliance with conservation of total momentum of the system.

In the low-energy limit, Compton scattering is observed in the form of classical Thomson scattering, which therefore signifies its low-energy boundary. Thomson scattering represents the elastic scattering experienced by electromagnetic waves under the influence of a free charged particle. In this scattering, the charged particle itself is accelerated. The acceleration is caused by the electric field of the incident wave. The charge thus accelerated emits dipolar electromagnetic radiation at the same frequency as the incident wave. In this way, the wave undergoes scattering. Thus there is no change in frequency or wavelength of the wave.

Crompton was awarded the Nobel Prize in physics in 1927 for his discovery of the Crompton effect in 1923, which showed that the electromagnetic radiation consisted of particles. He shared it with C.T. R. Wilson for his discovery of the cloud chamber showing the tracks of recoiling electrons.

2.6 de Broglie's Hypothesis

Louis-Victor Pierre Raymond de Broglie earned the Nobel Prize in physics in 1929 for discovering the wave nature of electrons (Broglie 1939). Figure 2.6 shows the electron moving as a wave in an atom.

Einstein's mass-energy equivalence states that any object having a mass m has an energy E related to its mass through the square of velocity of light (c^2) by the equation

$$E = mc^2 \tag{2.7}$$

Conversely any object having energy E has a corresponding mass m given by

$$m = \frac{E}{c^2} \tag{2.8}$$

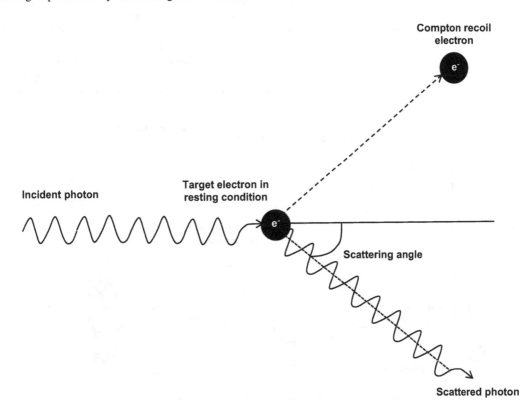

FIGURE 2.5 The Compton effect.

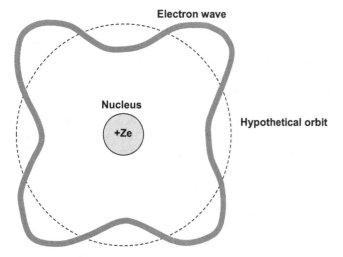

FIGURE 2.6 The electron wave in the atom.

According to Planck's hypothesis, energy E of radiation and its frequency ν are connected through the Planck's constant h as

$$E = h\nu \qquad (2.9)$$

Looking at eqs. (2.7) and (2.9), de Broglie wrote in 1924

$$mc^2 = h\nu \qquad (2.10)$$

de Broglie argued that real particles do not move with the velocity of light. So velocity of light must be substituted by the speed s of the particle giving

$$ms^2 = h\nu \qquad (2.11)$$

The frequency ν of radiation can be expressed in terms of speed s and wavelength λ as

$$\nu = \frac{s}{\lambda} \qquad (2.12)$$

Substituting for ν from eq. (2.11) in eq. (2.12) we get

$$ms^2 = h\frac{s}{\lambda} \qquad (2.13)$$

or

$$\lambda = \frac{h}{ms} = \frac{h}{p} \qquad (2.14)$$

where

$$p = ms \qquad (2.15)$$

is the momentum of the particle. The equation

$$\lambda = \frac{h}{p} \qquad (2.16)$$

relating the wavelength λ (a property of a wave) with momentum p (a property of a particle) is called de Broglie's equation.

2.7 Davisson-Germer Experiment on Electron Diffraction

During 1923–1927, American physicists Clinton Joseph Davisson and Lester Germer discovered experimentally that an electron beam is diffracted by a crystal. They designed a vacuum apparatus to measure the energies of electrons scattered by the surface of a metal. In this apparatus (Figure 2.7), the electron gun consists of a barium oxide-coated tungsten filament heated by a low voltage power supply. The electrons released from the gun by thermionic emission are accelerated by an anode to which a high voltage is applied. The anode is a cylinder with a narrow opening along its axis producing a fine collimated electron beam. This electron beam strikes a target of single-crystal nickel and the diffracted electrons fall on a Faraday box, used as an

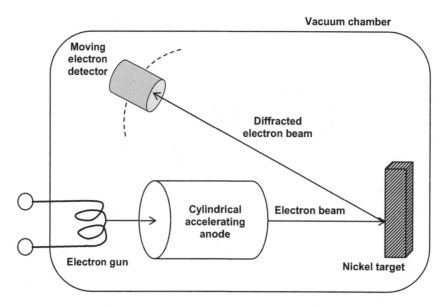

FIGURE 2.7 Setup of the Davisson-Germer experiment.

electron detector and connected to a galvanometer. The detector moves on a semi-circular arc so that the intensity of the electron beam can be measured for various values of the angle of scattering. When the accelerating voltage is raised from 44 to 68 V, a conspicuous peak in the intensity of scattered electrons is noticed. This peak is found at a voltage of 54 V (electron kinetic energy = 54 eV), and scattering angle $\phi = 50°$. It is interpreted in terms of the constructive interference of electron waves from the regularly spaced periodic arrangement of atoms in different planes of the nickel crystal, which acts as a reflective diffraction grating (Suzuki and Suzuki 2013).

Considering electrons as waves, the Bragg's law for X-ray diffraction is

$$n\lambda = 2d \sin\theta \qquad (2.17)$$

In this equation, n is an integer, λ is the wavelength of electron wave, d denotes the spacing between atomic planes in the nickel crystal, and θ represents the complement of the angle of incidence of the incoming electron beam. Since

$$\theta = 90° - \frac{\phi}{2} = 90° - \frac{50°}{2} = 90° - 25° = 65° \qquad (2.18)$$

from the known spacing of atoms in the Ni crystal $= 0.91 \times 10^{-10}$ m, taking $n = 1$, the wavelength of electron waves is found to be

$$\lambda = 2d \sin\theta = 2 \times 0.91 \times 10^{-10} \sin 65° \text{ m}$$

$$= 1.82 \times 10^{-10} \times 0.906 \text{ m} = 1.6489 \times 10^{-10} \text{ m}$$

$$= 0.165 \text{ nm} \qquad (2.19)$$

Clinton Joseph Davisson and George Paget Thomson jointly received the Nobel Prize in physics in 1937 for their experimental discovery of electron diffraction by crystals.

Controlled electron diffraction through a double slit was shown by Bach et al. in 2013. In their experimental setup (Figure 2.8), an electron beam produced by a thermionic tungsten filament and electrostatic lenses is collimated and passed through a wall having two slits. Probability distributions P_1, P_2, and P_{12} are observed for electrons passing through slit 1 open (with slit 2 closed), slit 2 open (with slit 1 closed), and through both slits 1 and 2 open. The patterns are magnified by a quadrupole lens. They are imaged on a 2-D microchannel plate and phosphor screen. Then they are recorded by a CCD camera. In this way, the wave properties of electrons are demonstrated.

Further, a diffraction pattern is constructed by recording single electron detection events for diffraction through a double slit. Hence the particle behavior of electrons is seen. Thus both wave and particle properties of electrons are established by this experiment.

2.8 Heisenberg's Uncertainty Principle

The German physicist Werner Karl Heisenberg asserted that a fundamental limit exists on the precision with which the pairs of certain properties of a particle such as momentum and position or energy and time can be simultaneously measured (Lindley 2008). The minimum value for the product of uncertainties of momentum (Δp) and position (Δx) is half the reduced Planck's constant $= \dfrac{\hbar}{2} = \dfrac{\frac{h}{2\pi}}{2} = \dfrac{h}{4\pi}$ where h is Planck's constant, so that

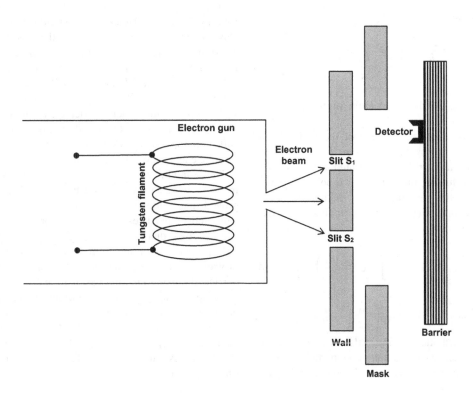

FIGURE 2.8 Controlled electron diffraction experiment using a double slit.

$$\Delta p \Delta x \geq \frac{\hbar}{2} \qquad (2.20)$$

and similarly

$$\Delta E \Delta t \geq \frac{\hbar}{2} \qquad (2.21)$$

By de Broglie's hypothesis, every particle is associated with a wave. How to find the position of the particle on this wave? The probability of finding the particle is maximum at a location of the wave where the undulations are most severe, i.e. accuracy of position is represented by intensity of wave undulations. But at the region of high intensity of undulations of the wave, its wavelength becomes more ill-defined. By de Broglie's equation, the wavelength = h/momentum of the particle. So the momentum of the particle is ill-defined. Thus we arrive at one part of uncertainty principle that exact determination of position leads to inexactness in finding the momentum.

Let us move the reverse way. At the position of the wave where undulations are less intense, wavelength and hence momentum are easily discernible. But the position of the particle is now not known precisely because the undulations being spread out, the chances of particle's presence become remote. So we get the second part of uncertainty principle in which exactness of momentum comes at the expense of inexactness of position.

Elaborating on the earlier thoughts, the particle is described by a wave function $\Psi(x)$, which for a single-moded plane wave of wave number k or momentum p is

$$\Psi(x) \propto \exp(ikx) = \exp\left(\frac{ipx}{\hbar}\right) \qquad (2.22)$$

because the wave number

$$k = \frac{2\pi}{\lambda} = \frac{2\pi}{\dfrac{h}{p}} = \frac{2\pi p}{h} = \frac{p}{\dfrac{h}{2\pi}} = \frac{p}{\hbar} \qquad (2.23)$$

by applying eq. (2.16). So the momentum of the particle is accurately known.

Let us now enquire about the position of the particle. The Born rule was put forward by Max Born in 1926. Max Born was a German physicist. His rule is enunciated as follows: The probability density of finding the particle at a point is proportional to the square of the wave function of the particle at that point. Hence the probability $P(x_1 < x < x_2)$ of finding the particle between points x_1 and x_2 is

$$P(x_1 < x < x_2) = \int_{x_1}^{x_2} |\Psi(x)|^2 \, dx = \int_{x_1}^{x_2} \left|\exp\left(\frac{ipx}{\hbar}\right)\right|^2 dx \quad (2.24)$$

Since the squared wave function $\left|\exp\left(\dfrac{ipx}{\hbar}\right)\right|^2$ represents a uniform or continuous distribution, all intervals of the same length are equally probable. Hence the position of the particle is highly uncertain because it can be found anywhere over all space.

Thus starting from accurately known momentum of the particle, we find ourselves unable to know its position exactly.

To know the position with greater accuracy, let us consider the wave function of a wave packet equivalent to a sum of waves written as

$$|\Psi(x)|_{\text{Packet}} \propto \sum_n P_n \exp\left(\frac{ip_n x}{\hbar}\right) \qquad (2.25)$$

where multiplicative factor P_n specifies the relative contribution of the mode p_n to the sum-total. At locations in the wave packet, where crests of two waves or troughs of two waves coincide, the amplitude is maximum (constructive interference) whereas at locations of crest-trough or trough-crest coincidence, the amplitude is zero (destructive interference). This interference pattern localizes the wave. It is possible to synthesize a wave packet from an infinite set of waves of different momenta such that constructive interference occurs only over a small region in space whereas in all the remaining regions, the interference is destructive, thus producing a short burst of localized wave action traveling as a whole unit. As the wave packet is extremely localized, the position of the particle becomes known with greater certainty. On the opposite side, the momentum of the wave packet obtained by the mixture of several waves of different momenta loses certainty. So the increase in accuracy of position is achieved at cost of losing accuracy in momentum.

Werner Karl Heisenberg was awarded the Nobel Prize in physics in 1932 for the creation of quantum mechanics.

2.9 Discussion and Conclusions

Young's experiment corroborated the wave nature of light refuting Newton's theory of corpuscular constitution of light. Davisson and Germer came up with the experimental proof that electrons, normally supposed to be particles, experienced diffraction like waves. Crompton's experiment unraveled that light consists of particles. Planck's hypothesized that radiation was emitted or absorbed discretely as quanta. Einstein too came up with the proposition about particulate nature of light explaining photoelectric emission by considering light as made of particles called photons. Bohr's atomic model was based on occupation of discrete energy levels by electrons.

These landmark experiments and path-breaking theories, many of which were honored by Nobel Prize, unequivocally proved that matter behaved like waves and radiation or light consisted of particles. This realization came to be known as wave-particle duality in nature. de Broglie connected the particle property of momentum with the wavelength parameter of waves. Heisenberg showed that a consequence of the wave-particle dualism in nature was that the product of uncertainties of complementary pairs of quantities such as momentum and position of a particle always exceeded a minimum limit which was Planck's constant divided by 4π. These findings mounted quantum theory on an unshakable throne from which it flourished flawlessly in the nanoworld in which nanoelectronics is a tiny kingdom.

Illustrative Exercises

2.1 The saturation velocity of electron in gallium arsenide is 1.2×10^7 cms^{-1}. Find the wavelength of the electron moving at this velocity.

$$\lambda = \frac{h}{ms} = \frac{6.626 \times 10^{-34} \text{ J s}}{9.109 \times 10^{-31} \text{ kg} \times \frac{1.2 \times 10^7}{100} \text{ m s}^{-1}}$$

$$= \frac{6.626 \times 10^{-8}}{9.109 \times 1.2} = 6.062 \times 10^{-9} \text{ m} \approx 6.1 \text{ nm} \quad (2.26)$$

2.2 A photon is a massless particle. Does it have a momentum? If yes, how much?

A photon has zero rest mass m_0. According to relativistic mechanics, the energy E of a photon is related to its momentum p as

$$E = \sqrt{m_0 c^2 + p^2 c^2} \quad (2.27)$$

where c is the velocity of light. Since $m_0 = 0$ for photon,

$$E = \sqrt{0 + p^2 c^2} = pc \quad (2.28)$$

or

$$p = \frac{E}{c} \quad (2.29)$$

But

$$E = h\nu = h\frac{c}{\lambda} \quad (2.30)$$

where ν is the frequency of photon and λ is its wavelength. So,

$$p = \frac{E}{c} = \frac{1}{c} \times h\frac{c}{\lambda} = \frac{h}{\lambda} \quad (2.31)$$

or

$$p = \frac{h}{\lambda} \quad (2.32)$$

i.e.,

$$\text{Momentum of the photon} = \frac{\text{Planck's constant}}{\text{Wavelength of the photon}} \quad (2.33)$$

2.3 How many green light photons of wavelength 550 nm will be required to produce 1 J energy? How many red light photons of wavelength 700 nm will be able to do the same?

Energy of a green light photon is

$$E_{\text{Green}} = \frac{hc}{\lambda} = \frac{6.626 \times 10^{-34} \times 3 \times 10^8}{550 \times 10^{-9}} = 3.614 \times 10^{-19} \text{ J} \quad (2.34)$$

Number of green light photons required is

$$N_{\text{Green}} = \frac{1 \text{ J}}{3.614 \times 10^{-19} \text{ J}} = 2.767 \times 10^{18} \quad (2.35)$$

Energy of a red light photon is

$$E_{\text{Green}} = \frac{hc}{\lambda} = \frac{6.626 \times 10^{-34} \times 3 \times 10^8}{700 \times 10^{-9}} = 2.8397 \times 10^{-19} \text{ J} \quad (2.36)$$

Number of red light photons required is

$$N_{\text{Green}} = \frac{1 \text{ J}}{2.8397 \times 10^{-19} \text{ J}} = 3.5215 \times 10^{18} \quad (2.37)$$

REFERENCES

Ananthaswamy, A. 2018. *Through Two Doors at Once: The Elegant Experiment that Captures the Enigma of Our Quantum Reality, Dutton*. An Imprint of Penguin Random House LLC: New York, 304 pp.

Bach, R., D. Pope, S.-H. Liou and H. Batelaan. 2013. Controlled double-slit electron diffraction. *New Journal of Physics* 15: 033018.

Broglie, L. D. 1939. *Matter and Light: The New Physics*. W. W. Norton & Co., Inc.: New York, 300 pp.

Kragh, H. 2012. *Niels Bohr and the Quantum Atom: The Bohr Model of Atomic Structure 1913–1925*, 1st Edition. Oxford University Press: Oxford, UK, 416 pp.

Lindley, D. 2008. *Uncertainty: Einstein, Heisenberg, Bohr, and the Struggle for the Soul of Science, Anchor Books*. Random House, Inc.: New York, 272 pp.

Planck, M. 2013. *The Theory of Heat Radiation*. Dover Publications, Inc.: New York, 256 pp.

Suzuki, M. and I. S. Suzuki. 2013 July. A proper understanding of the Davisson and Germer experiments for undergraduate modern physics course, arXiv:1307.6049v1 [physics.ed-ph], pp. 1–8.

Willet, E. 2005. *The Basics of Quantum Physics: Understanding the Photoelectric Effect and Line Spectra*. The Rosen Publishing Group, Inc, New York, 48 pp.

Wissmann, F. 2004. *Investigating the Structure of the Nucleons with Real Photons*. Springer-Verlag: Berlin, 156 pp.

3

The Schrodinger Wave Equation

3.1 Two Forms of Schrodinger Equation

Schrodinger's equation is a second order (involving second derivative as the highest derivative), linear (with unknown functions and their derivatives of first degree) partial differential equation (containing unknown multivariable functions and their partial derivatives) derived in 1925 and published in 1926 by Erwin Rudolf Josef Alexander Schrödinger, a theoretical physicist from Austria (Schrodinger 1926). Schrodinger was awarded the Nobel Prize in 1933 for this epoch-making work which is a fundamental equation of physics playing the same role in quantum mechanics as Newton's laws together with conservation of energy have in classical mechanics.

The equation is presented in two forms: the time-independent Schrodinger equation (TISE) and time-dependent Schrodinger equation (TDSE), although the two forms are not separate and the time-independent form is derivable from the time-dependent form (Schiff 1968, Schleich et al 2013). The TISE describes the allowed energies of a particle while the TDSE describes how the wave function of the particle evolves with respect to time. The wave function is a complex-valued function that describes the wave characteristics associated with a particle.

3.1.1 Time-Independent Schrodinger Equation (TISE)

Along the X-direction,

$$-\left(\frac{\hbar^2}{2m}\right)\frac{\partial^2}{\partial x^2}\Psi(x) + V(x)\Psi(x) = E\,\Psi(x) \tag{3.1}$$

where

$$\hbar = \frac{h}{2\pi} = \frac{\text{Planck's constant}}{2\pi} \tag{3.2}$$

m is the mass of the particle, $\Psi(x, t)$ is the wave function defined with respect to position x and time t, $V(x)$ is the potential energy of the particle at position x, and E is the total energy of the particle.

In three dimensions, the TISE is written using three partial spatial derivatives as

$$-\left(\frac{\hbar^2}{2m}\right)\left(\frac{\partial^2}{\partial x^2} + \frac{\partial^2}{\partial y^2} + \frac{\partial^2}{\partial z^2}\right)\Psi(x,y,z) + V(x,y,z)\Psi(x,y,z)$$

$$= E\Psi(x,y,z) \tag{3.3}$$

or

$$-\left(\frac{\hbar^2}{2m}\right)\nabla^2\Psi(x,y,z) + V(x,y,z)\Psi(x,y,z) = E\Psi(x,y,z) \tag{3.4}$$

where

$$\nabla^2 = \text{del squared} = \text{Laplacian operator} \tag{3.5}$$

3.1.2 Time-Dependent Schrodinger Equation (TDSE)

Along the X-direction,

$$-\left(\frac{\hbar^2}{2m}\right)\frac{\partial^2}{\partial x^2}\Psi(x, t) + V(x)\Psi(x, t) = i\hbar\frac{\partial}{\partial t}\Psi(x, t) \tag{3.6}$$

In three dimensions,

$$-\left(\frac{\hbar^2}{2m}\right)\nabla^2\Psi(x,y,z, t) + V(x,y,z)\Psi(x,y,z, t) = i\hbar\frac{\partial}{\partial t}\Psi(x,y,z, t)$$

$$\tag{3.7}$$

3.2 Formulation of Time-Independent Schrodinger Equation (TISE)

Schrodinger equation is axiomatic, self-evident and manifest, accepted and obvious. It cannot be derived. It is statement of principle of conservation of energy from the viewpoint of quantum mechanics (Barde et al 2015).

The total energy of the particle is

$$\text{Total energy}(E) = \text{Kinetic energy}(K) + \text{Potential energy}(V) \tag{3.8}$$

or

$$E = K + V = \left(\frac{1}{2}\right)mv^2 + V = \left(\frac{1}{2}\right)m \times \left(\frac{p}{m}\right)^2 + V = \frac{p^2}{2m} + V \tag{3.9}$$

$$\because p = mv \text{ or } v = \frac{p}{m} \tag{3.10}$$

The symbols v and p stand for the velocity of the particle and its momentum, respectively.

The particle is contemplated as a wave. The equation of this wave is written as

$$\Psi(x,t) = \cos(kx - \omega t) + i\sin(kx - \omega t) \qquad (3.11)$$

where k is the propagation constant or angular wave number or circular wave number, often simply called the wave number of the wave defined as

$$k = \frac{\text{Number of radians}}{\text{Distance}} = \frac{2\pi}{\lambda} \qquad (3.12)$$

and ω is the angular frequency

ω = Frequency expressed in radians per second

= $2\pi \times$ Regular or linear frequency in cycles per second (ν)

= $2\pi\nu$

$$\qquad (3.13)$$

By De Moivre's theorem,

$$\Psi(x,t) = \exp i(kx - \omega t) \qquad (3.14)$$

At this juncture, an obvious question arises. How do we obtain eq. (3.11) or eq. (3.14)? To answer this question, we note that in classical mechanics, the evolution of a wave is described by an equation of motion called the wave equation. It is a differential equation which yields the shape of the wave, e.g. by giving its displacement $y(x, t)$ in the Y-direction at every point x and every time instant t. The solution of this equation is known as the wave function. To clarify understanding, we shall derive the general equation of wave motion in classical mechanics in Section 3.5 and solve it in Section 3.6.

Wavelength associated with the particle is given by de Broglie's formula

$$\lambda = \frac{h}{p} \qquad (3.15)$$

$$\therefore k = \frac{2\pi}{\lambda} = \frac{2\pi}{h/p} = \frac{p}{h/(2\pi)} = \frac{p}{\hbar} \qquad (3.16)$$

$$\because \hbar = \frac{h}{2\pi} \qquad (3.17)$$

Taking the first partial derivative of the wave function $\Psi(x, t)$ with respect to x, keeping time t constant

$$\frac{\partial \Psi(x,t)}{\partial x} = \frac{\partial}{\partial x}\left\{ \exp i(kx - \omega t) \right\} = ik\exp i(kx - \omega t) \qquad (3.18)$$

Taking the second partial derivative of the wave function $\Psi(x, t)$, we have

$$\frac{\partial^2 \Psi(x,t)}{\partial x^2} = \frac{\partial}{\partial x}\left\{ ik\exp i(kx - \omega t)\right\} = ik \times ik \exp i(kx - \omega t)$$

$$= i^2 k^2 \Psi(x,t) = -k^2 \Psi(x,t) \qquad (3.19)$$

$$\because i^2 = \left(\sqrt{-1}\right)^2 = -1 \qquad (3.20)$$

Hence,

$$\frac{\partial^2 \Psi(x,t)}{\partial x^2} = -k^2 \Psi(x,t) = -\left(\frac{p}{\hbar}\right)^2 \Psi(x,t) = -\frac{p^2 \Psi(x,t)}{\hbar^2} \qquad (3.21)$$

From eq. (3.9)

$$E\Psi(x,t) = \left(\frac{p^2}{2m} + V\right)\Psi(x,t) = \frac{p^2 \Psi(x,t)}{2m} + V\Psi(x,t) \qquad (3.22)$$

or

$$\frac{p^2 \Psi(x,t)}{2m} = E\Psi(x,t) - V\Psi(x,t) = (E - V)\Psi(x,t) \qquad (3.23)$$

or

$$p^2 \Psi(x,t) = 2m(E - V)\Psi(x,t) \qquad (3.24)$$

Substituting for $p^2\Psi(x, t)$ from eq. (3.24) in eq. (3.21)

$$\frac{\partial^2 \Psi(x,t)}{\partial x^2} = -\frac{p^2 \Psi(x,t)}{\hbar^2} = -\frac{2m(E - V)\Psi(x,t)}{\hbar^2} \qquad (3.25)$$

or

$$\frac{\partial^2 \Psi(x,t)}{\partial x^2} + \frac{2m(E - V)\Psi(x,t)}{\hbar^2} = 0 \qquad (3.26)$$

Multiplying both sides of eq. (3.26) by $-\hbar^2/2m$,

$$\left(-\frac{\hbar^2}{2m}\right)\frac{\partial^2 \Psi(x,t)}{\partial x^2} + \left(\frac{-\hbar^2}{2m}\right)\frac{2m(E - V)\Psi(x,t)}{\hbar^2} = 0 \qquad (3.27)$$

or

$$\left(-\frac{\hbar^2}{2m}\right)\frac{\partial^2 \Psi(x,t)}{\partial x^2} - (E - V)\Psi(x,t) = 0 \qquad (3.28)$$

or

$$\left(-\frac{\hbar^2}{2m}\right)\frac{\partial^2 \Psi(x,t)}{\partial x^2} - E\Psi(x,t) + V\Psi(x,t) = 0 \qquad (3.29)$$

or

$$\left(-\frac{\hbar^2}{2m}\right)\frac{\partial^2 \Psi(x,t)}{\partial x^2} + V\Psi(x,t) = E\Psi(x,t) \qquad (3.30)$$

which is TISE.

3.3 Formulation of Time-Dependent Schrodinger Equation (TDSE)

Finding the partial derivative of wave function $\Psi(x, t)$ with respect to t keeping x constant, we get

$$\frac{\partial \Psi(x, t)}{\partial t} = \frac{\partial}{\partial t}\left\{\exp i(kx - \omega t)\right\} = -i\omega \exp i(kx - \omega t)$$

$$= -i\omega\, \Psi(x, t) \tag{3.31}$$

$$\because \exp i(kx - \omega t) = \Psi(x, t) \tag{3.32}$$

Energy of a photon of frequency ν is

$$E = h\nu \tag{3.33}$$

which can be written as

$$E = \frac{h}{2\pi} \times 2\pi\nu = \hbar\omega \tag{3.34}$$

$$\because \frac{h}{2\pi} = \hbar \text{ and } 2\pi\nu = \omega \tag{3.35}$$

Equation (3.34) gives

$$E\Psi(x, t) = \hbar\omega\Psi(x, t) \tag{3.36}$$

Multiplying both sides of eq. (3.36) by $-i/\hbar$

$$\left(-\frac{i}{\hbar}\right) \times E\Psi(x, t) = \left(-\frac{i}{\hbar}\right) \times \hbar\omega\Psi(x, t) = -i\omega\Psi(x, t) \tag{3.37}$$

Comparing eqs. (3.31) and (3.37)

$$\frac{\partial \Psi(x, t)}{\partial t} = \left(-\frac{i}{\hbar}\right) \times E\Psi(x, t) \tag{3.38}$$

or

$$\left(-\frac{\hbar}{i}\right)\frac{\partial \Psi(x, t)}{\partial t} = E\Psi(x, t) \tag{3.39}$$

or

$$\left(-\frac{\hbar}{i} \times \frac{i}{i}\right)\frac{\partial \Psi(x, t)}{\partial t} = E\Psi(x, t) \tag{3.40}$$

or

$$\left(-\frac{i\hbar}{i^2}\right)\frac{\partial \Psi(x, t)}{\partial t} = E\Psi(x, t) \tag{3.41}$$

or

$$\left(-\frac{i\hbar}{-1}\right)\frac{\partial \Psi(x, t)}{\partial t} = E\Psi(x, t) \tag{3.42}$$

or

$$i\hbar\frac{\partial \Psi(x, t)}{\partial t} = E\Psi(x, t) \tag{3.43}$$

But from TISE,

$$E\Psi(x, t) = -\left(\frac{\hbar^2}{2m}\right)\frac{\partial^2}{\partial x^2}\Psi(x, t) + V(x)\Psi(x, t) \tag{3.44}$$

where Ψ is taken as a function of x and t.

$$\therefore i\hbar\frac{\partial \Psi(x, t)}{\partial t} = -\left(\frac{\hbar^2}{2m}\right)\frac{\partial^2}{\partial x^2}\Psi(x, t) + V(x)\Psi(x, t) \tag{3.45}$$

which is TDSE.

3.4 TISE from TDSE: Method of Separation of Variables

Assuming the existence of separability of variables x and t in the wave function $\Psi(x, t)$, we can write

$$\Psi(x, t) = X(x)T(t) \tag{3.46}$$

where X is a function of x only and T is a function of t only.

Substituting for $\Psi(x, t)$ in TDSE (eq. (3.45)),

$$i\hbar\frac{\partial}{\partial t}\left\{X(x)T(t)\right\} = -\left(\frac{\hbar^2}{2m}\right)\frac{\partial^2}{\partial x^2}\left\{X(x)T(t)\right\}$$
$$+ V(x)\left\{X(x)T(t)\right\} \tag{3.47}$$

or

$$\left\{i\hbar\frac{\partial T(t)}{\partial t}\right\}X(x) = \left\{-\left(\frac{\hbar^2}{2m}\right)\frac{\partial^2 X(x)}{\partial x^2}\right\}T(t) + V(x)\left\{X(x)T(t)\right\}$$
$$= \left\{-\left(\frac{\hbar^2}{2m}\right)\frac{\partial^2 X(x)}{\partial x^2} + V(x)X(x)\right\}T(t) \tag{3.48}$$

We divide both sides of eq. (3.48) by $X(x)T(t)$. Also, we can replace the partial derivatives with respect to x and t with full derivatives because $X = X(x)$ and $T = T(t)$; hence

$$\left\{i\hbar\frac{dT(t)}{dt}\right\}\frac{X(x)}{X(x)T(t)} = \left\{-\left(\frac{\hbar^2}{2m}\right)\frac{d^2 X(x)}{dx^2} + V(x)X(x)\right\}$$
$$\times \frac{T(t)}{X(x)T(t)} \tag{3.49}$$

or

$$\frac{1}{T(t)}\left\{i\hbar\frac{dT(t)}{dt}\right\} = \frac{1}{X(x)}\left\{-\left(\frac{\hbar^2}{2m}\right)\frac{d^2X(x)}{dx^2} + V(x)X(x)\right\}$$

$$= -\frac{1}{X(x)}\left(\frac{\hbar^2}{2m}\right)\frac{d^2X(x)}{dx^2} + V(x) \qquad (3.50)$$

or

$$\frac{1}{T(t)}\left\{i\hbar\frac{dT(t)}{dt}\right\} = -\frac{1}{X(x)}\left(\frac{\hbar^2}{2m}\right)\frac{d^2X(x)}{dx^2} + V(x) \quad (3.51)$$

In this equation, the left-hand side (LHS) depends only on variable t whereas right-hand side (RHS) depends only on the position coordinate x. Therefore, the LHS and RHS must be equal to a constant, which is neither t-dependent nor x-dependent. Let us look at this equation dimensionally. In LHS, dimension of $1/T(t)$ outside the curly bracket is canceled by that of $T(t)$ in $dT(t)/dt$, $\hbar$ has the dimension [Energy × Time] while d/dt has the dimension [1/Time]. So, LHS has the dimension

$$[\text{LHS}] = [\text{Energy} \times \text{Time}] \times [1/\text{Time}] = [\text{Energy}] \qquad (3.52)$$

In the first term of RHS, dimension of $X(x)$ in the denominator will cancel with that of $X(x)$ in $d^2X(x)/dx^2$, the dimension of $\hbar$ is [Energy × Time], that of mass is [Mass] while $d^2X(x)/dx^2$ has the dimension [1/Length^2]. Therefore,

$$[\text{RHS}] = \frac{[\text{Energy} \times \text{Time}]^2}{[\text{Mass}]} \times \frac{1}{[\text{Length}^2]}$$

$$= \frac{[\text{Energy}]^2}{[\text{Mass}] \times \dfrac{[\text{Length}^2]}{[\text{Time}^2]}} = \frac{[\text{Energy}]^2}{[\text{Mass}] \times \dfrac{[\text{Length}]}{[\text{Time}^2]} \times [\text{Length}]}$$

$$= \frac{[\text{Energy}]^2}{[\text{Mass}] \times [\text{Acceleration}] \times [\text{Distance}]}$$

$$= \frac{[\text{Energy}]^2}{[\text{Force}] \times [\text{Distance}]} = \frac{[\text{Energy}]^2}{[\text{Energy}]} = [\text{Energy}] \qquad (3.53)$$

The second term of RHS is potential energy, which obviously has the dimension [Energy]. Therefore,

$$[\text{LHS}] = [\text{RHS}] = [\text{Energy}] \qquad (3.54)$$

Hence both sides of eq. (3.51) can be represented by the total energy E. So,

$$\frac{1}{T(t)}\left\{i\hbar\frac{dT(t)}{dt}\right\} = -\frac{1}{X(x)}\left(\frac{\hbar^2}{2m}\right)\frac{d^2X(x)}{dx^2} + V(x) = E \quad (3.55)$$

Consequently,

$$\frac{1}{T(t)}\left\{i\hbar\frac{dT(t)}{dt}\right\} = E \qquad (3.56)$$

or

$$i\hbar\frac{dT(t)}{dt} = ET(t) \qquad (3.57)$$

Also,

$$-\frac{1}{X(x)}\left(\frac{\hbar^2}{2m}\right)\frac{d^2X(x)}{dx^2} + V(x) = E \qquad (3.58)$$

or

$$-\left(\frac{\hbar^2}{2m}\right)\frac{d^2X(x)}{dx^2} + V(x)X(x) = EX(x) \qquad (3.59)$$

which is TISE with $X(x)$ = time-independent part of $\Psi(x, t) = \Psi(x)$. Thus

$$-\left(\frac{\hbar^2}{2m}\right)\frac{d^2\Psi(x)}{dx^2} + V(x)\Psi(x) = E\Psi(x) \qquad (3.60)$$

3.5 Deriving the General Equation for Wave Motion

To develop a mathematical description for wave motion, we note that the vertical displacement $y(x, t)$ of a particle in the Y-direction at an instant of time t is interconnected with its position x in the horizontal X-direction, i.e. $y(t)$ can be expressed as a function of x. Two possibilities arise: wave moving toward the right (positive X-axis) and wave moving toward the left (negative X-axis) (Figure 3.1).

A wave traveling in the positive direction along the X-axis is given by

$$y(x,t) = r(x - ct) \qquad (3.61)$$

where c is the phase velocity of wave, i.e. velocity of a particle of constant phase.

To understand why eq. (3.61) represents a wave moving in the positive direction along the X-axis, we note that after an instant of time Δt, the distance traveled by the wave is $c\Delta t$. Then the function becomes

$$y(x,t) = r\{x - c(t + \Delta t)\} = r(x - ct - c\Delta t) \qquad (3.62)$$

When the function changes, it no longer defines the same point on the wave. However, it can be restored to the previous value by applying a positive increment $+ c\Delta t$,

$$y(x,t) = r(x - ct - c\Delta t + c\Delta t) = r(x - ct) \qquad (3.63)$$

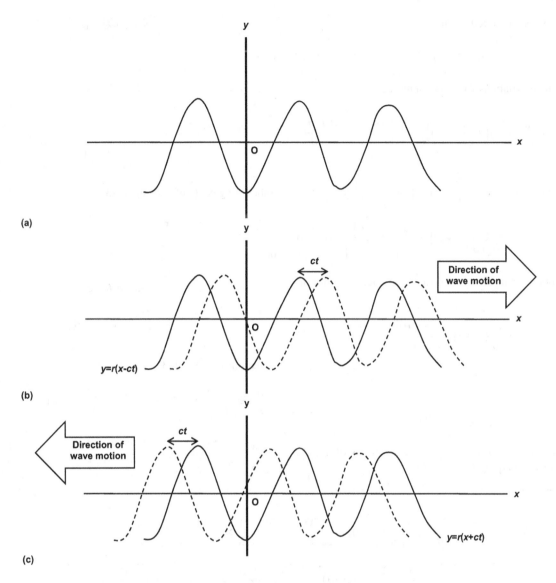

FIGURE 3.1 Depiction of wave motion: (a) initial position of the wave, (b) the wave displaced by moving toward the right with respect to initial position, and (c) the wave displaced by moving toward the left with respect to initial position.

In this way, we are always dealing with the same point on the wave as the x-coordinate is increasing.

Similarly, a wave advancing in the negative direction along the X-axis is expressed by the function

$$y(x,t) = l(x+ct) \qquad (3.64)$$

After a time interval Δt, the function changes to

$$y(x,t) = l\{x + c(t+\Delta t)\} = l(x + ct + c\Delta t) \qquad (3.65)$$

But when we add $-c\Delta t$, we get

$$y(x,t) = l(x + ct + c\Delta t - c\Delta t) = l(x + ct) \qquad (3.66)$$

which is the same as the original function. Therefore, our attention is restricted to the same point with decreasing x-coordinate.

A generalized function for the wave will include both the functions $r(x-ct)$ and $l(x+ct)$. The resultant wave function will be the combined effect of the waves moving rightward (denoted by r-function) and leftward (denoted by l-function) as given by the sum of the two functions

$$y(x,t) = r(x-ct) + l(x+ct) \qquad (3.67)$$

Let us introduce two variables θ and ξ as

$$x - ct = \theta \qquad (3.68a)$$

and

$$x + ct = \xi \qquad (3.68b)$$

Then the wave equation reduces to

$$y(x,t) = r(\theta) + l(\xi) \tag{3.69}$$

By partial differentiation of $y(x, t)$ with respect to θ

$$\frac{\partial}{\partial\theta}\{y(x,t)\} = \frac{\partial}{\partial\theta}\{r(\theta)\} + \frac{\partial}{\partial\theta}\{l(\xi)\} = \frac{\partial\{r(\theta)\}}{\partial\theta} + 0 = \frac{\partial\{r(\theta)\}}{\partial\theta} \tag{3.70}$$

By partial differentiation of $\dfrac{\partial}{\partial\theta}\{y(x,t)\}$ with respect to ξ

$$\frac{\partial}{\partial\xi}\left[\frac{\partial}{\partial\theta}\{y(x,t)\}\right] = \frac{\partial^2}{\partial\xi\partial\theta}\{y(x,t)\} = \frac{\partial}{\partial\xi}\left[\frac{\partial\{r(\theta)\}}{\partial\theta}\right] = 0 \tag{3.71}$$

Thus we obtain a pair of partial differential equations in terms of ξ and θ.

$$\frac{\partial}{\partial\theta}\{y(x,t)\} = \frac{\partial\{r(\theta)\}}{\partial\theta} \tag{3.72}$$

and

$$\frac{\partial^2}{\partial\xi\partial\theta}\{y(x,t)\} = 0 \tag{3.73}$$

These equations have to be written down in terms of x and t. By adding together eqs. (3.68a) and (3.68b),

$$x - ct + x + ct = \theta + \xi \tag{3.74}$$

or

$$2x = \theta + \xi \tag{3.75}$$

or

$$x = \frac{1}{2}(\theta + \xi) \tag{3.76}$$

Subtracting eq. (3.68b) from eq. (3.68a)

$$x - ct - x - ct = \theta - \xi \tag{3.77}$$

or

$$-2ct = \theta - \xi \tag{3.78}$$

$$\therefore t = -\frac{1}{2c}(\theta - \xi) = \frac{1}{2c}(\xi - \theta) \tag{3.79}$$

Equations (3.76) and (3.79) for x and t give

$$\frac{\partial x}{\partial\theta} = \frac{\partial}{\partial\theta}\left\{\frac{1}{2}(\theta + \xi)\right\} = \frac{1}{2}\frac{\partial\theta}{\partial\theta} + \frac{1}{2}\frac{\partial\xi}{\partial\theta} = \frac{1}{2}\times1 + 0 = \frac{1}{2} \tag{3.80}$$

$$\frac{\partial x}{\partial\xi} = \frac{\partial}{\partial\xi}\left\{\frac{1}{2}(\theta + \xi)\right\} = \frac{1}{2}\frac{\partial\theta}{\partial\xi} + \frac{1}{2}\frac{\partial\xi}{\partial\xi} = 0 + \frac{1}{2}\times1 = \frac{1}{2} \tag{3.81}$$

$$\frac{\partial t}{\partial\theta} = \frac{\partial}{\partial\theta}\left\{\frac{1}{2c}(\xi - \theta)\right\} = \frac{1}{2c}\frac{\partial\xi}{\partial\theta} - \frac{1}{2c}\frac{\partial\theta}{\partial\theta} = \frac{1}{2c}\times0 - \frac{1}{2c}\times1 = 0 - \frac{1}{2c} = -\frac{1}{2c} \tag{3.82}$$

$$\frac{\partial t}{\partial\xi} = \frac{\partial}{\partial\xi}\left\{\frac{1}{2c}(\xi - \theta)\right\} = \frac{1}{2c}\frac{\partial\xi}{\partial\xi} - \frac{1}{2c}\frac{\partial\theta}{\partial\xi} = \frac{1}{2c}\times1 - \frac{1}{2c}\times0 = \frac{1}{2c} \tag{3.83}$$

For transforming to x, t, we apply the partial differential chain rule

$$\frac{\partial y(x,t)}{\partial\theta} = \frac{\partial y(x,t)}{\partial x}\times\frac{\partial x}{\partial\theta} + \frac{\partial y(x,t)}{\partial t}\times\frac{\partial t}{\partial\theta} = \frac{\partial y(x,t)}{\partial x}\times\frac{1}{2} + \frac{\partial y(x,t)}{\partial t}\times-\frac{1}{2c} = \frac{1}{2}\frac{\partial y(x,t)}{\partial x} - \frac{1}{2c}\frac{\partial y(x,t)}{\partial t} \tag{3.84}$$

$$\frac{\partial y(x,t)}{\partial\xi} = \frac{\partial y(x,t)}{\partial x}\times\frac{\partial x}{\partial\xi} + \frac{\partial y(x,t)}{\partial t}\times\frac{\partial t}{\partial\xi} = \frac{\partial y(x,t)}{\partial x}\times\frac{1}{2} + \frac{\partial y(x,t)}{\partial t}\times\frac{1}{2c} = \frac{1}{2}\frac{\partial y(x,t)}{\partial x} + \frac{1}{2c}\frac{\partial y(x,t)}{\partial t} \tag{3.85}$$

Equations (3.84), (3.81), and (3.83) will be utilized to express the LHS of eq. (3.73) in terms of x and t as follows:

$$\frac{\partial^2}{\partial\xi\partial\theta}\{y(x,t)\} = \frac{\partial}{\partial\xi}\left\{\frac{\partial y(x,t)}{\partial\theta}\right\} = \frac{\partial}{\partial x}\left\{\frac{\partial y(x,t)}{\partial\theta}\right\}\times\frac{\partial x}{\partial\xi} + \frac{\partial}{\partial t}\left\{\frac{\partial y(x,t)}{\partial\theta}\right\}\times\frac{\partial t}{\partial\xi} = \frac{\partial}{\partial x}\left\{\frac{1}{2}\frac{\partial y(x,t)}{\partial x} - \frac{1}{2c}\frac{\partial y(x,t)}{\partial t}\right\}$$

$$\times\frac{\partial x}{\partial\xi} + \frac{\partial}{\partial t}\left\{\frac{1}{2}\frac{\partial y(x,t)}{\partial x} - \frac{1}{2c}\frac{\partial y(x,t)}{\partial t}\right\}\times\frac{\partial t}{\partial\xi} = \left[\frac{\partial}{\partial x}\left\{\frac{1}{2}\frac{\partial y(x,t)}{\partial x}\right\}\times\frac{\partial x}{\partial\xi}\right] - \left[\frac{\partial}{\partial x}\left\{\frac{1}{2c}\frac{\partial y(x,t)}{\partial t}\right\}\times\frac{\partial x}{\partial\xi}\right]$$

$$+ \left[\frac{\partial}{\partial t}\left\{\frac{1}{2}\frac{\partial y(x,t)}{\partial x}\right\}\times\frac{\partial t}{\partial\xi}\right] - \left[\frac{1}{2c}\frac{\partial}{\partial t}\left\{\frac{\partial y(x,t)}{\partial t}\right\}\times\frac{\partial t}{\partial\xi}\right] = \left\{\frac{1}{2}\frac{\partial^2 y(x,t)}{\partial x^2}\times\frac{\partial x}{\partial\xi}\right\} - \left\{\frac{1}{2c}\frac{\partial^2 y(x,t)}{\partial x\partial t}\times\frac{\partial x}{\partial\xi}\right\}$$

$$+\left\{\frac{1}{2}\frac{\partial^2 y(x,t)}{\partial t\,\partial x}\times\frac{\partial t}{\partial \xi}\right\}-\left\{\frac{1}{2c}\frac{\partial^2 y(x,t)}{\partial t^2}\times\frac{\partial t}{\partial \xi}\right\}=\left\{\frac{1}{2}\frac{\partial^2 y(x,t)}{\partial x^2}\times\frac{1}{2}\right\}-\left\{\frac{1}{2c}\frac{\partial^2 y(x,t)}{\partial x\,\partial t}\times\frac{1}{2}\right\}$$

$$+\left\{\frac{1}{2}\frac{\partial^2 y(x,t)}{\partial t\,\partial x}\times\frac{1}{2c}\right\}-\left\{\frac{1}{2c}\frac{\partial^2 y(x,t)}{\partial t^2}\times\frac{1}{2c}\right\}=\left\{\frac{1}{4}\frac{\partial^2 y(x,t)}{\partial x^2}\right\}-\left\{\frac{1}{4c}\frac{\partial^2 y(x,t)}{\partial x\,\partial t}\right\}+\left\{\frac{1}{4c}\frac{\partial^2 y(x,t)}{\partial t\,\partial x}\right\}-\left\{\frac{1}{4c^2}\frac{\partial^2 y(x,t)}{\partial t^2}\right\}$$

$$=\left\{\frac{1}{4}\frac{\partial^2 y(x,t)}{\partial x^2}\right\}-\left\{\frac{1}{4c^2}\frac{\partial^2 y(x,t)}{\partial t^2}\right\} \tag{3.86}$$

since

$$\left\{\frac{1}{4c}\frac{\partial^2 y(x,t)}{\partial x\,\partial t}\right\}=\left\{\frac{1}{4c}\frac{\partial^2 y(x,t)}{\partial t\,\partial x}\right\} \tag{3.87}$$

Thus

$$\frac{\partial^2}{\partial \xi\,\partial \theta}\{y(x,t)\}=\left\{\frac{1}{4}\frac{\partial^2 y(x,t)}{\partial x^2}\right\}-\left\{\frac{1}{4c^2}\frac{\partial^2 y(x,t)}{\partial t^2}\right\} \tag{3.88}$$

Equations (3.73) and (3.88) yield

$$\left\{\frac{1}{4}\frac{\partial^2 y(x,t)}{\partial x^2}\right\}-\left\{\frac{1}{4c^2}\frac{\partial^2 y(x,t)}{\partial t^2}\right\}=0 \tag{3.89}$$

or

$$\frac{1}{4}\frac{\partial^2 y(x,t)}{\partial x^2}=\frac{1}{4c^2}\frac{\partial^2 y(x,t)}{\partial t^2} \tag{3.90}$$

or

$$\frac{\partial^2 y(x,t)}{\partial x^2}=\frac{1}{c^2}\frac{\partial^2 y(x,t)}{\partial t^2} \tag{3.91}$$

This is the wave equation. It tells us that the second partial derivative (with respect to x) of vertical displacement $y(x,t)$ of a particle in the wave at a position x at time t equals $1/c^2$ times the second partial derivative (with respect to t) of vertical displacement y at the position x at time t.

3.6 Solving the General Equation for Wave Motion by Separating the Variables

Let us start with the assumption that the wave function can be written as the product of two functions: $X(x)$ which is purely a function of x and $T(t)$ which is purely a function of t:

$$y(x,t)=X(x)T(t) \tag{3.92}$$

This will lead to two equations, one containing x only and the other containing t only. Substituting for $y(x,t)$ in the wave equation (3.91),

Left-hand side of the wave equation $=\dfrac{\partial^2 y(x,t)}{\partial x^2}$

$$=\frac{\partial^2}{\partial x^2}\{X(x)T(t)\}=T(t)\frac{\partial^2 X(x)}{\partial x^2} \tag{3.93a}$$

and

Right-hand side of the wave equation $=\dfrac{1}{c^2}\dfrac{\partial^2 y(x,t)}{\partial t^2}$

$$=\frac{1}{c^2}\frac{\partial^2}{\partial t^2}\{X(x)T(t)\}=\frac{1}{c^2}X(x)\frac{\partial^2 T(t)}{\partial t^2} \tag{3.93b}$$

$$\therefore T(t)\frac{\partial^2 X(x)}{\partial x^2}=\frac{1}{c^2}X(x)\frac{\partial^2 T(t)}{\partial t^2} \tag{3.94}$$

Collecting together the 't' terms on one side and 'x' terms on the opposite side, we have

$$\frac{1}{X(x)}\frac{\partial^2 X(x)}{\partial x^2}=\frac{1}{c^2 T(t)}\frac{\partial^2 T(t)}{\partial t^2} \tag{3.95}$$

This equation means that LHS is a function of x only without any dependence on t. So, one can choose any value of x. Similarly, the RHS is a function of t only. Hence, one is free to pick any value of t. The variables x and t are independent of each other. They are not mathematically related. Because of freedom to choose x and t individualistically, we reach a puzzling situation. The only way these conditions can be true is when the two sides of eq. (3.95) are not functions of x and t, but are both equal to a constant. Then one can select any value of x or t. Those values will always result in a constant, $-k^2$ (suppose), irrespective of the values designated for x or t. Hence,

$$\frac{1}{X(x)}\frac{\partial^2 X(x)}{\partial x^2}=\frac{1}{c^2 T(t)}\frac{\partial^2 T(t)}{\partial t^2}=-k^2 \tag{3.96}$$

Thus the two sides of eq. (3.95), i.e. the x and t sides, can be separated to give two independent differential equations. For x, the equation is

$$\frac{1}{X(x)}\frac{\partial^2 X(x)}{\partial x^2}=-k^2 \tag{3.97}$$

Since X is a function of x only, we can write full derivative of $X(x)$ as

$$\frac{1}{X(x)}\frac{d^2 X(x)}{dx^2}=-k^2 \tag{3.98}$$

or

$$\frac{d^2 X(x)}{dx^2}=-k^2 X(x) \tag{3.99}$$

This is the equation for a simple harmonic oscillator. Let us search for a position function, which has the property that the second derivative of the position function is proportional to the negative of the position function. Both sine and cosine functions satisfy this property because if we take

$$X(x) = C\sin(kx) \tag{3.100}$$

then its first derivative is

$$\frac{dX(x)}{dx} = kC\cos(kx) \tag{3.101}$$

and the second derivative is

$$\frac{d^2X(x)}{dx^2} = \frac{d}{dx}\left(\frac{dX(x)}{dx}\right) = \frac{d}{dx}\{kC\cos(kx)\}$$
$$= -k^2C\sin(kx) = -k^2X(x) \tag{3.102}$$

complying with requirement that the second derivative of position function is proportional to negative of the position function.

Similarly, if we take

$$X(x) = C\cos(kx) \tag{3.103}$$

then its first derivative is

$$\frac{dX(x)}{dx} = -kC\sin(kx) \tag{3.104}$$

and the second derivative is

$$\frac{d^2X(x)}{dx^2} = \frac{d}{dx}\left(\frac{dX(x)}{dx}\right) = \frac{d}{dx}\{-kC\sin(kx)\}$$
$$= -k^2C\cos(kx) = -k^2X(x) \tag{3.105}$$

showing that it too meets the requirement.

Therefore, in a preliminary guess we can write

$$X(x) = C\sin(kx) \tag{3.106}$$

A problem with the proposed solution is that at $x=0$,

$$X(x) = C\sin(0) = 0 \tag{3.107}$$

which will be okay if we start from $x = 0$ but we may start from any value of x. So, it does not represent a general solution. To overcome this limitation, we need

$$X(x) = C\cos(kx) \tag{3.108}$$

Since the general solution must satisfy the requirement to start from any value of x, it is better to write it using phase angle ϕ

$$X(x) = C\sin(kx + \phi) = C\sin(kx)\cos\phi + C\cos(kx)\sin\phi \tag{3.109}$$

containing both sine and cosine terms, so that at $x = 0$

$$X(x) = C\sin(0)\cos\phi + C\cos(0)\sin\phi = 0 + C \times 1 \times \sin\phi = C\sin\phi \tag{3.110}$$

and for $x \neq 0$,

$$X(x) = C\sin(kx + \phi) \tag{3.111}$$

Now let us check whether the $X(x)$ given by eq. (3.111) can be a solution. We find the derivatives to verify this:

$$\frac{dX(x)}{dx} = \frac{d\{C\sin(kx+\phi)\}}{dx} = kC\cos(kx+\phi) \tag{3.112}$$

and

$$\frac{d^2X(x)}{dx^2} = \frac{d}{dx}\left(\frac{dX(x)}{dx}\right) = \frac{d}{dx}\{kC\cos(kx+\phi)\}$$
$$= -k \times kC\sin(kx+\phi) = -k^2X(x) \tag{3.113}$$

or

$$\frac{d^2X(x)}{dx^2} = -k^2X(x) \tag{3.114}$$

Yes, eq. (3.111) represents a solution because it satisfies eq. (3.99).

This solution can be put in a more elegant form using exponential functions. We write the sine of the sum $(kx + \phi)$ in full form as

$$X(x) = C\sin(kx+\phi) = C\sin kx\cos\phi + C\cos kx\sin\phi$$
$$= C\sin\phi \times \cos kx + C\cos\phi \times \sin kx \tag{3.115}$$

Put

$$C\sin\phi = A_x + B_x \tag{3.116}$$

and

$$C\cos\phi = i(A_x - B_x) \tag{3.117}$$

where A_x and B_x are constants.

Then

$$X(x) = (A_x + B_x)\cos(kx) + i(A_x - B_x)\sin(kx)$$
$$= A_x\cos(kx) + B_x\cos(kx) + iA_x\sin(kx) - iB_x\sin(kx)$$
$$= A_x\{\cos(kx) + i\sin(kx)\} + B_x\{\cos(kx) - i\sin(kx)\}$$
$$= A_x\exp(+ikx) + B_x\exp(-ikx) \tag{3.118}$$

or

$$X(x) = A_x\exp(+ikx) + B_x\exp(-ikx) \tag{3.119}$$

where Euler's formula has been applied.

After writing the solution for $X(x)$, let us follow the same procedure for the variable $T(t)$.

$$\frac{1}{c^2 T(t)} \frac{\partial^2 T(t)}{\partial t^2} = -k^2 \tag{3.120}$$

or

$$\frac{\partial^2 T(t)}{\partial t^2} = -c^2 k^2 T(t) \tag{3.121}$$

As earlier, we can rewrite the $T(t)$ equation using the full derivative

$$\frac{d^2 T(t)}{dt^2} = -c^2 k^2 T(t) \tag{3.122}$$

Putting

$$ck = \omega \tag{3.123}$$

we have

$$\frac{d^2 T(t)}{dt^2} = -\omega^2 T(t) \tag{3.124}$$

Exponential functions can be used to write the solution to this equation

$$T(t) = A_t \exp(+i\omega t) + B_t \exp(-i\omega t) \tag{3.125}$$

where A_t and B_t are constants.

We have solved separately for the X and T parts of the wave. Let us now combine these parts together to write the complete solution as the product XT

$$
\begin{aligned}
y(x, t) &= X(x)T(t) = \left\{ A_x \exp(+ikx) + B_x \exp(-ikx) \right\} \times \left\{ A_t \exp(+i\omega t) + B_t \exp(-i\omega t) \right\} \\
&= A_x A_t \exp(+ikx)\exp(+i\omega t) + A_x B_t \exp(+ikx)(-i\omega t) + B_x A_t \exp(-ikx)(+i\omega t) \\
&\quad + B_x B_t \exp(-ikx)\exp(-i\omega t) \\
&= A_x A_t \exp(+ikx)\exp(+i\omega t) + B_x B_t \exp(-ikx)\exp(-i\omega t) + A_x B_t \exp(+ikx)(-i\omega t) \\
&\quad + B_x A_t \exp(-ikx)(+i\omega t) \\
&= A_x A_t \exp\left\{ i(kx + \omega t) \right\} + B_x B_t \exp\left\{ -i(kx + \omega t) \right\} + A_x B_t \exp\left\{ i(kx - \omega t) \right\} \\
&\quad + B_x A_t \exp\left\{ -i(kx - \omega t) \right\} \\
&= A_\alpha \exp\left\{ i(kx + \omega t) \right\} + B_\alpha \exp\left\{ -i(kx + \omega t) \right\} + A_\beta \exp\left\{ i(kx - \omega t) \right\} + B_\beta \exp\left\{ -i(kx - \omega t) \right\}
\end{aligned} \tag{3.126}
$$

where

$$A_x A_t = A_\alpha, \quad B_x B_t = B_\alpha, \quad A_x B_t = A_\beta, \quad B_x A_t = B_\beta \tag{3.127}$$

Now we have to assign physical meanings to indiscriminately chosen symbols such as ω and k. For this purpose, we focus attention on a mechanical wave for which the real part of the solution only is relevant. So we separate the real and imaginary parts of eq. (3.126) as

$$
\begin{aligned}
y(x, t) &= A_\alpha \left\{ \cos(kx + \omega t) + i\sin(kx + \omega t) \right\} + B_\alpha \left\{ \cos(kx + \omega t) - i\sin(kx + \omega t) \right\} \\
&\quad + A_\beta \left\{ \cos(kx - \omega t) + i\sin(kx - \omega t) \right\} + B_\beta \left\{ \cos(kx - \omega t) - i\sin(kx - \omega t) \right\} \\
&= (A_\alpha + B_\alpha)\cos(kx + \omega t) + i(A_\alpha - B_\alpha)\sin(kx + \omega t) + (A_\beta + B_\beta)\cos(kx - \omega t) \\
&\quad + i(A_\beta - B_\beta)\sin(kx - \omega t) \\
&= (A_\alpha + B_\alpha)\cos(kx + \omega t) + (A_\beta + B_\beta)\cos(kx - \omega t) \\
&\quad + i\left\{ (A_\alpha - B_\alpha)\sin(kx + \omega t) + (A_\beta - B_\beta)\sin(kx - \omega t) \right\} \\
&= A\cos(kx + \omega t) + B\cos(kx - \omega t) + i\left\{ C\sin(kx + \omega t) + D\sin(kx - \omega t) \right\}
\end{aligned} \tag{3.128}
$$

where

$$A = A_\alpha + B_\alpha \tag{3.129}$$

$$B = A_\beta + B_\beta \qquad (3.130)$$

$$C = A_\alpha - B_\alpha \qquad (3.131)$$

and

$$D = A_\beta - B_\beta \qquad (3.132)$$

Constraining the solution to real values of $y(x,t)$, we can write the solution as the sum of two cosine terms

$$\mathrm{Re}\{y(x,\,t)\} = A\,\cos(kx + \omega t) + B\cos(kx - \omega t) \qquad (3.133)$$

where the phase term ϕ is ignored.

At $x = 0$, the solution for a mechanical wave is

$$y(0,\,t) = A\,\cos(+\omega t) + B\cos(-\omega t) \qquad (3.134)$$

The equation represents oscillatory motion with time period

$$T = \frac{2\pi}{\omega} \qquad (3.135)$$

where ω is the angular frequency.

Increasing the time from t to $t + T$, we find

$$y(0,\,t+T) = A\,\cos\{+\omega(t+T)\} + B\cos\{-\omega(t+T)\}$$

$$= A\,\cos\left\{+\omega\left(t+\frac{2\pi}{\omega}\right)\right\} + B\cos\left\{-\omega\left(t+\frac{2\pi}{\omega}\right)\right\}$$

$$= A\,\cos(+\omega t + 2\pi) + B\cos(-\omega t - 2\pi)$$

$$= A\,\cos(+\omega t) + B\cos(-\omega t) \qquad (3.136)$$

because the cosine function is periodic repeating itself after 2π intervals. So, this equation represents a point undergoing oscillation with angular frequency ω and the symbol ω denotes the angular frequency of oscillation of the point. Hence,

$$\omega = 2\pi\nu \qquad (3.137)$$

where ν is the linear frequency. During the derivation, we introduced another symbol $\omega = ck$. Therefore

$$\omega = 2\pi\nu = ck \qquad (3.138)$$

or

$$k = \frac{2\pi\nu}{c} \qquad (3.139)$$

But c being the phase velocity, we know that

$$c = \nu\lambda \qquad (3.140)$$

$$\therefore k = \frac{2\pi\nu}{c} = \frac{2\pi\nu}{\nu\lambda} = \frac{2\pi}{\lambda} \qquad (3.141)$$

which means that k is the wave number. Now putting $\omega = ck$ in eq. (3.133), we get

$$\mathrm{Re}\{y(x,\,t)\} = A\,\cos(kx + ckt) + B\cos(kx - ckt)$$

$$= A\,\cos\{k(x+ct)\} + B\cos\{k(x-ct)\} \qquad (3.142)$$

We can recognize that the first cosine term represents a wave propagating in the negative direction along the X-axis while the second cosine term indicates a wave traveling in the positive direction along the X-axis. Recalling our discussion at the beginning of the derivation, we had written two functions $r(x-ct)$ and $l(x + ct)$ for waves moving toward the right and left directions, respectively. Thus the solution is in concordance with our starting proposition. We now understand what the functions $r(x-\mathrm{ct})$ and $l(x + ct)$ actually signify. The general solution encompasses waves moving in both directions. The solution can be expressed including both real and imaginary parts as

$$y(x,\,t) = A\exp\{i(kx + \omega t)\} + B\exp\{i(kx - \omega t)\} \qquad (3.143)$$

the first term describing a wave moving toward $-X$-axis and the second term a wave traveling toward $+X$-axis.

3.7 Max Born's Physical Interpretation of the Wave Function Ψ

After learning about the Schrodinger equation and wave motion, a pertinent question is: What exactly is the physical significance of the wave function? The question was answered by the German-Jewish physicist Max Born who was awarded the Nobel Prize in physics in 1954 for his work in quantum mechanics, and particularly for the statistical interpretation of wave function offered by him (Born 1926).

According to Max Born (Figure 3.2), the wave function Ψ represents the probability amplitude. Being identical to the amplitude of a wave, it can acquire positive and negative values. The modulus of square of the wave function

$$|\Psi|^2\,dx = \Psi^*\Psi dx = P(x,t)dx \qquad (3.144)$$

is the probability density that the particle is found in the small region between x and $x + dx$ at a particular instant of time. Probability density is the probability per unit volume. The reason for taking the squared value is that the wave function can be positive or negative but the square of wave function is always positive (Figure 3.3). This is necessary to ascribe a physical meaning to the probability density, which must have positive values only. The reason for taking the product of the wave function with its complex conjugate is that the wave function is a complex number but the wave function multiplied by its conjugate is always real. In this way, imaginary values of probability density are avoided. This makes the definition of probability density physically acceptable because it is expected that the probability density will be always real.

Clearly demarcate the three terms in Born interpretation, namely, the probability amplitude, the probability density, and probability: the wave function is the probability amplitude, square of the modulus of the wave function is the probability density, and the product of probability density with

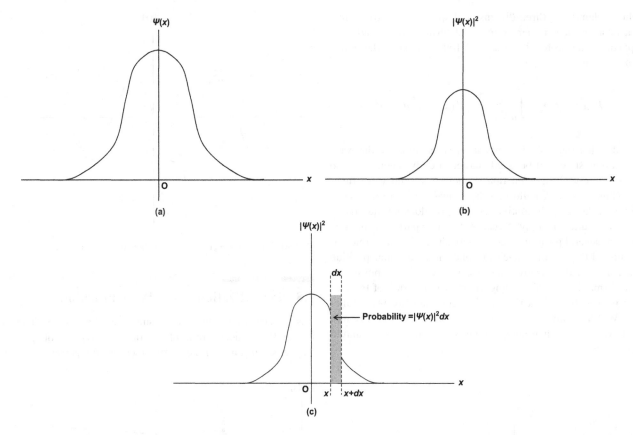

FIGURE 3.2 Born's interpretation of the wave function: (a) the wave function or probability amplitude, (b) the probability density, and (c) the probability of finding the particle in a narrow range between x and $x + dx$.

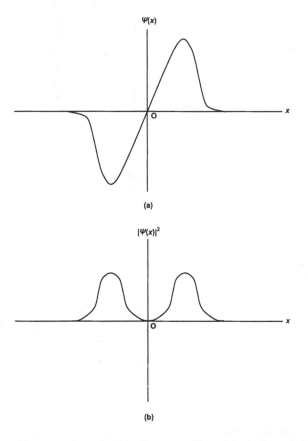

FIGURE 3.3 The wave function can be positive or negative, as in (a) but the square of its modulus is always positive, as in (b).

volume element in three-dimensional space is the probability. If a measurement is performed to determine the location of a particle, the probability that it is found between the points a, b is given by

$$P(a < x < b) = \int_{a_x}^{b_x} \int_{a_y}^{b_y} \int_{a_z}^{b_z} \left| \Psi(x, y, z) \right|^2 dx \, dy \, dz \quad (3.145)$$

In order that the Born interpretation is meaningful, the wave function must be well-behaved mathematically (Figure 3.4). So any randomly chosen function cannot be a valid wave function (Figure 3.5). A multiple-valued function or a function of infinite value or a discontinuous function does not qualify to be a wave function. Why? Because the same particle cannot be found at several positions; rather there should be an unambiguous probability of a particle's presence at a particular position and time. Further, an infinite probability value is physically absurd (maximum value being 1), and the location of the particle is ill-defined in a region of discontinuity of wave function. We shall enlist all the constraints on the wave function in the next chapter when laying down the postulates of quantum mechanics.

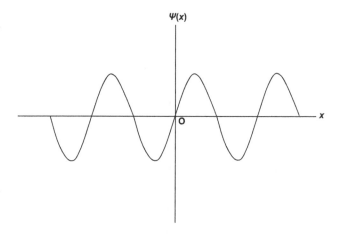

FIGURE 3.4   An acceptable wave function: the sine/cosine function.

3.8 Normalization of the Wave Function

It is physically expected that the sum-total of all probabilities of finding the particle within infinitesimal subdivisions of space at a specified instant will be unity; otherwise the particle will not

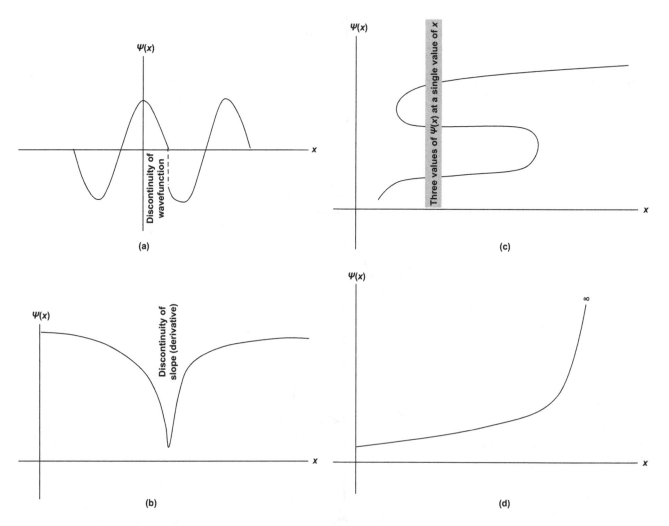

FIGURE 3.5   Disallowed wave functions due to discontinuity in the value of wave function (a), discontinuity in slope (b), not being single-valued (c), and having an infinite value over a finite region (d).

be found in the total volume under consideration. To circumvent such possibility of non-existence of particle, the total probability must equal unity, which is expressed by writing

$$\int_{-\infty}^{\infty} P(x,t)\,dx = \int_{-\infty}^{\infty} \Psi^*(x,t)\Psi(x,t)\,dx = 1 \qquad (3.146)$$

Probability is a real number. It lies between 0 and 1. Zero probability means an impossible result. Probability value of 1 indicates a certainty. Sometimes, it may be required to precede the integral in eq. (3.146) by a multiplicative constant factor to fulfill the condition of making the total probability equal to one; otherwise the statistical interpretation will become meaningless. This procedure of scaling is called normalization of the wave function, and the wave function after having undergone this procedure is said to be normalized. Equation (3.146), which is a statement of the unity value of total probability, is known as the normalization condition.

Consider the normalization of the wave function given by

$$\Psi(x) = C\exp(i\omega t)\exp\left(-\frac{x^2}{2\sigma^2}\right) \qquad (3.147)$$

where C, ω, and σ are real constants.

This wave function will be normalized if

$$\int_{-\infty}^{\infty} \Psi^*(x)\Psi(x)\,dx = 1 \qquad (3.148)$$

or

$$\int_{-\infty}^{\infty} C\exp(-i\omega t)\exp\left(-\frac{x^2}{2\sigma^2}\right) \times C\exp(i\omega t)\exp\left(-\frac{x^2}{2\sigma^2}\right)dx = 1 \qquad (3.149)$$

or

$$C^2 \int_{-\infty}^{\infty} \exp\left(-\frac{x^2}{\sigma^2}\right)dx = 1 \qquad (3.150)$$

Put

$$\frac{x}{\sigma} = u \qquad (3.151)$$

Then

$$\frac{dx}{\sigma} = du \qquad (3.152)$$

or

$$dx = \sigma\,du \qquad (3.153)$$

$$\therefore C^2 \int_{-\infty}^{\infty} \exp(-u^2)\sigma\,du = 1 \qquad (3.154)$$

or

$$\therefore C^2\sigma \int_{-\infty}^{\infty} \exp(-u^2)\,du = 1 \qquad (3.155)$$

or

$$C^2\sigma\sqrt{\pi} = 1 \qquad (3.156)$$

$$\because \int_{-\infty}^{\infty} \exp(-u^2)\,du = \sqrt{\pi} \qquad (3.157)$$

This integration is performed in Illustrative Exercise 3.3.

Thus

$$C^2 = \frac{1}{\sigma\sqrt{\pi}} \qquad (3.158)$$

or

$$C = \sqrt{\frac{1}{\sigma\sqrt{\pi}}} = \frac{1}{\sqrt{\sigma}\,\pi^{1/4}} \qquad (3.159)$$

Hence, putting the value of C in eq. (3.147), the normalized wave function is

$$\Psi(x) = \frac{1}{\sqrt{\sigma}\,\pi^{1/4}} \exp(i\omega t)\exp\left(-\frac{x^2}{2\sigma^2}\right) \qquad (3.160)$$

3.9 Discussion and Conclusions

Schrodinger equation is the core equation in quantum mechanics finding extensive applications in nanoelectronics. Solution to Schrodinger equation contains all the information about a quantum-mechanical system (Kilmister 1989). The time-independent wave equation considers the electrons as standing waves, and is applied to determine the shapes and sizes of the atomic orbitals, while the time-dependent wave equation treats them as traveling waves, and is applied to motion of free electrons. In an atom, the electrons are imagined to form standing waves, which fit into the atomic size. As a result, the atoms are confined within the atomic space. These standing waves are formed from trapping of traveling waves within the atom. Whenever traveling waves are trapped in a region of space, e.g. by the attraction of electrons to the proton in the atom, standing waves are formed. The time-independent equation can be solved analytically for many simple systems of electron motion being restricted within defined boundaries, thus providing useful information about them.

Born interpretation of wave function is of paramount importance in quantum mechanics because the wave function is a complex quantity, which is not observable and has no direct physical meaning. But the product of the wave function with its complex conjugate is a real quantity and the squared modulus of the wave function, being always real and positive, is proportional to the probability of finding the particle within a defined location. The Born interpretation imposes several requirements on the mathematical behavior of the wave function, which a wave function must fulfill for validity.

Illustrative Exercises

3.1 Which of the following functions can be separated into variables:

$$(i) \ \Psi(x,t) = \exp\{i(kx - \omega t)\} \qquad (3.161)$$

$$(ii) \ \Psi(x,t) = \cos(px)\sin(qt) \qquad (3.162)$$

$$(iii) \ \Psi(x,t) = \cos(kx + \omega t) \qquad (3.163)$$

$$(iv) \ \Psi(x,t) = \exp\{i(kx - \omega t)^3\} \qquad (3.164)$$

i. The given function is variable-separable because

$$\Psi(x,t) = \exp\{i(kx - \omega t)\} = \exp(ikx) \times \exp(-i\omega t)$$

$$= X(x) \times T(t) \qquad (3.165)$$

ii. The given function is variable-separable because

$$\Psi(x,t) = \cos(px)\sin(qt) = \cos(px) \times \sin(qt)$$

$$= X(x) \times T(t) \qquad (3.166)$$

iii. The given function is not variable-separable because

$$\Psi(x,t) = \cos(kx + \omega t) = \cos(kx)\cos(\omega t)$$

$$- \sin(kx)\sin(\omega t) \neq X(x) \times T(t) \qquad (3.167)$$

iv. The given function is not variable-separable because

$$\Psi(x,t) = \exp\{i(kx - \omega t)^3\}$$

$$= \exp\{i\left(k^3 x^3 - 3k^2 x^2 \omega t + 3kx\omega^2 t^2 - \omega^3 t^3\right)\} \neq X(x) \times T(t)$$

$$(3.168)$$

3.2 What is the difference between the waves described by the functions $\Psi = \cos kx$ and $\Psi = \cos(kx - \omega t)$ where $k = 2\pi/\lambda$?

Putting the value of $k = 2\pi/\lambda$, we can write,

$$\cos kx = \cos\left\{\left(\frac{2\pi}{\lambda}\right)x\right\} \qquad (3.169)$$

When $x = \lambda, 2\lambda, 3\lambda, 4\lambda, \ldots$

$$\cos kx = \cos\left\{\left(\frac{2\pi}{\lambda}\right)\lambda\right\}, \cos\left\{\left(\frac{2\pi}{\lambda}\right)2\lambda\right\},$$

$$\cos\left\{\left(\frac{2\pi}{\lambda}\right)3\lambda\right\}, \cos\left\{\left(\frac{2\pi}{\lambda}\right)4\lambda\right\}, \ldots$$

$$= \cos 2\pi, \cos 4\pi, \cos 6\pi, \cos 8\pi, \ldots \qquad (3.170)$$

We know that

$$\cos 2\pi = \cos 4\pi = \cos 6\pi = \cos 8\pi = 1 \qquad (3.171)$$

As we are increasing the position x in steps of λ, the function Ψ is attaining the same maximum value of 1 for each step. This kind of wave is called a standing wave with amplitude 1, as it does not move.

The wave $\Psi = \cos(kx - \omega t)$ has the maximum amplitude when

$$kx - \omega t = 0 \qquad (3.172)$$

because then

$$\cos(kx - \omega t) = \cos 0 = 1 \qquad (3.173)$$

From eq. (3.172),

$$kx = \omega t \qquad (3.174)$$

or

$$\frac{x}{t} = \frac{\omega}{k} \qquad (3.175)$$

But x/t = distance/time = velocity c. Hence,

$$c = \frac{\omega}{k} \qquad (3.176)$$

This equation represents a traveling wave with amplitude 1. This wave is moving with velocity $c = \omega/k$.

3.3 Prove that

$$\int_{-\infty}^{\infty} \exp(-u^2)\,du = \sqrt{\pi} \qquad (3.177)$$

We can write

$$\left\{\int_{-\infty}^{\infty} \exp(-u^2)\,du\right\}^2 = \left\{\int_{-\infty}^{\infty} \exp(-u^2)\,du\right\}$$

$$\times \left\{\int_{-\infty}^{\infty} \exp(-u^2)\,du\right\} \qquad (3.178)$$

The variable u is known as a variable of integration. It is also called a dummy variable. Its symbol is not important because that symbol integrates out. So, we can also use the symbol 'v' for 'u' in the second factor on the RHS. This trick enables us to transform eq. (3.178) into

$$\left\{\int_{-\infty}^{\infty} \exp(-u^2)\,du\right\}^2 = \left\{\int_{-\infty}^{\infty} \exp(-u^2)\,du\right\}$$

$$\times \left\{\int_{-\infty}^{\infty} \exp(-v^2)\,dv\right\} \qquad (3.179)$$

The two integrals can now be combined to form one double integral. Hence,

$$\left\{\int_{-\infty}^{\infty}\exp\left(-u^2\right)du\right\}^2 = \int_{-\infty}^{\infty}\int_{-\infty}^{\infty}\left\{\exp\left(-u^2\right)du\right\}\left\{\exp\left(-v^2\right)dv\right\}$$

$$= \int_{-\infty}^{\infty}\int_{-\infty}^{\infty}\exp\left(-u^2\right)\exp\left(-v^2\right)du\,dv$$

$$= \int_{-\infty}^{\infty}\int_{-\infty}^{\infty}\exp\left\{-\left(u^2+v^2\right)\right\}du\,dv \qquad (3.180)$$

Here $\exp\left(u^2+v^2\right)$ is being integrated over the full u–v plane. This integration can also be performed in plane polar coordinates, r and θ, where

$$u^2 + v^2 = r^2 \qquad (3.181)$$

and the area element

$$du\,dv = r\,d\theta\,dr \qquad (3.182)$$

This is evident from Figure 3.6.

While changing to plane polar coordinates, we alter the limits of integration as $0 \le \theta \le 2\pi$ and $0 \le r \le \infty$, because in this system, the point specified by the coordinates (r,θ) covers the entire plane with θ ranging from 0 to 2π and r ranging from 0 to ∞. Then

$$\left\{\int_{-\infty}^{\infty}\exp\left(-u^2\right)du\right\}^2 = \int_{-\infty}^{\infty}\int_{-\infty}^{\infty}\exp\left\{-\left(u^2+v^2\right)\right\}du\,dv$$

$$= \int_{0}^{2\pi}d\theta\int_{0}^{\infty}r\exp\left(-r^2\right)dr$$

$$= \left[\theta\right]_{0}^{2\pi}\int_{0}^{\infty}\exp\left(-r^2\right)r\,dr$$

$$= 2\pi\int_{0}^{\infty}\exp\left(-r^2\right)r\,dr \qquad (3.183)$$

Let us put $r^2 = v$. Then

$$2r\,dr = dv \qquad (3.184)$$

Hence,

$$\left\{\int_{0}^{\infty}\exp\left(-u^2\right)du\right\}^2 = \pi\int_{0}^{\infty}\exp(-v)dv \qquad (3.185)$$

Put $-v = w$. Then

$$-dv = dw \qquad (3.186)$$

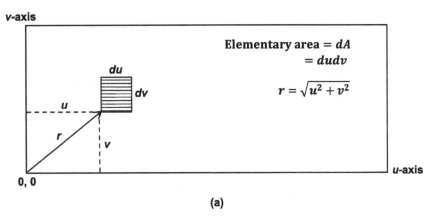

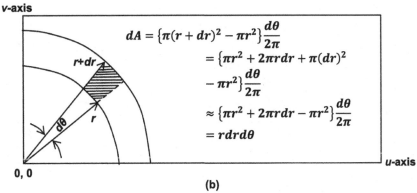

FIGURE 3.6 Calculation of differential of area in rectangular Cartesian and plane polar coordinates: (a) rectangular and (b) polar.

and

$$\left\{ \int_0^\infty \exp\left(-u^2\right) du \right\}^2 = \pi \int_0^\infty \exp(-v) dv = -\pi \int_0^\infty \exp(w) dw$$

$$= -\pi \left[\exp(w) \right]_0^\infty$$

$$= -\pi \left[\exp(-v) \right]_0^\infty$$

$$= -\pi \left\{ \exp(-\infty) - \exp(-0) \right\}$$

$$= -\pi (0-1) = -\pi \times -1 = \pi \qquad (3.187)$$

$$\therefore \int_0^\infty \exp\left(-u^2\right) du = \sqrt{\pi} \qquad (3.188)$$

REFERENCES

Barde, N. P., S. D. Patil, P. M. Kokne, P. P. Bardapurkar. 2015. Deriving time dependent Schrödinger equation from wave-mechanics, Schrödinger time independent equation, Classical and Hamilton-Jacobi. *Leonardo Electronic Journal of Practices and Technologies*, 26: 31–48.

Born, M. 1926. On the quantum mechanics of collision processes. *Zur Quantenmechanik der Stoßvorgänge, Zeitschrift für Physik*, 37: 863–867.

Kilmister, C. W. (Ed.) 1989. *Schrodinger: Centenary Celebration of a Polymath*. Cambridge University Press: London, 264 pp.

Schiff, L. I. 1968. *Quantum Mechanics*. McGraw Hill Book Company, Inc., New York, pp. 19–44.

Schleich, W. P., D. M. Greenbergera, D. H. Kobe and M. O. Scully. 2013. Schrödinger equation revisited. *PNAS*, 110(14): 5374–5379.

Schrodinger. 1926. An undulatory theory of the mechanics of atoms and molecules. *Physical Review*, 28(6): 1049–1070.

4

Operator Methods and Postulates of Quantum Mechanics

4.1 What Are Operators?

An operator, symbolized by the letter O with a hat as $\hat{O}$, is a mathematical command or instruction which acts on a function $f(x)$ to give another function $g(x)$; thus (Schechter 2003, Jordan 2006)

$$\hat{O} f(x) = g(x) \tag{4.1}$$

An operator is a rule for mapping one function to another function in the same way as a function

$$y = f(x) \tag{4.2}$$

is a relation between a set of input variables x and another set of output variables y such that each value of input variable x leads exactly to one value of output variable y. Action of an operator turns one function into another function just as action of a function turns one number into another number. Common examples of operators are:

$$\text{Square root operator}: \sqrt{f(x)} \tag{4.3}$$

$$\text{Differential operator}: \hat{D} = d/dx \tag{4.4}$$

and

$$\text{Integral operator}: \int f(x) \tag{4.5}$$

4.2 Main Operators in Quantum Mechanics

Let us start by introducing two fundamental operators, those for position and linear momentum, and move toward more operators because position and momentum are basic observables and other observables like energy and momentum can be derived from these two basic observables.

4.2.1 Position Operators

Position operator in the X-direction:

$$\hat{x}\Psi(x, y, z, t) = x\Psi(x, y, z, t) \tag{4.6}$$

Position operator in the Y-direction:

$$\hat{y}\Psi(x, y, z, t) = y\Psi(x, y, z, t) \tag{4.7}$$

and

Position operator in the Z-direction:

$$\hat{z}\Psi(x, y, z, t) = z\Psi(x, y, z, t) \tag{4.8}$$

4.2.2 Linear Momentum Operators

Linear momentum operator in the X-direction:

$$\hat{p}_x\Psi(x, y, z, t) = -i\hbar\frac{\partial}{\partial x}\Psi(x, y, z, t) \tag{4.9}$$

Linear momentum operator in the Y-direction:

$$\hat{p}_y\Psi(x, y, z, t) = -i\hbar\frac{\partial}{\partial y}\Psi(x, y, z, t) \tag{4.10}$$

and

Linear momentum operator in the Z-direction:

$$\hat{p}_z\Psi(x, y, z, t) = -i\hbar\frac{\partial}{\partial z}\Psi(x, y, z, t) \tag{4.11}$$

4.2.3 Kinetic Energy Operators

To write the operator for a property or observable, we write the classical expression for that property and substitute the earlier mentioned basic operators in that expression. To get the operator for kinetic energy T_x in the X-direction, we proceed as follows:

$$\text{Kinetic energy in the } X\text{-direction} = T_x = \frac{1}{2}mv_x^2 = \frac{p_x^2}{2m} \tag{4.12}$$

$$\therefore \text{Kinetic energy operator in the } X\text{-direction} = \hat{T}_x = \frac{\hat{p}_x^2}{2m}$$

$$= \frac{1}{2m}\left(-i\hbar\frac{\partial}{\partial x}\right)^2 = \frac{1}{2m}i^2\hbar^2\frac{\partial^2}{\partial x^2} = -\frac{\hbar^2}{2m}\frac{\partial^2}{\partial x^2} \tag{4.13}$$

Similarly,

$$\text{Kinetic energy operator in the } Y\text{-direction} = \hat{T}_y = -\frac{\hbar^2}{2m}\frac{\partial^2}{\partial y^2} \tag{4.14}$$

and

$$\text{Kinetic energy operator in the } Z\text{-direction} = \hat{T}_z = -\frac{\hbar^2}{2m}\frac{\partial^2}{\partial z^2} \tag{4.15}$$

4.2.4 Potential Energy Operators

Potential energy in the X-direction $= V_x = V(x)$ (4.16)

$\therefore$ Potential energy operator in the X-direction

$$= \hat{V}_x = \hat{V}(\hat{x}) = V(x) \tag{4.17}$$

Likewise,

Potential energy operator in the Y-direction

$$= \hat{V}_y = \hat{V}(\hat{y}) = V(y) \tag{4.18}$$

and

Potential energy operator in the Z-direction

$$= \hat{V}_z = \hat{V}(\hat{z}) = V(z) \tag{4.19}$$

4.2.5 Total Energy Operators

For total energy,

Total energy in X-direction

$$= E_x = T_x + V_x = \frac{p_x^2}{2m} + V(x) \tag{4.20}$$

$\therefore$ Total energy operator in X-direction

$$= \hat{E}_x = \frac{\hat{p}_x^2}{2m} + \hat{V}(\hat{x}) = -\frac{\hbar^2}{2m}\frac{\partial^2}{\partial x^2} + V(x) = \hat{H}_x \tag{4.21}$$

It is known as the Hamiltonian operator in the X-direction ($\hat{H}_x$). Identically,

Total energy operator in the Y-direction

$$= \hat{H}_y = -\frac{\hbar^2}{2m}\frac{\partial^2}{\partial y^2} + V(y) \tag{4.22}$$

Total energy operator in the Z-direction

$$= \hat{H}_z = -\frac{\hbar^2}{2m}\frac{\partial^2}{\partial z^2} + V(z) \tag{4.23}$$

and

Total energy operator in three dimensions

$$= \hat{H}_x + \hat{H}_y + \hat{H}_z = \hat{H} = \left\{ -\frac{\hbar^2}{2m}\frac{\partial^2}{\partial x^2} + V(x) \right\}$$

$$+ \left\{ -\frac{\hbar^2}{2m}\frac{\partial^2}{\partial y^2} + V(y) \right\} + \left\{ -\frac{\hbar^2}{2m}\frac{\partial^2}{\partial z^2} + V(z) \right\}$$

$$= -\frac{\hbar^2}{2m}\left(\frac{\partial^2}{\partial x^2} + \frac{\partial^2}{\partial y^2} + \frac{\partial^2}{\partial z^2} \right) + V(x, y, z)$$

$$= -\frac{\hbar^2}{2m}\nabla^2 + V(x, y, z) \tag{4.24}$$

4.2.6 Angular Momentum Operator

In classical mechanics, angular momentum $\mathbf{L}$ of a particle is the vector or cross product of the position $\mathbf{r}$ of the particle and its linear momentum $\mathbf{p}$ as

$$\mathbf{L} = \mathbf{r} \times \mathbf{p} \tag{4.25}$$

In three-dimensional rectangular Cartesian coordinates, $\mathbf{L}, \mathbf{r},$ and $\mathbf{p}$ can be expressed in terms of their X-, Y-, and Z-components as

$$\mathbf{L} = \hat{\mathbf{i}}L_x + \hat{\mathbf{j}}L_y + \hat{\mathbf{k}}L_z \tag{4.26}$$

$$\mathbf{r} = \hat{\mathbf{i}}x + \hat{\mathbf{j}}y + \hat{\mathbf{k}}z \tag{4.27}$$

and

$$\mathbf{p} = \hat{\mathbf{i}}p_x + \hat{\mathbf{j}}p_y + \hat{\mathbf{k}}p_z \tag{4.28}$$

where $\hat{\mathbf{i}}, \hat{\mathbf{j}}, \hat{\mathbf{k}}$ are the unit vectors in X, Y, Z directions, respectively.

Then left-hand side of eq. (4.25) is rewritten using eq. (4.26) while its right-hand side is recast with the help of eqs. (4.27) and (4.28). As a consequence, we obtain eq. (4.25) in terms of X-, Y-, Z-components of $\mathbf{L}, \mathbf{r},$ and $\mathbf{p}$ as

$$\hat{L} = \hat{\mathbf{i}}L_x + \hat{\mathbf{j}}L_y + \hat{\mathbf{k}}L_z = \left(\hat{\mathbf{i}}x + \hat{\mathbf{j}}y + \hat{\mathbf{k}}z \right) \times \left(\hat{\mathbf{i}}p_x + \hat{\mathbf{j}}p_y + \hat{\mathbf{k}}p_z \right)$$

$$= xp_x\left(\hat{\mathbf{i}} \times \hat{\mathbf{i}} \right) + xp_y\left(\hat{\mathbf{i}} \times \hat{\mathbf{j}} \right) + xp_z\left(\hat{\mathbf{i}} \times \hat{\mathbf{k}} \right)$$

$$+ yp_x\left(\hat{\mathbf{j}} \times \hat{\mathbf{i}} \right) + yp_y\left(\hat{\mathbf{j}} \times \hat{\mathbf{j}} \right) + yp_z\left(\hat{\mathbf{j}} \times \hat{\mathbf{k}} \right) + zp_x\left(\hat{\mathbf{k}} \times \hat{\mathbf{i}} \right)$$

$$+ zp_y\left(\hat{\mathbf{k}} \times \hat{\mathbf{j}} \right) + zp_z\left(\hat{\mathbf{k}} \times \hat{\mathbf{k}} \right)$$

$$= xp_x(0) + xp_y\left(\hat{\mathbf{k}} \right) + xp_z\left(-\hat{\mathbf{j}} \right) + yp_x\left(-\hat{\mathbf{k}} \right) + yp_y(0)$$

$$+ yp_z\left(\hat{\mathbf{i}} \right) + zp_x\left(\hat{\mathbf{j}} \right) + zp_y\left(-\hat{\mathbf{i}} \right) + zp_z(0)$$

$$= xp_y\left(\hat{\mathbf{k}} \right) + xp_z\left(-\hat{\mathbf{j}} \right) + yp_x\left(-\hat{\mathbf{k}} \right) + yp_z\left(\hat{\mathbf{i}} \right) + zp_x\left(\hat{\mathbf{j}} \right) + zp_y\left(-\hat{\mathbf{i}} \right)$$

$$= \hat{\mathbf{i}}\left(yp_z - zp_y \right) + \hat{\mathbf{j}}\left(zp_x - xp_z \right) + \hat{\mathbf{k}}\left(xp_y - yp_x \right) \tag{4.29}$$

from which

$$L_x = yp_z - zp_y \tag{4.30}$$

$$L_y = zp_x - xp_z \tag{4.31}$$

and

$$L_z = xp_y - yp_x \tag{4.32}$$

In operator form,

Angular momentum operator in X – direction

$$= \hat{L}_x = \hat{y}\hat{p}_z - \hat{z}\hat{p}_y = y\hat{p}_z - z\hat{p}_y$$

$$= y\left(-i\hbar\frac{\partial}{\partial z} \right) - z\left(-i\hbar\frac{\partial}{\partial y} \right) = -i\hbar\left(y\frac{\partial}{\partial z} - z\frac{\partial}{\partial y} \right) \tag{4.33}$$

Angular momentum operator in the Y-direction

$$= \hat{L}_y = \hat{z}\hat{p}_x - \hat{x}\hat{p}_z = z\hat{p}_x - x\hat{p}_z$$

$$= z\left(-i\hbar\frac{\partial}{\partial x}\right) - x\left(-i\hbar\frac{\partial}{\partial z}\right) = -i\hbar\left(z\frac{\partial}{\partial x} - x\frac{\partial}{\partial z}\right) \quad (4.34)$$

and

Angular momentum operator in the Z-direction

$$= \hat{L}_z = \hat{x}\hat{p}_y - \hat{y}\hat{p}_x = x\hat{p}_y - y\hat{p}_x$$

$$= x\left(-i\hbar\frac{\partial}{\partial y}\right) - y\left(-i\hbar\frac{\partial}{\partial x}\right) = -i\hbar\left(x\frac{\partial}{\partial y} - y\frac{\partial}{\partial x}\right) \quad (4.35)$$

4.3 Linear and Non-Linear Operators

An operator is said to be linear if it possesses the following properties:

i. The result obtained when the operator $\hat{O}$ acts on the sum of two functions $f(x)$ and $g(x)$ equals the sum of the results obtained when the operator $\hat{O}$ acts on $f(x)$ separately and when it acts on $g(x)$ separately, i.e.,

$$\hat{O}\{f(x) + g(x)\} = \hat{O}f(x) + \hat{O}g(x) \quad (4.36)$$

ii. If c is a coefficient given by

$$c = a + ib \quad (4.37)$$

the result of the product $\hat{O}\ cf(x)$ is the same as would be obtained by factoring out c, allowing the operator to act on the function $f(x)$, and multiplying the result with c.

$$\hat{O}\ cf(x) = c\ \hat{O}\ f(x) \quad (4.38)$$

The differential operator is a linear operator because

$$\frac{d}{dx}\{f(x) + g(x)\} = \frac{d}{dx}\{f(x)\} + \frac{d}{dx}\{g(x)\} \quad (4.39)$$

e.g.,

$$\frac{d}{dx}(7\sin x + 8\cos x) = \frac{d}{dx}(7\sin x) + \frac{d}{dx}(8\cos x)$$

$$= 7\frac{d}{dx}(\sin x) + 8\frac{d}{dx}(\cos x)$$

$$= 7\cos x - 8\sin x \quad (4.40)$$

However, the square root operator is a non-linear operator because

$$\sqrt{f(x) + g(x)} \neq \sqrt{f(x)} + \sqrt{g(x)} \quad (4.41)$$

e.g.,

$$\sqrt{4x^2 + 5x^2} = \sqrt{9x^2} = 3x \quad (4.42)$$

But

$$\sqrt{4x^2} + \sqrt{5x^2} = 2x + 2.236x = 4.236x \quad (4.43)$$

Almost all the operators in quantum mechanics, such as $\hat{x}$, $\hat{p}$, $\hat{H}$, $\hat{L}$, are linear. This statement can be verified by performing calculations with these operators.

4.4 Commutation of Operators and Its Implications

Consider an occasion in which we come across more than one operator acting upon a function. How to deal with such a situation? Can we interchange the order of operators? Operators cannot be always exchanged but in some circumstances it may be done.

Two operators for which the outcome remains same on changing their order are said to be commutative; such operators are said to commute with each other. Those operators which give a different result on changing order are known as non-commutative operators. For two non-commutative operators $\hat{P}, \hat{Q}$ acting on a function φ,

$$\hat{P}\hat{Q}\varphi \neq \hat{Q}\hat{P}\varphi \quad (4.44)$$

whereas for two commutative operators,

$$\hat{P}\hat{Q}\varphi = \hat{Q}\hat{P}\varphi \quad (4.45)$$

or

$$\hat{P}\hat{Q}\varphi - \hat{Q}\hat{P}\varphi = 0 \quad (4.46)$$

or

$$\left(\hat{P}\hat{Q} - \hat{Q}\hat{P}\right)\varphi = 0 \quad (4.47)$$

Since $\varphi \neq 0$,

$$\hat{P}\hat{Q} - \hat{Q}\hat{P} = 0 \quad (4.48)$$

which is written as

$$\left[\hat{P}, \hat{Q}\right] = 0 \quad (4.49)$$

Because each operator corresponds to an observable, commutation of two operators implies that the values of the observables associated with those operators can be simultaneously and exactly measured.

Let us check the operators for position and linear momentum in the X-direction. Let these operators act on a function $\varphi(x)$. To check these operators for commutation, we find whether

$$\left[\hat{x}, \hat{p}_x\right]\varphi(x) = 0 \text{ or } \left[\hat{x}, \hat{p}_x\right]\varphi(x) \neq 0 \quad (4.50)$$

and decide their behavior from this result. Now,

$$[\hat{x}, \hat{p}_x]\varphi(x) = [\hat{x}\hat{p}_x - \hat{p}_x\hat{x}]\varphi(x)$$

$$= \hat{x}\{\hat{p}_x\ \varphi(x)\} - \hat{p}_x\{\hat{x}\ \varphi(x)\}$$

$$= \hat{x}\left\{-i\hbar\frac{\partial}{\partial x}\ \varphi(x)\right\} - \left[-i\hbar\frac{\partial}{\partial x}\{\hat{x}\ \varphi(x)\}\right]$$

$$= \hat{x}\left\{-i\hbar\frac{\partial\varphi(x)}{\partial x}\ \right\} - \left[-i\hbar\frac{\partial\{x\ \varphi(x)\}}{\partial x}\right]$$

$$= x\left\{-i\hbar\frac{\partial\varphi(x)}{\partial x}\ \right\}$$

$$- \left\{-i\hbar\frac{\partial x}{\partial x}\times\varphi(x) - i\hbar x\frac{\partial\varphi(x)}{\partial x}\right\}$$

$$= -i\hbar x\frac{\partial\varphi(x)}{\partial x} + i\hbar\times 1\times\varphi(x) + i\hbar x\frac{\partial\varphi(x)}{\partial x}$$

$$= i\hbar\varphi(x) \neq 0 \tag{4.51}$$

Looking at the non-zero result, we learn that the operators $\hat{x}, \hat{p}_x$ are non-commuting, which obviously suggests the inability of their precise determination at the same time.

Let us now check whether the earlier remarks are applicable to position in the X-direction and linear momentum in the Y-direction by finding

$$[\hat{x}, \hat{p}_y]\varphi(x, y) = [\hat{x}\hat{p}_y - \hat{p}_y\hat{x}]\varphi(x, y)$$

$$= \hat{x}\{\hat{p}_y\ \varphi(x, y)\} - \hat{p}_y\{\hat{x}\ \varphi(x, y)\}$$

$$= \hat{x}\left\{-i\hbar\frac{\partial}{\partial y}\ \varphi(x, y)\right\} - \left[-i\hbar\frac{\partial}{\partial y}\{\hat{x}\ \varphi(x, y)\}\right]$$

$$= \hat{x}\left\{-i\hbar\frac{\partial\varphi(x, y)}{\partial y}\ \right\} - \left[-i\hbar\frac{\partial\{x\ \varphi(x, y)\}}{\partial y}\right]$$

$$= x\left\{-i\hbar\frac{\partial\varphi(x, y)}{\partial y}\ \right\}$$

$$- \left\{-i\hbar\frac{\partial x}{\partial y}\times\varphi(x, y) - i\hbar x\frac{\partial\varphi(x, y)}{\partial y}\right\}$$

$$= -i\hbar x\frac{\partial\varphi(x, y)}{\partial y} + i\hbar\times\frac{\partial x}{\partial y}\times\varphi(x, y)$$

$$+ i\hbar x\frac{\partial\varphi(x, y)}{\partial y}$$

$$= -i\hbar x\frac{\partial\varphi(x, y)}{\partial y} + 0 + i\hbar x\frac{\partial\varphi(x, y)}{\partial y} = 0 \tag{4.52}$$

The fact that the outcome is zero means that the operators $\hat{x}, \hat{p}_y$ are commutative. Hence, $\hat{x}, \hat{p}_y$ allow simultaneous exact measurement.

4.5 Eigenvalues and Eigenfunctions of an Operator

For an operator $\hat{O}$, the eigenvalues are those numbers a_j and the eigenfunctions are those functions φ_j which satisfy the equation

$$\hat{O}\varphi_j = a_j\varphi_j \tag{4.53}$$

where the subscript j is a label for the different eigenfunctions and the corresponding eigenvalues of the equation.

When $\hat{O}$ acts on one of its eigenfunctions, say φ_5, we have

$$\hat{O}\ \varphi_5 = a_5\varphi_5 = \text{Same eigenfunction } \varphi_5 \times \text{Eigenvalue } a_5 \tag{4.54}$$

The equation

$$\hat{O}\varphi = a\varphi \tag{4.55}$$

is called an eigenvalue equation in which φ is the eigenfunction with the eigenvalue a. As an example, the equation

$$\frac{d}{dx}\{\exp(kx)\} = k\exp(kx) \tag{4.56}$$

is an eigenvalue equation in which exp (kx) are the eigenfunctions with eigenvalues k having different values.

4.6 Schrodinger's Equation in Operator Form

Schrodinger's time-independent equation is written concisely and elegantly in operator form as

$$\hat{H}\Psi = E\Psi \tag{4.57}$$

where $\hat{H}$ is the Hamiltonian operator and E is the total energy of the system. The operator $\hat{H}$ acts on the wave function Ψ to give the product $E\Psi$. The Schrodinger equation is an eigenvalue equation in which the values of total energy E are the eigenvalues obtained when the Hamiltonian operator acts on the wave function Ψ. As this operation returns the original wave function multiplied by a constant E, the wave function Ψ is the eigenfunction.

4.7 Hermitian Operators

Since quantum mechanical operators are related to observable quantities, they cannot give imaginary values. Hence, the values obtained by applying a quantum-mechanical operator on a function are always real. Operators whose action leads to real values are known as Hermitian operators (Saleem 2015), i.e. Hermitian operators always yield real values. As an example, consider the operator $\hat{O}$ action on a function Ψ. Then the operator $\hat{O}$ will be a Hermitian operator if in the equation

$$\hat{O}\Psi = a\Psi \tag{4.58}$$

the value a is always real.

For the operator $\hat{O}$ acting on a function Ψ, multiplication by $\Psi*$ followed by integration gives

$$\int \Psi^* \hat{O} \Psi \, d\tau = \int \Psi^* a \Psi \, d\tau = a \int \Psi^* \Psi \, d\tau = a \times 1 = a \quad (4.59)$$

Taking the complex conjugate of eq. (4.58), we get

$$\hat{O} \Psi^* = a \Psi^* \quad (4.60)$$

Multiplying by Ψ and integrating, we have

$$\int \Psi \hat{O} \Psi^* \, d\tau = \int \Psi a \Psi^* \, d\tau = a \int \Psi \Psi^* \, d\tau = a \times 1 = a \quad (4.61)$$

Thus

$$\int \Psi^* \hat{O} \Psi \, d\tau = \int \Psi \hat{O} \Psi^* \, d\tau = a \quad (4.62)$$

This relationship is obeyed by Hermitian operators and any operator satisfying this equation is a Hermitian operator.

4.8 Expectation Value

If $\Psi(x, t)$ is a solution of the time-dependent Schrodinger equation (TDSE), then average position of the particle is given by

$$\bar{x} = \sum_i x_i \, P(x_i) \delta x \quad (4.63)$$

In the limit $\delta x \to 0$, the summation becomes an integration. The resulting value of position is called its expectation value, and is written as

$$x = \int_{-\infty}^{\infty} x P(x) dx = \int_{-\infty}^{\infty} x |\Psi(x, t)|^2 \, dx \quad (4.64)$$

The expectation value is the average value of all the possible results of a measurement taking into consideration their probability of occurrence, i.e. an average weighted by probabilities.

4.9 Postulates of Quantum Mechanics in Analogy to Classical Mechanics

The mathematical framework of quantum mechanics is constructed on postulates, which are not proven and cannot be derived but are taken to be true as a groundwork to formulate theories and make predictions. These predictions are tested by experimental investigations. If the predictions are verified experimentally, the postulates are confirmed to be correct, otherwise untrue.

Postulate I: State function: The state of a quantum mechanical system comprising a particle or several particles is completely specified by a state function $\Psi(x, y, z, t)$ called the wave function which depends on spatial coordinates x, y, z of the particle(s) and time t. The product

$$\Psi^*(x, y, z, t) \Psi(x, y, z, t) \quad (4.65)$$

i.e. complex conjugate of the wave function $\Psi^*(x, y, z, t)$ multiplied by the wave function $\Psi(x, y, z, t)$ itself represents the probability density of finding the particle at position (x, y, z) at time t, and

$$\Psi^*(x, y, z, t) \Psi(x, y, z, t) dx \, dy \, dz \quad (4.66)$$

is the probability of the particle being found in a small volume element $dx \, dy \, dz$ at position (x, y, z) at time t.

To articulate the absolute certainty or out-and-out sureness of existence of the particle somewhere in space, the normalization condition for the wave function over all space is written as

$$\iint \int_{-\infty}^{\infty} \Psi^*(x, y, z, t) \Psi(x, y, z, t) dx \, dy \, dz = 1 \quad (4.67)$$

or

$$\iint \int_{-\infty}^{\infty} \Psi^*(x, y, z, t) \Psi(x, y, z, t) d\tau = 1 \quad (4.68)$$

where

$$d\tau = dx \, dy \, dz \quad (4.69)$$

A wave function obeying the earlier equation (4.67) or (4.68) is called a normalized function. This equation is known as the normalization condition. An acceptable wave function:

i. must be single-valued because probability of finding the particle at a point cannot have multiple values;

ii. must be finite otherwise probability will become infinite;

iii. must be continuous and must have a continuous first derivative everywhere because $\partial^2 \Psi / \partial x^2$ in Schrodinger's equation will be finite everywhere if $\partial \Psi / \partial x$ and hence Ψ is continuous across a boundary.

iv. must be square-integrable so that the integral in probability equation is convergent. If such a wave function obeys the normalization condition at one instant then it satisfies the same at all later times.

In classical mechanics, the state of a system is characterized by dynamical variables, e.g. position (x, y, z) and momentum p in addition to static variables such as mass m, electronic charge. Thus $\Psi(x, y, z, t)$ is a replacement for dynamical variables of classical mechanics. Given the initial position and momentum of a particle and the forces acting on it, the trajectory of the particle is found. Classical mechanics is deterministic whereas quantum mechanics is probabilistic.

Postulate II: Observables and operators: To every physically observable or measurable property of the quantum-mechanical system corresponds a linear and Hermitian operator, which acts on the wave function according to a defined mathematical rule to give another function. Linearity of operators assures the applicability of linear or matrix algebra for their manipulation. Hermitian operators are chosen because they have real eigenvalues, not involving any imaginary numbers. Thus imaginary values are automatically excluded as being unrealistic. Variables like position, linear momentum, angular momentum, and kinetic and potential energies in classical mechanics are associated with individual operators in quantum mechanics.

Postulate III: Eigenvalues: When the action of an operator $\hat{O}$ on a function $\Psi(x, y, z, t)$ leads to an outcome, which is the original function $\Psi(x, y, z, t)$ itself scaled by a scalar constant A, the function $\Psi(x, y, z, t)$ is called the eigenfunction, the scalar constant A is known as the eigenvalue, and the equation describing the operation

$$\hat{O}\Psi(x, y, z, t) = A\Psi(x, y, z, t) \qquad (4.70)$$

is termed an eigenvalue equation. In quantum mechanics, the only allowed values of measurements of an observable are eigenvalues generated by the action of the operator for that observable on the wave function, and satisfying the eigenvalue equation. Thus the eigenvalues are the values of the observable. In classical mechanics, all values of dynamical variables are permitted.

Postulate IV: Equation of motion: The time evolution of the wave function is governed by Schrodinger's time-dependent equation

$$i\hbar \frac{\partial \Psi(x, y, z, t)}{\partial t} = \hat{H} \; \Psi(x, y, z, t) \qquad (4.71)$$

It is the equation of motion of the system. In classical mechanics, the same status and eminence is held by Newton's laws of motion. Thus Schrodinger's equation is the governing equation of quantum mechanics.

4.10 Discussion and Conclusions

The complete dynamical information about a quantum-mechanical system is contained inside a mathematical entity called the wave function. Operators are means of extracting this information from the wave function (Feynman et al 2011). Name any measurable property or observable and there is a corresponding operator acting on the wave function. Physically, the operator represents an observable property of the system such as position or momentum. Mathematically, an operator is a blackbox in which an input is fed to get an output.

In particular, if the wave function is an eigenfunction of an operator, then its eigenvalue is the value of the observable. An illustrious example of such a case is the Schrodinger equation. In this equation, $\hat{H}$ is the Hamiltonian operator of which Ψ is an eigenfunction. E is the eigenvalue of Ψ representing the total energy of the system. We know that observables are physical quantities that can be measured in an experiment. So it is necessary that the eigenvalues from which they are obtained must be real too. To assure that the eigenvalues are real, it is essential that the operators for the observables are Hermitian. Notwithstanding, in case the wave function is not an eigenfunction of the operator for a specific observable, it is still possible to calculate the expectation value of that property.

Quantum mechanics is an axiomatic theory, which builds up on few truths that cannot be further proved, being obvious by themselves. These postulates must be kept firmly instilled in mind.

Illustrative Exercises

4.1 Confirm the linearity of the integration operator.

For linearity of an operator $\hat{O}$ acting on a function $f(x)$, two conditions must be met:

$$\hat{O}\left\{f(x) + g(x)\right\} = \hat{O}f(x) + \hat{O}g(x) \qquad (4.72)$$

and

$$\hat{O}\, cf(x) = c\,\hat{O}\, f(x) \qquad (4.73)$$

where c is a coefficient.

Since

$$\int \left\{f(x) + g(x)\right\} dx = \int \left\{f(x)\right\} dx + \int \left\{g(x)\right\} dx \qquad (4.74)$$

the first condition is satisfied.

Further, since

$$\int c\left\{f(x)\right\} dx = c\int \left\{f(x)\right\} dx \qquad (4.75)$$

the second condition is also satisfied. Therefore, linearity of this operator is confirmed.

4.2 Prove that the Hamiltonian operator is linear.

The two linearity conditions can be combined together and jointly stated as

$$\hat{O}\left\{c_1 f(x) + c_2 g(x)\right\} = c_1 \hat{O}f(x) + c_2 \hat{O}g(x) \qquad (4.76)$$

To verify whether this condition holds for the Hamiltonian operator, we write

$$\left\{-\frac{\hbar^2}{2m}\frac{d^2}{dx^2} + V(x)\right\}\left\{c_1 f(x) + c_2 g(x)\right\}$$

$$= \left\{-\frac{\hbar^2}{2m}\frac{d^2}{dx^2} + V(x)\right\}\left\{c_1 f(x)\right\} + \left\{-\frac{\hbar^2}{2m}\frac{d^2}{dx^2} + V(x)\right\}\left\{c_2 g(x)\right\}$$

$$= \left(-\frac{\hbar^2}{2m}\frac{d^2}{dx^2}\right)\left\{c_1 f(x)\right\} + V(x)\left\{c_1 f(x)\right\}$$

$$+ \left(-\frac{\hbar^2}{2m}\frac{d^2}{dx^2}\right)\left\{c_2 g(x)\right\} + V(x)\left\{c_2 g(x)\right\}$$

$$= c_1\left(-\frac{\hbar^2}{2m}\frac{d^2}{dx^2}\right)f(x) + c_1\left\{V(x)\right\}f(x)$$

$$+ c_2\left(-\frac{\hbar^2}{2m}\frac{d^2}{dx^2}\right)g(x) + c_2\left\{V(x)\right\}g(x)$$

$$= c_1\left\{-\frac{\hbar^2}{2m}\frac{d^2}{dx^2} + V(x)\right\}f(x) + c_2\left\{-\frac{\hbar^2}{2m}\frac{d^2}{dx^2} + V(x)\right\}g(x)$$

$$\qquad (4.77)$$

so that the linearity condition is satisfied, establishing the linearity of Hamiltonian operator.

4.3 (a) Show that the position operator $\hat{x} = x$ is Hermitian.

(b) Is the potential energy operator $\hat{V}(\hat{x}) = V(x)$ a Hermitian operator?

(a) Condition for an operator to be Hermitian is

$$\int \left(\hat{O}\Psi\right)^* \Psi dx = \int \Psi^* \hat{O}\Psi dx \qquad (4.78)$$

For the position operator, we have to prove that

$$\int \left(\hat{x}\Psi\right)^* \Psi dx = \int \Psi^* \hat{x}\Psi dx \qquad (4.79)$$

Now

$$\int \left(\hat{x}\Psi\right)^* \Psi dx = \int \left(x\Psi\right)^* \Psi dx = \int \Psi^* x\Psi dx = \int \Psi^* \hat{x}\Psi dx \qquad (4.80)$$

Hence the position operator is Hermitian.

(b) Yes, the potential energy operator $\hat{V}(\hat{x}) = V(x)$ is a Hermitian operator because it is a function of the position operator, which itself is a Hermitian operator.

4.4 Show that the linear momentum operator $\hat{p}_x = -i\hbar \dfrac{\partial}{\partial x}$ is Hermitian.

We have to prove that

$$\int_{-\infty}^{+\infty} \left(\hat{p}_x\Psi\right)^* \Psi\, dx = \int_{-\infty}^{+\infty} \Psi^* \hat{p}_x\Psi\, dx \qquad (4.81)$$

Now

$$\int_{-\infty}^{+\infty} \left(\hat{p}_x\Psi\right)^* \Psi\, dx = \int_{-\infty}^{+\infty} \left(-i\hbar \frac{\partial\Psi}{\partial x}\right)^* \Psi\, dx$$

$$= \int_{-\infty}^{+\infty} (-i\hbar)^* \left(\frac{\partial\Psi}{\partial x}\right)^* \Psi\, dx$$

$$= \int_{-\infty}^{+\infty} (+i\hbar) \left(\frac{\partial\Psi}{\partial x}\right)^* \Psi\, dx$$

$$= i\hbar \int_{-\infty}^{+\infty} \left(\frac{\partial\Psi}{\partial x}\right)^* \Psi\, dx = i\hbar \int_{-\infty}^{+\infty} \Psi\, d\Psi^*$$

$$= i\hbar \left|\Psi\Psi^*\right|_{-\infty}^{+\infty} - i\hbar \int_{-\infty}^{+\infty} \Psi^*\, d\Psi \qquad (4.82)$$

after integrating by parts

$$\int u\, dv = uv - \int v\, du \qquad (4.83)$$

by taking

$$u = \Psi \text{ and } dv = d\Psi^* \qquad (4.84)$$

For a finite system or a confined particle, the wave function becomes zero at $+\infty$ or $-\infty$, i.e.,

$$\Psi\Psi^* \to 0 \text{ as } x \to \pm\infty \qquad (4.85)$$

Hence, the first term in eq. (4.82) is zero, so that

$$\int_{-\infty}^{+\infty} \left(\hat{p}_x\Psi\right)^* \Psi\, dx = -i\hbar \int_{-\infty}^{+\infty} \Psi^*\, d\Psi = -i\hbar \int_{-\infty}^{+\infty} \Psi^* \frac{d\Psi}{dx}\, dx$$

$$= \int_{-\infty}^{+\infty} \Psi^* \left(-i\hbar \frac{d\Psi}{dx}\right) dx = \int_{-\infty}^{+\infty} \Psi^* \hat{p}_x\Psi\, dx \qquad (4.86)$$

Hence proved.

4.5 Prove that the kinetic energy operator $\hat{T}_x = \dfrac{\hat{p}_x^2}{2m}$ and the Hamiltonian operator $\hat{H}_x = -\dfrac{\hbar^2}{2m} \dfrac{\hat{p}_x^2}{2m} + \hat{V}(\hat{x})$ are Hermitian.

To prove that the kinetic energy operator is Hermitian, we have to show that

$$\int \left(\hat{T}_x\Psi\right)^* \Psi\, dx = \int \Psi^* \hat{T}_x\Psi\, dx \qquad (4.87)$$

Now

$$\int \left(\hat{T}_x\Psi\right)^* \Psi\, dx = \int \left(\frac{\hat{p}_x^2}{2m}\Psi\right)^* \Psi\, dx = \frac{1}{2m} \int \left(\hat{p}_x^2\Psi\right)^* \Psi\, dx$$

$$= \frac{1}{2m} \int \left(\hat{p}_x\Psi\right)^* \hat{p}_x\Psi\, dx = \frac{1}{2m} \int \Psi^* \hat{p}_x^2\Psi\, dx$$

$$= \int \Psi^* \left(\frac{\hat{p}_x^2}{2m}\right)\Psi\, dx = \int \Psi^* \hat{T}_x\Psi\, dx \qquad (4.88)$$

Hence proved.

The Hamiltonian operator is

$$\hat{H}_x = -\hat{T}_x + \hat{V}(\hat{x}) \qquad (4.89)$$

As both the constituent operators in the Hamiltonian operator, namely the kinetic energy operator $\hat{T}_x$ and the potential energy operator $\hat{V}(\hat{x})$, are linear operators, as shown earlier, it readily follows that the Hamiltonian operator is also a linear operator.

REFERENCES

Feynman, R. P., R. B. Leighton, M. Sands. 2011. *The Feynman Lectures on Physics*, New Millennium edition. Basic Books: New York, vol. 3, pp. 20–1 to 20–7.

Jordan, T. F. 2006. *Linear Operators for Quantum Mechanics*. Dover Publications, Inc.: New York, 174 pp.

Saleem, M. 2015. *Quantum Mechanics*, Ch5: The Role of Hermitian Operators. IOP Publishing Ltd.: Bristol, pp. 5–1 to 5–39.

Schechter, M. 2003. *Operator Methods in Quantum Mechanics*. Dover Publications, Inc., New York, 368 pp.

5

Particle-in-a-Box and Related Problems

Although the particle in a box looks like a toy problem in quantum mechanics, it has a far-reaching impact on many advanced ideas (Shankar 1994, Griffiths 1995). We shall begin with an unshackled particle to appreciate how the particle responds on losing freedom.

5.1 Free particle in Quantum Mechanics

Like a classical free particle, a quantum-mechanical free particle is one unacted upon by any force, i.e. a particle in a field-free space or a region of uniform potential $V = 0$ everywhere.

5.1.1 Spatial Dependence

The one-dimensional time-independent Schrodinger equation (TISE)

$$-\frac{\hbar^2}{2m}\frac{\partial^2}{\partial x^2}\Psi(x)+V(x)\Psi(x)=E\Psi(x) \qquad (5.1)$$

reduces to

$$-\frac{\hbar^2}{2m}\frac{\partial^2}{\partial x^2}\Psi(x)+0\times\Psi(x)=E\Psi(x) \qquad (5.2)$$

or

$$-\frac{\hbar^2}{2m}\frac{\partial^2}{\partial x^2}\Psi(x)=E\Psi(x) \qquad (5.3)$$

or

$$\frac{\partial^2}{\partial x^2}\Psi(x)=-\frac{2mE}{\hbar^2}\,\Psi(x) \qquad (5.4)$$

or

$$\frac{\partial^2}{\partial x^2}\Psi(x)=-k^2\,\Psi(x) \qquad (5.5)$$

where

$$k^2=\frac{2mE}{\hbar^2},\text{ or }E=\frac{\hbar^2 k^2}{2m} \qquad (5.6)$$

Equation (5.5) is similar to the famous equation for a simple harmonic oscillator. It is a linear, homogeneous, second order differential equation with constant coefficients. We take a test function

$$\Psi(x)=\exp(rx) \qquad (5.7)$$

where r is the root. We put this solution in eq. (5.5) and find r. To do this, we determine the first and second partial derivatives of $\Psi(x)$:

$$\frac{\partial\Psi(x)}{\partial x}=\frac{\partial\{\exp(rx)\}}{\partial x}=r\exp(rx) \qquad (5.8a)$$

and

$$\frac{\partial^2\Psi(x)}{\partial x^2}=\frac{\partial}{\partial x}\left[\frac{\partial\{\exp(rx)\}}{\partial x}\right]=\frac{\partial}{\partial x}\{r\exp(rx)\}=r^2\exp(rx) \qquad (5.8b)$$

$$\therefore r^2\exp(rx)=-k^2\exp(rx) \qquad (5.9)$$

or

$$\left(r^2+k^2\right)\exp(rx)=0 \qquad (5.10)$$

We are interested in determining the value of r that will guarantee the validity of this equation for all values of x. We note that $\exp(rx)$ can be zero only when $x=-\infty$. Hence, this equation will be tenable if the factor

$$r^2+k^2=0 \qquad (5.11)$$

or

$$r^2=-k^2=i^2k^2 \qquad (5.12)$$

or

$$r=\pm ik \qquad (5.13)$$

Thus there are two functions: $\exp(+ikx)$ and $\exp(-ikx)$ that satisfy the differential equation (5.5). So, the most general solution of this equation describing all the solutions to it is a linear combination of these solutions, and can be expressed in the form

$$\Psi(x)=C_1\exp(ikx)+C_2\exp(-ikx) \qquad (5.14)$$

where C_1 and C_2 are constants.

Now, let us try

$$\Psi(x)=C_1\exp(ikx)+C_2\exp(-ikx) \qquad (5.15)$$

as a solution with constants C_1 and C_2. To verify that this is a solution, we find

$$\frac{\partial}{\partial x}\Psi(x)=\frac{\partial}{\partial x}\{C_1\exp(ikx)+C_2\exp(-ikx)\}$$

$$=ikC_1\exp(ikx)-ikC_2\exp(-ikx) \qquad (5.16)$$

and

$$\frac{\partial^2}{\partial x^2}\Psi(x) = \frac{\partial}{\partial x}\left\{\frac{\partial}{\partial x}\Psi(x)\right\}$$

$$= \frac{\partial}{\partial x}\left\{ikC_1\exp(ikx) - ikC_2\exp(-ikx)\right\}$$

$$= ik \times ik \times C_1\exp(ikx) - ik \times (-ik) \times C_2\exp(-ikx)$$

$$= i^2k^2C_1\exp(ikx) + i^2k^2C_2\exp(-ikx)$$

$$= i^2k^2\left\{C_1\exp(ikx) + C_2\exp(-ikx)\right\}$$

$$= -1 \times k^2 \times \Psi(x) = -k^2\Psi(x) \tag{5.17}$$

Since eq. (5.5) is satisfied by the trial solution, it is confirmed that the trial solution is actually a solution of eq. (5.5).

5.1.2 Time Dependence

The time-dependent Schrodinger equation (TDSE)

$$-\frac{\hbar^2}{2m}\frac{\partial^2}{\partial x^2}\Psi(x,t) + V(x)\Psi(x) = i\hbar\frac{\partial}{\partial t}\Psi(x,t) \tag{5.18}$$

simplifies to

$$-\frac{\hbar^2}{2m}\frac{\partial^2}{\partial x^2}\Psi(x,t) + 0 \times \Psi(x) = i\hbar\frac{\partial}{\partial t}\Psi(x,t) \tag{5.19}$$

or

$$-\frac{\hbar^2}{2m}\frac{\partial^2}{\partial x^2}\Psi(x,t) = i\hbar\frac{\partial}{\partial t}\Psi(x,t) \tag{5.20}$$

Since potential energy being zero is time-independent, we can separate the wave function in Schrodinger's equation into two parts: one part dependent on position only and the other part dependent on time only as

$$\Psi(x,t) = \psi(x)T(t) \tag{5.21}$$

where

$$\psi(x) = \Psi(x) = C_1\exp(ikx) + C_2\exp(-ikx) \tag{5.22}$$

Then TDSE is converted into the equation

$$-\frac{\hbar^2}{2m}\frac{\partial^2}{\partial x^2}\psi(x)T(t) = i\hbar\frac{\partial}{\partial t}\psi(x)T(t) \tag{5.23}$$

Since the partial derivative with respect to x acts on the function $\psi(x)$ only and does not act on $T(t)$ while the partial derivative with respect to t acts on the function $T(t)$ only and does not act on $X(x)$, we can rewrite the earlier eq. (5.23) as

$$\left\{-\frac{\hbar^2}{2m}\frac{\partial^2}{\partial x^2}\psi(x)\right\}T(t) = \left\{i\hbar\frac{\partial}{\partial t}T(t)\right\}\psi(x) \tag{5.24}$$

Dividing both sides of eq. (5.24) by $\psi(x)T(t)$, we have

$$\frac{1}{\psi(x)}\left\{-\frac{\hbar^2}{2m}\frac{\partial^2}{\partial x^2}\psi(x)\right\} = \frac{1}{T(t)}\left\{i\hbar\frac{\partial}{\partial t}T(t)\right\} \tag{5.25}$$

The left-hand side of this equation depends on position only whereas the right-hand side is only time-dependent. To ensure validity of this equation for all possible combinations of position and time, we set both sides of the equation = a constant η, arbitrarily chosen and undetermined. Then we get two equations

$$\frac{1}{\psi(x)}\left\{-\frac{\hbar^2}{2m}\frac{\partial^2}{\partial x^2}\psi(x)\right\} = \eta \tag{5.26}$$

and

$$\frac{1}{T(t)}\left\{i\hbar\frac{\partial}{\partial t}T(t)\right\} = \eta \tag{5.27}$$

Considering eq. (5.26),

$$[\text{Left-hand side}] = \frac{[\hbar^2]}{[m] \times [L^2]} = [\text{Energy}] = [E] \tag{5.28}$$

as we have seen in eqs. (3.51)–(3.55). For eq. (5.27),

$$[\text{Left-hand side}] = \frac{[\hbar]}{[T]} = [\text{Energy}] = [E] \tag{5.29}$$

Thus

$$\eta = E \tag{5.30}$$

We are interested in the time equation

$$i\hbar\frac{\partial}{\partial t}T(t) = ET(t) \tag{5.31}$$

or

$$\frac{\partial}{\partial t}T(t) = \frac{1}{i\hbar}ET(t) = \frac{i}{i^2\hbar}ET(t) = -\frac{i}{\hbar}ET(t) \tag{5.32}$$

or

$$\frac{\partial T(t)}{T(t)} = -\frac{iE}{\hbar}\partial t \tag{5.33}$$

or

$$\int\frac{\partial T(t)}{T(t)} = \int\left(-\frac{iE}{\hbar}\right)\partial t = -\frac{iE}{\hbar}\int\partial t \tag{5.34}$$

or

$$\ln T(t) = -\frac{iEt}{\hbar} + \ln K \tag{5.35}$$

giving

$$T(t) = K \exp\left(-\frac{iEt}{\hbar}\right) \tag{5.36}$$

where ln K is taken as the integration constant. Hence, from eqs. (5.21), (5.22), and (5.36)

$$\Psi(x,t) = \psi(x) T(t)$$

$$= \left\{ C_1 \exp(ikx) + C_2 \exp(-ikx) \right\} K \exp\left(-\frac{iEt}{\hbar}\right)$$

$$= C_1 K \exp(ikx) \exp\left(-\frac{iEt}{\hbar}\right) + C_2 K \exp(-ikx) \exp\left(-\frac{iEt}{\hbar}\right)$$

$$= C_1 K \exp\left(ikx - \frac{iEt}{\hbar}\right) + C_2 K \exp\left(-ikx - \frac{iEt}{\hbar}\right)$$

$$= C \exp\left\{\left(ikx - \frac{iEt}{\hbar}\right)\right\} + D \exp\left\{\left(-ikx - \frac{iEt}{\hbar}\right)\right\} \tag{5.37}$$

where

$$C_1 K = C; \text{ and } C_2 K = D \tag{5.38}$$

From eq. (5.6),

$$\frac{k^2}{2m} = \frac{E}{\hbar^2} = \frac{E}{\hbar \times \hbar} \tag{5.39}$$

or

$$\frac{E}{\hbar} = \frac{\hbar k^2}{2m} \tag{5.40}$$

Substituting for $E/\hbar$ from eq. (5.40) into eq. (5.37), we get

$$\Psi(x,t) = C \exp\left\{\left(ikx - \frac{iEt}{\hbar}\right)\right\} + D \exp\left\{-\left(ikx + \frac{iEt}{\hbar}\right)\right\}$$

$$= C \exp\left\{\left(ikx - \frac{i\hbar k^2 t}{2m}\right)\right\} + D \exp\left\{-\left(ikx + \frac{i\hbar k^2 t}{2m}\right)\right\}$$

$$= C \exp\left\{ik\left(x - \frac{\hbar k}{2m}t\right)\right\} + D \exp\left\{-ik\left(x + \frac{\hbar k}{2m}t\right)\right\} \tag{5.41}$$

We know that the graph of the complex exponential function $\exp(ikx)$ obtained by plotting the Taylor series expansion of $\exp(ikx)$ in the three-dimensional complex space with X-axis, real axis, and imaginary axis has the shape of a spiral or coil, and represents a wave moving along the X-axis (Figure 5.1). With change in time, the spiral translates, either in the $-X$-direction or $+X$-direction. Let us enquire how does the wave propagate?

Consider the first part of the wave function

First part of the wave function: $\Psi(x,t) = C \exp\left\{ik\left(x - \frac{\hbar k}{2m}t\right)\right\}$ \hfill (5.42)

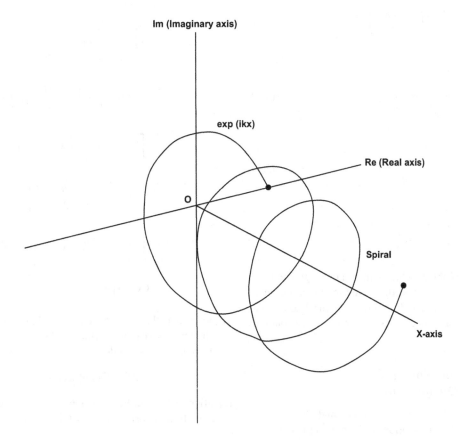

FIGURE 5.1 Graph of exponential function $\exp(ikx)$ in 3-D complex space representing a wave moving along the X-axis.

Let us concentrate our attention on a particular point of the wave described by this equation. To follow this point on the wave, we look at the argument of the wave function. It must remain constant otherwise the wave function will change, i.e.,

$$ik\left(x - \frac{\hbar k}{2m}t\right) = \text{Constant} \tag{5.43}$$

Let us take this constant to be zero, i.e. we are following the point for which

$$ik\left(x - \frac{\hbar k}{2m}t\right) = 0 \tag{5.44}$$

Since $ik \neq 0$,

$$x - \frac{\hbar k}{2m}t = 0 \tag{5.45}$$

or

$$x = \frac{\hbar k}{2m}t \tag{5.46}$$

This expression tells us the direction of wave motion. If t increases, x increases in the positive direction. Hence this part of the wave function represents a wave moving in the positive direction along the X-axis. On similar lines,

Second part of the wavefunction $\Psi(x,t)$

$$= D\exp\left\{-ik\left(x + \frac{\hbar k}{2m}t\right)\right\} \tag{5.47}$$

Let us now take

$$-ik\left(x + \frac{\hbar k}{2m}t\right) = 0 \tag{5.48}$$

or

$$x + \frac{\hbar k}{2m}t = 0 \tag{5.49}$$

or

$$x = -\frac{\hbar k}{2m}t \tag{5.50}$$

An increase in t brings about a change in x in the negative direction. Therefore, the second part of the wave function represents a wave advancing in the negative direction along the X-axis. Thus the two parts of the wave function are traveling wave solutions to TDSE.

5.1.3 Boundary Conditions and Absence of Quantization

For the free particle, $V(x) = 0$ everywhere and there are no boundaries within which the motion of the particle is confined. Hence, there are no boundary conditions for the free particle.

This implies that the parameter k can assume any possible value and therefore energy E also can acquire any value. Hence, no quantization takes place for a free particle in quantum mechanics. The eigen energies for the free particle are given by eq. (5.6), which allows continuous energy values.

5.1.4 Non-Normalization of Wave Function

Another major question arises when we examine the normalization of the wave function. The normalization condition is

$$\int_{-\infty}^{\infty} \Psi^*(x, t)\Psi(x, t)dx = 1 \tag{5.51}$$

For waves moving in the positive direction along the X-axis, the wave function can be written as

$$\Psi(x,t) = C\exp\left\{ik\left(x - \frac{\hbar k}{2m}t\right)\right\} \tag{5.52}$$

The complex conjugate of $\Psi(x, t)$ is

$$\Psi^*(x,t) = C^*\exp\left\{-ik\left(x - \frac{\hbar k}{2m}t\right)\right\} \tag{5.53}$$

Putting the values of $\Psi(x, t)$ and $\Psi^*(x, t)$ on the left-hand side in eq. (5.51), we get

$$\text{Left-hand side} = \int_{-\infty}^{\infty} \Psi^*(x, t)\Psi(x, t)dx$$

$$= \int_{-\infty}^{\infty}\left[C\exp\left\{ik\left(x - \frac{\hbar k}{2m}t\right)\right\}\right.$$

$$\left.\times C^*\exp\left\{-ik\left(x - \frac{\hbar k}{2m}t\right)\right\}\right]dx$$

$$= \int_{-\infty}^{\infty}CC^*\left[\exp\left\{ik\left(x - \frac{\hbar k}{2m}t\right)\right\}\right.$$

$$\left.\times\exp\left\{-ik\left(x - \frac{\hbar k}{2m}t\right)\right\}\right]dx$$

$$= \int_{-\infty}^{\infty}CC^* \times 1 \times dx = CC^*\int_{-\infty}^{\infty}dx = CC^* \times [x]_{-\infty}^{\infty}$$

$$= CC^* \times [\infty - (-\infty)] = \infty \tag{5.54}$$

To normalize the wave function, we have to multiply by $1/\infty$, which means we cannot normalize the wave function.

At $t = 0$, eq. (5.41) reduces to

$$\Psi(x,0) = C\exp(ikx) + D\exp(-ikx) \tag{5.55}$$

Equation (5.55) can be rewritten in a more convenient form by applying Euler's formula. Using this formula,

$$\exp(ikx) = \cos(kx) + i\sin(kx) \tag{5.56}$$

and

$$\exp(-ikx) = \cos(kx) - i\sin(kx) \tag{5.57}$$

and writing sine term first followed by cosine term, we get

$$\Psi(x,0) = C\left\{i\sin(kx) + \cos(kx)\right\} + D\left\{-i\sin(kx) + \cos(kx)\right\}$$

$$= Ci\sin(kx) - Di\sin(kx) + C\cos(kx) + D\cos(kx)$$

$$= i(C-D)\sin(kx) + (C+D)\cos(kx)$$

$$= A\sin(kx) + B\cos(kx) \tag{5.58}$$

where

$$i(C-D) = A \tag{5.59}$$

and

$$C+D = B \tag{5.60}$$

A moving free particle with zero potential energy is represented by a wave function $\Psi(x, 0)$ involving the wave number k and hence the precisely known momentum p of the particle through the relation

$$k = p/\hbar \tag{5.61}$$

as given by eq. (2.23).

If $B = 0$, eq. (5.58) for wave function simplifies to

$$\Psi(x,0) = A\sin(kx) \tag{5.62}$$

Plotting $\Psi(x, 0)$ along the Y-axis and x along the X-axis, we get a sinusoidal wave called the plane wave, which stretches indefinitely in both directions (Figure 5.2). Now, we emphasize that error in momentum $\Delta p = 0$. So, according to Heisenberg's uncertainty principle, it is impossible to determine the exact location of the particle. The free particle is spread out infinitely in space. It behaves as a wave extending from $-\infty$ to $+\infty$. Hence, the uncertainty in position is

$$\Delta x = \infty \tag{5.63}$$

Because of extension of plane waves to infinity in both directions, these waves cannot represent a particle with a non-zero wave function in a certain spatial region. Therefore, we utilize the framework of superposition in which traveling wave solutions are superimposed to form wave packets by writing

$$\Psi(x, t = 0) = \text{Superposition (overlapping and interaction)}$$

$$\text{of travelling waves} \tag{5.64}$$

5.1.5 Concept of Wave Packets

If we wish to describe a particle most likely to be found within a confined region of space, we must try to produce a localized wave. This can be done by superimposition of an infinite number of plane waves (Figure 5.3). In this superimposed wave, each wave differs in wavelength λ and hence momentum p from the preceding wave by a small amount because from eq. (2.16)

$$\lambda = h/p \tag{5.65}$$

The localized region is formed at a place where the waves interfere constructively. In the remaining space, destructive interference occurs up to $+\infty$ and $-\infty$. This superimposed wave is known as a wave packet. It behaves as a wave spread over a small region Δx in which there is likelihood of finding the particle. Thus a wave packet is a single wave formed by the superimposition of an infinite number of plane waves and propagating as a small localized disturbance over a length Δx representing the likely position of the particle. This length Δx is the uncertainty in position. It has a finite value unlike the infinite value for the wave function of a free particle. For the wave packet, the momentum being the combined effect of slightly different momenta is no longer exact. It has a finite uncertainty Δp. Thus the position and momentum of the particle described by the wave packet are known with uncertainties Δx, Δp, respectively, and they are interrelated in accordance with the uncertainty principle. None of them is infinite.

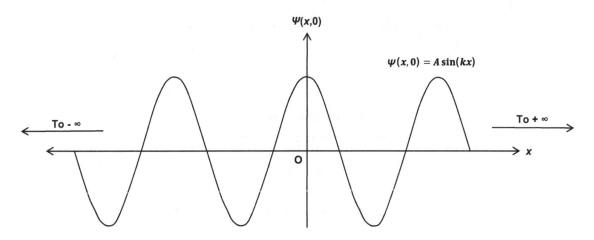

FIGURE 5.2 Plane wave stretching indefinitely in both directions.

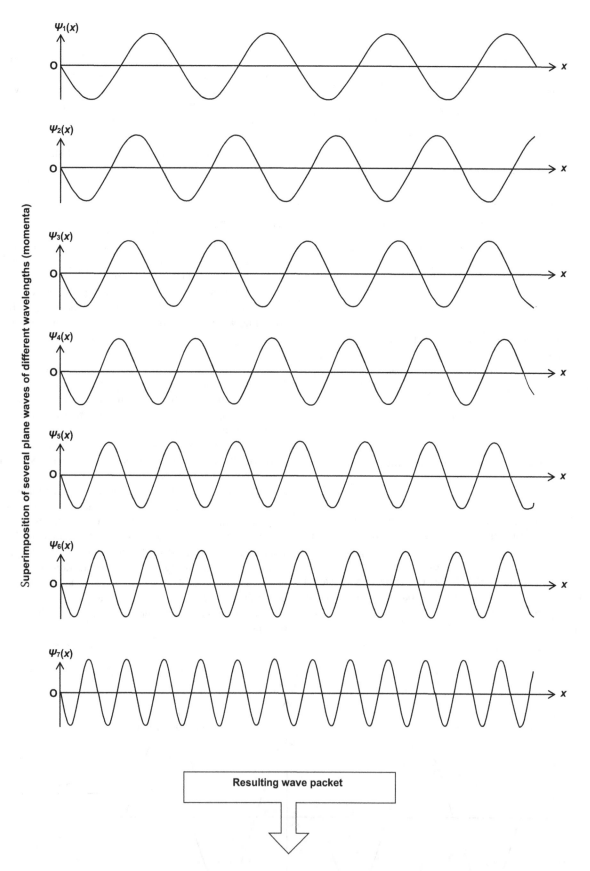

FIGURE 5.3 Formation of spatially localized wave packet by linear superimposition of plane waves of different wavelengths.

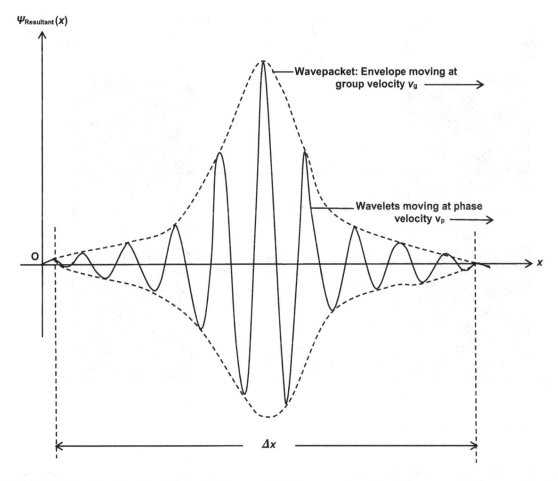

$\Psi_{\text{Resultant}}(x)$

Wavepacket: Envelope moving at group velocity v_g →

Wavelets moving at phase velocity v_p →

O

Δx

FIGURE 5.3 (CONTINUED) Formation of spatially localized wave packet by linear superimposition of plane waves of different wavelengths.

5.2 Particle in a Box or Infinite Potential Well

This problem is also called the study of particle in a rigid box, particle in an infinite potential well, or particle in infinite square well (Powell and Crasemann 2015). Here, a particle of mass m such as an electron is confined in a well of infinite depth (Figure 5.4). The particle is placed at the bottom of the well and has zero potential energy $V(x) = 0$. But it has a finite kinetic energy T so that its total energy $E = T + V$ is finite. By virtue of its kinetic energy, it can move to and fro at the bottom of the well but cannot escape from the captivity of the well whose walls are infinitely high. It is assumed that the collisions of the particle with the walls of the well are perfectly elastic, i.e. both energy and momentum are conserved during these collisions. The problem, often the first one a student is confronted with in quantum mechanics, consists in exploring the motion of the particle leftward and rightward inside the well along its width placed on the horizontal X-axis. The problem represents a hypothetical situation, not generally encountered in nature but is applicable to many complex circumstances. It is intended to familiarize the student with ideas like quantization of energy. The mathematics of the wave function is exquisitely illustrated by this problem without any approximations.

5.2.1 The Solution

The solution consists of the following steps:

i. Definition of potential energy $V(x)$:
   The left wall of the well coincides with the Y-axis and the right wall is at a distance L from the left wall so that the width of the well is L. Potential energy inside the well is zero and that outside the well is infinitely high, as defined by the equations

$$V(x) = 0 \text{ if } 0 \le x \le L \tag{5.66}$$

and

$$V(x) = \infty \text{ for } x \langle 0 \text{ or } x \rangle L \tag{5.67}$$

This means that if the particle is inside the well it will have zero potential energy but if it is found outside the well, its potential energy will be infinite.

ii. Solution of Schrodinger's equation to get the wave function inside the well:
   Because $V(x) = 0$ inside the well, the TISE is

$$-\frac{\hbar^2}{2m}\frac{\partial^2}{\partial x^2}\Psi(x) = E\Psi(x) \tag{5.68}$$

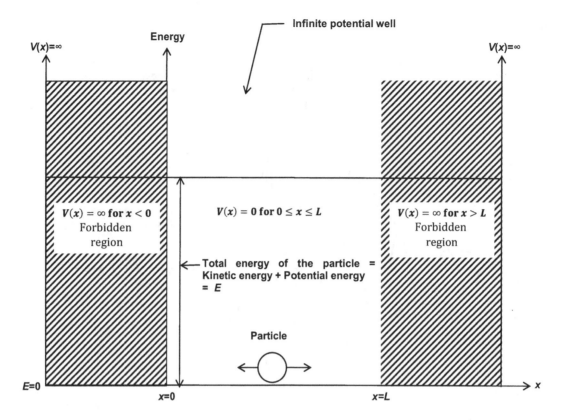

FIGURE 5.4   The particle in an infinite potential well.

The equation has been solved earlier (see eq. (5.58)) for a free particle, giving the solution

$$\Psi(x) = A\sin(kx) + B\cos(kx) \quad (5.69)$$

In this equation, there are three unknown constants. To know the wave function, the constants A, B, and k must be determined.

iii. Definition of the wave function by applying the boundary conditions:

5.2.2 Determination of B

For continuity of $\Psi(x)$,

$$\Psi(x) \text{ at } x = 0, \Psi(0) = 0 \quad (5.70)$$

and

$$\Psi(x) \text{ at } x = L, \Psi(L) = 0 \quad (5.71)$$

These equations mean that the probability of finding the particle either at $x = 0$ or at $x = L$ is zero. We can also argue like this: At $x = 0$, $V(x) = \infty$. Suppose $\Psi(x)$ has a finite value at $x = 0$, i.e. $\Psi(0)$ is a finite number. Then $V(x)\Psi(0) = \infty$. Putting this value in TISE we find

$$-\frac{\hbar^2}{2m}\frac{\partial^2}{\partial x^2}\Psi(x) + \infty = E\,\Psi(x) \quad (5.72)$$

Since $\Psi(x)$ has a finite value and is $\neq\infty$, E must be $= \infty$, which contradicts our starting assumption that the particle has a finite energy. So, $\Psi(x)$ cannot have a finite value at $x = 0$. The only way by which the earlier contradiction can be avoided is by putting $\Psi(0) = 0$.

When $x = 0$,

$$\Psi(0) = A\sin(0) + B\cos(0)$$

$$= A \times 0 + B\cos\Psi(0) = B\cos(0) \quad (5.73)$$

Since $\Psi(0) = 0$,

$$B\cos(0) = 0 \quad (5.74)$$

We know that $\cos(0) = 1$; hence $B = 0$. Thus the wave function will have no cosine term and becomes

$$\Psi(x) = A\sin(kx) + 0 \times \cos(kx) = A\sin(kx) \quad (5.75)$$

5.2.3 Determination of k

We can determine k by working backward. Since

$$\Psi(x) = A\sin(kx) \quad (5.76)$$

Differentiating twice,

$$\frac{\partial \Psi(x)}{\partial x} = kA\cos(kx) \quad (5.77)$$

and

$$\frac{\partial^2 \Psi(x)}{\partial x} = \frac{\partial}{\partial x}\left\{\frac{\partial \Psi(x)}{\partial x}\right\} = \frac{\partial}{\partial x}\left\{kA\cos(kx)\right\}$$

$$= kA \times k \times -\sin(kx) = -k^2 A\sin(kx)$$

$$= -k^2 \Psi(x) \qquad (5.78)$$

From TISE eq. (5.4),

$$\frac{\partial^2 \Psi(x)}{\partial x^2} = -\frac{2mE}{\hbar^2}\Psi(x) \qquad (5.79)$$

Comparing eqs. (5.78) and (5.79),

$$k^2 = \frac{2mE}{\hbar^2} \qquad (5.80)$$

or

$$k = \frac{\sqrt{2mE}}{\hbar} \qquad (5.81)$$

Putting the value of k in eq. (5.76),

$$\Psi(x) = A\sin(kx) = A\sin\left(\frac{\sqrt{2mE}}{\hbar}x\right) \qquad (5.82)$$

When $x = L$, $\Psi(x) = 0$

$$\therefore A\sin\left(\frac{\sqrt{2mE}}{\hbar}L\right) = 0 \qquad (5.83)$$

which is true only when

$$kL = \frac{\sqrt{2mE}}{\hbar}L = n\pi \qquad (5.84)$$

where n is an integer $= 1, 2, 3, \ldots$ because

$$\sin 0° = \sin 180° = \sin 360° = \sin 540° = \cdots = 0 \qquad (5.85)$$

or

$$\sin 0° = \sin\pi = \sin 2\pi = \sin 3\pi = \cdots = 0 \qquad (5.86)$$

From eq. (5.84),

$$\frac{\sqrt{2mE}}{\hbar} = \frac{n\pi}{L} \qquad (5.87)$$

Equations (5.82) and (5.87) give

$$\Psi(x) = A\sin\left(\frac{\sqrt{2mE}}{\hbar}x\right) = A\sin\left(\frac{n\pi}{L}x\right) \qquad (5.88)$$

Further, squaring both sides of eq. (5.87)

$$\left(\frac{\sqrt{2mE}}{\hbar}\right)^2 = \left(\frac{n\pi}{L}\right)^2 \qquad (5.89)$$

or

$$\frac{2mE}{\hbar^2} = \frac{n^2\pi^2}{L^2} \qquad (5.90)$$

In this equation, m, $\hbar$, L are all constants. Of these, m, L are properties of the system. The only variable that we can solve for is the energy E.

$$E = \frac{n^2\pi^2\hbar^2}{2mL^2} \qquad (5.91)$$

This equation contains n, which can take only integral values and is called the quantum number. The idea of energy quantization appears at this point. Starting from a system which was continuous, we find that only a discrete set of energies are allowed for different integral values of $n = 1, 2, 3, \ldots$. Why? It happens because the wave function is a sinusoidal function. It has a value of zero after each step of 180°. In the intervening angles, the sine function is non-zero. So, we find that the wave nature of the particle led to the Schrodinger equation. The equation when solved and subjected to boundary condition produced a discrete or quantum solution. This is the reason of quantization. Thus the wave function equation can be written with subscript n as

$$\Psi_n(x) = A\sin\left(\frac{n\pi}{L}x\right) \qquad (5.92)$$

The energy equation can also be written in the same fashion as

$$E_n = \frac{n^2\pi^2\hbar^2}{2mL^2} \qquad (5.93)$$

to indicate that only values corresponding to integral n are permitted.

5.2.4 Determination of A

The normalization condition is

$$1 = \int_0^L \Psi_n^*(x)\Psi_n(x)\,dx$$

$$= \int_0^L A\sin\left(\frac{n\pi}{L}x\right) \times A\sin\left(\frac{n\pi}{L}x\right)dx$$

$$= \int_0^L A^2 \sin^2\left(\frac{n\pi}{L}x\right)dx$$

$$= \frac{A^2}{2}\int_0^L 2\sin^2\left(\frac{n\pi}{L}x\right)dx = \frac{A^2}{2}\int_0^L\left\{1-\cos\left(\frac{2n\pi}{L}x\right)\right\}dx$$

$$= \frac{A^2}{2}\int_0^L dx - \frac{A^2}{2}\int_0^L \cos\left(\frac{2n\pi}{L}x\right)dx$$

$$= \frac{A^2}{2}[x]_0^L - \frac{A^2}{2}\left[\frac{\sin\left(\frac{2n\pi}{L}x\right)}{2}\right]_0^L$$

$$= \frac{A^2}{2}(L-0) - \frac{A^2}{2}\left\{ \frac{\sin\left(\frac{2n\pi}{L}L\right)}{2} - \frac{\sin\left(\frac{2n\pi}{L}\times 0\right)}{2} \right\}$$

$$= \frac{A^2 L}{2} - \frac{A^2}{2}\left\{ \frac{\sin(2n\pi)}{2} - \frac{\sin(0)}{2} \right\} = \frac{A^2 L}{2} - \frac{A^2}{2}\left(\frac{0}{2} - \frac{0}{2} \right)$$

$$= \frac{A^2 L}{2} - \frac{A^2}{2}\times 0 = \frac{A^2 L}{2} - 0 = \frac{A^2 L}{2} \qquad (5.94)$$

or

$$\frac{A^2 L}{2} = 1 \qquad (5.95)$$

or

$$A^2 = \frac{2}{L} \qquad (5.96)$$

$$\therefore A = \sqrt{\frac{2}{L}} \qquad (5.97)$$

Putting the value of A in eq. (5.92),

$$\Psi_n(x) = \sqrt{\frac{2}{L}}\sin\left(\frac{n\pi}{L}x\right) \qquad (5.98)$$

5.2.5 Plotting E_n, Ψ_n, $|\Psi_n|^2$ versus x Graphs

As the solutions to the particle-in-a-box problem we have obtained the wave functions given by

$$\Psi_n(x) = \sqrt{\frac{2}{L}}\sin\left(\frac{n\pi}{L}x\right) \qquad (5.99)$$

These wave functions are associated with discontinuous energy levels given by

$$E_n = \frac{n^2\pi^2\hbar^2}{2mL^2} \qquad (5.100)$$

Put $n = 1$. Then

$$E_1 = \frac{\pi^2\hbar^2}{2mL^2} \qquad (5.101)$$

and

$$\Psi_1(x) = \sqrt{\frac{2}{L}}\sin\left(\frac{\pi}{L}x\right) \qquad (5.102)$$

We plot the position x on the X-axis. On the Y-axis, we have energy E_1 (Figure 5.5) and wave function $\Psi(x)$ (Figure 5.6). The graph of E_1 is a straight line because E_1 remains constant as x changes.

The graph of $\Psi_1(x)$ contains only one half cycle of the sine wave so that we can write

$$L = \lambda/2 \qquad (5.103)$$

or

$$\lambda = 2L \qquad (5.104)$$

This happens because for $x = L$, the sine factor in eq. (5.102) becomes $\sin(\pi)$, and the angular span (0 to π) covers only half of the sine curve with the full curve extending over 2π.

For $n = 2$,

$$E_2 = \frac{4\pi^2\hbar^2}{2mL^2} \qquad (5.105)$$

and

$$\Psi_2(x) = \sqrt{\frac{2}{L}}\sin\left(\frac{2\pi}{L}x\right) \qquad (5.106)$$

It is found that

$$E_2 = 4E_1 \qquad (5.107)$$

The graph of $\Psi_2(x)$ contains two half cycles or one full cycle of the sine wave; hence

$$L = 2\frac{\lambda}{2} = \lambda \qquad (5.108)$$

or

$$\lambda = L \qquad (5.109)$$

Putting $n = 3$

$$E_3 = \frac{9\pi^2\hbar^2}{2mL^2} \qquad (5.110)$$

and

$$\Psi_3(x) = \sqrt{\frac{2}{L}}\sin\left(\frac{3\pi}{L}x\right) \qquad (5.111)$$

Note that

$$E_3 = 9E_1 \qquad (5.112)$$

The graph of $\Psi_3(x)$ contains three half cycles of the sine wave; hence

$$L = \frac{3\lambda}{2} \qquad (5.113)$$

or

$$\lambda = \frac{2L}{3} \qquad (5.114)$$

When $n = 4$

$$E_4 = \frac{16\pi^2\hbar^2}{2mL^2} \qquad (5.115)$$

and

$$\Psi_4(x) = \sqrt{\frac{2}{L}}\sin\left(\frac{4\pi}{L}x\right) \qquad (5.116)$$

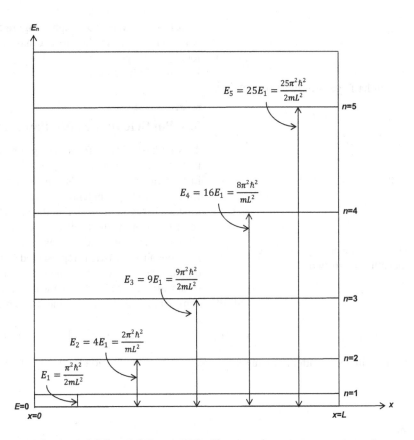

FIGURE 5.5 Energy-distance graph for the particle in an infinite potential well.

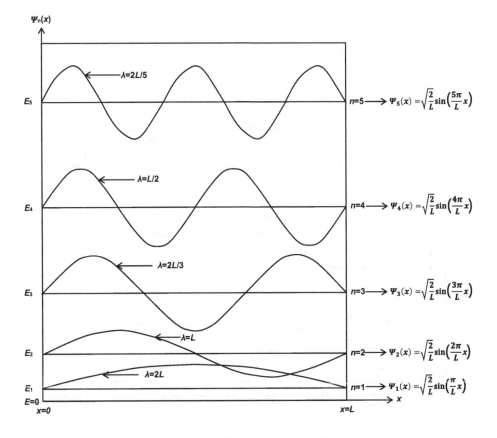

FIGURE 5.6 Wave functions of the particle in an infinite potential well. E is the energy, n is the quantum number.

We get

$$E_4 = 16E_1 \qquad (5.117)$$

The graph of $\Psi_4(x)$ contains four half cycles or two full cycles of the sine wave; hence

$$L = \frac{4\lambda}{2} = 2\lambda \qquad (5.118)$$

or

$$\lambda = \frac{L}{2} \qquad (5.119)$$

In general, for the different modes, the energy levels are E_1, $4E_1$, $9E_1$, $16E_1$, ... and the wavelengths are given by

$$\lambda = \frac{2L}{n} \qquad (5.120)$$

$$\text{for } n = 1, 2, 3, \ldots \qquad (5.121)$$

The energy separation is found to increase with increasing n.

In the $|\Psi_n|^2$ versus x graphs (Figure 5.7), there are points at which $|\Psi_n|^2 = 0$; so the particle has zero probability of being found. Such points are called nodes. The number of nodes increases with rising n.

5.3 Particle in a Finite Potential Well

This problem differs from the rigid box problem in the respect that here the potential energy at the walls of the box is finite instead of an infinite value in the previous case (Figure 5.8). Inside the well, the potential energy of the particle is zero but its kinetic energy is finite so that the particle is in a state of motion (Zettili 2009). At the location of the walls, the potential energy climbs up suddenly and reaches a value $V_0 \# \infty$. In Figure 5.8, the position of the particle is represented on the X-axis and its energy on the Y-axis. Left wall of the well is at a position $x = 0$ and the right wall at $x = L$. So, L is the width of the well. The space under consideration is subdivided into three regions:

$$\text{Region I}: x \leq 0 : V(x) = V_0 \qquad (5.122)$$

$$\text{Region II}: 0 < x < L : V(x) = 0 \qquad (5.123)$$

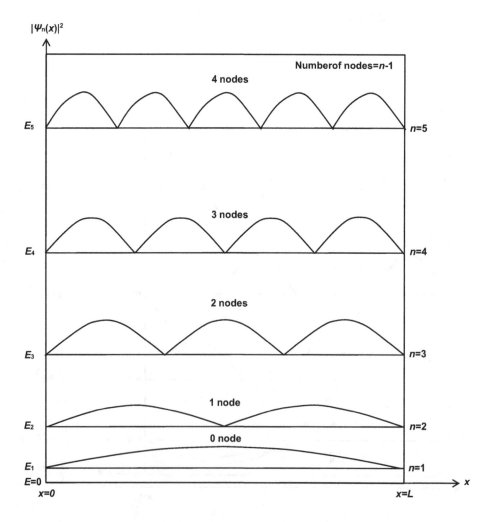

FIGURE 5.7 Probability function plots of the particle in an infinite potential well. E is the energy, n is the quantum number.

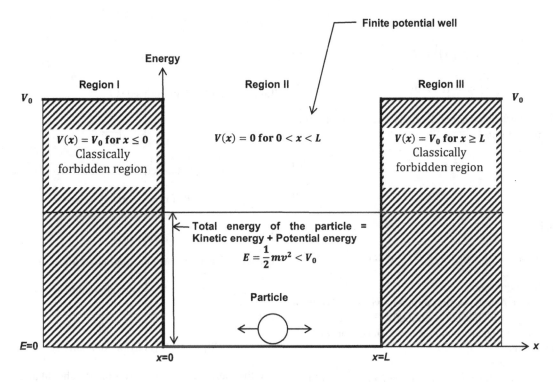

FIGURE 5.8 The particle in a finite potential well of depth V_0. The particle has an energy E less than V_0.

and

$$\text{Region III}: x \geq L : V(x) = V_0 \qquad (5.124)$$

It is assumed that the total energy E of the particle is $< V_0$, the height of the barrier. We can look upon V_0 as the potential energy that the particle must have if it is outside the well. Classical mechanics forbids any chance of particle's existence outside the well. Let us examine the possibilities in quantum mechanics. To find this, the wave function describing the motion of the particle in this potential well is to be determined.

5.3.1 Regions I and III

For this case, the TISE is

$$-\frac{\hbar^2}{2m}\frac{\partial^2}{\partial x^2}\Psi_{\text{I,III}}(x) + V_0(x)\Psi_{\text{I,III}}(x) = E\Psi_{\text{I,III}}(x) \qquad (5.125)$$

or

$$-\frac{\hbar^2}{2m}\frac{\partial^2}{\partial x^2}\Psi_{\text{I,III}}(x) + V_0(x)\Psi_{\text{I,III}}(x) - E\Psi_{\text{I,III}}(x) = 0 \qquad (5.126)$$

or

$$\frac{\hbar^2}{2m}\frac{\partial^2}{\partial x^2}\Psi_{\text{I,III}}(x) - V_0(x)\Psi_{\text{I,III}}(x) + E\Psi_{\text{I,III}}(x) = 0 \qquad (5.127)$$

or

$$\frac{\hbar^2}{2m}\frac{\partial^2}{\partial x^2}\Psi_{\text{I,III}}(x) - \{V_0(x) - E\}\Psi_{\text{I,III}}(x) = 0 \qquad (5.128)$$

Dividing both sides of eq. (5.128) by $\hbar^2/(2m)$

$$\frac{\partial^2}{\partial x^2}\Psi_{\text{I,III}}(x) - \frac{2m}{\hbar^2}\{V_0(x) - E\}\Psi_{\text{I,III}}(x) = 0 \qquad (5.129)$$

Putting

$$\frac{2m}{\hbar^2}\{V_0(x) - E\} = \alpha^2 \qquad (5.130)$$

and noting that the quantity $(V_0 - E)$ is positive since $E < V_0$, and also

$$\alpha = \sqrt{\frac{2m}{\hbar^2}\{V_0(x) - E\}} \qquad (5.131)$$

is a real number
we get

$$\frac{\partial^2}{\partial x^2}\Psi_{\text{I,III}}(x) - \alpha^2\Psi_{\text{I,III}}(x) = 0 \qquad (5.132)$$

Let us assume the solution to be of the form $\exp(\gamma x)$. Plugging the solution into the differential equation (5.129)

$$\frac{\partial^2}{\partial x^2}\{\exp(\gamma x)\} - \alpha^2\{\exp(\gamma x)\} = 0 \qquad (5.133)$$

or

$$\gamma^2\{\exp(\gamma x)\} - \alpha^2\{\exp(\gamma x)\} = 0 \qquad (5.134)$$

or

$$\{\exp(\gamma x)\}(\gamma^2 - \alpha^2) = 0 \qquad (5.135)$$

Since,

$$\{\exp(\gamma x)\} \neq 0 \qquad (5.136)$$

$$\therefore \gamma^2 - \alpha^2 = 0 \qquad (5.137)$$

This is the characteristic equation. Its roots give the solutions to the differential equation.

$$\because \gamma^2 = \alpha^2 \qquad (5.138)$$

$$\therefore \gamma = \pm\alpha \qquad (5.139)$$

The solution is

$$\Psi_{I,III}(x) = C \exp(\alpha x) + D \exp(-\alpha x) \qquad (5.140)$$

where C and D are constants.

In region I, $x < 0$. As $x \to -\infty$, first term $\to 0$ while second term $\to \infty$. Then $\Psi(x) \to \infty$, which is not meaningful. To make it meaningful, we take $D = 0$. Then second term $= 0$. Hence for region I, the wave function is

$$\Psi_I(x) = C \exp(\alpha x) + 0 = C \exp(\alpha x) \qquad (5.141)$$

We argue on similar lines for region III. In region III, $x > 0$. As $x \to \infty$, first term $\to \infty$; second term $\to 0$. Then $\Psi(x) \to \infty$, which is meaningless. To make it meaningful, we take $C = 0$. Then first term $= 0$. Hence for region III, the wave function is

$$\Psi_{III}(x) = 0 + D \exp(-\alpha x) = D \exp(-\alpha x) \qquad (5.142)$$

5.3.2 Region II

This case has already been examined for a free particle with $V(x) = 0$. The Schrodinger's equation is

$$-\frac{\hbar^2}{2m}\frac{\partial^2}{\partial x^2}\Psi_{II}(x) + 0 = E\Psi_{II}(x) \qquad (5.143)$$

or

$$-\frac{\hbar^2}{2m}\frac{\partial^2}{\partial x^2}\Psi_{II}(x) = E\Psi_{II}(x) \qquad (5.144)$$

or

$$-\frac{\partial^2}{\partial x^2}\Psi_{II}(x) = \frac{2m}{\hbar^2}E\Psi_{II}(x) \qquad (5.145)$$

or

$$\frac{\partial^2}{\partial x^2}\Psi_{II}(x) + \frac{2m}{\hbar^2}E\Psi_{II}(x) = 0 \qquad (5.146)$$

or

$$\frac{\partial^2}{\partial x^2}\Psi_{II}(x) + k^2\Psi_{II}(x) = 0 \qquad (5.147)$$

where

$$k^2 = \frac{2m}{\hbar^2}E \qquad (5.148)$$

Taking square root of both sides

$$k = \sqrt{\frac{2m}{\hbar^2}E} \qquad (5.149)$$

The wave function obtained is the same as derived earlier, eq. (5.58)

$$\Psi_{II}(x) = A \sin(kx) + B \cos(kx) \qquad (5.150)$$

5.3.3 Boundary Conditions

The wave functions must be continuous all throughout giving

$$\Psi_I(0) = \Psi_{II}(0) \text{ for continuity at } x = 0 \qquad (5.151)$$

and

$$\Psi_{II}(L) = \Psi_{III}(L) \text{ for continuity at } x = L \qquad (5.152)$$

The first derivative of the wave function should also be continuous. Hence,

$$\frac{\partial}{\partial x}\Psi_I(0) = \frac{\partial}{\partial x}\Psi_{II}(0) \text{ for continuity at } x = 0 \qquad (5.153)$$

and

$$\frac{\partial}{\partial x}\Psi_{II}(L) = \frac{\partial}{\partial x}\Psi_{III}(L) \text{ for continuity at } x = L \qquad (5.154)$$

We take these equations one by one. At $x = 0$,

$$\Psi_I(0) = \Psi_{II}(0) \qquad (5.155)$$

or

$$[C \exp(\alpha x)]_{x=0} = [A \sin(kx) + B \cos(kx)]_{x=0} \qquad (5.156)$$

or

$$C \exp(\alpha \times 0) = A \sin(k \times 0) + B \cos(k \times 0) \qquad (5.157)$$

or

$$C \exp(0) = A \sin(0) + B \cos(0) \qquad (5.158)$$

or

$$C \times 1 = A \times 0 + B \times 1 = 0 + B = B \qquad (5.159)$$

or

$$C = B \qquad (5.160)$$

At $x = L$,

$$\Psi_{II}(L) = \Psi_{III}(L) \qquad (5.161)$$

or

$$-A\sin(kx) + B\cos(kx)\big]_{x=L} = \big[D\exp(-\alpha x)\big]_{x=L} \quad (5.162)$$

or

$$A\sin(kL) + B\cos(kL) = D\exp(-\alpha L) \quad (5.163)$$

At $x = 0$,

$$\frac{\partial}{\partial x}\Psi_{\mathrm{I}}(0) = \frac{\partial}{\partial x}\Psi_{\mathrm{II}}(0) \quad (5.164)$$

or

$$\left[\frac{\partial}{\partial x}\{C\exp(\alpha x)\}\right]_{x=0} = \left[\frac{\partial}{\partial x}\{A\sin(kx) + B\cos(kx)\}\right]_{x=0} \quad (5.165)$$

or

$$\big[C\alpha\exp(\alpha x)\big]_{x=0} = \big[Ak\cos(kx) - Bk\sin(kx)\big]_{x=0} \quad (5.166)$$

or

$$C\alpha\exp(\alpha \times 0) = Ak\cos(k \times 0) - Bk\sin(k \times 0) \quad (5.167)$$

or

$$C\alpha\exp(0) = Ak\cos(0) - Bk\sin(0) \quad (5.168)$$

or

$$C\alpha \times 1 = Ak \times 1 - Bk \times 0 \quad (5.169)$$

or

$$C\alpha = Ak \quad (5.170)$$

At $x = L$,

$$\frac{\partial}{\partial x}\Psi_{\mathrm{II}}(L) = \frac{\partial}{\partial x}\Psi_{\mathrm{III}}(L) \quad (5.171)$$

or

$$\left[\frac{\partial}{\partial x}\{A\sin(kx) + B\cos(kx)\}\right]_{x=L} = \left[\frac{\partial}{\partial x}\{D\exp(-\alpha x)\}\right]_{x=L} \quad (5.172)$$

or

$$-Ak\cos(kx) - Bk\sin(kx)\big]_{x=L} = \big[-D\alpha\exp(-\alpha x)\big]_{x=L} \quad (5.173)$$

or

$$Ak\cos(kL) - Bk\sin(kL) = -D\alpha\exp(-\alpha L) \quad (5.174)$$

By applying the boundary conditions, we obtained four equations, which are collected together in the following order:

$$C = B \quad (5.175)$$

$$C\alpha = Ak \quad (5.176)$$

$$A\sin(kL) + B\cos(kL) = D\exp(-\alpha L) \quad (5.177)$$

and

$$Ak\cos(kL) - Bk\sin(kL) = -D\alpha\exp(-\alpha L) \quad (5.178)$$

From eqs. (5.175) and (5.176),

$$B\alpha = Ak \quad (5.179)$$

From eq. (5.177),

$$D = \{A\sin(kL) + B\cos(kL)\}\exp(\alpha L) \quad (5.180)$$

Putting this value of D in eq. (5.178)

$$Ak\cos(kL) - Bk\sin(kL)$$
$$= -\{A\sin(kL) + B\cos(kL)\}\exp(\alpha L) \times \alpha\exp(-\alpha L)$$
$$= -\alpha\{A\sin(kL) + B\cos(kL)\} \quad (5.181)$$

or

$$Ak\cos(kL) - Bk\sin(kL) = -\alpha A\sin(kL) - \alpha B\cos(kL) \quad (5.182)$$

or

$$Ak\cos(kL) + \alpha B\cos(kL) = Bk\sin(kL) - \alpha A\sin(kL) \quad (5.183)$$

or

$$(Bk - \alpha A)\sin(kL) = (Ak + \alpha B)\cos(kL) \quad (5.184)$$

or

$$\frac{\sin(kL)}{\cos(kL)} = \frac{Ak + \alpha B}{Bk - \alpha A} \quad (5.185)$$

or

$$\tan(kL) = \frac{Ak + \alpha B}{Bk - \alpha A} = \frac{\alpha B + \alpha B}{Bk - \alpha A} = \frac{2\alpha B}{B\left(k - \dfrac{\alpha A}{B}\right)} = \frac{2\alpha B}{B\left(k - \dfrac{\alpha A}{\dfrac{Ak}{\alpha}}\right)}$$

$$= \frac{2\alpha B}{B\left(k - \alpha A \times \dfrac{\alpha}{Ak}\right)} = \frac{2\alpha}{k - \alpha \times \dfrac{\alpha}{k}} = \frac{2\alpha}{\dfrac{k^2 - \alpha \times \alpha}{k}}$$

$$= \frac{2\alpha k}{k^2 - \alpha^2} \quad (5.186)$$

where eq. (5.179) has been applied. Thus

$$\tan(kL) = \frac{2\alpha k}{k^2 - \alpha^2} \quad (5.187)$$

Division of the numerator and denominator of the right-hand side of eq. (5.187) by k^2 gives us

$$\tan(kL) = \frac{\dfrac{2\alpha k}{k^2}}{\dfrac{k^2 - \alpha^2}{k^2}} = \frac{2\left(\dfrac{\alpha}{k}\right)}{1 - \dfrac{\alpha^2}{k^2}} \qquad (5.188)$$

or

$$\tan\left\{2\left(\frac{1}{2}kL\right)\right\} = \frac{2\left(\dfrac{\alpha}{k}\right)}{1 - \dfrac{\alpha^2}{k^2}} \qquad (5.189)$$

Comparison of eq. (5.189) with the equation

$$\tan(2\theta) = \frac{2\tan\theta}{1 - \tan^2\theta} \qquad (5.190)$$

enables us to write, $\tan\theta = \dfrac{\alpha}{k}$ and $\theta = \dfrac{1}{2}kL$, so that

$$\tan\left(\frac{1}{2}kL\right) = \frac{\alpha}{k} \qquad (5.191)$$

or

$$k\tan\left(\frac{1}{2}kL\right) = \alpha \qquad (5.192)$$

Similarly by comparing eq. (5.189) with

$$\tan(2\theta) = \frac{2\cot(-\theta)}{1 - \cot^2(-\theta)} \qquad (5.193)$$

we have, $\cot(-\theta) = \dfrac{\alpha}{k}$ and $\theta = \dfrac{1}{2}kL$, whence it follows that

$$\cot\left(-\frac{1}{2}kL\right) = \frac{\alpha}{k} \qquad (5.194)$$

or

$$k\cot\left(-\frac{1}{2}kL\right) = \alpha \qquad (5.195)$$

or

$$-k\cot\left(\frac{1}{2}kL\right) = \alpha \qquad (5.196)$$

Since from eq. (5.6)

$$E = \frac{\hbar^2 k^2}{2m} \qquad (5.197)$$

From eqs. (5.131) and (5.197), α can be expressed as

$$\alpha = \sqrt{\frac{2m}{\hbar^2}\{V_0(x) - E\}} = \sqrt{\frac{2m}{\hbar^2}\left\{V_0(x) - \frac{\hbar^2 k^2}{2m}\right\}}$$

$$= \sqrt{\frac{2m}{\hbar^2}V_0(x) - k^2} \qquad (5.198)$$

Putting

$$\frac{2m}{\hbar^2}V_0(x) = k_0^2 \qquad (5.199)$$

we get

$$\alpha = \sqrt{k_0^2 - k^2} \qquad (5.200)$$

Then eq. (5.192) converts to

$$k\tan\left(\frac{1}{2}kL\right) = \sqrt{k_0^2 - k^2} \qquad (5.201)$$

and eq. (5.196) turns to

$$-k\cot\left(\frac{1}{2}kL\right) = \sqrt{k_0^2 - k^2} \qquad (5.202)$$

Let us define new dimensionless variables

$$u = \frac{1}{2}kL \text{ and } u_0 = \frac{1}{2}k_0 L \qquad (5.203)$$

Then

$$k = \frac{2u}{L} \text{ and } k_0 = \frac{2u_0}{L} \qquad (5.204)$$

So,

$$\sqrt{k_0^2 - k^2} = \sqrt{\left(\frac{2u_0}{L}\right)^2 - \left(\frac{2u}{L}\right)^2} = \sqrt{\frac{4u_0^2}{L^2} - \frac{4u^2}{L^2}}$$

$$= \sqrt{\frac{4}{L^2}(u_0^2 - u^2)} = \frac{2}{L}\sqrt{u_0^2 - u^2} \qquad (5.205)$$

Substituting for $\sqrt{k_0^2 - k^2}$ from eq. (5.205) and using eqs. (5.203) and (5.204), eq. (5.201) reduces to

$$k\tan(u) = \frac{2}{L}\sqrt{u_0^2 - u^2} \qquad (5.206)$$

or

$$\frac{2u}{L} \times \tan(u) = \frac{2}{L}\sqrt{u_0^2 - u^2} \qquad (5.207)$$

Dividing both sides by $2/L$

$$u\tan(u) = \sqrt{u_0^2 - u^2} \qquad (5.208)$$

Similarly, substituting for $\sqrt{k_0^2 - k^2}$ from eq. (5.205) and using eqs. (5.203) and (5.204), eq. (5.202) reduces to

$$-k \cot(u) = \frac{2}{L}\sqrt{u_0^2 - u^2} \qquad (5.209)$$

or

$$-\frac{2u}{L} \times \cot(u) = \frac{2}{L}\sqrt{u_0^2 - u^2} \qquad (5.210)$$

or

$$-u \cot(u) = \sqrt{u_0^2 - u^2} \qquad (5.211)$$

We collect together the master equations for discussion:

$$u \tan(u) = \sqrt{u_0^2 - u^2} \qquad (5.212)$$

and

$$-u \cot(u) = \sqrt{u_0^2 - u^2} \qquad (5.213)$$

In these equations, u is the parameter containing the energy E of the particle through parameter k and u_0 is the parameter involving the potential V_0. So, by solving these equations we can get the energy values E that the particle can possess in relation to the potential V_0. But both these equations are transcendental equations, i.e. equations containing functions, e.g. trigonometric functions, which transcend algebra and often do not have solutions in closed form. To solve these equations, we use numerical or graphical methods. We describe the graphical method here. First consider eq. (5.212). We plot u-values on the X-axis. On the Y-axis, we plot the left-hand side $u \tan(u)$ for different energy values E_1, E_2, E_3, ... Thus we get several curves corresponding to energies E_1, E_2, E_3, ... On the Y-axis, we also plot the right-hand side expression $\sqrt{(u_0^2 - u^2)}$. The intersection points of the curves for the left-hand side and right-hand side variables are the solutions. These intersections determine k and hence energy values E. It is found that the number of intersections increases as the V_0 value taken in u_0 equation increases, and therefore there are greater number of energy levels. Thus it is shown graphically that only discrete energy values pertaining to the intersection points are possible, not continuous values. In other words, energy is quantized.

Proceeding on the same lines, eq. (5.213) shows that another discrete set of energies is permissible. All these energies are concealed in the equation for k so that from eq. (5.197)

$$E_n = \frac{\hbar^2 k^2}{2m} \qquad (5.214)$$

and are obtained for different values of V_0 hiding in the equation for k_0, such that

$$V_0 = \frac{\hbar^2 k_0^2}{2m} \qquad (5.215)$$

from eq. (5.199).

In a nutshell, we solved the Schrodinger equation for the particle in regions I, III outside the one-dimensional box. The solution contains two constants C and D. We also solved the Schrodinger equation for the particle in region II inside the box, when two further constant A, B arose. To find the constants, we specified the boundary conditions that the wave functions must be continuous and continuously differentiable. Applying the continuity conditions, we obtained one equation in tangent form and another in cotangent form, both containing the parameters k, α related to energy E of the particle and the potential V_0. These continuity equations are not satisfied by any arbitrary value of energy. Only specific values of energy which are solutions to either or both the equations are permissible, indicating quantization of energy. Further, these equations being transcendental in nature, we applied the graphical method of solution. We found that there were limited points of intersection on the graph representing the solutions, whereby the energy quantization is shown pictorially. We present the details of the graphical solution in the next section.

5.3.4 Graphical Solution of Finite Potential Well Equations

The calculation procedure for the graph in Figure 5.9 is described below:

i. To get the first circular arc, take

$$V_0(x)L^2 = 4\frac{\hbar^2}{2m} \qquad (5.216)$$

Then from eqs. (5.203) and (5.215)

$$u_0 = \frac{1}{2}k_0 L = \frac{L}{2}\sqrt{\frac{2m}{\hbar^2}V_0(x)} = \frac{1}{2}\sqrt{\frac{2m}{\hbar^2}V_0(x)L^2}$$

$$= \frac{1}{2}\sqrt{\frac{2m}{\hbar^2}\frac{4\hbar^2}{2m}} = \frac{2}{2} = 1 \qquad (5.217)$$

Calculate $\sqrt{u_0^2 - u^2}$ for different values of u corresponding to various energies, as shown in Table 5.1.

ii. To get the second circular arc, proceed as follows:
Take

$$V_0(x)L^2 = 16\frac{\hbar^2}{2m} \qquad (5.218)$$

Then from eqs. (5.203) and (5.215)

$$u_0 = \frac{1}{2}k_0 L = \frac{L}{2}\sqrt{\frac{2m}{\hbar^2}V_0(x)} = \frac{1}{2}\sqrt{\frac{2m}{\hbar^2}V_0(x)L^2}$$

$$= \frac{1}{2}\sqrt{\frac{2m}{\hbar^2}\frac{16\hbar^2}{2m}} = \frac{4}{2} = 2 \qquad (5.219)$$

Calculate $\sqrt{u_0^2 - u^2}$ for different values of u corresponding to various energies as shown in Table 5.2.

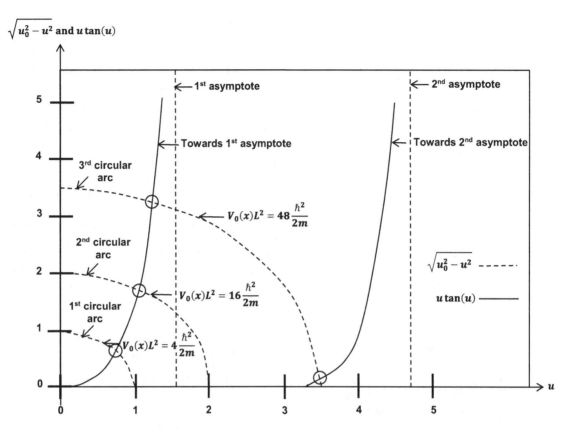

FIGURE 5.9 Graphical solution of equation $u \tan(u) = \sqrt{u_0^2 - u^2}$ for three different values of $V_0(x)L^2$.

TABLE 5.1

Calculation of $\sqrt{u_0^2 - u^2}$ for Different Values of u for the First Circular Arc

Sl. No.	Value of u	Value of $\sqrt{u_0^2 - u^2}$
1.	0	$\sqrt{1-0} = 1$
2.	0.157	$\sqrt{1-0.0246} = 0.988$
3.	0.314	$\sqrt{1-0.099} = 0.949$
4.	0.3925	$\sqrt{1-0.154} = 0.92$
5.	0.47	$\sqrt{1-0.221} = 0.88$
6.	0.496	$\sqrt{1-0.246} = 0.87$
7.	0.499	$\sqrt{1-0.249} = 0.866$
8.	0.628	$\sqrt{1-0.394} = 0.78$
9.	0.72	$\sqrt{1-0.5184} = 0.694$
10.	0.85	$\sqrt{1-0.7225} = 0.527$
11.	0.942	$\sqrt{1-0.887} = 0.336$
12.	0.9985	$\sqrt{1-0.997} = 0.055$

TABLE 5.2

Calculation of $\sqrt{u_0^2 - u^2}$ for Different Values of u for the Second Circular Arc

Sl. No.	Value of u	Value of $\sqrt{u_0^2 - u^2}$
1.	0	$\sqrt{4-0} = 2$
2.	0.3925	$\sqrt{4-0.154} = 1.96$
3.	0.499	$\sqrt{4-0.249} = 1.94$
4.	0.85	$\sqrt{4-0.7225} = 1.81$
5.	1.256	$\sqrt{4-1.576} = 1.56$
6.	1.507	$\sqrt{4-2.271} = 1.32$
7.	1.8	$\sqrt{4-3.24} = 0.872$
8.	1.9	$\sqrt{4-3.61} = 0.62$
9.	1.95	$\sqrt{4-3.8} = 0.45$
10.	1.99	$\sqrt{4-3.96} = 0.2$
11.	1.995	$\sqrt{4-3.98} = 0.14$
12.	1.999	$\sqrt{4-3.996} = 0.063$

iii. To get the third circular arc, work as follows:
Take

$$V_0(x)L^2 = 48\frac{\hbar^2}{2m} \qquad (5.220)$$

Then from eqs. (5.203) and (5.215)

$$u_0 = \frac{1}{2}k_0 L = \frac{L}{2}\sqrt{\frac{2m}{\hbar^2}V_0(x)} = \frac{1}{2}\sqrt{\frac{2m}{\hbar^2}V_0(x)L^2}$$

$$= \frac{1}{2}\sqrt{\frac{2m}{\hbar^2}\frac{48\hbar^2}{2m}} = \frac{6.928}{2} = 3.464 \qquad (5.221)$$

The calculated values are tabulated in Table 5.3.

TABLE 5.3

Calculation of $\sqrt{u_0^2 - u^2}$ for Different Values of u for the Third Circular Arc

Sl. No.	Value of u	Value of $\sqrt{u_0^2 - u^2}$
1.	0	$\sqrt{11.999 - 0} = 3.464$
2.	0.499	$\sqrt{11.999 - 0.249} = 3.43$
3.	1.507	$\sqrt{11.999 - 2.271} = 3.12$
4.	1.95	$\sqrt{11.999 - 3.8} = 2.86$
5.	1.999	$\sqrt{11.999 - 3.996} = 2.83$
6.	2.5	$\sqrt{11.999 - 6.25} = 2.4$
7.	3.14	$\sqrt{11.999 - 9.8596} = 1.46$
8.	3.45	$\sqrt{11.999 - 11.9} = 0.31$
9.	3.459	$\sqrt{11.999 - 11.965} = 0.18$
10.	3.463	$\sqrt{11.999 - 11.992369} = 0.081$

We note that the first and second circular arcs intersect the tangent curves at one point each. Hence, there is one solution for each of these arcs. The third circular arc cuts both the tangent curves. Hence there are two solutions for this arc.

iv. To get the tangent graphs toward the two asymptotes, note that the asymptotes in tangent graph are repeated after π units. In Table 5.4, in the second column, the values of u refer to the tangent graph toward the first asymptote. In the third column, the values of u are given after addition of $\pi = 3.14$; these values are used for plotting the tangent graph toward the second asymptote. Method of determination of u numerically (in numbers)

and in degrees is illustrated in second column, second row, and shown directly further.

The calculation procedure for the cotangent graphs in Figure 5.10 is similar to that for the tangent graphs already drawn, and is explained below:

The first, second, and third circular arcs are the same as for Figure 5.9. To get the cotangent graphs, we note that values of $\cot(u)$ for u from 0 to $\pi/2$ are positive; so the resulting $-u \cot(u)$ function will be negative which is unacceptable. Therefore, we start after $u = \pi/2$ and work as shown in Table 5.5.

We find that the first circular arc does not intersect the cotangent curve at any point. So there is no solution for this arc. The second circular arc cuts the cotangent curve at one point yielding one solution. Likewise, the third circular arc cuts the cotangent curve at one point leading to one solution.

Considering both Figures 5.9 and 5.10 for the tangent and cotangent cases, it is evident that altogether there is one solution from the first circular arc (the tangent curve), two solutions from the second circular arc (tangent and cotangent curves), and three solutions from the third circular arc (tangent and cotangent curves).

5.4 Alternative Method: Determining Solutions for Even (Symmetric) and Odd (Antisymmetric) Wave Functions

A wave function (in fact any function) is said to be even or symmetric and has positive parity if

$$\Psi(x) = \Psi(-x) \tag{5.222}$$

TABLE 5.4

Calculation of $u \tan(u)$ for Different Values of u

Sl. No.	Value of u for First Asymptote	Value of u for Second Asymptote	Value of $u \tan(u)$
1.	$0\pi = 0$	$0 + 3.14 = 3.14$	$0 \times 0 = 0$
2.	$0.05\pi = 0.05 \times 3.14 = 0.157$ $0.05\pi = 0.05 \times 180° = 9°$	$0.57 + 3.14 = 3.3$	$0.157 \times \tan(9°) = 0.157 \times 0.158 = 0.025$
3.	$0.1\pi = 0.314 = 18°$	$0.314 + 3.14 = 3.45$	$0.314 \times \tan(18°) = 0.314 \times 0.325 = 0.1$
4.	$0.125\pi = 0.3925 = 22.5°$	$03925 + 3.14 = 3.5$	$0.3925 \times \tan(22.5°) = 0.3925 \times 0.41 = 0.16$
5.	$0.15\pi = 0.47 = 27°$	$0.47 + 3.14 = 3.6$	$0.47 \times \tan(27°) = 0.47 \times 0.51 = 0.24$
6.	$0.158\pi = 0.496 = 28.44°$	$0.496 + 3.14 = 3.64$	$0.496 \times \tan(28.44°) = 0.496 \times 0.54 = 0.27$
7.	$0.159\pi = 0.499 = 28.62°$	$0.499 + 3.14 = 3.64$	$0.499 \times \tan(28.62°) = 0.499 \times 0.55 = 0.274$
8.	$0.2\pi = 0.628 = 36°$	$0.628 + 3.14 = 3.77$	$0.628 \times \tan(36°) = 0.628 \times 0.73 = 0.46$
9.	$0.23\pi = 0.72 = 41.4°$	$0.72 + 3.14 = 3.9$	$0.72 \times \tan(41.4°) = 0.72 \times 0.882 = 0.635$
10.	$0.27\pi = 0.85 = 48.6°$	$0.85 + 3.14 = 4$	$0.85 \times \tan(48.6°) = 0.85 \times 1.134 = 0.964$
11.	$0.3\pi = 0.942 = 54°$	$0.942 + 3.14 = 4.1$	$0.942 \times \tan(54°) = 0.942 \times 1.38 = 1.3$
12.	$0.318\pi = 0.9985 = 57.24°$	$0.9985 + 3.14 = 4.14$	$0.9985 \times \tan(57.84°) = 0.9985 \times 1.59 = 1.59$
13.	$0.4\pi = 1.256 = 72°$	$1.256 + 3.14 = 4.4$	$1.256 \times \tan(72°) = 1.256 \times 3.078 = 3.87$
14.	$0.419\pi = 1.316 = 75.42°$	$1.316 + 3.14 = 4.5$	$1.316 \times \tan(75.42°) = 1.316 \times 3.84 = 5.05$
15.	$0.45\pi = 1.413 = 81°$	$1.413 + 3.14 = 4.6$	$1.413 \times \tan(81°) = 1.413 \times 6.3 = 8.9$
16.	$0.48\pi = 1.507 = 86.4°$	$1.507 + 3.14 = 4.657$	$1.507 \times \tan(86.4°) = 1.507 \times 15.89 = 23.9$
17.	$0.5\pi = 1.57 = 90°$	$1.57 + 3.14 = 4.7$	$1.57 \times \tan(90°) = \infty$

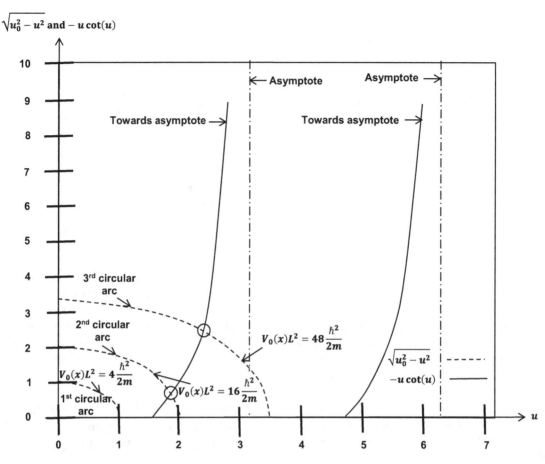

FIGURE 5.10   Graphical solution of equation $-u \cot(u) = \sqrt{u_0^2 - u^2}$ for three different values of $V_0(x)L^2$.

TABLE 5.5

Calculation of $-u \cot(u)$ for Different Values of u

Sl. No.	Value of u for First Asymptote	Value of u for Second Asymptote	Value of $u \cot(u)$
1.	$0.51\pi = 1.6014 = 91.8°$	$1.6014 + 3.14 = 4.7414$	$-1.6014 \times \cot(91.8°) = -1.6014 \times -0.03143 = 0.05$
2.	$0.6\pi = 1.884 = 108°$	$1.884 + 3.14 = 5.024$	$-1.884 \times \cot(108°) = -1.884 \times -0.3249 = 0.612$
3.	$0.7\pi = 2.198 = 126°$	$2.198 + 3.14 = 5.338$	$-2.198 \times \cot(126°) = -2.198 \times -0.7265 = 1.5969$
4.	$0.8\pi = 2.512 = 144°$	$2.512 + 3.14 = 5.652$	$-2.512 \times \cot(144°) = -2.512 \times -1.376 = 3.4575$
5.	$0.9\pi = 2.826 = 162°$	$2.826 + 3.14 = 5.966$	$-2.826 \times \cot(162°) = -2.826 \times -3.0777 = 8.6975$

It is said to be odd or anti-symmetric and has negative parity if

$$\Psi(x) = -\Psi(-x) \quad (5.223)$$

5.4.1 Even Solution Boundary Conditions

For an even function for second region wave function $\Psi_{II}(x)$, any sine term is inadmissible because $\sin(kx)$ is an odd function since

$$\sin(kx) = -\sin(-kx) \quad (5.224)$$

For exclusion of sine term $A = 0$; hence from eq. (5.150)

$$\Psi_{II}(x) = A\sin(kx) + B\cos(kx) = 0 \times \sin(kx) + B\cos(kx)$$

$$= 0 + B\cos(kx) = B\cos(kx) \quad (5.225)$$

Further, for an even solution, the wave functions for first and third regions must be interrelated. So, in the equations for $\Psi_I(x)$ and $\Psi_{III}(x)$, we must have the coefficients equal as $C = D$ in eqs. (5.141) and (5.142), e.g. D in eq. (5.141) and again D in eq. (5.142). Applying the condition of continuity of the wave function and its derivative at the boundary between second and third regions, we get from eqs. (5.142) and (5.225)

$$\left| D\exp(-\alpha x) \right|_{x=L} = \left| B\cos(kx) \right|_{x=L} \quad (5.226)$$

or

$$D\exp(-\alpha L) = B\cos(kL) \quad (5.227)$$

Also,

$$\left|\frac{\partial}{\partial x}\left\{D\exp(-\alpha x)\right\}\right|_{x=L} = \left|\frac{\partial}{\partial x}\left\{B\cos(kx)\right\}\right|_{x=L} \quad (5.228)$$

or

$$\left|-\alpha D\exp(-\alpha x)\right|_{x=L} = \left|-Bk\sin(kx)\right|_{x=L} \quad (5.229)$$

or

$$-\alpha D\exp(-\alpha L) = -Bk\sin(kL) \quad (5.230)$$

Dividing eq. (5.230) by eq. (5.227), we have

$$-\frac{\alpha D\exp(-\alpha L)}{D\exp(-\alpha L)} = -\frac{Bk\sin(kL)}{B\cos(kL)} \quad (5.231)$$

or

$$-\alpha = -k\frac{\sin(kL)}{\cos(kL)} = -k\tan(kL) \quad (5.232)$$

$$\therefore k\tan(kL) = \alpha \quad (5.233)$$

5.4.2 Odd Solution Boundary Conditions

In case the second region wave function $\Psi_{II}(x)$ is an odd function, any cosine term is disallowed because $\cos(kx)$ is an even function following the equation

$$\cos(kx) = \cos(-kx) \quad (5.234)$$

For prohibition of cosine term $B = 0$; hence from eq. (5.150)

$$\Psi_{II}(x) = A\sin(kx) + B\cos(kx) = A\sin(kx) + 0\times\cos(kx)$$

$$= A\sin(kx) + 0 = A\sin(kx) \quad (5.235)$$

Another requirement for an odd solution to the wave functions for first and third regions, is that in the equations for $\Psi_I(x)$ and $\Psi_{III}(x)$, we must have the coefficients equal in magnitude and opposite in sign as $C = -D$ in eqs. (5.141) and (5.142), e.g. D in eq. (5.141) and $-D$ in eq. (5.142). For continuity of the wave function at the boundary between second and third regions, eqs. (5.142) and (5.235) give

$$\left|-D\exp(-\alpha x)\right|_{x=L} = \left|A\sin(kx)\right|_{x=L} \quad (5.236)$$

or

$$-D\exp(-\alpha L) = A\sin(kL) \quad (5.237)$$

Continuity condition for first derivative of wave function yields

$$\left|\frac{\partial}{\partial x}\left\{-D\exp(-\alpha x)\right\}\right|_{x=L} = \left|\frac{\partial}{\partial x}\left\{A\sin(kx)\right\}\right|_{x=L} \quad (5.238)$$

or

$$\left|\alpha D\exp(-\alpha x)\right|_{x=L} = \left|Ak\cos(kx)\right|_{x=L} \quad (5.239)$$

or

$$\alpha D\exp(-\alpha L) = Ak\cos(kL) \quad (5.240)$$

Dividing eq. (5.240) by eq. (5.237), we have

$$-\frac{\alpha D\exp(-\alpha L)}{D\exp(-\alpha L)} = \frac{Ak\cos(kL)}{A\sin(kL)} \quad (5.241)$$

or

$$-\alpha = k\frac{\cos(kL)}{\sin(kL)} = k\cot(kL) \quad (5.242)$$

$$\therefore -k\cot(kL) = \alpha \quad (5.243)$$

5.4.3 Normalization of Even Wave Functions

For normalization we write with a pre-factor 2 for taking limits of integration between 0 and ∞ instead of $-\infty$ to $+\infty$. From eqs. (5.142) and (5.225),

$$1 = 2\int_0^\infty \left|\Psi^2(x)\right|dx = 2\int_0^L \left|\Psi_{II}^2(x)\right|dx + 2\int_L^\infty \left|\Psi_{III}^2(x)\right|dx$$

$$= 2\int_0^L \left|B\cos(kx)\right|^2 dx + 2\int_L^\infty \left|D\exp(-\alpha x)\right|^2 dx$$

$$= 2\int_0^L \left|B\right|^2\cos^2(kx)dx + 2\int_L^\infty \left|D\right|^2\exp(-2\alpha x)dx$$

$$= 2\left|B\right|^2\int_0^L\cos^2(kx)dx + 2\left|D\right|^2\int_L^\infty\exp(-2\alpha x)dx \quad (5.244)$$

Here,

$$\int_0^L\cos^2(kx)dx = \int_0^L\frac{1}{2}\left\{2\cos^2(kx)\right\}dx = \frac{1}{2}\int_0^L(1+\cos 2kx)dx$$

$$= \frac{1}{2}\int_0^L dx + \frac{1}{2}\int_0^L(\cos 2kx)dx$$

$$= \frac{1}{2}\left|x\right|_0^L + \frac{1}{2}\int_0^L(\cos 2kx)dx$$

$$= \frac{L}{2} + \frac{1}{2}\int_0^L(\cos 2kx)dx \quad (5.245)$$

Put

$$2kx = w \quad (5.246)$$

Then

$$2kdx = dw \quad (5.247)$$

or

$$kdx = \frac{dw}{2} \quad (5.248)$$

Hence,

$$\frac{1}{2}\int_0^L (\cos 2kx)\,dx = \frac{1}{2k}\int_0^L (\cos 2kx)\,k\,dx = \frac{1}{2k}\int_0^L (\cos w)\frac{dw}{2}$$

$$= \frac{1}{2k}\left|\frac{\sin w}{2}\right|_0^L = \frac{1}{2k}\left|\frac{\sin(2kx)}{2}\right|_0^L$$

$$= \frac{1}{2k}\left(\frac{\sin(2kL)}{2} - \frac{\sin(2k\times 0)}{2}\right)$$

$$= \frac{1}{2k}\left\{\frac{\sin(2kL)}{2} - \sin 0\right\}$$

$$= \frac{1}{2k}\left\{\frac{\sin(2kL)}{2} - 0\right\} = \frac{\sin 2kL}{4k} \qquad (5.249)$$

To determine the second term in eq. (5.244), put

$$v = -2\alpha x \qquad (5.250)$$

Then

$$dv = -2\alpha\,dx \qquad (5.251)$$

or

$$dx = -\frac{dv}{2\alpha} \qquad (5.252)$$

So,

$$\int_L^\infty \exp(-2\alpha x)\,dx = \int_L^\infty \exp(v)\times -\frac{dv}{2\alpha} = -\frac{1}{2\alpha}\int_L^\infty \exp(v)\,dv$$

$$= \left|-\frac{\exp(v)}{2\alpha}\right|_L^\infty = \left|-\frac{\exp(-2\alpha x)}{2\alpha}\right|_L^\infty$$

$$= -\frac{\exp(-2\alpha\times\infty)}{2\alpha} - \left\{-\frac{\exp(-2\alpha L)}{2\alpha}\right\}$$

$$= 0 + \frac{\exp(-2\alpha L)}{2\alpha} \qquad (5.253)$$

With the help of eqs. (5.245), (5.249), and (5.253), eq. (5.244) becomes

$$1 = 2|B|^2 \times \left\{\frac{L}{2} + \frac{\sin(2kL)}{4k}\right\} + 2|D|^2 \times \frac{\exp(-2\alpha L)}{2\alpha}$$

$$= 2|B|^2 \times \frac{1}{2}\left\{L + \frac{\sin(2kL)}{2k}\right\} + |D|^2 \times \frac{\exp(-2\alpha L)}{\alpha}$$

$$= |B|^2 \times \left\{L + \frac{\sin(2kL)}{2k}\right\} + |D|^2 \times \frac{\exp(-2\alpha L)}{\alpha} \qquad (5.254)$$

But from eq. (5.227)

$$\frac{D}{\exp(\alpha L)} = B\cos(kL) \qquad (5.255)$$

or

$$D = B\exp(\alpha L)\cos(kL) \qquad (5.256)$$

Hence, eq. (5.254) is converted into

$$1 = |B|^2 \times \left\{L + \frac{\sin(2kL)}{2k}\right\} + \{|B|\exp(\alpha L)\cos(kL)\}^2$$

$$\times \frac{\exp(-2\alpha L)}{\alpha}$$

$$= |B|^2\left\{L + \frac{\sin(2kL)}{2k}\right\} + |B|^2\exp(2\alpha L)\cos^2(kL)\times\frac{\exp(-2\alpha L)}{\alpha}$$

$$= |B|^2\left\{L + \frac{\sin(2kL)}{2k}\right\} + |B|^2\cos^2(kL)\times\frac{1}{\alpha}$$

$$= |B|^2\left(L + \frac{\sin 2kL}{2k} + \frac{\cos^2(kL)}{\alpha}\right)$$

$$= |B|^2\left\{L + \frac{\sin(2kL)}{2k} + \frac{\cos^2(kL)}{k\tan(kL)}\right\}$$

$$= |B|^2\left\{L + \frac{2\sin(kL)\cos(kL)}{2k} + \frac{\cos^2(kL)}{k\frac{\sin(kL)}{\cos(kL)}}\right\}$$

$$= |B|^2\left\{L + \frac{\sin(kL)\cos(kL)}{k} + \frac{\cos^3(kL)}{k\sin(kL)}\right\} \qquad (5.257)$$

$$\because \sin(2kL) = 2\sin(kL)\cos(kL) \text{ and } \alpha = k\tan(kL) \qquad (5.258)$$

from eq. (5.233).
   Now,

$$1 = |B|^2\left\{L + \frac{\sin^2(kL)\cos(kL)}{k\sin(kL)} + \frac{\cos^3(kL)}{k\sin(kL)}\right\}$$

$$= |B|^2\left[L + \frac{\cos(kL)}{k\sin(kL)}\{\sin^2(kL) + \cos^2(kL)\}\right]$$

$$= |B|^2\left[L + \frac{\cos(kL)}{k\sin(kL)}(1)\right] = |B|^2\left[L + \frac{1}{k\frac{\sin(kL)}{\cos(kL)}}\right]$$

$$= |B|^2\left[L + \frac{1}{k\tan(kL)}\right] = |B|^2\left(L + \frac{1}{\alpha}\right) \qquad (5.259)$$

$$\because k\tan(kL) = \alpha \qquad (5.260)$$

Thus

$$1 = |B|^2\left(L + \frac{1}{\alpha}\right) \qquad (5.261)$$

or

$$B = \frac{1}{\sqrt{L + \dfrac{1}{\alpha}}} \qquad (5.262)$$

and from eqs. (5.256) and (5.262)

$$D = B \exp(\alpha L) \cos(kL) = \frac{\exp(\alpha L) \cos(kL)}{\sqrt{L + \dfrac{1}{\alpha}}} \qquad (5.263)$$

From eqs. (5.141), (5.142), using the condition of equal coefficients, both in magnitude and sign, $C = D$, along with eq. (5.225), the even wave functions are given by

$$\Psi_I(x) = D \exp(\alpha x) \qquad (5.264)$$

$$\Psi_{II}(x) = B \cos(kx) \qquad (5.265)$$

and

$$\Psi_{III}(x) = D \exp(-\alpha x) \qquad (5.266)$$

where the constants D and B are obtained from eqs. (5.263) and (5.262).

5.4.4 Normalization of Odd Wave Functions

As for even wave functions, the normalization integral is written from eqs. (5.142) with $-D$, and (5.235) as

$$1 = 2 \int_0^\infty \left| \Psi^2(x) \right| dx = 2 \int_0^L \left| \Psi_{II}^2(x) \right| dx + 2 \int_L^\infty \left| \Psi_{III}^2(x) \right| dx$$

$$= 2 \int_0^L \left| A \sin(kx) \right|^2 dx + 2 \int_L^\infty \left| -D \exp(-\alpha x) \right|^2 dx$$

$$= 2 \int_0^L |A|^2 \sin^2(kx) dx + 2 \int_L^\infty |D|^2 \exp(-2\alpha x) dx$$

$$= 2|A|^2 \int_0^L \sin^2(kx) dx + 2|D|^2 \int_L^\infty \exp(-2\alpha x) dx \qquad (5.267)$$

Here,

$$\int_0^L \sin^2(kx) dx = \int_0^L \frac{1}{2} \left\{ 2 - 2\cos^2(kx) \right\} dx$$

$$= \frac{1}{2} \int_0^L \left\{ 2 - (1 + \cos 2kx) \right\} dx$$

$$= \frac{1}{2} \int_0^L (1 - \cos 2kx) dx$$

$$= \frac{1}{2} \int_0^L dx - \frac{1}{2} \int_0^L (\cos 2kx) dx$$

$$= \frac{1}{2} |x|_0^L - \frac{1}{2} \int_0^L (\cos 2kx) dx$$

$$= \frac{L}{2} - \frac{1}{2} \int_0^L (\cos 2kx) dx \qquad (5.268)$$

Applying eq. (5.268), eq. (5.249) and eq. (5.253), eq. (5.267) becomes

$$1 = 2|A|^2 \times \left\{ \frac{L}{2} - \frac{\sin(2kL)}{4k} \right\} + 2|D|^2 \times \frac{\exp(-2\alpha L)}{2\alpha}$$

$$= 2|A|^2 \times \frac{1}{2} \left\{ L - \frac{\sin(2kL)}{2k} \right\} + |D|^2 \times \frac{\exp(-2\alpha L)}{\alpha}$$

$$= |A|^2 \times \left\{ L - \frac{\sin(2kL)}{2k} \right\} + |D|^2 \times \frac{\exp(-2\alpha L)}{\alpha} \qquad (5.269)$$

But from eq. (5.237)

$$-D \exp(-\alpha L) = A \sin(kL) \qquad (5.270)$$

or

$$D = -A \exp(\alpha L) \sin(kL) \qquad (5.271)$$

Putting the value of D from eq. (5.271) in eq. (5.269), we get

$$1 = |A|^2 \times \left\{ L - \frac{\sin(2kL)}{2k} \right\} + \left\{ -A \exp(\alpha L) \sin(kL) \right\}^2$$

$$\times \frac{\exp(-2\alpha L)}{\alpha}$$

$$= |A|^2 \left\{ L - \frac{\sin(2kL)}{2k} \right\} + |-A|^2 \exp(2\alpha L) \sin^2(kL)$$

$$\times \frac{\exp(-2\alpha L)}{\alpha}$$

$$= |A|^2 \left\{ L - \frac{\sin(2kL)}{2k} \right\} + |-A|^2 \sin^2(kL) \times \frac{1}{\alpha}$$

$$= |A|^2 \left(L - \frac{\sin 2kL}{2k} + \frac{\sin^2(kL)}{\alpha} \right)$$

$$= |A|^2 \left\{ L - \frac{\sin(2kL)}{2k} + \frac{\sin^2(kL)}{-k\cot(kL)} \right\}$$

$$= |A|^2 \left\{ L - \frac{2\sin(kL)\cos(kL)}{2k} + \frac{\sin^2(kL)}{-\dfrac{k\cos(kL)}{\sin(kL)}} \right\}$$

$$= |A|^2 \left\{ L - \frac{\sin(kL)\cos(kL)}{k} - \frac{\sin^3(kL)}{k\cos(kL)} \right\} \qquad (5.272)$$

$$\because \sin(2kL) = 2\sin(kL)\cos(kL) \text{ and } \alpha = -k\cot(kL) \qquad (5.273)$$

from eq. (5.243).

Now using $\cot(kL) = -\alpha/k$, eq. (5.243)

$$1 = |A|^2 \left\{ L - \frac{\sin(kL)\cos(kL)}{k} - \frac{\sin^3(kL)}{k\cos(kL)} \right\}$$

$$= |A|^2 \left[L - \frac{\sin(kL)}{k\cos(kL)} \left\{ \cos^2(kL) + \sin^2(kL) \right\} \right]$$

$$= |A|^2 \left[L - \frac{1}{\frac{k\cos(kL)}{\sin(kL)}} \left\{ \cos^2(kL) + \sin^2(kL) \right\} \right]$$

$$= |A|^2 \left[L - \frac{1}{k\cot(kL)}(1) \right]$$

$$= |A|^2 \left[L - \frac{1}{k \times -\frac{\alpha}{k}}(1) \right] = |A|^2 \left(L + \frac{1}{\alpha} \right) \qquad (5.274)$$

or

$$A = \frac{1}{\sqrt{L + \frac{1}{\alpha}}} \qquad (5.275)$$

Plugging this value of A in eq. (5.271) begets

$$D = -A\exp(\alpha L)\sin(kL) = -\frac{1}{\sqrt{L + \frac{1}{\alpha}}}\exp(\alpha L)\sin(kL)$$

$$= -\frac{\exp(\alpha L)\sin(kL)}{\sqrt{L + \frac{1}{\alpha}}} \qquad (5.276)$$

From eqs. (5.141), (5.142), and using the condition of equality in magnitude but opposite in sign of coefficients C, D, along with eq. (5.235), the odd wave functions are given by

$$\Psi_I(x) = D\exp(\alpha x) \qquad (5.277)$$

$$\Psi_{II}(x) = A\sin(kx) \qquad (5.278)$$

and

$$\Psi_{III}(x) = -D\exp(-\alpha x) \qquad (5.279)$$

where the constants D and A are obtained from eqs. (5.276) and (5.275), respectively.

5.4.5 Plotting the Wave Functions: Quantum-Mechanical Tunneling

When the wave functions $\Psi_I(x)$, $\Psi_{II}(x)$, $\Psi_{III}(x)$ described by eqs. (5.264–5.266) and (5.277–5.279) are plotted for the ground or lowest energy state and higher energies, continuous curves are

obtained, as sketched in Figure 5.11; the corresponding curves for infinite potential well are drawn alongside to reveal the differences between the two cases. As mentioned in the diagram, the first wave function is an even wave function, the second wave function is an odd wave function, and again the third wave function is an even wave function. The odd and even wave functions keep alternating. Moreover, the number of nodes is zero, one, and two, respectively.

Look at the graph prudently. The graph has a noteworthy feature. The wave function decreases exponentially outside the walls of the finite box. It does not suddenly decrease to zero outside the walls, disparately from the case of an infinite well. The non-zero values of the wave function point toward interesting possibilities. They divulge that there are non-zero probabilities for existence of the particle in these regions. This is an unequivocal difference from classical mechanics. Newtonian mechanics affirmatively states that the particle cannot be found here. Thus according to quantum mechanics, there is a distinct probability that the particle can momentarily penetrate through the walls of a potential V_0 despite the fact that its energy is less than V_0.

To reiterate, the non-zero value of wave function in classically prohibited regions is of purely quantum-mechanical origin and is ascribed to the phenomenon of quantum-mechanical tunneling through the barrier. This possibility leads to a transitory violation of the principle of conservation of energy but can be explained in terms of the time-energy uncertainty principle (eq. (2.21))

$$\Delta E \Delta t \geq \frac{\hbar}{2} \qquad (5.280)$$

representing a fundamental limit to the simultaneous exact measurement of the pair of complementary variables, energy and time.

5.4.6 Unbound States

Solutions obtained by solving the TISE for energies of the particle $E > V_0$ are found to be oscillatory in nature and are non-normalizable, meaning that the particle has a continuous energy spectrum for $E > V_0$.

5.5 Derivation of Tunneling Probability Equation

An electron having an energy E strikes a potential barrier of height V_0 and width L. Let $E < V_0$. Both V_0 and L are finite. The electron behavior is studied by partitioning the space under consideration into three regions I, II and III representing the space on the left side of the barrier from where the electron wave approaches the barrier, the space of electron journey inside the barrier, and the space of the electron emerging out of the barrier, if possible, on the right side of the barrier (Figure 5.12). The equations describing a potential barrier of finite height and finite width are:

$$\text{Region I}: x < 0 \text{ or, } -\infty < x < 0 : V(x) = 0 \qquad (5.281)$$

$$\text{Region II}: 0 \leq x \leq L : V(x) = V_0 \qquad (5.282)$$

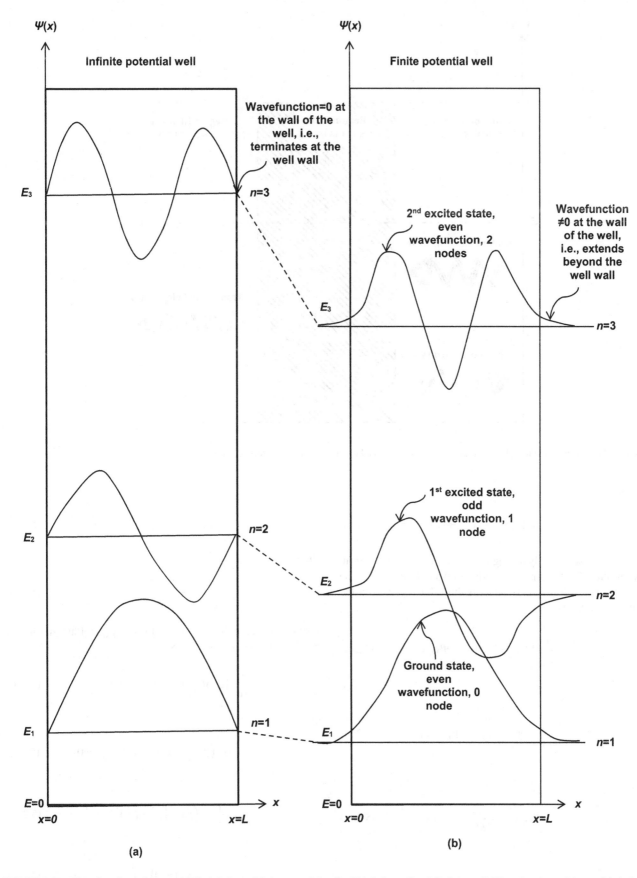

FIGURE 5.11 Wave functions of the particle in infinite and finite potential wells: (a) infinite well and (b) finite well. Wave functions of the particle in finite potential well spread out into classically forbidden regions while those of the particle in infinite well are restricted to well width; also the energy spacing between different discrete levels is less in the finite well than in the infinite well. E is the energy, n is the quantum number.

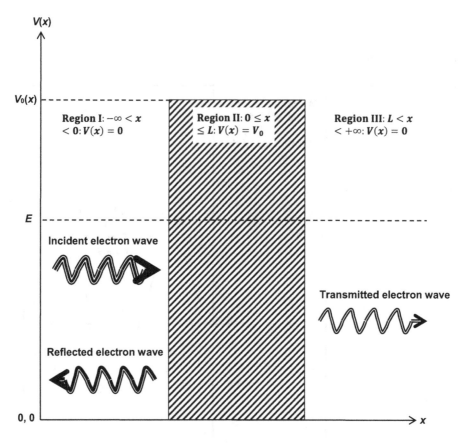

FIGURE 5.12 Creation of three physical regions by a barrier of height $V_0(x)$ to an incident electron wave.

and

$$\text{Region III} : x > L \text{ or, } L < x < +\infty : V(x) = 0 \quad (5.283)$$

The reader must carefully distinguish these three equations from those written for the finite potential well in Section 5.3. The TISE for regions I and III is same as those for region II in the finite well case, i.e.,

$$-\frac{\hbar^2}{2m}\frac{\partial^2}{\partial x^2}\Psi_{I,III}(x) + 0 = E\,\Psi_{I,III}(x) \quad (5.284)$$

or

$$-\frac{\hbar^2}{2m}\frac{\partial^2}{\partial x^2}\Psi_{I,III}(x) = E\Psi_{I,III}(x) \quad (5.285)$$

or

$$-\frac{\partial^2}{\partial x^2}\Psi_{I,III}(x) = \frac{2m}{\hbar^2}E\Psi_{I,III}(x) \quad (5.286)$$

or

$$\frac{\partial^2}{\partial x^2}\Psi_{I,III}(x) + \frac{2m}{\hbar^2}E\Psi_{I,III}(x) = 0 \quad (5.287)$$

or

$$\frac{\partial^2}{\partial x^2}\Psi_{I,III}(x) + k^2\Psi_{I,III}(x) = 0 \quad (5.288)$$

where

$$k^2 = \frac{2m}{\hbar^2}E \quad (5.289)$$

Assume the solution to be of the form $\exp(\gamma x)$. Plugging it into the differential equation (5.288),

$$\frac{\partial^2}{\partial x^2}\{\exp(\gamma x)\} + k^2\{\exp(\gamma x)\} = 0 \quad (5.290)$$

or

$$\gamma^2\{\exp(\gamma x)\} - (ik)^2\{\exp(\gamma x)\} = 0 \quad (5.291)$$

or

$$\{\exp(\gamma x)\}\{\gamma^2 - (ik)^2\} \quad (5.292)$$

Since,

$$\{\exp(\gamma x)\} \neq 0 \quad (5.293)$$

$$\gamma^2 - (ik)^2 = 0 \quad (5.294)$$

This is the characteristic equation. Its roots give the solutions to the differential equation (5.288).

$$\because \gamma^2 = (ik)^2 \qquad (5.295)$$

$$\therefore \gamma = \pm ik \qquad (5.296)$$

The solutions for $\Psi_I(x)$ and $\Psi_{III}(x)$ may be written as

$$\Psi_I(x) = A\exp(ikx) + B\exp(-ikx) \qquad (5.297)$$

and

$$\Psi_{III}(x) = F\exp(ikx) + G\exp(-ikx) \qquad (5.298)$$

where A, B, F, and G are constants. We take $G = 0$ because it is the coefficient of an oppositely directed incident wave, which is obviously not physical. Then

$$\Psi_{III}(x) = F\exp(ikx) \qquad (5.299)$$

Similarly, the TISE for region II is same as those for regions I and III for the finite well case:

$$-\frac{\hbar^2}{2m}\frac{\partial^2}{\partial x^2}\Psi_{II}(x) + V_0(x)\Psi_{II}(x) = E\Psi_{II}(x) \qquad (5.300)$$

or

$$-\frac{\hbar^2}{2m}\frac{\partial^2}{\partial x^2}\Psi_{II}(x) + V_0(x)\Psi_{II}(x) - E\Psi_{II}(x) = 0 \qquad (5.301)$$

or

$$\frac{\hbar^2}{2m}\frac{\partial^2}{\partial x^2}\Psi_{II}(x) - V_0(x)\Psi_{II}(x) + E\Psi_{II}(x) = 0 \qquad (5.302)$$

or

$$\frac{\hbar^2}{2m}\frac{\partial^2}{\partial x^2}\Psi_{II}(x) - \left\{V_0(x) - E\right\}\Psi_{II}(x) = 0 \qquad (5.303)$$

Dividing both sides of eq. (5.303) by $\hbar^2/(2m)$ we get

$$\frac{\partial^2}{\partial x^2}\Psi_{II}(x) - \frac{2m}{\hbar^2}\left\{V_0(x) - E\right\}\Psi_{II}(x) = 0 \qquad (5.304)$$

As before for finite well, putting

$$\frac{2m}{\hbar^2}\left\{V_0(x) - E\right\} = \alpha^2 \qquad (5.305)$$

and taking note of the fact that the quantity $(V_0 - E)$ is positive since $E < V_0$, and also

$$\alpha = \sqrt{\frac{2m}{\hbar^2}\left\{V_0(x) - E\right\}} \qquad (5.306)$$

is a real number

we get

$$\frac{\partial^2}{\partial x^2}\Psi_{II}(x) - \alpha^2\Psi_{II}(x) = 0 \qquad (5.307)$$

Let us presume the solution to be of the form $\exp(\beta x)$. Plugging the guessed solution into the differential equation (5.307), we have

$$\frac{\partial^2}{\partial x^2}\left\{\exp(\beta x)\right\} - \alpha^2\left\{\exp(\beta x)\right\} = 0 \qquad (5.308)$$

or

$$\beta^2\left\{\exp(\beta x)\right\} - \alpha^2\left\{\exp(\beta x)\right\} = 0 \qquad (5.309)$$

or

$$\left\{\exp(\beta x)\right\}\left(\beta^2 - \alpha^2\right) = 0 \qquad (5.310)$$

Since,

$$\left\{\exp(\beta x)\right\} \neq 0 \qquad (5.311)$$

$$\therefore \beta^2 - \alpha^2 = 0 \qquad (5.312)$$

This is the characteristic equation. Its roots give the solutions to the differential equation (5.307)

$$\because \beta^2 = \alpha^2 \qquad (5.313)$$

$$\therefore \beta = \pm\alpha \qquad (5.314)$$

The solution is

$$\Psi_{II}(x) = C\exp(\alpha x) + D\exp(-\alpha x) \qquad (5.315)$$

where C and D are constants.

We collect together the equations for $\Psi_I(x)$, $\Psi_{II}(x)$, $\Psi_{III}(x)$:

$$\Psi_I(x) = A\exp(ikx) + B\exp(-ikx) \qquad (5.316)$$

$$\Psi_{II}(x) = C\exp(\alpha x) + D\exp(-\alpha x) \qquad (5.317)$$

and

$$\Psi_{III}(x) = F\exp(ikx) \qquad (5.318)$$

The continuity condition at the boundary between regions I and II is

$$\Psi_I(0) = \Psi_{II}(0) \qquad (5.319)$$

so that from eqs. (5.316) and (5.317)

$$A\exp(0) + B\exp(-0) = C\exp(0) + D\exp(-0) \qquad (5.320)$$

or

$$A + B = C + D \qquad (5.321)$$

The continuity condition at the boundary between regions II and III is

$$\Psi_{\mathrm{II}}(L) = \Psi_{\mathrm{III}}(L) \tag{5.322}$$

Hence, from eqs. (5.317) and (5.318), we have

$$C \exp(\alpha L) + D \exp(-\alpha L) = F \exp(ikL) \tag{5.323}$$

The smoothness condition requires that the first derivative of the solution must be continuous at the boundary between regions I and II, i.e.,

$$\left| \frac{d\Psi_{\mathrm{I}}(x)}{dx} \right|_{x=0} = \left| \frac{d\Psi_{\mathrm{II}}(x)}{dx} \right|_{x=0} \tag{5.324}$$

or

$$\left| \frac{d}{dx} \{ A \exp(ikx) + B \exp(-ikx) \} \right|_{x=0}$$
$$= \left| \frac{d}{dx} \{ C \exp(\alpha x) + D \exp(-\alpha x) \} \right|_{x=0} \tag{5.325}$$

or

$$\left| ikA \exp(ikx) - ikB \exp(-ikx) \right|_{x=0}$$
$$= \left| \alpha C \exp(\alpha x) - \alpha D \exp(-\alpha x) \right|_{x=0} \tag{5.326}$$

or

$$ikA \exp(0) - ikB \exp(-0) = \alpha C \exp(0) - \alpha D \exp(-0) \tag{5.327}$$

or

$$ikA - ikB = \alpha C - \alpha D \tag{5.328}$$

or

$$ik(A - B) = \alpha(C - D) \tag{5.329}$$

From the smoothness condition requirement that the first derivative of the solution must be continuous at the boundary between regions II and III,

$$\left| \frac{d\Psi_{\mathrm{II}}(x)}{dx} \right|_{x=L} = \left| \frac{d\Psi_{\mathrm{III}}(x)}{dx} \right|_{x=L} \tag{5.330}$$

or

$$\left| \frac{d}{dx} \{ C \exp(\alpha x) + D \exp(-\alpha x) \} \right|_{x=L} = \left| \frac{d}{dx} \{ F \exp(ikx) \} \right|_{x=L} \tag{5.331}$$

or

$$\left| \alpha C \exp(\alpha x) - \alpha D \exp(-\alpha x) \right|_{x=L} = \left| ikF \exp(ikx) \right|_{x=L} \tag{5.332}$$

or

$$\alpha C \exp(\alpha L) - \alpha D \exp(-\alpha L) = ikF \exp(ikL) \tag{5.333}$$

Thus we have four equations compiled as follows:

$$A + B = C + D \tag{5.334}$$

$$C \exp(\alpha L) + D \exp(-\alpha L) = F \exp(ikL) \tag{5.335}$$

$$ik(A - B) = \alpha(C - D) \tag{5.336}$$

and

$$\alpha C \exp(\alpha L) - \alpha D \exp(-\alpha L) = ikF \exp(ikL) \tag{5.337}$$

Inopportunely, these four equations contain five unknowns, viz. *A*, *B*, *C*, *D*, and *F*. Obviously, we need five equations to find all the unknowns and we are short of one equation. But there is a ray of hope when we focus only on the unknown whose value is sought. We notice that we have to determine the transmission coefficient $T(L, E)$ for a particle of energy *E* incident on a barrier of width *L* defined as

$$T(L, E) = \frac{\text{Transmitted intensity} |F|^2}{\text{Incident intensity} |A|^2} = \frac{|F|^2}{|A|^2} = \left| \frac{F}{A} \right|^2 \tag{5.338}$$

So we play a trick. Each of the four equations is divided by *A* yielding four equations in terms of fractions *B/A*, *C/A*, *D/A*, and *F/A*. The fractions are denoted as:

$$\frac{B}{A} = X, \frac{C}{A} = Y, \frac{D}{A} = Z, \frac{F}{A} = W \tag{5.339}$$

Application of the trick has resulted in producing four unknown fractions, viz. *X*, *Y*, *Z*, and *W*. Also, there are four equations from which to evaluate these fractions. So by this division artifice, the number of unknowns has been cut down to the same number as the number of equations, facilitating our task. It is required to determine the fraction $F/A = W$. Proceeding to execute this plan of action, eq. (5.334) becomes

$$\frac{A}{A} + \frac{B}{A} = \frac{C}{A} + \frac{D}{A} \tag{5.340}$$

or

$$1 + \frac{B}{A} = \frac{C}{A} + \frac{D}{A} \tag{5.341}$$

or

$$1 + X = Y + Z \tag{5.342}$$

Also, from eq. (5.335)

$$\frac{C}{A} \exp(\alpha L) + \frac{D}{A} \exp(-\alpha L) = \frac{F}{A} \exp(ikL) \tag{5.343}$$

or

$$Y \exp(\alpha L) + Z \exp(-\alpha L) = W \exp(ikL) \qquad (5.344)$$

Moreover, from eq. (5.336)

$$ik\left(\frac{A}{A} - \frac{B}{A}\right) = \alpha\left(\frac{C}{A} - \frac{D}{A}\right) \qquad (5.345)$$

or

$$ik(1 - X) = \alpha(Y - Z) \qquad (5.346)$$

and from eq. (5.337)

$$\alpha\frac{C}{A}\exp(\alpha L) - \alpha\frac{D}{A}\exp(-\alpha L) = ik\frac{F}{A}\exp(ikL) \quad (5.347)$$

or

$$\alpha Y \exp(\alpha L) - \alpha Z \exp(-\alpha L) = ikW \exp(ikL) \quad (5.348)$$

We put the relevant equations together

$$1 + X = Y + Z \qquad (5.349)$$

$$Y \exp(\alpha L) + Z \exp(-\alpha L) = W \exp(ikL) \qquad (5.350)$$

$$ik(1 - X) = \alpha(Y - Z) \qquad (5.351)$$

and

$$\alpha Y \exp(\alpha L) - \alpha Z \exp(-\alpha L) = ikW \exp(ikL) \quad (5.352)$$

Let us first eliminate X. From eq. (5.349)

$$X = Y + Z - 1 \qquad (5.353)$$

Substituting for X from eq. (5.353) in eq. (5.351) we get

$$ik\{1 - (Y + Z - 1)\} = \alpha(Y - Z) \qquad (5.354)$$

or

$$ik(2 - Y - Z) = \alpha(Y - Z) \qquad (5.355)$$

Now X is eliminated and we have three equations in Y, Z, and W:

$$ik(2 - Y - Z) = \alpha(Y - Z) \qquad (5.356)$$

$$Y \exp(\alpha L) + Z \exp(-\alpha L) = W \exp(ikL) \qquad (5.357)$$

and

$$\alpha Y \exp(\alpha L) - \alpha Z \exp(-\alpha L) = ikW \exp(ikL) \quad (5.358)$$

Let us now eliminate Y. From eq. (5.356),

$$2ik - ikY - ikZ = \alpha Y - \alpha Z \qquad (5.359)$$

or

$$-ikY - \alpha Y = -\alpha Z + ikZ - 2ik \qquad (5.360)$$

or

$$-Y(ik + \alpha) = -\alpha Z + ikZ - 2ik \qquad (5.361)$$

or

$$Y = \frac{\alpha Z - ikZ + 2ik}{ik + \alpha} \qquad (5.362)$$

Substituting for Y from eq. (5.362) in eqs. (5.357) and (5.358), we get

$$\frac{\alpha Z - ikZ + 2ik}{ik + \alpha}\exp(\alpha L) + Z \exp(-\alpha L) = W \exp(ikL) \quad (5.363)$$

and

$$\alpha\frac{\alpha Z - ikZ + 2ik}{ik + \alpha}\exp(\alpha L) - \alpha Z \exp(-\alpha L) = ikW \exp(ikL)$$

$$(5.364)$$

Thus Y is eliminated and we are left with two equations containing two unknowns W and Z. We next eliminate Z. From eq. (5.363)

$$Z \exp(-\alpha L) = W \exp(ikL) - \frac{\alpha Z - ikZ + 2ik}{ik + \alpha}\exp(\alpha L) \quad (5.365)$$

or

$$Z(ik + \alpha)\exp(-\alpha L) = W \exp(ikL)(ik + \alpha)$$
$$- (\alpha Z - ikZ + 2ik)\exp(\alpha L) \qquad (5.366)$$

or

$$Z(ik + \alpha)\exp(-\alpha L) + Z(\alpha - ik)\exp(\alpha L)$$
$$= W \exp(ikL)(ik + \alpha) - 2ik \exp(\alpha L) \qquad (5.367)$$

or

$$Z = \frac{W \exp(ikL)(ik + \alpha) - 2ik \exp(\alpha L)}{(ik + \alpha)\exp(-\alpha L) + (\alpha - ik)\exp(\alpha L)} \qquad (5.368)$$

Rearranging eq. (5.364) and substituting for Z from eq. (5.368) in eq. (5.364),

$$\alpha\frac{(\alpha - ik)Z + 2ik}{ik + \alpha}\exp(\alpha L) - \alpha Z \exp(-\alpha L) = ikW \exp(ikL)$$

$$(5.369)$$

or

$$\frac{\alpha(\alpha - ik)Z \exp(\alpha L)}{ik + \alpha} - \alpha Z \exp(-\alpha L)$$

$$= ikW \exp(ikL) - \frac{2ik\alpha}{ik + \alpha}\exp(\alpha L) \qquad (5.370)$$

or

$$\alpha(\alpha - ik)Z\exp(\alpha L) - \alpha(ik + \alpha)Z\exp(-\alpha L)$$
$$= ikW(ik + \alpha)\exp(ikL) - 2ik\alpha\exp(\alpha L) \qquad (5.371)$$

or

$$\{\alpha(\alpha - ik)\exp(\alpha L) - \alpha(ik + \alpha)\exp(-\alpha L)\}Z$$
$$= ikW(ik + \alpha)\exp(ikL) - 2ik\alpha\exp(\alpha L) \qquad (5.372)$$

or

$$\{\alpha(\alpha - ik)\exp(\alpha L) - \alpha(ik + \alpha)\exp(-\alpha L)\}$$
$$\times \frac{W\exp(ikL)(ik + \alpha) - 2ik\exp(\alpha L)}{(ik + \alpha)\exp(-\alpha L) + (\alpha - ik)\exp(\alpha L)}$$
$$= ikW(ik + \alpha)\exp(ikL) - 2ik\alpha\exp(\alpha L) \qquad (5.373)$$

or

$$W\exp(ikL)(ik + \alpha)\frac{\alpha\{(\alpha - ik)\exp(\alpha L) - (ik + \alpha)\exp(-\alpha L)\}}{(ik + \alpha)\exp(-\alpha L) + (\alpha - ik)\exp(\alpha L)}$$
$$- 2ik\exp(\alpha L)\frac{\alpha\{(\alpha - ik)\exp(\alpha L) - (ik + \alpha)\exp(-\alpha L)\}}{(ik + \alpha)\exp(-\alpha L) + (\alpha - ik)\exp(\alpha L)}$$
$$= ikW(ik + \alpha)\exp(ikL) - 2ik\alpha\exp(\alpha L) \qquad (5.374)$$

or

$$W\alpha\exp(ikL)(ik + \alpha)\{(\alpha - ik)\exp(\alpha L) - (ik + \alpha)\exp(-\alpha L)\}$$
$$- 2ik\alpha\exp(\alpha L)\{(\alpha - ik)\exp(\alpha L) - (ik + \alpha)\exp(-\alpha L)\}$$
$$= ikW(ik + \alpha)\exp(ikL)\{(ik + \alpha)\exp(-\alpha L) + (\alpha - ik)\exp(\alpha L)\}$$
$$- 2ik\alpha\exp(\alpha L)\{(ik + \alpha)\exp(-\alpha L) + (\alpha - ik)\exp(\alpha L)\}$$
$$\qquad (5.375)$$

or

$$W\alpha\exp(ikL)(ik + \alpha)\{(\alpha - ik)\exp(\alpha L) - (ik + \alpha)\exp(-\alpha L)\}$$
$$- ikW(ik + \alpha)\exp(ikL)\{(ik + \alpha)\exp(-\alpha L) + (\alpha - ik)\exp(\alpha L)\}$$
$$= 2ik\alpha\exp(\alpha L)\{(\alpha - ik)\exp(\alpha L) - (ik + \alpha)\exp(-\alpha L)\}$$
$$- 2ik\alpha\exp(\alpha L)\{(ik + \alpha)\exp(-\alpha L) + (\alpha - ik)\exp(\alpha L)\}$$
$$= 2ik\alpha\exp(\alpha L)\{(\alpha - ik)\exp(\alpha L) - (ik + \alpha)\exp(-\alpha L)$$
$$- (ik + \alpha)\exp(-\alpha L) - (\alpha - ik)\exp(\alpha L)\}$$
$$= 2ik\alpha\exp(\alpha L)\times -2(ik + \alpha)\exp(-\alpha L) = -4ik\alpha(ik + \alpha)$$
$$\qquad (5.376)$$

Dividing both sides of eq. (5.376) by $-4ik\alpha\,(ik + \alpha)$,

$$-\frac{W\exp(ikL)\{(\alpha - ik)\exp(\alpha L) - (ik + \alpha)\exp(-\alpha L)\}}{4ik}$$
$$+\frac{W\exp(ikL)\{(ik + \alpha)\exp(-\alpha L) + (\alpha - ik)\exp(\alpha L)\}}{4\alpha} = 1$$
$$\qquad (5.377)$$

or

$$W = \frac{\exp(-ikL)}{-\dfrac{\{(\alpha - ik)\exp(\alpha L) - (ik + \alpha)\exp(-\alpha L)\}}{4ik} + \dfrac{\{(ik + \alpha)\exp(-\alpha L) + (\alpha - ik)\exp(\alpha L)\}}{4\alpha}}$$

$$= \frac{\exp(-ikL)}{-\dfrac{1}{4}\left\{\left(\dfrac{\alpha}{ik} - 1\right)\exp(\alpha L) - \left(1 + \dfrac{\alpha}{ik}\right)\exp(-\alpha L)\right\} + \dfrac{1}{4}\left\{\left(\dfrac{ik}{\alpha} + 1\right)\exp(-\alpha L) + \left(1 - \dfrac{ik}{\alpha}\right)\exp(\alpha L)\right\}}$$

$$= \frac{\exp(-ikL)}{-\dfrac{1}{4}\left\{\dfrac{\alpha}{ik}\exp(\alpha L) - \exp(\alpha L) - \exp(-\alpha L) - \dfrac{\alpha}{ik}\exp(-\alpha L)\right\} + \dfrac{1}{4}\left\{\dfrac{ik}{\alpha}\exp(-\alpha L) + \exp(-\alpha L) + \exp(\alpha L) - \dfrac{ik}{\alpha}\exp(\alpha L)\right\}}$$

$$= \frac{\exp(-ikL)}{-\dfrac{1}{4}\left\{\dfrac{\alpha}{ik}\exp(\alpha L) - \dfrac{\alpha}{ik}\exp(-\alpha L) - \exp(\alpha L) - \exp(-\alpha L)\right\} + \dfrac{1}{4}\left\{\dfrac{ik}{\alpha}\exp(-\alpha L) - \dfrac{ik}{\alpha}\exp(\alpha L) + \exp(-\alpha L) + \exp(\alpha L)\right\}}$$

$$= \frac{\exp(-ikL)}{-\dfrac{1}{2}\left\{\dfrac{\alpha}{ik}\dfrac{\{\exp(\alpha L) - \exp(-\alpha L)\}}{2} - \dfrac{\exp(\alpha L) + \exp(-\alpha L)}{2}\right\} + \dfrac{1}{2}\left\{-\dfrac{ik}{\alpha}\dfrac{\exp(\alpha L) - \exp(-\alpha L)}{2} + \dfrac{\exp(\alpha L) + \exp(-\alpha L)}{2}\right\}}$$

$$= \frac{\exp(-ikL)}{-\frac{1}{2}\left\{\frac{\alpha}{ik}\sinh(\alpha L) - \cosh(\alpha L)\right\} + \frac{1}{2}\left\{-\frac{ik}{\alpha}\sinh(\alpha L) + \cosh(\alpha L)\right\}}$$

$$= \frac{\exp(-ikL)}{\frac{1}{2}\cosh(\alpha L) - \frac{1}{2}\frac{\alpha}{ik}\sinh(\alpha L) + \frac{1}{2}\cosh(\alpha L) - \frac{1}{2}\frac{ik}{\alpha}\sinh(\alpha L)}$$

$$= \frac{\exp(-ikL)}{\frac{1}{2}\cosh(\alpha L) + \frac{1}{2}\cosh(\alpha L) - \frac{1}{2}\frac{\alpha}{ik}\sinh(\alpha L) - \frac{1}{2}\frac{ik}{\alpha}\sinh(\alpha L)}$$

$$= \frac{\exp(-ikL)}{\cosh(\alpha L) - \frac{1}{2}\frac{\alpha}{ik}\sinh(\alpha L) - \frac{1}{2}\frac{ik}{\alpha}\sinh(\alpha L)} = \frac{\exp(-ikL)}{\cosh(\alpha L) - \frac{1}{2}i\left\{\frac{\alpha}{i^2 k}\sinh(\alpha L) + \frac{k}{\alpha}\sinh(\alpha L)\right\}}$$

$$= \frac{\exp(-ikL)}{\cosh(\alpha L) - \frac{1}{2}i\left\{-\frac{\alpha}{k}\sinh(\alpha L) + \frac{k}{\alpha}\sinh(\alpha L)\right\}} = \frac{\exp(-ikL)}{\cosh(\alpha L) + \frac{1}{2}i\left\{+\frac{\alpha}{k}\sinh(\alpha L) - \frac{k}{\alpha}\sinh(\alpha L)\right\}}$$

$$= \frac{\exp(-ikL)}{\cosh(\alpha L) + \frac{1}{2}i\left(\frac{\alpha}{k} - \frac{k}{\alpha}\right)\sinh(\alpha L)} = \frac{\exp(-ikL)}{\cosh(\alpha L) + \frac{1}{2}i\xi\sinh(\alpha L)} \tag{5.378}$$

where

$$\xi = \frac{\alpha}{k} - \frac{k}{\alpha} \tag{5.379}$$

The equation for tunneling probability or transmission coefficient is obtained as

$$T(L, E) = \left|\frac{F}{A}\right|^2 = W^2 = W \cdot W^* = \frac{\exp(-ikL)}{\cosh(\alpha L) + \frac{1}{2}i\xi\sinh(\alpha L)} \times \frac{\exp(+ikL)}{\cosh(\alpha L) - \frac{1}{2}i\xi\sinh(\alpha L)}$$

$$= \frac{1}{\cosh^2(\alpha L) - \left\{\frac{1}{2}i\xi\sinh(\alpha L)\right\}^2} = \frac{1}{\cosh^2(\alpha L) - \frac{1}{4}\times -1\times\xi^2\sinh^2(\alpha L)}$$

$$= \frac{1}{\cosh^2(\alpha L) + \frac{1}{4}\xi^2\sinh^2(\alpha L)} \tag{5.380}$$

where we use eqs. (5.289) and (5.306) to get

$$\frac{1}{4}\xi^2 = \frac{1}{4}\left(\frac{\alpha}{k} - \frac{k}{\alpha}\right)^2 = \frac{1}{4}\left(\frac{\sqrt{\frac{2m}{\hbar^2}\{V_0(x) - E\}}}{\sqrt{\frac{2mE}{\hbar^2}}} - \frac{\sqrt{\frac{2mE}{\hbar^2}}}{\sqrt{\frac{2m}{\hbar^2}\{V_0(x) - E\}}}\right)^2 = \frac{1}{4}\left[\frac{\frac{2m}{\hbar^2}\{V_0(x) - E\}}{\frac{2mE}{\hbar^2}} + \frac{\frac{2mE}{\hbar^2}}{\frac{2m}{\hbar^2}\{V_0(x) - E\}}\right.$$

$$\left. - 2\frac{\sqrt{\frac{2m}{\hbar^2}\{V_0(x) - E\}}}{\sqrt{\frac{2mE}{\hbar^2}}} \times \frac{\sqrt{\frac{2mE}{\hbar^2}}}{\sqrt{\frac{2m}{\hbar^2}\{V_0(x) - E\}}}\right] = \frac{1}{4}\left[\frac{\frac{2m}{\hbar^2}\{V_0(x) - E\}}{\frac{2mE}{\hbar^2}} + \frac{\frac{2mE}{\hbar^2}}{\frac{2m}{\hbar^2}\{V_0(x) - E\}} - 2\right]$$

$$= \frac{1}{4}\left[\frac{\{V_0(x) - E\}}{E} + \frac{E}{\{V_0(x) - E\}} - 2\right] = \frac{1}{4}\left[\frac{\left\{1 - \frac{E}{V_0(x)}\right\}}{\frac{E}{V_0(x)}} + \frac{\frac{E}{V_0(x)}}{\left\{1 - \frac{E}{V_0(x)}\right\}} - 2\right] \tag{5.381}$$

Alternatively, we apply the relation

$$\cosh^2(\alpha L) = 1 + \sinh^2(\alpha L) \tag{5.382}$$

in eq. (5.380) to get

$$T(L, E) = \frac{1}{1 + \sinh^2(\alpha L) + \dfrac{1}{4}\xi^2 \sinh^2(\alpha L)} = \frac{1}{1 + \left(1 + \dfrac{1}{4}\xi^2\right)\sinh^2(\alpha L)} \tag{5.383}$$

Here, using eq. (5.381)

$$\left(1 + \frac{1}{4}\xi^2\right)\sinh^2(\alpha L) = \left[1 + \frac{\{V_0(x) - E\}^2 + E^2 - 2E\{V_0(x) - E\}}{4E\{V_0(x) - E\}}\right]\sinh^2(\alpha L)$$

$$= \left[\frac{4E\{V_0(x) - E\} + \{V_0(x)\}^2 + E^2 - 2V_0(x)E + E^2 - 2E\{V_0(x) - E\}}{4E\{V_0(x) - E\}}\right]\sinh^2(\alpha L)$$

$$\tag{5.384}$$

$$= \left[\frac{4EV_0(x) - 4E^2 + \{V_0(x)\}^2 + E^2 - 2V_0(x)E + E^2 - 2EV_0(x) + 2E^2}{4E\{V_0(x) - E\}}\right]\sinh^2(\alpha L)$$

$$= \frac{\{V_0(x)\}^2}{4E\{V_0(x) - E\}}\sinh^2(\alpha L)$$

From eqs. (5.383) and (5.384), we see that

$$T(L, E) = \frac{1}{1 + \dfrac{\{V_0(x)\}^2}{4E\{V_0(x) - E\}}\sinh^2(\alpha L)} \tag{5.385}$$

This equation can be approximated as

$$T(L, E) \approx \frac{1}{\dfrac{\{V_0(x)\}^2}{4E\{V_0(x) - E\}}\sinh^2(\alpha L)} = \frac{1}{\dfrac{\{V_0(x)\}^2}{4E\{V_0(x) - E\}}\left\{\dfrac{\exp(\alpha L) - \exp(-\alpha L)}{2}\right\}^2} \tag{5.386}$$

When the barrier is wide and high,

$$\alpha L \gg 1, \quad \therefore \exp(\alpha L) \gg \exp(-\alpha L) \tag{5.387}$$

so that eq. (5.386) reduces to the form

$$T(L, E) = \frac{1}{\dfrac{\{V_0(x)\}^2}{4E\{V_0(x) - E\}}\left\{\dfrac{\exp(\alpha L) - \exp(-\alpha L)}{2}\right\}^2} \approx \frac{1}{\dfrac{\{V_0(x)\}^2}{4E\{V_0(x) - E\}}\left\{\dfrac{\exp(\alpha L)}{2}\right\}^2}$$

$$= \frac{1}{\dfrac{\{V_0(x)\}^2}{4E\{V_0(x) - E\}} \times \dfrac{\exp(2\alpha L)}{4}} = \frac{\exp(-2\alpha L)}{\dfrac{\{V_0(x)\}^2}{16E\{V_0(x) - E\}}} = \frac{16E\{V_0(x) - E\}\exp(-2\alpha L)}{\{V_0(x)\}^2}$$

$$= 16\left\{\frac{E}{V_0(x)}\right\}\left\{1 - \frac{E}{V_0(x)}\right\}\exp(-2\alpha L) \tag{5.388}$$

5.6 Discussion and Conclusions

i. A quantum-mechanical free particle can have any value of energy. The probability of finding the particle at a point x is spread out evenly in space so that position of the particle is uncertain within $\pm\infty$. The ability to represent a localized particle by plane waves is overcome by creating a non-zero wave function in a limited region of space which goes to zero far away in either direction. This wave function called a wave packet is formed by addition or superimposition of a large number of plane waves, which must individually be solutions of Schrodinger equation. As the number of waves added increases, the wave packet becomes more localized. Thus the free particle is described by a wave packet perspective.

ii. (a) For the particle in an infinite potential well, the energy is discontinuous or quantized and the lowest energy (zero point energy) has a non-zero value (Zettili 2009). With increasing quantum number, the energy separation becomes larger and the number of nodes (regions of zero probability of particle's presence) rises. So the main difference from classical mechanical solution is that the energy of the particle is quantized.

(b) The particle in an infinite well is analogous to an electron orbiting the atomic nucleus. The electron wave shows a standing wave pattern much like the standing wave pattern produced on a string comprising nodes (no displacement or destructive interference points) and antinodes (maximum amplitude or constructive interference points).

iii. (a) The particle in a finite potential well shows two major differences from classical mechanics:

 i. quantization of energy of the particle, and

 ii. likelihood of the particle being found in a region where its presence is outright impermissible in classical mechanics. This likelihood is the pointer toward the well-known tunneling phenomenon in quantum mechanics.

(b) The finite well has unbound states for $E > V_0$; infinite well has no such states.

iv. The occurrence of tunneling phenomenon for the particle in a finite well is expounded and the tunneling probability equation is obtained in full and approximate versions.

Illustrative Exercises

5.1 A particle is confined in a rigid box of width 10 nm? What is the probability of finding the particle in the location between $x = 4$ nm and $x = 6$ nm assuming that the particle is in ground state?

Probability of finding the particle in the location from $x = 4$ nm to $x = 6$ nm is from eq. (5.99)

$$\int_{x=4 \text{ nm}}^{x=6 \text{ nm}} \left| \Psi_n(x) \right|^2 dx = \int_{x=4 \text{ nm}}^{x=6 \text{ nm}} \left\{ \sqrt{\frac{2}{L}} \sin\left(\frac{n\pi}{L} x\right) \right\}^2 dx$$

$$= \int_{x=4 \text{ nm}}^{x=6 \text{ nm}} \frac{2}{L} \sin^2\left(\frac{n\pi}{L} x\right) dx$$

$$= \frac{1}{L} \int_{x=4 \text{ nm}}^{x=6 \text{ nm}} 2\sin^2\left(\frac{n\pi}{L} x\right) dx$$

$$= \frac{1}{L} \int_{x=4 \text{ nm}}^{x=6 \text{ nm}} \left\{ 1 - \cos\left(\frac{2n\pi}{L} x\right) \right\} dx$$

$$= \frac{1}{L} \int_{x=4 \text{ nm}}^{x=6 \text{ nm}} dx - \frac{1}{L} \int_{x=4 \text{ nm}}^{x=6 \text{ nm}} \cos\left(\frac{2n\pi}{L} x\right) dx$$

$$= \frac{1}{L} |x|_{x=4 \text{ nm}}^{x=6 \text{ nm}} - \frac{1}{L} \times \frac{1}{\frac{2n\pi}{L}} \left| \sin\frac{2n\pi}{L} x \right|_{x=4 \text{ nm}}^{x=6 \text{ nm}}$$

$$= \frac{1}{L}(6-4) - \frac{1}{2n\pi} \left(\sin\frac{2n\pi}{L} 6 \right)$$

$$+ \frac{1}{2n\pi} \left(\sin\frac{2n\pi}{L} 4 \right) \qquad (5.389)$$

where we have used the integral

$$\int \cos ax \, dx = \frac{1}{a} \sin ax \qquad (5.390)$$

Putting $n = 1$ for ground state and $L = 10$ nm, we get from eq. (5.389)

$$\int_{x=4 \text{ nm}}^{x=6 \text{ nm}} \left| \Psi_n(x) \right|^2 dx = \frac{1}{10}(6-4) - \frac{1}{2\pi}\left(\sin\frac{2\pi}{10}6\right) + \frac{1}{2\pi}\left(\sin\frac{2\pi}{10}4\right)$$

$$= \frac{2}{10} - \frac{1}{2 \times 3.14} \sin(360° \times 0.6)$$

$$+ \frac{1}{2 \times 3.14} \sin(360° \times 0.4)$$

$$= 0.2 - \frac{1}{6.28} \sin 216° + \frac{1}{6.28} \sin 144°$$

$$= 0.2 + \frac{0.5878}{6.28} + \frac{0.5878}{6.28}$$

$$= 0.2 + 0.0.0936 + 0.0936 = 0.3872 \quad (5.391)$$

5.2 For the particle in Illustrative Exercise 5.1, calculate the probability of finding the particle in the: (a) left half of the box and (b) right half of the box.

Applying arguments similar to eqs. (5.389)–(5.391),

(a)

$$\int_{x=0 \text{ nm}}^{x=5 \text{ nm}} \left| \Psi_n(x) \right|^2 dx = \frac{1}{10}(5-0) - \frac{1}{2\pi}\left(\sin\frac{2\pi}{10}5\right) + \frac{1}{2\pi}\left(\sin\frac{2\pi}{10}0\right)$$

$$= \frac{5}{10} - \frac{1}{6.28} \sin 180° + \frac{1}{6.28} \sin 0°$$

$$= 0.5 - \frac{0}{6.28} + \frac{0}{6.28} = 0.5 \qquad (5.392)$$

(b)

$$\int_{x=5\text{ nm}}^{x=10\text{ nm}} |\Psi_n(x)|^2\, dx = \frac{1}{10}(10-5) - \frac{1}{2\pi}\left(\sin\frac{2\pi}{10}10\right)$$

$$+ \frac{1}{2\pi}\left(\sin\frac{2\pi}{10}5\right)$$

$$= \frac{5}{10} - \frac{1}{6.28}\sin 360° + \frac{1}{6.28}\sin 180°$$

$$= 0.5 - \frac{0}{6.28} + \frac{0}{6.28} = 0.5 \tag{5.393}$$

5.3   What is the energy of an electron in the ground state in an infinite well of width 6.7 nm?
From eq. (5.93),

$$E_n = \frac{n^2\pi^2\hbar^2}{2mL^2} \tag{5.394}$$

Here $n = 1$, $L = 6.7$ nm. Hence

$$E_n = \frac{1^2\pi^2\hbar^2}{2mL^2} = \frac{\pi^2\hbar^2}{2mL^2} = \frac{(3.14)^2 \times (1.05457\times 10^{-34})^2}{2\times 9.109\times 10^{-31}\times (6.7\times 10^{-9})^2}\ \text{J}$$

$$= \frac{(3.14)^2 \times (1.05457\times 10^{-34})^2}{2\times 9.109\times 10^{-31}\times (6.7\times 10^{-9})^2} \times \frac{1}{1.602\times 10^{-19}}\ \text{eV}$$

$$= \frac{(3.14)^2 \times (1.05457)^2}{2\times 9.109\times (6.7)^2\times 1.602} \times \frac{10^{-68}}{10^{-31}\times 10^{-18}\times 10^{-19}}\ \text{eV}$$

$$= 0.008369\ \text{eV} \tag{5.395}$$

5.4   (a) Find the ground state energy of an electron in a finite well of depth 25 eV and width 500 pm.
(b) What will be the ground state energy if the well is of infinite depth?

(a) The allowed energies of the electron are contained in the equation (5.191)

$$\tan\left(\frac{1}{2}kL\right) = \frac{\alpha}{k} \tag{5.396}$$

where from eqs. (5.149) and (5.198)

$$k = \sqrt{\frac{2mE}{\hbar^2}} \tag{5.397}$$

$$\alpha = \sqrt{\frac{2m}{\hbar^2}\{V_0(x) - E\}} \tag{5.398}$$

$$\frac{\alpha}{k} = \sqrt{\frac{V_0(x) - E}{E}} = \sqrt{\frac{V_0(x)}{E} - 1} \tag{5.399}$$

Combining eqs. (5.396), (5.397), and (5.399) we get the transcendental equation

$$\tan\left(\frac{L}{2}\sqrt{\frac{2mE}{\hbar^2}}\right) = \sqrt{\frac{V_0(x)}{E} - 1} \tag{5.400}$$

which when solved numerically yields the solution $E = 1.123$ eV, as verified below

$$\tan\left(\frac{L}{2}\sqrt{\frac{2mE}{\hbar^2}}\right) = \tan\left(\frac{L}{2}\frac{\sqrt{2mE}}{\hbar}\right)$$

$$= \tan\left(\frac{500\times 10^{-12}}{2}\frac{\sqrt{2\times 9.109\times 10^{-31}\times 1.123\times 1.602\times 10^{-19}}}{\frac{h}{2\pi}}\right)$$

$$= \tan\left(\frac{500\times 10^{-12}}{2}\frac{\sqrt{2\times 9.109\times 1.123\times 1.602\times 10^{-50}}}{\frac{h}{2\pi}}\right)$$

$$= \tan\left(500\times 10^{-12}\times \frac{5.725\times 10^{-25}}{h}\pi\right)$$

$$= \tan\left(\frac{2.8625\times 10^{-34}}{6.626\times 10^{-34}}\times 180°\right)$$

$$= \tan\left(\frac{2.8625\times 10^{-34}}{6.626\times 10^{-34}}\times 180°\right) = \tan(77.762) = 4.61 \tag{5.401}$$

and

$$\sqrt{\frac{V_0(x)}{E} - 1} = \sqrt{\frac{25}{1.123} - 1} = 4.611 \tag{5.402}$$

Since the left-hand side equals the right-hand side, the solution $E = 1.123$ eV is true.

(b) From eq. (5.93),

$$E_1 = \frac{1^2\pi^2\hbar^2}{2mL^2} = \frac{(3.14159)^2\left(\frac{6.626\times 10^{-34}}{2\pi}\right)^2}{2\times 9.109\times 10^{-31}\times (500\times 10^{-12})^2}$$

$$= \frac{9.87\times\left(\frac{6.626\times 10^{-34}}{2\times 3.14}\right)^2}{4.55\times 10^{-49}} = \frac{1.099\times 10^{-67}}{4.55\times 10^{-49}}$$

$$= 2.42\times 10^{-19}\ \text{J} = \frac{2.42\times 10^{-19}}{1.602\times 10^{-19}}\ \text{eV} = 1.511\ \text{eV} \tag{5.403}$$

Note: The ground state energy of the electron in the infinite well (1.511 eV) is larger than the ground state energy of electron in the finite well (1.123 eV).

REFERENCES

Griffiths, D. J. 1995. *Introduction to Quantum Mechanics*. Prentice-Hall: Upper Saddle River, NJ, pp. 20–74.

Powell, J. L. and Crasemann, B. 2015. *Quantum Mechanics*, Dover edition. Addison-Wesley Publishing Company, Inc.: Reading, MA, pp. 102–157.

Shankar, R. 1994. *Principles of Quantum Mechanics*. Springer Science + Business Media, LLC: New York, pp. 151–178.

Zettili, N. 2009. *Quantum Mechanics: Concepts and Applications*. John Wiley & Sons: Chichester, UK, pp. 215–282.

6

The Hydrogen Atom

A problem of utmost importance in the field of atomic and molecular structure is the problem of structure of the hydrogen atom, not only because it affords a simpler analytical treatment but also because it lays down the foundation for dealing with complex atomic systems (Haken and Wolf 1994, Thakkar 2017). Historically too it has played a pivotal role in the development of physical and chemical theories keeping in view the seminal contributions of Bohr's model and Schrodinger's solution of the wave equation for the hydrogen atom.

The hydrogen atom is the simplest atom comprising a nucleus containing a single proton and an electron revolving around the nucleus. The oppositely charged electron and the proton are held together by the Coulombic electrostatic attraction. The potential energy of the system is

$$V(r) = -\frac{e^2}{4\pi\varepsilon_0 r} \tag{6.1}$$

where e is the electronic charge $= 1.6 \times 10^{-19}$C, r is the distance between the electron (charge $-e$) and the proton (charge $+e$), and ε_0 is permittivity of free space $= 8.85 \times 10^{-12}$ F m^{-1}.

6.1 Extension of the Schrodinger Wave Equation for Describing Two-Particle Motion

For dealing with the hydrogen atom, the Schrodinger wave equation describing the motion of a single particle in an external field of force must be extended to two particles (Gradinetti 2019), the proton (p) and the electron (e), which are mutually attracted towards each other by a force dependent on the distance between them. In place of three rectangular coordinates (x_e, y_e, z_e) for a single particle of mass m_e, we use six rectangular coordinates $(x_e, y_e, z_e, x_p, y_p, z_p)$ for the two particles of masses m_e and m_p, to write the time-independent Schrodinger equation as

$$\left\{ -\frac{\hbar^2}{2m_e}\left(\frac{\partial^2}{\partial x_e^2} + \frac{\partial^2}{\partial y_e^2} + \frac{\partial^2}{\partial z_e^2} \right) - \frac{\hbar^2}{2m_p}\left(\frac{\partial^2}{\partial x_p^2} + \frac{\partial^2}{\partial y_p^2} + \frac{\partial^2}{\partial z_p^2} \right) \right.$$

$$\left. + V\left(x_e, y_e, z_e, x_p, y_p, z_p \right) \right\} \Psi\left(x_e, y_e, z_e, x_p, y_p, z_p \right)$$

$$= E\Psi\left(x_e, y_e, z_e, x_p, y_p, z_p \right) \tag{6.2}$$

where the potential energy has been assumed to depend in an arbitrary fashion on all the six coordinates. Two coordinate systems are defined:

   i. relative coordinates (x, y, z) and
  ii. center-of-mass coordinates (X, Y, Z).

$$x = x_e - x_p, \; y = y_e - y_p, z = z_e - z_p \tag{6.3}$$

$$X = \frac{m_e x_e + m_p x_p}{M}, Y = \frac{m_e y_e + m_p y_p}{M}, Z = \frac{m_e z_e + m_p z_p}{M} \tag{6.4}$$

and

$$M = m_e + m_p = \text{Total mass of the two} - \text{particle system} \tag{6.5}$$

According to the multivariable chain rule, for a function

$$z = f\left(x(t), y(t) \right) \tag{6.6}$$

differentiable at t, we have

$$\frac{dz}{dt} = \frac{\partial}{\partial x} f\left(x(t), y(t) \right)\frac{dx}{dt} + \frac{\partial}{\partial y} f\left(x(t), y(t) \right)\frac{dy}{dt} \tag{6.7}$$

Then the first derivatives of $\Psi(x, X)$ in the X-direction are

$$\frac{\partial\Psi(x,X)}{\partial x_e} = \frac{\partial\Psi}{\partial x}\frac{\partial x}{\partial x_e} + \frac{\partial\Psi}{\partial X}\frac{\partial X}{\partial x_e}$$

$$= \frac{\partial\Psi}{\partial x}(1-0) + \frac{\partial\Psi}{\partial X}\frac{\partial}{\partial x_e}\left(\frac{m_e x_e + m_p x_p}{M} \right)$$

$$= \frac{\partial\Psi}{\partial x} + \frac{\partial\Psi}{\partial X}\left(\frac{m_e + 0}{M} \right) = \frac{\partial\Psi}{\partial x} + \frac{m_e}{M}\frac{\partial\Psi}{\partial X} \tag{6.8}$$

$$\frac{\partial\Psi(x,X)}{\partial x_p} = \frac{\partial\Psi}{\partial x}\frac{\partial x}{\partial x_p} + \frac{\partial\Psi}{\partial X}\frac{\partial X}{\partial x_p}$$

$$= \frac{\partial\Psi}{\partial x}(0-1) + \frac{\partial\Psi}{\partial X}\frac{\partial}{\partial x_p}\left(\frac{m_e x_e + m_p x_p}{M} \right)$$

$$= -\frac{\partial\Psi}{\partial x} + \frac{\partial\Psi}{\partial X}\left(\frac{0 + m_p}{M} \right) = -\frac{\partial\Psi}{\partial x} + \frac{m_p}{M}\frac{\partial\Psi}{\partial X} \tag{6.9}$$

where eqs. (6.3) and (6.4) have been applied. Second derivatives of $\Psi(x, X)$ in the X-direction are

$$\frac{\partial}{\partial x_e}\left(\frac{\partial \Psi}{\partial x_e}\right) = \frac{\partial}{\partial x_e}\left(\frac{\partial \Psi}{\partial x} + \frac{m_e}{M}\frac{\partial \Psi}{\partial X}\right)$$

$$= \frac{\partial}{\partial x}\left(\frac{\partial \Psi}{\partial x} + \frac{m_e}{M}\frac{\partial \Psi}{\partial X}\right)\frac{\partial x}{\partial x_e} + \frac{\partial}{\partial X}\left(\frac{\partial \Psi}{\partial x} + \frac{m_e}{M}\frac{\partial \Psi}{\partial X}\right)\frac{\partial X}{\partial x_e}$$

$$= \left(\frac{\partial^2 \Psi}{\partial x^2} + 0\right)\frac{\partial x}{\partial x_e} + \left(0 + \frac{m_e}{M}\frac{\partial^2 \Psi}{\partial X^2}\right)\frac{\partial X}{\partial x_e}$$

$$= \frac{\partial^2 \Psi}{\partial x^2}\frac{\partial x}{\partial x_e} + \frac{m_e}{M}\frac{\partial^2 \Psi}{\partial X^2}\frac{\partial X}{\partial x_e}$$

$$= \frac{\partial^2 \Psi}{\partial x^2}(1-0) + \frac{m_e}{M}\frac{\partial^2 \Psi}{\partial X^2}\frac{\partial}{\partial x_e}\left(\frac{m_e x_e + m_p x_p}{M}\right)$$

$$= \frac{\partial^2 \Psi}{\partial x^2} + \frac{m_e}{M}\frac{\partial^2 \Psi}{\partial X^2} \times \frac{m_e}{M} = \frac{\partial^2 \Psi}{\partial x^2} + \left(\frac{m_e}{M}\right)^2\frac{\partial^2 \Psi}{\partial X^2}$$

$$\tag{6.10}$$

or

$$\frac{\partial^2 \Psi}{\partial x_e^2} = \frac{\partial^2 \Psi}{\partial x^2} + \left(\frac{m_e}{M}\right)^2\frac{\partial^2 \Psi}{\partial X^2} \tag{6.11}$$

or

$$\frac{1}{m_e}\frac{\partial^2 \Psi}{\partial x_e^2} = \frac{1}{m_e}\frac{\partial^2 \Psi}{\partial x^2} + \frac{m_e}{M^2}\frac{\partial^2 \Psi}{\partial X^2} \tag{6.12}$$

Similarly,

$$\frac{\partial}{\partial x_p}\left(\frac{\partial \Psi}{\partial x_p}\right) = \frac{\partial}{\partial x_p}\left(-\frac{\partial \Psi}{\partial x} + \frac{m_p}{M}\frac{\partial \Psi}{\partial X}\right)$$

$$= \frac{\partial}{\partial x}\left(-\frac{\partial \Psi}{\partial x} + \frac{m_p}{M}\frac{\partial \Psi}{\partial X}\right)\frac{\partial x}{\partial x_p}$$

$$+ \frac{\partial}{\partial X}\left(-\frac{\partial \Psi}{\partial x} + \frac{m_p}{M}\frac{\partial \Psi}{\partial X}\right)\frac{\partial X}{\partial x_p}$$

$$= \left(-\frac{\partial^2 \Psi}{\partial x^2} + 0\right)\frac{\partial x}{\partial x_p} + \left(0 + \frac{m_p}{M}\frac{\partial^2 \Psi}{\partial X^2}\right)\frac{\partial X}{\partial x_p}$$

$$= \left(-\frac{\partial^2 \Psi}{\partial x^2}\right)\frac{\partial x}{\partial x_p} + \frac{m_p}{M}\frac{\partial^2 \Psi}{\partial X^2}\frac{\partial X}{\partial x_p}$$

$$= \left(-\frac{\partial^2 \Psi}{\partial x^2}\right)(0-1)$$

$$+ \frac{m_p}{M}\frac{\partial^2 \Psi}{\partial X^2}\left\{\frac{\partial}{\partial x_p}\left(\frac{m_e x_e + m_p x_p}{M}\right)\right\}$$

$$= \frac{\partial^2 \Psi}{\partial x^2} + \frac{m_p}{M}\frac{\partial^2 \Psi}{\partial X^2} \times \frac{m_p}{M}$$

or

$$= \frac{\partial^2 \Psi}{\partial x^2} + \left(\frac{m_p}{M}\right)^2\frac{\partial^2 \Psi}{\partial X^2} \tag{6.13}$$

$$\frac{\partial^2 \Psi}{\partial x_p^2} = \frac{\partial^2 \Psi}{\partial x^2} + \left(\frac{m_p}{M}\right)^2\frac{\partial^2 \Psi}{\partial X^2} \tag{6.14}$$

or

$$\frac{1}{m_p}\frac{\partial^2 \Psi}{\partial x_p^2} = \frac{1}{m_p}\frac{\partial^2 \Psi}{\partial x^2} + \frac{m_p}{M^2}\frac{\partial^2 \Psi}{\partial X^2} \tag{6.15}$$

Adding together eqs. (6.12) and (6.15) for the X-direction,

$$\frac{1}{m_e}\frac{\partial^2 \Psi}{\partial x_e^2} + \frac{1}{m_p}\frac{\partial^2 \Psi}{\partial x_p^2} = \frac{1}{m_e}\frac{\partial^2 \Psi}{\partial x^2} + \frac{m_e}{M^2}\frac{\partial^2 \Psi}{\partial X^2} + \frac{1}{m_p}\frac{\partial^2 \Psi}{\partial x^2} + \frac{m_p}{M^2}\frac{\partial^2 \Psi}{\partial X^2}$$

$$= \left(\frac{1}{m_e}\frac{\partial^2 \Psi}{\partial x^2} + \frac{1}{m_p}\frac{\partial^2 \Psi}{\partial x^2}\right)$$

$$+ \left(\frac{m_e}{M^2}\frac{\partial^2 \Psi}{\partial X^2} + \frac{m_p}{M^2}\frac{\partial^2 \Psi}{\partial X^2}\right)$$

$$= \left(\frac{1}{m_e} + \frac{1}{m_p}\right)\frac{\partial^2 \Psi}{\partial x^2} + \left(\frac{m_e + m_p}{M^2}\right)\frac{\partial^2 \Psi}{\partial X^2}$$

$$= \frac{m_e + m_p}{m_e m_p}\frac{\partial^2 \Psi}{\partial x^2} + \frac{M}{M^2}\frac{\partial^2 \Psi}{\partial X^2}$$

$$= \frac{1}{\frac{m_e m_p}{m_e + m_p}}\frac{\partial^2 \Psi}{\partial x^2} + \frac{1}{M}\frac{\partial^2 \Psi}{\partial X^2}$$

$$= \frac{1}{\mu}\frac{\partial^2 \Psi}{\partial x^2} + \frac{1}{M}\frac{\partial^2 \Psi}{\partial X^2} \tag{6.16}$$

where μ is the reduced mass of the two-particle system given by

$$\mu = \frac{m_e m_p}{m_e + m_p} \tag{6.17}$$

Multiplying both sides of eq. (6.16) by $\hbar^2/2$

$$\frac{\hbar^2}{2m_e}\frac{\partial^2 \Psi}{\partial x_e^2} + \frac{\hbar^2}{2m_p}\frac{\partial^2 \Psi}{\partial x_p^2} = \frac{\hbar^2}{2\mu}\frac{\partial^2 \Psi}{\partial x^2} + \frac{\hbar^2}{2M}\frac{\partial^2 \Psi}{\partial X^2} \tag{6.18}$$

After calculating the first and second partial derivatives of $\Psi(y, Y)$ and $\Psi(z, Z)$ we are able to write

$$\frac{\hbar^2}{2m_e}\frac{\partial^2 \Psi}{\partial y_e^2} + \frac{\hbar^2}{2m_p}\frac{\partial^2 \Psi}{\partial y_p^2} = \frac{\hbar^2}{2\mu}\frac{\partial^2 \Psi}{\partial y^2} + \frac{\hbar^2}{2M}\frac{\partial^2 \Psi}{\partial Y^2} \tag{6.19}$$

and

$$\frac{\hbar^2}{2m_e}\frac{\partial^2 \Psi}{\partial z_e^2} + \frac{\hbar^2}{2m_p}\frac{\partial^2 \Psi}{\partial z_p^2} = \frac{\hbar^2}{2\mu}\frac{\partial^2 \Psi}{\partial z^2} + \frac{\hbar^2}{2M}\frac{\partial^2 \Psi}{\partial Z^2} \tag{6.20}$$

If it is assumed that the potential energy depends only on the relative coordinates x, y, z, eq. (6.2) is amended to the form

$$\left\{ -\left(\frac{\hbar^2}{2m_e} \frac{\partial^2 \Psi}{\partial x_e^2} + \frac{\hbar^2}{2m_p} \frac{\partial^2 \Psi}{\partial x_p^2} \right) - \left(\frac{\hbar^2}{2m_e} \frac{\partial^2 \Psi}{\partial y_e^2} + \frac{\hbar^2}{2m_p} \frac{\partial^2 \Psi}{\partial y_p^2} \right) \right.$$

$$\left. -\left(\frac{\hbar^2}{2m_e} \frac{\partial^2 \Psi}{\partial z_e^2} + \frac{\hbar^2}{2m_p} \frac{\partial^2 \Psi}{\partial z_p^2} \right) + V(x, y, z)\Psi \right\} = E\Psi \qquad (6.21)$$

From eq. (6.21) along with eqs. (6.18), (6.19), and (6.20), we see that

$$\left\{ -\left(\frac{\hbar^2}{2\mu} \frac{\partial^2 \Psi}{\partial x^2} + \frac{\hbar^2}{2M} \frac{\partial^2 \Psi}{\partial X^2} \right) - \left(\frac{\hbar^2}{2\mu} \frac{\partial^2 \Psi}{\partial y^2} + \frac{\hbar^2}{2M} \frac{\partial^2 \Psi}{\partial Y^2} \right) \right.$$

$$\left. -\left(\frac{\hbar^2}{2\mu} \frac{\partial^2 \Psi}{\partial z^2} + \frac{\hbar^2}{2M} \frac{\partial^2 \Psi}{\partial Z^2} \right) + V(x, y, z)\Psi \right\} = E\Psi \qquad (6.22)$$

We collect together terms of partial derivatives in X, Y, Z and x, y, z coordinate systems in eq. (6.22). Then this equation is converted into

$$-\frac{\hbar^2}{2M}\left(\frac{\partial^2 \Psi}{\partial X^2} + \frac{\partial^2 \Psi}{\partial Y^2} + \frac{\partial^2 \Psi}{\partial Z^2} \right) - \frac{\hbar^2}{2\mu}\left(\frac{\partial^2 \Psi}{\partial x^2} + \frac{\partial^2 \Psi}{\partial y^2} + \frac{\partial^2 \Psi}{\partial z^2} \right)$$

$$+ V(x, y, z)\Psi = E\Psi \qquad (6.23)$$

Let us pronounce the symbolization bearing uppercase 'R' in subscript with Laplacian operator as:

$$\frac{\partial^2 \Psi}{\partial X^2} + \frac{\partial^2 \Psi}{\partial Y^2} + \frac{\partial^2 \Psi}{\partial Z^2} \equiv \nabla_R^2 \Psi \qquad (6.24)$$

when we are performing partial differentiation in the center-of-mass system of coordinates (X, Y, Z), and having lowercase 'r' in subscript with Laplacian operator as:

$$\frac{\partial^2 \Psi}{\partial x^2} + \frac{\partial^2 \Psi}{\partial y^2} + \frac{\partial^2 \Psi}{\partial z^2} \equiv \nabla_r^2 \Psi \qquad (6.25)$$

when we are carrying out partial differentiation in the relative system of coordinates (x, y, z). Then eq. (6.23) gives

$$-\frac{\hbar^2}{2M}\nabla_R^2 \Psi(\mathbf{R}, \mathbf{r}) - \frac{\hbar^2}{2\mu}\nabla_r^2 \Psi(\mathbf{R}, \mathbf{r}) + V(\mathbf{r})\Psi(\mathbf{R}, \mathbf{r}) = E\Psi(\mathbf{R}, \mathbf{r})$$

$$(6.26)$$

where

$$\mathbf{R} = X\mathbf{i} + Y\mathbf{j} + Z\mathbf{k} \qquad (6.27)$$

and

$$\mathbf{r} = x\mathbf{i}' + y\mathbf{j}' + z\mathbf{k}' \qquad (6.28)$$

$\mathbf{i}, \mathbf{j}, \mathbf{k}, \mathbf{i}', \mathbf{j}', \mathbf{k}'$ are unit vectors along X, Y, Z, x, y, z directions.

6.2 Splitting the Schrodinger Equation into Equation for the Hydrogen Atom as a Whole and Equation for Its Internal States

Suppose that the wave function can be written as

$$\Psi(\mathbf{R}, \mathbf{r}) = S(\mathbf{R})U(\mathbf{r}) \qquad (6.29)$$

Putting $\Psi(\mathbf{R}, \mathbf{r})$ from eq. (6.29) in eq. (6.26),

$$-\frac{\hbar^2}{2M}\nabla_R^2\{S(\mathbf{R})U(\mathbf{r})\} - \frac{\hbar^2}{2\mu}\nabla_r^2\{S(\mathbf{R})U(\mathbf{r})\}$$

$$+ V(\mathbf{r})\{S(\mathbf{R})U(\mathbf{r})\} = E\{S(\mathbf{R})U(\mathbf{r})\} \qquad (6.30)$$

or

$$-U(\mathbf{r})\frac{\hbar^2}{2M}\nabla_R^2\{S(\mathbf{R})\} - S(\mathbf{R})\frac{\hbar^2}{2\mu}\nabla_r^2\{U(\mathbf{r})\}$$

$$+ V(\mathbf{r})\{S(\mathbf{R})U(\mathbf{r})\} = E\{S(\mathbf{R})U(\mathbf{r})\} \qquad (6.31)$$

or

$$-U(\mathbf{r})\frac{\hbar^2}{2M}\nabla_R^2\{S(\mathbf{R})\} + S(\mathbf{R})\left\{ -\frac{\hbar^2}{2\mu}\nabla_r^2 + V(\mathbf{r}) \right\}\{U(\mathbf{r})\}$$

$$= E\{S(\mathbf{R})U(\mathbf{r})\} \qquad (6.32)$$

Dividing both sides of eq. (6.32) by $S(\mathbf{R})U(\mathbf{r})$,

$$-\frac{1}{S(\mathbf{R})}\frac{\hbar^2}{2M}\nabla_R^2\{S(\mathbf{R})\} + \frac{1}{U(\mathbf{r})}\left\{ -\frac{\hbar^2}{2\mu}\nabla_r^2 + V(\mathbf{r}) \right\}\{U(\mathbf{r})\} = E$$

$$(6.33)$$

or

$$-\frac{1}{S(\mathbf{R})}\frac{\hbar^2}{2M}\nabla_R^2\{S(\mathbf{R})\} = E - \frac{1}{U(\mathbf{r})}\left\{ -\frac{\hbar^2}{2\mu}\nabla_r^2 + V(\mathbf{r}) \right\}\{U(\mathbf{r})\}$$

$$(6.34)$$

The left-hand side of this equation depends only on $\mathbf{R}$ while the right-hand side depends only on $\mathbf{r}$, which is only possible if both sides = a constant = $E_{Constant}$. Hence, we get two separated equations as follows:

$$-\frac{1}{S(\mathbf{R})}\frac{\hbar^2}{2M}\nabla_R^2\{S(\mathbf{R})\} = E_{Constant} \qquad (6.35)$$

or

$$-\frac{\hbar^2}{2M}\nabla_R^2\{S(\mathbf{R})\} = E_{Constant}\, S(\mathbf{R}) \qquad (6.36)$$

and

$$E - \frac{1}{U(\mathbf{r})}\left\{ -\frac{\hbar^2}{2\mu}\nabla_r^2 + V(\mathbf{r}) \right\}\{U(\mathbf{r})\} = E_{\text{Constant}} \quad (6.37)$$

or

$$-\frac{1}{U(\mathbf{r})}\left\{ -\frac{\hbar^2}{2\mu}\nabla_r^2 + V(\mathbf{r}) \right\}\{U(\mathbf{r})\} = E_{\text{Constant}} - E = E_H \quad (6.38)$$

These equations can be solved separately. Equation (6.36) states that the center of mass of the two particles, electron and proton, moves like a free particle of mass M, the sum-total of the masses of the particles. It is the wave equation for a free particle of mass M, which we have already solved in Section 5.1. Its solutions are the wave functions (eq. (5.14))

$$S(\mathbf{R}) = \exp(i\mathbf{K} \cdot \mathbf{R}) \quad (6.39)$$

and eigen energies (eq. (5.6))

$$E_{\text{Constant}} = \frac{\hbar^2 K^2}{2M} \quad (6.40)$$

Thus eq. (6.36) describes the motion of the hydrogen atom as a whole as a particle of mass M. The other eq. (6.38) describes the internal relative motion of the electron and proton as a particle of reduced mass μ in a potential energy $V(\mathbf{r})$. Solutions of this equation will give the internal states, i.e. the orbitals and energies of the hydrogen atom. We shall be interested in the energy levels associated with the relative motion. The protonic mass being much larger than the electronic mass, the reduced mass μ is only slightly smaller than the electronic mass m_e.

6.3 Writing the Schrodinger Equation for Internal Relative Motion of the Electron and Proton

We begin by rewriting the Schrodinger's equation in three dimensions as

$$\frac{\partial^2 \Psi(x, y, z, t)}{\partial x^2} + \frac{\partial^2 \Psi(x, y, z, t)}{\partial y^2}$$

$$+ \frac{\partial^2 \Psi(x, y, z, t)}{\partial z^2} + \frac{2\mu}{\hbar^2}(E - V)\Psi(x, y, z, t) = 0 \quad (6.41)$$

where

$$\mu = \text{Reduced mass of the electron}$$

$$- \text{nucleus system} = \frac{m_{\text{Electron}} M_{\text{Nucleus}}}{m_{\text{Electron}} + M_{\text{Nucleus}}} \quad (6.42)$$

m_{Electron} and m_{Nucleus} are the masses of electron and nucleus, respectively.

Substituting for $V(r)$ from eq. (6.1) in eq. (6.41),

$$\frac{\partial^2 \Psi(x, y, z, t)}{\partial x^2} + \frac{\partial^2 \Psi(x, y, z, t)}{\partial y^2}$$

$$+ \frac{\partial^2 \Psi(x, y, z, t)}{\partial z^2} + \frac{2\mu}{\hbar^2}\left(E + \frac{e^2}{4\pi\varepsilon_0 r} \right)\Psi(x, y, z, t) = 0 \quad (6.43)$$

The three coordinates in spherical polar coordinates are (r, θ, ϕ). They are related to rectangular Cartesian coordinates (x, y, z) as (Figure 6.1)

$r = $ Length of the radius vector from the origin to the point (x, y, z)

$$= \sqrt{x^2 + y^2 + z^2} \quad (6.44)$$

$\theta = $ Angle between radius vector and $+ Z$-axis

$$= \cos^{-1}\left(\frac{z}{\sqrt{x^2 + y^2 + z^2}} \right) \quad (6.45)$$

and

$\phi = $ Angle between projection of radius vector on XY-plane and

$$+ X\text{-axis} = \tan^{-1}\left(\frac{y}{x} \right) \quad (6.46)$$

Conversely,

$$x = r\sin\theta\cos\phi \quad (6.47)$$

$$y = r\sin\theta\sin\phi \quad (6.48)$$

and

$$z = r\cos\theta \quad (6.49)$$

In spherical polar coordinates, the Laplacian

$$\frac{\partial^2 \Psi(x, y, z, t)}{\partial x^2} + \frac{\partial^2 \Psi(x, y, z, t)}{\partial y^2} + \frac{\partial^2 \Psi(x, y, z, t)}{\partial z^2}$$

$$= \frac{1}{r^2}\frac{\partial}{\partial r}\left(r^2 \frac{\partial \Psi}{\partial r} \right) + \frac{1}{r^2 \sin\theta}\frac{\partial}{\partial \theta}\left(\sin\theta \frac{\partial \Psi}{\partial \theta} \right) + \frac{1}{r^2 \sin^2\theta}\frac{\partial^2 \Psi}{\partial \phi^2} \quad (6.50)$$

So, Schrodinger's equation becomes

$$\frac{1}{r^2}\frac{\partial}{\partial r}\left(r^2 \frac{\partial \Psi}{\partial r} \right) + \frac{1}{r^2 \sin\theta}\frac{\partial}{\partial \theta}\left(\sin\theta \frac{\partial \Psi}{\partial \theta} \right)$$

$$+ \frac{1}{r^2 \sin^2\theta}\frac{\partial^2 \Psi}{\partial \phi^2} + \frac{2\mu}{\hbar^2}\left(E + \frac{e^2}{4\pi\varepsilon_0 r} \right)\Psi = 0 \quad (6.51)$$

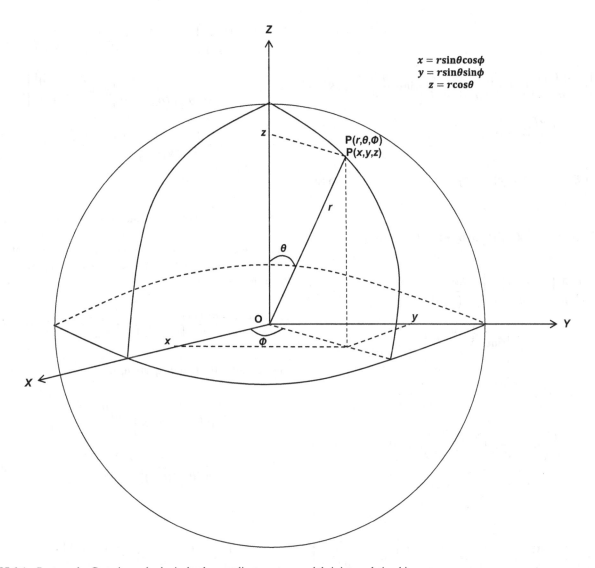

FIGURE 6.1 Rectangular Cartesian and spherical polar coordinate systems and their inter-relationship.

6.4 Separation of Variables in the Schrodinger Equation to Form Radial and Angular Equations

By solving the Schrodinger's equation for Ψ, we can obtain the full information regarding the hydrogen atom. This equation is solved by using a method called 'separation of variables' involving splitting of a wave function Ψ into a radial part $R(r)$ which is a function of r only and an angular part $Y(\theta, \phi)$ which is a function of θ and ϕ only

$$\Psi(r,\theta,\phi) = \text{Radial part } R(r) \times \text{Angular part } Y(\theta,\phi)$$

$$= R(r)Y(\theta,\phi) \tag{6.52}$$

and inserting the separated Ψ(r, θ, ϕ) into eq. (6.51) as

$$\frac{1}{r^2}\frac{\partial}{\partial r}\left[r^2\frac{\partial\{R(r)Y(\theta,\phi)\}}{\partial r}\right] + \frac{1}{r^2\sin\theta}\frac{\partial}{\partial\theta}\left[\sin\theta\frac{\partial\{R(r)Y(\theta,\phi)\}}{\partial\theta}\right]$$

$$+ \frac{1}{r^2\sin^2\theta}\frac{\partial^2\{R(r)Y(\theta,\phi)\}}{\partial\phi^2}$$

$$+ \frac{2\mu}{\hbar^2}\left(E + \frac{e^2}{4\pi\varepsilon_0 r}\right)\{R(r)Y(\theta,\phi)\} = 0 \tag{6.53}$$

or

$$\frac{1}{r^2}\frac{\partial}{\partial r}\left[r^2\frac{\partial R(r)}{\partial r}\{Y(\theta,\phi)\}\right] + \frac{1}{r^2\sin\theta}\frac{\partial}{\partial\theta}\left[\sin\theta\frac{\partial Y(\theta,\phi)}{\partial\theta}\{R(r)\}\right]$$

$$+ \frac{1}{r^2\sin^2\theta}\frac{\partial^2 Y(\theta,\phi)}{\partial\phi^2}\{R(r)\}$$

$$+ \frac{2\mu}{\hbar^2}\left(E + \frac{e^2}{4\pi\varepsilon_0 r}\right)\{R(r)Y(\theta,\phi)\} = 0 \tag{6.54}$$

$$x = r\sin\theta\cos\phi$$
$$y = r\sin\theta\sin\phi$$
$$z = r\cos\theta$$

or

$$\frac{1}{r^2}\frac{\partial}{\partial r}\left\{r^2\frac{\partial R(r)}{\partial r}\right\}\{Y(\theta,\phi)\}+\frac{1}{r^2\sin\theta}\frac{\partial}{\partial\theta}\left\{\sin\theta\frac{\partial Y(\theta,\phi)}{\partial\theta}\right\}\{R(r)\}$$

$$+\frac{1}{r^2\sin^2\theta}\frac{\partial^2 Y(\theta,\phi)}{\partial\phi^2}\{R(r)\}$$

$$+\frac{2\mu}{\hbar^2}\left(E+\frac{e^2}{4\pi\varepsilon_0 r}\right)\{R(r)Y(\theta,\phi)\}=0 \tag{6.55}$$

Dividing both sides of eq. (6.55) by $R(r)\{Y(\theta,\phi)\}$ we get

$$\frac{1}{r^2}\frac{\partial}{\partial r}\left\{r^2\frac{\partial R(r)}{\partial r}\right\}\frac{\{Y(\theta,\phi)\}}{R(r)\{Y(\theta,\phi)\}}$$

$$+\frac{1}{r^2\sin\theta}\frac{\partial}{\partial\theta}\left\{\sin\theta\frac{\partial\{Y(\theta,\phi)\}}{\partial\theta}\right\}\frac{\{R(r)\}}{R(r)\{Y(\theta,\phi)\}}$$

$$+\frac{1}{r^2\sin^2\theta}\frac{\partial^2\{Y(\theta,\phi)\}}{\partial\phi^2}\frac{\{R(r)\}}{R(r)\{Y(\theta,\phi)\}}$$

$$+\frac{2\mu}{\hbar^2}\left(E+\frac{e^2}{4\pi\varepsilon_0 r}\right)\frac{R(r)\{Y(\theta,\phi)\}}{R(r)\{Y(\theta,\phi)\}}=0 \tag{6.56}$$

or

$$\frac{1}{r^2}\frac{1}{R(r)}\frac{\partial}{\partial r}\left\{r^2\frac{\partial R(r)}{\partial r}\right\}+\frac{1}{r^2\sin\theta}\frac{1}{\{Y(\theta,\phi)\}}\frac{\partial}{\partial\theta}\left\{\sin\theta\frac{\partial\{Y(\theta,\phi)\}}{\partial\theta}\right\}$$

$$+\frac{1}{r^2\sin^2\theta}\frac{1}{\{Y(\theta,\phi)\}}\frac{\partial^2\{Y(\theta,\phi)\}}{\partial\phi^2}+\frac{2\mu}{\hbar^2}\left(E+\frac{e^2}{4\pi\varepsilon_0 r}\right)=0 \tag{6.57}$$

or

$$\frac{1}{r^2}\frac{1}{R(r)}\frac{\partial}{\partial r}\left\{r^2\frac{\partial R(r)}{\partial r}\right\}+\frac{2\mu}{\hbar^2}\left(E+\frac{e^2}{4\pi\varepsilon_0 r}\right)$$

$$=-\frac{1}{r^2\sin\theta}\frac{1}{\{Y(\theta,\phi)\}}\frac{\partial}{\partial\theta}\left\{\sin\theta\frac{\partial\{Y(\theta,\phi)\}}{\partial\theta}\right\}$$

$$-\frac{1}{r^2\sin^2\theta}\frac{1}{\{Y(\theta,\phi)\}}\frac{\partial^2\{Y(\theta,\phi)\}}{\partial\phi^2} \tag{6.58}$$

Multiplying both sides of eq. (6.58) by r^2,

$$\frac{1}{R(r)}\frac{\partial}{\partial r}\left\{r^2\frac{\partial R(r)}{\partial r}\right\}+\frac{2\mu r^2}{\hbar^2}\left(E+\frac{e^2}{4\pi\varepsilon_0 r}\right)$$

$$=-\frac{1}{\sin\theta}\frac{1}{\{Y(\theta,\phi)\}}\frac{\partial}{\partial\theta}\left\{\sin\theta\frac{\partial\{Y(\theta,\phi)\}}{\partial\theta}\right\}$$

$$-\frac{1}{\sin^2\theta}\frac{1}{\{Y(\theta,\phi)\}}\frac{\partial^2\{Y(\theta,\phi)\}}{\partial\phi^2} \tag{6.59}$$

or

$$\text{Function of }(r) = \text{Function of }(\theta,\phi)\text{ only} \tag{6.60}$$

This equation is satisfied only if both sides of the equation are equal to a constant, independent of (r, θ, ϕ). Suppose this constant is $l(l+1)$. Thus

$$\frac{1}{R(r)}\frac{\partial}{\partial r}\left\{r^2\frac{\partial R(r)}{\partial r}\right\}+\frac{2\mu r^2}{\hbar^2}\left(E+\frac{e^2}{4\pi\varepsilon_0 r}\right)=l(l+1) \tag{6.61}$$

which is called the radial equation, and

$$-\frac{1}{\sin\theta}\frac{1}{\{Y(\theta,\phi)\}}\frac{\partial}{\partial\theta}\left\{\sin\theta\frac{\partial\{Y(\theta,\phi)\}}{\partial\theta}\right\}$$

$$-\frac{1}{\sin^2\theta}\frac{1}{\{Y(\theta,\phi)\}}\frac{\partial^2\{Y(\theta,\phi)\}}{\partial\phi^2}=l(l+1) \tag{6.62}$$

or

$$\frac{1}{\sin\theta}\frac{1}{\{Y(\theta,\phi)\}}\frac{\partial}{\partial\theta}\left\{\sin\theta\frac{\partial\{Y(\theta,\phi)\}}{\partial\theta}\right\}$$

$$+\frac{1}{\sin^2\theta}\frac{1}{\{Y(\theta,\phi)\}}\frac{\partial^2\{Y(\theta,\phi)\}}{\partial\phi^2}=-l(l+1) \tag{6.63}$$

which is known as the angular equation. The quantity l in the separation constant $l(l+1)$ is the angular momentum quantum number. This is where the angular momentum quantum number is introduced in the analysis. It appears as a separation constant factor when the radial and angular parts of the Schrodinger equation are separated.

6.5 Separation of Variables in the Angular Equation to Form Polar and Azimuthal Angle Equations

Writing

$$Y(\theta,\phi)=\Theta(\theta)\Phi(\phi) \tag{6.64}$$

the angular equation is written as

$$\frac{1}{\sin\theta}\frac{1}{\{\Theta(\theta)\Phi(\phi)\}}\frac{\partial}{\partial\theta}\left\{\sin\theta\frac{\partial\{\Theta(\theta)\Phi(\phi)\}}{\partial\theta}\right\}$$

$$+\frac{1}{\sin^2\theta}\frac{1}{\{\Theta(\theta)\Phi(\phi)\}}\frac{\partial^2\{\Theta(\theta)\Phi(\phi)\}}{\partial\phi^2}=-l(l+1) \quad (6.65)$$

or

$$\frac{1}{\sin\theta}\frac{1}{\{\Theta(\theta)\Phi(\phi)\}}\frac{\partial}{\partial\theta}\left\{\sin\theta\frac{\partial\{\Theta(\theta)\}}{\partial\theta}\right\}\Phi(\phi)$$

$$+\frac{1}{\sin^2\theta}\frac{1}{\{\Theta(\theta)\Phi(\phi)\}}\frac{\partial^2\{\Phi(\phi)\}}{\partial\phi^2}\Theta(\theta)=-l(l+1) \quad (6.66)$$

or

$$\frac{1}{\sin\theta}\frac{1}{\{\Theta(\theta)\}}\frac{\partial}{\partial\theta}\left\{\sin\theta\frac{\partial\{\Theta(\theta)\}}{\partial\theta}\right\}$$

$$+\frac{1}{\sin^2\theta}\frac{1}{\{\Phi(\phi)\}}\frac{\partial^2\{\Phi(\phi)\}}{\partial\phi^2}=-l(l+1) \quad (6.67)$$

Multiplying both sides of eq. (6.67) by $\sin^2\theta$

$$\frac{\sin^2\theta}{\sin\theta}\frac{1}{\{\Theta(\theta)\}}\frac{\partial}{\partial\theta}\left\{\sin\theta\frac{\partial\{\Theta(\theta)\}}{\partial\theta}\right\}$$

$$+\frac{\sin^2\theta}{\sin^2\theta}\frac{1}{\{\Phi(\phi)\}}\frac{\partial^2\{\Phi(\phi)\}}{\partial\phi^2}=-l(l+1)\sin^2\theta \quad (6.68)$$

or

$$\frac{\sin\theta}{\{\Theta(\theta)\}}\frac{\partial}{\partial\theta}\left\{\sin\theta\frac{\partial\{\Theta(\theta)\}}{\partial\theta}\right\}$$

$$+\frac{1}{\{\Phi(\phi)\}}\frac{\partial^2\{\Phi(\phi)\}}{\partial\phi^2}=-l(l+1)\sin^2\theta \quad (6.69)$$

or

$$\frac{\sin\theta}{\{\Theta(\theta)\}}\frac{\partial}{\partial\theta}\left\{\sin\theta\frac{\partial\{\Theta(\theta)\}}{\partial\theta}\right\}$$

$$+l(l+1)\sin^2\theta=-\frac{1}{\{\Phi(\phi)\}}\frac{\partial^2\{\Phi(\phi)\}}{\partial\phi^2} \quad (6.70)$$

Arguing as earlier regarding the exclusive dependence of the left-hand side on θ only and that of the right-hand side on ϕ only, and observing that this is possible only when

Left-hand side = Right-hand side = constant = m_l^2 (6.71)

we have

$$\frac{\sin\theta}{\{\Theta(\theta)\}}\frac{\partial}{\partial\theta}\left\{\sin\theta\frac{\partial\{\Theta(\theta)\}}{\partial\theta}\right\}+l(l+1)\sin^2\theta=m_l^2 \quad (6.72a)$$

and

$$-\frac{1}{\{\Phi(\phi)\}}\frac{\partial^2\{\Phi(\phi)\}}{\partial\phi^2}=m_l^2 \quad (6.72b)$$

or

$$\frac{1}{\Phi(\phi)}\frac{\partial^2\Phi(\phi)}{\partial\phi^2}=-m_l^2 \quad (6.73)$$

The Θ equation is called the polar angle equation and the Φ equation is known as the azimuthal angle equation. The quantity m_l is the magnetic quantum number. Note that it is introduced at this stage. It appears as a separation constant when the Θ and Φ equations are separated.

6.6 Solution of the Radial Equation

The radial equation (6.61) is written in full derivatives as

$$\frac{1}{R(r)}\frac{d}{dr}\left\{r^2\frac{dR(r)}{dr}\right\}+\frac{2\mu r^2}{\hbar^2}\left(E+\frac{e^2}{4\pi\varepsilon_0 r}\right)=l(l+1) \quad (6.74)$$

or

$$\frac{1}{R(r)}\frac{d}{dr}\left\{r^2\frac{dR(r)}{dr}\right\}+\frac{2\mu r^2}{\hbar^2}\left(E+\frac{e^2}{4\pi\varepsilon_0 r}\right)-l(l+1)=0 \quad (6.75)$$

or

$$\frac{d}{dr}\left\{r^2\frac{dR(r)}{dr}\right\}+\frac{2\mu r^2}{\hbar^2}\left(E+\frac{e^2}{4\pi\varepsilon_0 r}\right)R(r)-l(l+1)R(r)=0$$

$$(6.76)$$

or

$$r^2\frac{d^2R(r)}{dr^2}+2r\frac{dR(r)}{dr}$$

$$+\left\{\frac{2\mu r^2 E}{\hbar^2}+\frac{2\mu r^2}{\hbar^2}\times\frac{e^2}{4\pi\varepsilon_0 r}-l(l+1)\right\}R(r)=0 \quad (6.77)$$

or

$$r^2\frac{d^2R(r)}{dr^2}+2r\frac{dR(r)}{dr}+\left\{\frac{2\mu r^2 E}{\hbar^2}+\frac{\mu e^2 r}{2\pi\varepsilon_0\hbar^2}-l(l+1)\right\}R(r)=0$$

$$(6.78)$$

Putting

$$\sqrt{-\frac{2\mu E}{\hbar^2}} = \kappa \qquad (6.79)$$

Equation (6.78) is converted into

$$r^2 \frac{d^2 R(r)}{dr^2} + 2r \frac{dR(r)}{dr}$$

$$+ \left\{ -\left(\sqrt{-\frac{2\mu E}{\hbar^2}}\right)^2 r^2 + \frac{\mu e^2 r}{2\pi\varepsilon_0 \hbar^2} - l(l+1) \right\} R(r) = 0 \qquad (6.80)$$

or

$$r^2 \frac{d^2 R(r)}{dr^2} + 2r \frac{dR(r)}{dr} + \left\{ -k^2 r^2 + \frac{\mu e^2 r}{2\pi\varepsilon_0 \hbar^2} - l(l+1) \right\} R(r) = 0$$

$$(6.81)$$

Since

$$\kappa = \frac{\sqrt{-2\mu E}}{\hbar} = \frac{\sqrt{\text{kg} \times \text{kg} \times \frac{\text{m}}{\text{s}^2} \times \text{m}}}{\text{m}^2 \times \frac{\text{kg}}{\text{s}^2} \times \text{s}} = \frac{\sqrt{\text{kg}^2 \times \frac{\text{m}^2}{\text{s}^2}}}{\text{m}^2 \times \frac{\text{kg}}{\text{s}}}$$

$$= \frac{\text{kg} \times \frac{\text{m}}{\text{s}}}{\text{m}^2 \times \frac{\text{kg}}{\text{s}}} = \text{kg} \times \frac{\text{m}}{\text{s}} \times \frac{1}{\text{m}^2} \times \frac{\text{s}}{\text{kg}} = \frac{1}{\text{m}} = \text{m}^{-1} \qquad (6.82)$$

κ has the unit of m^{-1}, i.e. dimensions of 1/length. Let us construct a unit-less quantity ρ. To achieve this, let us make a substitution

$$\rho = 2r\kappa \qquad (6.83)$$

or

$$r = \frac{\rho}{2\kappa} \qquad (6.84)$$

This substitution will alter the equation to a more convenient form. We note that

$$\frac{dR(\rho)}{dr} = \frac{dR(\rho)}{d\rho} \times \frac{d\rho}{dr} = \frac{dR(\rho)}{d\rho} \times \frac{d(2rk)}{dr} = 2k \frac{dR(\rho)}{d\rho} \qquad (6.85)$$

and

$$\frac{d^2 R(\rho)}{dr^2} = \frac{d}{dr} \left\{ \frac{dR(\rho)}{dr} \right\} = \frac{d}{dr} \left\{ 2k \frac{dR(\rho)}{d\rho} \right\} = \frac{d}{d\rho} \times \frac{d\rho}{dr} \left\{ 2k \frac{dR(\rho)}{d\rho} \right\}$$

$$= \frac{d}{d\rho} \times \frac{d(2rk)}{dr} \left\{ 2k \frac{dR(\rho)}{d\rho} \right\} = 2k \times \frac{d}{d\rho} \left\{ 2k \frac{dR(\rho)}{d\rho} \right\}$$

$$= 4\kappa^2 \frac{d}{d\rho} \left\{ \frac{dR(\rho)}{d\rho} \right\} = 4\kappa^2 \frac{d^2 R(\rho)}{d\rho^2} \qquad (6.86)$$

Eqs. (6.81), (6.85), and (6.86) give

$$r^2 \times 4\kappa^2 \frac{d^2 R(\rho)}{d\rho^2} + 2r \times 2k \frac{dR(\rho)}{d\rho}$$

$$+ \left\{ -k^2 r^2 + \frac{\mu e^2 r}{2\pi\varepsilon_0 \hbar^2} - l(l+1) \right\} R(\rho) = 0 \qquad (6.87)$$

or

$$4k^2 r^2 \frac{d^2 R(\rho)}{d\rho^2} + 4kr \frac{dR(\rho)}{d\rho}$$

$$+ \left\{ -k^2 r^2 + \frac{\mu e^2 r}{2\pi\varepsilon_0 \hbar^2} - l(l+1) \right\} R(\rho) = 0 \qquad (6.88)$$

or

$$(2kr)^2 \frac{d^2 R(\rho)}{d\rho^2} + 2 \times 2kr \frac{dR(\rho)}{d\rho}$$

$$+ \left\{ -\left(\frac{2kr}{2}\right)^2 + \frac{\mu e^2 r}{2\pi\varepsilon_0 \hbar^2} - l(l+1) \right\} R(\rho) = 0 \qquad (6.89)$$

or

$$\rho^2 \frac{d^2 R(\rho)}{d\rho^2} + 2\rho \frac{dR(\rho)}{d\rho} + \left\{ -\frac{\rho^2}{4} + \frac{\mu e^2 r}{2\pi\varepsilon_0 \hbar^2} - l(l+1) \right\} R(\rho) = 0$$

$$(6.90)$$

Define

$$u(\rho) = \rho R(\rho) \qquad (6.91)$$

Then

$$R(\rho) = \frac{u(\rho)}{\rho} \qquad (6.92)$$

$$\frac{dR(\rho)}{d\rho} = \frac{d}{d\rho} \left\{ \frac{u(\rho)}{\rho} \right\} = \frac{\rho \frac{du(\rho)}{d\rho} - 1 \times u(\rho)}{\rho^2} = \left(\frac{1}{\rho}\right) \frac{du(\rho)}{d\rho} - \frac{u(\rho)}{\rho^2}$$

$$(6.93)$$

and

$$\frac{d^2 R(\rho)}{d\rho^2} = \frac{d}{d\rho} \left\{ \frac{dR(\rho)}{d\rho} \right\} = \frac{d}{d\rho} \left\{ \left(\frac{1}{\rho}\right) \frac{du(\rho)}{d\rho} - \frac{u(\rho)}{\rho^2} \right\}$$

$$= \frac{d}{d\rho} \left\{ \rho^{-1} \frac{du(\rho)}{d\rho} - \rho^{-2} u(\rho) \right\}$$

$$= -\rho^{-2} \frac{du(\rho)}{d\rho} + \rho^{-1} \frac{d^2 u(\rho)}{d\rho^2} + 2\rho^{-3} u(\rho) - \rho^{-2} \frac{du(\rho)}{d\rho}$$

$$(6.94)$$

So, using eqs. (6.92)–(6.94), eq. (6.90) is changed to

$$\rho^2\left\{-\rho^{-2}\frac{du(\rho)}{d\rho}+\rho^{-1}\frac{d^2u(\rho)}{d\rho^2}+2\rho^{-3}u(\rho)-\rho^{-2}\frac{du(\rho)}{d\rho}\right\}$$

$$+2\rho\left\{\left(\frac{1}{\rho}\right)\frac{du(\rho)}{d\rho}-\frac{u(\rho)}{\rho^2}\right\}$$

$$+\left\{-\frac{\rho^2}{4}+\frac{\mu e^2 r}{2\pi\varepsilon_0\hbar^2}-l(l+1)\right\}\frac{u(\rho)}{\rho}=0 \qquad (6.95)$$

or

$$-\frac{du(\rho)}{d\rho}+\rho\frac{d^2u(\rho)}{d\rho^2}+2\rho^{-1}u(\rho)-\frac{du(\rho)}{d\rho}$$

$$+2\frac{du(\rho)}{d\rho}-2\frac{u(\rho)}{\rho}+\left\{-\frac{\rho^2}{4}+\frac{\mu e^2 r}{2\pi\varepsilon_0\hbar^2}-l(l+1)\right\}\frac{u(\rho)}{\rho}=0$$

$$(6.96)$$

or

$$\rho\frac{d^2u(\rho)}{d\rho^2}+\left\{-\frac{\rho^2}{4}+\frac{\mu e^2 r}{2\pi\varepsilon_0\hbar^2}-l(l+1)\right\}\frac{u(\rho)}{\rho}=0 \qquad (6.97)$$

or

$$\rho^2\frac{d^2u(\rho)}{d\rho^2}+\left\{-\frac{\rho^2}{4}+\frac{\mu e^2 r}{2\pi\varepsilon_0\hbar^2}-l(l+1)\right\}u(\rho)=0 \qquad (6.98)$$

or

$$\rho^2\frac{d^2u(\rho)}{d\rho^2}=-\left\{-\frac{\rho^2}{4}+\frac{\mu e^2 r}{2\pi\varepsilon_0\hbar^2}-l(l+1)\right\}u(\rho)$$

$$=\left\{\frac{\rho^2}{4}-\frac{\mu e^2 r}{2\pi\varepsilon_0\hbar^2}+l(l+1)\right\}u(\rho) \qquad (6.99)$$

Dividing both sides of eq. (6.99) by ρ^2 we get

$$\frac{d^2u(\rho)}{d\rho^2}=\left\{\frac{1}{4}-\frac{\mu e^2 r}{2\pi\varepsilon_0\hbar^2\rho^2}+\frac{l(l+1)}{\rho^2}\right\}u(\rho) \qquad (6.100)$$

Put

$$\frac{\mu e^2}{2\pi\varepsilon_0\kappa\hbar^2}=\rho_0 \qquad (6.101)$$

so that

$$\frac{\mu e^2 r}{2\pi\varepsilon_0\hbar^2\rho^2}=\frac{\mu e^2}{2\pi\varepsilon_0\kappa\hbar^2}\times\frac{kr}{\rho^2}=\rho_0\times\frac{2kr}{2\rho^2}=\rho_0\times\frac{\rho}{2\rho^2}=\frac{\rho_0}{2\rho}$$

$$(6.102)$$

and eq. (6.100) acquires the form

$$\frac{d^2u(\rho)}{d\rho^2}=\left\{\frac{1}{4}-\frac{\rho_0}{2\rho}+\frac{l(l+1)}{\rho^2}\right\}u(\rho) \qquad (6.103)$$

Let us probe into the two limiting cases, first when $\rho\to0$ and second when $\rho\to\infty$ where ρ is given by eq. (6.83). When $\rho\to0$, the term with ρ^2 in the denominator is dominant and eq. (6.103) reduces to

$$\frac{d^2u(\rho)}{d\rho^2}=\left\{\frac{l(l+1)}{\rho^2}\right\}u_1(\rho) \qquad (6.104)$$

where $u_1(\rho)$ is the solution for $\rho\to0$. To solve this equation, we consider a generic polynomial of the form

$$u_1(\rho)=a\rho^p \qquad (6.105)$$

Then

$$\frac{du_1(\rho)}{d\rho}=\frac{d}{d\rho}(a\rho^p)=ap\rho^{p-1} \qquad (6.106)$$

and

$$\frac{d^2u_1(\rho)}{d\rho^2}=\frac{d}{d\rho}(ap\rho^{p-1})=ap(p-1)\rho^{p-2} \qquad (6.107)$$

Hence,

$$ap(p-1)\rho^{p-2}=\left\{\frac{l(l+1)}{\rho^2}\right\}a\rho^p=al(l+1)\rho^{p-2} \qquad (6.108)$$

which means that there will be solution for

$$p(p-1)=l(l+1) \qquad (6.109)$$

or

$$p^2-p-l(l+1)=0 \qquad (6.110)$$

or

$$p^2-(l+1)p+lp-l(l+1)=0 \qquad (6.111)$$

or

$$p\{p-(l+1)\}+l\{p-(l+1)\}=0 \qquad (6.112)$$

or

$$(p+l)\{p-(l+1)\}=0 \qquad (6.113)$$

Either

$$p+l=0; \text{ hence } p=-l \qquad (6.114)$$

or

$$p - (l+1) = 0; \text{ hence } p = l+1 \tag{6.115}$$

The general solution is expressed as a linear combination using constants C_1 and C_2

$$u_1(\rho) = C_1 \rho^{-l} + C_2 \rho^{l+1} \tag{6.116}$$

Since ρ is near zero, $C_1 = 0$; otherwise this term will indefinitely increase and the function will be non-normalizable. So,

$$u_1(\rho) = 0 + C_2 \rho^{l+1} = C_2 \rho^{l+1} \tag{6.117}$$

When $\rho \to \infty$, eq. (6.103) is

$$\frac{d^2 u_2(\rho)}{d\rho^2} = \left\{ \frac{1}{4} - \frac{\rho_0}{2\infty} + \frac{l(l+1)}{\infty^2} \right\} u_2(\rho)$$

$$= \left\{ \frac{1}{4} - 0 + 0 \right\} u_2(\rho) = \frac{1}{4} u_2(\rho) \tag{6.118}$$

where $u_2(\rho)$ is the solution for this case.

This equation has a solution of the form

$$u_2(\rho) = A \exp\left(\frac{\rho}{2} \right) + B \exp\left(-\frac{\rho}{2} \right) \tag{6.119}$$

where A and B are constants. We can check whether this solution satisfies the given equation. It is easy to see that

$$\frac{du_2(\rho)}{d\rho} = \frac{d}{d\rho} \left\{ A \exp\left(\frac{\rho}{2} \right) + B \exp\left(-\frac{\rho}{2} \right) \right\}$$

$$= \frac{d}{d\rho} \left\{ A \exp\left(\frac{\rho}{2} \right) \right\} + \frac{d}{d\rho} \left\{ B \exp\left(-\frac{\rho}{2} \right) \right\}$$

$$= \frac{1}{2} A \exp\left(\frac{\rho}{2} \right) - \frac{1}{2} B \exp\left(-\frac{\rho}{2} \right) \tag{6.120}$$

and

$$\frac{d^2 u_2(\rho)}{d\rho^2} = \frac{d}{d\rho} \left\{ \frac{1}{2} A \exp\left(\frac{\rho}{2} \right) - \frac{1}{2} B \exp\left(-\frac{\rho}{2} \right) \right\}$$

$$= \frac{d}{d\rho} \left\{ \frac{1}{2} A \exp\left(\frac{\rho}{2} \right) \right\} - \frac{d}{d\rho} \left\{ \frac{1}{2} B \exp\left(-\frac{\rho}{2} \right) \right\}$$

$$= \frac{1}{2} \times \frac{1}{2} A \exp\left(\frac{\rho}{2} \right) - \frac{1}{2} \times \left(-\frac{1}{2} \right) \left\{ B \exp\left(-\frac{\rho}{2} \right) \right\}$$

$$= \frac{1}{4} A \exp\left(\frac{\rho}{2} \right) + \frac{1}{4} B \exp\left(-\frac{\rho}{2} \right) = \frac{1}{4} u_2(\rho) \tag{6.121}$$

Thus the solution is true. But the growing exponential $A \exp\left(\frac{\rho}{2} \right)$ will not be normalizable and is therefore rejected, yielding

$$u_2(\rho) = 0 + B \exp\left(-\frac{\rho}{2} \right) = B \exp\left(-\frac{\rho}{2} \right) \tag{6.122}$$

as the solution.

Taking both the regimes $\rho \to 0$ and $\rho \to \infty$, we write the full solution in terms of a function $L(\rho)$ of ρ as

$$u(\rho) = \rho^{l+1} \exp\left(-\frac{\rho}{2} \right) L(\rho) \tag{6.123}$$

which behaves correctly in the limits $\rho \to 0$ and $\rho \to \infty$.

Then

$$\frac{du(\rho)}{d\rho} = \frac{d}{d\rho} \left\{ \rho^{l+1} \exp\left(-\frac{\rho}{2} \right) L(\rho) \right\}$$

$$= \left[\frac{d}{d\rho} \{ \rho^{l+1} \} \right] \exp\left(-\frac{\rho}{2} \right) L(\rho) + \left[\frac{d}{d\rho} \left\{ \exp\left(-\frac{\rho}{2} \right) \right\} \right] \{ \rho^{l+1} L(\rho) \} + \left[\frac{d}{d\rho} \{ L(\rho) \} \right] \left\{ \rho^{l+1} \exp\left(-\frac{\rho}{2} \right) \right\}$$

$$= (l+1) \rho^{l+1-1} \exp\left(-\frac{\rho}{2} \right) L(\rho) + \left\{ \left(-\frac{1}{2} \right) \exp\left(-\frac{\rho}{2} \right) \right\} \{ \rho^{l+1} L(\rho) \} + \frac{d\{ L(\rho) \}}{d\rho} \left\{ \rho^{l+1} \exp\left(-\frac{\rho}{2} \right) \right\}$$

$$= (l+1) \rho^{l} \exp\left(-\frac{\rho}{2} \right) L(\rho) - \frac{1}{2} \exp\left(-\frac{\rho}{2} \right) \{ \rho^{l+1} L(\rho) \} + \frac{d\{ L(\rho) \}}{d\rho} \left\{ \rho^{l+1} \exp\left(-\frac{\rho}{2} \right) \right\}$$

$$= \left\{ \rho^{l+1} \exp\left(-\frac{\rho}{2} \right) \right\} \left[(l+1) \frac{L(\rho)}{\rho} - \frac{1}{2} L(\rho) + \frac{d\{ L(\rho) \}}{d\rho} \right] \tag{6.124}$$

$$\frac{d^2u(\rho)}{d\rho^2} = \left[\frac{d}{d\rho}\{\rho^{l+1}\}\right]\exp\left(-\frac{\rho}{2}\right)\left[(l+1)\frac{L(\rho)}{\rho} - \frac{1}{2}L(\rho) + \frac{d\{L(\rho)\}}{d\rho}\right]$$

$$+ \left[\frac{d}{d\rho}\left\{\exp\left(-\frac{\rho}{2}\right)\right\}\right]\left[(l+1)\frac{L(\rho)}{\rho} - \frac{1}{2}L(\rho) + \frac{d\{L(\rho)\}}{d\rho}\right]\rho^{l+1} + \left\{\rho^{l+1}\exp\left(-\frac{\rho}{2}\right)\right\}\times\frac{d}{d\rho}\left[(l+1)\frac{L(\rho)}{\rho} - \frac{1}{2}L(\rho) + \frac{d\{L(\rho)\}}{d\rho}\right]$$

$$= (l+1)\rho^{l+1-1}\exp\left(-\frac{\rho}{2}\right)\left[(l+1)\frac{L(\rho)}{\rho} - \frac{1}{2}L(\rho) + \frac{d\{L(\rho)\}}{d\rho}\right] - \frac{1}{2}\left\{\exp\left(-\frac{\rho}{2}\right)\right\}\rho^{l+1}\left[(l+1)\frac{L(\rho)}{\rho} - \frac{1}{2}L(\rho) + \frac{d\{L(\rho)\}}{d\rho}\right]$$

$$+ \left\{\rho^{l+1}\exp\left(-\frac{\rho}{2}\right)\right\}\times\left[(l+1)\frac{\left\{\dfrac{d\{L(\rho)\}}{d\rho}\times\rho - L(\rho)\times 1\right\}}{\rho^2} - \frac{1}{2}\frac{dL(\rho)}{d\rho} + \frac{d^2\{L(\rho)\}}{d\rho^2}\right]$$

$$= \left\{\exp\left(-\frac{\rho}{2}\right)\right\}\rho^{l+1}\left[\frac{(l+1)^2}{\rho}\times\frac{L(\rho)}{\rho} - \frac{1}{2}\frac{(l+1)}{\rho}\times L(\rho) + \frac{(l+1)}{\rho}\times\frac{d\{L(\rho)\}}{d\rho} - \frac{1}{2}(l+1)\frac{L(\rho)}{\rho} + \frac{1}{4}L(\rho)\right.$$

$$\left. - \frac{1}{2}\frac{d\{L(\rho)\}}{d\rho} + \frac{(l+1)}{\rho}\frac{d\{L(\rho)\}}{d\rho} - \frac{(l+1)}{\rho^2}L(\rho) - \frac{1}{2}\frac{dL(\rho)}{d\rho} + \frac{d^2\{L(\rho)\}}{d\rho^2}\right]$$

$$= \left\{\exp\left(-\frac{\rho}{2}\right)\right\}\rho^{l+1}\left[\frac{d^2\{L(\rho)\}}{d\rho^2} + \frac{(l+1)}{\rho}\times\frac{d\{L(\rho)\}}{d\rho} - \frac{1}{2}\frac{d\{L(\rho)\}}{d\rho} + \frac{(l+1)}{\rho}\frac{d\{L(\rho)\}}{d\rho} - \frac{1}{2}\frac{dL(\rho)}{d\rho} + \frac{(l+1)^2}{\rho}\times\frac{L(\rho)}{\rho}\right.$$

$$\left. - \frac{1}{2}\frac{(l+1)}{\rho}\times L(\rho) - \frac{1}{2}\frac{(l+1)L(\rho)}{\rho} + \frac{1}{4}L(\rho) - \frac{(l+1)}{\rho^2}L(\rho)\right]$$

$$= \left\{\exp\left(-\frac{\rho}{2}\right)\right\}\rho^{l+1}\left[\frac{d^2\{L(\rho)\}}{d\rho^2} + \left\{\frac{(l+1)}{\rho} - \frac{1}{2} + \frac{(l+1)}{\rho} - \frac{1}{2}\right\}\frac{d\{L(\rho)\}}{d\rho} + \left\{\frac{(l+1)^2}{\rho}\times\frac{1}{\rho} - \frac{1}{2}\frac{(l+1)}{\rho} - \frac{1}{2}\frac{(l+1)}{\rho} + \frac{1}{4} - \frac{(l+1)}{\rho^2}\right\}L(\rho)\right]$$

$$= \left\{\exp\left(-\frac{\rho}{2}\right)\right\}\rho^{l+1}\left[\frac{d^2\{L(\rho)\}}{d\rho^2} + \left\{\frac{2(l+1)-\rho+2(l+1)-\rho}{2\rho}\right\}\frac{d\{L(\rho)\}}{d\rho} + \left\{\frac{4(l+1)^2 - 2\rho(l+1) - 2\rho(l+1) + \rho^2 - 4(l+1)}{4\rho^2}\right\}L(\rho)\right]$$

$$= \left\{\exp\left(-\frac{\rho}{2}\right)\right\}\rho^{l+1}\left[\frac{d^2\{L(\rho)\}}{d\rho^2} + \left(\frac{4l+4-2\rho}{2\rho}\right)\frac{d\{L(\rho)\}}{d\rho} + \left\{\frac{4(l+1)^2 - 4\rho(l+1) + \rho^2 - 4(l+1)}{4\rho^2}\right\}L(\rho)\right] \tag{6.125}$$

Substituting for $d^2u(\rho)/d\rho^2$ from eq. (6.125) in eq. (6.103), and putting the value of $u(\rho)$ from eq. (6.123), we get

$$\left\{\exp\left(-\frac{\rho}{2}\right)\right\}\rho^{l+1}\left[\frac{d^2\{L(\rho)\}}{d\rho^2} + \left(\frac{2l+2-\rho}{\rho}\right)\frac{d\{L(\rho)\}}{d\rho} + \left\{\frac{4(l+1)^2 - 4\rho(l+1) + \rho^2 - 4(l+1)}{4\rho^2}\right\}L(\rho)\right]$$

$$= \left\{\frac{1}{4} - \frac{\rho_0}{2\rho} + \frac{l(l+1)}{\rho^2}\right\}u(\rho) = \left\{\frac{1}{4} - \frac{\rho_0}{2\rho} + \frac{l(l+1)}{\rho^2}\right\}\rho^{l+1}\exp\left(-\frac{\rho}{2}\right)L(\rho) \tag{6.126}$$

Dividing both sides of eq. (6.126) by $\rho^{l+1} \exp(-\rho/2)$

$$\frac{d^2\{L(\rho)\}}{d\rho^2} + \left\{\frac{2(l+1)-\rho}{\rho}\right\}\frac{d\{L(\rho)\}}{d\rho}$$

$$+ \left\{\frac{4(l+1)^2 - 4\rho(l+1) + \rho^2 - 4(l+1)}{4\rho^2}\right\}L(\rho)$$

$$= \left\{\frac{1}{4} - \frac{\rho_0}{2\rho} + \frac{l(l+1)}{\rho^2}\right\}L(\rho) \qquad (6.127)$$

or

$$\frac{d^2\{L(\rho)\}}{d\rho^2} + \left\{\frac{2(l+1)-\rho}{\rho}\right\}\frac{d\{L(\rho)\}}{d\rho}$$

$$+ \left\{\frac{4(l+1)^2 - 4\rho(l+1) + \rho^2 - 4(l+1)}{4\rho^2}\right\}L(\rho)$$

$$- \left\{\frac{1}{4} - \frac{\rho_0}{2\rho} + \frac{l(l+1)}{\rho^2}\right\}L(\rho) = 0 \qquad (6.128)$$

or

$$\frac{d^2\{L(\rho)\}}{d\rho^2} + \left\{\frac{2(l+1)-\rho}{\rho}\right\}\frac{d\{L(\rho)\}}{d\rho}$$

$$+ \left\{\frac{4(l+1)^2 - 4\rho(l+1) + \rho^2 - 4(l+1)}{4\rho^2}\right\}L(\rho)$$

$$- \left\{\frac{\rho^2 - 2\rho\rho_0 + 4l(l+1)}{4\rho^2}\right\}L(\rho) = 0 \qquad (6.129)$$

or

$$\frac{d^2\{L(\rho)\}}{d\rho^2} + \left\{\frac{2(l+1)-\rho}{\rho}\right\}\frac{d\{L(\rho)\}}{d\rho}$$

$$+ \left\{\frac{4(l+1)^2 - 4\rho(l+1) + \rho^2 - 4(l+1)}{4\rho^2}\right.$$

$$\left. - \frac{\rho^2 - 2\rho\rho_0 + 4l(l+1)}{4\rho^2}\right\}L(\rho) = 0 \qquad (6.130)$$

or

$$\frac{d^2\{L(\rho)\}}{d\rho^2} + \left\{\frac{2(l+1)-\rho}{\rho}\right\}\frac{d\{L(\rho)\}}{d\rho}$$

$$+ \left\{\frac{4(l+1)^2 - 4\rho(l+1) + \rho^2 - 4(l+1) - \rho^2 + 2\rho\rho_0 - 4l(l+1)}{4\rho^2}\right\}$$

$$\times L(\rho) = 0$$

$$(6.131)$$

or

$$\frac{d^2\{L(\rho)\}}{d\rho^2} + \left\{\frac{2(l+1)-\rho}{\rho}\right\}\frac{d\{L(\rho)\}}{d\rho}$$

$$+ \left\{\frac{4(l+1)^2 - 4\rho(l+1) + 2\rho\rho_0 - 4(l+1)(1+l)}{4\rho^2}\right\}L(\rho) = 0$$

$$(6.132)$$

or

$$\frac{d^2\{L(\rho)\}}{d\rho^2} + \left\{\frac{2(l+1)-\rho}{\rho}\right\}\frac{d\{L(\rho)\}}{d\rho}$$

$$+ \left\{\frac{4(l+1)^2 - 4\rho(l+1) + 2\rho\rho_0 - 4(l+1)^2}{4\rho^2}\right\}L(\rho) = 0$$

$$(6.133)$$

or

$$\frac{d^2\{L(\rho)\}}{d\rho^2} + \left\{\frac{2(l+1)-\rho}{\rho}\right\}\frac{d\{L(\rho)\}}{d\rho}$$

$$+ \left\{\frac{-4\rho(l+1) + 2\rho\rho_0}{4\rho^2}\right\}L(\rho) = 0 \qquad (6.134)$$

or

$$\frac{d^2\{L(\rho)\}}{d\rho^2} + \left\{\frac{2(l+1)-\rho}{\rho}\right\}\frac{d\{L(\rho)\}}{d\rho}$$

$$+ \left\{\frac{-4(l+1) + 2\rho_0}{4\rho}\right\}L(\rho) = 0 \qquad (6.135)$$

Multiplying both sides of eq. (6.135) by ρ,

$$\rho\frac{d^2\{L(\rho)\}}{d\rho^2} + \{2(l+1)-\rho\}\frac{d\{L(\rho)\}}{d\rho}$$

$$+ \left\{\frac{\rho_0}{2} - (l+1)\right\}L(\rho) = 0 \qquad (6.136)$$

Let us put

$$2l + 1 = v \qquad (6.137)$$

Then

$$2(l+1) = 2l + 2 = 2l + 1 + 1 = v + 1 \qquad (6.138)$$

Also, let us put

$$\frac{\rho_0}{2} - (l+1) = \eta \qquad (6.139)$$

Then eq. (6.136) is transformed to

$$\rho\left\{\frac{d^2 L(\rho)}{d\rho^2}\right\} + \{v+1-\rho\}\frac{d\{L(\rho)\}}{d\rho} + \eta L(\rho) = 0 \qquad (6.140)$$

Let us compare this equation with the associated Laguerre equation

$$x\left\{\frac{d^2 L_\eta^v(x)}{dx^2}\right\} + (v+1-x)\left\{\frac{dL_\eta^v(x)}{dx}\right\} + \eta L_\eta^v(x) = 0 \qquad (6.141)$$

where v and η are real numbers. The associated Laguerre equation has non-singular solution when $\eta = $ non-negative integer $= 0$, 1, 2, 3, ... These solutions are denoted by $L_\eta^v(\rho)$. They are called associated Laguerre polynomials.

We found that eq. (6.140) is identical to the associated Laguerre equation. Moving further, since the solutions to the associated Laguerre equation are the associated Laguerre polynomials $L_n^v(x)$, the solutions to eq. (6.140) must also be similar polynomials given by

$$L(\rho) = L_\eta^v(\rho) = L_{\frac{1}{2}\rho_0 -(l+1)}^{2l+1}(\rho) = L_{n-(l+1)}^{2l+1}(\rho) \qquad (6.142)$$

by defining

$$\frac{1}{2}\rho_0 = \text{an integer} = n = 1, 2, 3, \dots \qquad (6.143)$$

where n is the principal quantum number. Note the stage of its introduction in the analysis. It appears when the radial part of the Schrodinger equation is solved. We have said that eq. (6.141) has non-singular solution only for positive values of η. Since

$$\eta = \frac{\rho_0}{2} - (l+1) = n - (l+1) \qquad (6.144)$$

it will be positive only when

$$n \geq (l+1) \qquad (6.145)$$

Another way to argue is that it is necessary to guarantee a convergent series to represent the non-asymptotic part of the radial wave function. Equation (6.140) or (6.141) is solved by proposing a power series in ρ or x. The condition for the power series to terminate at some power is that n (see eqs. (6.139) and (6.143)) is an integer and

$$n \geq (l+1) \qquad (6.146)$$

Hence the restriction on value of l with respect to that of n is explained. The wave function with $n = 1$, $l = 0$ is called 1s state, that with $n = 2$, $l = 1$ is known as 2p state.

Since ρ_0 is related with κ and hence energy, integral values for $(1/2)\rho_0$ indicate that energy is quantized. From eqs. (6.143) (6.101) and (6.79),

$$n = \frac{1}{2}\rho_0 = \frac{1}{2} \times \frac{\mu e^2}{2\pi\varepsilon_0\kappa\hbar^2} = \frac{\mu e^2}{4\pi\varepsilon_0\sqrt{-\frac{2\mu E}{\hbar^2}}\hbar^2} \qquad (6.147)$$

Squaring both sides of eq. (6.147),

$$n^2 = \left(\frac{\mu e^2}{4\pi\varepsilon_0\sqrt{-\frac{2\mu E}{\hbar^2}}\hbar^2}\right)^2 = \frac{\mu^2 e^4}{16\pi^2\varepsilon_0^2\left(-\frac{2\mu E}{\hbar^2}\right)\hbar^4}$$

$$= -\frac{\mu^2 e^4}{32\pi^2\varepsilon_0^2\hbar^2\mu E} = -\frac{\mu e^4}{32\pi^2\varepsilon_0^2\hbar^2 E} \qquad (6.148)$$

or

$$E = -\frac{\mu e^4}{32\pi^2\varepsilon_0^2\hbar^2 n^2} = -\left(\frac{\mu e^4}{32\pi^2\hbar^2\varepsilon_0^2}\right)\frac{1}{n^2} = -\text{Ry} \times \frac{1}{n^2} = -\frac{\text{Ry}}{n^2}$$

$$(6.149)$$

where

$$\text{Ry} = \text{Rydberg constant} = \frac{\mu e^4}{32\pi^2\hbar^2\varepsilon_0^2} = \frac{\left(\frac{m_e m_p}{m_e + m_p}\right)e^4}{32\pi^2\hbar^2\varepsilon_0^2}$$

$$= \frac{\left(\frac{9.109\times 10^{-31}\ \text{kg}\times 1.6726\times 10^{-27}\ \text{kg}}{9.109\times 10^{-31}\ \text{kg} + 1.6726\times 10^{-27}\ \text{kg}}\right)(1.602\times 10^{-19}\ \text{C})^4}{32(3.1415)^2(1.05457\times 10^{-34}\ \text{Js})^2(8.854\times 10^{-12}\ \text{Fm}^{-1})^2}$$

$$= \frac{\frac{15.2357\times 10^{-58}}{1.6735\times 10^{-27}}\times 6.5864\times 10^{-76}}{32\times 9.869\times 1.1121\times 10^{-68}\times 78.3933\times 10^{-24}}$$

$$= \frac{9.1041\times 10^{-31}\times 6.5864\times 10^{-76}}{27532.5169\times 10^{-92}}$$

$$= \frac{59.963\times 10^{-107}}{27532.5169\times 10^{-92}} = 2.17789\times 10^{-18}\ \text{J}$$

$$= \frac{2.17789\times 10^{-18}}{1.602\times 10^{-19}}\ \text{eV} = 13.5948\ \text{eV} \qquad (6.150)$$

$$\therefore E = -\frac{\text{Ry}}{n^2} = -\frac{13.59}{n^2}\text{eV} \qquad (6.151)$$

The reduced mass μ of the electron and proton system is 9.1041×10^{-31} kg, which differs by a very small amount from the electron rest mass (9.109×10^{-31} kg). The negative sign of energy implies that energy has to be expended against the electric field to separate the electron from the binding force of the proton in the atomic nucleus. Thus arises the concept of energy quantization. Equation (6.151) is the same as derived in the primitive model of Bohr.

6.7 Construction and Normalization of the Radial Wave Function

Let us go back and combine eqs. (6.123) and (6.142) to write the radial wave function using appropriate subscripts n, l as

$$u_{n,l}\left(\rho_n\right) = \rho_n^{l+1} \exp\left(-\frac{\rho_n}{2}\right) L\left(\rho_n\right) = \rho_n^{l+1} \exp\left(-\frac{\rho_n}{2}\right) L_{n-(l+1)}^{2l+1}\left(\rho_n\right) \tag{6.152}$$

where from eq. (6.83)

$$\rho_n = 2k_n r \tag{6.153}$$

But from eq. (6.92)

$$R_{n,l}\left(\rho_n\right) = \frac{u_{n,l}\left(\rho_n\right)}{\rho_n} = \frac{\rho_n^{l+1} \exp\left(-\dfrac{\rho_n}{2}\right) L_{n-(l+1)}^{2l+1}\left(\rho_n\right)}{\rho_n}$$

$$= \rho_n^l \exp\left(-\frac{\rho_n}{2}\right) L_{n-(l+1)}^{2l+1}\left(\rho_n\right) \tag{6.154}$$

For the normalization of the radial part we write

$$R_{n,l}\left(\rho_n\right) = A_{n,l}\ \rho_n^l \exp\left(-\frac{\rho_n}{2}\right) L_{n-(l+1)}^{2l+1}\left(\rho_n\right) \tag{6.155}$$

or

$$R_{n,l}\left(2k_n r\right) = A_{n,l}\left(2k_n r\right)^l \exp\left(-\frac{2k_n r}{2}\right) L_{n-(l+1)}^{2l+1}\left(2k_n r\right) \tag{6.156}$$

For convenience of notation, we write

$$2k_n r = x \tag{6.157}$$

Hence

$$R_{n,l}\left(x\right) = A_{n,l} x^l \exp\left(-\frac{x}{2}\right) L_{n-(l+1)}^{2l+1}\left(x\right) \tag{6.158}$$

and so

$$R_{n,l}^*\left(x\right) = A_{n,l} x^l \exp\left(-\frac{x}{2}\right) L_{n-(l+1)}^{2l+1}\left(x\right) \tag{6.159}$$

The normalization integral is expressed as

$$\int_0^\infty R_{n,l}^*\left(r\right) R_{n,l}\left(r\right) r^2\ dr = 1 \tag{6.160}$$

or

$$\int_0^\infty R_{n,l}^*\left(x\right) R_{n,l}\left(x\right) \left(\frac{x}{2k_n}\right)^2 d\left(\frac{x}{2k_n}\right) = 1 \tag{6.161}$$

or

$$\int_0^\infty R_{n,l}^*\left(x\right) R_{n,l}\left(x\right) \frac{x^2}{4k_n^2} \times \frac{1}{2k_n}\ dx = 1 \tag{6.162}$$

or

$$\frac{1}{8k_n^3} \int_0^\infty R_{n,l}^*\left(x\right) R_{n,l}\left(x\right) x^2\ dx = 1 \tag{6.163}$$

Substituting for $R_{n,l}^*\left(x\right)$ and $R_{n,l}\left(x\right)$ from eqs. (6.158) and (6.159) in eq. (6.163) we have

$$\frac{1}{8k_n^3} \int_0^\infty \left\{ A_{n,l} x^l \exp\left(-\frac{x}{2}\right) L_{n-(l+1)}^{2l+1}\left(x\right) \right.$$

$$\left. \times A_{n,l} x^l \exp\left(-\frac{x}{2}\right) L_{n-(l+1)}^{2l+1}\left(x\right) \times x^2 dx \right\} = 1 \tag{6.164}$$

or

$$\frac{A_{n,l}^2}{8k_n^3} \int_0^\infty x^l L_{n-(l+1)}^{2l+1}\left(x\right) \times x^l \exp\left(-\frac{x}{2}-\frac{x}{2}\right) L_{n-(l+1)}^{2l+1}\left(x\right) \times x^2\ dx = 1 \tag{6.165}$$

or

$$\frac{A_{n,l}^2}{8k_n^3} \int_0^\infty x^l L_{n-(l+1)}^{2l+1}\left(x\right) \times x^l \exp(-x) L_{n-(l+1)}^{2l+1}\left(x\right) \times x^2\ dx = 1 \tag{6.166}$$

or

$$\frac{A_{n,l}^2}{8k_n^3} \int_0^\infty x^{2l} \left\{ L_{n-(l+1)}^{2l+1}\left(x\right) \right\}^2 \exp(-x) \times x^2\ dx = 1 \tag{6.167}$$

or

$$\frac{A_{n,l}^2}{8k_n^3} \int_0^\infty x^{2l+2} \left\{ L_{n-(l+1)}^{2l+1}\left(x\right) \right\}^2 \exp(-x)\ dx = 1 \tag{6.168}$$

We apply the general integral for associated Laguerre polynomial

$$\int_0^\infty x^{k+1} \left\{ L_j^k\left(x\right) \right\}^2 \exp(-x)\ dx = \frac{(2j+k+1)(j+k)!}{j!} \tag{6.169}$$

to get

$$\int_0^\infty x^{(2l+1)+1}\left\{L_{n-(l+1)}^{2l+1}(x)\right\}^2 \exp(-x)dx$$

$$= \frac{\left[2\{n-(l+1)\}+2l+1+1\right]\times\{n-(l+1)+2l+1\}!}{\{n-(l+1)\}!}$$

$$= \frac{[2n-2l-2+2l+1+1]\times\{n-l-1+2l+1\}!}{\{n-(l+1)\}!}$$

$$= \frac{[2n-2l-2+2l+2]\times\{n-l-1+2l+1\}!}{\{n-l-1\}!} = \frac{(2n)(n+l)!}{(n-l-1)!}$$

$$(6.170)$$

Substituting the value of the integral from eq. (6.170) in eq. (6.168) we have

$$\frac{A_{n,l}^2}{8k_n^3}\times\frac{(2n)(n+l)!}{(n-l-1)!} = 1 \qquad (6.171)$$

or

$$A_{n,l}^2 = \frac{\left(8k_n^3\right)(n-l-1)!}{(2n)(n+l)!} \qquad (6.172)$$

or

$$A_{n,l} = \sqrt{\frac{\left(8k_n^3\right)(n-l-1)!}{(2n)(n+l)!}} = (2k_n)^{3/2}\sqrt{\frac{(n-l-1)!}{(2n)(n+l)!}} \qquad (6.173)$$

Putting the value of the normalization constant from eq. (6.173) in eq. (6.158)

$$R_{n,l}(2k_nr) = (2k_n)^{3/2}\sqrt{\frac{(n-l-1)!}{(2n)(n+l)!}}(2k_nr)^l$$

$$\exp\left(-\frac{2k_nr}{2}\right)L_{n-(l+1)}^{2l+1}(2k_nr)$$

$$= (2k_n)^{l+3/2}\sqrt{\frac{(n-l-1)!}{(2n)(n+l)!}}\exp(-k_nr)r^l L_{n-(l+1)}^{2l+1}(2k_nr)$$

$$(6.174)$$

which may be simply written as

$$R_{n,l}(r) = (2k_n)^{l+3/2}\sqrt{\frac{(n-l-1)!}{(2n)(n+l)!}}\exp(-k_nr)r^l L_{n-(l+1)}^{2l+1}(2k_nr)$$

$$(6.175)$$

6.8 Solution of the Polar Equation

The polar equation (6.72a) can be written in full derivatives as

$$\sin\theta\frac{d}{d\theta}\left\{\sin\theta\frac{d\{\Theta(\theta)\}}{d\theta}\right\}+l(l+1)\sin^2\theta\{\Theta(\theta)\} = m_l^2\{\Theta(\theta)\}$$

$$(6.176)$$

or

$$\sin\theta\frac{d}{d\theta}\left\{\sin\theta\frac{d\{\Theta(\theta)\}}{d\theta}\right\}+l(l+1)\sin^2\theta\{\Theta(\theta)\}-m_l^2\{\Theta(\theta)\} = 0$$

$$(6.177)$$

$$\text{First term} = \sin\theta\frac{d}{d\theta}\left\{\sin\theta\frac{d\{\Theta(\theta)\}}{d\theta}\right\}$$

$$= \sin\theta\left[\frac{d}{d\theta}(\sin\theta)\times\frac{d\{\Theta(\theta)\}}{d\theta}+\frac{d}{d\theta}\left\{\frac{d\{\Theta(\theta)\}}{d\theta}\right\}\times\sin\theta\right]$$

$$= \sin\theta\left[\cos\theta\times\frac{d\{\Theta(\theta)\}}{d\theta}+\left\{\frac{d^2\{\Theta(\theta)\}}{d\theta^2}\right\}\times\sin\theta\right]$$

$$= \sin\theta\cos\theta\frac{d\{\Theta(\theta)\}}{d\theta}+\sin^2\theta\frac{d^2\{\Theta(\theta)\}}{d\theta^2} \qquad (6.178)$$

Putting the first term as obtained in eq. (6.178) into eq. (6.177) we get

$$\sin^2\theta\frac{d^2\{\Theta(\theta)\}}{d\theta^2}+\sin\theta\cos\theta\frac{d\{\Theta(\theta)\}}{d\theta}$$

$$+l(l+1)\sin^2\theta\{\Theta(\theta)\}-m_l^2\{\Theta(\theta)\} = 0 \qquad (6.179)$$

We change variables using

$$x = \cos\theta \qquad (6.180)$$

Then

$$dx = -\sin\theta d\theta \qquad (6.181)$$

or

$$\frac{dx}{d\theta} = -\sin\theta \qquad (6.182)$$

Now,

$$\frac{d\{\Theta(\theta)\}}{d\theta} = \frac{d\{\Theta(x)\}}{dx} \times \frac{dx}{d\theta}$$

$$= \frac{d\{\Theta(x)\}}{dx} \times -\sin\theta = -\sin\theta \frac{d\{\Theta(x)\}}{dx} \qquad (6.183)$$

and

$$\frac{d^2\{\Theta(\theta)\}}{d\theta^2} = \frac{d}{d\theta}\left[\frac{d\{\Theta(\theta)\}}{d\theta}\right] = \frac{d}{d\theta}\left[-\sin\theta\frac{d\{\Theta(x)\}}{dx}\right]$$

$$= \frac{d}{d\theta}(-\sin\theta) \times \frac{d\{\Theta(x)\}}{dx} + \frac{d}{d\theta}\left[\frac{d\{\Theta(x)\}}{dx}\right] \times (-\sin\theta)$$

$$= -\cos\theta\frac{d\{\Theta(x)\}}{dx} - \sin\theta\frac{d}{dx}\left[\frac{d\{\Theta(x)\}}{dx}\right] \times \frac{dx}{d\theta}$$

$$= -\cos\theta\frac{d\{\Theta(x)\}}{dx} - \sin\theta\left[\frac{d^2\{\Theta(x)\}}{dx^2}\right] \times (-\sin\theta)$$

$$= -\cos\theta\frac{d\{\Theta(x)\}}{dx} + \sin^2\theta\frac{d^2\{\Theta(x)\}}{dx^2} \qquad (6.184)$$

Substituting for $d/d\{\Theta(\theta)\}$ and $d^2/d\theta^2\{\Theta(\theta)\}$ from eqs. (6.183) and (6.184) into eq. (6.179)

$$\sin^2\theta\left[-\cos\theta\frac{d\{\Theta(x)\}}{dx} + \sin^2\theta\frac{d^2\{\Theta(x)\}}{dx^2}\right]$$

$$+ \sin\theta\cos\theta\left[-\sin\theta\frac{d\{\Theta(x)\}}{dx}\right]$$

$$+ l(l+1)\sin^2\theta\{\Theta(x)\} - m_l^2\{\Theta(x)\} = 0 \qquad (6.185)$$

or

$$\sin^4\theta\frac{d^2\{\Theta(x)\}}{dx^2} - \sin^2\theta\cos\theta\frac{d\{\Theta(x)\}}{dx}$$

$$- \sin^2\theta\cos\theta\frac{d\{\Theta(x)\}}{dx} + l(l+1)\sin^2\theta\{\Theta(x)\} - m_l^2\{\Theta(x)\} = 0$$

$$(6.186)$$

Dividing both sides of eq. (6.186) by $\sin^2\theta$,

$$\sin^2\theta\frac{d^2\{\Theta(x)\}}{dx^2} - \cos\theta\frac{d\{\Theta(x)\}}{dx}$$

$$- \cos\theta\frac{d\{\Theta(x)\}}{dx} + l(l+1)\{\Theta(x)\} - \frac{m_l^2\{\Theta(x)\}}{\sin^2\theta} = 0 \qquad (6.187)$$

By putting

$$\cos\theta = x \qquad (6.188)$$

$$\sin^2\theta = 1 - \cos^2\theta = 1 - x^2 \qquad (6.189)$$

we obtain

$$(1-x^2)\frac{d^2\{\Theta(x)\}}{dx^2} - x\frac{d\{\Theta(x)\}}{dx}$$

$$- x\frac{d\{\Theta(x)\}}{dx} + l(l+1)\{\Theta(x)\} - \frac{m_l^2\{\Theta(x)\}}{1-x^2} = 0 \qquad (6.190)$$

or

$$(1-x^2)\frac{d^2\{\Theta(x)\}}{dx^2} - 2x\frac{d\{\Theta(x)\}}{dx}$$

$$+ l(l+1)\{\Theta(x)\} - \frac{m_l^2\{\Theta(x)\}}{1-x^2} = 0 \qquad (6.191)$$

Let us compare eq. (6.191) with the associated Legendre differential equation

$$(1-x^2)\frac{d^2y}{dx^2} - 2x\frac{dy}{dx} + l(l+1)y - \frac{m^2 y}{1-x^2} = 0 \qquad (6.192)$$

where l is a non-negative integer and m is an integer in the range $-l \leq m \leq +l$. We find that eq. (6.191) is the associated Legendre differential equation. But since we have m_l in eq. (6.191) instead of m in eq. (6.192), we write the limits on m_l as $-l \leq m_l \leq +l$. The quantum number l with the above restriction of values is needed in order that the power series in the Legendre polynomial converges to yield a physically sensible solution. It may be noted here that the associated Legendre differential equation is a generalized form of the Legendre differential equation

$$(1-x^2)\frac{d^2y}{dx^2} - 2x\frac{dy}{dx} + l(l+1)y = 0 \qquad (6.193)$$

In the Legendre differential equation, $m = 0$.

The solutions to the associated Legendre equation are the associated Legendre polynomials $P_{l,m}(x)$ which are generated from Legendre polynomials $P_l(x)$ using the equation

$$P_l^m(x) = (-1)^m \sqrt{(1-x^2)^m} \frac{d^m}{dx^m} P_l(x) \qquad (6.194)$$

and the Legendre polynomials themselves are generated from the equation

$$P_l(x) = \frac{(-1)^l}{2^l \, l!} \frac{d^l}{dx^l}(1-x^2)^l \qquad (6.195)$$

Thus solution of the polar equation is

$$\Theta(\theta) = P_l^m(\cos\theta) \qquad (6.196)$$

where P_l^m is an associated Legendre polynomial.

6.9 Solution of the Azimuthal Equation

Since Φ is a function of ϕ only, we can write the full derivative for Φ in eq. (6.73):

$$\frac{1}{\Phi}\frac{d^2\Phi}{d\phi^2} = -m_l^2 \qquad (6.197)$$

or

$$\frac{d^2\Phi}{d\phi^2} = -m_l^2\Phi \qquad (6.198)$$

It is a second order ordinary differential equation (ODE) with constant coefficients. Its solution is

$$\Phi = C_1\exp(im_l\phi) + C_2\exp(-im_l\phi) \qquad (6.199)$$

Allowing m_l to be positive or negative, we can take only one solution

$$\Phi = C_1\exp(im_l\phi) \qquad (6.200)$$

When ϕ is allowed to cycle through 2π, the function returns to the same position

$$\Phi(\phi) = \Phi(\phi + 2\pi) \qquad (6.201)$$

Since ϕ and $\phi + 2\pi$ represent the same point in space,

$$\exp(im_l\phi) = \exp\{im_l(\phi + 2\pi)\} \qquad (6.202)$$

or

$$\frac{\exp\{im_l(\phi + 2\pi)\}}{\exp(im_l\phi)} = 1 \qquad (6.203)$$

or

$$\exp\left[\{im_l(\phi + 2\pi)\} - im\phi\right] = 1 \qquad (6.204)$$

or

$$\exp\left[im_l\phi + im_l 2\pi - im_l\phi\right] = 1 \qquad (6.205)$$

or

$$\exp(im_l 2\pi) = 1 \qquad (6.206)$$

or

$$\cos(2\pi m_l) + i\sin(2\pi m_l) = 1 \qquad (6.207)$$

which is true if m_l is zero or an integer, i.e. 0, ±1, ±2, ±3, ... This quantum number arises from the need for the azimuthal wave function Φ to return to its initial value after 2π rotation of ϕ, i.e. when ϕ is cycled through ϕ, the azimuthal angle wave function returns to the same position.

Thus the solutions $\Phi(\phi)$ to the azimuthal angle equation are exponentials involving m_l, and

$$m_l = 0, \ \pm 1, \pm 2, \pm 3, \ldots \qquad (6.208)$$

Since we have already mentioned the limits on m_l in Section 6.8 as $-l \le m_l \le +l$, the last value of m_l in eq. (6.208) will be $\pm l$. Hence, eq. (6.208) becomes

$$m_l = 0, \ \pm 1, \pm 2, \pm 3, \ldots, \pm l \qquad (6.209)$$

6.10 Combining the Angular Partial Solutions

We write from eqs. (6.196) and (6.200),

$$\Psi(\theta,\phi) = N\Theta(\theta)\Phi(\phi) = NP_l^m(\cos\theta)\exp(im_l\phi) \qquad (6.210)$$

where N is the unknown normalization constant. The normalized product of the polar and azimuthal angular solutions is called spherical harmonics and is written as

$$Y_l^m(\theta,\phi) = NP_l^m(\cos\theta)\exp(im_l\phi) \qquad (6.211)$$

The normalization of spherical harmonics will ensure

$$\int_0^{2\pi}\int_0^{\pi}\left\{Y_l^m(\theta,\phi)\right\}^*\left\{Y_l^m(\theta,\phi)\right\}\sin\theta\,d\theta\,d\phi = 1 \qquad (6.212)$$

or

$$\int_0^{2\pi}\int_0^{\pi}\left\{NP_l^m(\cos\theta)\exp(im_l\phi)\right\}^*$$
$$\times\left\{NP_l^m(\cos\theta)\exp(im_l\phi)\right\}\sin\theta\,d\theta\,d\phi = 1 \qquad (6.213)$$

or

$$N^2\int_0^{2\pi}\int_0^{\pi}\left\{P_l^m(\cos\theta)\right\}^*\exp(-im_l\phi)$$
$$\times\left\{P_l^m(\cos\theta)\right\}\exp(im_l\phi)\sin\theta\,d\theta\,d\phi = 1 \qquad (6.214)$$

or

$$N^2\int_0^{2\pi}\int_0^{\pi}\left[\left\{P_l^m(\cos\theta)\right\}^*\left\{P_l^m(\cos\theta)\right\}\sin\theta\,d\theta\right]$$
$$\times\left\{\exp(-im_l\phi)\exp(im_l\phi)\,d\phi\right\} = 1 \qquad (6.215)$$

or

$$N^2 \int_0^{2\pi} d\phi \times \int_0^{\pi} \left\{ P_l^m (\cos\theta) \right\}^* \left\{ P_l^m (\cos\theta) \right\} \sin\theta \, d\theta = 1 \quad (6.216)$$

or

$$N^2 \int_0^{2\pi} d\phi \times \int_0^{\pi} \left\{ P_l^m (\cos\theta) \right\} \left\{ P_l^m (\cos\theta) \right\} \sin\theta \, d\theta = 1 \quad (6.217)$$

But

$$\int_0^{2\pi} d\phi = |\phi|_0^{2\pi} = 2\pi - 0 = 2\pi \quad (6.218)$$

The associated Legendre polynomials satisfy the orthogonality condition for fixed m

$$\int_0^{\pi} P_k^m (\cos\theta) \left\{ P_l^m (\cos\theta) \right\} \sin\theta \, d\theta = \frac{2}{2l+1} \frac{(l+|m|)!}{(l-|m|)!} \delta_{k,l} \quad (6.219)$$

where $\delta_{k,l}$ is the Kronecker delta = 0 if $k \neq l$ and =1 if $k = l$. Here $k = l$; hence

$$\int_0^{\pi} P_l^m (\cos\theta) \left\{ P_l^m (\cos\theta) \right\} \sin\theta \, d\theta = \frac{2}{2l+1} \frac{(l+|m|)!}{(l-|m|)!} \delta_{l,l}$$

$$= \frac{2}{2l+1} \frac{(l+|m|)!}{(l-|m|)!} \times 1 = \frac{2}{2l+1} \frac{(l+|m|)!}{(l-|m|)!} \quad (6.220)$$

From eqs. (6.217), (6.218), and (6.220),

$$N^2 \times 2\pi \times \frac{2}{2l+1} \frac{(l+|m|)!}{(l-|m|)!} = 1 \quad (6.221)$$

or

$$N^2 = \frac{2l+1}{4\pi} \frac{(l-|m|)!}{(l+|m|)!} \quad (6.222)$$

or

$$N = \sqrt{\frac{2l+1}{4\pi} \frac{(l-|m|)!}{(l+|m|)!}} \quad (6.223)$$

Substituting for N from eq. (6.223) in eq. (6.211)

$$Y_{l,m} (\theta, \phi) = \varepsilon \sqrt{\frac{2l+1}{4\pi} \frac{(l-|m|)!}{(l+|m|)!}} P_{l,m} (\cos\theta) \exp(im_l\phi) \quad (6.224)$$

where

$$\varepsilon = (-1)^{\xi} \text{ when } \xi \geq 0, \text{ and } \varepsilon = 1 \text{ when } \xi \leq 0 \quad (6.225)$$

6.11 Putting together the Complete Wave Function

From eqs. (6.52), (6.175), and (6.224)

$$\Psi(r, \theta, \phi) = \text{Radial part } R_{n,l}(r) \times \text{Angular part } Y_{l,m}(\theta, \phi)$$

$$= (2k_n)^{l+3/2} \sqrt{\frac{(n-l-1)!}{(2n)(n+l)!}} \exp(-k_n r) r^l L_{n-(l+1)}^{2l+1} (2k_n r)$$

$$\times \varepsilon \sqrt{\frac{2l+1}{4\pi} \frac{(l-|m|)!}{(l+|m|)!}} P_{l,m} (\cos\theta) \exp(im_l\phi)$$

$$(6.226)$$

6.12 Discussion and Conclusions

The hydrogen atom occupies a coveted position in quantum mechanics because it is a two-body problem, which has provided several analytical closed-form solutions (Tang 2005).

The Schrodinger equation for the hydrogen atom in spherical coordinates is split into the equation for the hydrogen atom as a whole and equation for its internal states. To solve the Schrodinger equation for internal relative motion of the electron and proton, it is broken down into radial and angular components. Further from the angular component, azimuthal and polar equations are formed. Then three equations, namely, radial, polar, and azimuthal equations, for the three spatial variables R, Θ, Φ are solved. While solving these equations, three quantum numbers originate. The angular momentum quantum number l is familiarized in the analysis as a separation constant factor when the Schrodinger equation is decomposed into radial and angular parts. The magnetic quantum number m_l comes as a separation constant when the Θ and Φ equations are disconnected. The principal quantum number n appears when the radial part of the Schrodinger equation is solved. During the solution of the three equations, it is further found that n can take integral values 1, 2, 3, ... while $l = 0$, 1, 3, ... $(n-1)$ and $m_l = 0, \pm1, \pm2, \pm3, ... \pm m_l$. The combined eigenfunction is the product of the associated Laguerre polynomial for the radial part, the associated Legendre polynomial for the polar part, and the exponential function for the azimuthal part. The second and third factors jointly represent spherical harmonics so that the combined eigenfunction can also be expressed as the product of the associated Laguerre polynomial for the radial part and spherical harmonics for the angular part. The eigenvalues are obtained during the solution of the radial part. The energy levels for the hydrogen atom are found to be the same as in Bohr's model. Thus a clear picture of the appearance of quantum numbers, the allowed values of the quantum numbers, and the energy levels of the hydrogen atom emerges from the solution of the Schrodinger equation.

Illustrative Exercises

6.1 Calculate the energy of the photon emitted when an electron falls from energy level $n = 2$ to energy level $n = 1$ in a hydrogen atom.

The energies of the states for different n values are given by

$$E_n = -\frac{13.6}{n^2} \text{ eV} \tag{6.227}$$

$$E_{\text{Initial}} = E_2 = -\frac{13.6}{2^2} \text{ eV} = -3.4 \text{ eV} \tag{6.228}$$

$$E_{\text{Final}} = E_1 = -\frac{13.6}{1^2} \text{ eV} = -13.6 \text{ eV} \tag{6.229}$$

The photon energy for the transition between two energy states is

$$\Delta E = E_{\text{Initial}} - E_{\text{Final}} = E_2 - E_1$$
$$= -3.4 - (-13.6) = -3.4 + 13.6 = 10.2 \text{ eV} \tag{6.230}$$

6.2 An electron fell down from an unknown higher energy orbit to $n = 2$ energy level in a hydrogen atom emitting a radiation of wavelength 434 nm. From what energy level did the electron fall down?

$$E_{\text{Initial}} = -\frac{13.6}{n_{\text{Initial}}^2} \text{ eV} \tag{6.231}$$

$$E_{\text{Final}} = -\frac{13.6}{n_{\text{Final}}^2} \text{ eV} \tag{6.232}$$

$$\therefore \Delta E = E_{\text{Initial}} - E_{\text{Final}} = -\frac{13.6}{n_{\text{Initial}}^2} \text{ eV} - \left(-\frac{13.6}{n_{\text{Final}}^2} \text{ eV}\right)$$

$$= 13.6 \text{ eV} \left(-\frac{1}{n_{\text{Initial}}^2} + \frac{1}{n_{\text{Final}}^2}\right) = 13.6 \left(\frac{1}{n_{\text{Final}}^2} - \frac{1}{n_{\text{Initial}}^2}\right) \text{ eV} \tag{6.233}$$

or

$$\frac{hc}{\lambda} = 13.6 \left(\frac{1}{n_{\text{Final}}^2} - \frac{1}{n_{\text{Initial}}^2}\right) \text{ eV} \tag{6.234}$$

where h is Planck's constant, c is the velocity of light, and λ is the wavelength of radiation emitted; here 434 nm. From eq. (6.234),

$$\frac{1}{\lambda} = \frac{13.6}{hc} \left(\frac{1}{n_{\text{Final}}^2} - \frac{1}{n_{\text{Initial}}^2}\right) \text{ eV}$$

$$= \frac{13.6 \text{ eV}}{4.1357 \times 10^{-15} \text{ eV s} \times 3 \times 10^8 \text{ m s}^{-1}} \left(\frac{1}{n_{\text{Final}}^2} - \frac{1}{n_{\text{Initial}}^2}\right)$$

$$= 1.096 \times 10^7 \left(\frac{1}{n_{\text{Final}}^2} - \frac{1}{n_{\text{Initial}}^2}\right) \text{ m}^{-1} \tag{6.235}$$

Hence,

$$\frac{1}{434 \times 10^{-9} \text{ m}} = 1.096 \times 10^7 \left(\frac{1}{2^2} - \frac{1}{n_{\text{Initial}}^2}\right) \text{ m}^{-1} \tag{6.236}$$

or

$$2.30 \times 10^6 = \frac{1.096 \times 10^7}{4} - \frac{1.096 \times 10^7}{n_{\text{Initial}}^2} \tag{6.237}$$

or

$$2.3 \times 10^6 - 2.74 \times 10^6 = -\frac{1.096 \times 10^7}{n_{\text{Initial}}^2} \tag{6.238}$$

or

$$-0.44 \times 10^6 = -\frac{1.096 \times 10^7}{n_{\text{Initial}}^2} \tag{6.239}$$

or

$$n_{\text{Initial}}^2 = \frac{1.096 \times 10^7}{0.44 \times 10^6} = 24.91 \approx 25 \tag{6.240}$$

$$\therefore n_{\text{Initial}} = 5 \tag{6.241}$$

REFERENCES

Gradinetti, P. J. 2019. The Hydrogen Atom, https://www.grandinetti.org/resources/Teaching/Chem4300/LectureCh18.pdf.

Haken, H. and H. C. Wolf. 1994. Ch.10: Quantum mechanics of the hydrogen atom. In: *The Physics of Atoms and Quanta*. Springer: Berlin, Heidelberg, pp.149–166.

Tang, C. 2005. The hydrogen atom. In: *Fundamentals of Quantum Mechanics: For Solid-State Electronics and Optics*. Cambridge University Press: Cambridge, UK, pp. 89–109.

Thakkar, A. J. 2017. Ch.6: The hydrogen atom. In: *Quantum Chemistry: A Concise Introduction for Students of Physics, Chemistry, Biochemistry and Materials Science*. Morgan & Claypool Publishers: San Rafael, CA, USA, (A Morgan & Claypool publication as part of IOP Concise Physics), pp. 6-1 to 6-10.

Part II

Condensed Matter Physics for Nanoelectronics

7

Drude-Lorentz Free Electron Model

7.1 Condensed Matter Physics

It is the physics of condensed states of matter and by condensed states we mean solids and liquids. 'Condensed' is the antonym of 'dilute'. Condensed matter consists of a large number of particles, lying so close together that their interactions are no longer ignorable. Condensed matter serves as an ideal workshop for quantum and statistical mechanics along with electromagnetics, offering opportunities to design new materials and nanostructures. Striking examples are the crystalline lattices of atoms, the electrons in metals, the ferromagnetic and antiferromagnetic materials, and the superconducting phase.

In condensed mater physics, the diversity of properties of materials is investigated experimentally and theoretical physics is applied to develop mathematical models for explaining these properties. Experiments are supported by theories and conversely. Theories are validated and checked for accuracy and consistency. If necessary, they are updated for conformance with experimental data. Otherwise new theories supersede older theories. The Drude model for metals, proposed long back in 1900, is an illustrious theoretical development which still continues to be applied as a first-cut tool for easy visualization and approximate calculations, albeit with several improvements. Drude explained electrical conduction in metals by applying the kinetic theory of gases to metals by considering them to be made of a gas of electrons (Sólyom 2009, Di Sia 2012). Drude derived the electron velocity by applying Boltzmann's equipartition theorem (Drude 1900a,b). Lorentz introduced the concept of Maxwell-Boltzmann distribution of velocities into the explanation of electron motion in metals (Lorentz 1909).

7.2 From Kinetic Theory of Gases to the Drude Model

The kinetic theory of gases is a fairly successful theory to explain the macroscopic properties such as pressure and temperature of gases. It postulates that gases consist of a large number of submicroscopic molecules moving randomly in all directions such that their separation is much larger than their size (Figure 7.1). The molecules of a gas do not experience any attractive or repulsive forces among themselves. They undergo perfectly elastic collisions with each other, i.e. collisions in which they do not undergo any gain or loss of energy.

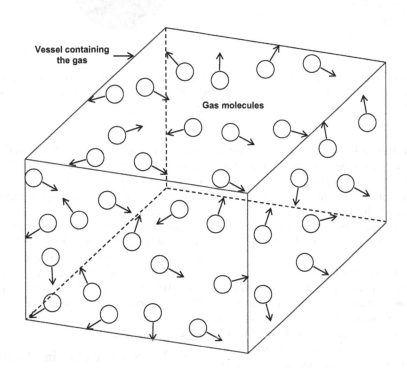

FIGURE 7.1   Chaotic motion of gas molecules in a container.

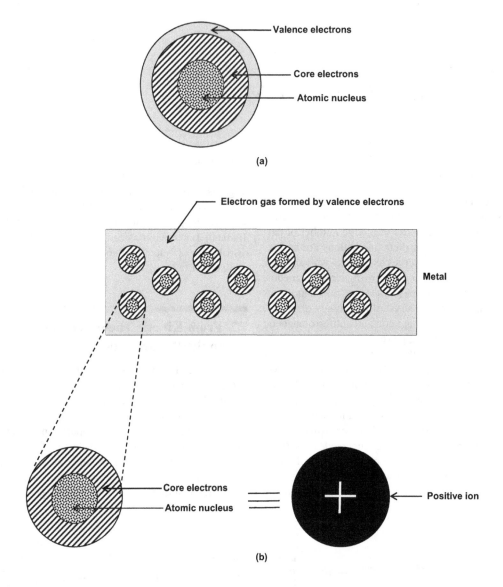

FIGURE 7.2 A metallic element: (a) single atom along with its electron shell consisting of core electrons and valence electrons, and (b) the electron gas formed from detached valence electrons when several atoms combine together. The nucleus and the core electrons of each atom remain unaffected and together they constitute a positive ion.

Molecules do not interact with each other in any way between collisions. They also collide against the walls of the containing vessel. The average kinetic energy of molecules is directly proportional to absolute temperature. The molecular motion follows Newton's laws and the gravitational forces acting on them are negligible. Relativistic and quantum-mechanical effects are ignored. Thus molecular dynamics is amenable to a classical and statistical treatment.

Like the kinetic theory of gases, the Drude model is a classical model which is able to explain the dynamics of charged particles particularly for simple alkaline metals, and is successful in general under many circumstances, despite being founded on several oversimplified assumptions. The central idea is that in a metal, the electric field of the atomic nucleus is decreased by the screening or shielding effect of electrons in inner shells whereby electrons in the valence shell are loosely bound to the nucleus. At normal temperatures, the valence electrons are liberated forming

a gas of free or conduction electrons (Figure 7.2). Therefore, a metal consists of positive ions formed by detachment of valence electrons from neutral atoms along with the free electron gas. This free electron gas can be studied like an ideal gas dealt with in the kinetic theory.

Interactions between charged particles are obviously different from that between neutral molecules in gases. Electrons are negatively charged particles moving in a sea of positively charged ions in metals, and are liable to repel each other or be attracted toward the positive ions. Let us see the difficulties in extension of kinetic theory ideas to electron motion.

7.3 Electron Densities in Metals

In a metal, the free electron concentration per unit volume (n) is given by

$$n = \frac{\text{Density}\,(\rho) \times \text{Avogadro's number }(N_{\mathrm{A}}) \times \text{Number of valence electrons per atom}\,(N_{\mathrm{V}})}{\text{Atomic mass }(M_{\mathrm{A}})} \tag{7.1}$$

Since $N_{\mathrm{A}} = 6.022 \times 10^{23}/\text{mol}$ and for cesium (Cs), $\rho = 1.879\,\text{g cm}^{-3}$, $M_{\mathrm{A}} = 132.9\,\text{g mol}^{-1}$, $N_{\mathrm{V}} = +1$,

$$\therefore n = \frac{1.879 \times 6.022 \times 10^{23} \times 1}{132.9} = 0.085 \times 10^{23}$$

$$= 8.5 \times 10^{21}\ \text{electrons/cm}^3 \tag{7.2}$$

For copper (Cu), $\rho = 8.96\,\text{g cm}^{-3}$, $M_{\mathrm{A}} = 63.546\,\text{g mol}^{-1}$, $N_{\mathrm{V}} = +1, +2$,

$$\therefore n = \frac{8.96 \times 6.022 \times 10^{23} \times 1}{63.546} = 0.849 \times 10^{23}\ \text{electrons/cm}^3$$

$$= 8.49 \times 10^{22}\ \text{electrons/cm}^3 \tag{7.3}$$

for valence 1 and 1.698×10^{23} electrons/cm^3 for valence 2.

For beryllium (Be), $\rho = 1.85\,\text{g cm}^{-3}$, $M_{\mathrm{A}} = 9.012\,\text{g mol}^{-1}$, $N_{\mathrm{V}} = +2$,

$$\therefore n = \frac{1.85 \times 6.022 \times 10^{23} \times 2}{9.012} = 2.47 \times 10^{23}\ \text{electrons/cm}^3 \tag{7.4}$$

For comparison, the number of molecules per unit volume in a gas is

$$n = \frac{\text{Density}\,(\rho) \times \text{Avogadro's number }(N_{\mathrm{A}})}{\text{Molar mass }(M_{\mathrm{M}})} \tag{7.5}$$

Taking the example of nitrogen (N), $\rho = 1.2506\,\text{g L}^{-1} = 1.2506/1000\,\text{g cm}^{-3}$ at standard temperature and pressure (STP), $M_{\mathrm{M}} = 28.014\,\text{g mol}^{-1}$,

$$n = \frac{1.2506 \times 6.022 \times 10^{23}}{1000 \times 28.014} = 0.2688 \times 10^{20}$$

$$= 2.688 \times 10^{19}\ \text{molecules / cm}^3 \tag{7.6}$$

So the number of molecules per unit volume in a gas at STP is several orders of magnitude less than in a metal, and the application of kinetic theory concepts to metals is debatable.

7.4 Separation between Electrons

The average distance r between any two electrons in a metal is obtained by considering that:

n electrons are present in unit volume

So, 1 electron is present in $1/n$ volume; hence

$$\frac{1}{n} = \frac{4}{3}\pi r^3 \tag{7.7}$$

where the right-hand side is the volume of a sphere per conduction electron.

or

$$r^3 = \frac{3}{4\pi n} \tag{7.8}$$

or

$$r = \left(\frac{3}{4\pi n}\right)^{1/3} \tag{7.9}$$

For Cs,

$$r = \left(\frac{3}{4 \times 3.14159 \times 8.5 \times 10^{21}}\right)^{1/3} = 3.0397 \times 10^{-8}\ \text{cm} \tag{7.10}$$

For Cu,

$$r = \left(\frac{3}{4 \times 3.14159 \times 8.49 \times 10^{22}}\right)^{1/3} = 1.411 \times 10^{-8}\ \text{cm} \tag{7.11}$$

For Be,

$$r = \left(\frac{3}{4 \times 3.14159 \times 2.47 \times 10^{23}}\right)^{1/3} = 9.887 \times 10^{-9}\ \text{cm} \tag{7.12}$$

For a gas like nitrogen at STP,

$$r = \left(\frac{3}{4 \times 3.14159 \times 2.688 \times 10^{19}}\right)^{1/3} = 2.07 \times 10^{-7}\ \text{cm} \tag{7.13}$$

The electron-electron separation is much smaller than intermolecular distance in gases. Hence straightway extension of kinetic theory seems unreasonable. Still Drude very boldly proposed that the electron gas in metals is similar to the ideal gas in kinetic theory of gases, and came forward with encouraging predictions.

7.5 Assumptions of the Drude Model

The electric field of the atomic nucleus is decreased by the screening or shielding effect of electrons in inner shells whereby electrons in the valence shell are loosely bound to the nucleus. At normal temperatures, the valence electrons are liberated forming a gas of free or conduction electrons. Therefore a metal consists of positive ions formed by detachment of valence electrons from neutral atoms along with the free electron gas.

i. Free electron approximation: Except for collisions, the electrons do not interact with the ions. So between any two collisions, the electron motion is same irrespective of existence or non-existence of ions.

ii. Independent electron approximation: Electron-electron interactions are overlooked. Whereas in the Drude model, the electrons do not interact with each other, in kinetic theory of gases, collisions between molecules of a gas help in attaining equilibrium.

iii. Collisions and relaxation time: Collisions of electrons with ions are considered to be instantaneous events. They change the velocity of the electron. The probability per unit time of an electron to undergo collision is $1/\tau$ where the parameter τ is called the relaxation time or momentum relaxation time. It represents the average time between collisions. So, it is also known as the mean free time.

iv. Collisions and thermal equilibrium: Collisions of electrons with ions, referred to as scattering of electrons, act as a means by which the electrons attain thermal equilibrium with their surroundings. After a collision, the electron loses memory of its pre-collision velocity and the new velocity is acquired by thermal equilibrium at the collision location. In statistical mechanics, the law of equipartition of energy relates the temperature of a system to its average energy. It states that in thermal equilibrium at absolute temperature T, each quadratic term in the equation for energy of a system of particles will contribute $1/2\, k_B T$ to the average energy; here k_B is the Boltzmann constant. Degree of freedom of a system is the minimum number of independent parameters such as coordinates needed to specify its configuration. The motion of an electron has three degrees of freedom pertaining to the x, y, and z components of its momentum. Owing to the quadratic appearance of these momenta in the equation for kinetic energy of the electron, each electron is ascribed a kinetic energy of $(3/2)\, k_B T$. Hence, the post-collision velocity v of the electron of mass m is determined by the equality

$$\frac{1}{2}mv^2 = \frac{3}{2}k_B T \qquad (7.14)$$

where T is the temperature of the location on Kelvin scale. Thus hotter the collision location, faster is the emergent electron.

The implications of these postulates need elaboration. The electrons are classical particles complying with Maxwell-Boltzmann distribution of velocities at a particular temperature. They are treated as identical and distinguishable meaning that exchanging electrons between two energy levels with the identical number of electrons in those energy levels leads to a new state. Pauli's exclusion principle is not applicable to electrons so

that any number of electrons can populate an energy level, i.e. no upper boundary is imposed on the number of electrons in an energy level. Further, there is no interaction of electrons with ions except for collisions. Also, electrons do not exert any influence on each other. These declarations mean that the potential within the solid is uniform and constant averting any accumulation of electrons at a favored location.

7.6 Thermal Velocity of Electrons

From eq. (7.14)

$$v^2 = \frac{3k_B T}{m} \qquad (7.15)$$

or

$$v = v_{\text{Thermal}} = \sqrt{\frac{3k_B T}{m}} \qquad (7.16)$$

is the mean thermal velocity of electrons. At $T = 300$ K,

$$v_{\text{Thermal}} = \sqrt{\frac{3 \times 1.3806\,\text{J}\,\text{K}^{-1} \times 10^{-23} \times 300\,\text{K}}{9.109 \times 10^{-31}\,\text{kg}}}$$

$$= \sqrt{136.4079 \times 10^8} = 11.679 \times 10^4\ \text{m}\,\text{s}^{-1} = 1.17 \times 10^5\ \text{m}\,\text{s}^{-1} \qquad (7.17)$$

The random motion of electrons taking place at this velocity does not lead to any flow of current in a given direction. So, this velocity is a scalar quantity.

7.7 DC Electrical Conductivity of Metals

Let us take a metal wire of cross-sectional area A (Figure 7.3). Let the electron number density, i.e. number of electrons per unit

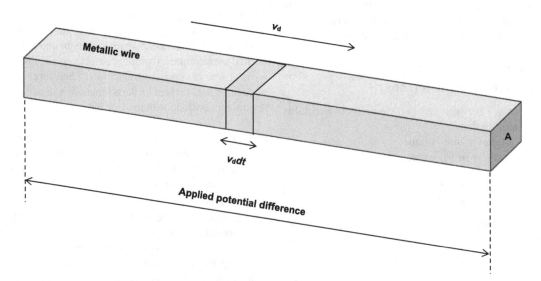

FIGURE 7.3 Current flowing in a metallic wire due to electrons moving with a velocity v_d.

volume in the wire be n. Suppose the electrons are moving in the wire with a uniform velocity v_d under the influence of an applied potential difference. Then

Distance traveled by electrons in time dt is

$$= \frac{\text{Distance}}{\text{Time}} \times \text{Time} = \text{Velocity} \times \text{Time} = v_d dt \quad (7.18)$$

Number of electrons crossing unit area in time dt

$$= \text{Number of electrons per unit volume} \times v_d dt = nv_d dt \quad (7.19)$$

So,

Number of electrons crossing area A in time dt $= nv_d A dt$ (7.20)

Further,

Charge carried by these electrons

$= \text{Charge on the electron} \times nv_d A dt$

$$= -e \times nv_d A dt = -nev_d A dt \quad (7.21)$$

Now

$$\text{Current carried by these electrons} = \frac{\text{Total charge}}{\text{Time}}$$

$$= -\frac{nev_d A dt}{dt} = -nev_d A \quad (7.22)$$

$$\therefore \text{Current density} = \mathbf{J} = \frac{\text{Current}}{\text{Area}} = -\frac{nev_d A}{A} = -nev_d \quad (7.23)$$

Suppose the voltage is applied to the metal wire for a time interval t. Then the electric field $\mathbf{E}$ in the wire (=applied voltage/length of the wire) accelerates the electrons.

$$E = \frac{\text{Force}}{\text{Charge}} = \frac{\mathbf{F}}{-e} \quad (7.24)$$

or

$$\mathbf{F} = -e\mathbf{E} \quad (7.25)$$

By Newton's second law of motion,

Force = Mass of the electron $(m) \times$ its acceleration (7.26)

or

$$\mathbf{F} = m\mathbf{a} = m\frac{dv_d}{dt} \quad (7.27)$$

From eqs. (7.25) and (7.27)

$$-e\mathbf{E} = m\frac{dv_d}{dt} \quad (7.28)$$

or

$$dv_d = -\frac{e\mathbf{E}}{m}dt \quad (7.29)$$

This equation is integrated to get v_d. However, the time limits of integration must be carefully selected. We cannot arbitrarily integrate from $t = 0$ to $t = \infty$. Obviously, we will get erroneous result because that will mean that the electron velocity is increasing indefinitely. The physical picture is different. As electrons move, they collide with the ions and the velocity after collision becomes zero. The new velocity is the velocity gained from zero value till the next collision. So the upper limit is prescribed by the time between collisions, which we shall recall is the relaxation time τ. Setting the time limits as $t = 0$ to $t = \tau$, we get

$$\int dv_d = -\int_0^\tau \frac{e\mathbf{E}}{m}dt = -\frac{e\mathbf{E}}{m}\int_0^\tau dt = -\frac{e\mathbf{E}}{m}|t|_0^\tau = -\frac{e\mathbf{E}\tau}{m} \quad (7.30)$$

or

$$v_d = -\frac{e\mathbf{E}\tau}{m} \quad (7.31)$$

During integration, the electric field $\mathbf{E}$ is assumed to be time-independent. Equation (7.31) is rewritten as

$$v_d = -\left(\frac{e\tau}{m}\right)\mathbf{E} = -\mu_n\mathbf{E} \quad (7.32)$$

where

$$\mu_n = \frac{e\tau}{m} \quad (7.33)$$

is the electron mobility. Since from eq. (7.32)

$$\mu_n = -\frac{v_d}{\mathbf{E}} \quad (7.34)$$

mobility of a charge carrier is defined as the drift velocity per unit applied electric field. Furthermore, eq. (7.23) together with eq. (7.31) yields the current density as

$$\mathbf{J} = -nev_d = -ne\left(-\frac{e\mathbf{E}\tau}{m}\right) = \frac{ne^2\tau}{m}\mathbf{E} = \sigma\mathbf{E} \quad (7.35)$$

where

$$\sigma = \frac{ne^2\tau}{m} \quad (7.36)$$

is the electrical conductivity. Equation (7.35) is written as

$$\mathbf{J} = \sigma\mathbf{E} \quad (7.37)$$

It represents the familiar Ohm's law.

Equation (7.36) is applied to semiconductors by taking into consideration the roles played by both electrons and holes in current flow. If m_n^* and m_p^* denote the effective masses of electrons and holes, n_n and n_h are their concentrations per unit volume, and τ_n and τ_p are their relaxation times, the conductivity σ_n due to electrons is

$$\sigma_n = \frac{n_n e^2 \tau_n}{m_n^*} \tag{7.38}$$

and the conductivity σ_p due to holes is

$$\sigma_p = \frac{n_p e^2 \tau_p}{m_p^*} \tag{7.39}$$

so that total conductivity σ of the semiconductor is

$$\sigma = \sigma_n + \sigma_p = \frac{n_n e^2 \tau_n}{m_n^*} + \frac{n_p e^2 \tau_p}{m_p^*} = e^2 \left(\frac{n_n \tau_n}{m_n^*} + \frac{n_p \tau_p}{m_p^*} \right) \tag{7.40}$$

From eq. (7.36)

$$\tau = \frac{m\sigma}{ne^2} \tag{7.41}$$

Conductivity of pure annealed copper is 5.8×10^7 S m^{-1}. From eq. (7.3), $n = 8.49 \times 10^{22}$ electrons/cm^3 = 8.49×10^{28} electrons/m^3. Hence, the relaxation time

$$\tau = \frac{m\sigma}{ne^2} = \frac{9.109 \times 10^{-31} \text{ kg} \times 5.8 \times 10^7 \text{ Sm}^{-1}}{8.49 \times 10^{28} \text{ electrons/m}^3 \times \left(1.602 \times 10^{-19} \text{ C} \right)^2}$$

$$= \frac{52.8322 \times 10^{-24}}{21.7888 \times 10^{10}} = 2.425 \times 10^{-14} \text{ s} \tag{7.42}$$

A related term is mean free path l of electrons

$$l = \text{Distance between two consecutive collisions} = v_{\text{Thermal}} \tau \tag{7.43}$$

For copper, eqs. (7.17), (7.42) and (7.43) give

$$l = 1.17 \times 10^5 \text{ ms}^{-1} \times 2.425 \times 10^{-14} \text{ s} = 2.837 \times 10^{-9} \text{ m} \tag{7.44}$$

where eq. (7.43) is applied. The value of l being comparable to interatomic distances appears to be reasonable.

Also, from eqs. (7.34) and (7.31)

$$\mu_n = -\frac{v_d}{E} = -\frac{-\dfrac{eE\tau}{m}}{E} = \frac{e\tau}{m} = \frac{1.602 \times 10^{-19} \text{ C} \times 2.425 \times 10^{-14} \text{ s}}{9.109 \times 10^{-31} \text{ kg}}$$

$$= \frac{3.8849 \times 10^{-33}}{9.109 \times 10^{-31}} = 0.4265 \times 10^{-2}$$

$$= 4.265 \times 10^{-3} \text{ m}^2 \text{ V}^{-1} \text{s}^{-1} \tag{7.45}$$

and from eq. (7.34), the drift velocity in an electric field of 1 V m^{-1} is

$$v_d = -\mu_n E = -4.265 \times 10^{-3} \text{ m}^2 \text{ V}^{-1} \text{s}^{-1} \times 1 \text{ V m}^{-1}$$

$$= -4.265 \times 10^{-3} \text{ ms}^{-1} \ll v_{\text{Thermal}} \left(1.17 \times 10^5 \text{ ms}^{-1} \right) \tag{7.46}$$

using eq. (7.17).

Thus the scenario inside the metallic conductor is contemplated as the superimposition of a much slower directed motion of electrons at velocity ~10^{-3}m s^{-1} over a much faster chaotic directionless motion of electrons at speed ~10^5m s^{-1} (Figure 7.4). The random speed is higher than directed velocity by a factor of 10^8. Then how does an electric bulb light instantaneously when switched on? The reason is the association of electromagnetic wave propagation with electron flow. The high velocity electromagnetic waves accompanying an electrical signal traversing a highly conducting wire propagate at the speed of light in the external medium, either air or some dielectric, while inside the wire the electrons travel at a much slower pace.

7.8 Drude's Equation of Motion of an Electron in an Electric Field

If the average velocity of the electrons at time t is $\mathbf{v}(t)$,

Average momentum of the electrons at time t is $\mathbf{p}(t) = m\mathbf{v}(t)$

$$\tag{7.47}$$

Let us calculate the average momentum of electrons after an infinitesimal time dt has elapsed. During the time interval dt, some of the electrons will experience collision, others will not be colliding. So the electrons are divided into two classes.

Since

Probability per unit time of electrons suffering collision is $\dfrac{1}{\tau}$

$$\tag{7.48}$$

hence,

Probability of electrons suffering collision during time interval dt is $\dfrac{dt}{\tau}$

$$\tag{7.49}$$

and

Probability of electrons not suffering collision during time interval dt is $\left(1 - \dfrac{dt}{\tau} \right)$

$$\tag{7.50}$$

Electrons of the first group, namely the ones encountering collision(s), will lose their momenta and emerge with new momenta in random directions. So, the total momentum resulting from this group of electrons sums up to zero and hence the average value is zero. This allows us to neglect the contribution of this group of electrons.

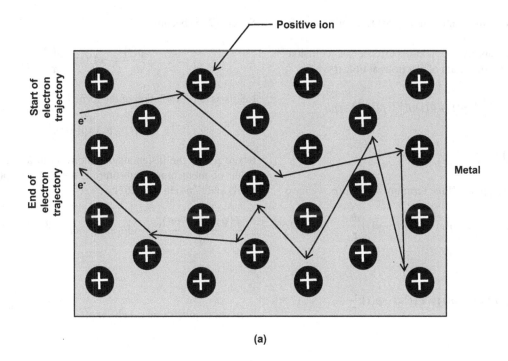

(a)

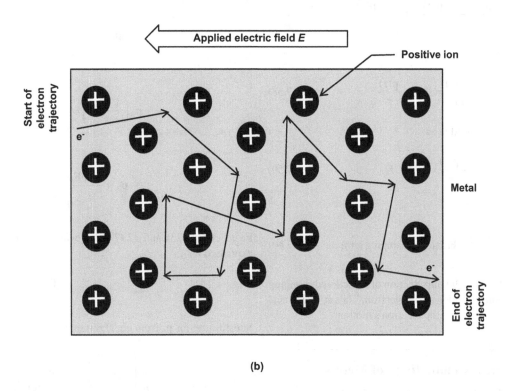

(b)

FIGURE 7.4 Trajectory of a conduction electron in the metallic solid: (a) in absence of an electric field (random motion only) and (b) under applied electric field (random + directed motion).

Electrons of the second group which survive up to time span dt without any collision gain velocity under the influence of the force $\mathbf{F}(t)$ due to the field. Since

$$\mathbf{F}(t) = \frac{\text{Change of momentum}}{\text{Time}} = \frac{\Delta \mathbf{p}(t)}{dt} \qquad (7.51)$$

or

$$\Delta \mathbf{p}(t) = \mathbf{F}(t)dt \qquad (7.52)$$

Consequently,

Momentum of an electron after time interval dt is

$$= \mathbf{p}(t) + \Delta \mathbf{p}(t) = \mathbf{p}(t) + \mathbf{F}(t)dt + 0 = \mathbf{p}(t) + \mathbf{F}(t)dt \qquad (7.53)$$

where zero value of momentum in the last term originates from electrons of the first group.

Since the probability of non-occurrence of collision is $(1 - dt/\tau)$, average momentum of electrons at time $(t+dt)$ is

$$\mathbf{p}(t+dt)=\left(1-\frac{dt}{\tau}\right)\{\mathbf{p}(t)+\mathbf{F}(t)dt\}=\mathbf{p}(t)+\mathbf{F}(t)dt$$

$$-\mathbf{p}(t)\frac{dt}{\tau}-\mathbf{F}(t)\frac{(dt)^2}{\tau} \quad (7.54)$$

Neglecting and dropping off the term involving $(dt)^2$, we have

$$\mathbf{p}(t+dt)=\mathbf{p}(t)+\mathbf{F}(t)dt-\mathbf{p}(t)\frac{dt}{\tau} \quad (7.55)$$

or

$$\mathbf{p}(t+dt)-\mathbf{p}(t)=\mathbf{F}(t)dt-\mathbf{p}(t)\frac{dt}{\tau} \quad (7.56)$$

or

$$\frac{\mathbf{p}(t+dt)-\mathbf{p}(t)}{dt}=\mathbf{F}(t)-\frac{\mathbf{p}(t)}{\tau} \quad (7.57)$$

In the limit $dt\to 0$,

$$\frac{d\mathbf{p}(t)}{dt}=\mathbf{F}(t)-\frac{\mathbf{p}(t)}{\tau} \quad (7.58)$$

This equation may be written as

$$\frac{d\mathbf{p}(t)}{dt}=\mathbf{F}(t)-\mathbf{F}_f \quad (7.59)$$

where

$$\mathbf{F}_f=\frac{\mathbf{p}(t)}{\tau}=\text{Frictional damping term} \quad (7.60)$$

The effect of collision is the introduction of a frictional damping term in the equation of motion of the electron. Thus collision acts as a frictional force damping the electron's motion.

7.9 AC Electrical Conductivity of Metals

If the applied field $\mathbf{E}$ is a time-dependent AC field with angular frequency ω,

$$\mathbf{E}(t)=\mathbf{E}_0\exp(-i\omega t) \quad (7.61)$$

where $\mathbf{E}(t)$ is the instantaneous electric field and $\mathbf{E}_0$ is the peak value of electric field.

Applying eq. (7.25), the time-varying force $\mathbf{F}(t)$ is written in terms of field $\mathbf{E}(t)$ as

$$\mathbf{F}(t)=-e\mathbf{E}(t) \quad (7.62)$$

So, eq. (7.58) becomes

$$\frac{d\mathbf{p}(t)}{dt}=-e\mathbf{E}(t)-\frac{\mathbf{p}(t)}{\tau} \quad (7.63)$$

Let us assume a solution of the form

$$\mathbf{p}(t)=\mathbf{p}_0\exp(-i\omega t) \quad (7.64)$$

where $\mathbf{p}(t)$ is the instantaneous momentum and $\mathbf{p}_0$ is the peak value of momentum. Substituting for $\mathbf{E}(t)$ and $\mathbf{p}(t)$ from eqs. (7.61) and (7.64) in eq. (7.63) we get

$$\frac{d\{\mathbf{p}_0\exp(-i\omega t)\}}{dt}=-e\mathbf{E}_0\exp(-i\omega t)-\frac{\mathbf{p}_0\exp(-i\omega t)}{\tau} \quad (7.65)$$

or

$$\mathbf{p}_0\exp(-i\omega t)\times-i\omega=-e\mathbf{E}_0\exp(-i\omega t)-\frac{\mathbf{p}_0\exp(-i\omega t)}{\tau} \quad (7.66)$$

or

$$-i\omega\mathbf{p}_0+\frac{\mathbf{p}_0}{\tau}=-e\mathbf{E}_0 \quad (7.67)$$

or

$$\mathbf{p}_0\left(\frac{1}{\tau}-i\omega\right)=-e\mathbf{E}_0 \quad (7.68)$$

or

$$\mathbf{p}_0=-\frac{e\mathbf{E}_0}{\dfrac{1}{\tau}-i\omega} \quad (7.69)$$

From eqs. (7.23) and (7.47), the peak current density can be expressed as

$$\mathbf{J}_0=-nev_d=-\frac{ne\mathbf{p}_0}{m} \quad (7.70)$$

Substituting for $\mathbf{p}_0$ from eq. (7.69) in eq. (7.70), we have

$$\mathbf{J}_0=\frac{ne}{m}\left(\frac{e\mathbf{E}_0}{\dfrac{1}{\tau}-i\omega}\right)=\frac{ne^2\mathbf{E}_0}{m\left(\dfrac{1-i\omega\tau}{\tau}\right)}$$

$$=\left(\frac{ne^2\tau}{m}\right)\frac{\mathbf{E}_0}{1-i\omega\tau}=\sigma_0\frac{\mathbf{E}_0}{1-i\omega\tau} \quad (7.71)$$

where we have put

$$\sigma_0=\text{DC conductivity}=\frac{ne^2\tau}{m} \quad (7.72)$$

from eq. (7.36) for σ. Equation (7.71) can be written as

$$\frac{\mathbf{J}_0}{\mathbf{E}_0} = \frac{\sigma_0}{1-i\omega\tau} \tag{7.73}$$

or

$$\sigma = \frac{\sigma_0}{1-i\omega\tau} \tag{7.74}$$

taking

$$\frac{\mathbf{J}_0}{\mathbf{E}_0} = \sigma \tag{7.75}$$

In the limit $\omega \to 0$, eq. (7.74) reduces to

$$\sigma = \frac{\sigma_0}{1-0} = \sigma_0 = \frac{ne^2\tau}{m} \tag{7.76}$$

as for the DC case.

7.10 Discussion and Conclusions

The road to the Drude model (1900) had several milestones en route, viz. the discovery of Ohm's law (1827), the Joule law of electrical heating (1841), the equipartition theorem (1845), and most importantly the discovery of electron in 1897. The Drude model explains fairly well the DC and AC conductivity of metals at room temperature and the thermal conductivity of metals. Reasonable resistivity values are obtained at room temperature. It works well particularly with the alkali and alkaline earth metals (Levi 2016). However electrical conductivities of alloys, semiconductors, and insulators are difficult to explain.

It partly elucidates the Wiedemann-Franz law according to which, for metals the ratio

$$\frac{\text{Electrical component of thermal conductivity}(K)}{\text{Thermal conductivity}(\sigma_{\text{Th}})}$$

$$= \text{Lorentz number}(L) \times \text{Absolute temperature}(T) \tag{7.77}$$

or

$$\frac{K}{\sigma_{\text{Th}}} = LT = \text{Constant} \tag{7.78}$$

The predicted value of the Lorentz number by the Drude model is half the experimental value. At low temperature due to difference in variation of electrical and thermal conductivities, the ratio is not constant whereas it should be constant by the Drude model.

It grossly over-estimates the electronic heat capacities of metals. According to the Drude model, insulators having no free electrons should have low heat capacity. But practically, their heat capacities are found to be of the same order as metals.

The Drude model confirms Hall's observation that magnetoresistance, the ratio of electric field E_x along the conductor to the current density j_x in the same direction, is independent of magnetic field H_z. More careful experiments have shown that

the magnetoresistance depends on magnetic field in some metals. Further, in Hall effect, the Hall coefficient determines the sign of charge carriers. By the Drude model, the Hall coefficient can only be negative. Positive Hall coefficient R_H shown by some metals is not explained by the Drude model.

Seebeck effect, ferromagnetism, blackbody radiation, photoelectric effect, and Crompton effect are not expounded. Theoretical paramagnetic susceptibility is higher than experimental value. Observed value of thermoelectric power at room temperature is 100 times smaller than that predicted by the Drude model. In Drude's calculation, electronic specific heat capacity c_v was 0.01 times off the actual value while mean square electron velocity v^2 was 100 times off, leading to annulment of errors from these parameters occurring in the expression for the ratio in Wiedemann-Franz law.

The electron mean free path calculated by the Drude model is typically the distance equal to lattice spacing showing that electrons scatter off positive ion cores. In practice, it can be several orders of magnitude higher ~10^8 times the lattice constant. The electrons are delocalized and are scattered by defects. They do not bump from the ion cores.

The Drude model is applicable where exact comprehension of scattering mechanism is not necessary, especially for electrical conductivity of spatially homogeneous time-dependent electric fields and static magnetic fields which are spatially uniform.

The reasons for the failure of the Drude model are the non-obeyance of Maxwell-Boltzmann statistics by electrons and non-inclusion of the interactions of electrons with themselves, with the lattice, and with the defects and impurities in the material. The Drude model is a free electron model and an independent electron model.

Discrepancies of the Drude model with experiments begin to dwindle when models are framed taking cognizance of these caveats (Dresselhaus et al 2018).

Illustrative Exercises

7.1 Where does the relation between average molecular kinetic energy and temperature come from? Apply it to calculate the thermal velocity of argon atoms at 27°C. Compare the thermal velocity of argon atoms with electron thermal velocity.

It comes from the kinetic theory of gases.

For argon, the molar mass is 39.948 g mol^{-1}. Since 1 mol of argon contains 6.022×10^{23} atoms, the

$$\text{Mass of 1 atom of argon} = \frac{39.948}{6.022 \times 10^{23}} = 6.63 \times 10^{-23} \text{ g}$$

$$= 6.63 \times 10^{-20} \text{ kg} \tag{7.79}$$

$$(v_{\text{Thermal}})_{\text{Argon}} = \sqrt{\frac{3k_B T}{m}} = \sqrt{\frac{3 \times 1.3806 \times 10^{-23} \times 300}{6.63 \times 10^{-20}}}$$

$$= \sqrt{187.41 \times 10^{-3}} = \sqrt{0.18741} = 0.433 \text{ m s}^{-1} \tag{7.80}$$

so that from eqs. (7.17) and (7.80), we have

$$\frac{\left(v_{\text{Thermal}}\right)_{\text{Argon}}}{\left(v_{\text{Thermal}}\right)_{\text{Electron}}} = \frac{0.433 \text{ m s}^{-1}}{1.17 \times 10^5 \text{ m s}^{-1}} = 3.7 \times 10^{-6} \quad (7.81)$$

times lower than electron thermal velocity, and this is expected to be so because of the relatively larger mass of argon atoms as compared to electrons.

7.2 Find (a) the number of electrons per unit volume, (b) relaxation time, (c) drift velocity in an applied electric field of 10 V m⁻¹, and (d) mobility of electrons in aluminum having atomic weight =26.981539 amu, valency = +3, density = 2.7 g cm⁻³, and electrical conductivity = 3.5 × 10⁷ S m⁻¹.

(a) The number of electrons per unit volume is

$$n = \frac{2.7 \times 6.022 \times 10^{23} \times 3}{26.981539} = \frac{2.7 \times 6.022 \times 3}{26.981539} \times 10^{23}$$

$$= 1.8078 \times 10^{23} \text{ cm}^{-3} = 1.8078 \times 10^{29} \text{ m}^{-3} \quad (7.82)$$

(b) Relaxation time is

$$\tau = \frac{m\sigma}{ne^2} = \frac{9.109 \times 10^{-31} \times 3.5 \times 10^7}{1.8078 \times 10^{29} \times \left(1.602 \times 10^{-19}\right)^2}$$

$$= \frac{9.109 \times 3.5 \times 10^{-24}}{1.8078 \times \left(1.602\right)^2 \times 10^{-9}} = 6.872 \times 10^{-15} \text{ s} \quad (7.83)$$

(c) Drift velocity is

$$v_{\text{d}} = -\frac{e\mathbf{E}\tau}{m} = -\frac{1.602 \times 10^{-19} \times 10 \times 6.872 \times 10^{-15}}{9.109 \times 10^{-31}}$$

$$= -\frac{1.602 \times 6.872}{9.109} \times 10^{-2} = 1.2086 \times 10^{-2} \text{ m s}^{-1} \quad (7.84)$$

(d) Mobility is

$$\mu_{\mathbf{n}} = \frac{e\tau}{m} = \frac{1.602 \times 10^{-19} \times 6.872 \times 10^{-15}}{9.109 \times 10^{-31}}$$

$$= 1.2086 \times 10^{-3} \text{ m}^2 \text{ V}^{-1} \text{ s}^{-1} \quad (7.85)$$

REFERENCES

Di Sia, P. 2012. Modelling at nanoscale. In: K. Y. Kim (Ed.), *Plasmonics - Principles and Applications*. IntechOpen: London, UK, pp. 1–22. https://www.intechopen.com/books/plasmonics-principles-and-applications/modelling-at-nanoscale.

Dresselhaus, M., G. Dresselhaus, S. B. Cronin and A. Gomes Souza Filho. 2018. Drude theory–Free carrier contribution to the optical properties. In: *Solid State Properties: From Bulk to Nano. Graduate Texts in Physics*. Springer: Berlin, Heidelberg, pp. 329–324.

Drude, P. 1900a. Zur elektronentheorie der metalle. *Annalen der Physik* 306(3): 566–613.

Drude, P. 1900b. Zur elektronentheorie der metalle; II. Teil. galvanomagnetische und thermomagnetische effecte. *Annalen der Physik* 308(11): 369–402.

Levi, A. F. J. 2016. Ch6: The Drude model. In: *Essential Classical Mechanics for Device Physics*. Morgan & Claypool Publishers: San Rafael, CA, Morgan & Claypool publication as part of IOP Concise Physics, pp. 6–1 to 6–20.

Lorentz H. A. 1909. *The Theory of Electrons and its Applications to the Phenomena of Light and Radiant Heat*. Columbia University Press: New York, pp. 356.

Sólyom, J. 2009. Free-electron model of metals. In: Sólyom, J. (Eds.), *Fundamentals of the Physics of Solids*, Vol. 2: Electronic Properties. Springer: Berlin, Heidelberg, pp. 1–76.

8

Sommerfeld Free Electron Fermi Gas Model

8.1 Strengthening Drude Model with Quantum Mechanics

Sommerfeld retained many ideas of the Drude model, mainly its free electron picture which assumes that a constant potential is maintained in the solid under which the electrons wander untied to any particular atom (Simon 2016). But he argued that electrons are not classical particles. They follow the laws of quantum mechanics and therefore must be addressed quantum mechanically, rather than classically. Pauli's exclusion principle strictly applies. So the Maxwell-Boltzmann distribution of electron velocities is invalidated. Instead, Fermi-Dirac statistics must be used so that the number of electrons in any energy level is prescribed by the Fermi-Dirac distribution function. The electrons are identical and indistinguishable so that swapping few electrons between two energy levels filled with equal number of electrons does not produce a new state. The electron population of an energy level is not unlimited but restricted by the number of states available at that energy level. Particles conforming to Fermi-Dirac statistics are called fermions, and so are electrons too. Electrons possess half-integral spins of ±1/2. Thus the Sommerfeld model is essentially the Drude model strengthened with quantum-mechanical Fermi-Dirac statistics.

8.2 Assumptions of the Sommerfeld Model

The main assumptions of the Sommerfeld model are as follows (Chambers 1990, Singleton 2012):

i. The free electron approximation is continued, in which electron-ion interaction is only through scattering, which is not necessarily via collision. But between any two scattering events, there is no other mechanism involved.

ii. The independent electron approximation is also continued, as in the Drude model.

iii. The relaxation time approximation: The electron-ion scattering probability is inversely proportional to the relaxation time τ denoting the average time between two such events.

iv. Pauli's exclusion principle must be adhered to, which is taken due care of by using Fermi-Dirac statistics.

8.3 Behavior of a Free Electron Gas in One Dimension

8.3.1 Wave Function with Box Boundary Conditions

When we think of a free electron, we imagine that the electron is not affected by external forces, which, in turn, indicates that the electron is under a constant potential V. We can consider this constant potential as zero, i.e. set $V = 0$ for simplification of analysis and treat the electron of mass m confined on a line of length L by infinite potential barriers. This problem is already solved in Section 5.2, using the fixed or box boundary condition

$$\Psi_n(0) = \Psi_n(L) = 0 \tag{8.1}$$

The wave functions are given by (eqs. (5.98), (5.99))

$$\Psi_n(x) = \sqrt{\frac{2}{L}} \sin\left(\frac{n\pi x}{L}\right) \tag{8.2}$$

and the eigenvalues by (eq. (5.100))

$$E_n = \frac{n^2 \pi^2 \hbar^2}{2mL^2} \tag{8.3}$$

where $n = 1, 2, 3, \ldots$

Since from eq. (5.84)

$$k = \frac{n\pi}{L} \tag{8.4}$$

and

$$k = \frac{2\pi}{\lambda} \tag{8.5}$$

where λ is the wavelength of the electron wave, so

$$\frac{2\pi}{\lambda} = \frac{n\pi}{L} \tag{8.6}$$

or

$$n = \frac{2L}{\lambda} = \frac{L}{\frac{\lambda}{2}} \tag{8.7}$$

meaning that the quantum number n expresses the number of half-wavelengths in the wave function

Also,

$$k^2 = \frac{n^2\pi^2}{L^2} \qquad (8.8)$$

$$\therefore E_n = \frac{\hbar^2 k^2}{2m} \qquad (8.9)$$

from eq. (8.3). The wave functions are standing waves carrying no current and all the permitted values of k are positive.

8.3.2 Wave Function with Periodic Boundary Conditions

It is more expedient to treat free electrons as traveling waves in place of standing waves by applying periodic or Born-von Karman boundary conditions pronouncing smoothly matching wave function at $x = L$ with $x = 0$:

$$\Psi(0) = \Psi(L) \text{ and } \left|\frac{\partial \Psi}{\partial x}\right|_{x=0} = \left|\frac{\partial \Psi}{\partial x}\right|_{x=L} \qquad (8.10)$$

Here exponential waves are used instead of sine waves satisfying

$$\exp(ik0) = \exp(ikL) = 1 = \exp(i2\pi n) \qquad (8.11)$$

so that

$$ikL = i2\pi n \qquad (8.12)$$

or

$$k = \frac{2\pi n}{L}; n = 0, \pm 1, \pm 2, \pm 3, \ldots \qquad (8.13)$$

giving

$$k = 0, \pm\frac{2\pi}{L}, \pm, \ldots \qquad (8.14)$$

Both signs of k are allowed leading to two degenerate states at each energy level carrying opposite signs of k, with the exception of $k = 0$. In the limit $L \to \infty$, the calculation is unaffected by the choice of boundary conditions, whether fixed or periodic.

In place of eq. (5.88)

$$\Psi_n(x) = A\sin\left(\frac{n\pi}{L}x\right) \qquad (8.15)$$

of Section 5.2.3, we write

$$\Psi_n(x) = P\exp\left(i\frac{2\pi n}{L}x\right) \qquad (8.16)$$

where P is a constant to be determined from the normalization condition.

8.3.3 Normalization of the Wave Function

The condition for normalization is

$$1 = \int_0^L \Psi_n^*(x)\Psi_n(x)dx$$

$$= \int_0^L P\exp\left(-i\frac{2\pi n}{L}x\right) \times P\exp\left(i\frac{2\pi n}{L}x\right)dx$$

$$= \int_0^L P^2\exp(0)dx = P^2\int_0^L dx = P^2(L-0) = P^2L \qquad (8.17)$$

or

$$1 = P^2L \qquad (8.18)$$

or

$$P^2 = \frac{1}{L} \qquad (8.19)$$

or

$$P = \frac{1}{\sqrt{L}} \qquad (8.20)$$

Equations (8.16) and (8.20) enable us to write

$$\Psi_n(x) = \frac{1}{\sqrt{L}}\exp\left(i\frac{2\pi n}{L}x\right) \qquad (8.21)$$

8.3.4 Quantum Numbers and Filling of Energy States

The solution of the wave equation for a system of one electron represents an orbital. In a linear solid, the orbital of a free electron is characterized by two quantum numbers: principal quantum number n, a positive integer and magnetic quantum number m_s, which has two possible values $\pm 1/2$ depending on the spin orientation. By Pauli's exclusion principle, no two fermions in a quantum state of a system can have all their quantum numbers identical. Therefore, each orbital can accommodate a maximum of two electrons, one with spin up orientation and the other with spin down orientation. Suppose N valence electrons are to be accommodated in the given quantum states. We start from the bottom state with lowest energy and proceed upward to reach the topmost state of highest energy. Due to the two-electron limit laid down by Pauli's exclusion principle, the topmost state is reached when

$$n = n_F = \frac{N}{2} \qquad (8.22)$$

where n_F is the value of n for the topmost energy state.

8.3.5 Fermi Energy

The idea of Fermi energy (not Fermi level) is a concept defined at absolute zero temperature as the difference of energy between the highest filled and the lowest filled energy states with the lowest state for a Fermi gas being taken as one having zero kinetic energy. With this definition, the Fermi energy ε_F is obtained from

$$\varepsilon_F = E_n \text{ for } n = n_F = \frac{N}{2} \qquad (8.23)$$

Hence, from eq. (8.3)

$$\varepsilon_F = \frac{n_F^2 \pi^2 \hbar^2}{2mL^2} = \frac{\left(\frac{N}{2}\right)^2 \pi^2 \hbar^2}{2mL^2} = \frac{\hbar^2}{2m}\left(\frac{N\pi}{2L}\right)^2 \qquad (8.24)$$

8.4 Free Electron Gas in Three Dimensions

8.4.1 Potential inside the Box, and the Schrodinger Equation

For an electron in a three-dimensional box of dimensions L_x, L_y, L_z, with potential $V = 0$ everywhere inside the box and infinite outside, we have

$$V(x,y,z) = 0 \quad \text{for } 0 < x < L_x, 0 < x < L_y, 0 < x < L_z \quad (8.25)$$

and

$$V(x,y,z) = \infty \text{ otherwise} \qquad (8.26)$$

The three-dimensional time-independent Schrodinger equation for the electron inside the box is

$$-\frac{\hbar^2}{2m}\left(\frac{\partial^2}{\partial x^2} + \frac{\partial^2}{\partial y^2} + \frac{\partial^2}{\partial z^2}\right)\Psi(x,y,z) + V\Psi(x,y,z)$$
$$= E\Psi(x,y,z) \qquad (8.27)$$

or

$$-\frac{\hbar^2}{2m}\left(\frac{\partial^2}{\partial x^2} + \frac{\partial^2}{\partial y^2} + \frac{\partial^2}{\partial z^2}\right)\Psi(x,y,z) + +0\times\Psi(x,y,z)$$
$$= E\Psi(x,y,z) \qquad (8.28)$$

or

$$-\frac{\hbar^2}{2m}\left(\frac{\partial^2}{\partial x^2} + \frac{\partial^2}{\partial y^2} + \frac{\partial^2}{\partial z^2}\right)\Psi(x,y,z) = E\Psi(x,y,z) \quad (8.29)$$

or

$$\left(\frac{\partial^2}{\partial x^2} + \frac{\partial^2}{\partial y^2} + \frac{\partial^2}{\partial z^2}\right)\Psi(x,y,z) = -\frac{2m}{\hbar^2}E\Psi(x,y,z) \quad (8.30)$$

or

$$\left(\frac{\partial^2}{\partial x^2} + \frac{\partial^2}{\partial y^2} + \frac{\partial^2}{\partial z^2}\right)\Psi(x,y,z) = -k^2\Psi(x,y,z) \quad (8.31)$$

where

$$k^2 = \frac{2mE}{\hbar^2} \qquad (8.32)$$

8.4.2 Factoring the Wave Function

If we speculate that the wave function $\Psi(x, y, z)$ is factorable into three independent components along the three spatial coordinates x, y, z, we have

$$\Psi(x,y,z) = X(x)Y(y)Z(z) \qquad (8.33)$$

so that eq. (8.31) reduces to

$$\left(\frac{\partial^2}{\partial x^2} + \frac{\partial^2}{\partial y^2} + \frac{\partial^2}{\partial z^2}\right)X(x)Y(y)Z(z)$$
$$= -k^2 X(x)Y(y)Z(z) \qquad (8.34)$$

or

$$\frac{\partial^2 X(x)}{\partial x^2}Y(y)Z(z) + \frac{\partial^2 Y(y)}{\partial y^2}X(x)Z(z)$$
$$+ \frac{\partial^2 Z(z)}{\partial z^2}X(x)Y(y) = -k^2 X(x)Y(y)Z(z) \qquad (8.35)$$

Dividing both sides of eq. (8.35) by $X(x)Y(y)Z(z)$, we get

$$\frac{1}{X(x)}\frac{\partial^2 X(x)}{\partial x^2} + \frac{1}{Y(y)}\frac{\partial^2 Y(y)}{\partial y^2} + \frac{1}{Z(z)}\frac{\partial^2 Z(z)}{\partial z^2} = -k^2 \quad (8.36)$$

Letting

$$k^2 = k_x^2 + k_y^2 + k_z^2 \qquad (8.37)$$

eq. (8.36) becomes

$$\frac{1}{X(x)}\frac{\partial^2 X(x)}{\partial x^2} + \frac{1}{Y(y)}\frac{\partial^2 Y(y)}{\partial y^2} + \frac{1}{Z(z)}\frac{\partial^2 Z(z)}{\partial z^2}$$
$$= -k_x^2 - k_y^2 - k_z^2 \qquad (8.38)$$

or

$$\left\{\frac{1}{X(x)}\frac{\partial^2 X(x)}{\partial x^2} + k_x^2\right\} + \left\{\frac{1}{Y(y)}\frac{\partial^2 Y(y)}{\partial y^2} + k_y^2\right\}$$
$$+ \left\{\frac{1}{Z(z)}\frac{\partial^2 Z(z)}{\partial z^2} + k_z^2\right\} = 0 \qquad (8.39)$$

Validity of this equation for all values of x, y, and z requires that each of the bracketed terms is independently zero leading to three independent equations in x, y, and z

$$\frac{1}{X(x)}\frac{\partial^2 X(x)}{\partial x^2} + k_x^2 = 0 \text{ where } k_x^2 = \frac{2mE_x}{\hbar^2} \qquad (8.40)$$

$$\frac{1}{Y(y)}\frac{\partial^2 Y(y)}{\partial y^2} + k_y^2 = 0 \text{ where } k_y^2 = \frac{2mE_y}{\hbar^2} \qquad (8.41)$$

and

$$\frac{1}{Z(z)}\frac{\partial^2 Z(z)}{\partial z^2}+k_z^2=0 \text{ where } k_z^2=\frac{2mE_z}{\hbar^2} \qquad (8.42)$$

Also,

$$E=E_x+E_y+E_z \qquad (8.43)$$

Solutions of the three differential equations (8.40), (8.41), and (8.42) are

$$X(x)=A_x\sin(k_x x)+B_x\cos(k_x x) \qquad (8.44)$$

$$Y(y)=A_y\sin(k_y y)+B_y\cos(k_y y) \qquad (8.45)$$

and

$$Z(z)=A_z\sin(k_z z)+B_z\cos(k_z z) \qquad (8.46)$$

where the constants A_x, B_x, A_y, B_y, A_z, and B_z are determined from the boundary conditions.

Considering eq. (8.44),

$$X(x)=0 \text{ at } x=0 \qquad (8.47)$$

we have

$$0=A_x\sin(k_x 0)+B_x\cos(k_x 0) \qquad (8.48)$$

Since $\cos 0 = 1$, B_x must be zero. Hence,

$$X(x)=A_x\sin(k_x x) \qquad (8.49)$$

Furthermore,

$$X(x)=0 \text{ at } x=L_x \qquad (8.50)$$

So,

$$A_x\sin(k_x L_x)=0 \qquad (8.51)$$

or

$$\sin(k_x L_x)=0 \qquad (8.52)$$

which holds when

$$k_x L_x=n_x\pi \qquad (8.53)$$

or

$$k_x=\frac{n_x\pi}{L_x} \text{ where } n_x=1,2,3,... \qquad (8.54)$$

8.4.3 Determination of Constants by Normalization of Wave Function

The constant A_x is determined by applying the normalization condition, as discussed in Section 5.2.4, eq. (5.97), obtaining

$$A_x=\sqrt{\frac{2}{L_x}} \qquad (8.55)$$

Thus from eqs. (8.49), (8.54) and (8.55)

$$X(x)=\sqrt{\frac{2}{L_x}}\sin\left(\frac{n_x\pi x}{L_x}\right) \qquad (8.56)$$

The boundary conditions for eq. (8.45) are

$$Y(y)=0 \text{ at } y=0 \qquad (8.57)$$

and

$$Y(y)=0 \text{ at } y=L_y \qquad (8.58)$$

giving together with the normalization condition

$$Y(y)=\sqrt{\frac{2}{L_y}}\sin\left(\frac{n_y\pi y}{L_y}\right) \text{ where } n_y=1,2,3,... \qquad (8.59)$$

The boundary conditions for eq. (8.46) are

$$Z(z)=0 \text{ at } z=0 \qquad (8.60)$$

and

$$Z(z)=0 \text{ at } z=L_z \qquad (8.61)$$

giving together with the normalization condition

$$Z(z)=\sqrt{\frac{2}{L_z}}\sin\left(\frac{n_z\pi z}{L_z}\right) \text{ where } n_z=1,2,3,... \qquad (8.62)$$

8.4.4 The Complete Wave Function

Combining eqs. (8.56), (8.59), and (8.62) with eq. (8.33), the wave function is found to be

$$\Psi(x,y,z)=\sqrt{\frac{2}{L_x}}\sin\left(\frac{n_x\pi x}{L_x}\right)\times\sqrt{\frac{2}{L_y}}\sin\left(\frac{n_y\pi y}{L_y}\right)\times\sqrt{\frac{2}{L_z}}\sin\left(\frac{n_z\pi z}{L_z}\right)$$

$$=\sqrt{\frac{2\times2\times2}{L_xL_yL_z}}\sin\left(\frac{n_x\pi x}{L_x}\right)\times\sin\left(\frac{n_y\pi y}{L_y}\right)\times\sin\left(\frac{n_z\pi z}{L_z}\right)$$

$$=\sqrt{\frac{8}{L_xL_yL_z}}\sin\left(\frac{n_x\pi x}{L_x}\right)\sin\left(\frac{n_y\pi y}{L_y}\right)\sin\left(\frac{n_z\pi z}{L_z}\right)$$

where $n_x=1,2,3,...$; $n_y=1,2,3,...$; $n_z=1,2,3,...$

$$(8.63)$$

which represents a standing wave. Further from Section 8.3.1, eq. (8.3), the eigenvalues are:
the x-component of E

$$E_x=\frac{n_x^2\pi^2\hbar^2}{2mL_x^2} \qquad (8.64)$$

the y-component of E

$$E_y = \frac{n_y^2 \pi^2 \hbar^2}{2mL_y^2} \tag{8.65}$$

and the z-component of E

$$E_z = \frac{n_z^2 \pi^2 \hbar^2}{2mL_z^2} \tag{8.66}$$

Combining eqs. (8.64), (8.65), and (8.66) with eq. (8.43)

$$E = \frac{n_x^2 \pi^2 \hbar^2}{2mL_x^2} + \frac{n_y^2 \pi^2 \hbar^2}{2mL_y^2} + \frac{n_z^2 \pi^2 \hbar^2}{2mL_z^2}$$

$$= \frac{\pi^2 \hbar^2}{2m} \left(\frac{n_x^2}{L_x^2} + \frac{n_y^2}{L_y^2} + \frac{n_z^2}{L_z^2} \right) \tag{8.67}$$

This equation may also be expressed as

$$E = \frac{\hbar^2}{2m} \left(\frac{\pi^2 n_x^2}{L_x^2} + \frac{\pi^2 n_y^2}{L_y^2} + \frac{\pi^2 n_z^2}{L_z^2} \right) = \frac{\hbar^2}{2m} \left(k_x^2 + k_y^2 + k_z^2 \right) \tag{8.68}$$

noting from eq. (8.4) that

$$\frac{\pi^2 n_x^2}{L_x^2} = k_x^2 \tag{8.69}$$

and similarly for k_y^2, k_z^2. Equation (8.68) can be stated as

$$E = \frac{\hbar^2}{2m} \left(k_x^2 + k_y^2 + k_z^2 \right) = \frac{\hbar^2 k^2}{2m} \tag{8.70}$$

8.4.5 Wave Function with Periodic Boundary Conditions

As for the one-dimensional case, it is customary to replace the box boundary conditions with periodic boundary conditions satisfying

$$\exp(ik_x 0) = \exp(ik_x L_x) = 1 \tag{8.71}$$

Since

$$\exp(i2\pi n_x) = \cos(2\pi n_x) + i\sin(2\pi n_x) = 1 \tag{8.72}$$

eq. (8.71) gives

$$\exp(ik_x L_x) = \exp(i2\pi n_x) \tag{8.73}$$

or

$$ik_x L_x = i2\pi n_x \tag{8.74}$$

or

$$k_x = \frac{2\pi n_x}{L_x}; n_x = 0, \pm 1, \pm 2, \pm 3, \ldots \tag{8.75}$$

In analogy to eq. (8.21), we can write

$$X(x) = P\exp(ik_x x) = P\exp\left(i\frac{2\pi n_x}{L_x} x \right) \tag{8.76}$$

where $P = \frac{1}{\sqrt{L_x}}$, $n_x = 0, \pm 1, \pm 2, \pm 3, \ldots$

$$Y(y) = Q\exp(ik_y y) = Q\exp\left(i\frac{2\pi n_y}{L_y} y \right) \tag{8.77}$$

where $Q = \frac{1}{\sqrt{L_y}}$, $n_y = 0, \pm 1, \pm 2, \pm 3, \ldots$

and

$$Z(z) = R\exp(ik_z z) = R\exp\left(i\frac{2\pi n_z}{L_z} z \right) \tag{8.78}$$

where $R = \frac{1}{\sqrt{L_z}}$, $n_z = 0, \pm 1, \pm 2, \pm 3, \ldots$

The solution to eq. (8.27) or its modified form, eq. (8.31) may be written as

$$\Psi(x,y,z) = P\exp\left(i\frac{2\pi n_x}{L_x} x \right) \times Q\exp\left(i\frac{2\pi n_y}{L_y} y \right) \times R\exp\left(i\frac{2\pi n_z}{L_z} z \right)$$

$$= PQR\exp\left(i\frac{2\pi n_x}{L_x} x \right)\exp\left(i\frac{2\pi n_y}{L_y} y \right)\exp\left(i\frac{2\pi n_z}{L_z} z \right)$$

$$= \frac{1}{\sqrt{L_x}}\frac{1}{\sqrt{L_y}}\frac{1}{\sqrt{L_z}}\exp\left(i\frac{2\pi n_x}{L_x} x \right)\exp\left(i\frac{2\pi n_y}{L_y} y \right)\exp\left(i\frac{2\pi n_z}{L_z} z \right)$$

$$= \frac{1}{\sqrt{L_x L_y L_z}}\exp\left(i\frac{2\pi n_x}{L_x} x \right)\exp\left(i\frac{2\pi n_y}{L_y} y \right)\exp\left(i\frac{2\pi n_z}{L_z} z \right)$$

$$= \frac{1}{\sqrt{\Omega}}\exp\left\{ i2\pi\left(\frac{n_x}{L_x} x + \frac{n_y}{L_y} y + \frac{n_z}{L_z} z \right) \right\} \tag{8.79}$$

where Ω is the volume of the box.

$$\Omega = L_x L_y L_z \tag{8.80}$$

The solution may be written in altered form in terms of k as

$$\Psi(x,y,z) = P\exp(ik_x x) \times Q\exp(ik_y y) \times R\exp(ik_z z)$$

$$= PQR\exp(ik_x x)\exp(ik_y y)\exp(ik_z z)$$

$$= \frac{1}{\sqrt{L_x L_y L_z}}\exp\left\{ i\left(k_x x + k_y y + k_z z \right) \right\}$$

$$= \frac{1}{\sqrt{\Omega}}\exp\left\{ i\left(k_x x + k_y y + k_z z \right) \right\}$$

$$= \frac{1}{\sqrt{\Omega}}\exp(i\mathbf{k}\cdot\mathbf{r}) \tag{8.81}$$

or

$$\Psi_k(\mathbf{r}) = \frac{1}{\sqrt{\Omega}} \exp(i\mathbf{k} \cdot \mathbf{r}) \qquad (8.82)$$

Since the linear momentum operator

$$\mathbf{p} = -i\hbar\nabla \qquad (8.83)$$

we can write from eq. (8.82)

$$\mathbf{p}\Psi_k(\mathbf{r}) = \mathbf{p}\left\{\frac{1}{\sqrt{\Omega}}\exp(i\mathbf{k}\cdot\mathbf{r})\right\} = -i\hbar\nabla\left\{\frac{1}{\sqrt{\Omega}}\exp(i\mathbf{k}\cdot\mathbf{r})\right\}$$

$$= -\frac{i\hbar}{\sqrt{\Omega}}\nabla\left\{\exp(i\mathbf{k}\cdot\mathbf{r})\right\} = -\frac{i\hbar}{\sqrt{\Omega}} \times i\mathbf{k}\exp(i\mathbf{k}\cdot\mathbf{r})$$

$$= -\frac{i^2\hbar}{\sqrt{\Omega}}\mathbf{k}\exp(i\mathbf{k}\cdot\mathbf{r}) = -\frac{(-1)\hbar}{\sqrt{\Omega}}\mathbf{k}\exp(i\mathbf{k}\cdot\mathbf{r})$$

$$= \hbar\mathbf{k} \times \frac{1}{\sqrt{\Omega}}\exp(i\mathbf{k}\cdot\mathbf{r}) = \hbar\mathbf{k}\Psi_k(\mathbf{r}) \qquad (8.84)$$

So, the plane wave $\Psi_k(\mathbf{r})$ is an eigenfunction of the linear momentum $\mathbf{p}$ with the eigenvalue $\hbar\mathbf{k}$.

8.4.6 The Momentum Space

The allowed values of $\mathbf{k}$ in the three directions can be combined to form three-dimensional wave vectors having components given by

$$k = \left(\frac{2\pi n_x}{L_x}, \frac{2\pi n_y}{L_y}, \frac{2\pi n_z}{L_z}\right) \text{ where } n_x, n_y, n_z = 0, \pm 1, \pm 2, \pm 3, \ldots$$

$$(8.85)$$

These points are plotted in a three-dimensional space. Three mutually perpendicular axes k_x, k_y, and k_z are chosen. This choice of axes forms a k-space. As the momentum $p = \hbar k$, this space is often referred to as the momentum space. In the ground state, a system of N electrons at 0 K occupies states that have lowest possible energies. As the states are filled with electrons according to the rules for fermions, the energy levels with lowest energies are first filled and this filling moves toward higher energies. The last available state is filled with the electron having the highest energy.

8.4.7 Fermi Energy and Related Terms

The energy corresponding to the last highest energy state thus filled up is the Fermi energy ε_F. An electron having an energy = Fermi energy propagates with a velocity known as the Fermi velocity v_F. The Fermi temperature T_F is defined as the temperature obtained by equating the thermal energy $k_B T_F$ with Fermi energy where k_B is the Boltzmann constant.

These filled states lie inside a sphere. The size of the sphere is determined by the highest energy electron state. For higher energy electrons, the radius of the sphere is larger. The radius of the sphere generated by filled states in k-space at 0 K is denoted by k_F. The k_F is called the Fermi wave vector. This sphere in k-space is termed the Fermi sphere (Figure 8.1) and its surface is

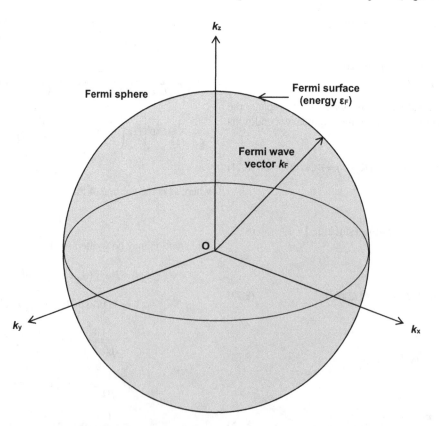

FIGURE 8.1   The Fermi sphere in the momentum space with axes k_x, k_y, k_z.

called the Fermi surface. So Fermi surface is the surface of the Fermi sphere in k-space defining the Fermi energy. The Fermi surface lies at the energy ε_F. Looking at the overall picture, the Fermi sphere is associated with a Fermi surface, Fermi energy ε_F, Fermi velocity v_F, and Fermi temperature T_F.

8.4.8 Filling of Energy States

The filling of the states in the sphere by electrons is like filling it with cells having dimensions of the three-dimensional box. The volume occupied by a cell of dimensions L_x, L_y, L_z in k-space is

$$V_{State} = \frac{2\pi}{L_x} \times \frac{2\pi}{L_y} \times \frac{2\pi}{L_z} = \frac{(2\pi)^3}{L_x L_y L_z} = \frac{(2\pi)^3}{\Omega} \quad (8.86)$$

Number of cells accommodated in a space of unit volume ($=1$) is

$$N_{Cell} = \frac{\text{Volume available}}{\text{Volume occupied by a state}} = \frac{1}{V_{State}}$$

$$= \frac{1}{\frac{(2\pi)^3}{\Omega}} = \frac{\Omega}{(2\pi)^3} \quad (8.87)$$

Since there are two values of spin quantum number for one allowed wave vector represented by the triplet set k_x, k_y, k_z, we must multiply by 2 to obtain the

Number of allowed states per unit volume $= 2N_{Cell} = 2\dfrac{\Omega}{(2\pi)^3}$

$$\quad (8.88)$$

To find the number of states in the Fermi sphere of radius k_F, we note that

$$\text{Volume of the Fermi sphere} = \frac{4}{3}\pi k_F^3 \quad (8.89)$$

Therefore,

Number of allowed states in the Fermi sphere

$$= 2\frac{\Omega}{(2\pi)^3} \times \text{Volume of the Fermi sphere}$$

$$= 2\frac{\Omega}{(2\pi)^3} \times \frac{4}{3}\pi k_F^3 = 2 \times \frac{\Omega}{8\pi^3} \times \frac{4}{3}\pi k_F^3 = \frac{\Omega}{3\pi^2} k_F^3 \quad (8.90)$$

For a system of N electrons,

$$\frac{\Omega}{3\pi^2} k_F^3 = N \quad (8.91)$$

or

$$k_F^3 = \frac{3\pi^2 N}{\Omega} \quad (8.92)$$

$$\therefore k_F = \left(\frac{3\pi^2 N}{\Omega}\right)^{1/3} \quad (8.93)$$

At the Fermi energy ε_F, eq. (8.70) is modified by putting $E = \varepsilon_F$ and $k = k_F$ to get

$$\varepsilon_F = \frac{\hbar^2 k_F^2}{2m} \quad (8.94)$$

Substituting for k_F from eq. (8.93) in eq. (8.94)

$$\varepsilon_F = \frac{\hbar^2}{2m} \left\{ \left(\frac{3\pi^2 N}{\Omega}\right)^{1/3} \right\}^2 = \frac{\hbar^2}{2m} \left(\frac{3\pi^2 N}{\Omega}\right)^{2/3} \quad (8.95)$$

For finding the Fermi temperature, we set

$$\varepsilon_F = k_B T_F \quad (8.96)$$

getting

$$T_F = \frac{\varepsilon_F}{k_B} = \frac{\hbar^2}{2mk_B} \left(\frac{3\pi^2 N}{\Omega}\right)^{2/3} \quad (8.97)$$

In quantum mechanics, an electron of energy ε represents a wave of linear frequency ν or angular frequency ω as

$$\varepsilon = h\nu = \frac{h}{2\pi} \times 2\pi\nu = \hbar \times 2\pi\nu = \hbar\omega \quad (8.98)$$

giving

$$\omega = \frac{\varepsilon}{\hbar} \quad (8.99)$$

In a dispersive medium, the angular frequency ω depends on $\mathbf{k}$. A wave packet is formed by superimposition of waves having different $\mathbf{k}$ values close to a given ω value. This wave packet has a velocity called the group velocity v_g. The group velocity is defined as

$$v_g = \frac{\partial\omega}{\partial k} \quad (8.100)$$

Substituting for ω from eq. (8.99) in eq. (8.100)

$$v_g = \frac{\partial}{\partial k}\left(\frac{\varepsilon}{\hbar}\right) = \frac{1}{\hbar}\frac{\partial\varepsilon}{\partial k} \quad (8.101)$$

Applying eq. (5.6) for energy of free electron

$$v_g = \frac{1}{\hbar}\frac{\partial}{\partial k}\left(\frac{\hbar^2 k^2}{2m}\right) = \frac{\hbar^2}{2m\hbar}\frac{\partial}{\partial k}(k^2) = \frac{\hbar^2}{2m\hbar}(2k) = \frac{\hbar k}{m} \quad (8.102)$$

To determine the Fermi velocity v_F, we put $k = k_F$ in eq. (8.102) to get

$$v_F = \frac{\hbar k_F}{m} \quad (8.103)$$

Substituting for k_F from eq. (8.93) in eq. (8.103),

$$v_F = \frac{\hbar}{m}\left(\frac{3\pi^2 N}{\Omega}\right)^{1/3} \quad (8.104)$$

8.4.9 Density of States

Density of states is defined as the number of states enclosed in unit volume per unit energy. It is found by writing eq. (8.95) as

$$\varepsilon = \frac{\hbar^2}{2m}\left(\frac{3\pi^2 N}{\Omega}\right)^{2/3} \qquad (8.105)$$

Rearranging eq. (8.105) and raising both sides to 3/2 power

$$\left(\frac{2m\varepsilon}{\hbar^2}\right)^{3/2} = \frac{3\pi^2 N}{\Omega} \qquad (8.106)$$

or

$$N = \frac{\Omega}{3\pi^2}\left(\frac{2m\varepsilon}{\hbar^2}\right)^{3/2} \qquad (8.107)$$

Then the density of states is

$$(N)_{3D} \equiv \frac{dN}{d\varepsilon} = \frac{d}{d\varepsilon}\left\{\frac{\Omega}{3\pi^2}\left(\frac{2m\varepsilon}{\hbar^2}\right)^{3/2}\right\} = \frac{\Omega}{3\pi^2}\left(\frac{2m}{\hbar^2}\right)^{3/2}\frac{d}{d\varepsilon}(\varepsilon)^{3/2}$$

$$= \frac{\Omega}{3\pi^2}\left(\frac{2m}{\hbar^2}\right)^{3/2}\times\frac{3}{2}(\varepsilon)^{3/2-1} = \frac{\Omega}{2\pi^2}\left(\frac{2m}{\hbar^2}\right)^{3/2}\varepsilon^{1/2} \qquad (8.108)$$

Figure 8.2 displays the graph between density of states $(N)_{3D}$ of a free electron gas and electron energy ε.

Another useful formula for density of states is derived by taking natural logarithm of both sides of eq. (8.107)

$$\ln(N) = \ln\left\{\frac{\Omega}{3\pi^2}\left(\frac{2m\varepsilon}{\hbar^2}\right)^{3/2}\right\} = \ln\left\{\frac{\Omega}{3\pi^2}\left(\frac{2m}{\hbar^2}\right)^{3/2}\varepsilon^{3/2}\right\}$$

$$= \ln\left\{\frac{\Omega}{3\pi^2}\left(\frac{2m}{\hbar^2}\right)^{3/2}\right\} + \ln\left(\varepsilon^{3/2}\right) = \text{Constant} + \ln\left(\varepsilon^{3/2}\right) \qquad (8.109)$$

By differentiation with respect to energy,

$$\frac{d}{d\varepsilon}\{\ln(N)\} = \frac{d}{d\varepsilon}\left\{\text{Constant} + \ln\left(\varepsilon^{3/2}\right)\right\} = 0 + \frac{d}{d\varepsilon}\left\{\ln\left(\varepsilon^{3/2}\right)\right\} \qquad (8.110)$$

or

$$\frac{d}{dN}\{\ln(N)\}\times\frac{dN}{d\varepsilon} = \left(\frac{3}{2}\right)\frac{1}{\varepsilon^{3/2}}\times\varepsilon^{3/2-1}$$

$$= \left(\frac{3}{2}\right)\frac{1}{\varepsilon^{3/2}}\times\varepsilon^{1/2} = \left(\frac{3}{2}\right)\frac{1}{\varepsilon^{3/2-1/2}} = \frac{3}{2\varepsilon} \qquad (8.111)$$

or

$$\frac{1}{N}\times\frac{dN}{d\varepsilon} = \frac{3}{2\varepsilon} \qquad (8.112)$$

giving

$$\frac{dN}{d\varepsilon} = \frac{3N}{2\varepsilon} \qquad (8.113)$$

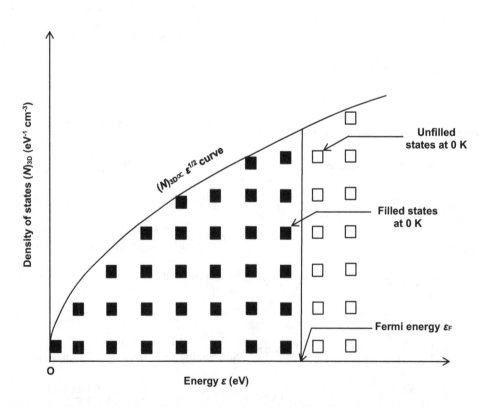

FIGURE 8.2 Plot of the density of states as a function of energy for a free electron gas in three dimensions.

Hence,

$$(N)_{3D} = \frac{dN}{d\varepsilon} = \frac{3N}{2\varepsilon} \qquad (8.114)$$

In particular, at Fermi energy ε_F,

$$\left|(N)_{3D}\right|_{\text{Fermi}} = \frac{3N}{2\varepsilon_F} \qquad (8.115)$$

where N is the three-dimensional electron density of the system. Thus

$$\text{Density of states at Fermi energy} = \frac{3}{2} \times \frac{\text{Electron density}}{\text{Fermi energy}}$$

$$(8.116)$$

8.5 Fermi Velocity versus Drift Velocity

8.5.1 Higher Value of Fermi Velocity than Random Thermal Velocity

The Fermi velocity in the Sommerfeld Fermi free electron gas model is equivalent to the thermal velocity in the Drude free electron model. Corresponding to any occupied momentum state **p** in the Fermi sphere, there exists a momentum state −**p**, so that the net momentum is zero, i.e. there is no resultant motion from the random motion of electrons, as in the concept of thermal velocity.

The Fermi velocity for copper is calculated in Illustrative Exercise 8.2 as 1.57×10^6 m s^{-1}. Comparing the Fermi velocity for copper = 1.57×10^6 m s^{-1} of the Sommerfeld model with the electron thermal velocity, determined in Section 7.6, viz., 1.17×10^5 m s^{-1}, of the Drude model, we note that the Fermi velocity is an order of magnitude higher than the thermal velocity. One can contemplate that in the absence of an applied electric field, the electrons in a metal are moving randomly in different directions at enormous velocities with an average value equal to the Fermi velocity.

8.5.2 Continuation of the Drift Velocity Concept

When an electric field is applied, the electrons start moving opposite to the direction of the field. The drift velocity concept describing the accelerated motion of electrons during a period τ and the related equation still holds although it was derived in the Drude model from classical concepts. The reason is obvious because τ was a measured parameter in the model. Recall from Section 7.7, eq. (7.42), that τ was calculated from the measured conductivity of the metal. Further, as determined previously, the drift velocity in copper is 4.265×10^{-3} m s^{-1}. So, the Fermi velocity is 3.68×10^8 times higher than drift velocity.

8.5.3 Relaxation Time Re-Interpretation

The different interpretations of τ in the two models need to be emphasized. In the Drude model, all the electrons contribute toward conduction. So, τ is the mean relaxation time considering all the electrons. However, in the Sommerfeld model, the most significant role in conduction is played by the electrons at the Fermi surface, and therefore τ is the relaxation time of electrons at the Fermi surface, not the averaged value for all electrons.

8.5.4 Dominant Role of the Small Number of Electrons near the Fermi Surface

If any current measurement is done as a function of energy at low temperature, it will be revealed that the net current at a particular instant is non-zero only within a few $k_B T$ of the Fermi energy at that instant, the quasi-Fermi energy. This revelation points to the fact that we need not be concerned about the dynamics of the full sea of conduction electrons. It suffices to look only into the dynamics of electrons in vicinity of the Fermi energy.

To understand why the current flows through electrons within a few $k_B T$ of the quasi-Fermi energy, we note that at equilibrium the Fermi sphere is symmetric about the origin (Figure 8.3). States toward the left and right sides of the origin, i.e. in the −x or +x directions, are filled with electrons up to the same energy. So the net current is zero.

When a field is applied, the Fermi sphere is shifted opposite to the direction of the field (Figure 8.4). The region deep inside the Fermi sphere is not affected. Assuming that the applied field is small, only the states in a distance +δk_F toward the direction of drift of the Fermi sphere, which were previously empty, are now filled with electrons. Likewise, the states in a distance −δk_F away from the direction of drift of the Fermi sphere, which were originally occupied, are now emptied of electrons. As a consequence, states in the region near +δk_F carrying current toward the direction of drift of the Fermi sphere are filled up to a higher energy ε^+. States in the region near −δk_F carrying current away from the direction of drift of the Fermi sphere are filled up to a lower energy ε^-. Equilibrium is disturbed and current flows in the energy range between ε^+ and ε^-. Moreover, only a small number of electrons adjoining ±δk_F have their energies altered. The electric field affects only the electrons within ±δk_F.

From the earlier discussion it is logical to state that although the electron is described by an averaged drift velocity v_d, in reality the electric field only moves a small number of electrons near −k_F and +k_F with Fermi velocity v_F. The current density equation (7.23) may therefore be modified as

$$J = e\left(n \frac{v_d}{v_F}\right) v_F \qquad (8.117)$$

where $n v_d / v_F$ is the total number of electrons moving with the Fermi velocity. Note that the total number of electrons has been decreased by the factor v_d / v_F. Although the current density equation reduces to the same as in the Drude model, its physical explanation is different. So the formulae for conductivity and mobility are also same.

8.5.5 Electron Mean Free Path Determination

For copper, the relaxation time was determined as 2.425×10^{-14} s and the mean free path was 2.837×10^{-9} m as determined in Section 7.7, eqs. (7.42) and (7.44) from the Drude model. This mean free path was calculated using thermal

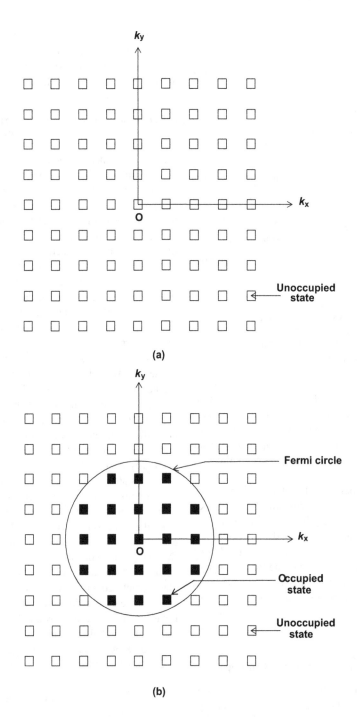

FIGURE 8.3 Electronic states and the Fermi circle: (a) representation of empty electronic states in a bounded region of the two-dimensional momentum space by unfilled squares, (b) drawing the Fermi circle at equilibrium (in the absence of electric field) when it is symmetrically placed around the origin and the states within the circle are filled at 0 K and those outside the circle are empty.

velocity $= 1.17 \times 10^5$ m s^{-1}. Instead of thermal velocity, we use the Fermi velocity (Illustrative exercise 8.2(e)) to get the mean free path

$$l = v_F \tau = 1.573 \times 10^6 \text{ m s}^{-1} \times 2.425 \times 10^{-14} \text{ s} = 3.81 \times 10^{-8} \text{ m} \tag{8.118}$$

Since the lattice constant of copper is 3.6147×10^{-10} m

$$l_{\text{Drude}} = \frac{2.837 \times 10^{-9} \text{ m}}{3.6147 \times 10^{-10} \text{ m}} = 7.85 \text{ lattice constants} \tag{8.119}$$

$$l_{\text{Sommerfeld}} = \frac{3.81 \times 10^{-8} \text{ m}}{3.6147 \times 10^{-10} \text{ m}} = 105.4 \text{ lattice constants} \tag{8.120}$$

These calculations show that using the Drude model, the collisions of electrons with lattice are very frequent due to the shorter mean free path, i.e. 8 lattice constants but when the Sommerfeld model is applied, they occur less often owing to the longer mean free path, i.e. 105 lattice constants. This gives a more realistic picture

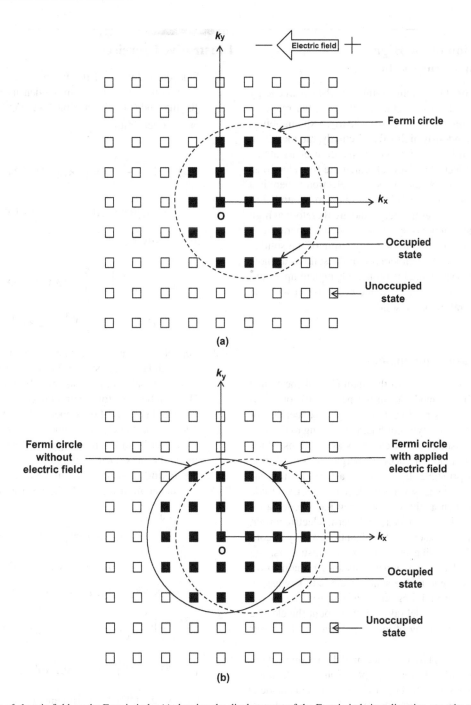

FIGURE 8.4 Effect of electric field on the Fermi circle: (a) showing the displacement of the Fermi circle in a direction opposite to that of electric field together with its filled states and (b) showing the relative offset of Fermi circle positions in the presence of electric field with respect to its position in the absence of electric field which was drawn in Figure 8.3(b).

of the electron scattering because scattering does not take place in a perfect lattice but at lattice imperfections, due to impurities or lattice vibrations (phonons). Quantum-mechanically, we shall see that the electron can propagate unscattered through a perfect periodic lattice. The effect of lattice is duly accounted for by using an effective mass in place of the actual mass of electron. From these discussions, a longer mean free path for electrons is closer toward reality than the shorter one.

Electron mobility is highly temperature-dependent. The higher the temperature, the greater the phonon scattering. The phonon scattering can be suppressed by lowering the temperature.

At liquid helium temperature (4.2 K), the electron relaxation time in copper diminishes to 1×10^{-9} s, so from eq. (8.118)

$$l_{\text{Sommerfeld}} = 1.573 \times 10^6 \text{ ms}^{-1} \times 1 \times 10^{-9} \text{ s} = 1.573 \times 10^{-3} \text{ m}$$

$$= 0.0016 \text{ m} = \frac{0.0016 \text{ m}}{3.6147 \times 10^{-10} \text{ m}} \text{ lattice constants}$$

$$= 4.43 \times 10^6 \text{ lattice constants}$$

$$(8.121)$$

8.6 Reconciliation of the High Fermi Temperature Value

A material is at 0 K or at room temperature and the Fermi energy is a few eV, but the Fermi temperature can be in the range of thousands or even tens of thousands. Clearly, explanation is needed for this tremendously high value. Actually, the measured temperature of a solid is the average temperature of its atoms and the constituent electrons, a few of which are free electrons, the focus of interest here. Among these free electrons, there is a distribution of energies. So a small fraction of the total number of free electrons possess the Fermi energy and are therefore at high temperature. But the number of electrons at high temperature is very small and so their effect on the temperature of the solid is insignificant. So, the average temperature of the solid as a whole is still low. From this viewpoint, the immensely high temperature value does not appear to be amazing, and is consistent with the low measured temperature of the solid.

8.7 Discussion and Conclusions

The Sommerfeld model adheres to the main core of the Drude model but relooks at the model from the perspective of quantum mechanics. It applies modifications such as conception of electrons in a particle-a-box type confinement leading to discrete energy levels and consideration of electrons as fermions which obey Fermi-Dirac statistics and occupy energy states according to Pauli's exclusion principle. By replacing a free electron gas by a free electron Fermi gas, several inadequacies of the Drude model are removed such as those concerning the Wiedemann-Franz law, the electronic specific heat, and thermoelectric power. At the same time glaring issues like the gigantic diversity in conductivity of materials behaving as conductors and insulators, and magnetic field dependence of magnetoresistance remain unresolved. This strongly suggests that some crucial and key aspect of the phenomenon is still missing, and this concerns the contribution of positive ions. A model taking into account the electric field produced by a static array of ions is expected to be closer to reality.

The drift velocity concept of the Drude model is retained in the Sommerfeld model with different interpretations of relaxation time as that of electrons near the Fermi surface. The current density equation is also same as in the Drude model but has a different physical explanation. Mobility and conductivity formulae are identical in the two models. But the calculation of relaxation time using the Fermi velocity instead of thermal velocity provides a more realistic picture of electron scattering in which the scattering events are appreciably less frequent than in the Drude model with correspondingly longer mean free paths.

Justification for the monumentally high value of Fermi temperature is that only an extremely small number of electrons possess Fermi energy and the resultant high temperature so that the average temperature value remains reasonable.

Illustrative Exercises

8.1 Find the radius k_F of the Fermi sphere for: (a) beryllium with free electron number density $(N/\Omega) = 24.7 \times 10^{28}\,\mathrm{m}^{-3}$ and (b) silver having $N/\Omega = 5.86 \times 10^{28}\,\mathrm{m}^{-3}$.

(a) For beryllium,

$$(k_F)_{Be} = \left(\frac{3\pi^2 N}{\Omega}\right)^{1/3} = \left\{3 \times (3.14)^2 \times 24.7 \times 10^{28}\right\}^{1/3}$$

$$= \left(7.30596 \times 10^{30}\right)^{1/3} = 1.94 \times 10^{10}\,\mathrm{m}^{-1} \tag{8.122}$$

(b) For silver

$$(k_F)_{Be} = \left(\frac{3\pi^2 N}{\Omega}\right)^{1/3} = \left\{3 \times (3.14)^2 \times 5.86 \times 10^{28}\right\}^{1/3}$$

$$= \left(1.73332 \times 10^{30}\right)^{1/3} = 1.2 \times 10^{10}\,\mathrm{m}^{-1} \tag{8.123}$$

8.2 Copper has a face-centered cubic (FCC) crystal structure with lattice constant = $3.6147 \times 10^{-10}\,\mathrm{m}$ (see reference: General, atomic, and crystallographic...copper). The number of atoms per unit cell is 4. It has two valence states: 2, 1. For copper with the valence state 1, calculate (a) the electron concentration, (b) the Fermi wave vector, (c) Fermi energy, (d) Fermi temperature, and (e) Fermi velocity.

(a) Taking 1 valence electron per atom, since there are four atoms in a unit cell, the electron concentration is

$$\frac{N}{\Omega} = \frac{1 \times 4}{\left(3.6147 \times 10^{-10}\right)^3} = \frac{4}{47.23 \times 10^{-30}}$$

$$= 0.08469 \times 10^{30} = 8.469 \times 10^{28}\,\mathrm{m}^{-3} \tag{8.124}$$

(b) By eq. (8.93), the Fermi wave vector is

$$k_F = \left(\frac{3\pi^2 N}{\Omega}\right)^{1/3} = \left\{3 \times (3.14159)^2 \times 8.469 \times 10^{28}\right\}^{1/3}$$

$$= \left(2.50757 \times 10^{30}\right)^{1/3} = 1.3586 \times 10^{10}\,\mathrm{m}^{-1} \tag{8.125}$$

(c) By eq. (8.94), the Fermi energy is

$$\varepsilon_F = \frac{\hbar^2 k_F^2}{2m} = \frac{\left(1.05457 \times 10^{-34}\,\mathrm{Js}\right)^2 \times \left(1.3586 \times 10^{10}\,\mathrm{m}^{-1}\right)^2}{2 \times 9.109 \times 10^{-31}\,\mathrm{kg}}$$

$$= \frac{1.112 \times 10^{-68} \times 1.846 \times 10^{20}}{1.822 \times 10^{-30}} = 1.1266 \times 10^{-18}\,\mathrm{J}$$

$$= \frac{1.1266 \times 10^{-18}}{1.602 \times 10^{-19}}\,\mathrm{eV} = \frac{11.266}{1.602} = 7.032\,\mathrm{eV}$$

$$\tag{8.126}$$

(d) By eq. (8.97), the Fermi temperature is

$$T_F = \frac{\varepsilon_F}{k_B} = \frac{7.032 \text{ eV}}{8.617 \times 10^{-5} \text{ eV K}^{-1}} = 0.81606 \times 10^5 \text{K}$$

$$= 81606 \text{ K} \qquad (8.127)$$

(e) By eq. (8.103), the Fermi velocity is

$$v_F = \frac{\hbar k_F}{m} = \frac{1.05457 \times 10^{-34} \text{ Js} \times 1.3586 \times 10^{10} \text{ m}^{-1}}{9.109 \times 10^{-31} \text{ kg}}$$

$$= 0.1573 \times 10^7 \text{ m s}^{-1} = 1.573 \times 10^6 \text{ m s}^{-1} \qquad (8.128)$$

REFERENCES

Chambers, R. G. 1990. *Electrons in Metals and Semiconductors.* Springer: Netherlands, pp. 1–15.

General, Atomic and Crystallographic Properties and Features of Copper. (Feb. 1992). Properties of Copper and Copper Alloys at Cryogenic Temperatures, by N. J. Simon, E. S. Drexler, and R. P. Reed, National Institute of Standards and Technology Monograph 177 U.S. Government Printing Office, Washington, DC, 850 pages, https://www.copper.org/resources/properties/atomic_properties.html. Accessed 19 March 2019

Simon, S. H. 2016. *The Oxford Solid State Basics.* Oxford University Press: Oxford, UK, pp. 1–38.

Singleton, J. 2012. *Band Theory and Electronic Properties of Solids.* Oxford University Press: Oxford, UK, pp. 1–15.

9

Kronig-Penney Periodic Potential Model

9.1 From Particle-in-a-Box to Particle-in-a-Periodic Lattice

In the previous two chapters, electron transport models in solids were studied, the first model based on purely classical concepts and the second model on semi-classical concepts because in the second model quantum mechanics was applied but still there were some lacunae, which needed to be bridged. Actually, both the models assumed that the potential in the solid was uniform and constant. When we considered a particle-in-a-box model, we assumed that potential inside the box is uniform and equal to zero. Here lies the crux of the problem. An electron traveling in a solid approaches a positive ion, crosses the ion, and recedes away from it; then it moves across another positive ion in a similar way, and this process is repeatedly followed. Thus instead of a particle in a box, a closer analogy to reality is a particle in a periodic lattice or particle in an array of potential barriers/wells.

The problem is complex and three-dimensional (3D). The crystal potential is created by the positive ions in the lattice. When the electron is near the positive ion, it is in the well region where the potential is assumed as zero. But when the electron is in a region between two ions, it is in the barrier region where the potential has a finite value. The width of the well equals the lattice spacing, which can be taken to extend from an ion placed at the center of one well to the ion placed at the center of next well.

9.2 Simplification of the Problem

In a gross oversimplification, all the wells and barriers may be assumed to be equispaced. All the wells may be assumed to be of the same width and depth. All the barriers may too be assumed to be having the same width and height. The 3D problem may be analyzed by considering a one-dimensional periodic potential. The Kronig-Penney model is a simple, one-dimensional idealized representation of the periodic potential in a crystal lattice in the form of an infinite array of square potential barriers formed by a linear array of atoms (Kronig and Penney 1931, Kittel 2005). The advantage derived by simplicity is that the solution of the Schrodinger equation becomes tractable and the clear quantum-mechanical picture demonstrating the formation of energy bands and forbidden energy gaps in solids emerges. Electrons and ions are assumed to interact purely by Coulombic forces. Thermal vibration of lattice atoms is ignored. Electron-electron interactions are not taken into account. So are the electron collisions with lattice or effects of presence of impurities in the solid.

The Kronig-Penney model serves as a very useful pedagogical or instructive tool to show the characteristic behavior of electrons in a periodic potential.

9.3 The Periodic Potential

The potential assumed in the model is expressed as (Mcquarrie 1996)

$$V(x) = 0 \quad \text{for } 0 < x < a \text{ (Region I)} \tag{9.1}$$

and

$$V(x) = V_0 \quad \text{for} -b < x < 0 \text{ (Region II)} \tag{9.2}$$

The potential is shown in Figure 9.1.

Since the well region extends up to a distance a and the barrier region extends up to a distance $-b$, the period of the potential is

$$c = a - (-b) = a + b \tag{9.3}$$

9.4 Schrodinger Equations for Regions I and II, and Their Solutions

The time-independent Schrodinger equation

$$-\frac{\hbar^2}{2m}\frac{\partial^2 \Psi(x)}{\partial x^2} + V(x)\Psi(x) = E\Psi(x) \tag{9.4}$$

is written for regions I and II. In region I, $V(x)=0$; so

$$-\frac{\hbar^2}{2m}\frac{\partial^2 \Psi(x)}{\partial x^2} + 0 \times \Psi(x) = E\Psi(x) \tag{9.5}$$

or

$$-\frac{\partial^2 \Psi(x)}{\partial x^2} = \frac{2m}{\hbar^2}E\Psi(x) \tag{9.6}$$

or

$$\frac{\partial^2 \Psi(x)}{\partial x^2} = -\frac{2mE}{\hbar^2}\{\Psi(x)\} \tag{9.7}$$

Put

$$\frac{2mE}{\hbar^2} = \alpha^2 \tag{9.8}$$

or

$$\alpha = \sqrt{\frac{2mE}{\hbar^2}} \tag{9.9}$$

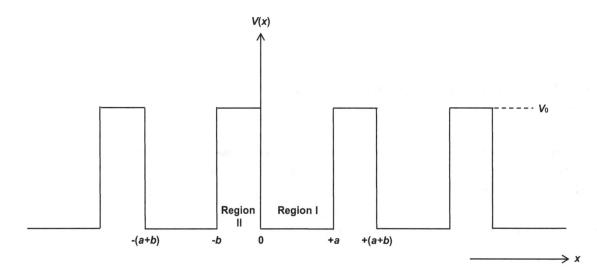

FIGURE 9.1   The periodic potential assumed in the Kronig-Penney model.

Then

$$\frac{\partial^2 \Psi(x)}{\partial x^2} = -\alpha^2 \{\Psi(x)\} \qquad (9.10)$$

This is the same as eq. (5.5) and has the solution (eq. (5.14))

$$\Psi_{\mathrm{I}}(x) = A\exp(i\alpha x) + B\exp(-i\alpha x) \qquad (9.11)$$

where A and B are constants.

In region II, $V(x) = V_0$; so the Schrodinger equation is

$$-\frac{\hbar^2}{2m}\frac{\partial^2 \Psi(x)}{\partial x^2} + V_0\Psi(x) = E\Psi(x) \qquad (9.12)$$

or

$$-\frac{\partial^2 \Psi(x)}{\partial x^2} + \frac{2mV_0}{\hbar^2}\Psi(x) = \frac{2mE}{\hbar^2}\Psi(x) \qquad (9.13)$$

or

$$-\frac{\partial^2 \Psi(x)}{\partial x^2} + \frac{2mV_0}{\hbar^2}\Psi(x) - \frac{2mE}{\hbar^2}\Psi(x) = 0 \qquad (9.14)$$

or

$$-\frac{\partial^2 \Psi(x)}{\partial x^2} + \frac{2m}{\hbar^2}\{V_0\Psi(x) - E\Psi(x)\} = 0 \qquad (9.15)$$

or

$$\frac{\partial^2 \Psi(x)}{\partial x^2} - \frac{2m}{\hbar^2}\{V_0\Psi(x) - E\Psi(x)\} = 0 \qquad (9.16)$$

or

$$\frac{\partial^2 \Psi(x)}{\partial x^2} - \frac{2m(V_0 - E)}{\hbar^2}\Psi(x) = 0 \qquad (9.17)$$

Put

$$\frac{2m(V_0 - E)}{\hbar^2} = \beta^2 \qquad (9.18)$$

or

$$\beta = \sqrt{\frac{2m(V_0 - E)}{\hbar^2}} \qquad (9.19)$$

It is assumed that $E < V_0$ so that β is real. Then

$$\frac{\partial^2 \Psi(x)}{\partial x^2} - \beta^2\Psi(x) = 0 \qquad (9.20)$$

which is the same as eq. (5.132) and has the solution (eq. (5.140))

$$\Psi_{\mathrm{II}}(x) = C\exp(\beta x) + D\exp(-\beta x) \qquad (9.21)$$

where C and D are constants.

9.5 Introducing Bloch Theorem by Symmetry Analysis

For solving the wave equation in a periodic crystal lattice, it is necessary to introduce Bloch's theorem. This can be done either by the more simple and intuitive symmetry analysis or the more difficult Fourier analysis, which is albeit convertible to matrix format and therefore amenable to computer simulations. We pursue the symmetry analysis path.

A translation operator $\hat{T}$ is defined that shifts a function $y(x)$ by c to convert it into $y(x+c)$:

$$\hat{T}y(x) = y(x+c) \qquad (9.22)$$

When we apply the translation operator to the Schrodinger equation, we get

$$\hat{T}\hat{H}\Psi(x) = -\frac{\hbar^2}{2m}\frac{\partial^2 \Psi(x+c)}{\partial(x+c)^2} + V(x+c)\Psi(x+c) = E\Psi(x+c)$$

$$(9.23)$$

So we can write

$$\hat{T}\hat{H}\Psi(x) = \hat{H}\Psi(x+c) \qquad (9.24)$$

But from eq. (9.22)

$$\hat{T}\Psi(x) = \Psi(x+c) \qquad (9.25)$$

Substituting for $\Psi(x+c)$ from eq. (9.25) in eq. (9.24), we have

$$\hat{T}\hat{H}\Psi(x) = \hat{H}\Psi(x+c) = \hat{H}\hat{T}\Psi(x) \qquad (9.26)$$

or

$$\hat{T}\hat{H}\Psi(x) = \hat{H}\hat{T}\Psi(x) \qquad (9.27)$$

or

$$\left(\hat{T}\hat{H} - \hat{H}\hat{T}\right)\Psi(x) = 0 \qquad (9.28)$$

which means that the operators $\hat{T}, \hat{H}$ have the same eigenfunction $\Psi(x)$. This allows us to use the eigenfunction of the Hamiltonian operator $\hat{H}$ to determine the eigenvalue of the translation operator $\hat{T}$. Suppose the eigenvalue of the translation operator is γ. Then we can write using eq. (9.22)

$$\hat{T}\Psi(x) = \Psi(x+c) = \gamma\Psi(x) \qquad (9.29)$$

Applying the translation operator on the function $\Psi(x)$ two times we get

$$\hat{T}^2\Psi(x) = \hat{T}\hat{T}\Psi(x) = \hat{T}\Psi(x+c) = \Psi(x+2c) = \gamma^2\Psi(x) \quad (9.30)$$

By n times application of the translation operator on the function $\Psi(x)$ we have

$$\hat{T}^n\Psi(x) = \Psi(x+nc) = \gamma^n\Psi(x) \qquad (9.31)$$

The Born-von Karman boundary condition is applied. Here the lattice is imagined as a one-dimensional ring of N atoms. Then

$$\Psi(x+Nc) = \Psi(x) \qquad (9.32)$$

From eq. (9.31),

$$\Psi(x+Nc) = \gamma^N\Psi(x) \qquad (9.33)$$

Looking at eqs. (9.32) and (9.33), we have

$$\gamma^N\Psi(x) = \Psi(x) \qquad (9.34)$$

giving

$$\gamma^N = 1 \qquad (9.35)$$

or

$$\gamma = (1)^{1/N} \qquad (9.36)$$

The root of unity is a complex number which yields unity when raised to the power N, where N is an integer. It is called a de Moivre number. To find γ, we write

$$\gamma^N = 1 = \cos 0 + i\sin 0 = \cos 2l\pi + i\sin 2l\pi \text{ where } l \text{ is an integer} \qquad (9.37)$$

Taking $1/N$th power of both sides

$$\left(\gamma^N\right)^{1/N} = \left(\cos 2l\pi + i\sin 2l\pi\right)^{1/N} \qquad (9.38)$$

or

$$\gamma = \left(\cos 2l\pi + i\sin 2l\pi\right)^{1/N} \qquad (9.39)$$

By de Moivre's theorem

$$\gamma = \cos\frac{2l\pi}{N} + i\sin\frac{2l\pi}{N} \text{ where } l = 0, 1, 2, 3, \dots (N-1) \qquad (9.40)$$

But

$$\cos\frac{2l\pi}{N} + i\sin\frac{2l\pi}{N} = \exp\left(i\frac{2l\pi}{N}\right) \qquad (9.41)$$

So,

$$\gamma = \exp\left(i\frac{2l\pi}{N}\right) \qquad (9.42)$$

Let us define

$$k = \frac{2l\pi}{Nc} \qquad (9.43)$$

Then

$$i\frac{2l\pi}{N} = \frac{2l\pi}{Nc} \times ic = ikc \qquad (9.44)$$

Hence,

$$\gamma = \exp(ikc) \qquad (9.45)$$

From eqs. (9.29) and (9.45), we straightway write

$$\Psi(x+c) = \gamma\Psi(x) = \exp(ikc)\Psi(x) \qquad (9.46)$$

which gives us an important result

$$\Psi(x+c) = \exp(ikc)\Psi(x) \qquad (9.47)$$

Equation (9.47) is referred to as the Bloch condition.

Now any function satisfying eq. (9.47) can be expressed in the form

$$\Psi(x) = \exp(ikx).u(x) \qquad (9.48)$$

Here $u(x)$ has the same periodicity as the lattice, i.e. the function $u(x)$ is a periodic function of x having period c,

$$u(x) = u(x+c) \qquad (9.49)$$

For verification of truth of this statement, we write

$$\Psi(x+c) = \exp\{ik(x+c)\} \cdot u(x+c) = \exp(ikx) \cdot \exp(ikc)u(x+c)$$

$$= \exp(ikc) \cdot \exp(ikx)u(x+c) = \exp(ikc) \cdot \exp(ikx)u(x) \tag{9.50}$$

by applying eq. (9.49) relating $u(x)$ and $u(x + c)$. Then from eq. (9.48),

$$\Psi(x+c) = \exp(ikc)\Psi(x) \tag{9.51}$$

which is the Bloch condition given in eq. (9.47). Equation (9.48) is referred as the Bloch theorem. It is a fundamental theorem in solid-state physics.

9.6 Boundary Conditions

i. Continuity of the wave function

$$\Psi_I(x=0) = \Psi_{II}(x=0) \tag{9.52}$$

ii. Continuity of the first derivative of the wave function

$$\left.\frac{d\Psi_I}{dx}\right|_{x=0} = \left.\frac{d\Psi_{II}}{dx}\right|_{x=0} \tag{9.53}$$

iii. Periodicity of the function $u(x)$

$$u(x=-b) = u(x=a) \tag{9.54}$$

Since

$$\Psi_{II}(x=-b) = \exp(-ikb) \cdot u(-b) \tag{9.55}$$

$$\therefore u(-b) = \exp(+ikb)\Psi_{II}(x=-b) \tag{9.56}$$

Also, since

$$\Psi_I(x=a) = \exp(ika) \cdot u(a) \tag{9.57}$$

$$\therefore u(a) = \exp(-ika)\Psi_I(x=a) \tag{9.58}$$

From eqs. (9.54), (9.56), and (9.58)

$$\exp(-ika)\Psi_I(x=a) = \exp(+ikb)\Psi_{II}(x=-b) \tag{9.59}$$

or

$$\Psi_I(x=a) = \frac{\exp(+ikb)\Psi_{II}(x=-b)}{\exp(-ika)} = \exp(+ika)\exp(+ikb)\Psi_{II}(x=-b) \tag{9.60}$$

or

$$\Psi_I(x=a) = \exp ik(a+b)\Psi_{II}(x=-b) \tag{9.61}$$

iv. Periodicity of the first derivative of function $u(x)$

$$\left.\frac{du}{dx}\right|_{x=-b} = \left.\frac{du}{dx}\right|_{x=a} \tag{9.62}$$

Since from eq. (9.55)

$$\left.\frac{d\Psi_{II}}{dx}\right|_{x=-b} = \frac{d}{dx}\{\exp(-ikb) \cdot u(-b)\} = \exp(-ikb)\left.\frac{du}{dx}\right|_{x=-b} \tag{9.63}$$

$$\therefore \left.\frac{du}{dx}\right|_{x=-b} = \exp(+ikb)\left.\frac{d\Psi_{II}}{dx}\right|_{x=-b} \tag{9.64}$$

and since from eq. (9.57)

$$\left.\frac{d\Psi_I}{dx}\right|_{x=a} = \frac{d}{dx}\{\exp(ika) \cdot u(a)\} = \exp(ika)\left.\frac{du}{dx}\right|_{x=a} \tag{9.65}$$

$$\therefore \left.\frac{du}{dx}\right|_{x=a} = \exp(-ika)\left.\frac{d\Psi_I}{dx}\right|_{x=a} \tag{9.66}$$

From eqs. (9.62), (9.64), and (9.66)

$$\exp(-ika)\left.\frac{d\Psi_I}{dx}\right|_{x=a} = \exp(+ikb)\left.\frac{d\Psi_{II}}{dx}\right|_{x=-b} \tag{9.67}$$

or

$$\left.\frac{d\Psi_I}{dx}\right|_{x=a} = \frac{\exp(+ikb)}{\exp(-ika)}\left.\frac{d\Psi_{II}}{dx}\right|_{x=-b} = \exp(+ika)\exp(+ikb)\left.\frac{d\Psi_{II}}{dx}\right|_{x=-b} \tag{9.68}$$

or

$$\left.\frac{d\Psi_I}{dx}\right|_{x=a} = \exp ik(a+b)\left.\frac{d\Psi_{II}}{dx}\right|_{x=-b} \tag{9.69}$$

9.7 Application of Boundary Conditions

i. Since from eq. (9.11)

$$\left.|\Psi_I(x)|\right|_{x=0} = \left.|A\exp(i\alpha x) + B\exp(-i\alpha x)|\right|_{x=0}$$

$$= A\exp(i\alpha 0) + B\exp(-i\alpha 0) = A\exp(0) + B\exp(-0) = A+B \tag{9.70}$$

and from eq. (9.21)

$$\left.|\Psi_{II}(x)|\right|_{x=0} = \left.|C\exp(\beta x) + D\exp(-\beta x)|\right|_{x=0}$$

$$= C\exp(\beta 0) + D\exp(-\beta 0) = C\exp(0) + D\exp(-0) = C+D \tag{9.71}$$

$$\therefore A+B = C+D \tag{9.72}$$

or

$$A + B - C - D = 0 \tag{9.73}$$

ii. Since from eq. (9.11)

$$\left| \frac{d\Psi_{\mathrm{I}}}{dx} \right|_{x=0} = \left| \frac{d}{dx} \left\{ A\exp(i\alpha x) + B\exp(-i\alpha x) \right\} \right|_{x=0}$$

$$= \left| i\alpha A\exp(i\alpha x) - i\alpha B\exp(-i\alpha x) \right|_{x=0}$$

$$= i\alpha A\exp(i\alpha 0) - i\alpha B\exp(-i\alpha 0)$$

$$= i\alpha A\exp(0) - i\alpha B\exp(-0) = i\alpha A - i\alpha B \tag{9.74}$$

and from eq. (9.21)

$$\left| \frac{d\Psi_{\mathrm{II}}}{dx} \right|_{x=0} = \left| \frac{d}{dx} \left\{ C\exp(\beta x) + D\exp(-\beta x) \right\} \right|_{x=0}$$

$$= \left| \beta C\exp(\beta x) - \beta D\exp(-\beta x) \right|_{x=0}$$

$$= \beta C\exp(\beta 0) - \beta D\exp(-\beta 0) = \beta C - \beta D \tag{9.75}$$

$$\therefore i\alpha A - i\alpha B = \beta C - \beta D \tag{9.76}$$

or

$$i\alpha(A - B) = \beta(C - D) \tag{9.77}$$

or

$$A - B = \frac{\beta}{i\alpha}(C - D) = \frac{i\beta}{i^2\alpha}(C - D) = \frac{i\beta}{-1\times\alpha}(C - D)$$

$$= -\frac{i\beta}{\alpha}(C - D) = -\frac{i\beta}{\alpha}C + \frac{i\beta}{\alpha}D \tag{9.78}$$

or

$$A - B + \frac{i\beta}{\alpha}C - \frac{i\beta}{\alpha}D = 0 \tag{9.79}$$

iii. Equation (9.61) will be applied. Since from eq. (9.11)

$$\Psi_{\mathrm{I}}(x = a) = \left| A\exp(i\alpha x) + B\exp(-i\alpha x) \right|_{x=a}$$

$$= A\exp(i\alpha a) + B\exp(-i\alpha a) \tag{9.80}$$

and from eq. (9.21)

$$\Psi_{\mathrm{II}}(x = -b) = \left| C\exp(\beta x) + D\exp(-\beta x) \right|_{x=-b}$$

$$= C\exp(-\beta b) + D\exp(+\beta b) \tag{9.81}$$

$$\therefore A\exp(i\alpha a) + B\exp(-i\alpha a)$$

$$= \exp ik(a + b)\left\{ C\exp(-\beta b) + D\exp(+\beta b) \right\} \tag{9.82}$$

or

$$\exp(i\alpha a)A + \exp(-i\alpha a)B = \exp(-\beta b)\exp\left\{ ik(a + b) \right\}C$$

$$+ \exp(+\beta b)\exp\left\{ ik(a + b) \right\}D \tag{9.83}$$

or

$$\exp(i\alpha a)A + \exp(-i\alpha a)B - \exp(-\beta b)\exp\left\{ ik(a + b) \right\}C$$

$$- \exp(+\beta b)\exp\left\{ ik(a + b) \right\}D = 0 \tag{9.84}$$

iv. Equation (9.69) will be applied, i.e.,

$$\left| \frac{d\Psi_{\mathrm{I}}}{dx} \right|_{x=a} = \exp ik(a + b)\left| \frac{d\Psi_{\mathrm{II}}}{dx} \right|_{x=-b} \tag{9.85}$$

Since from eq. (9.11)

$$\left| \frac{d\Psi_{\mathrm{I}}}{dx} \right|_{x=a} = \left| \frac{d}{dx} \left\{ A\exp(i\alpha x) + B\exp(-i\alpha x) \right\} \right|_{x=a}$$

$$= \left| i\alpha A\exp(i\alpha x) - i\alpha B\exp(-i\alpha x) \right|_{x=a}$$

$$= i\alpha A\exp(i\alpha a) - i\alpha B\exp(-i\alpha a) \tag{9.86}$$

and from eq. (9.21)

$$\left| \frac{d\Psi_{\mathrm{II}}}{dx} \right|_{x=-b} = \left| \frac{d}{dx} \left\{ C\exp(\beta x) + D\exp(-\beta x) \right\} \right|_{x=-b}$$

$$= \left| \beta C\exp(\beta x) - \beta D\exp(-\beta x) \right|_{x=-b}$$

$$= \beta C\exp(-\beta b) - \beta D\exp(+\beta b) \tag{9.87}$$

$$\therefore i\alpha A\exp(i\alpha a) - i\alpha B\exp(-i\alpha a)$$

$$= \exp ik(a + b)\left\{ \beta C\exp(-\beta b) - \beta D\exp(+\beta b) \right\} \tag{9.88}$$

or

$$i\alpha\exp(i\alpha a)A - i\alpha\exp(-i\alpha a)B$$

$$= \beta\exp(-\beta b)\exp ik(a + b)C - \beta\exp(+\beta b)\exp ik(a + b)D \tag{9.89}$$

or

$$\exp(i\alpha a)A - \exp(-i\alpha a)B = \frac{\beta}{i\alpha}\exp(-\beta b)\exp ik(a + b)C$$

$$- \frac{\beta}{i\alpha}\exp(+\beta b)\exp ik(a + b)D \tag{9.90}$$

or

$$\exp(i\alpha a)A - \exp(-i\alpha a)B = \frac{i\beta}{i^2\alpha}\exp(-\beta b)\exp ik(a + b)C$$

$$- \frac{i\beta}{i^2\alpha}\exp(+\beta b)\exp ik(a + b)D \tag{9.91}$$

or

$$\exp(i\alpha a)A - \exp(-i\alpha a)B = \frac{i\beta}{-1\times\alpha}\exp(-\beta b)\exp ik(a + b)C$$

$$- \frac{i\beta}{-1\times\alpha}\exp(+\beta b)\exp ik(a + b)D \tag{9.92}$$

or

$$\exp(i\alpha a)A - \exp(-i\alpha a)B = -\frac{i\beta}{\alpha}\exp(-\beta b)\exp ik(a+b)C$$

$$+\frac{i\beta}{\alpha}\exp(+\beta b)\exp ik(a+b)D \qquad (9.93)$$

or

$$\exp(i\alpha a)A - \exp(-i\alpha a)B + \frac{i\beta}{\alpha}\exp(-\beta b)\exp ik(a+b)C$$

$$-\frac{i\beta}{\alpha}\exp(+\beta b)\exp ik(a+b)D = 0 \qquad (9.94)$$

Equations (9.73), (9.79), (9.84), and (9.94) may be stated as

$$A + B - C - D = 0 \qquad (9.95)$$

$$A - B + \frac{i\beta}{\alpha}C - \frac{i\beta}{\alpha}D = 0 \qquad (9.96)$$

$$\exp(i\alpha a)A + \exp(-i\alpha a)B - \exp(-\beta b)\exp\{ik(a+b)\}C$$

$$-\exp(+\beta b)\exp\{ik(a+b)\}D = 0 \qquad (9.97)$$

and

$$\exp(i\alpha a)A - \exp(-i\alpha a)B + \frac{i\beta}{\alpha}\exp(-\beta b)\exp ik(a+b)C$$

$$-\frac{i\beta}{\alpha}\exp(+\beta b)\exp ik(a+b)D = 0 \qquad (9.98)$$

The 4×4 matrix representing the earlier homogeneous system of linear equations is

9.8 Calculation of the Determinant

The determinant is calculated in Illustrative Exercise 9.3. This calculation results in the following equation:

$$\left(\frac{\beta^2 - \alpha^2}{2\alpha\beta}\right)\sinh(\beta b)\sin(\alpha a) + \cosh(\beta b)\cos(\alpha a) = \cos\{k(a+b)\}$$

$$(9.101)$$

Let us decrease the barrier width b to zero and increase the barrier height V_0 to infinity in such a way that the product $bV_0 = $ constant. Then the potential is converted into a delta function train at $x = a$. It is repeated with a period $= a$, so that

$$\text{Delta function train} = bV_0\delta(x - b - na) \qquad (9.102)$$

where n denotes an integer.

$$\text{As } b \to 0, \sinh(\beta b) \to \beta b \text{ and } \cosh(\beta b) \to 1 \qquad (9.103)$$

and from eqs. (9.8) and (9.18)

$$\beta^2 - \alpha^2 = \frac{2m(V_0 - E)}{\hbar^2} - \frac{2mE}{\hbar^2} = \frac{2m}{\hbar^2}(V_0 - 2E) = \frac{2m}{\hbar^2}(\infty - 2E) = \infty$$

$$(9.104)$$

$$\therefore \beta^2 \gg \alpha^2 \qquad (9.105)$$

So, eq. (9.101) reduces to

$$\left(\frac{\beta^2}{2\alpha\beta}\right)\beta b\sin(\alpha a) + 1\times\cos(\alpha a) = \cos\{k(a+0)\} \qquad (9.106)$$

or

$$\frac{\beta^2 b}{2\alpha}\sin(\alpha a) + \cos(\alpha a) = \cos(ka) \qquad (9.107)$$

$$M = \begin{bmatrix} 1 & 1 & -1 & -1 \\ 1 & -1 & \dfrac{i\beta}{\alpha} & -\dfrac{i\beta}{\alpha} \\ \exp(i\alpha a) & \exp(-i\alpha a) & -\exp(-\beta b)\exp\{ik(a+b)\} & -\exp(\beta b)\exp\{ik(a+b)\} \\ \exp(i\alpha a) & -\exp(-i\alpha a) & \dfrac{i\beta}{\alpha}\exp(-\beta b)\exp ik(a+b) & -\dfrac{i\beta}{\alpha}\exp(\beta b)\exp ik(a+b) \end{bmatrix} \qquad (9.99)$$

The equations have a non-trivial solution if the determinant of the previous matrix is zero, i.e.,

$$D = \begin{vmatrix} 1 & 1 & -1 & -1 \\ 1 & -1 & \dfrac{i\beta}{\alpha} & -\dfrac{i\beta}{\alpha} \\ \exp(i\alpha a) & \exp(-i\alpha a) & -\exp(-\beta b)\exp\{ik(a+b)\} & -\exp(\beta b)\exp\{ik(a+b)\} \\ \exp(i\alpha a) & -\exp(-i\alpha a) & \dfrac{i\beta}{\alpha}\exp(-\beta b)\exp\{ik(a+b)\} & -\dfrac{i\beta}{\alpha}\exp(\beta b)\exp\{ik(a+b)\} \end{vmatrix} = 0 \qquad (9.100)$$

Putting

$$\frac{\beta^2 ba}{2} = P \tag{9.108}$$

we have

$$\frac{\beta^2 b}{2\alpha} = \frac{\beta^2 ba}{2} \times \frac{1}{\alpha a} = \frac{P}{\alpha a} \tag{9.109}$$

and eq. (9.107) becomes

$$\frac{P}{\alpha a} \sin(\alpha a) + \cos(\alpha a) = \cos(ka) \tag{9.110}$$

9.9 Protrayal of Bandgaps in Tabular and Graphical Formats

We can write the left-hand side of eq. (9.110) as a function F of αa:

$$F(\alpha a) = \frac{P}{\alpha a} \sin(\alpha a) + \cos(\alpha a) = \cos(ka) \tag{9.111}$$

Also, from eq. (9.9), we can write

$$\alpha a = \sqrt{\frac{2mE}{\hbar^2}} a \tag{9.112}$$

or

$$(\alpha a)^2 = \frac{2mE}{\hbar^2} a^2 \tag{9.113}$$

or

$$E = \frac{\hbar^2}{2ma^2} (\alpha a)^2 \tag{9.114}$$

or

$$\frac{E}{\frac{\hbar^2}{2ma^2}} = (\alpha a)^2 \tag{9.115}$$

or

$$\frac{E}{\xi} = (\alpha a)^2 \tag{9.116}$$

where

$$\xi = \frac{\hbar^2}{2ma^2} = \text{constant} \tag{9.117}$$

We shall use eqs. (9.111) and (9.116) to compile a table (Table 9.1) and prepare graphs (to be described and shown in Figures 9.2 and 9.3) by taking

$$P = \frac{3\pi}{2} \text{ and } \xi = 0.1 \tag{9.118}$$

showing that a continuous range of energies is not allowed for electrons, and there are energy ranges that are clearly prohibited.

First we calculate $F(\alpha a)$ for different values of $\alpha a/\pi$. Starting from rows 2 and 3 of Table 9.1, we complete the remaining rows. Using eq. (9.111), for

$$\frac{\alpha a}{\pi} = 0 \tag{9.119}$$

$$F(\alpha a) = \frac{3\pi}{2} \left\{ \frac{\sin(0)}{0} \right\} + \cos(0) \tag{9.120}$$

To evaluate sin (0)/0, we recall the Taylor series expansion of $\sin x$:

$$\sin x = x - \frac{x^3}{3!} + \frac{x^5}{5!} - \frac{x^7}{7!} + \frac{x^9}{9!} - \cdots \tag{9.121}$$

$$\therefore \frac{\sin x}{x} = 1 - \frac{x^2}{3!} + \frac{x^4}{5!} - \frac{x^6}{7!} + \frac{x^8}{9!} - \cdots \tag{9.122}$$

and

$$\frac{\sin(0)}{0} = 1 - 0 + 0 - 0 + 0 - \cdots = 1 \tag{9.123}$$

Hence from eq. (9.120)

$$F(\alpha a) = \frac{3\pi}{2} \times 1 + 1 = \frac{3 \times 3.14}{2} \times 1 + 1 = 4.71 + 1 = 5.71 \tag{9.124}$$

For

$$\frac{\alpha a}{\pi} = -0.25 \tag{9.125}$$

$$F(\alpha a) = \frac{\frac{3\pi}{2}}{-0.25\pi} \sin(-0.25\pi) + \cos(-0.25\pi)$$

$$= -6 \times -0.707 + 0.707 = 4.242 + 0.707 = +4.949 \tag{9.126}$$

For

$$\frac{\alpha a}{\pi} = -0.5 \tag{9.127}$$

$$F(\alpha a) = \frac{\frac{3\pi}{2}}{-0.5\pi} \sin(-0.5\pi) + \cos(-0.5\pi) = -3 \times -1 + 0 = 3 + 0 = +3 \tag{9.128}$$

Similar calculations are done for

$$\frac{\alpha a}{\pi} = -0.7, -0.72, -0.75, -1.0, -1.25, -1.5, -1.75, -2.0, -2.25,$$

$$-2.365, -2.37, -2.38, -2.4, -2.5, -2.75,$$

$$-3.0, -3.25, -3.5, -3.75, -4.0, \ldots \tag{9.129}$$

TABLE 9.1

Calculation Chart for Kronig-Penney Model

Row 0	E/ξ	0	0.616	2.465	4.831	5.111	5.546	9.86	15.41	22.184	29.851	39.44	49.91
1.	$(\alpha a)^2$	0	0.616	2.465	4.831	5.111	5.546	9.86	15.41	22.184	29.851	39.44	49.91
2.	$\alpha a/\pi$	0	-0.25π	-0.5π	-0.7π	-0.72π	-0.75π	$-\pi$	-1.25π	-1.5π	-1.75π	-2π	-2.25π
3.	$F(\alpha a)$	+5.7	+4.949	+3	+1.097	+0.968	+0.707	−1	+1.555	−1	+0.101	+1	+1.181
4.	$\cos^{-1}\{F(\alpha a)\}$ or ka	No number	No number	No number	No number	14.53	45	180	No number	180	84.2	0	No number
5.	ka/π	Not applicable	Not applicable	Not applicable	Not applicable	0.08	0.25	1	Not applicable	1	0.47	0	Not applicable
6.	Net ka/π					0		π		π		2π	

Row 0	E/ξ	55.147	55.38	55.85	56.79	61.62	74.56	88.74	104.14	120.78	138.65	157.75
1.	$(\alpha a)^2$	55.147	55.38	55.85	56.79	61.62	74.56	88.74	104.14	120.78	138.65	157.75
2.	$\alpha a/\pi$	-2.365π	-2.37π	-2.38π	-2.40π	-2.5π	-2.75π	-3π	-3.25π	-3.5π	-3.75π	-4.0π
3.	$F(\alpha a)$	0.989	0.978	0.954	0.903	0.6	−0.321	−1	−1.032	−0.4286	0.4242	1
4.	$\cos^{-1}\{F(\alpha a)\}$ or ka	8.51	12.04	17.45	25.44	53.13	108.72	180	No number	115.38	64.9	0
5.	ka/π	0.047	0.067	0.097	0.141	0.295	0.604	1	Not applicable	0.641	0.361	0
6.	Net ka/π	2π						3π		3π		4π

The cosine of an angle is a trigonometrical ratio whose maximum/ minimum values are ±1. We know that for any angle θ, at $\theta=0$, 2π, ... $\cos\theta=+1$, maximum value of $\cos\theta$ is +1. At $\theta=\pi$, ... $\cos\theta=-1$, minimum value of $\cos\theta$ is −1. Hence in eq. (9.111), for the right-hand side

$$-1 < \cos(ka) < +1 \qquad (9.130)$$

The same must be true for left-hand side of eq. (9.111), i.e.,

$$-1 < F(\alpha a) < +1 \qquad (9.131)$$

But by looking at Table 9.1, we notice that many values of $F(\alpha a)$ in Table 9.1 are within bounds ±1, e.g.

$$F(\alpha a) = +0.968, +0.707, -1, -1, +0.101, +1,...\text{for } \alpha a$$
$$= -0.72\pi, -0.75\pi, -\pi, -1.5\pi, -1.75\pi, -2\pi,... \qquad (9.132)$$

These are the allowed values for $F(\alpha a)$. But there are also values that exceed ±1, e.g.

$$F(\alpha a) = +5.7, 4.949, +3, +1.097, -1.555, +1.181, -1.032...\text{for } \alpha a$$
$$= 0, -0.25\pi, -0.5\pi, -0.7\pi, -1.25\pi, -2.25\pi, -3.25\pi... \qquad (9.133)$$

Obviously, these values of $F(\alpha a)$ are not permitted. But from eq. (9.9) we know that α is proportional to energy, i.e.,

$$\alpha \propto E \qquad (9.134)$$

Using eq. (9.116), we have calculated E/ξ in row 0. Hence, the range of energies E/ξ corresponding to α values in eq. (9.132) is allowed; these fall in the allowed energy band. Also, the range of energies corresponding to α values in eq. (9.133) is forbidden; these fall in the forbidden bandgap.

The earlier remarks can be pictorially illustrated by plotting two graphs. The first graph is between $F(\alpha a)$ on the Y-axis and E/ξ on the X-axis (Figure 9.2). In this graph, the only allowed values of $F(\alpha a)$ are those falling between the two boundary lines for +1 and −1. Values of $F(\alpha a)$ that fall beyond these boundary, either above or below, are disallowed. From the fourth row of Table 9.1, we can see that there are no real number values for $\cos^{-1}\{F(ka)\}$ at the values of $F(\alpha a)$ crossing the boundaries. The second graph is drawn between E/ξ on the Y-axis and net ka/π on the X-axis (Figure 9.3). In this graph, there are breaks in the E/ξ values at $ka/\pi=1, 2, 3$ or $\alpha a/\pi=1.25, 2.25, 3.25$. These breaks are the indicators of energy gaps. The E/ξ differences on the Y-axis are the bandgaps. Thus a series of disallowed energy ranges or bandgaps is evident separating the allowed energy ranges or bands.

9.10 Different Schemes for Drawing Energy-Band Diagrams

In Figure 9.3, we have shown one of the many ways of drawing energy-band diagrams. Figure 9.4 shows the E-k plot for free electrons. It is this plot which is amended by the periodic potential in a solid. So, it is helpful in understanding the alterations. We keep this plot as a reference in our mind and look at the several ways in which the changes to this plot are represented. We list below some common methods of representation:

i. Extended zone representation: The plot of energy E versus wave vector k starting from a common origin and spreading across lattice points, as given in Figure 9.5, is called extended zone representation.

ii. Flat band diagram: A common representation is in the form of energy E versus distance x. Here flat horizontal lines are drawn from the allowed and disallowed energy ranges to show the two types of bands, as in Figure 9.6.

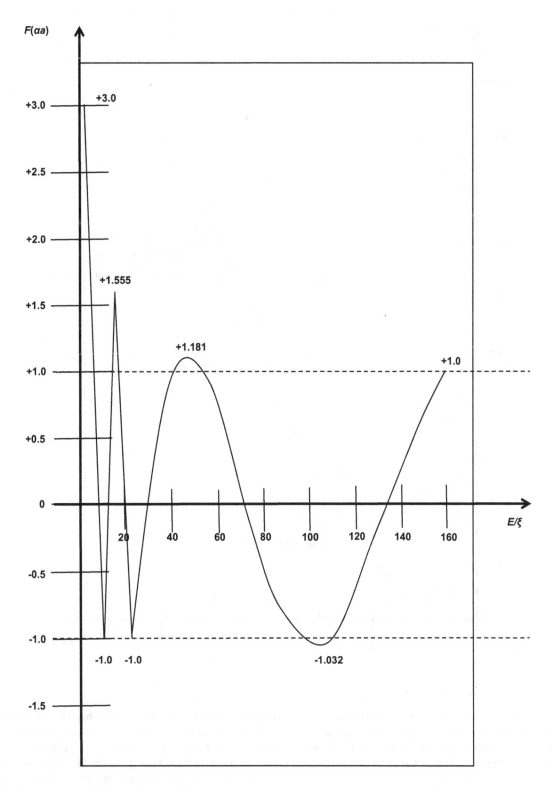

FIGURE 9.2 Graph constructed between $F(\alpha a)$ and E/ξ.

(iii) Repeated zone representation: Selection of the origin of k space is arbitrary. From crystal symmetry, each point of the lattice is equivalent to any other point. By drawing the E-k diagram at every lattice point, a more complete pictorial representation of the energy bands can be made. This kind of representation results in the repeated zone representation (Figure 9.7).

(iv) Reduced zone scheme: The region in k-space, also called momentum space or reciprocal space, between $+\pi/a$ and $-\pi/a$, i.e. within $\pm\pi/a$ boundaries, is called the first Brillouin zone. Recalling crystal terminology, we note that the first Brillouin zone is a uniquely defined primitive cell in reciprocal lattice. The subdivision of a reciprocal lattice into Brillouin zones in reciprocal

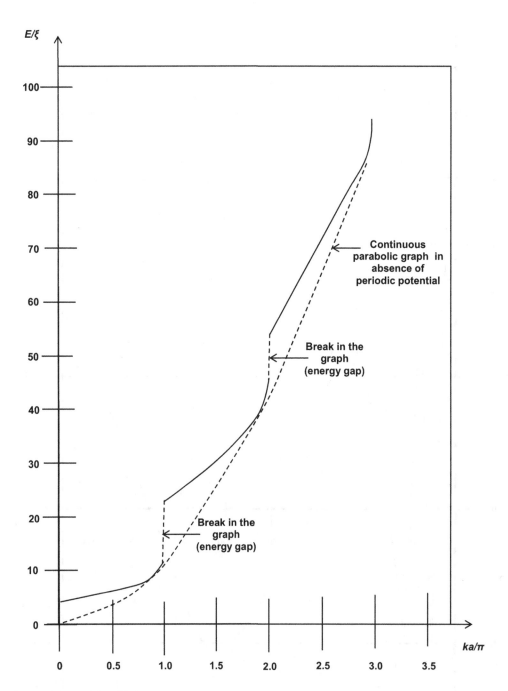

FIGURE 9.3   Graph drawn between E/ξ and ka/π.

space is analogous to the subdivision of a Bravais lattice into Wigner-Seitz cells in real or physical space. So, the Brillouin zone in reciprocal space is the counterpart of the Wigner-Seitz cell in real space.

It appears logical to cramp all the information pertaining to allowed/disallowed energy bands within the first Brillouin zone $-\pi/a < k < +\pi/a$, for a concise representation, when the bandgaps are seen at $k = 0$ and $k = \pm\pi/a$. This E-k representation is referred to as the reduced zone scheme (Figure 9.8). Figure 9.9 shows how the flat energy-band model follows from the reduced zone scheme.

In all these schemes, there is a method of naming the energy bands. The energy band below the bandgap is called

the valence band. The other energy band above the bandgap is known as the conduction band. The band structures of conductors, insulators, and semiconductors are different. In a conductor, the valence band is either partially filled or overlaps the conduction band. The pertinent question is: How does conduction take place? Look at the existence of a large number of empty states in the partially filled valence band. These are the states that make conduction possible on applying an electric field. Then the valence band itself becomes the conduction band. Similar effect is observed on overlapping valence and conduction bands.

The band structure of an insulator is an opposite extreme to that of a conductor. In an insulator, the valence band is completely filled, while the conduction band is nearly empty. In

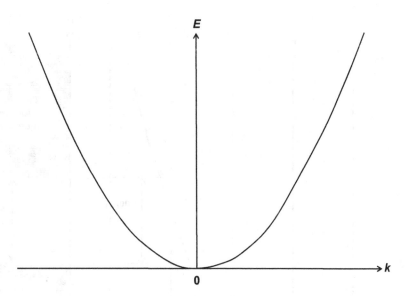

FIGURE 9.4 Parabolic *E-k* plot for free electrons in a solid.

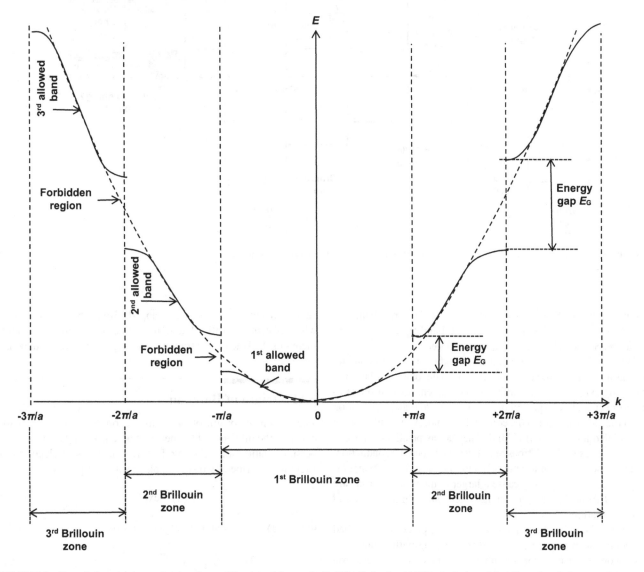

FIGURE 9.5 Extended zone scheme showing the modification of the parabolic *E-k* relation in a solid by inclusion of the effect of periodic potential.

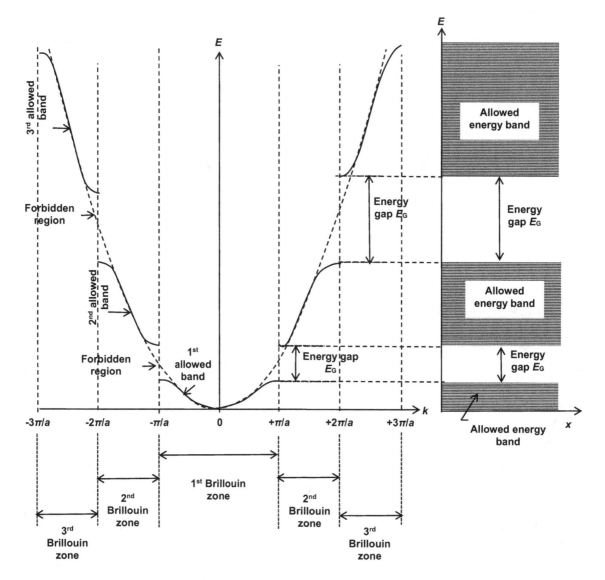

FIGURE 9.6 Drawing the equivalent flat energy-band diagram from the extended zone scheme.

addition, the bandgap is large. Because of the large bandgap, very few electrons can move across the bandgap at room temperature. So electrical conduction is inhibited.

What happens in the case of a semiconductor? A semiconductor has an energy-band diagram similar to that of an insulator but its bandgap is relatively smaller. So thermal energy at room temperature enables a large number of electrons to make a transition from the valence band to the conduction band. As a result, vacant sites called holes are left behind. They act as positively charged carriers. Thus both electrons and holes contribute to conduction.

The conductivity of a semiconductor increases as the temperature is elevated. This is because a larger number of electrons are liberated from the chemical bonds increasing the electron and hole populations.

The conductivity of a semiconductor can be precisely controlled by doping with impurities. For silicon, doping with pentavalent phosphorous, arsenic, or antimony increases the free electron concentration leading to conductivity enhancement. The resulting

semiconductor is said to be N-doped. Similarly doping silicon with trivalent impurities like boron, indium, gallium, aluminum, increases the hole concentration rendering the material P-doped.

9.11 Origin of Bandgaps

The reason for origin of bandgaps becomes obvious when we consider the influence of the periodic potential on the energy dispersion relation for electrons. For free electrons or electrons in vacuum, described by traveling plane waves (eq. (5.14))

$$\Psi(x) = \exp(ikx) \qquad (9.135)$$

the energy dispersion relation is (eq. (5.6))

$$E(k) = \frac{\hbar^2 k^2}{2m_0} = \frac{p^2}{2m_0} \qquad (9.136)$$

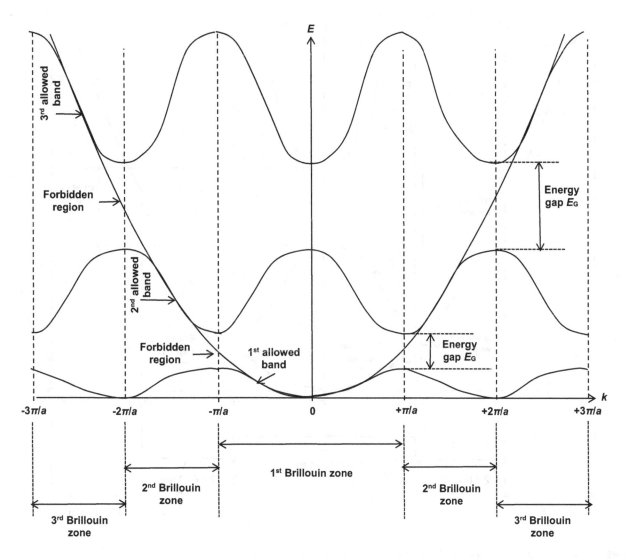

FIGURE 9.7   Repeated zone scheme of energy-band diagram representation.

where m_0 is the mass of the free electron, k its wave vector, and p its momentum. The E-k plot for the free electrons is a continuous parabola as shown in Figure 9.4. The same plot for the electrons in a periodic potential is a discontinuous parabola (Figure 9.5), and the discontinuities called the bandgaps arise from the interaction of electrons with the periodic potential. Electrons having a wavelength widely different from the lattice spacing a are able to propagate undisturbed through the crystal like the traveling waves of free electrons. The continuous ranges of allowed energies correspond to this undisturbed motion of electrons. For electrons having a wavelength twice the lattice spacing i.e. for $k = \pi/a$, the motion is perturbed by the periodic potential. It is these electrons whose interaction with periodic potential leads to the creation of bandgaps. These electrons are reflected by the periodic potential. Through such Bragg reflections, standing waves are formed. So we have no longer traveling waves like free electrons. The standing waves are formed by linear combinations of plane waves with $k = \pm\,\pi/a$. The two standing waves are labeled as $\Psi_{+\text{Bloch}}$ and $\Psi_{-\text{Bloch}}$. They are similar to each other but are mutually displaced along the X-axis by a distance of $a/2$. The first wave concentrates the probability of electron location at the ionic sites whereas the second wave concentrates the electron

probability in the region between ionic sites. In the first kind of event, the potential energy is lowered while in the second kind of event, the potential energy is raised. The consequence of this difference in electronic charge distribution is that the two standing waves have different energies at the same value of k, which appears as an energy bandgap in the dispersion relation.

The multiplicity of bandgaps arises from the periodicity of the crystal lattice due to which Bragg reflections can take place at several values of k given by

$$k = n\frac{\pi}{a} \qquad (9.137)$$

where n is an integer. Thus bandgaps occur at different k values.

9.12 Concepts of Effective Mass and Hole

How do we account for the interaction of the electron with the periodic potential? The effective mass concept enables us to do so. An electron in a crystal may move slowly or faster than one in vacuum. In a broad sense, an effective mass larger than the rest

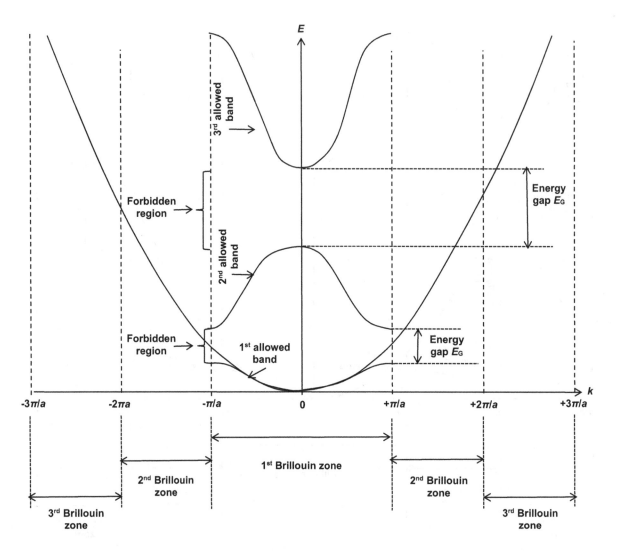

FIGURE 9.8 Reduced zone scheme in which all the information is represented in the first Brillouin zone.

mass of free electron reflects less agile electrons while an effective mass smaller than the free electron rest mass reflects higher electron agility.

As already stated, a free electron obeys the same energy-wave vector or energy-momentum relationship as a classical particle (eq. (9.136)). The velocity of the electron is the group velocity v_g of the wave packet given by

$$v_g = \frac{d\omega}{dk} = \frac{d\omega}{dE} \times \frac{dE}{dk} = \frac{d}{dE}\left(\frac{E}{\hbar}\right) \times \frac{dE}{dk} = \frac{1}{\hbar}\frac{dE}{dk} \quad (9.138)$$

since from Planck-Einstein formula

$$E = h\nu = \frac{h}{2\pi} \times 2\pi\nu = \hbar\omega \quad (9.139)$$

or

$$\omega = \frac{E}{\hbar} \quad (9.140)$$

Here ω is the angular frequency, ν is the linear frequency, E is the energy, h is the Planck's constant, and $\hbar$ is the reduced Planck's constant.

On applying an external force F on the wave packet (apart from the one due to periodic potential which is already taken into consideration in the wave function solution), the work dE done by the external force on the wave packet is given by

$$dE = \text{Force} \times \text{Distance} = F dx \quad (9.141)$$

But

$$dx = \text{Velocity} \times \text{Time} = v_g dt \quad (9.142)$$

$$\therefore dE = F v_g dt \quad (9.143)$$

or

$$F = \frac{1}{v_g}\frac{dE}{dt} \quad (9.144)$$

Substituting for v_g from eq. (9.138) in eq. (9.144)

$$F = \frac{1}{v_g}\frac{dE}{dt} = \frac{1}{\frac{1}{\hbar}\frac{dE}{dk}}\frac{dE}{dt} = \hbar\frac{dk}{dE}\frac{dE}{dt} = \hbar\frac{dk}{dt} \quad (9.145)$$

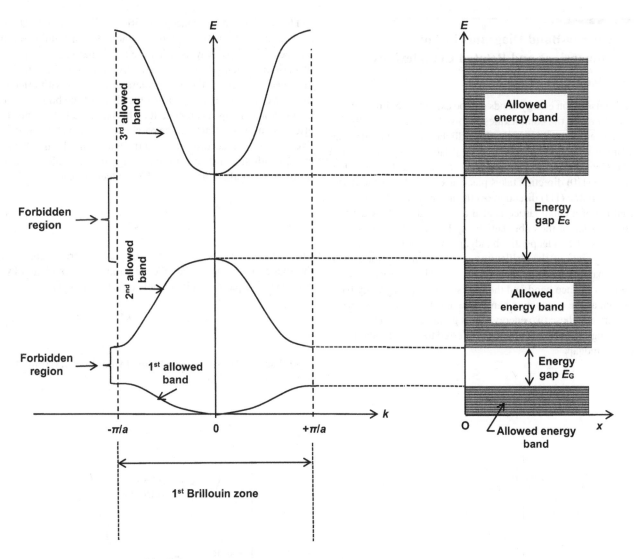

FIGURE 9.9 Flat energy-band model from the reduced zone scheme.

Acceleration a of the electron can be derived by differentiating the group velocity with respect to time:

$$a = \frac{dv_g}{dt} \qquad (9.146)$$

Substituting for v_g from eq. (9.138) in eq. (9.146),

$$a = \frac{d}{dt}\left(\frac{1}{\hbar}\frac{dE}{dk}\right) = \frac{d}{dk}\left(\frac{1}{\hbar}\frac{dE}{dk}\right) \times \frac{dk}{dt} = \frac{1}{\hbar}\left(\frac{d^2E}{dk^2}\right)\frac{dk}{dt} \qquad (9.147)$$

Now

Force = Effective mass of electron × Acceleration of the electron (9.148)

or

$$F = m^* a = m^* \times \frac{1}{\hbar}\left(\frac{d^2E}{dk^2}\right)\frac{dk}{dt} \qquad (9.149)$$

by applying eq. (9.147). Equation (9.145) helps us to write

$$\hbar\frac{dk}{dt} = m^* \times \frac{1}{\hbar}\left(\frac{d^2E}{dk^2}\right)\frac{dk}{dt} \qquad (9.150)$$

from which

$$m^* = \frac{1}{\dfrac{1}{\hbar^2}\left(\dfrac{d^2E}{dk^2}\right)} \qquad (9.151)$$

The second derivative of E with respect to k tells us about the curvature of the E-k graph. If the second derivative is positive, m^* is positive and the graph is concave upward. Contrarily, a negative second derivative means m^* is negative and the graph is concave downward. Such a particle with a negative effective mass and positive charge equal in magnitude to the electronic charge is ascribed to an electron vacancy from a chemical bond from which an electron is removed. It is called a hole. While the electron has a negative charge e^-, spin 1/2, and a positive effective mass m_e^*, the hole has a positive charge e^+, spin 1/2, and a negative effective mass m_h^* or m_p^*.

9.13 Energy-Band Diagrams in Three Dimensions and Related Complexities

9.13.1 Multitude of Bandgaps

The discussion on energy bands can be extended to three dimensions. However, the complexities involved must be carefully followed. The periodicities in the three directions may be different. So, Bragg reflections in the 3D crystal lattice vary (Koole et al 2014). Considering a primitive cubic lattice, the Bragg reflection in the (110) direction takes place at $k = \sqrt{2}\pi/a$ but the Bragg reflection in the (111) direction occurs at $k = \sqrt{3}\pi/a$. The reciprocal lattice of a cubic lattice is also a cubic lattice. It is the first Brillouin zone of the cubic lattice. As Bragg reflections are the originators of bandgaps, the bandgaps will not be the same at different points of the Brillouin zone. In such situations when several bandgaps arise, the highest occupied energy band is called the valence band and the lowest unoccupied energy band is known as the conduction band. The intervening energy difference between these two energy bands giving the smallest energy gap amongst the various options is referred to as the fundamental energy bandgap.

The effective mass concept greatly facilitates our treatment of electron motion. By applying it, the electron motion can be treated in a simple way as if it were a free particle. The effects of periodic potential are automatically taken into account. We only need to replace the actual mass of electron with effective mass. We work by approximating the shape of the bottom of the conduction band by a parabola. It is this parabola from which the effective mass is extracted. The parabola gives the electron effective mass m_e^* in accordance with the equation relating electron energy with the wave vector k and energy of the conduction band minimum E_C called the edge of the conduction band

$$E \cong E_C + \frac{\hbar^2 k^2}{2m_e^*} \qquad (9.152)$$

The motion of holes is amenable to identical treatment. Here, the shape of the top of the valence band is approximated by a parabola. The parabola gives a hole effective mass m_h^*:

$$E \cong E_V - \frac{\hbar^2 k^2}{2m_h^*} \qquad (9.153)$$

Note that the effective mass obtained from the curvature of the parabola is a constant.

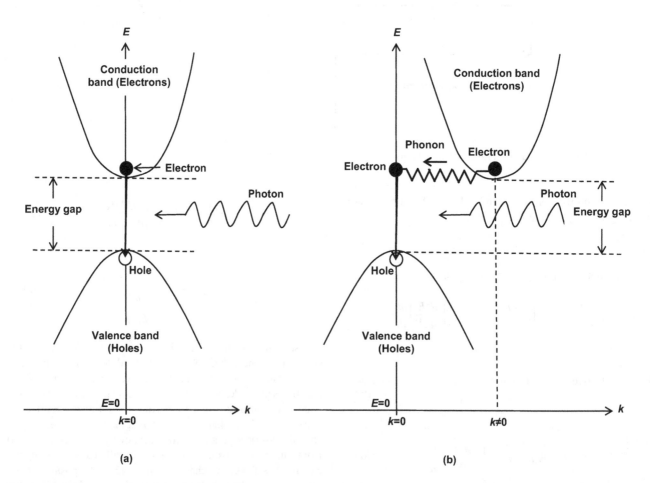

(a)   **(b)**

FIGURE 9.10   Reduced zone E-k diagrams of semiconductors: (a) direct bandgap and (b) indirect bandgap. Processes initiated by the absorption of a photon in the two types of materials are shown.

9.13.2 Direct- and Indirect-Bandgap Semiconductors

It is not necessary that the valence band maximum and the conduction band minimum should be located at the value of k. If they are situated on the same k value, the semiconductor is said to be a direct bandgap material (Figure 9.10). Otherwise, it is an indirect bandgap material. Hence, in a direct bandgap semiconductor, the wave vector or momentum of the highest energy state in the valence band and the lowest energy state in the conduction band are identical while in an indirect bandgap semiconductor, they differ. Notable examples of direct bandgap semiconductors include GaAs, InAs, InSb, InN, GaN, ZnO, ZnS, CdSe. Examples of indirect bandgap semiconductors are Si, Ge, diamond (C), GaP.

The difference between direct and indirect nature of bandgap is particularly important for optical optoelectronic devices such as light-emitting diodes and laser diodes. On absorption of a photon having energy greater than the bandgap, an electron-hole pair is generated in a direct bandgap material. Photon alone is sufficient for this process. However, in an indirect bandgap material, a change of momentum is also to be produced. So, in addition to photon, the electron must interact with a phonon or quantum of vibrational energy of the lattice. The requirement of interaction amongst three particles (electron, photon, and phonon) makes the process slower. The reverse process of emission of light by recombination of an electron with a hole is also difficult in indirect bandgap semiconductors because of the need of mediation by a phonon.

9.13.3 Silicon Band Structure and Two Types of Effective Masses of Carriers

In the E-k diagram for silicon (Figure 9.11), there are three valence band maxima, all at $k=0$. Two of these maxima are located at $E=0$ eV. These are the light and heavy hole bands. The third maxima is at $E=0.044$ eV. This is the split-off hole band. From the curvatures of the bands,

$$\text{Light hole effective mass} = m_{lh}^* = 0.16 m_0 \quad (9.154)$$

$$\text{Heavy hole effective mass} = m_{hh}^* = 0.46 m_0 \quad (9.155)$$

$$\text{Split-off hole effective mass} = m_{V,so}^* = 0.29 m_0 \quad (9.156)$$

where m_0 is the rest mass of electron $= 9.11 \times 10^{-31}$ kg.

There is one conduction band minimum at $k = 0$ at an energy $E_{C,\ Direct}$ much larger than the bandgap. Besides, there are six equivalent conduction band minima which are at $k \neq 0$ and energy $E_{C,\ Indirect} = 1.12$ eV. So, taking the lower-energy conduction band minima, the bandgap is $E_g = 1.12$ eV $- 0$ eV $= 1.12$ eV. The conduction band minima are characterized by a longitudinal mass along the (100) direction and two transverse masses in the plane orthogonal to the longitudinal direction.

$$\text{Longitudinal electron effective mass} = m_{el}^* = 0.98 m_0 \quad (9.157)$$

$$\text{Transverse electron effective mass} = m_{et}^* = 0.19 m_0 \quad (9.158)$$

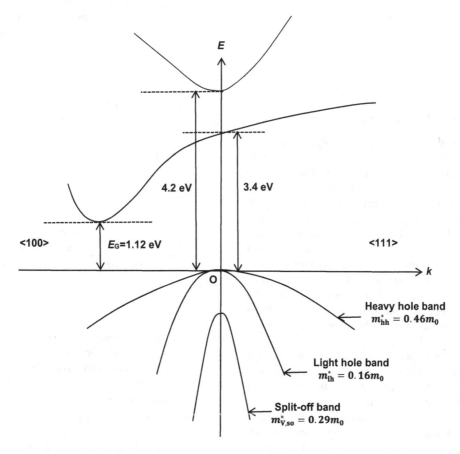

FIGURE 9.11 Simplified energy-band diagram of silicon.

Two types of effective masses are defined. These are intended to be used for different purposes. As electrical conductivity arising from several band minima/maxima is proportional to the sum of the inverse of the individual masses, an effective mass for conductivity calculations, also called conductivity effective mass, is introduced. For anisotropic minima containing one longitudinal and two transverse masses, the mass to be used for calculations involving mobilities and diffusion constants is given by

$$m_{el, Conductivity}^* = \frac{3}{\frac{1}{m_{el}^*} + \frac{1}{m_{et}^*} + \frac{1}{m_{et}^*}} = \frac{3}{\frac{1}{m_{el}^*} + \frac{2}{m_{et}^*}} = \frac{3}{\frac{1}{0.98m_0} + \frac{2}{0.19m_0}}$$

$$= \frac{3m_0}{\frac{1}{0.98} + \frac{2}{0.19}} = \frac{3m_0}{1.02 + 10.526} = \frac{3m_0}{11.546} = 0.26m_0$$

$$(9.159)$$

For a single band minimum containing one longitudinal mass and two transverse masses, the effective mass for density-of-states calculations, also called the density-of-states effective mass, is taken as the geometric mean of the three masses as

$$m_{el, DOS}^* = (\text{Number of equivalent band minima})^{2/3} \times \left(m_{el}^* m_{et}^* m_{et}^*\right)^{1/3}$$

$$= (\text{Number of equivalent band minima})^{2/3} \times \left(m_{el}^*\right)^{1/3} \times \left(m_{et}^*\right)^{2/3}$$

$$= (6)^{2/3} \times (0.98m_0)^{1/3} \times (0.19m_0)^{2/3}$$

$$= (6)^{2/3} \times (0.98)^{1/3} \times (0.19)^{2/3} \times m_0^{1/3} \times m_0^{2/3}$$

$$= 3.322 \times 0.9933 \times 0.3287 \times m_0 = 1.08m_0 \qquad (9.160)$$

9.14 Discussion and Conclusions

The Kronig-Penney model is a simplified prototype of an electron moving in a one-dimensional rectangular periodic potential. It offers the opportunity of analytical determination of energy eigenvalues and eigenfunctions. It also provides an eloquent demonstration of the existence of energy bands and bandgaps in solids, thereby systematically accounting for the differences in properties exhibited by metals, semiconductors, and insulators. The electronic band structure emerging from this model shares many features with sophisticated models requiring elaborate numerical calculations. Reshodko et al. (2019) obtained an analytical solution of the finite Kronig-Penney model with delta scatterers of random heights located at arbitrary points within a box that can be used to treat many problems hitherto requiring numerical techniques.

The energy-band diagrams can be constructed in various ways but the flat band diagrams and the reduced zone scheme have been immensely popular. A striking feature of the energy-band analysis is that the effect of the periodic potential can be taken into consideration in a simple way through the concept of effective mass. This concept helps in using free electron models with the electron mass substituted by effective mass. The artifice makes mathematical formulation incredibly easier. The effective

mass can be looked upon as a construct designed to make mathematics for electron motion through a solid easier, as if obeying Newton's laws and assuming that the mobile charge carriers are near the band edges. Besides the conductivity effective mass for calculations using models such as the Drude model, another effective mass, the density-of-states effective mass, is necessary for density-of-states calculations.

Illustrative Exercises

9.1 The conductivity effective mass in silicon is $0.26m_0$, in germanium $0.12m_0$, and in gallium arsenide $0.067m_0$. The electron mobilities are $1400\,cm^2 V\ s^{-1}$, $3900\,cm^2 V\ s^{-1}$, and $8500\,cm^2 V\ s^{-1}$. Find the mean free times in the three semiconductors.

From eq. (7.33)

$$\mu = \frac{e\tau}{m^*} \qquad (9.161)$$

or

$$\tau = \frac{m^*\mu}{e} \qquad (9.162)$$

Hence,

$$\tau_{Si} = \frac{0.26 \times 9.109 \times 10^{-31} \times \frac{1400}{10000}}{1.602 \times 10^{-19}} = \frac{0.26 \times 9.109 \times 10^{-31} \times 0.14}{1.602 \times 10^{-19}}$$

$$= 2.07 \times 10^{-13}\ s \qquad (9.163)$$

$$\tau_{Ge} = \frac{0.12 \times 9.109 \times 10^{-31} \times \frac{3900}{10000}}{1.602 \times 10^{-19}} = \frac{0.12 \times 9.109 \times 10^{-31} \times 0.39}{1.602 \times 10^{-19}}$$

$$= 2.66 \times 10^{-13}\ s \qquad (9.164)$$

and

$$\tau_{GaAs} = \frac{0.067 \times 9.109 \times 10^{-31} \times \frac{8500}{10000}}{1.602 \times 10^{-19}} = \frac{0.067 \times 9.109 \times 10^{-31} \times 0.85}{1.602 \times 10^{-19}}$$

$$= 3.238 \times 10^{-13}\ s \qquad (9.165)$$

9.2 (a) A direct bandgap semiconductor has a bandgap E_G of 1.42 eV. An electron is to be raised from the top of the valence band to the bottom of the conduction band. Find the energy and wave number of the photon necessary for the task.

(b) An indirect bandgap semiconductor has a bandgap E_G of 1.12 eV. A photon of energy 0.72 eV strikes it. What additional phonon energy is necessary to lift an electron from valence band top to bottom of the conduction band? If the valence band maximum occurs at $k = 0$ and the conduction band minimum at $k = k_m$, what should be the wave number of the phonon?

(a) Energy E of photon necessary to raise an electron from the top of the valence band to the bottom of the conduction band = Bandgap of the semiconductor = $E_G = 1.42\,\text{eV}$.

To find the wave number of the photon, we note that

$$E = h\nu = h\frac{c}{\lambda} \tag{9.166}$$

where h is Planck's constant; ν and λ are frequency and wavelength of the photon, respectively. From eq. (9.166)

$$\lambda = \frac{hc}{E} \tag{9.167}$$

and the wave number is

$$k = \frac{2\pi}{\lambda} = \frac{2\pi}{\frac{hc}{E}} = \frac{2\pi E}{hc} = \frac{2 \times 3.14 \times 1.42\,\text{eV}}{4.13567 \times 10^{-15}\,\text{eV s} \times 3 \times 10^8\,\text{m s}^{-1}}$$

$$= 0.71875 \times 10^7 \approx 7.2 \times 10^6\,\text{m}^{-1} \tag{9.168}$$

(b) Energy of phonon = Energy bandgap − Energy of photon = 1.12 eV − 0.72 eV = 0.40 eV

The wave number of phonon = k_m

9.3 Show how the determinant in eq. (9.100) leads to the equation

$$\left(\frac{\beta^2 - \alpha^2}{2\alpha\beta}\right)\sinh(\beta b)\sin(\alpha a) + \cosh(\beta b)\cos(\alpha a) = \cos\{k(a+b)\} \tag{9.169}$$

We begin with rewriting the determinant in eq. (9.100)

$$D = \begin{vmatrix} 1 & 1 & -1 & -1 \\ 1 & -1 & \dfrac{i\beta}{\alpha} & -\dfrac{i\beta}{\alpha} \\ \exp(i\alpha a) & \exp(-i\alpha a) & -\exp(-\beta b)\exp\{ik(a+b)\} & -\exp(\beta b)\exp\{ik(a+b)\} \\ \exp(i\alpha a) & -\exp(-i\alpha a) & \dfrac{i\beta}{\alpha}\exp(-\beta b)\exp\{ik(a+b)\} & -\dfrac{i\beta}{\alpha}\exp(\beta b)\exp\{ik(a+b)\} \end{vmatrix} = 0 \tag{9.170}$$

The simplification of the determinant is based on many elementary column or row operations on determinants such as columns or rows of the determinant can be interchanged, each element in a column or row of the determinant can be multiplied by a non-zero number, a column or row of the determinant can be multiplied by a non-zero number, and the multiplication result can be added to another column or row.

In eq. (9.170), replace column 1 by (column 1 + column 2) and replace column 3 by (column 3 + column 4)

$$\begin{vmatrix} 2 & 1 & -2 & -1 \\ 0 & -1 & 0 & -\dfrac{i\beta}{\alpha} \\ 2\cos(\alpha a) & \exp(-i\alpha a) & -2\cosh(\beta b)\exp\{ik(a+b)\} & -\exp(\beta b)\exp\{ik(a+b)\} \\ 2i\sin(\alpha a) & -\exp(-i\alpha a) & -2\dfrac{i\beta}{\alpha}\sinh(\beta b)\exp\{ik(a+b)\} & -\dfrac{i\beta}{\alpha}\exp(\beta b)\exp\{ik(a+b)\} \end{vmatrix} = 0 \tag{9.171}$$

because for column 1

$$1 + 1 = 2 \tag{9.172}$$

$$1 - 1 = 0 \tag{9.173}$$

$$\exp(i\alpha a) + \exp(-i\alpha a) = 2\cos(\alpha a) \text{ since}$$

$$\cos x = \frac{\exp(ix) + \exp(-ix)}{2} \tag{9.174}$$

and

$$\exp(i\alpha a) - \exp(-i\alpha a) = 2i\sin(\alpha a) \text{ since}$$

$$\sin x = \frac{\exp(ix) - \exp(-ix)}{2i} \tag{9.175}$$

and for column 3

$$-1 - 1 = -2 \tag{9.176}$$

$$\frac{i\beta}{\alpha} - \frac{i\beta}{\alpha} = 0 \tag{9.177}$$

$$-\exp(-\beta b)\exp\{ik(a+b)\} - \exp(\beta b)\exp\{ik(a+b)\}$$

$$= -\exp\{ik(a+b)\}\left[\exp(\beta b) + \exp(-\beta b)\right] = -\exp\{ik(a+b)\} \times 2\cosh(\beta b)$$

$$= -2\cosh(\beta b)\exp\{ik(a+b)\} \text{ since } \cosh x = \frac{\exp(x) + \exp(-x)}{2} \tag{9.178}$$

and

$$\frac{i\beta}{\alpha}\exp(-\beta b)\exp\{ik(a+b)\} - \frac{i\beta}{\alpha}\exp(\beta b)\exp\{ik(a+b)\}$$

$$= -\frac{i\beta}{\alpha}\exp\{ik(a+b)\}\{\exp(\beta b) - \exp(-\beta b)\} = -\frac{i\beta}{\alpha}\exp\{ik(a+b)\}2\sinh(\beta b)$$

$$= -2\frac{i\beta}{\alpha}\sinh(\beta b)\exp\{ik(a+b)\} \text{ since } \sinh x = \frac{\exp(x) - \exp(-x)}{2} \tag{9.179}$$

In eq. (9.171), multiply column 1 by 1/2, column 3 by – 1/2, and column 4 by –1

$$\begin{vmatrix} 1 & 1 & 1 & 1 \\ 0 & -1 & 0 & \dfrac{i\beta}{\alpha} \\ \cos(\alpha a) & \exp(-i\alpha a) & \cosh(\beta b)\exp\{ik(a+b)\} & \exp(\beta b)\exp\{ik(a+b)\} \\ i\sin(\alpha a) & -\exp(-i\alpha a) & \dfrac{i\beta}{\alpha}\sinh(\beta b)\exp\{ik(a+b)\} & \dfrac{i\beta}{\alpha}\exp(\beta b)\exp\{ik(a+b)\} \end{vmatrix} = 0 \tag{9.180}$$

In eq. (9.180), multiply column 4 by $-i\alpha/\beta$

$$\begin{vmatrix} 1 & 1 & 1 & -\dfrac{i\alpha}{\beta} \\ 0 & -1 & 0 & 1 \\ \cos(\alpha a) & \exp(-i\alpha a) & \cosh(\beta b)\exp\{ik(a+b)\} & -\dfrac{i\alpha}{\beta}\exp(\beta b)\exp\{ik(a+b)\} \\ i\sin(\alpha a) & -\exp(-i\alpha a) & \dfrac{i\beta}{\alpha}\sinh(\beta b)\exp\{ik(a+b)\} & \exp(\beta b)\exp\{ik(a+b)\} \end{vmatrix} = 0 \tag{9.181}$$

because

$$1 \times -\frac{i\alpha}{\beta} = -\frac{i\alpha}{\beta} \tag{9.182}$$

$$\frac{i\beta}{\alpha} \times -\frac{i\alpha}{\beta} = -i^2 = -\left(\sqrt{-1}\right)^2 = -(-1) = 1 \tag{9.183}$$

$$\exp(\beta b)\exp\{ik(a+b)\} \times -\frac{i\alpha}{\beta} = -\frac{i\alpha}{\beta}\exp(\beta b)\exp\{ik(a+b)\}$$

$$\tag{9.184}$$

and

$$\frac{i\beta}{\alpha}\exp(\beta b)\exp\{ik(a+b)\} \times -\frac{i\alpha}{\beta}$$

$$= -i^2\exp(\beta b)\exp\{ik(a+b)\} = -\left(\sqrt{-1}\right)^2\exp(\beta b)\exp\{ik(a+b)\}$$

$$= -(-1)\exp(\beta b)\exp\{ik(a+b)\} = \exp(\beta b)\exp\{ik(a+b)\}$$

$$\tag{9.185}$$

In eq. (9.181), replacing column 4 by (column 2+column 4)

$$\begin{vmatrix} 1 & 1 & 1 & 1-\dfrac{i\alpha}{\beta} \\[2mm] 0 & -1 & 0 & 0 \\[2mm] \cos(\alpha a) & \exp(-i\alpha a) & \cosh(\beta b)\exp\{ik(a+b)\} & \exp(-i\alpha a)-\dfrac{i\alpha}{\beta}\exp(\beta b)\exp\{ik(a+b)\} \\[2mm] i\sin(\alpha a) & -\exp(-i\alpha a) & \dfrac{i\beta}{\alpha}\sinh(\beta b)\exp\{ik(a+b)\} & -\exp(-i\alpha a)+\exp(\beta b)\exp\{ik(a+b)\} \end{vmatrix}=0 \tag{9.186}$$

Expanding the determinant about the second row which has the maximum number of zeroes, we get

$$+0\begin{vmatrix} 1 & 1 & 1-\dfrac{i\alpha}{\beta} \\[2mm] \exp(-i\alpha a) & \cosh(\beta b)\exp\{ik(a+b)\} & \exp(-i\alpha a)-\dfrac{i\alpha}{\beta}\exp(\beta b)\exp\{ik(a+b)\} \\[2mm] -\exp(-i\alpha a) & \dfrac{i\beta}{\alpha}\sinh(\beta b)\exp\{ik(a+b)\} & -\exp(-i\alpha a)+\exp(\beta b)\exp\{ik(a+b)\} \end{vmatrix}$$

$$-(-1)\begin{vmatrix} 1 & 1 & 1-\dfrac{i\alpha}{\beta} \\[2mm] \cos(\alpha a) & \cosh(\beta b)\exp\{ik(a+b)\} & \exp(-i\alpha a)-\dfrac{i\alpha}{\beta}\exp(\beta b)\exp\{ik(a+b)\} \\[2mm] i\sin(\alpha a) & \dfrac{i\beta}{\alpha}\sinh(\beta b)\exp\{ik(a+b)\} & -\exp(-i\alpha a)+\exp(\beta b)\exp\{ik(a+b)\} \end{vmatrix}$$

$$+0\begin{vmatrix} 1 & 1 & 1-\dfrac{i\alpha}{\beta} \\[2mm] \cos(\alpha a) & \exp(-i\alpha a) & \exp(-i\alpha a)-\dfrac{i\alpha}{\beta}\exp(\beta b)\exp\{ik(a+b)\} \\[2mm] i\sin(\alpha a) & -\exp(-i\alpha a) & -\exp(-i\alpha a)+\exp(\beta b)\exp\{ik(a+b)\} \end{vmatrix}$$

$$-0\begin{vmatrix} 1 & 1 & 1 \\[2mm] \cos(\alpha a) & \exp(-i\alpha a) & \cosh(\beta b)\exp\{ik(a+b)\} \\[2mm] i\sin(\alpha a) & -\exp(-i\alpha a) & \dfrac{i\beta}{\alpha}\sinh(\beta b)\exp\{ik(a+b)\} \end{vmatrix}=0 \tag{9.187}$$

or

$$\begin{vmatrix} 1 & 1 & 1-\dfrac{i\alpha}{\beta} \\[2mm] \cos(\alpha a) & \cosh(\beta b)\exp\{ik(a+b)\} & \exp(-i\alpha a)-\dfrac{i\alpha}{\beta}\exp(\beta b)\exp\{ik(a+b)\} \\[2mm] i\sin(\alpha a) & \dfrac{i\beta}{\alpha}\sinh(\beta b)\exp\{ik(a+b)\} & -\exp(-i\alpha a)+\exp(\beta b)\exp\{ik(a+b)\} \end{vmatrix}=0 \tag{9.188}$$

In eq. (9.188), multiplying column 3 by $\beta/(\beta-i\alpha)$

$$\begin{vmatrix} 1 & 1 & 1 \\ \cos(\alpha a) & \cosh(\beta b)\exp\{ik(a+b)\} & \dfrac{\beta}{\beta-i\alpha}\exp(-i\alpha a)-\dfrac{i\alpha}{\beta-i\alpha}\exp(\beta b)\exp\{ik(a+b)\} \\ i\sin(\alpha a) & \dfrac{i\beta}{\alpha}\sinh(\beta b)\exp\{ik(a+b)\} & -\dfrac{\beta}{\beta-i\alpha}\exp(-i\alpha a)+\dfrac{\beta}{\beta-i\alpha}\exp(\beta b)\exp\{ik(a+b)\} \end{vmatrix} = 0 \qquad (9.189)$$

because

$$\left(1-\frac{i\alpha}{\beta}\right)\times\frac{\beta}{\beta-i\alpha}=\frac{\beta-i\alpha}{\beta}\times\frac{\beta}{\beta-i\alpha}=1$$

$$\left[\exp(-i\alpha a)-\frac{i\alpha}{\beta}\exp(\beta b)\exp\{ik(a+b)\}\right]\times\frac{\beta}{\beta-i\alpha}$$

$$=\frac{\beta}{\beta-i\alpha}\exp(-i\alpha a)-\frac{\beta}{\beta-i\alpha}\times\frac{i\alpha}{\beta}\exp(\beta b)\exp\{ik(a+b)\}$$

$$=\frac{\beta}{\beta-i\alpha}\exp(-i\alpha a)-\frac{i\alpha}{\beta-i\alpha}\exp(\beta b)\exp\{ik(a+b)\}$$

$$\left[-\exp(-i\alpha a)+\exp(\beta b)\exp\{ik(a+b)\}\right]\times\frac{\beta}{\beta-i\alpha}$$

$$=-\frac{\beta}{\beta-i\alpha}\exp(-i\alpha a)+\frac{\beta}{\beta-i\alpha}\exp(\beta b)\exp\{ik(a+b)\} \qquad (9.190)$$

In eq. (9.189), replacing column 2 by (column 2−column 1) and column 3 by (column 3−column 1)

$$\begin{vmatrix} 1 & 0 & 0 \\ \cos(\alpha a) & \cosh(\beta b)\exp\{ik(a+b)\}-\cos(\alpha a) & \dfrac{\beta}{\beta-i\alpha}\exp(-i\alpha a)-\dfrac{i\alpha}{\beta-i\alpha}\exp(\beta b)\exp\{ik(a+b)\}-\cos(\alpha a) \\ i\sin(\alpha a) & \dfrac{i\beta}{\alpha}\sinh(\beta b)\exp\{ik(a+b)\}-i\sin(\alpha a) & -\dfrac{\beta}{\beta-i\alpha}\exp(-i\alpha a)+\dfrac{\beta}{\beta-i\alpha}\exp(\beta b)\exp\{ik(a+b)\}-i\sin(\alpha a) \end{vmatrix} = 0$$

$$(9.191)$$

Expanding this determinant about the first row which has the maximum number of zeroes,

$$+1\begin{vmatrix} \cosh(\beta b)\exp\{ik(a+b)\}-\cos(\alpha a) & \dfrac{\beta}{\beta-i\alpha}\exp(-i\alpha a)-\dfrac{i\alpha}{\beta-i\alpha}\exp(\beta b)\exp\{ik(a+b)\}-\cos(\alpha a) \\ \dfrac{i\beta}{\alpha}\sinh(\beta b)\exp\{ik(a+b)\}-i\sin(\alpha a) & -\dfrac{\beta}{\beta-i\alpha}\exp(-i\alpha a)+\dfrac{\beta}{\beta-i\alpha}\exp(\beta b)\exp\{ik(a+b)\}-i\sin(\alpha a) \end{vmatrix}$$

$$-0\begin{vmatrix} \cos(\alpha a) & \dfrac{\beta}{\beta-i\alpha}\exp(-i\alpha a)-\dfrac{i\alpha}{\beta-i\alpha}\exp(\beta b)\exp\{ik(a+b)\}-\cos(\alpha a) \\ i\sin(\alpha a) & -\dfrac{\beta}{\beta-i\alpha}\exp(-i\alpha a)+\dfrac{\beta}{\beta-i\alpha}\exp(\beta b)\exp\{ik(a+b)\}-i\sin(\alpha a) \end{vmatrix}$$

$$+0\begin{vmatrix} \cos(\alpha a) & \cosh(\beta b)\exp\{ik(a+b)\}-\cos(\alpha a) \\ i\sin(\alpha a) & \dfrac{i\beta}{\alpha}\sinh(\beta b)\exp\{ik(a+b)\}-i\sin(\alpha a) \end{vmatrix} = 0 \qquad (9.192)$$

or

$$\begin{vmatrix} \cosh(\beta b)\exp\{ik(a+b)\}-\cos(\alpha a) & \dfrac{\beta}{\beta-i\alpha}\exp(-i\alpha a)-\dfrac{i\alpha}{\beta-i\alpha}\exp(\beta b)\exp\{ik(a+b)\}-\cos(\alpha a) \\[2ex] \dfrac{i\beta}{\alpha}\sinh(\beta b)\exp\{ik(a+b)\}-i\sin(\alpha a) & -\dfrac{\beta}{\beta-i\alpha}\exp(-i\alpha a)+\dfrac{\beta}{\beta-i\alpha}\exp(\beta b)\exp\{ik(a+b)\}-i\sin(\alpha a) \end{vmatrix}=0 \qquad (9.193)$$

Since

$$\exp(-ix)=\cos x-i\sin x \qquad (9.194)$$

$$\frac{\beta}{\beta-i\alpha}\exp(-i\alpha a)-\frac{i\alpha}{\beta-i\alpha}\exp(\beta b)\exp\{ik(a+b)\}-\cos(\alpha a)$$

$$=\frac{\beta}{\beta-i\alpha}\cos(\alpha a)-\frac{\beta}{\beta-i\alpha}i\sin(\alpha a)-\frac{i\alpha}{\beta-i\alpha}\exp(\beta b)\exp\{ik(a+b)\}-\cos(\alpha a)$$

$$=\frac{\beta}{\beta-i\alpha}\cos(\alpha a)-\cos(\alpha a)-\frac{i\alpha}{\beta-i\alpha}\exp(\beta b)\exp\{ik(a+b)\}$$

$$-\frac{i\beta}{\beta-i\alpha}\sin(\alpha a)=\frac{\{\beta-(\beta-i\alpha)\}\cos(\alpha a)}{\beta-i\alpha}-\frac{i\alpha}{\beta-i\alpha}\exp(\beta b)\exp\{ik(a+b)\}-\frac{i\beta}{\beta-i\alpha}\sin(\alpha a)$$

$$=\frac{i\alpha}{\beta-i\alpha}\cos(\alpha a)-\frac{i\alpha}{\beta-i\alpha}\exp(\beta b)\exp\{ik(a+b)\}-\frac{i\beta}{\beta-i\alpha}\sin(\alpha a) \qquad (9.195)$$

and

$$-\frac{\beta}{\beta-i\alpha}\exp(-i\alpha a)+\frac{\beta}{\beta-i\alpha}\exp(\beta b)\exp\{ik(a+b)\}-i\sin(\alpha a)$$

$$=-\frac{\beta}{\beta-i\alpha}\cos(\alpha a)+\frac{\beta}{\beta-i\alpha}i\sin(\alpha a)+\frac{\beta}{\beta-i\alpha}\exp(\beta b)\exp\{ik(a+b)\}-i\sin(\alpha a)$$

$$=\frac{\beta}{\beta-i\alpha}i\sin(\alpha a)-i\sin(\alpha a)+\frac{\beta}{\beta-i\alpha}\exp(\beta b)\exp\{ik(a+b)\}-\frac{\beta}{\beta-i\alpha}\cos(\alpha a)$$

$$=\frac{\{i\beta-i(\beta-i\alpha)\}\sin(\alpha a)}{\beta-i\alpha}+\frac{\beta}{\beta-i\alpha}\exp(\beta b)\exp\{ik(a+b)\}-\frac{\beta}{\beta-i\alpha}\cos(\alpha a) \qquad (9.196)$$

$$=\frac{i^{2}\alpha\sin(\alpha a)}{\beta-i\alpha}+\frac{\beta}{\beta-i\alpha}\exp(\beta b)\exp\{ik(a+b)\}-\frac{\beta}{\beta-i\alpha}\cos(\alpha a)$$

$$=-\frac{\alpha}{\beta-i\alpha}\sin(\alpha a)+\frac{\beta}{\beta-i\alpha}\exp(\beta b)\exp\{ik(a+b)\}-\frac{\beta}{\beta-i\alpha}\cos(\alpha a)$$

Hence, eq. (193) reduces to

$$\begin{vmatrix} \cosh(\beta b)\exp\{ik(a+b)\}-\cos(\alpha a) & \dfrac{i\alpha}{\beta-i\alpha}\cos(\alpha a)-\dfrac{i\alpha}{\beta-i\alpha}\exp(\beta b)\exp\{ik(a+b)\}-\dfrac{i\beta}{\beta-i\alpha}\sin(\alpha a) \\[2ex] \dfrac{i\beta}{\alpha}\sinh(\beta b)\exp\{ik(a+b)\}-i\sin(\alpha a) & -\dfrac{\alpha}{\beta-i\alpha}\sin(\alpha a)+\dfrac{\beta}{\beta-i\alpha}\exp(\beta b)\exp\{ik(a+b)\}-\dfrac{\beta}{\beta-i\alpha}\cos(\alpha a) \end{vmatrix}=0 \qquad (9.197)$$

or

$$\left[\cosh(\beta b)\exp\{ik(a+b)\} - \cos(\alpha a) \right]$$

$$\times \left[-\frac{\alpha}{\beta - i\alpha}\sin(\alpha a) + \frac{\beta}{\beta - i\alpha}\exp(\beta b)\exp\{ik(a+b)\} - \frac{\beta}{\beta - i\alpha}\cos(\alpha a) \right]$$

$$-\left[\frac{i\beta}{\alpha}\sinh(\beta b)\exp\{ik(a+b)\} - i\sin(\alpha a) \right]$$

$$\times \left[\frac{i\alpha}{\beta - i\alpha}\cos(\alpha a) - \frac{i\alpha}{\beta - i\alpha}\exp(\beta b)\exp\{ik(a+b)\} - \frac{i\beta}{\beta - i\alpha}\sin(\alpha a) \right] = 0 \tag{9.198}$$

or

$$-\frac{\alpha}{\beta - i\alpha}\sin(\alpha a)\cosh(\beta b)\exp\{ik(a+b)\} - \frac{\alpha}{\beta - i\alpha}\sin(\alpha a)\times -\cos(\alpha a)$$

$$+\frac{\beta}{\beta - i\alpha}\exp(\beta b)\exp\{ik(a+b)\}\times\cosh(\beta b)\exp\{ik(a+b)\}$$

$$+\frac{\beta}{\beta - i\alpha}\exp(\beta b)\exp\{ik(a+b)\}\times -\cos(\alpha a)$$

$$-\frac{\beta}{\beta - i\alpha}\cos(\alpha a)\cosh(\beta b)\exp\{ik(a+b)\} - \frac{\beta}{\beta - i\alpha}\cos(\alpha a)\times -\cos(\alpha a)$$

$$-\frac{i\beta}{\alpha}\sinh(\beta b)\exp\{ik(a+b)\}\times\frac{i\alpha}{\beta - i\alpha}\cos(\alpha a) - \frac{i\beta}{\alpha}\sinh(\beta b)\exp\{ik(a+b)\}$$

$$\times -\frac{i\alpha}{\beta - i\alpha}\exp(\beta b)\exp\{ik(a+b)\}$$

$$-\frac{i\beta}{\alpha}\sinh(\beta b)\exp\{ik(a+b)\}\times -\frac{i\beta}{\beta - i\alpha}\sin(\alpha a) + i\sin(\alpha a)\times\frac{i\alpha}{\beta - i\alpha}\cos(\alpha a) + i\sin(\alpha a)$$

$$\times -\frac{i\alpha}{\beta - i\alpha}\exp(\beta b)\exp\{ik(a+b)\} + i\sin(\alpha a)\times -\frac{i\beta}{\beta - i\alpha}\sin(\alpha a) = 0 \tag{9.199}$$

or

$$-\frac{\alpha}{\beta - i\alpha}\sin(\alpha a)\cosh(\beta b)\exp\{ik(a+b)\} - \frac{i\beta}{\alpha}\sinh(\beta b)\exp\{ik(a+b)\}$$

$$\times -\frac{i\beta}{\beta - i\alpha}\sin(\alpha a) + i\sin(\alpha a)\times -\frac{i\alpha}{\beta - i\alpha}\exp(\beta b)\exp\{ik(a+b)\} + \frac{\beta}{\beta - i\alpha}\exp(\beta b)\exp\{ik(a+b)\}$$

$$\times -\cos(\alpha a) - \frac{\beta}{\beta - i\alpha}\cos(\alpha a)\cosh(\beta b)\exp\{ik(a+b)\} - \frac{i\beta}{\alpha}\sinh(\beta b)\exp\{ik(a+b)\}$$

$$\times \frac{i\alpha}{\beta - i\alpha}\cos(\alpha a) - \frac{\alpha}{\beta - i\alpha}\sin(\alpha a)\times -\cos(\alpha a) + i\sin(\alpha a)\times\frac{i\alpha}{\beta - i\alpha}\cos(\alpha a) + i\sin(\alpha a)$$

$$\times -\frac{i\beta}{\beta - i\alpha}\sin(\alpha a) - \frac{\beta}{\beta - i\alpha}\cos(\alpha a)\times -\cos(\alpha a) + \frac{\beta}{\beta - i\alpha}\exp(\beta b)\exp\{ik(a+b)\}$$

$$\times \cosh(\beta b)\exp\{ik(a+b)\} - \frac{i\beta}{\alpha}\sinh(\beta b)\exp\{ik(a+b)\}$$

$$\times -\frac{i\alpha}{\beta - i\alpha}\exp(\beta b)\exp\{ik(a+b)\} = 0 \tag{9.200}$$

or

$$\left[-\frac{\alpha}{\beta-i\alpha}\cosh(\beta b)-\frac{i\beta}{\alpha}\sinh(\beta b)\times-\frac{i\beta}{\beta-i\alpha}+i\times-\frac{i\alpha}{\beta-i\alpha}\exp(\beta b)\right]\exp\{ik(a+b)\}\sin(\alpha a)$$

$$+\left[-\frac{\beta}{\beta-i\alpha}\exp(\beta b)-\frac{\beta}{\beta-i\alpha}\cosh(\beta b)-\frac{i\beta}{\alpha}\sinh(\beta b)\times\frac{i\alpha}{\beta-i\alpha}\right]\exp\{ik(a+b)\}\cos(\alpha a)$$

$$+\left[-\frac{\alpha}{\beta-i\alpha}\times-1+i\times\frac{i\alpha}{\beta-i\alpha}\right]\sin(\alpha a)\cos(\alpha a)+i\times-\frac{i\beta}{\beta-i\alpha}\sin^2(\alpha a)-\frac{\beta}{\beta-i\alpha}\cos^2(\alpha a)\times-1$$

$$+\frac{\beta}{\beta-i\alpha}\exp(\beta b)\cosh(\beta b)\exp 2\{ik(a+b)\}-\frac{i\beta}{\alpha}\times-\frac{i\alpha}{\beta-i\alpha}\exp(\beta b)\sinh(\beta b)\exp 2\{ik(a+b)\}=0 \qquad (9.201)$$

or

$$\left[-\frac{\alpha}{\beta-i\alpha}\cosh(\beta b)-\frac{\beta^2}{\alpha(\beta-i\alpha)}\sinh(\beta b)+\frac{\alpha}{\beta-i\alpha}\exp(\beta b)\right]\exp\{ik(a+b)\}\sin(\alpha a)$$

$$-\left[\frac{\beta}{\beta-i\alpha}\exp(\beta b)+\frac{\beta}{\beta-i\alpha}\cosh(\beta b)-\frac{\beta}{\beta-i\alpha}\sinh(\beta b)\right]\exp\{ik(a+b)\}\cos(\alpha a)$$

$$+\left[\frac{\alpha}{\beta-i\alpha}-\frac{\alpha}{\beta-i\alpha}\right]\sin(\alpha a)\cos(\alpha a)+\frac{\beta}{\beta-i\alpha}\sin^2(\alpha a)+\frac{\beta}{\beta-i\alpha}\cos^2(\alpha a)$$

$$+\frac{\beta}{\beta-i\alpha}\exp(\beta b)\cosh(\beta b)\exp 2\{ik(a+b)\}-\frac{\beta}{\beta-i\alpha}\exp(\beta b)\sinh(\beta b)\exp 2\{ik(a+b)\}=0 \qquad (9.202)$$

or

$$\left[-\frac{\alpha}{\beta-i\alpha}\cosh(\beta b)-\frac{\beta^2}{\alpha(\beta-i\alpha)}\sinh(\beta b)+\frac{\alpha}{\beta-i\alpha}\exp(\beta b)\right]\exp\{ik(a+b)\}\sin(\alpha a)$$

$$-\left[\frac{\beta}{\beta-i\alpha}\exp(\beta b)+\frac{\beta}{\beta-i\alpha}\cosh(\beta b)-\frac{\beta}{\beta-i\alpha}\sinh(\beta b)\right]\exp\{ik(a+b)\}\cos(\alpha a)$$

$$+0\times\sin(\alpha a)\cos(\alpha a)+\frac{\beta}{\beta-i\alpha}\sin^2(\alpha a)+\frac{\beta}{\beta-i\alpha}\cos^2(\alpha a)$$

$$+\frac{\beta}{\beta-i\alpha}\exp(\beta b)\cosh(\beta b)\exp 2\{ik(a+b)\}-\frac{\beta}{\beta-i\alpha}\exp(\beta b)\sinh(\beta b)\exp 2\{ik(a+b)\}=0 \qquad (9.203)$$

or

$$\left[-\frac{\alpha}{\beta-i\alpha}\cosh(\beta b)-\frac{\beta^2}{\alpha(\beta-i\alpha)}\sinh(\beta b)+\frac{\alpha}{\beta-i\alpha}\{\cosh(\beta b)+\sinh(\beta b)\}\right]\exp\{ik(a+b)\}\sin(\alpha a)$$

$$-\left[\frac{\beta}{\beta-i\alpha}\{\cosh(\beta b)+\sinh(\beta b)\}+\frac{\beta}{\beta-i\alpha}\cosh(\beta b)-\frac{\beta}{\beta-i\alpha}\sinh(\beta b)\right]\exp\{ik(a+b)\}\cos(\alpha a)+0$$

$$+\frac{\beta}{\beta-i\alpha}\{\sin^2(\alpha a)+\cos^2(\alpha a)\}+\frac{\beta}{\beta-i\alpha}\{\cosh(\beta b)-\sinh(\beta b)\}\exp(\beta b)\exp 2\{ik(a+b)\}=0 \qquad (9.204)$$

since

$$\exp x=\cosh x+\sinh x \qquad (9.205)$$

Equation (9.204) reduces to

$$\left[-\frac{\alpha}{\beta-i\alpha}\cosh(\beta b)-\frac{\beta^2}{\alpha(\beta-i\alpha)}\sinh(\beta b)+\frac{\alpha}{\beta-i\alpha}\cosh(\beta b)+\frac{\alpha}{\beta-i\alpha}\sinh(\beta b)\right]\exp\{ik(a+b)\}\sin(\alpha a)$$

$$-\left[\frac{\beta}{\beta-i\alpha}\cosh(\beta b)+\frac{\beta}{\beta-i\alpha}\sinh(\beta b)+\frac{\beta}{\beta-i\alpha}\cosh(\beta b)-\frac{\beta}{\beta-i\alpha}\sinh(\beta b)\right]\exp\{ik(a+b)\}\cos(\alpha a)+\frac{\beta}{\beta-i\alpha}\times1$$

$$+\frac{\beta}{\beta-i\alpha}\{\cosh(\beta b)-\sinh(\beta b)\}\exp(\beta b)\exp2\{ik(a+b)\}=0 \tag{9.206}$$

or

$$\left[-\frac{\beta^2}{\alpha(\beta-i\alpha)}\sinh(\beta b)+\frac{\alpha}{\beta-i\alpha}\sinh(\beta b)\right]\exp\{ik(a+b)\}\sin(\alpha a)$$

$$-\left[\frac{\beta}{\beta-i\alpha}\cosh(\beta b)+\frac{\beta}{\beta-i\alpha}\cosh(\beta b)\right]\exp\{ik(a+b)\}\cos(\alpha a)$$

$$+\frac{\beta}{\beta-i\alpha}+\frac{\beta}{\beta-i\alpha}\{\cosh(\beta b)-\sinh(\beta b)\}\exp(\beta b)$$

$$\times\exp2\{ik(a+b)\}=0 \tag{9.207}$$

or

$$\left[-\frac{\beta^2}{\alpha(\beta-i\alpha)}+\frac{\alpha}{\beta-i\alpha}\right]\exp\{ik(a+b)\}\sinh(\beta b)\sin(\alpha a)$$

$$-\left[\frac{\beta}{\beta-i\alpha}+\frac{\beta}{\beta-i\alpha}\right]\exp\{ik(a+b)\}\cosh(\beta b)\cos(\alpha a)$$

$$+\frac{\beta}{\beta-i\alpha}+\frac{\beta}{\beta-i\alpha}\{\cosh(\beta b)-\sinh(\beta b)\}\exp(\beta b)\exp2$$

$$\times\{ik(a+b)\}=0 \tag{9.208}$$

or

$$\frac{\beta}{\beta-i\alpha}\left[-\frac{\beta}{\alpha}+\frac{\alpha}{\beta}\right]\exp\{ik(a+b)\}\sinh(\beta b)\sin(\alpha a)$$

$$-\frac{\beta}{\beta-i\alpha}(1+1)\exp\{ik(a+b)\}\cosh(\beta b)\cos(\alpha a)+\frac{\beta}{\beta-i\alpha}$$

$$+\frac{\beta}{\beta-i\alpha}\{\cosh(\beta b)-\sinh(\beta b)\}\exp(\beta b)\exp2\{ik(a+b)\}=0 \tag{9.209}$$

Dividing both sides of eq. (9.209) by $\beta/(\beta-i\alpha)$

$$\left[-\frac{\beta}{\alpha}+\frac{\alpha}{\beta}\right]\exp\{ik(a+b)\}\sinh(\beta b)\sin(\alpha a)$$

$$-(1+1)\exp\{ik(a+b)\}\cosh(\beta b)\cos(\alpha a)+1$$

$$+\{\cosh(\beta b)-\sinh(\beta b)\}\exp(\beta b)\exp2\{ik(a+b)\}=0 \tag{9.210}$$

or

$$\exp\{ik(a+b)\}\left[\begin{array}{l}\left(\dfrac{\alpha^2-\beta^2}{\alpha\beta}\right)\sinh(\beta b)\sin(\alpha a)-2\cosh(\beta b)\cos(\alpha a)\\ +\exp\{-ik(a+b)\}+\{\cosh(\beta b)-\sinh(\beta b)\}\\ \times\exp(\beta b)\exp\{ik(a+b)\}\end{array}\right]=0 \tag{9.211}$$

or

$$\left(\frac{\alpha^2-\beta^2}{\alpha\beta}\right)\sinh(\beta b)\sin(\alpha a)-2\cosh(\beta b)\cos(\alpha a)$$

$$+\exp\{-ik(a+b)\}$$

$$+\{\cosh(\beta b)-\sinh(\beta b)\}\exp(\beta b)\exp\{ik(a+b)\}=0 \tag{9.212}$$

or

$$\left(\frac{\alpha^2-\beta^2}{\alpha\beta}\right)\sinh(\beta b)\sin(\alpha a)-2\cosh(\beta b)\cos(\alpha a)$$

$$+\exp\{-ik(a+b)\}+\exp(-\beta b)\exp(\beta b)\exp\{ik(a+b)\}=0 \tag{9.213}$$

since

$$\cosh x-\sinh x=\exp(-x) \tag{9.214}$$

Equation (9.213) can be written as

$$\left(\frac{\alpha^2 - \beta^2}{\alpha\beta}\right)\sinh(\beta b)\sin(\alpha a) - 2\cosh(\beta b)\cos(\alpha a)$$

$$+ \exp\{-ik(a+b)\} + \exp\{ik(a+b)\} = 0 \qquad (9.215)$$

or

$$\left(\frac{\alpha^2 - \beta^2}{\alpha\beta}\right)\sinh(\beta b)\sin(\alpha a) - 2\cosh(\beta b)\cos(\alpha a)$$

$$+ 2\cos\{k(a+b)\} = 0 \qquad (9.216)$$

since

$$\frac{\exp(x) + \exp(-x)}{2} = \cos x \qquad (9.217)$$

Equation (9.216) is relooked upon:

$$\left(\frac{\alpha^2 - \beta^2}{\alpha\beta}\right)\sinh(\beta b)\sin(\alpha a) - 2\cosh(\beta b)\cos(\alpha a)$$

$$+ 2\cos\{k(a+b)\} = 0 \qquad (9.218)$$

Dividing throughout by -2

$$\left(\frac{\alpha^2 - \beta^2}{-2\alpha\beta}\right)\sinh(\beta b)\sin(\alpha a) - \frac{2\cosh(\beta b)\cos(\alpha a)}{-2} + \frac{2\cos\{k(a+b)\}}{-2} = 0$$

$$(9.219)$$

or

$$\left(\frac{\beta^2 - \alpha^2}{2\alpha\beta}\right)\sinh(\beta b)\sin(\alpha a) + \cosh(\beta b)\cos(\alpha a) - \cos\{k(a+b)\} = 0$$

$$(9.220)$$

or

$$\left(\frac{\beta^2 - \alpha^2}{2\alpha\beta}\right)\sinh(\beta b)\sin(\alpha a) + \cosh(\beta b)\cos(\alpha a) = \cos\{k(a+b)\}$$

$$(9.221)$$

which is eq. (9.169).

REFERENCES

Kittel, C. 2005. *Introduction to Solid State Physics*. John Wiley & Sons, Inc.: Hoboken, NJ, pp. 161–184.

Koole, R., E. Groeneveld, D. Vanmaekelbergh, A. Meijerink and C. de Mello Donegá. 2014. Chapter 2: Size effects on semiconductor nanoparticles. In: C. de Mello Donegá (Ed.), *Nanoparticles: Workhorses of Nanoscience*. Springer-Verlag: Berlin Heidelberg, pp. 13–51.

Kronig, R. de L. and W. G. Penney. 1931 Quantum mechanics of electrons in crystal lattices. *Proceedings of the Royal Society of London A* 130: 499–513.

Mcquarrie, D. A. 1996. The Kronig–Penney model: A single lecture illustrating the band structure of solids. *The Chemical Educator* 1(1): 1–10.

Reshodko, I., A. Benseny, J. Romhányi and T. Busch. 2019. Topological states in the Kronig–Penney model with arbitrary scattering potentials. *New Journal of Physics* 21: 013010.

Part III

Electron Behavior in Nanostructures

10

Quantum Confinement and Electronic Structure of Quantum Dots

10.1 Length Scale for Quantum Confinement

As already defined in Chapter 1, quantum confinement is concerned with the group of phenomena, notably the modification of free-particle dispersion relation and consequent effects, e.g. discretization of energy states and enlarging of bandgap, arising from the spatial confinement of electrons and holes in a nanostructure or nanomaterial having dimensions of the length scale of de Broglie wavelength of electrons. Zero-dimensional quantum dots exhibit quantum confinement in three dimensions (3D confinement). Likewise, two-dimensional (2D) confinement is observed in one-dimensional (1D) quantum wires while 2D quantum wells show 1D confinement.

We are curious to know: How do we specify this spatial confinement? What is the characteristic length to observe quantum effect? It is described with reference to the spatial extension of exciton (excited electron and accompanying hole) expressed in terms of the exciton Bohr radius. In semiconductor materials of interest, the relevant ultra-small length scale for occurrence of quantum confinement stretches over 2–50 nm because the exciton Bohr radius in these materials lies in this range. When the electron wavelength becomes the same or is nearby the exciton Bohr radius in a material, quantum confinement is inevitable.

10.2 Recapitulation of the Bohr Radius

In atomic physics, the Bohr radius is the radius of the ground state of a hydrogen atom calculated from a combination of classical and quantum mechanics on a simple atomic model postulating that the electrons revolve in concentric circular orbits around the nucleus containing protons and neutrons bound by the Coulomb attractive force between the negatively charged electrons and the positively charged nucleus. The model is founded on two basic assumptions (Section 2.4):

i. The first assumption is regarding the forces responsible for uniform circular motion of the electron around the nucleus. This motion takes place with the center-seeking or centripetal force = the attractive electrostatic force between the electron and the nucleus. If m_e is the mass of the electron and e is the electronic charge, and electron is moving with an orbital velocity v in a circular orbit of radius r,

$$\frac{m_e v^2}{r} = \frac{Ze^2}{4\pi\varepsilon_0 r^2} = \frac{e^2}{4\pi\varepsilon_0 r^2} \tag{10.1}$$

where Z is the atomic number of the element (=1 for hydrogen) and ε_0 is the permittivity of free space.

ii. The second assumption relates to the condition that must be fulfilled by the angular momentum of the electron. The angular momentum is discrete or quantized changing discontinuously or in a stepped manner governed by the rule

$$m_e vr = \frac{nh}{2\pi} \tag{10.2}$$

where n is the principal quantum number which can acquire only integral values and h is Planck's constant.

Solving for the velocity v we get

$$v = \frac{nh}{2\pi m_e r} \tag{10.3}$$

Substituting for v from eq. (10.3) in eq. (10.1) we have

$$\frac{m_e}{r}\left(\frac{nh}{2\pi m_e r}\right)^2 = \frac{e^2}{4\pi\varepsilon_0 r^2} \tag{10.4}$$

or

$$m_e n^2 h^2 \times 4\pi\varepsilon_0 r^2 = e^2 \times r \times 4\pi^2 m_e^2 r^2 \tag{10.5}$$

$$\therefore r = \frac{m_e n^2 h^2 \times 4\pi\varepsilon_0}{e^2 \times 4\pi^2 m_e^2} = \frac{\varepsilon_0 n^2 h^2}{\pi m_e e^2} \tag{10.6}$$

For the ground state $n = 1$; hence

$$r = \frac{\varepsilon_0 h^2}{\pi m_e e^2} \tag{10.7}$$

Replacing h by the reduced Planck's constant in eq. (10.7)

$$\hbar = \frac{h}{2\pi} \tag{10.8}$$

or

$$h = 2\pi\hbar \tag{10.9}$$

we may write eq. (10.7) as

$$r = \frac{\varepsilon_0 (2\pi\hbar)^2}{\pi m_e e^2} = \frac{4\pi\varepsilon_0 \hbar^2}{m_e e^2} \tag{10.10}$$

So, the Bohr radius is

$$a_0 = \frac{4\pi\varepsilon_0\hbar^2}{m_e e^2}$$

$$= \frac{4\times 3.1415\times 8.854\times 10^{-12}\ \mathrm{F\,m^{-1}}\times\left(1.05457\times 10^{-34}\ \mathrm{J\,s}\right)^2}{9.109\times 10^{-31}\ \mathrm{kg}\times\left(1.602\times 10^{-19}\ \mathrm{C}\right)^2}$$

$$= \frac{123.7335\times 10^{-80}}{23.3774\times 10^{-69}} = 5.293\times 10^{-11}\ \mathrm{m} = 0.05293\times 10^{-9}\ \mathrm{m}$$

$$= 0.0529\ \mathrm{nm} \tag{10.11}$$

10.3 Exciton, Exciton Bohr Radius, and Exciton Binding Energy

10.3.1 Exciton

An exciton is an electrically quasi-neutral particle in a semiconductor or insulator representing an electron excited from the valence to conduction band, e.g. by absorption of a photon, but still bound to its associated hole left behind in the valence band by the Coulombic force of electrostatic attraction (Figure 10.1). This coupled electron-hole pair is a mobile entity or concentration of energy in the crystal. Ultimately, the electron will recombine with the hole resulting in the decay of the exciton but after a finite lifetime.

The exciton may be compared with the hydrogen atom in which the electron orbits around the proton but it is strikingly dissimilar to the hydrogen atom because the electron is much lighter than the proton and therefore the radius of the exciton is much larger than that of the hydrogen atom.

In a high-permittivity semiconductor material, the Coulomb force between electron and hole is weak. Consequently, the electron-hole binding energy is low ~0.01 eV and hence the exciton radius is larger. This exciton is called a Wannier-Mott exciton.

In a low-permittivity insulating material, the Coulomb force between the electron and the hole is large. As a result, the electron-hole binding energy is high (~0.1–1 eV) and exciton radius is small. This exciton is known as a Frenkel exciton.

Three types of excitons are found (Figure 10.2): Wannier-Mott exciton, Frenkel exciton, and charge-transfer excitons. Wannier-Mott excitons are delocalized over a large number of unit cells of the lattice. They can move freely through the crystal and hence are also called free excitons. Their binding energy is very low (~0.01 eV). The binding energy for GaAs is 0.004 eV. The radius, i.e. mean electron-to-hole distance (~40–100 Å), is an order of magnitude larger than the intermolecular distance. It is 10–100 times the lattice constant. Frenkel excitons are localized on a single molecule. They are found in insulating solid rare gases, alkali halide crystals, and molecular organic crystals. Radius ≤ 5 Å. So, the radius of Wannier-Mott excitons ≫ the radius of Frenkel excitons. The binding energy of Frenkel excitons is very high (~0.1–1 eV), and hence these excitons are also known as tightly bound excitons. They move by hopping from one atom to another. Charge-transfer excitons are produced when the electron and hole occupy adjoining molecules, e.g. in an organic semiconductor. Radius = 1–2 × nearest neighbor intermolecular separation.

10.3.2 Exciton Bohr Radius

The exciton Bohr radius is the average distance between an electron and hole in an exciton in the ground state. It is denoted by a_X. In SI units,

$$a_X = \frac{4\pi\varepsilon_0\varepsilon_r\hbar^2}{\mu e^2} \tag{10.12}$$

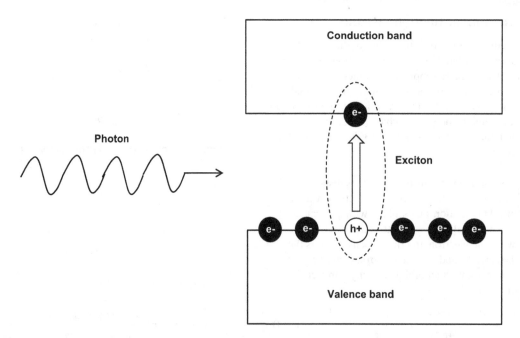

FIGURE 10.1 Formation of an exciton by absorption of a photon by an atom and excitation of an electron (e⁻) from valence to conduction band leaving behind a hole (h⁺) in the valence band.

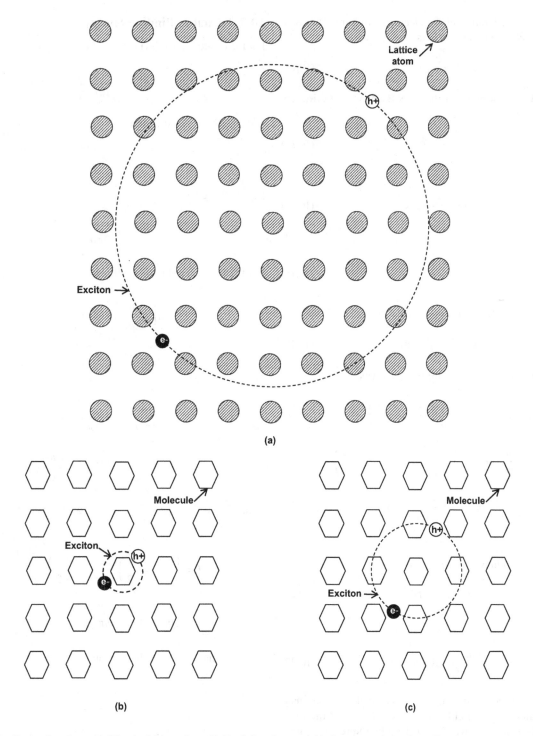

FIGURE 10.2 Types of excitons: (a) Wannier-Mott exciton, (b) Frenkel exciton, and (c) charge-transfer exciton. Wannier-Mott excitons appear in high dielectric constant inorganic semiconductors.

where μ is the reduced mass

$$\mu = \frac{m_e^* m_h^*}{m_e^* + m_h^*} \qquad (10.13)$$

with m_e^* and m_h^* being the effective masses of electron and hole, respectively; ε_r is the dielectric constant of the semiconductor.

To derive the equation for the exciton Bohr radius, let us apply Bohr's model to the exciton. The exciton is visualized as an electron and a hole orbiting around each other. Instead of the electron alone, we are now faced with a two-body problem in which the two bodies are the electron and the hole. The two-body problem must be reduced to a one-body problem to apply Bohr's model. By Newton's second law of motion, the force exerted by the hole on the electron is

$$\mathbf{F}_{eh} = m_e^* \mathbf{a}_e \qquad (10.14)$$

where $\mathbf{a}_e$ is the acceleration of the electron. Similarly, the force exerted by the electron on the hole is

$$\mathbf{F}_{he} = m_h^* \mathbf{a}_h \tag{10.15}$$

where $\mathbf{a}_h$ is the acceleration of the hole. By Newton's third law of motion,

$$\mathbf{F}_{eh} = -\mathbf{F}_{he} \tag{10.16}$$

or

$$m_e^* \mathbf{a}_e = -m_h^* \mathbf{a}_h \tag{10.17}$$

or

$$\mathbf{a}_h = -\frac{m_e^* \mathbf{a}_e}{m_h^*} \tag{10.18}$$

The relative acceleration between the electron and the hole is

$$\mathbf{a}_{\text{Relative}} = \mathbf{a}_e - \mathbf{a}_h \tag{10.19}$$

Substituting for $\mathbf{a}_h$ from eq. (10.18) in eq. (10.19),

$$\mathbf{a}_{\text{Relative}} = \mathbf{a}_e - \mathbf{a}_h = \mathbf{a}_e - \left(-\frac{m_e^* \mathbf{a}_e}{m_h^*}\right) = \mathbf{a}_e + \frac{m_e^* \mathbf{a}_e}{m_h^*}$$

$$= \left(1 + \frac{m_e^*}{m_h^*}\right)\mathbf{a}_e = \left(\frac{m_h^* + m_e^*}{m_h^*}\right)\mathbf{a}_e = \left(\frac{m_h^* + m_e^*}{m_h^* m_e^*}\right)m_e^* \mathbf{a}_e$$

$$= \left(\frac{m_e^* + m_h^*}{m_e^* m_h^*}\right)m_e^* \mathbf{a}_e = \left(\frac{m_e^* + m_h^*}{m_e^* m_h^*}\right)\mathbf{F}_{eh} \tag{10.20}$$

by applying eq. (10.14).

Equation (10.20) may be rewritten as

$$\mathbf{a}_{\text{Relative}} = \mu \mathbf{F}_{eh} \tag{10.21}$$

where

$$\mu = \frac{m_e^* m_h^*}{m_e^* + m_h^*} = \text{Reduced mass} \tag{10.22}$$

allows looking at the electron and hole as a single body of mass μ. It is the same as eq. (10.13). Thus in eq. (10.7) for the Bohr radius, the electron mass will be replaced by the reduced mass of the electron-hole system when we are dealing with the exciton. Another change is that the electron and the hole are moving inside a semiconductor material so that the free space permittivity ε_0 will be replaced by the dielectric scaling factor $\varepsilon_0 \varepsilon_r$ where ε_r is the relative permittivity of the semiconductor material. On making these two changes, eq. (10.10) becomes

$$a_X = \frac{4\pi\varepsilon_0 \varepsilon_r \hbar^2}{\mu e^2} \tag{10.23}$$

which is the same as eq. (10.12).

10.3.3 Exciton Binding Energy

For the exciton, eq. (10.1) is altered to

$$\frac{\mu v^2}{r} = \frac{e^2}{4\pi\varepsilon_0 \varepsilon_r r^2} \tag{10.24}$$

and eq. (10.2) is changed to

$$\mu v r = \frac{nh}{2\pi} \tag{10.25}$$

Eq. (10.25) gives

$$v = \frac{nh}{2\pi\mu r} \tag{10.26}$$

Putting the value of v in eq. (10.24)

$$\frac{\mu\left(\dfrac{nh}{2\pi\mu r}\right)^2}{r} = \frac{e^2}{4\pi\varepsilon_0 \varepsilon_r r^2} \tag{10.27}$$

or

$$\frac{\mu n^2 h^2}{4\pi^2 \mu^2 r^3} = \frac{e^2}{4\pi\varepsilon_0 \varepsilon_r r^2} \tag{10.28}$$

$$\therefore r = \frac{\mu n^2 h^2 \times 4\pi\varepsilon_0 \varepsilon_r}{4\pi^2 \mu^2 e^2} = \frac{n^2 h^2 \varepsilon_0 \varepsilon_r}{\pi\mu e^2} \tag{10.29}$$

Using eq. (10.25), the kinetic energy of the exciton is

$$T = \frac{1}{2}\mu v^2 = \left(\frac{1}{2}\right)\frac{(\mu v r)^2}{\mu r^2} = \left(\frac{1}{2}\right)\frac{\left(\dfrac{nh}{2\pi}\right)^2}{\mu r^2}$$

$$= \frac{n^2 h^2}{2\mu r^2 \times 4\pi^2} = \frac{n^2 h^2}{8\pi^2 \mu r^2} \tag{10.30}$$

The potential energy of the exciton is

$$U = -\frac{e^2}{4\pi\varepsilon_0 \varepsilon_r r} \tag{10.31}$$

From eqs. (10.30) and (10.31), the total energy of the exciton is

$$E = T + U = \frac{n^2 h^2}{8\pi^2 \mu r^2} - \frac{e^2}{4\pi\varepsilon_0 \varepsilon_r r} = \frac{n^2 h^2}{8\pi^2 \mu \left(\dfrac{n^2 h^2 \varepsilon_0 \varepsilon_r}{\pi\mu e^2}\right)^2}$$

$$- \frac{e^2}{4\pi\varepsilon_0 \varepsilon_r \left(\dfrac{n^2 h^2 \varepsilon_0 \varepsilon_r}{\pi\mu e^2}\right)} \tag{10.32}$$

where r has been substituted from eq. (10.29). Hence,

$$E = \frac{n^2 h^2 \pi^2 \mu^2 e^4}{8\pi^2 \mu n^4 h^4 \varepsilon_0^2 \varepsilon_r^2} - \frac{\mu e^4}{4n^2 h^2 \varepsilon_0^2 \varepsilon_r^2} = \frac{\mu e^4}{8n^2 h^2 \varepsilon_0^2 \varepsilon_r^2}$$

$$-\frac{\mu e^4}{4n^2 h^2 \varepsilon_0^2 \varepsilon_r^2} = \frac{\mu e^4 - 2\mu e^4}{8n^2 h^2 \varepsilon_0^2 \varepsilon_r^2} = -\frac{\mu e^4}{8n^2 h^2 \varepsilon_0^2 \varepsilon_r^2}$$

$$= -\frac{\mu e^4}{8n^2 (2\pi\hbar)^2 \varepsilon_0^2 \varepsilon_r^2} = -\frac{\mu e^4}{32\pi^2 n^2 \hbar^2 \varepsilon_0^2 \varepsilon_r^2} \quad (10.33)$$

The negative energy indicates that work must be done against the electric field to separate the charges. Comparing it with eq. (6.149) for hydrogen atom derived from the Schrodinger equation bears eloquent testimony to the success of Bohr's model.

The Rydberg unit of energy is

$$1\text{Ry} = \frac{m_e e^4}{8h^2 \varepsilon_0^2} = \frac{m_e e^4}{8(2\pi\hbar)^2 \varepsilon_0^2} = \frac{m_e e^4}{32\pi^2 \hbar^2 \varepsilon_0^2} \quad (10.34)$$

Like the hydrogen atom, the exciton Rydberg energy is

$$1\text{Ry}^* = \frac{\mu e^4}{32\pi^2 \hbar^2 \varepsilon_0^2 \varepsilon_r^2} \quad (10.35)$$

by replacing m_e by μ.

$$\therefore E = -\frac{1}{n^2} \text{Ry}^* \quad (10.36)$$

by applying eq. (10.33). To express E in terms of Ry, we note from eqs. (10.33) and (10.34) that

$$E = -\frac{\mu e^4}{32\pi^2 n^2 \hbar^2 \varepsilon_0^2 \varepsilon_r^2} = -\left(\frac{\mu}{m_e}\right)\frac{m_e e^4}{32\pi^2 n^2 \hbar^2 \varepsilon_0^2 \varepsilon_r^2}$$

$$= -\left(\frac{\mu}{m_e}\right)\left(\frac{m_e e^4}{32\pi^2 \hbar^2 \varepsilon_0^2}\right)\left(\frac{1}{\varepsilon_r^2 n^2}\right) = -\left(\frac{\mu}{m_e}\right)\left(\frac{1}{\varepsilon_r^2 n^2}\right)\text{Ry} \quad (10.37)$$

But from eqs. (10.11) and (10.12)

$$\frac{a_0}{a_X} = \frac{4\pi\varepsilon_0\hbar^2}{m_e e^2} \times \frac{\mu e^2}{4\pi\varepsilon_0\varepsilon_r\hbar^2} = \frac{\mu}{m_e\varepsilon_r} \quad (10.38)$$

$$\therefore \frac{\mu}{m_e} = \left(\frac{a_0}{a_X}\right)\varepsilon_r \quad (10.39)$$

Substituting for μ/m_e from eq. (10.39) in eq. (10.37),

$$E = -\left(\frac{a_0}{a_X}\right)\varepsilon_r\left(\frac{1}{\varepsilon_r^2 n^2}\right)\text{Ry} = -\frac{1}{\varepsilon_r}\left(\frac{a_0}{a_X}\right)\frac{\text{Ry}}{n^2} \quad (10.40)$$

10.4 Weak, Moderate, and Strong Quantum Confinement

10.4.1 Exciton, Electron, and Hole Bohr Radii

We have derived the equation for the exciton Bohr radius previously, but to delve deeply into the quantum confinement, we need to extend the analysis further, and need similar parameters

for electrons and holes. Recalling the exciton Bohr radius, we note that

$$a_X = \frac{4\pi\varepsilon_0\varepsilon_r\hbar^2}{\mu e^2} = \frac{4\pi\varepsilon_0\hbar^2}{e^2} \times \frac{\varepsilon_r}{\mu}$$

$$= \frac{4 \times 3.14159 \times 8.854 \times 10^{-12}\,\text{Fm}^{-1} \times (1.05457 \times 10^{-34}\,\text{J}-\text{s})^2}{(1.602 \times 10^{-19}\,\text{C})^2}$$

$$\times \frac{\varepsilon_r}{\mu} = \frac{123.7371 \times 10^{-80}}{2.5664 \times 10^{-38}}\left(\frac{\varepsilon_r}{\mu}\right) = 48.214 \times 10^{-42}\left(\frac{\varepsilon_r}{\mu}\right)\text{m}$$

$$= 4.82 \times 10^{-41}\left(\frac{\varepsilon_r}{\mu}\right)\text{m} \quad (10.41)$$

Let us now turn our attention to the two other important parameters, viz. the Bohr radii for electrons and holes.

The Bohr radius of an electron, symbolized as a_{Electron}, is a fundamental physical quantity representing the most probable distance between the electron and the nucleus in a hydrogen atom in the ground or lowest energy state according to Bohr's model. It is given by

$$a_{\text{Electron}} = \frac{4\pi\varepsilon_0\varepsilon_r\hbar^2}{m_e^* e^2} = 4.82 \times 10^{-41}\left(\frac{\varepsilon_r}{m_e^*}\right)\text{m} \quad (10.42)$$

where ε_0 is the permittivity of free space, ε_r is the relative permittivity of the medium, $\hbar$ is reduced Planck's constant, and e is the elementary charge; m stands for meter (not mass).

The Bohr radius of a hole, symbolized as a_{Hole}, is

$$a_{\text{Hole}} = \frac{4\pi\varepsilon_0\varepsilon_r\hbar^2}{m_h^* e^2} = 4.82 \times 10^{-41}\left(\frac{\varepsilon_r}{m_h^*}\right)\text{m} \quad (10.43)$$

10.4.2 Explaining the Three Confinement Regimes

If R is the radius of the nanostructure, three regimes are distinguished as follows (Yoffe 1993, Barbagiovanni et al 2012): (i) Weak confinement: $R > a_X$, $R > a_{\text{Electron}}$, $R > a_{\text{Hole}}$. (ii) Moderate confinement: $R \sim a_X$, $a_{\text{Electron}} > R > a_{\text{Hole}}$. Only electrons experience confinement. (iii) Strong confinement: $R < a_X$, $R < a_{\text{Electron}}$, $R < a_{\text{Hole}}$.

10.5 Dispersion Relation for Excitons

Like the hydrogen atom (Sections 6.1 and 6.2), the interaction between electron and hole, according to Coulomb's law, to form the exciton state is described by a two-particle Schrodinger equation

$$\left\{-\frac{\hbar^2}{2m_e}\left(\frac{\partial^2}{\partial x_e^2} + \frac{\partial^2}{\partial y_e^2} + \frac{\partial^2}{\partial z_e^2}\right) - \frac{\hbar^2}{2m_h}\left(\frac{\partial^2}{\partial x_h^2} + \frac{\partial^2}{\partial y_h^2} + \frac{\partial^2}{\partial z_h^2}\right)\right.$$

$$\left. + V(x_e, y_e, z_e, x_h, y_h, z_h)\right\}\Psi(x_e, y_e, z_e, x_h, y_h, z_h)$$

$$= E\Psi(x_e, y_e, z_e, x_h, y_h, z_h) \quad (10.44)$$

where the subscript p for proton is replaced by h for hole. Proceeding further, like the hydrogen atom, we can write the modified wave equation for the exciton as

$$-\frac{\hbar^2}{2M}\left(\frac{\partial^2\Psi}{\partial X^2}+\frac{\partial^2\Psi}{\partial Y^2}+\frac{\partial^2\Psi}{\partial Z^2}\right)-\frac{\hbar^2}{2\mu}\left(\frac{\partial^2\Psi}{\partial x^2}+\frac{\partial^2\Psi}{\partial y^2}+\frac{\partial^2\Psi}{\partial z^2}\right)$$

$$+V(x,y,z)\Psi=E\Psi \qquad (10.45)$$

where

$$x=x_e-x_h,\, y=y_e-y_h,\, z=z_e-z_h \qquad (10.46)$$

$$X=\frac{m_e x_e+m_h x_h}{M},\, Y=\frac{m_e y_e+m_h y_h}{M},\, Z=\frac{m_e z_e+m_h z_h}{M} \qquad (10.47)$$

$$M=m_e+m_h \qquad (10.48)$$

Equation (10.45) is further modified to give

$$-\frac{\hbar^2}{2M}\nabla_R^2\Psi(\mathbf{R},\mathbf{r})-\frac{\hbar^2}{2\mu}\nabla_r^2\Psi(\mathbf{R},\mathbf{r})+V(\mathbf{r})\Psi(\mathbf{R},\mathbf{r})=E\Psi(\mathbf{R},\mathbf{r})$$

$$(10.49)$$

where the symbols have the same meanings as before. Equation (10.49) is broken into two parts with

$$-\frac{1}{S(\mathbf{R})}\frac{\hbar^2}{2M}\nabla_R^2\{S(\mathbf{R})\}=E_{\text{Constant}} \qquad (10.50)$$

as the equation for a free particle of mass *M*, describing the translational center-of-mass motion of the electron-hole system, and the following equation:

$$-\frac{1}{U(\mathbf{r})}\left\{-\frac{\hbar^2}{2\mu}\nabla_r^2+V(\mathbf{r})\right\}\{U(\mathbf{r})\}=E_{\text{Constant}}-E=E_H \quad (10.51)$$

describing the internal states of the electron-hole pair. Equation (10.50) leads to the kinetic energy of translational motion

$$E_{\text{Constant}}=\frac{\hbar^2\mathbf{K}^2}{2M} \qquad (10.52)$$

where **K** is the wave vector of the exciton.

Equation (10.51) gives the hydrogen-like discrete energy levels numbered as $\bar{n}$

$$E_n=-\frac{\mu e^4}{32\pi^2\varepsilon_0{}^2\varepsilon_r{}^2\hbar^2\bar{n}^2}=-\left(\frac{\mu e^4}{32\pi^2\hbar^2\varepsilon_0^2\varepsilon_r{}^2}\right)\frac{1}{\bar{n}^2}$$

$$=-\text{Ry}^*\times\frac{1}{\bar{n}^2}=-\frac{\text{Ry}^*}{\bar{n}^2} \qquad (10.53)$$

where from eq. (10.35)

$$\frac{\mu e^4}{32\pi^2\hbar^2\varepsilon_0^2\varepsilon_r^2}=\text{Ry}^*=\text{Exciton Rydberg energy} \qquad (10.54)$$

From eqs. (10.52) and (10.53), the dispersion relation relating the exciton energy with wave vector **K** is written as

$$E_n(\mathbf{K})=E_G-\frac{\text{Ry}^*}{\bar{n}^2}+\frac{\hbar^2\mathbf{K}^2}{2M} \qquad (10.55)$$

where E_G is the energy gap of the semiconductor. To understand how the band structure of a semiconductor is influenced by quantum confinement, we calculate the energy of electrons and holes in such an environment by finding the energy of the particle in a spherical box.

10.6 Confinement Energies of Electrons and Holes

The band structure of bulk semiconductors is described by considering an infinite crystal. While extending this description to a semiconductor nanocrystal such as a quantum dot, a correction factor is applied to the Bloch wave function $\Psi_{\text{Bloch}}(x)$ of the bulk semiconductor. This factor has the form of an envelope function $\phi_{\text{Envelope}}(x)$. Thus the corrected wave function $\Psi_{\text{Corrected}}(x)$ is expressed as the product

Corrected wave function $\Psi_{\text{Corrected}}(x)$

= Bloch wave function describing the bulk semiconductor

$\{\Psi_{\text{Bloch}}(x)\}\times$ Envelope function describing the effects of

confinement of electrons, holes and excitons $\{\phi_{\text{Envelope}}(x)\}$

$$(10.56)$$

or

$$\Psi_{\text{Corrected}}(x)=\Psi_{\text{Bloch}}(x)\times\phi_{\text{Envelope}}(x) \qquad (10.57)$$

The envelope function $\phi_{\text{Envelope}}(x)$ is the solution of the Schrodinger equation for a particle in a spherically symmetric potential with an infinite barrier (Figure 10.3):

$$V(r)=0 \quad \text{for } 0\le r\le a \qquad (10.58)$$

and

$$V(r)=\infty \quad \text{for } r>a \qquad (10.59)$$

Within the defined region, the function must be well-behaved. It should be square-integrable at $r=0$ and $=0$ at $r=a$, the radius of the sphere. Essentially, the small piece of semiconductor material in which the exciton is confined is treated as the problem of a particle in a spherical box.

Inside the spherical box $V(r)=0$ so that the Schrodinger equation in spherical polar coordinates (Sections 6.3-6.6)

$$\frac{1}{r^2}\frac{\partial}{\partial r}\left(r^2\frac{\partial\Psi}{\partial r}\right)+\frac{1}{r^2\sin\theta}\frac{\partial}{\partial\theta}\left(\sin\theta\frac{\partial\Psi}{\partial\theta}\right)+\frac{1}{r^2\sin^2\theta}\frac{\partial^2\Psi}{\partial\phi^2}$$

$$+\frac{2m_e^*}{\hbar^2}\{E+V(r)\}\Psi=0 \qquad (10.60)$$

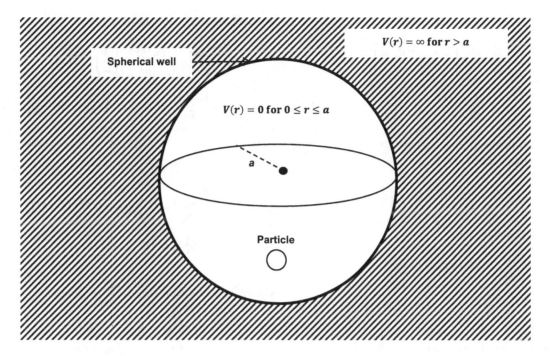

FIGURE 10.3 Particle in an infinite spherical potential well of radius a.

becomes

$$\frac{1}{r^2}\frac{\partial}{\partial r}\left(r^2\frac{\partial\Psi}{\partial r}\right)+\frac{1}{r^2\sin\theta}\frac{\partial}{\partial\theta}\left(\sin\theta\frac{\partial\Psi}{\partial\theta}\right)$$
$$+\frac{1}{r^2\sin^2\theta}\frac{\partial^2\Psi}{\partial\phi^2}+\frac{2m_e^*}{\hbar^2}E\Psi=0 \quad (10.61)$$

where m_e^* is the effective mass of electrons.

Within the region $0<r<a$, the wave function Ψ is non-zero and square-integrable at $r=0$ and $\Psi=0$ at $r=a$. Separating the wave function $\Psi(r,\theta,\phi)$ into radial $R_{n,l}(r)$ and angular $Y_{l,m}(\theta,\phi)$ parts

$$\Psi(r,\theta,\phi)=R_{n,l}(r)\times Y_{l,m}(\theta,\phi) \quad (10.62)$$

where n, l, and m are quantum numbers.

In the region $0<r<a$, the radial function $R_{n,l}$ satisfies the equation (see eq. (6.74))

$$\frac{1}{R_{n,l}(r)}\frac{\partial}{\partial r}\left\{r^2\frac{\partial R_{n,l}(r)}{\partial r}\right\}+\frac{2m_e^*r^2E}{\hbar^2}=l(l+1) \quad (10.63)$$

or

$$\frac{r^2}{R_{n,l}(r)}\frac{\partial^2 R_{n,l}(r)}{\partial r^2}+\frac{2r}{R_{n,l}(r)}\frac{\partial R_{n,l}(r)}{\partial r}+\frac{2m_e^*r^2E}{\hbar^2}=l(l+1)$$
$$(10.64)$$

Multiplying both sides by $R_{n,l}(r)$,

$$r^2\frac{\partial^2 R_{n,l}(r)}{\partial r^2}+2r\frac{\partial R_{n,l}(r)}{\partial r}+\frac{2m_e^*r^2E}{\hbar^2}R_{n,l}(r)=l(l+1)R_{n,l}(r)$$
$$(10.65)$$

or

$$r^2\frac{\partial^2 R_{n,l}(r)}{\partial r^2}+2r\frac{\partial R_{n,l}(r)}{\partial r}+\left\{\frac{2m_e^*r^2E}{\hbar^2}-l(l+1)\right\}R_{n,l}(r)=0$$
$$(10.66)$$

or

$$r^2\frac{\partial^2 R_{n,l}(r)}{\partial r^2}+2r\frac{\partial R_{n,l}(r)}{\partial r}+\left\{\left(\sqrt{\frac{2m_e^*E}{\hbar^2}}\right)^2 r^2-l(l+1)\right\}$$
$$\times R_{n,l}(r)=0 \quad (10.67)$$

or

$$r^2\frac{\partial^2 R_{n,l}(r)}{\partial r^2}+2r\frac{\partial R_{n,l}(r)}{\partial r}+\left\{k^2r^2-l(l+1)\right\}R_{n,l}(r)=0 \quad (10.68)$$

where

$$k=\sqrt{\frac{2m_e^*E}{\hbar^2}} \quad (10.69)$$

Equation (10.68) is the spherical Bessel differential equation. Introducing the scaled variable

$$x=kr \quad (10.70)$$

we have

$$r\frac{\partial R_{n,l}(r)}{\partial r}=\left(\frac{kr}{k}\right)\frac{\partial R_{n,l}(r)}{\partial r}=kr\frac{\partial R_{n,l}(r)}{\partial(kr)}=x\frac{\partial R_{n,l}(r)}{\partial x} \quad (10.71)$$

and

$$r^2 \frac{\partial^2 R_{n,l}(r)}{\partial r^2} = \left(\frac{k^2 r^2}{k^2}\right) \frac{\partial^2 R_{n,l}(r)}{\partial r^2}$$

$$= k^2 r^2 \frac{\partial^2 R_{n,l}(r)}{\partial (k^2 r^2)} = x^2 \frac{\partial^2 R_{n,l}(r)}{\partial x^2} \qquad (10.72)$$

By making substitutions from eqs. (10.71) and (10.72) in eq. (10.68), it is transformed to

$$x^2 \frac{\partial^2 R_{n,l}(r)}{\partial x^2} + 2x \frac{\partial R_{n,l}(r)}{\partial x} + \left\{x^2 - l(l+1)\right\} R_{n,l}(r) = 0 \quad (10.73)$$

Let us try a solution of the form

$$R_{n,l}(r) = \xi(x) x^{-1/2} \qquad (10.74)$$

Then

$$\frac{\partial R_{n,l}(r)}{\partial x} = \frac{\partial}{\partial x}\left\{\xi(x) x^{-\frac{1}{2}}\right\} = x^{-\frac{1}{2}} \frac{\partial \xi(x)}{\partial x} + \xi(x)\left(-\frac{1}{2}\right) x^{-\frac{1}{2}-1}$$

$$= x^{-\frac{1}{2}} \frac{\partial \xi(x)}{\partial x} - \frac{1}{2}\xi(x) x^{-\frac{3}{2}} \qquad (10.75)$$

and

$$\frac{\partial^2 R_{n,l}(r)}{\partial x^2} = \frac{\partial}{\partial x}\left\{x^{-\frac{1}{2}} \frac{\partial \xi(x)}{\partial x} - \frac{1}{2}\xi(x) x^{-\frac{3}{2}}\right\}$$

$$= x^{-\frac{1}{2}} \frac{\partial \xi^2(x)}{\partial x^2} - \left(\frac{1}{2}\right) x^{-\frac{1}{2}-1} \frac{\partial \xi(x)}{\partial x} - \left(\frac{1}{2}\right) \frac{\partial \xi(x)}{\partial x} x^{-\frac{3}{2}}$$

$$- \left(\frac{1}{2}\right)\xi(x) \times \left\{-\frac{3}{2}\right\} x^{-\frac{3}{2}-1} = x^{-\frac{1}{2}} \frac{\partial \xi^2(x)}{\partial x^2}$$

$$- \left(\frac{1}{2}\right) x^{-\frac{3}{2}} \frac{\partial \xi(x)}{\partial x} - \left(\frac{1}{2}\right) \frac{\partial \xi(x)}{\partial x} x^{-\frac{3}{2}}$$

$$+ \left(\frac{1}{2} \times \frac{3}{2}\right)\xi(x) x^{-\frac{5}{2}} = x^{-\frac{1}{2}} \frac{\partial \xi^2(x)}{\partial x^2} - x^{-\frac{3}{2}} \frac{\partial \xi(x)}{\partial x}$$

$$+ \left(\frac{3}{4}\right) x^{-\frac{5}{2}}\xi(x)$$

$$(10.76)$$

Combining eqs. (10.73), (10.75), and (10.76),

$$x^2 \left\{x^{-\frac{1}{2}} \frac{\partial \xi^2(x)}{\partial x^2} - x^{-\frac{3}{2}} \frac{\partial \xi(x)}{\partial x} + \left(\frac{3}{4}\right) x^{-\frac{5}{2}}\xi(x)\right\}$$

$$+ 2x\left\{x^{-\frac{1}{2}} \frac{\partial \xi(x)}{\partial x} - \frac{1}{2}\xi(x) x^{-\frac{3}{2}}\right\} + \left\{x^2 - l(l+1)\right\}\xi(x) x^{-\frac{1}{2}} = 0$$

$$(10.77)$$

Dividing eq. (10.77) throughout by $x^{-1/2}$,

$$x^2 \left\{\frac{\partial \xi^2(x)}{\partial x^2} - x^{-1} \frac{\partial \xi(x)}{\partial x} + \left(\frac{3}{4}\right) x^{-2}\xi(x)\right\}$$

$$+ 2x\left\{\frac{\partial \xi(x)}{\partial x} - \frac{1}{2}\xi(x) x^{-1}\right\} + \left\{x^2 - l(l+1)\right\}\xi(x) = 0$$

$$(10.78)$$

or

$$x^2 \frac{\partial \xi^2(x)}{\partial x^2} - x \frac{\partial \xi(x)}{\partial x} + \left(\frac{3}{4}\right)\xi(x) + 2x \frac{\partial \xi(x)}{\partial x} - \xi(x)$$

$$+ \left\{x^2 - l(l+1)\right\}\xi(x) = 0 \qquad (10.79)$$

Collecting terms with $\frac{\partial \xi^2(x)}{\partial x^2}, \frac{\partial \xi(x)}{\partial x}, \xi(x)$

$$x^2 \frac{\partial \xi^2(x)}{\partial x^2} + (-x + 2x) \frac{\partial \xi(x)}{\partial x}$$

$$+ \left\{\left(\frac{3}{4}\right) - 1 + x^2 - l(l+1)\right\}\xi(x) = 0 \qquad (10.80)$$

or

$$x^2 \frac{\partial \xi^2(x)}{\partial x^2} + x \frac{\partial \xi(x)}{\partial x} + \left\{-\frac{1}{4} + x^2 - l(l+1)\right\}\xi(x) = 0 \qquad (10.81)$$

or

$$x^2 \frac{\partial \xi^2(x)}{\partial x^2} + x \frac{\partial \xi(x)}{\partial x} + \left\{-\frac{1}{4} + x^2 - l^2 - l\right\}\xi(x) = 0 \qquad (10.82)$$

or

$$x^2 \frac{\partial \xi^2(x)}{\partial x^2} + x \frac{\partial \xi(x)}{\partial x} + \left\{x^2 - l^2 - l - \frac{1}{4}\right\}\xi(x) = 0 \qquad (10.83)$$

or

$$x^2 \frac{\partial \xi^2(x)}{\partial x^2} + x \frac{\partial \xi(x)}{\partial x} + \left\{x^2 - \left(l^2 + l + \frac{1}{4}\right)\right\}\xi(x) = 0 \qquad (10.84)$$

or

$$x^2 \frac{\partial \xi^2(x)}{\partial x^2} + x \frac{\partial \xi(x)}{\partial x} + \left[x^2 - \left\{l^2 + \left(2l \times \frac{1}{2}\right) + \left(\frac{1}{2}\right)^2\right\}\right]\xi(x) = 0$$

$$(10.85)$$

or

$$x^2 \frac{\partial \xi^2(x)}{\partial x^2} + x \frac{\partial \xi(x)}{\partial x} + \left\{x^2 - \left(l + \frac{1}{2}\right)^2\right\}\xi(x) = 0 \quad (10.86)$$

The solutions to this well-known second order differential equation are Bessel functions of half-integral order.

$$\xi(x) = AJ_{l+1/2}(x) + BY_{l+1/2}(x) \tag{10.87}$$

where A and B are constants. Therefore, from eq. (10.74), the solutions to the original equation are written as

$$R_{n,l}(r) = \frac{\xi(x)}{\sqrt{x}} = \frac{AJ_{l+1/2}(x) + BY_{l+1/2}(x)}{\sqrt{x}}$$

$$= \frac{AJ_{l+1/2}(x)}{\sqrt{x}} + \frac{BY_{l+1/2}(x)}{\sqrt{x}} = A\frac{J_{l+1/2}(kr)}{\sqrt{kr}} + B\frac{Y_{l+1/2}(kr)}{\sqrt{kr}}$$

$$\tag{10.88}$$

The two independent solutions are contained in the first and second terms. Using the symbols

$$j_l(kr) = \text{Bessel function of the first kind} = \sqrt{\frac{\pi}{2}}\frac{J_{l+1/2}(kr)}{\sqrt{kr}} \tag{10.89}$$

and

$$y_l(kr) = \text{Bessel function of the second kind} = \sqrt{\frac{\pi}{2}}\frac{Y_{l+1/2}(kr)}{\sqrt{kr}}$$

$$\tag{10.90}$$

we write the general solution as

$$R_{n,l}(r) = A\sqrt{\frac{2}{\pi}}j_l(kr) + B\sqrt{\frac{2}{\pi}}y_l(kr) = Cj_l(kr) + Dy_l(kr) \tag{10.91}$$

where

$$C = A\sqrt{\frac{2}{\pi}} \text{ and } D = B\sqrt{\frac{2}{\pi}} \tag{10.92}$$

The functions $j_l(kr)$ and $y_l(kr)$ are referred to as spherical Bessel functions. The first few Bessel functions of the first kind are:

$$j_0(kr) = \frac{\sin(kr)}{kr} \tag{10.93}$$

$$j_1(kr) = \frac{\sin(kr)}{(kr)^2} - \frac{\cos(kr)}{kr} \tag{10.94}$$

and

$$j_2(kr) = \left\{\frac{3}{(kr)^3} - \frac{1}{(kr)}\right\}\sin(kr) - \frac{3}{(kr)^2}\cos(kr) \tag{10.95}$$

The first few Bessel functions of the second kind are:

$$y_0(kr) = -\frac{\cos(kr)}{kr} \tag{10.96}$$

$$y_1(kr) = -\frac{\cos(kr)}{(kr)^2} - \frac{\sin(kr)}{kr} \tag{10.97}$$

and

$$y_2(kr) = -\left\{\frac{3}{(kr)^3} - \frac{1}{(kr)}\right\}\cos(kr) - \frac{3}{(kr)^2}\sin(kr) \tag{10.98}$$

On plotting the Bessel functions of the first and second kinds versus kr, i.e. $j_l(kr)$ against kr and $y_l(kr)$ against kr, we find that both the functions are oscillatory in nature (Figures 10.4 and 10.5). The plotted curves pass through zero several times. Functions $y_l(kr)$ are found to be misbehaved. They are not square-integrable at $kr = 0$ and so physically inadmissible. But functions $j_l(kr)$ are well-behaved.

To satisfy the boundary condition

$$\text{At } r = a, R_{n,l}(r) = R_{n,l}(a) = 0 \tag{10.99}$$

the value of k must be selected in such a way that $x = kr = ka$ corresponds to one of the zeroes of the Bessel function $j_l(x)$. Denoting the nth zero of $j_l(x)$ as $x_{n,l}$, we may write

$$ka = x_{n,l} \quad \text{for } n = 1,2,3,\ldots \tag{10.100}$$

or

$$k = \frac{x_{n,l}}{a} \tag{10.101}$$

Hence from eq. (10.69) for k

$$\sqrt{\frac{2m_e^*E}{\hbar^2}} = \frac{x_{n,l}}{a} \tag{10.102}$$

or

$$\frac{2m_e^*E}{\hbar^2} = \frac{x_{n,l}^2}{a^2} \tag{10.103}$$

or

$$E = \frac{x_{n,l}^2\hbar^2}{2m_e^*a^2} \tag{10.104}$$

Therefore, the allowed energy levels of the electrons are represented as

$$E_{n,l} = \frac{x_{n,l}^2\hbar^2}{2m_e^*a^2} \tag{10.105}$$

where $x_{n,l}$ are the roots of the Bessel function; n, the number of the root of Bessel function, is the principal quantum number; and l, the order of the function, is the azimuthal quantum number. The energy levels of the electron in a small piece of solid called a quantum dot become discrete like those in isolated atoms. The s, p, d, f atomic orbitals may be defined for the quantum dot in the same manner as for an atom. The resemblance of the behavior of envelope wave functions with atom-like wave functions leads to the perception of quantum dots as artificial atoms. The continuum of energy levels in energy bands is not applicable for the quantum dots.

The energy obtained by imagining an electron enclosed in a small spatial dimension being the consequence of quantum confinement of the electron is known as the confinement energy of electrons. A similar equation may be written for the hole confinement energy

$$E_{n,l} = \frac{x_{n,l}^2\hbar^2}{2m_h^*a^2} \tag{10.106}$$

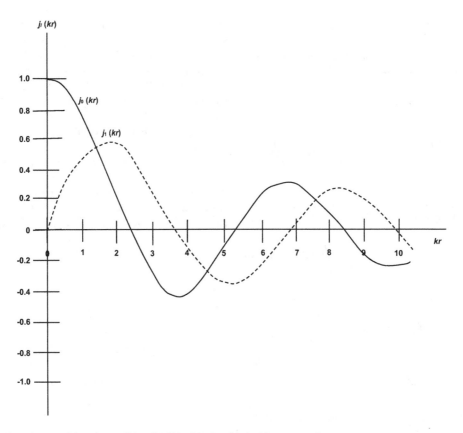

FIGURE 10.4   Plot of first two Bessel functions j_0 (*kr*) and j_1 (*kr*) of the first kind with respect to *kr*.

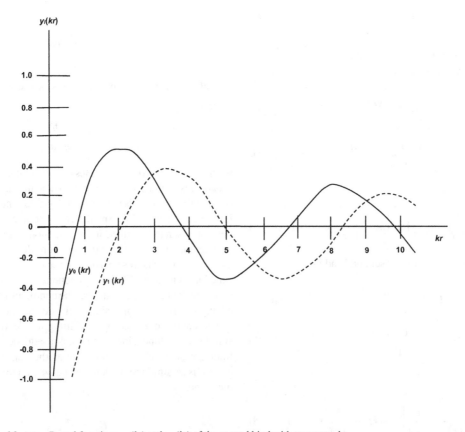

FIGURE 10.5   Plot of first two Bessel functions y_0 (*kr*) and y_1 (*kr*) of the second kind with respect to *kr*.

where m_h^* is the effective mass of holes.

Apart from discreteness of energy levels with solitary atom-like properties, another prominent observation is that if we consider the energy difference between any two consecutive energy levels of electrons or holes, suppose nth and $(n-1)$th energy levels,

$$E_{n,l} - E_{n-1,l} = \frac{x_{n,l}^2 \hbar^2}{2m_e^* a^2} - \frac{x_{n-1,l}^2 \hbar^2}{2m_e^* a^2} = \frac{\hbar^2}{2m_e^* a^2}\left(x_{n,l}^2 - x_{n-1,l}^2\right) \quad (10.107)$$

it is found that the energy difference is inversely proportional to the square of the radius of the quantum dot. Hence, the smaller the dot, the larger the energy difference. This means that quantum confinement increases the separation between the two states accounting for the widening of the energy bandgap, and this effect is accentuated by taking quantum dots of smaller sizes. Conversely, the energy gap decreases on taking quantum dots of larger size.

The meanings of the three quantum numbers n, m, and l need to be elucidated. The quantum number n is used for the internal states of the exciton while the quantum numbers m and l are used for its external state. The internal states originate from the Coulomb electrostatic interaction between the electron and the hole. These states will be labeled by capital letters: 1S, 2S, 2P, 3S, 3P, 3D, …. The external states arise from the external potential barrier in the form of a spherical box. Lowercase letters will be used for labeling them: 1s, 1p, 1d, … 2s, 2p, 2d, … Combined letters S(s), P(p), D(d), F(f) … will represent the states with $l = 0, 1, 2, 3, …$

The quantum numbers for the lowest state of the exciton are: $n = 1$, $m = 1$, and $l = 0$. This is the 1S1s state. The Bessel function root is

$$x_{10} = \pi \quad (10.110)$$

Putting this value in eq. (10.109),

$$
\begin{aligned}
E_{\text{1S1s}} &= E_G - \frac{\text{Ry}^*}{1^2} + \frac{x_{10}^2 \hbar^2}{2Ma^2} = E_G - \text{Ry}^* + \frac{\pi^2 \hbar^2}{2Ma^2} = E_G - \text{Ry}^*\left(1 - \frac{1}{\text{Ry}^*}\frac{\pi^2 \hbar^2}{2Ma^2}\right) = E_G - \text{Ry}^*\left\{1 - \frac{\mu}{M}\left(\frac{\pi^2 \hbar^2}{2\mu a^2} \times \frac{32\pi^2 \hbar^2 \varepsilon_0^2 \varepsilon_r^2}{\mu e^4}\right)\right\} \\
&= E_G - \text{Ry}^*\left\{1 - \frac{\mu}{M}\left(\frac{16\pi^4 \hbar^4 \varepsilon_0^2 \varepsilon_r^2}{\mu^2 e^4 a^2}\right)\right\} = E_G - \text{Ry}^*\left\{1 - \frac{\mu}{M}\left(\frac{\pi^2}{a^2} \times \frac{16\pi^2 \hbar^4 \varepsilon_0^2 \varepsilon_r^2}{\mu^2 e^4}\right)\right\} = E_G - \text{Ry}^*\left[1 - \frac{\mu}{M}\left\{\frac{\pi^2}{a^2} \times \left(\frac{4\pi \varepsilon_0 \varepsilon_r \hbar^2}{\mu e^2}\right)^2\right\}\right] \\
&= E_G - \text{Ry}^*\left\{1 - \frac{\mu}{M}\left(\frac{\pi^2}{a^2} \times a_X^2\right)\right\} = E_G - \text{Ry}^*\left\{1 - \frac{\mu}{M}\left(\frac{\pi a_X}{a}\right)^2\right\} = E_G - \text{Ry}^* + \text{Ry}^* \frac{\mu}{M}\left(\frac{\pi a_X}{a}\right)^2
\end{aligned}
$$

$$(10.111)$$

10.7 Energy Equation of Excitons in Weak Confinement

For large quantum dots having radius several times the exciton Bohr radius, we replace the kinetic energy term by the solution for the spherical box as

$$E_{\text{Constant}} = \frac{\hbar^2 K^2}{2M} = \frac{x_{n,l}^2 \hbar^2}{2Ma^2} \quad (10.108)$$

so that the dispersion relation in eq. (10.55) becomes

$$E_{nml}(K) = E_G - \frac{\text{Ry}^*}{\bar{n}^2} + \frac{\hbar^2 K^2}{2M} = E_G - \frac{\text{Ry}^*}{\bar{n}^2} + \frac{x_{m,l}^2 \hbar^2}{2Ma^2} \quad (10.109)$$

The substitution of the spherical box solution means that the motion of the center of mass of the exciton, which was so far like a free particle, has now become quantized due to the confinement. $x_{m,l}$ are the roots of the spherical Bessel function and M is the sum of electron and hole effective masses. The subscript m, l is used in the third term to distinguish between quantum numbers n related to reduced mass motion and m concerned with center-of-mass motion.

by applying eqs. (10.54) and (10.12) for R_y^* and a_X, respectively. Hence the shift in energy for the 1S1s state is

$$\Delta E_{\text{Shift}} = +\text{Ry}^* \frac{\mu}{M}\left(\frac{\pi a_X}{a}\right)^2 \quad (10.112)$$

Since $a_X \ll a$, the energy shift expressing the effect of quantum confinement is smaller than the Rydberg energy. This explains why the confinement is said to be weak in this case.

10.8 Energy Equations of Electrons and Holes in Strong Confinement

To a first approximation, the Coulomb interaction between the electron and the hole, although much larger than in a bulk crystal with the charges enclosed in a smaller volume, is presumed to be insufficient to keep them in a bound exciton state. The kinetic energies of the electrons and holes overcome the attractive electrostatic force. Hence the electrons and holes may be treated as independent particles instead of their existence as an exciton. Consequently, the energy spectra for these particles are described by separate equations. Let us consider the energy level of the top of the valence band as the zero energy point,

$E_V = 0$. With reference to $E_V = 0$ as the origin, and denoting the energy gap of material by E_G, the energy spectrum for electrons is expressed by the following equation:

$$E_{ml}^e = E_V + E_G + \frac{x_{m,l}^2\hbar^2}{2m_e^*a^2} = 0 + E_G + \frac{x_{m,l}^2\hbar^2}{2m_e^*a^2} = E_G + \frac{x_{m,l}^2\hbar^2}{2m_e^*a^2}$$

(10.113)

Similarly, the energy spectrum for holes is written as

$$E_{ml}^h = E_V - \frac{x_{m,l}^2\hbar^2}{2m_h^*a^2} = 0 - \frac{x_{m,l}^2\hbar^2}{2m_h^*a^2} = -\frac{x_{m,l}^2\hbar^2}{2m_h^*a^2}$$

(10.114)

Since the laws of conservation of energy and momentum lead to selection rules that permit optical transitions between electron and hole states having the same principal and azimuthal quantum numbers n and l, the absorption spectrum reduces to a set of discrete bands with energy peaks at

$$E_{n,l} = E_G + \frac{x_{n,l}^2\hbar^2}{2m_e^*a^2} - \left(-\frac{x_{n,l}^2\hbar^2}{2m_h^*a^2}\right) = E_G + \frac{x_{n,l}^2\hbar^2}{2m_e^*a^2} + \frac{x_{n,l}^2\hbar^2}{2m_h^*a^2}$$

$$= E_G + \frac{x_{n,l}^2\hbar^2}{2a^2}\left(\frac{1}{m_e^*} + \frac{1}{m_h^*}\right) = E_G + \frac{x_{n,l}^2\hbar^2}{2a^2}\left(\frac{m_h^* + m_e^*}{m_e^*m_h^*}\right)$$

$$= E_G + \frac{x_{n,l}^2\hbar^2}{2a^2}\frac{1}{\dfrac{m_e^*m_h^*}{m_e^* + m_h^*}} = E_G + \frac{x_{n,l}^2\hbar^2}{2a^2}\frac{1}{\mu}$$

$$= E_G + \frac{x_{n,l}^2\hbar^2}{2\mu a^2}$$

(10.115)

The valence band moves downward and the conduction band moves upward. The net result is an enlargement of the energy bandgap. For the 1S1s state of the system, using eq. (10.110) and expressing the result in terms of Ry* (eq. (10.54)) and a_X (eq. (10.12))

$$\frac{x_{n,l}^2\hbar^2}{2\mu a^2} = \frac{x_{10}^2\hbar^2}{2\mu a^2} = \frac{Ry^*}{Ry^*}\frac{\pi^2\hbar^2}{2\mu a^2} = \frac{\pi^2\hbar^2}{2\mu a^2 \times \dfrac{\mu e^4}{32\pi^2\hbar^2\varepsilon_0^2\varepsilon_r^2}}Ry^*$$

$$= \frac{\pi^2\hbar^2 \times 32\pi^2\hbar^2\varepsilon_0^2\varepsilon_r^2}{2\mu a^2 \times \mu e^4}Ry^*$$

$$= \frac{16\pi^4\hbar^4\varepsilon_0^2\varepsilon_r^2}{\mu^2 e^4 a^2}Ry^* = \frac{\pi^2}{a^2}\left(\frac{4\pi\hbar^2\varepsilon_0\varepsilon_r}{\mu e^2}\right)^2 Ry^*$$

$$= \frac{\pi^2}{a^2}(a_X)^2 Ry^* = \left(\frac{\pi a_X}{a}\right)^2 Ry^*$$

(10.116)

Combining eqs. (10.115) and (10.116),

$$E_{n,l} = E_G + \left(\frac{\pi a_X}{a}\right)^2 Ry^*$$

(10.117)

Since $a_X \gg a$, the zero-point kinetic energy of the two-particle system is much larger than the Rydberg energy. Hence the confinement is strong.

In the preliminary analysis, the Coulomb interaction between the electron and the hole is ignored. However, restriction of the electron and hole within the boundaries of a volume smaller than the exciton Bohr radius and assumption of the absence of any interaction between them seem to give an erroneous and unrealistic picture of the phenomenon. Furthermore, spatial correlation between electron and hole has to be accounted for. More generally, the energy of the lowest energy state of the exciton can be written as

$E_{1S1s} =$ Infinite crystal bandgap of the semiconductor

+ Infinite spherical well contribution to bandgap

+ Coulomb interaction correction

+ Spatial correlation correction

$$= E_G + \frac{\pi^2\hbar^2}{2\mu a^2} - \frac{1.786e^2}{4\pi\varepsilon_0\varepsilon_r a} - 0.248 Ry^*$$

(10.118)

It may be noted that the second term on the right-hand side is inversely proportional to the square of crystal radius whereas the third term varies as first power of crystal radius only. The second term owes its origin to Brus (1983, 1984, 1986) and the third term to Kayunama (1986, 1988).

10.9 Bottom-Up Approach to the Evolution of the Energy Band Structure of Nanocrystalline Solids

Hitherto, our attention has been focused on the top-down view of the band structure of crystals. Now we reverse our thinking. In the bottom-up approach, the nanocrystal is built up starting from individual atoms. Molecular orbital theory proposes that individual atomic orbitals combine to form molecular orbitals. The unification of atomic orbitals is accomplished by linearly conjoining the atomic orbitals. This approach is referred to as linear combination of atomic orbitals (LCAO). Suppose the atomic orbital of an atom A is described by the wave function Ψ_A while that of atom B is assigned the wave function Ψ_B. Let the electron clouds of the atoms A and B overlap. Then two molecular orbitals are formed:

i. One molecular orbital is formed by the addition of wave functions of atoms A and B.

ii. Another molecular orbital is obtained by subtraction of these wave functions.

Thus, the wave function of the molecular orbital constructed from atomic orbitals A and B is expressed as

$$\Psi_{\text{Molecular orbital}} = \Psi_{\text{Atom A}} \pm \Psi_{\text{Atom B}}$$

(10.119)

The molecular orbital formed by the addition of wave functions Ψ_A, Ψ_B of constituent atoms A, B is called a bonding molecular orbital (BMO):

$$\Psi_{\text{BMO}} = \Psi_{\text{Atom A}} + \Psi_{\text{Atom B}}$$

(10.120)

The molecular orbital obtained by subtraction of wave functions Ψ_A, Ψ_B is known as an antibonding molecular orbital (ABMO):

$$\Psi_{ABMO} = \Psi_{Atom\ A} - \Psi_{Atom\ B} \qquad (10.121)$$

The orbitals BMO and ABMO differ in their relative energies. The BMO is lower in energy than the energies of the interacting atomic orbitals while the ABMO is higher in energy than the energies of atomic orbitals. The higher energy results from the increase in energy due to attraction between nuclei and electrons of both the atoms undergoing combination whereas the lower energy is caused by electron movement away from the nuclei being in a repulsive state.

The simplest multielectron molecule is exemplified by the diatomic hydrogen molecule. The hydrogen molecule is formed by the combination of two atomic orbitals to form two molecular orbitals, a BMO and an ABMO (Figure 10.6). These molecular orbitals are spread over both the participating hydrogen atoms. Accommodation of electrons in the molecular orbital takes place in such a way that the potential energy of the molecule is minimized. Before combination, each hydrogen atom has an electron in the 1s atomic orbital. After combination, these 1s electrons occupy the BMO. So the ABMO is left vacant. The highest occupied molecular orbital is called HOMO while the lowest unoccupied molecular orbital is known as the LUMO and the intervening gap between them is the HOMO-LUMO energy gap.

Larger molecules, clusters of molecules, and the bulk solids are formed in the identical manner from atomic orbitals (Figure 10.7) With the molecule growing in size, the number of atomic orbitals overlapping with each other for molecular orbital formation increases. Each molecular orbital has a distinct energy level. So, a large number of energy levels are produced. During this molecular orbital creation, there is a greater tendency toward formation of molecular orbitals of intermediate energy values than those with maximum or minimum energies, i.e. fully bonding or antibonding types of molecular orbitals. Naturally, the molecular orbital states are denser at intermediate energy values than at the extreme values, whether highest or lowest. In the limiting case of formation of a bulk solid, the number of overlapping atomic orbitals is very high. Therefore, the number of energy levels of molecular orbitals becomes excessively large. Then the spacing between adjoining energy levels becomes exceeding small and almost disappears resulting in a quasi-continuum of energy levels which is referred to as an energy band. As there are two types of molecular orbitals, the BMO and ABMO, so there must be two energy bands. These are referred to as the valence and conduction bands in the energy band model. The HOMO level is recognized as the top of the valence band and the LUMO level as the bottom of the conduction band with the HOMO-LUMO energy gap as the energy bandgap of the material. Further, recall that the greater tendency of production of molecular orbitals of intermediate energies causes a crowding of energy states at intermediate energies and the states become scantier toward the edges, i.e. near the valence band top or conduction band bottom boundaries. So, the density of energy levels decreases from the energy levels having intermediate energy values and becomes lesser as one moves proximate to the band edges where the levels become discrete.

The difference in width of the energy bandgap between a nanocrystal and a bulk solid arises from the difference in the number of atomic orbitals that are available for forming the molecular orbitals. In a nanocrystal, the number of atoms is low. So, the number of atomic orbitals is small. Obviously, the number of BMOs and ABMOs produced by the overlapping of a meager number of atomic orbitals is also paltry. As more atomic materials become available for overlapping, greater number of BMOs and ABMOs are produced and the HOMO-LUMO gap

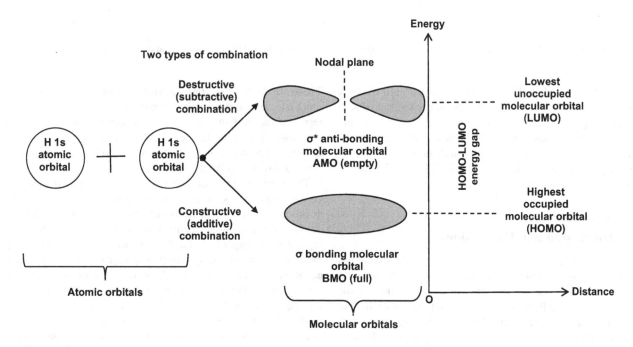

FIGURE 10.6 Formation of antibonding and bonding molecular orbitals from atomic orbitals of hydrogen, creation of LUMO and HOMO energy levels, and the HOMO-LUMO energy gap.

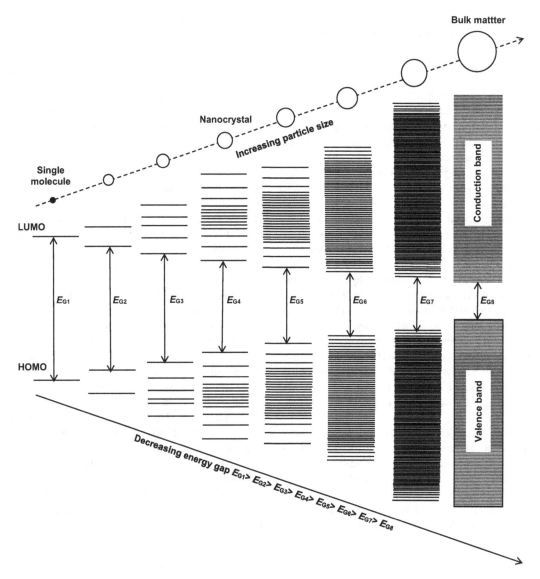

FIGURE 10.7   Evolving the energy band structure by bottom-up approach: from single molecule through nanocrystal to bulk solid.

progressively diminishes. This trend is continued until the bulk solid state is reached. Then the number of energy levels produced by overlapping of atomic orbitals has reached the pinnacle and the energy gap too has attained the lowest value. Greater the number of atoms and hence atomic orbitals taking part to produce a crystal, smaller is the resultant HOMO-LUMO energy gap between molecular orbitals.

10.10 Discussion and Conclusions

It is emphasized that the deterministic length scale for quantum confinement is the exciton Bohr radius. For deriving the equation for the exciton Bohr radius, we first obtain the equation for the Bohr radius from the Bohr atomic model and apply it to find the exciton Bohr radius. We also find the exciton binding energy getting an equation similar to the one for hydrogen atom. Besides the exciton Bohr radius, two other parameters are necessary to clearly demarcate the three regimes of quantum confinement.

These are the Bohr radii for electrons and holes. Then the weak, moderate, and strong confinement regimes are distinguished in terms of the exciton Bohr radius and Bohr radii for electrons and holes. Following in the footsteps of hydrogen atom solution described in Chapter 6, the dispersion relation for exciton is derived showing hydrogen-like discrete energy levels.

The main property describing quantum confinement is the confinement energy obtained by imagining an electron to be enclosed within a small region in space. The equation for the confinement energy of electrons and holes is derived by solving the Schrodinger equation for the particle-in-a-spherical box problem. The resulting equation is a spherical Bessel differential equation. Its solutions are Bessel functions of half-integral order. The general solution contains Bessel functions of the first kind and the second kind. But Bessel functions of the second kind are misbehaved. So the allowed energy levels are expressed in terms of roots of the Bessel function of the first kind. It is found that the s, p, d, f atomic orbitals may be defined for the quantum dot as if it was an atom. Hence it may be called an artificial atom. The energy spectrum is a line spectrum instead of being a continuous band and the energy difference corresponding to the bandgap varies inversely with the square of the radius of the quantum dot meaning that reduction of the size of the dot rapidly increases the energy difference or the bandgap.

The energy equation for weak confinement is written down by utilizing the equation for the confinement energy. The three quantum numbers n, l and m are explained. It is found that the shift in energy for the 1S1s state is less than the Rydberg energy; hence the confinement is weak. For analyzing the strong confinement regime, the electrons and holes are considered as independent particles for which separate equations for energy spectrum are written down, and the bandgap widening is interpreted. For the 1S1s state, the zero-point kinetic energy is appreciably larger than the Rydberg energy justifying the term 'strong confinement'. A Coulomb interaction term owing its origin to Brus and a spatial correction term owing its origin to Kayunama are added to the energy equation.

Breaking away from the earlier top-down approach, the bottom-up approach is pursued to evolve the energy band structure of a quantum dot. The atomic orbitals can be linearly combined starting from individual atoms. In this line of thinking, the HOMO-LUMO gap is the equivalent of the forbidden energy bandgap in the top-down approach. The wider HOMO-LUMO gap obtained in a quantum dot arises from the smaller number of atomic orbitals due to the scanter number of atoms in the nanocrystal. The gap width becomes larger as smaller number of atomic orbitals become available.

Illustrative Exercises

10.1 Find a_X for Si (dielectric constant $\varepsilon_r = 11.7$) given $m_e^* = 0.26m_0$ and $m_h^* = 0.36m_0$ where $m_0 = 9.11 \times 10^{-31}$ kg, $\varepsilon_0 = 8.85 \times 10^{-12}$ F m^{-1}, $e = 1.6 \times 10^{-19}$ C, $\hbar = 1.05457 \times 10^{-34}$ Js.

By applying eq. (10.13)

$$\mu = \frac{m_e^* m_h^*}{m_e^* + m_h^*} = \frac{0.26m_0 \times 0.36m_0}{0.26m_0 + 0.36m_0}$$

$$= \frac{0.26 \times 0.36 m_0^2}{(0.26 + 0.36)m_0} = \frac{0.0936}{0.62}m_0 = 0.151m_0$$

$$= 0.151 \times 9.11 \times 10^{-31} \text{ kg} = 1.3756 \times 10^{-31} \text{ kg} \quad (10.122)$$

By eq. (10.12)

$$a_X = \frac{4\pi\varepsilon_0\varepsilon_r\hbar^2}{\mu e^2} = \frac{4 \times 3.14 \times 8.85 \times 10^{-12} \times 11.7 \times (1.05457 \times 10^{-34})^2}{1.3756 \times 10^{-31} \times (1.6 \times 10^{-19})^2}$$

$$= \frac{1446.337 \times 10^{-80}}{3.52 \times 10^{-69}} = 410.89 \times 10^{-11} \text{ m}$$

$$= 4.11 \times 10^{-9} \text{ m} = 4.11 \text{ nm} \quad (10.123)$$

10.2 In Ge (dielectric constant = 16.2), $m_e^* = 0.12m_0$ and $m_h^* = 0.21m_0$; $m_0 = 9.11 \times 10^{-31}$ kg. Calculate a_X for Ge. $\varepsilon_0 = 8.85 \times 10^{-12}$ F m^{-1}, $e = 1.6 \times 10^{-19}$ C, $\hbar = 1.05457 \times 10^{-34}$ Js.

By applying eq. (10.13)

$$\mu = \frac{m_e^* m_h^*}{m_e^* + m_h^*} = \frac{0.12m_0 \times 0.21m_0}{0.12m_0 + 0.21m_0} = \frac{0.12 \times 0.21m_0^2}{(0.12 + 0.21)m_0}$$

$$= \frac{0.0252}{0.33}m_0 = 0.076m_0 = 0.076 \times 9.11 \times 10^{-31} \text{ kg}$$

$$= 0.692 \times 10^{-31} \text{ kg} \quad (10.124)$$

By eq. (10.12)

$$a_X = \frac{4\pi\varepsilon_0\varepsilon_r\hbar^2}{\mu e^2}$$

$$= \frac{4 \times 3.14 \times 8.85 \times 10^{-12} \times 16.2 \times (1.05457 \times 10^{-34})^2}{0.692 \times 10^{-31} \times (1.6 \times 10^{-19})^2}$$

$$= \frac{2002.621 \times 10^{-80}}{1.77 \times 10^{-69}} = 1131.424 \times 10^{-11} \text{ m} = 11.31 \times 10^{-9} \text{ m}$$

$$= 11.31 \text{ nm}$$

$$\quad (10.125)$$

REFERENCES

Barbagiovanni, E. G., D. J. Lockwood, P. J. Simpson and L. V. Goncharova. 2012. Quantum confinement in Si and Ge nanostructures. *Journal of Applied Physics* 111: 034307-1–034307-9.

Brus, L. E. 1983. A simple model for the ionization potential, electron affinity, and aqueous redox potentials of small semiconductor crystallites. *Journal of Chemical Physics* 79(11): 5566–5571.

Brus, L. E. 1984. Electron-electron and electron-hole interactions in small semiconductor crystallites, the size dependence of the lowest excited electronic state. *Journal of Chemical Physics.* 80(9): 4403–4409.

Brus, L. E. 1986. Electronic wave functions in semiconductor cluster, experiment and theory. *Journal of Physical Chemistry* 90(12): 2555–2560.

Kayunama, Y. 1986. Wannier exciton in microcrystals. *Solid State Communications* 59: 405–408.

Kayunama, Y. 1988. Quantum-size effects of interacting electrons and holes in semiconductor microcrystals with spherical shape. *Physical Review B* 38: 9797–9805.

Yoffe, A. D. 1993 Low-dimensional systems: quantum size effects and electronic properties of semiconductor microcrystallites (zero-dimensional systems) and some quasi-two-dimensional systems. *Advances in Physics* 42(2): 173–262.

11

Electrons in Quantum Wires and Landauer-Büttiker Formalism

11.1 Two-Terminal Quantum Wire with Macroscopic Contacts

A quantum wire is a wire having diameter/cross section in the nanoscale with a magnitude of the order of the de Broglie wavelength of the electron. In such a wire, the carrier transport properties are controlled by quantum mechanics. Hence it is called a quantum wire. It is a one-dimensional (1D) conductor.

The center of attention is the current flowing through a two-terminal quantum wire device (Figure 11.1). It comprises a thin wire terminated at both ends with thick macroscopic contact regions, and subjected to a bias. The contact regions are reservoirs of electrons acting as infinite sources and sinks of electrons. The wire has a rectangular cross section. The current is flowing in the arbitrarily chosen X-direction. The dimensions along the Y- and Z-directions are L_y and L_z respectively. Along the X-direction, the wire has a length L extending from $x = 0$ to $x = L$, and it is assumed that $L \ll L_m$ (mean free path = distance traversed by electron before an elastic collision), L_ϕ (phase coherence length = distance covered by the electron before an inelastic collision, i.e. over which the wave function of the electron retains its coherence). The electron journey from one contact to another is hurdle-free. As the electrons traverse this small distance, no collisions are encountered enroute. Electron transport under this condition is said to be ballistic. Ultra-short channel nanoscale field-effect transistors (FETs) and carbon nanotube (CNT) FETs are practical realizations of this concept.

The carriers flow freely longitudinally along the X-direction while their motion is constrained along the other two directions (Y- and Z-directions), which are the transverse directions. As a consequence of restriction of the transverse motion, only one degree of freedom is available for propagation of an electron wave. This motion is characterized by the wave vector k_x. The electron energy in the other two confinement directions is discretized and represented as ε_{n_y, n_z} because it depends on integral quantum numbers n_y, n_z. The discretization of electron energy results in the formation of sub-bands or channels for electron flow, which are known as the transverse modes. Therefore, it is utmost important to know the wave functions and the energy levels of the electrons.

There are several ingredients to the problem of determination of the current flowing in the wire. We are enthused to know the distribution function for electrons in the quantum wire, and the number of modes, both at zero and higher temperatures. To account for the electron reflection at the interface between the contact and the quantum wire, we are inquisitive to evaluate the transmission coefficient for electron movement across the interface.

The simplest case is that of a single-mode wire at zero temperature without any scattering. This case can be generalized to the situation of a wire with multiple modes and exposed to arbitrary temperatures.

11.2 Solution of the Schrodinger Equation for the Quantum Wire

The charge carriers in the quantum wire are under the influence of a potential $W(y, z)$ confining the carriers in the YZ-plane, hence called the confinement potential. The total potential $V(x, y, z)$ can be written as the sum of the potential $W(y, z)$ and a potential $V(x)$ along the axis of the wire oriented along the X-axis. So,

$$V(x, y, z) = V(x) + W(y, z) \tag{11.1}$$

In accordance with the definition of potential $V(x, y, z)$, the wave function of the electrons $\Psi(x, y, z)$ is the product of two functions

$$\Psi(x, y, z) = \Psi(x)\Psi(y, z) \tag{11.2}$$

So, the time-independent Schrodinger equation can be written in the form

$$\left\{ -\frac{\hbar^2}{2m}\left(\frac{\partial^2}{\partial x^2} + \frac{\partial^2}{\partial y^2} + \frac{\partial^2}{\partial z^2} \right) + V(x) + W(y,z) \right\} \Psi(x)\Psi(y, z)$$
$$= E\Psi(x)\Psi(y, z) \tag{11.3}$$

or

$$\left\{ -\left(\frac{\hbar^2}{2m} \right)\frac{\partial^2}{\partial x^2} - \left(\frac{\hbar^2}{2m} \right)\frac{\partial^2}{\partial y^2} - \left(\frac{\hbar^2}{2m} \right)\frac{\partial^2}{\partial z^2} + V(x) + W(y,z) \right\}$$
$$\Psi(x)\Psi(y, z) = E\Psi(x)\Psi(y, z) \tag{11.4}$$

or

$$-\left(\frac{\hbar^2}{2m} \right)\frac{\partial^2}{\partial x^2}\{\Psi(x)\Psi(y, z)\} - \left(\frac{\hbar^2}{2m} \right)\frac{\partial^2}{\partial y^2}\{\Psi(x)\Psi(y, z)\}$$

$$-\left(\frac{\hbar^2}{2m} \right)\frac{\partial^2}{\partial z^2}\{\Psi(x)\Psi(y, z)\} + V(x)\Psi(x)\Psi(y, z)$$

$$+ W(y,z)\Psi(x)\Psi(y, z) = E\Psi(x)\Psi(y, z) \tag{11.5}$$

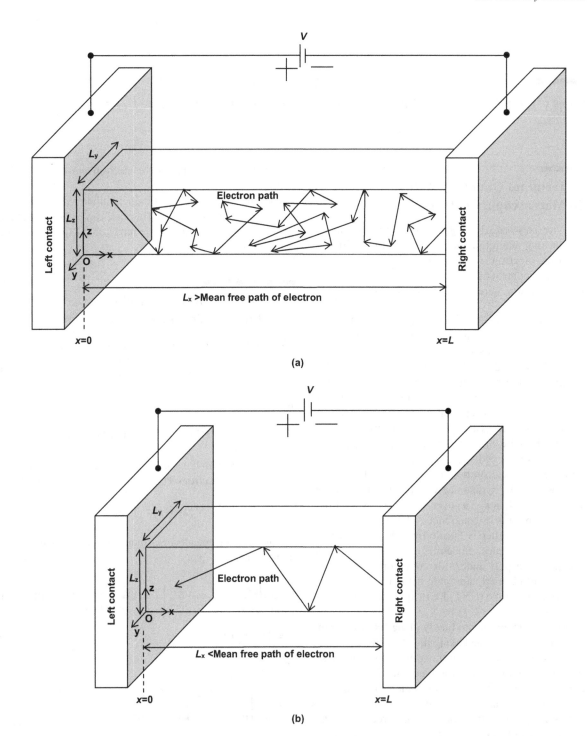

FIGURE 11.1   Electron transport mechanisms in nanowires: (a) diffusive transport and (b) ballistic transport. In (b), reflection from the walls is likely but no scattering.

or

$$-\left(\frac{\hbar^2}{2m}\right)\Psi(y,z)\frac{\partial^2\Psi(x)}{\partial x^2}-\left(\frac{\hbar^2}{2m}\right)\Psi(x)\frac{\partial^2\Psi(y,z)}{\partial y^2}$$

$$-\left(\frac{\hbar^2}{2m}\right)\Psi(x)\frac{\partial^2\Psi(y,z)}{\partial z^2}+V(x)\Psi(x)\Psi(y,z)$$

$$+W(y,z)\Psi(x)\Psi(y,z)=E\Psi(x)\Psi(y,z) \qquad (11.6)$$

Dividing both sides of eq. (11.6) by $\Psi(x)\,\Psi(y,z)$,

$$-\left(\frac{\hbar^2}{2m}\right)\frac{1}{\Psi(x)}\frac{\partial^2\Psi(x)}{\partial x^2}-\left(\frac{\hbar^2}{2m}\right)\frac{1}{\Psi(y,z)}\frac{\partial^2\Psi(y,z)}{\partial y^2}$$

$$-\left(\frac{\hbar^2}{2m}\right)\frac{1}{\Psi(y,z)}\frac{\partial^2\Psi(y,z)}{\partial z^2}$$

$$+V(x)+W(y,z)=E \qquad (11.7)$$

or

$$-\left(\frac{\hbar^2}{2m}\right)\frac{1}{\Psi(x)}\frac{\partial^2\Psi(x)}{\partial x^2}+V(x)$$

$$=E-W(y,z)+\left(\frac{\hbar^2}{2m}\right)\frac{1}{\Psi(y,z)}\frac{\partial^2\Psi(y,z)}{\partial y^2}$$

$$+\left(\frac{\hbar^2}{2m}\right)\frac{1}{\Psi(y,z)}\frac{\partial^2\Psi(y,z)}{\partial z^2} \tag{11.8}$$

This equation implies that

A function of x only = A function of (y,z) only   (11.9)

which is only possible if both sides = a constant E_V. Hence,

$$-\left(\frac{\hbar^2}{2m}\right)\frac{1}{\Psi(x)}\frac{\partial^2\Psi(x)}{\partial x^2}+V(x)=E_V \tag{11.10}$$

or

$$-\left(\frac{\hbar^2}{2m}\right)\frac{\partial^2\Psi(x)}{\partial x^2}+V(x)\Psi(x)=E_V\,\Psi(x) \tag{11.11}$$

and

$$E-W(y,z)+\left(\frac{\hbar^2}{2m}\right)\frac{1}{\Psi(y,z)}\frac{\partial^2\Psi(y,z)}{\partial y^2}$$

$$+\left(\frac{\hbar^2}{2m}\right)\frac{1}{\Psi(y,z)}\frac{\partial^2\Psi(y,z)}{\partial z^2}=E_V \tag{11.12}$$

or

$$-\left(\frac{\hbar^2}{2m}\right)\frac{1}{\Psi(y,z)}\frac{\partial^2\Psi(y,z)}{\partial y^2}-\left(\frac{\hbar^2}{2m}\right)\frac{1}{\Psi(y,z)}\frac{\partial^2\Psi(y,z)}{\partial z^2}$$

$$+W(y,z)=E-E_V=E_W \tag{11.13}$$

or

$$-\left(\frac{\hbar^2}{2m}\right)\frac{1}{\Psi(y,z)}\left\{\frac{\partial^2\Psi(y,z)}{\partial y^2}+\frac{\partial^2\Psi(y,z)}{\partial z^2}\right\}+W(y,z)=E_W$$

$$\tag{11.14}$$

or

$$-\left(\frac{\hbar^2}{2m}\right)\left(\frac{\partial^2}{\partial y^2}+\frac{\partial^2}{\partial z^2}\right)\Psi(y,z)+W(y,z)\Psi(y,z)=E_W\Psi(y,z)$$

$$\tag{11.15}$$

By separating the original equation into two parts, the particle motion is decoupled into two motions:

   i. The motion along the X-direction in which there is no confinement potential; hence $V(x)=0$.

   ii. The motion along the Y- and Z-directions taking place under the confinement potential $W(y,z)$.

So eq. (11.11) is reduced to

$$-\left(\frac{\hbar^2}{2m}\right)\frac{\partial^2\Psi(x)}{\partial x^2}+0\times\Psi(x)=E_V\,\Psi(x) \tag{11.16}$$

or

$$-\left(\frac{\hbar^2}{2m}\right)\frac{\partial^2\Psi(x)}{\partial x^2}=E_V\,\Psi(x) \tag{11.17}$$

This is the Schrodinger equation for a particle in free space. It is solved in Section 5.1. There is no energy quantization. The energies can acquire continuous values. The dispersion relation is (eq. (5.6))

$$E_V=\frac{\hbar^2 k_x^2}{2m} \tag{11.18}$$

where

$$k_x=\sqrt{\frac{2mE_V}{\hbar^2}} \tag{11.19}$$

The wave functions are plane waves of the form given by eq. (5.14)

$$\Psi_k(x)=\exp(\pm ik_x) \tag{11.20}$$

Now let us direct our attention to the potential $W(y,z)$.

$$W(y,z)=0 \quad \text{for } 0<x<L_x \text{ and } 0<y<L_y \tag{11.21}$$

and

$$W(y,z)=\infty \text{ otherwise} \tag{11.22}$$

where L_y and L_z denote the dimensions of the quantum wire in Y- and Z-directions, respectively. The earlier equations can be stated in words as:

   i. inside a rectangular box of dimensions L_y and L_z, the potential $W(y,z)$ is everywhere zero, while

   ii. outside this box, it is infinite.

Inside the rectangular box $W(y,z)=0$. Hence, eq. (11.15) becomes

$$-\left(\frac{\hbar^2}{2m}\right)\left(\frac{\partial^2}{\partial y^2}+\frac{\partial^2}{\partial z^2}\right)\Psi(y,z)+0\times\Psi(y,z)=E_W\Psi(y,z)$$

$$\tag{11.23}$$

or

$$-\left(\frac{\hbar^2}{2m}\right)\left\{\frac{\partial^2\Psi(y,z)}{\partial y^2}+\frac{\partial^2\Psi(y,z)}{\partial z^2}\right\}=E_W\Psi(y,z) \tag{11.24}$$

The wave function $\Psi(y,z)$ can be written as a product of the wave functions $\Psi(y)$ and $\Psi(z)$ in the Y- and Z-directions, respectively:

$$\Psi(y,z)=\Psi(y)\Psi(z) \tag{11.25}$$

Making substitution for $\Psi(y, z)$ from eq. (11.25) in eq. (11.24),

$$-\left(\frac{\hbar^2}{2m}\right)\left[\frac{\partial^2\{\Psi(y)\Psi(z)\}}{\partial y^2}+\frac{\partial^2\{\Psi(y)\Psi(z)\}}{\partial z^2}\right]=E_W\Psi(y)\Psi(z)$$

(11.26)

or

$$-\left(\frac{\hbar^2}{2m}\right)\Psi(z)\frac{\partial^2\Psi(y)}{\partial y^2}-\left(\frac{\hbar^2}{2m}\right)\Psi(y)\frac{\partial^2\Psi(z)}{\partial z^2}$$

$$=E_W\Psi(y)\Psi(z)$$

(11.27)

Dividing both sides of eq. (11.27) by $\Psi(y)\Psi(z)$,

$$-\left(\frac{\hbar^2}{2m}\right)\frac{1}{\Psi(y)}\frac{\partial^2\Psi(y)}{\partial y^2}-\left(\frac{\hbar^2}{2m}\right)\frac{1}{\Psi(z)}\frac{\partial^2\Psi(z)}{\partial z^2}=E_W$$

(11.28)

or

$$-\left(\frac{\hbar^2}{2m}\right)\frac{1}{\Psi(y)}\frac{\partial^2\Psi(y)}{\partial y^2}=E_W+\left(\frac{\hbar^2}{2m}\right)\frac{1}{\Psi(z)}\frac{\partial^2\Psi(z)}{\partial z^2}$$

(11.29)

As in eq. (11.8), introspection of left and right-hand sides of eq. (11.29) shows that in this equation each side is a function of a single variable, and these variables are different for the two sides, so that the equation means that

A function of y only = A function of z only   (11.30)

which is legitimate if both sides = a constant = E_{Wy} (say). Then

$$-\left(\frac{\hbar^2}{2m}\right)\frac{1}{\Psi(y)}\frac{\partial^2\Psi(y)}{\partial y^2}=E_{Wy}$$

(11.31)

or

$$-\left(\frac{\hbar^2}{2m}\right)\frac{\partial^2\Psi(y)}{\partial y^2}=E_{Wy}\Psi(y)$$

(11.32)

and

$$E_W+\left(\frac{\hbar^2}{2m}\right)\frac{1}{\Psi(z)}\frac{\partial^2\Psi(z)}{\partial z^2}=E_{Wy}$$

(11.33)

or

$$-\left(\frac{\hbar^2}{2m}\right)\frac{1}{\Psi(z)}\frac{\partial^2\Psi(z)}{\partial z^2}=E_W-E_{Wy}=E_{Wz}$$

(11.34)

or

$$-\left(\frac{\hbar^2}{2m}\right)\frac{\partial^2\Psi(z)}{\partial z^2}=E_{Wz}\Psi(z)$$

(11.35)

Thus we are confronted with the problem of a particle in an infinite potential well which was solved in Section 5.2. The wave function for eq. (11.32) is (eqs. (5.76) and (5.98))

$$\Psi(y)=\sqrt{\frac{2}{L_y}}\sin(k_y y)=\sqrt{\frac{2}{L_y}}\sin\left(\frac{n_y\pi y}{L_y}\right)\text{where }n_y=1,2,3,\dots$$

(11.36)

and the wave function for eq. (11.35) is

$$\Psi(z)=\sqrt{\frac{2}{L_y}}\sin(k_z z)=\sqrt{\frac{2}{L_z}}\sin\left(\frac{n_z\pi z}{L_z}\right)\text{where }n_z=1,2,3,\dots$$

(11.37)

$$k_y=\frac{n_y\pi}{L_y}\text{ and }k_z=\frac{n_z\pi}{L_z}$$

(11.38)

where n_y and n_z are the principal quantum numbers.

So, from eqs. (11.25), (11.36), and (11.37)

$$\Psi(y,z)=\Psi(y)\Psi(z)=\sqrt{\frac{2}{L_y}}\sin\left(\frac{n_y\pi y}{L_y}\right)\times\sqrt{\frac{2}{L_z}}\sin\left(\frac{n_z\pi z}{L_z}\right)$$

$$=\sqrt{\frac{4}{L_yL_z}}\sin\left(\frac{n_y\pi y}{L_y}\right)\sin\left(\frac{n_z\pi z}{L_z}\right)\text{where}$$

$$n_y=1,2,3,\dots;n_z=1,2,3,\dots$$

(11.39)

The energy levels are (eq. (5.91))

$$E_{n_y}=\frac{\hbar^2}{2m}\left(\frac{n_y\pi}{L_y}\right)^2\text{where }n_y=1,2,3,\dots$$

(11.40)

and

$$E_{n_z}=\frac{\hbar^2}{2m}\left(\frac{n_z\pi}{L_z}\right)^2\text{where }n_z=1,2,3,\dots$$

(11.41)

From eqs. (11.20) and (11.39), the wave function of the electron is written as

$$\Psi(x,y,z)=\exp(\pm ikx)\sqrt{\frac{4}{L_yL_z}}\sin\left(\frac{n_y\pi y}{L_y}\right)\sin\left(\frac{n_z\pi z}{L_z}\right)$$

= Wave function describing the translational motion

of electrons as plane waves in the X-direction

× Wave function describing the distribution of

electrons in the transverse Y, Z directions   (11.42)

The energy spectrum of the electron consists of a series of sub-bands in one dimension represented as paraboloids in the energy versus k_x space. The energies are

$$E_{k_x,n_y,n_z}=\frac{\hbar^2k_x^2}{2m^*}+\frac{\hbar^2}{2m^*}\left(\frac{n_y\pi}{L_y}\right)^2+\frac{\hbar^2}{2m^*}\left(\frac{n_z\pi}{L_z}\right)^2$$

$$=\frac{\hbar^2k_x^2}{2m^*}+\frac{\hbar^2}{2m^*}\left\{\left(\frac{n_y\pi}{L_y}\right)^2+\left(\frac{n_z\pi}{L_z}\right)^2\right\}=\frac{\hbar^2k_x^2}{2m^*}+\varepsilon_{n_y,n_z}$$

= Energy for continuous motion + Energy of discrete levels

(11.43)

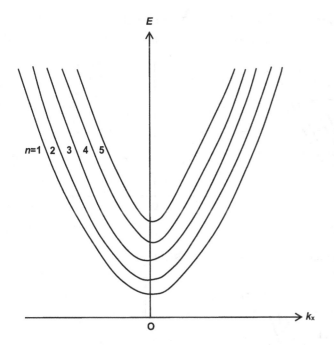

FIGURE 11.2 Modification of the 1-D dispersion relation by energies of transverse modes.

n_x and n_y are integers, and k_x is a 1D wave vector directed along the axis of the quantum wire, the direction of propagation of the current. This equation for energies represents the 1D dispersion relation

$$E_{k_x} = \frac{\hbar^2 k_x^2}{2m^*} \qquad (11.44)$$

offset by energies of the transverse modes or conduction channels (Figure 11.2). Current flow is restricted only to those modes or channels for which the energies are smaller than the Fermi energy level E_F of the contacts. As shown in Figure 11.3(a), the dispersion relation graph at equilibrium cuts the Fermi level E_F and only those energies fulfilling the earlier condition are allowed. Note that we retain the symbol m for mass in general, and use m^* for effective mass. We shall also use the symbol m in subscript for quantum number as per common usage.

11.3 Distribution Function of Electrons in the Quantum Wire under Bias

Since the energy of electrons is related to the applied voltage V as

$$E = -eV \qquad (11.45)$$

the left side contact of the quantum wire on which higher potential is applied has lower Fermi energy and the right side contact at lower potential has higher Fermi energy. The Fermi level on the higher potential side shifts downward by 1/2 eV and becomes $E_F - 1/2$ eV while that on the lower potential side shifts upward by 1/2 eV becoming $E_F + 1/2$ eV. The quasi-Fermi levels in the wide electrodes remain uninfluenced by the bias because there are a large number of available states, as indicated by dense hatching but those in the ballistic conductor are affected because of the presence of few transverse modes.

From the earlier arguments, quasi-Fermi levels will be established on the two contacts of the quantum wire, as shown in Figure 11.3(b). On the positive side contact, the quasi-Fermi level is shown at the position $E_{FL} = E_F - 1/2$ eV (Figure 11.3c). On the negative side contact, the quasi-Fermi level is shown at the position $E_{FR} = E_F + 1/2$ eV. Electronic states with $+k$ values originate from the right contact having higher Fermi energy (lower electrical potential) while those with $-k$ values arise from the left contact having lower Fermi energy (higher electrical potential). The responsibility of the current transport is borne only by states having $+k$ values lying in the region between quasi-Fermi levels E_{FR} and E_{FL}. Below the energy E_{FL}, all states with $+k$ values and those with $-k$ values are filled leading to zero current.

Applying equation (11.43), the Fermi-Dirac distribution function for electron energies on the left side contact is represented as

$$f_L(E) = f\left\{ \frac{\hbar^2 k_x^2}{2m^*} + \varepsilon_{n_y,n_z} - \left(E_F - 1/2\,\text{eV}\right) \right\}$$

$$= f\left(\frac{\hbar^2 k_x^2}{2m^*} + \varepsilon_{n_y,n_z} - E_F + 1/2\,\text{eV} \right)$$

$$= f\left(\frac{\hbar^2 k_x^2}{2m^*} + \varepsilon_{n_y,n_z} + 1/2\,\text{eV} - E_F \right)$$

$$= f\left(E_{k_x,n_y,n_z} + 1/2\,\text{eV} - E_F \right)$$

$$= \frac{1}{1 + \exp\left(\dfrac{E_{k_x,n_y,n_z} + 1/2\,\text{eV} - E_F}{k_B T} \right)} \qquad (11.46)$$

where k_B is the Boltzmann constant.

Similarly, the electron energy distribution function on the right side contact is represented as

$$f_R(E) = f\left\{ \frac{\hbar^2 k_x^2}{2m^*} + \varepsilon_{n_y,n_z} - \left(E_F + 1/2\,\text{eV}\right) \right\}$$

$$= f\left(\frac{\hbar^2 k_x^2}{2m^*} + \varepsilon_{n_y,n_z} - E_F - 1/2\,\text{eV} \right)$$

$$= f\left(\frac{\hbar^2 k_x^2}{2m^*} + \varepsilon_{n_y,n_z} - 1/2\,\text{eV} - E_F \right)$$

$$= f\left(E_{k_x,n_y,n_z} - 1/2\,\text{eV} - E_F \right)$$

$$= \frac{1}{1 + \exp\left(\dfrac{E_{k_x,n_y,n_z} - 1/2\,\text{eV} - E_F}{k_B T} \right)} \qquad (11.47)$$

For convenience, we write

$$f_L\left(E_{k_x,n,m} \right) = f\left(E_{k_x,n,m} + 1/2\,\text{eV} - E_F \right) \qquad (11.48)$$

$$f_R\left(E_{k_x,n,m} \right) = f\left(E_{k_x,n,m} - 1/2\,\text{eV} - E_F \right) \qquad (11.49)$$

where n and m are the transverse quantum numbers. They represent the wave vectors k_y and k_z defined in eq. (11.38).

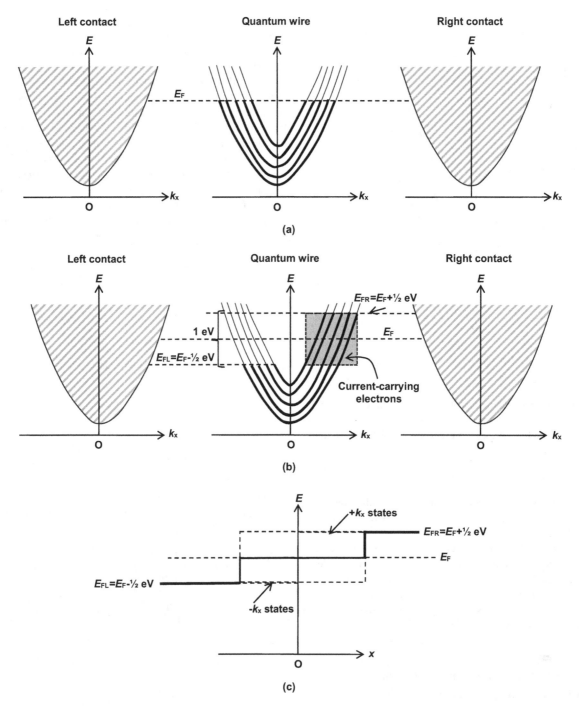

FIGURE 11.3 Ballistic conductor with wide contacts: (a) in the unbiased state and (b) in the biased condition by a voltage V; (c) variation of the quasi-Fermi level from left contact to right contact.

In the zero temperature limit, the Fermi-Dirac function becomes the Heaviside step function (Figure 11.4). So

$$\lim_{T \to 0} f_L \left(E_{k_x,n,m} \right) = \theta \left\{ E_F^0 - \left(E_{k_x,n_y,n_z} + 1/2 \text{ eV} \right) \right\} \quad (11.50)$$

$$\lim_{T \to 0} f_R \left(E_{k_x,n,m} \right) = \theta \left\{ E_F^0 - \left(E_{k_x,n_y,n_z} - 1/2 \text{ eV} \right) \right\} \quad (11.51)$$

where E_F^0 is the Fermi level at $T = 0$. For notational convenience, we can write

$$f_L \left(E, T = 0 \right) = \theta \left\{ E_F^0 - \left(E_{k_x,n,m} + 1/2 \text{ eV} \right) \right\} \quad (11.52)$$

$$f_R \left(E, T = 0 \right) = \theta \left\{ E_F^0 - \left(E_{k_x,n,m} - 1/2 \text{ eV} \right) \right\} \quad (11.53)$$

We shall need these equations in Section 11.6.

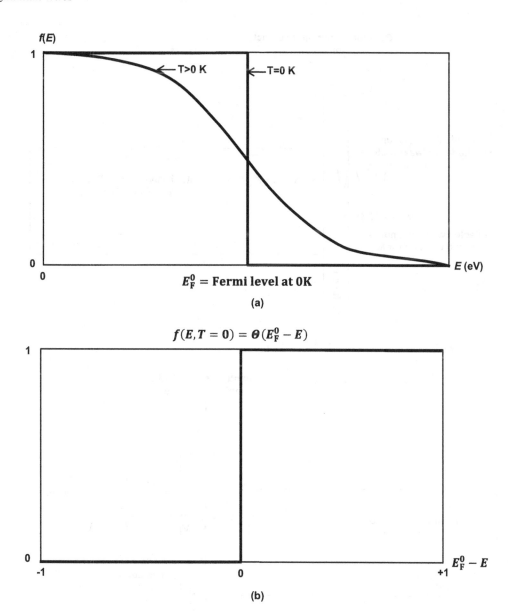

FIGURE 11.4 Fermi-Dirac distribution function: (a) at 0 K and higher temperature and (b) representation of its behavior at 0 K by the Heaviside step function $H(x) = f(E, T = 0) = \Theta(x) = \Theta(E_F^0 - E) = 0$ when $x = E_F^0 - E < 0$ or $E_F^0 < E$ or $E > E_F^0$, which is found to be true from (a); and $H(x) = f(E) = \Theta(x) = \Theta(E_F^0 - E) = 1$ when $x = E_F^0 - E > 0$ or $E_F^0 > E$ or $E < E_F^0$, which is again verified as true from (a).

11.4 Transmission Coefficient of Electrons across the Contact/Quantum Wire Interface

Here, electron transport along the length of the quantum wire is under consideration. Therefore, the relevant wave function is $\Psi_k(x)$. Consider the interface between the left contact and the quantum wire (Figure 11.5). Let an electron wave of unit amplitude and wave vector k_L moving from left to right fall on the interface between the left contact and the quantum wire. Suppose the interface is located at $x = 0$. The wave undergoes both reflection and transmission at the interface. After suffering reflection at the interface, let the amplitude change to to r_L. Thus the reflected wave differs from the incident wave in its amplitude and direction. Suppose the transmitted wave has an amplitude t_R and wave vector k_R. Obviously, its direction is the same as that of the incident wave.

The wave functions can be defined for the two regions: $x \leq 0$ and $x \geq 0$ as

$\Psi_L(x) =$ Incident wave of unit amplitude and wave vector k_L moving from left to right along the positive X-direction

+ Reflected wave of amplitude r_L and wave vector k_L moving from right to left along the negative X-direction

$= 1 \cdot \exp(+ik_L x) + r_L \exp(-ik_L x) = \exp(+ik_L x) + r_L \exp(-ik_L x)$ for $x \leq 0$ (11.54)

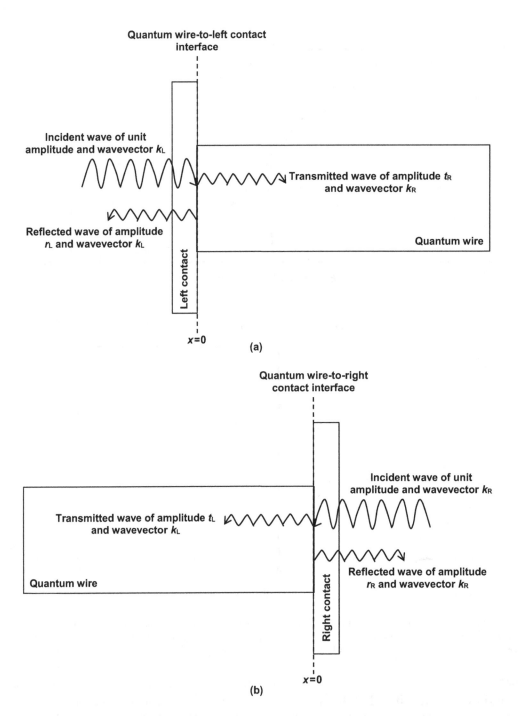

FIGURE 11.5   Transmission of electron waves across the interfaces between: (a) left contact and quantum wire, and (b) right contact and quantum wire.

and

$\Psi_R(x)$ = Trasmitted wave of amplitude t_R and wave vector k_R

    moving from left to right along the positive X-direction

$$= t_R \exp(+ik_R x) \text{ for } x \geq 0 \qquad (11.55)$$

Applying the continuity condition on $\Psi_L(x)$ at $x = 0$

$$\exp(+ik_L x) + r_L \exp(-ik_L x) = t_R \exp(+ik_R x) \qquad (11.56)$$

or

$$\exp(+ik_L \times 0) + r_L \exp(-ik_L \times 0) = t_R \exp(+ik_R \times 0) \qquad (11.57)$$

or

$$1 + r_L \times 1 = t_R \times 1 \qquad (11.58)$$

or

$$1 + r_L = t_R \qquad (11.59)$$

To apply the continuity condition on $d\Psi_L(x)/dx$, we note that

$$\left\{\frac{d}{dx}\exp(+ik_L x)\right\} + \frac{d}{dx}\left\{r_L \exp(-ik_L x)\right\} = \frac{d}{dx}\left\{t_R \exp(+ik_R x)\right\}$$

(11.60)

or

$$\exp(+ik_L x) \times (+ik_L) + r_L \exp(-ik_L x) \times (-ik_L)$$
$$= t_R \exp(+ik_R x) \times (+ik_R)$$

(11.61)

At $x = 0$,

$$\exp(+ik_L \times 0) \times (+ik_L) + r_L \exp(-ik_L \times 0) \times (-ik_L)$$
$$= t_R \exp(+ik_R \times 0) \times (+ik_R)$$

(11.62)

or

$$1 \times (+ik_L) + r_L \times 1 \times (-ik_L) = t_R \times 1 \times (+ik_R)$$

(11.63)

or

$$ik_L - ik_L r_L = ik_R t_R$$

(11.64)

or

$$(1 - r_L)ik_L = ik_R t_R$$

(11.65)

or

$$(1 - r_L)k_L = k_R t_R$$

(11.66)

From eq. (11.59)

$$r_L = t_R - 1$$

(11.67)

Substituting for r_L from eq. (11.67) in eq. (11.66),

$$\{1 - (t_R - 1)\}k_L = k_R t_R$$

(11.68)

or

$$(2 - t_R)k_L = k_R t_R$$

(11.69)

or

$$2k_L - t_R k_L = k_R t_R$$

(11.70)

or

$$t_R(k_L + k_R) = 2k_L$$

(11.71)

$$\therefore t_R = \frac{2k_L}{k_L + k_R}$$

(11.72)

Substituting for t_R from eq. (11.72) in eq. (11.67),

$$r_L = t_R - 1 = \frac{2k_L}{k_L + k_R} - 1 = \frac{2k_L - (k_L + k_R)}{k_L + k_R} = \frac{k_L - k_R}{k_L + k_R}$$

(11.73)

Now, particle flux is the number of particles passing a specified point per second = square of the amplitude of the wave (representing the average number of particles per unit length) × velocity = $|A|^2$ × velocity = $|A|^2$ × momentum/mass = $|A|^2 \times \hbar k/m$ from eq. (2.23). Hence,

Flux of incident electrons

$$= |\text{Amplitude of incident electrons}|^2 \frac{\hbar k_L}{m^*} = |1|^2 \frac{\hbar k_L}{m^*} = \frac{\hbar k_L}{m^*}$$

(11.74)

and

Flux of reflected electrons

$$= |\text{Amplitude of reflected electrons}|^2 \frac{\hbar k_L}{m^*} = |r_L|^2 \frac{\hbar k_L}{m^*}$$

∴ Reflection coefficient of electrons moving from left to right

$$= R_{L \to R}$$

$$= \frac{\text{Flux of reflected electrons}}{\text{Flux of incident electrons}} = \frac{|r_L|^2 \frac{\hbar k_L}{m^*}}{\frac{\hbar k_L}{m^*}} = |r_L|^2 = \left|\frac{k_L - k_R}{k_L + k_R}\right|^2$$

using equation (11.73).

(11.75)

Since

Flux of transmitted electrons

$$= |\text{Amplitude of transmitted electrons}|^2 \frac{\hbar k_R}{m^*} = |t_R|^2 \frac{\hbar k_R}{m^*}$$

(11.76)

∴ Transmission coefficient of electrons moving from left to right

$$= T_{L \to R} = \frac{\text{Flux of transmitted electrons}}{\text{Flux of incident electrons}}$$

$$= \frac{|t_R|^2 \frac{\hbar k_R}{m^*}}{\frac{\hbar k_L}{m^*}} = |t_R|^2 \frac{k_R}{k_L} = \left|\frac{2k_L}{k_L + k_R}\right|^2 \frac{k_R}{k_L}$$

(11.77)

using equation (11.72).
Also,

Flux of incident electrons – Flux of reflected electrons

(11.78)

$$= \text{Flux of transmitted electrons}$$

or

$$\frac{\hbar k_L}{m^*} - |r_L|^2 \frac{\hbar k_L}{m^*} = |t_R|^2 \frac{\hbar k_R}{m^*}$$

(11.79)

or

$$\left(1 - |r_L|^2\right)\frac{\hbar k_L}{m^*} = |t_R|^2 \frac{\hbar k_R}{m^*}$$

(11.80)

or

$$\left(1-\left|r_{\mathrm{L}}\right|^2\right)k_{\mathrm{L}} = \left|t_{\mathrm{R}}\right|^2 k_{\mathrm{R}} \tag{11.81}$$

or

$$\left(1-\left|r_{\mathrm{L}}\right|^2\right)k_{\mathrm{L}} k_{\mathrm{R}} = \left|t_{\mathrm{R}}\right|^2 k_{\mathrm{R}}^2 \tag{11.82}$$

Now consider electrons moving from right to left. An electron wave of unit amplitude and wave vector k_{R} moving from right to left strikes the interface between the right contact and the quantum wire. After reflection at the interface, let the amplitude change to r_{R}. Suppose the transmitted wave has an amplitude t_{L} and wave vector k_{L}. As earlier, the wave functions are defined as

$\Psi_{\mathrm{R}}(x) =$ Incident wave of unit amplitude and wave vector k_{R}

moving from right to left along the negative X-direction

$+$ Reflected wave of amplitude r_{R} and wave vector k_{R}

moving from left to right along the positive X-direction

$= 1 \cdot \exp(-ik_{\mathrm{R}}x) + r_{\mathrm{R}} \exp(+ik_{\mathrm{R}}x)$

$= \exp(-ik_{\mathrm{R}}x) + r_{\mathrm{R}} \exp(+ik_{\mathrm{R}}x)$ for $x \geq 0$ $\tag{11.83}$

and

$\Psi_{\mathrm{R}}(x) =$ Trasmitted wave of amplitude t_{L} and wave vector k_{L}

moving from right to left along the negative X-direction

$= t_{\mathrm{L}} \exp(-ik_{\mathrm{L}}x)$ for $x \leq 0$ $\tag{11.84}$

For application of the continuity condition on $\Psi_{\mathrm{R}}(x)$ at $x = 0$, we write

$$\exp(-ik_{\mathrm{R}}x) + r_{\mathrm{R}} \exp(+ik_{\mathrm{R}}x) = t_{\mathrm{L}} \exp(-ik_{\mathrm{L}}x) \tag{11.85}$$

At $x = 0$,

$$\exp(-ik_{\mathrm{R}} \times 0) + r_{\mathrm{R}} \exp(+ik_{\mathrm{R}} \times 0) = t_{\mathrm{L}} \exp(-ik_{\mathrm{L}} \times 0) \tag{11.86}$$

or

$$1 + r_{\mathrm{R}} \times 1 = t_{\mathrm{L}} \times 1 \tag{11.87}$$

or

$$1 + r_{\mathrm{R}} = t_{\mathrm{L}} \tag{11.88}$$

To apply the continuity condition on $d\Psi_{\mathrm{R}}(x)/dx$, the equality condition is

$$\frac{d}{dx}\left\{\exp(-ik_{\mathrm{R}}x)\right\} + \frac{d}{dx}\left\{r_{\mathrm{R}} \exp(+ik_{\mathrm{R}}x)\right\} = \frac{d}{dx}\left\{t_{\mathrm{L}} \exp(-ik_{\mathrm{L}}x)\right\} \tag{11.89}$$

or

$$\exp(-ik_{\mathrm{R}}x) \times (-ik_{\mathrm{R}}) + r_{\mathrm{R}} \exp(+ik_{\mathrm{R}}x) \times +ik_{\mathrm{R}}$$
$$= \left\{t_{\mathrm{L}} \exp(-ik_{\mathrm{L}}x)\right\} \times (-ik_{\mathrm{L}}) \tag{11.90}$$

At $x = 0$,

$$\exp(-ik_{\mathrm{R}} \times 0) \times (-ik_{\mathrm{R}}) + r_{\mathrm{R}} \exp(+ik_{\mathrm{R}} \times 0) \times +ik_{\mathrm{R}}$$
$$= \left\{t_{\mathrm{L}} \exp(-ik_{\mathrm{L}} \times 0)\right\} \times (-ik_{\mathrm{L}}) \tag{11.91}$$

or

$$1 \times (-ik_{\mathrm{R}}) + r_{\mathrm{R}} \times 1 \times +ik_{\mathrm{R}} = (t_{\mathrm{L}} \times 1) \times (-ik_{\mathrm{L}}) \tag{11.92}$$

or

$$-ik_{\mathrm{R}} + ik_{\mathrm{R}} r_{\mathrm{R}} = -ik_{\mathrm{L}} t_{\mathrm{L}} \tag{11.93}$$

or

$$+k_{\mathrm{R}} - k_{\mathrm{R}} r_{\mathrm{R}} = +k_{\mathrm{L}} t_{\mathrm{L}} \tag{11.94}$$

or

$$(1 - r_{\mathrm{R}})k_{\mathrm{R}} = k_{\mathrm{L}} t_{\mathrm{L}} \tag{11.95}$$

From eq. (11.88)

$$r_{\mathrm{R}} = t_{\mathrm{L}} - 1 \tag{11.96}$$

Substituting for r_{R} from eq. (11.96) in eq. (11.95),

$$\left\{1 - (t_{\mathrm{L}} - 1)\right\}k_{\mathrm{R}} = k_{\mathrm{L}} t_{\mathrm{L}} \tag{11.97}$$

or

$$(2 - t_{\mathrm{L}})k_{\mathrm{R}} = k_{\mathrm{L}} t_{\mathrm{L}} \tag{11.98}$$

or

$$2k_{\mathrm{R}} - t_{\mathrm{L}} k_{\mathrm{R}} = k_{\mathrm{L}} t_{\mathrm{L}} \tag{11.99}$$

or

$$t_{\mathrm{L}}(k_{\mathrm{R}} + k_{\mathrm{L}}) = 2k_{\mathrm{R}} \tag{11.100}$$

$$\therefore t_{\mathrm{L}} = \frac{2k_{\mathrm{R}}}{k_{\mathrm{R}} + k_{\mathrm{L}}} \tag{11.101}$$

From eqs. (11.96) and (11.101),

$$r_{\mathrm{R}} = t_{\mathrm{L}} - 1 = \frac{2k_{\mathrm{R}}}{k_{\mathrm{L}} + k_{\mathrm{R}}} - 1 = \frac{2k_{\mathrm{R}} - (k_{\mathrm{L}} + k_{\mathrm{R}})}{k_{\mathrm{L}} + k_{\mathrm{R}}} = \frac{k_{\mathrm{R}} - k_{\mathrm{L}}}{k_{\mathrm{R}} + k_{\mathrm{L}}} \tag{11.102}$$

The following equations can be written down:

Flux of incident electrons

$$= \left|\text{Amplitude of incident electrons}\right|^2 \frac{\hbar k_{\mathrm{R}}}{m^*} = \left|1\right|^2 \frac{\hbar k_{\mathrm{R}}}{m^*} = \frac{\hbar k_{\mathrm{R}}}{m^*} \tag{11.103}$$

and

Flux of reflected electrons

$$= \left|\text{Amplitude of reflected electrons}\right|^2 \frac{\hbar k_{\mathrm{R}}}{m^*} = \left|r_{\mathrm{R}}\right|^2 \frac{\hbar k_{\mathrm{R}}}{m^*} \tag{11.104}$$

$\therefore$ Reflection coefficient of electrons moving from right to left

$$= R_{R \to L} = \frac{\text{Flux of reflected electrons}}{\text{Flux of incident electrons}}$$

$$= \frac{|r_R|^2 \dfrac{\hbar k_R}{m^*}}{\dfrac{\hbar k_R}{m^*}} = |r_R|^2 = \left| \frac{k_R - k_L}{k_R + k_L} \right|^2 \qquad (11.105)$$

using equation (11.102).

Since

Flux of transmitted electrons

$$= \left| \text{Amplitude of transmitted electrons} \right|^2 \frac{\hbar k_L}{m^*} = |t_L|^2 \frac{\hbar k_L}{m^*} \qquad (11.106)$$

$\therefore$ Transmission coefficient of electrons moving from right to left

$$= T_{R \to L} = \frac{\text{Flux of transmitted electrons}}{\text{Flux of incident electrons}}$$

$$= \frac{|t_L|^2 \dfrac{\hbar k_L}{m^*}}{\dfrac{\hbar k_R}{m^*}} = |t_L|^2 \frac{k_L}{k_R} = \left| \frac{2k_R}{k_R + k_L} \right|^2 \frac{k_L}{k_R} \qquad (11.107)$$

using equation (11.101).

Comparing eqs. (11.73) and (11.102), we find

$$|r_L|^2 = |r_R|^2 \qquad (11.108)$$

Further,

Flux of transmitted electrons = Flux of incident electrons

$$- \text{Flux of reflected electrons} \qquad (11.109)$$

or

$$|t_L|^2 \frac{\hbar k_L}{m^*} = \frac{\hbar k_R}{m^*} - |r_R|^2 \frac{\hbar k_R}{m^*} \qquad (11.110)$$

or

$$|t_L|^2 \frac{\hbar k_L}{m^*} = \left(1 - |r_R|^2\right) \frac{\hbar k_R}{m^*} \qquad (11.111)$$

or

$$|t_L|^2 k_L = \left(1 - |r_R|^2\right) k_R \qquad (11.112)$$

or

$$|t_L|^2 k_L^2 = \left(1 - |r_R|^2\right) k_R k_L \qquad (11.113)$$

or

$$|t_L|^2 k_L^2 = \left(1 - |r_L|^2\right) k_R k_L \qquad (11.114)$$

from eq. (11.108).

Comparison of eqs. (11.82) and (11.114) unveils that

$$|t_L|^2 k_L^2 = |t_R|^2 k_R^2 \qquad (11.115)$$

We found that

$$T_{L \to R} = |t_R|^2 \frac{k_R}{k_L} \qquad (11.116)$$

from (eq. (11.77)), and

$$T_{R \to L} = |t_L|^2 \frac{k_L}{k_R} = |t_L|^2 \frac{k_L^2}{k_R} \times \frac{1}{k_L} = \frac{|t_R|^2 k_R^2}{k_R k_L} = |t_R|^2 \frac{k_R}{k_L} \qquad (11.117)$$

by applying eq. (11.115). Thus from equations (11.116) and (11.117),

$$T_{L \to R} = T_{R \to L} \qquad (11.118)$$

Expressing transmission coefficient as a function of kinetic energy in the *X*-direction

$$T_{L \to R}(E_x) = T_{R \to L}(E_x) \qquad (11.119)$$

11.5 Current Propagation through the Quantum Wire

Let us confine our attention to electrons in a single energy state in the left contact lead. This energy state has quantum numbers k_x, n, and m. Then

Number of electrons in the single energy state

$$= 2 \times f\left(E_{k_x, n, m} + 1/2 \text{ eV} - E_F\right) \qquad (11.120)$$

where factor 2 accounts for the spin degeneracy, i.e. there are two electrons corresponding to spin quantum numbers +1/2 and −1/2.

Since the quantum wire has a length L,

Number of electrons in the single energy state per unit length

$$= \frac{2f\left(E_{k_x, n, m} + 1/2 \text{ eV} - E_F\right)}{L} \qquad (11.121)$$

and electric current I_L due to electrons entering the quantum wire from left contact is

$$I_L = \text{Electron charge} \times \sum_{n, \, m} \sum_{k_x > 0} \frac{2f\left(E_{k_x, n, m} + 1/2 \text{ eV} - E_F\right)}{L}$$

$$\times \text{ Transmission coefficient in the } X\text{-direction}$$

$$\times \text{ Electron velocity in the } X\text{-direction}$$

$$= (-e) \times \sum_{n, \, m} \sum_{k_x} \frac{2f\left(E_{k_x, n, m} + 1/2 \text{ eV} - E_F\right)}{L} \times T(E_x) \times v(x)$$

$$\qquad (11.122)$$

because

Current = Charge carried by one electron

$$\times \frac{\text{Number of electrons available}}{\text{Length}}$$

$$\times \text{Fraction of electrons transmitted} \times \frac{\text{Length}}{\text{Time}} \quad (11.123)$$

Hence,

$$I_L = \left(-\frac{2e}{L}\right) \sum_{n,\,m} \sum_{k_x > 0} v(x) T(E_x) f\left(E_{k_x,n,\,m} + 1/2 \text{ eV} - E_F\right)$$

$$(11.124)$$

Similarly, the electric current I_R caused by electrons entering the quantum wire from the right contact is

$$I_R = \left(-\frac{2e}{L}\right) \sum_{n,\,m} \sum_{k_x > 0} v(x) T(E_x) f\left(E_{k_x,n,\,m} - 1/2 \text{ eV} - E_F\right)$$

$$(11.125)$$

So, the total current flowing in the quantum wire is

$$I = I_L - I_R = \left(-\frac{2e}{L}\right) \sum_{n,\,m} \sum_{k_x > 0} v(x) T(E_x) f\left(E_{k_x,n,\,m} + 1/2 \text{ eV} - E_F\right)$$

$$-\left\{\left(-\frac{2e}{L}\right) \sum_{n,\,m} \sum_{k_x > 0} v(x) T(E_x) f\left(E_{k_x,n,\,m} - 1/2 \text{ eV} - E_F\right)\right\}$$

$$(11.126)$$

Since the electron velocity and transmission coefficient do not depend on the quantum numbers n and m, eq. (11.126) can be written as

$$I = \left(-\frac{2e}{L}\right) \sum_{n,\,m} \sum_{k_x > 0} v(x) T(E_x) \left\{f\left(E_{k_x,n,\,m} + 1/2 \text{ eV} - E_F\right)\right.$$

$$\left. - f\left(E_{k_x,n,\,m} - 1/2 \text{ eV} - E_F\right)\right\} \quad (11.127)$$

Further onward, we shall be concerned with the variation of longitudinal part of energy along the X-direction. i.e. the kinetic energy of longitudinal motion. Since the energy distribution function explicitly contains the longitudinal $\left(E_{k_x}\right)$ and transverse $\left(E_{n,m}\right)$ energies as

$$f\left(E_{k_x,n,\,m} - E_F\right) = \frac{1}{1 + \exp\left(\dfrac{E_{k_x,n,\,m} - E_F}{k_B T}\right)}$$

$$= \frac{1}{1 + \exp\left(\dfrac{E_{k_x} + E_{n,m} - E_F}{k_B T}\right)} \quad (11.128)$$

we can write in eq. (11.127)

$$2 \sum_{n,\,m} \sum_{k_x > 0} f\left(E_{k_x,n,\,m} - E_F\right) = \sum_{k_x > 0} \sum_{n,\,m} \frac{2}{1 + \exp\left(\dfrac{E_{k_x} + E_{n,m} - E_F}{k_B T}\right)}$$

$$= \sum_{k_x > 0} f\left(E_{k_x} - E_F\right) \quad (11.129)$$

Consequently eq. (11.127) takes the simple form

$$I = \left(-\frac{e}{L}\right) \sum_{k_x > 0} v(x) T(E_x) \left\{f\left(E_{k_x} + 1/2 \text{ eV} - E_F\right)\right.$$

$$\left. - f\left(E_{k_x} - 1/2 \text{ eV} - E_F\right)\right\} \quad (11.130)$$

The summation over k_x may be replaced by an integration over k_x through the relation

$$\sum_{k_x > 0} \{\ldots\ldots\} = L \int \frac{dk_x}{2\pi} \{\ldots\ldots\} \quad (11.131)$$

This conversion relationship arises by recalling the derivation of equation for 1D density of states as follows:

The Schrodinger equation for electron confined in one dimension is

$$-\frac{\hbar^2}{2m} \frac{\partial^2 \Psi(x)}{\partial x^2} = E\Psi(x) \quad (11.132)$$

or

$$-\frac{\partial^2 \Psi(x)}{\partial x^2} = \frac{2m}{\hbar^2} E\Psi(x) \quad (11.133)$$

or

$$\frac{\partial^2 \Psi(x)}{\partial x^2} + \frac{2m}{\hbar^2} E\Psi(x) = 0 \quad (11.134)$$

or

$$\frac{\partial^2 \Psi(x)}{\partial x^2} + k_x^2 \Psi(x) = 0 \quad (11.135)$$

where

$$k_x = \sqrt{\frac{2mE}{\hbar^2}} \quad (11.136)$$

The solutions to eq. (11.135) consist of sine and cosine functions (eq. (5.69)) written as

$$\Psi(x) = A\cos(k_x x) + B\sin(k_x x) \quad (11.137)$$

where A and B are constants. Only the sine functions are valid because the wave function vanishes at the infinite barriers of the line. Hence, from eqs. (5.76) and (5.88)

$$\Psi(x) = A\sin(k_x x) \text{ where } k_x = n_x \frac{\pi}{L} \text{ for } n_x = \pm 1, 2, 3, \ldots$$

$$(11.138)$$

Now, the length of the single state in k-space is

$$L_{\text{Single state}} = \frac{\pi}{L} \tag{11.139}$$

But the length of the line in k-space is

$$L_{\text{Line}} = k_x \tag{11.140}$$

Hence, the number of filled states in a line is

$$N = \frac{L_{\text{Line}}}{L_{\text{Single state}}} = \frac{k_x}{\frac{\pi}{L}} \times 2 \times \frac{1}{2} = \frac{k_x L}{\pi} \tag{11.141}$$

Multiplication by 2 is done for two spin states and division by 2 to correct for counting identical states $\pm n_x$ twice.

$$\therefore dN = \frac{d}{dk_x}\left(\frac{kL}{\pi}\right)dk_x = \frac{L}{\pi}dk_x = 2\left(\frac{L}{2\pi}\right)dk_x \tag{11.142}$$

Returning to eq. (11.130), the replacement of summation is done by integration according to eqs. (11.131) and (11.142) as

$$\left(-\frac{e}{L}\right)\sum_{k_x>0} v(x)T(E_x)\left\{f\left(E_{k_x} + 1/2 \text{ eV} - E_F\right)\right.$$

$$\left. -f\left(E_{k_x} - 1/2 \text{ eV} - E_F\right)\right\}$$

$$= \left(-\frac{e}{L}\right)L\int \frac{dk_x}{2\pi}v(x)T(E_x)\left\{f\left(E_{k_x} + 1/2 \text{ eV} - E_F\right)\right.$$

$$\left. -f\left(E_{k_x} - 1/2 \text{ eV} - E_F\right)\right\} \tag{11.143}$$

Note that the group velocity (eq. (9.138))

$$v(x) = \frac{1}{\hbar}\frac{dE_{k_x}}{dk_x} \tag{11.144}$$

$$\therefore I = -\frac{e}{2\pi}\int dk_x \times \frac{1}{\hbar}\frac{dE_{k_x}}{dk_x} \times T(E_x)\left\{f\left(E_{k_x} + 1/2 \text{ eV} - E_F\right)\right.$$

$$\left. -f\left(E_{k_x} - 1/2 \text{ eV} - E_F\right)\right\}$$

$$= -\frac{e}{2\pi\hbar}\int T(E_x)\left\{f\left(E_{k_x} + 1/2 \text{ eV} - E_F\right)\right.$$

$$\left. -f\left(E_{k_x} - 1/2 \text{ eV} - E_F\right)\right\}dE_{k_x} \tag{11.145}$$

where the integration is performed over E_{k_x}.

Let us consider the replacement of summation by integration for

$$\sum_{k_x>0}\sum_{n,\,m}\frac{1}{1 + \exp\left(\dfrac{E_{k_x} + E_{n,m} - E_F}{k_BT}\right)} \tag{11.146}$$

For this replacement, we write the equation for two-dimensional density of states to be derived in Section 12.4.4, eq. (12.75)

$$\{g(E)\}_{2D} = \frac{m^*}{\pi\hbar^2} \tag{11.147}$$

Applying this equation, the density of electrons per unit area is

$$n_{2D} = \int \{g(E)\}_{2D}\, f\left(E_{k_x} - E_F\right)dE_{k_x} = \frac{m^*}{\pi\hbar^2}\int f\left(E_{k_x} - E_F\right)dE_{k_x} \tag{11.148}$$

In view of this equation,

$$\sum_{k_x>0}\left\{f\left(E_{k_x} + 1/2 \text{ eV} - E_F\right) - f\left(E_{k_x} - 1/2 \text{ eV} - E_F\right)\right\} = \frac{m^*}{\pi\hbar^2}\left[\int\left\{f\left(E_{k_x} + 1/2 \text{ eV} - E_F\right) - f\left(E_{k_x} - 1/2 \text{ eV} - E_F\right)\right\}dE_{k_x}\right] \tag{11.148a}$$

$$\therefore I = -\frac{e}{2\pi\hbar}\int T(E_x)dE_{k_x} \times \frac{m^*}{\pi\hbar^2}\left[\int\left\{f\left(E_{k_x} + 1/2 \text{ eV} - E_F\right) - f\left(E_{k_x} - 1/2 \text{ eV} - E_F\right)\right\}dE_{k_x}\right]$$

$$= -\frac{em^*}{2\pi^2\hbar^3}\int T(E_x)dE_{k_x}\left[\int\left\{f\left(E_{k_x} + 1/2 \text{ eV} - E_F\right) - f\left(E_{k_x} - 1/2 \text{ eV} - E_F\right)\right\}dE_{k_x}\right] \tag{11.149}$$

from equation (11.145).

In eq. (11.149),

$$\int\left\{f\left(E_{k_x} + 1/2 \text{ eV} - E_F\right) - f\left(E_{k_x} - 1/2 \text{ eV} - E_F\right)\right\}dE_{k_x} = \int f\left(E_{k_x} + 1/2 \text{ eV} - E_F\right)dE_{k_x} - \int f\left(E_{k_x} - 1/2 \text{ eV} - E_F\right)dE_{k_x}$$

$$= \int \frac{dE_{k_x}}{1 + \exp\left(\dfrac{E_{k_x} + E_{n,m} + 1/2 \text{ eV} - E_F}{k_BT}\right)} - \int \frac{dE_{k_x}}{1 + \exp\left(\dfrac{E_{k_x} + E_{n,m} - 1/2 \text{ eV} - E_F}{k_BT}\right)} \tag{11.150}$$

$$\text{First term} = \int \frac{dE_{k_x}}{\exp\left(\dfrac{E_{k_x} + E_{n,m} + 1/2 \text{ eV} - E_F}{k_BT}\right)\left[\exp\left\{-\left(\dfrac{E_{k_x} + E_{n,m} + 1/2 \text{ eV} - E_F}{k_BT}\right)\right\} + 1\right]} \tag{11.151}$$

Put

$$\eta = \exp\left\{-\left(\frac{E_{k_x} + E_{n,m} + 1/2 \text{ eV} - E_F}{k_B T}\right)\right\} \qquad (11.152)$$

Taking derivative, we obtain

$$\frac{d\eta}{dE_{k_x}} = \frac{d}{dE_{k_x}}\left[\exp\left\{-\left(\frac{E_{k_x} + E_{n,m} + 1/2 \text{ eV} - E_F}{k_B T}\right)\right\}\right]$$

$$= \exp\left\{-\left(\frac{E_{k_x} + E_{n,m} + 1/2 \text{ eV} - E_F}{k_B T}\right)\right\} \times \left(-\frac{1}{k_B T}\right)$$

$$= \eta \times \left(-\frac{1}{k_B T}\right) = -\frac{\eta}{k_B T} \qquad (11.153)$$

or

$$dE_{k_x} = -(k_B T)\frac{d\eta}{\eta} \qquad (11.154)$$

and

$$\text{First term} = \int \frac{-(k_B T)\dfrac{d\eta}{\eta}}{\eta^{-1}(\eta+1)} = -(k_B T)\int \frac{d\eta}{\eta+1} \qquad (11.155)$$

Put

$$\eta + 1 = \gamma \qquad (11.156)$$

Then

$$d\eta = d\gamma \qquad (11.157)$$

and

$$\text{First term} = -(k_B T)\int \frac{d\gamma}{\gamma} = -(k_B T)\ln\gamma$$

$$= -(k_B T)\ln(\eta+1)$$

$$= -(k_B T)\ln\left[\exp\left\{-\left(\frac{E_{k_x} + E_{n,m} + 1/2 \text{ eV} - E_F}{k_B T}\right)\right\} + 1\right]$$

$$= -(k_B T)\ln\left[1 + \exp\left\{-\left(\frac{E_{k_x} + E_{n,m} + 1/2 \text{ eV} - E_F}{k_B T}\right)\right\}\right] \qquad (11.158)$$

Proceeding in the same way

Second term

$$= -(k_B T)\ln\left[1 + \exp\left\{-\left(\frac{E_{k_x} + E_{n,m} - 1/2 \text{ eV} - E_F}{k_B T}\right)\right\}\right] \qquad (11.159)$$

Substituting the expressions for the first and second terms determined in eqs. (11.158) and (11.159) into eq. (11.150),

$$\int \left\{f\left(E_{k_x} + 1/2 \text{ eV} - E_F\right) - f\left(E_{k_x} - 1/2 \text{ eV} - E_F\right)\right\}dE_{k_x}$$

$$= -(k_B T)\ln\left[1 + \exp\left\{-\left(\frac{E_{k_x} + E_{n,m} + 1/2 \text{ eV} - E_F}{k_B T}\right)\right\}\right]$$

$$+ (k_B T)\ln\left[1 + \exp\left\{-\left(\frac{E_{k_x} + E_{n,m} - 1/2 \text{ eV} - E_F}{k_B T}\right)\right\}\right] \qquad (11.160)$$

From eqs. (11.149) and (11.160),

$$I = -\frac{em^*}{2\pi^2\hbar^3}\int T(E_x)\,dE_{k_x}$$

$$\times \left(-(k_B T)\ln\left[1 + \exp\left\{-\left(\frac{E_{k_x} + E_{n,m} + 1/2 \text{ eV} - E_F}{k_B T}\right)\right\}\right]\right.$$

$$\left. + (k_B T)\ln\left[1 + \exp\left\{-\left(\frac{E_{k_x} + E_{n,m} - 1/2 \text{ eV} - E_F}{k_B T}\right)\right\}\right]\right)$$

$$= -\frac{em^*}{2\pi^2\hbar^3}\int T(E_x)\,dE_{k_x}(k_B T)$$

$$\times \left(-\ln\left[1 + \exp\left\{-\left(\frac{E_{k_x} + E_{n,m} + 1/2 \text{ eV} - E_F}{k_B T}\right)\right\}\right]\right.$$

$$\left. + \ln\left[1 + \exp\left\{-\left(\frac{E_{k_x} + E_{n,m} - 1/2 \text{ eV} - E_F}{k_B T}\right)\right\}\right]\right)$$

$$= -\frac{em^* k_B T}{2\pi^2\hbar^3}\int T(E_x)\,dE_{k_x}$$

$$\times \ln\left[\frac{1 + \exp\left\{-\left(\dfrac{E_{k_x} + E_{n,m} - 1/2 \text{ eV} - E_F}{k_B T}\right)\right\}}{1 + \exp\left\{-\left(\dfrac{E_{k_x} + E_{n,m} + 1/2 \text{ eV} - E_F}{k_B T}\right)\right\}}\right] \qquad (11.161)$$

This equation expresses the dependence of the current flowing in a quantum wire as a function of electron concentration and temperature.

11.6 Determination of Conductance in the Zero Temperature Limit

Following eqs. (11.143)–(11.145), the current equation is written in a more general form including all the three quantum numbers k_x, n, and m

$$I = -\frac{e}{\hbar} \int T(E_x) \frac{dE}{2\pi} \Big[f\big(E(k_x, n, m) + 1/2\ eV - E_F\big) - f\big(E(k_x, n, m) - 1/2\ eV - E_F\big) \Big]$$

$$= -\frac{e}{2\pi\hbar} \int T(E_x) \Big[f\big(E(k_x, n, m) + 1/2\ eV - E_F\big) - f\big(E(k_x, n, m) - 1/2\ eV - E_F\big) \Big] dE \qquad (11.162)$$

$$= -\frac{e}{h} \int T(E_x) \Big[f\big(E(k_x, n, m) + 1/2\ eV - E_F\big) - f\big(E(k_x, n, m) - 1/2\ eV - E_F\big) \Big] dE$$

because

$$2\pi\hbar = h \qquad (11.163)$$

Let us write

$$f\big(E(k_x, n, m) + 1/2\ eV - E_F\big) = f_L \qquad (11.164)$$

$$f\big(E(k_x, n, m) - 1/2\ eV - E_F\big) = f_R \qquad (11.165)$$

The outcome is

$$I = -\frac{e}{h} \int T(E_x)(f_L - f_R)\, dE \qquad (11.166)$$

When both the contacts are at the same potential

$$f_L = f_R \text{ and } I = 0 \qquad (11.167)$$

Let us disturb the equilibrium condition by applying a small bias, i.e. let the bias V be small. Then the current is proportional to the voltage and we can write

$$\delta I = -\frac{e}{h} \int \Big[\{T(E_x)\}_{\text{Eq.}}\, \delta(f_L - f_R) + (f_L - f_R)_{\text{Eq.}}\, \delta\{T(E_x)\} \Big] dE \qquad (11.168)$$

where the subscript "Eq." refers to the equilibrium condition. Since

$$(f_L - f_R)_{\text{Eq.}} = 0 \qquad (11.169)$$

the second term is zero leading to

$$\delta I = -\frac{e}{h} \int \Big[\{T(E_x)\}_{\text{Eq.}}\, \delta(f_L - f_R) \Big] dE \qquad (11.170)$$

By a Taylor series expansion

$$f(x) - f(x_0) = f'(x_0)(x - x_0) + \dots \qquad (11.171)$$

so that

$$\delta(f_L - f_R) = \left(\frac{\partial f}{\partial V}\right)_{\text{Eq.}} (V_L - V_R) = \left(\frac{\partial f}{\partial V}\right)_{\text{Eq.}} \left\{ +\frac{1}{2} V - \left(-\frac{1}{2} V\right) \right\}$$

$$= -\left(\frac{\partial f}{\partial E}\right)_{\text{Eq.}} \left\{ +\frac{1}{2} eV - \left(-\frac{1}{2} eV\right) \right\} = -\left(\frac{\partial f_0}{\partial E}\right) eV \qquad (11.172)$$

where $f_0(E)$ is the Fermi-Dirac distribution function at equilibrium

$$f_0(E) = \frac{1}{1 + \exp\left\{\dfrac{E(k_x, n, m) - E_F}{k_B T}\right\}} \qquad (11.173)$$

Substituting for $\delta(f_L - f_R)$ from eq. (11.172) in eq. (11.170)

$$\delta I = -\frac{e}{h} \int \left[\{T(E_x)\}_{\text{Eq.}} \left\{ -\left(\frac{\partial f_0}{\partial E}\right) eV \right\} \right] dE$$

$$= \frac{e^2 V}{h} \int \left[\{T(E_x)\}_{\text{Eq.}} \left\{ \left(\frac{\partial f_0}{\partial E}\right) \right\} \right] dE \qquad (11.174)$$

For $E(k_x, n, m) - E_F < 0$, when $T \to 0$, the power of the exponential term in the Fermi-Dirac distribution function approaches $-\infty$. So, the exponential term becomes zero and the Fermi-Dirac function equals unity. For $E(k_x, n, m) - E_F > 0$, when $T \to 0$, the power of the exponential term in the Fermi-Dirac distribution function approaches $+\infty$. As a result, the exponential term becomes infinity and the Fermi-Dirac function equals zero. Thus at absolute zero temperature, all the states below the Fermi level are filled and those above are vacant. The Fermi level represents the highest filled energy level and all other filled levels lie below it.

Recall from Section 11.3 that at very low temperatures, the Fermi-Dirac function becomes the Heaviside step function

$$\lim_{T \to 0} f\{E_F - E(k_x, n, m)\} = \theta\{E_F - E(k_x, n, m)\} \qquad (11.175)$$

so that

$$\theta\{E_F - E(k_x, n, m)\} = 1 \text{ for } E_F - E(k_x, n, m) > 0 \text{ or}$$

$$E_F > E(k_x, n, m) \qquad (11.176)$$

and

$$\theta\{E_F - E(k_x, n, m)\} = 0 \text{ for } E_F - E(k_x, n, m) < 0 \text{ or}$$

$$E_F < E(k_x, n, m) \tag{11.177}$$

The derivative of the Heaviside function is the unit impulse function, also known as the Dirac delta function.

$$\therefore \frac{\partial\left[\theta\{E_F - E(k_x, n, m)\}\right]}{\partial E} = \delta\{E_F - E(k_x, n, m)\} \tag{11.178}$$

so that

$$\delta\{E_F - E(k_x, n, m)\} = 0 \text{ for } E_F - E(k_x, n, m) \neq 0$$

$$\text{or } E_F \neq E(k_x, n, m) \tag{11.179}$$

$$\delta\{E_F - E(k_x, n, m)\} = \text{Undefined for } E_F - E(k_x, n, m)$$

$$= 0 \text{ or } E_F = E(k_x, n, m) \tag{11.180}$$

and

$$\int_{-\infty}^{+E_{F0}} \delta\{E_F - E(k_x, n, m)\} dE = 1 \tag{11.181}$$

where E_{F0} is the Fermi level at $T = 0$ K.

From eqs. (11.174), (11.178), and (11.181),

$$I = \frac{e}{h} \times eV \int T(E_F, n, m) \int_{-\infty}^{+E_{F0}} \delta\{E_F - E(k_x, n, m)\} dE$$

$$= \frac{e^2 V}{h} \int T(E_F, n, m) \times 1 \tag{11.182}$$

Writing the integral as a sum

$$I = \frac{e^2 V}{h} \sum_{n,m,s} T(E_F, n, m) \rightarrow I = \frac{2e^2 V}{h} \sum_{n,m} T(E_F, n, m) \tag{11.183}$$

where the summation is carried out only over electron states (n, m) having energies $< E_F$. Prefactor 2 is included for the two spin states of the electron.

Therefore the conductance of the quantum wire is

$$G = \frac{I}{V} = \frac{\dfrac{2e^2 V}{h} \sum_{n,m} T(E_F, n, m)}{V} = \frac{2e^2}{h} \sum_{n,m} T(E_F, n, m) \tag{11.184}$$

Equation (11.184) is known as Landauer's formula (Landauer 1957, 1992). Rolf William Landauer was a German-American physicist. As the conductance equation does not contain L, the conductance of the quantum wire is independent of its length.

Assuming reflectionless contacts

$$\sum_{n,m} T(E_F, n, m) = 1 \tag{11.185}$$

Hence, at zero temperature and without any reflection of electron waves at the contact/quantum wire interface

$$G = \frac{2e^2}{h} \tag{11.186}$$

It is already assumed that the length of the quantum wire is so short that any chances of internal scattering are excluded.

The quantity

$$G_0 = \frac{2e^2}{h} = \frac{2(1.602 \times 10^{-19} \text{ C})^2}{6.62607 \times 10^{-34} \text{ J s}} = \frac{2 \times 2.5664 \times 10^{-38}}{6.62607 \times 10^{-34}}$$

$$= 0.7746 \times 10^{-4} \text{ S} = 77.46 \times 10^{-6} \text{ S} = 77.46 \text{ μS} \tag{11.187}$$

is called the quantum of conductance. Its reciprocal has the value 12.91 kΩ.

11.7 Discussion on Landauer's Formula and the Length Scale of Validity of Ohm's Law

Landauer's formula has great consequences. It leads to the ideas of the quantization of conductance and the statement of minimum limiting quantum value of conductance that occurs in the idealized assumed situation of zero temperature, with neither any scattering in the short-length quantum wire nor any reflections at the quantum wire/contact interface (transmission coefficient of unity). In such an idealistic situation, the conductance is expected to be infinite according to Ohm's law. Ohm's law expressing the conductance G of a conductor in terms of its length L, width W, and conductivity σ is given by

$$G = \frac{\sigma W}{L} \tag{11.188}$$

When the length of a conductor connected across two large contacts is decreased below the distance at which an electron encounters no scattering, the conductor should show infinite conductance or zero resistance. But experimentally it still has a finite conductance or resistance value. This non-zero resistance value arises from interface resistance between the conductor and each contact. An explanation for the interface resistance, commonly called contact resistance, can be offered on the basis of the difference in the number of transverse modes in the contacts and the quantum wire.

As already mentioned, in a narrow conductor, electronic states evolve from electric sub-bands arising from electrostatic confinement. The sub-bands are distinctly separated in energy. Such a conductor is often referred to as an electron waveguide. The sub-bands are also called transverse modes in analogy to the modes of a waveguide for electromagnetic waves.

The current transport in the contacts takes place through infinitely many electric transverse modes whereas in the conductor, it occurs via only a few modes. The interface between each contact and quantum wire is the boundary between a region with infinite modes and infinite electron population with all electrons in equilibrium and a region with a small number of electrons that are not in equilibrium. The interfacial resistance is produced by the redistribution of current at the interface among the infinitely many transverse modes of the contacts and the relatively few modes of the conductor of the quantum wire. The observed conductance due to these interface effects becomes independent of the length of the conductor.

Another deviation from Ohm's law is that the conductance no longer decreases linearly with the width of the conductor. Instead, it falls in discrete steps in accordance with the number of transverse modes in the conductor/contacts. Thus there are two corrections, one observed in the form of additional interface resistance and the other seen as the discretization of conductance changes. Both the corrections, namely interface resistance and discretization, are incorporated in the Landauer formula for conductance.

A conductor obeying Ohm's law is said to be ohmic. The law is valid as long as the dimensions of the conductor are much greater than each of the four characteristic lengths (Datta 1999): (i) the de Broglie wavelength = Planck's constant/Momentum of the charge-carrying particle; (ii) the mean free path: average distance traversed by the electron before its initial momentum becomes zero; (iii) the phase relaxation length: distance traveled by the electron before destruction of its initial phase; and (iv) the Debye screening length: distance over which mobile carriers such as electrons shield the effect of electric field in plasma or other conductors.

11.8 Alternative Approaches to Landauer's Formula

Landauer's formula is often derived by considering current carried by a single energy level in a single-mode quantum wire at 0 K, then generalizing to a quantum wire with multiple modes, further to multiple energy levels and a wire containing a scattering site at a higher temperature.

11.8.1 Landauer's Formula for Current Carried by a Single Energy Level in a Single-Mode/Multimode Quantum Wire at Zero Temperature

i. Current flowing through a ballistic conductor with reflectionless contacts

Let us consider a conducting wire connected between two large contacts L and R (Figure 11.6). The assumption of zero temperature assures that current flow is completely restricted to the energy range between V_L and V_R. Energy distributions of the incident electrons in the two leads are step functions. Transport takes place from contact L to contact R only. There is no transport in the reverse direction from contact R to contact L. Note that contrary to Figures 11.1 and 11.3, the left contact is taken at a negative potential (higher energy) and the right contact at positive potential (lower energy) in Figure 11.6. This will enable us to understand both situations.

The contacts are assumed to be reflectionless implying that the electrons from the conductor can enter a contact without suffering any reflections. The situation is considerably simplified for the case of reflectionless contacts. The $+k_x$ states in the conductor are only occupied by electrons that originate in the contact L. This is

because electrons from the left contact fill the $+k_x$ states and migrate to the right contact without suffering any reflection. Similarly, the $-k_x$ states in the conductor are only occupied by electrons that are created in the contact R. This is because electrons from the right contact fill the $-k_x$ states and migrate to the left contact without undergoing any reflection.

We can now foresee that the quasi-Fermi level of the $+k_x$ states is always E_{FL} even under the application of bias because if we keep both contacts at the same potential, then the Fermi level for the $+k_x$ states is obviously E_{FL} and if we change the potential of the right contact to E_{FR}, then the quasi-Fermi level for the $+k_x$ states still remains undisturbed owing to the non-existence of a causal relation between the right contact and $+k_x$ states, i.e. no electron arising from the right contact is able to reach any $+k_x$ state. The earlier reasoning also holds for the right contact and we find that the quasi-Fermi level of the $-k_x$ states is always E_{FR}. Thus at low temperature, electrons lying in the $+k_x$ states between E_{FL} and E_{FR} levels act as current carriers.

The current I in the wire can be expressed in terms of the number N of uncompensated electrons and the transit time τ of electrons from left to right contact as

$$I = -e \frac{N}{\tau} \tag{11.189}$$

The current can be obtained from the knowledge of N and τ. For N, we recall from the calculation of 1D density of states,

$$k\text{-space occupied by each state} = \Delta k_x = \frac{2\pi}{L} \tag{11.190}$$

Then

N = Number of electrons in states in the k_x interval

$k_{xL} > k_x > k_{xR}$ or energy interval $E_{FL} > E_x > E_{FR}$

$$= 2 \int_{k_{xR}}^{k_{xL}} \frac{dk_x}{\frac{2\pi}{L}} = 2 \int_{k_{xR}}^{k_{xL}} \frac{L dk_x}{2\pi} \tag{11.191}$$

where factor 2 accounts for two electrons per k-state. The transit time τ is

$$\tau = \frac{\text{Length of the wire} (L)}{\text{Group velocity of electrons} (v_g)}$$

$$= \frac{L}{\frac{1}{\hbar} \frac{dE_x}{dk_x}} = L\hbar \frac{dk_x}{dE_x} \tag{11.192}$$

using eq. (9.138).

Combining eqs. (11.189), (11.191), and (11.192)

$$I = -2e \int_{k_{xR}}^{k_{xL}} \frac{L dk_x}{2\pi} \Big/ \left(L\hbar \frac{dk_x}{dE_x} \right) = -2e \int_{k_{xR}}^{k_{xL}} \frac{L dk_x}{2\pi} \times \frac{1}{L\hbar} \frac{dE_x}{dk_x}$$

$$= -\frac{2e}{2\pi\hbar} \int_{k_{xR}}^{k_{xL}} dE_x \tag{11.193}$$

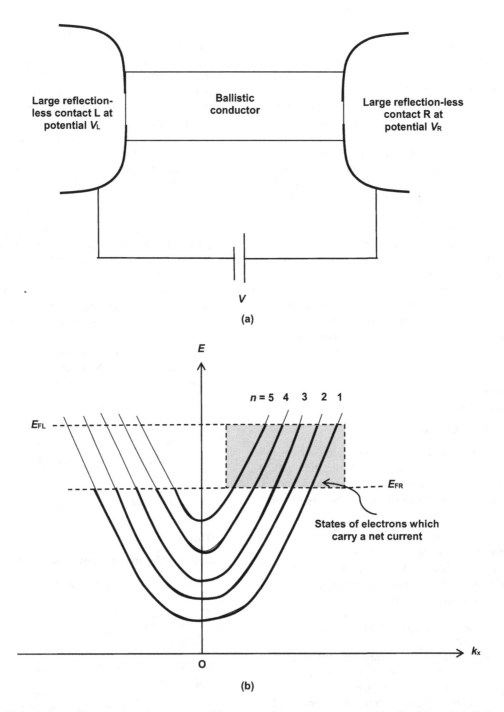

FIGURE 11.6 Ballistic conductor: (a) a narrow conductor connected between two large contacts across which a bias V is applied. (b) Plot of the dispersion relation for transverse modes in the conductor.

Substituting

$$\hbar = \frac{h}{2\pi} \qquad (11.194)$$

and changing the variable of integration as

$$k_{xR} \rightarrow E_{FR} \text{ and } k_{xL} \rightarrow E_{FL} \qquad (11.195)$$

we can write

$$I = -\frac{2e}{2\pi \dfrac{h}{2\pi}} \int_{E_{FR}}^{E_{FL}} dE_x = -\frac{2e}{h} \int_{E_{FR}}^{E_{FL}} dE_x$$

$$= -\frac{2e}{h} \left| E_x \right|_{E_{FR}}^{E_{FL}} = -\frac{2e}{h} \left(E_{FL} - E_{FR} \right) \qquad (11.196)$$

A multimode wire comprising M modes is equivalent to a parallel combination of M single-mode wires. Since parallel currents add up, the current in the multimode wire is

$$I = -\frac{2eM}{h}(E_{FL} - E_{FR}) \qquad (11.197)$$

ii. Current flowing through a conductor having transmission probability T connected to large contacts by ballistic conductor leads

Referring to Figure 11.7, electron influx from lead L into the wire is

$$I_L^+ = -\frac{2eM}{h}(E_{FL} - E_{FR}) \qquad (11.198)$$

Electron outflux from lead R into contact R is

$$I_R^+ = \text{Electron influx from lead } L \text{ into the wire } \left(I_L^+\right)$$

$$\times \text{Transmission probability}(T) = -\frac{2eM}{h}(E_{FL} - E_{FR}) \times T$$

$$= -\frac{2eMT}{h}(E_{FL} - E_{FR}) \qquad (11.199)$$

where T is the transmission probability per mode at the Fermi energy. It is assumed to be constant over the range $E_{FL} < E_F < E_{FR}$.

Electron flux reflected back to contact L is

$$I_L^- = -\frac{2eM}{h}(E_{FL} - E_{FR}) \times (1 - T)$$

$$= -\frac{2eM(1-T)}{h}(E_{FL} - E_{FR}) \qquad (11.200)$$

The total current flowing at any point either internally in the wire or in the external circuit is

$$I = I_L^+ - I_L^- = -\frac{2eM}{h}(E_{FL} - E_{FR}) - \left\{ -\frac{2eM(1-T)}{h}(E_{FL} - E_{FR}) \right\}$$

$$= -\frac{2eM}{h}(E_{FL} - E_{FR})\{1 - (1-T)\}$$

$$= -\frac{2eM}{h}(E_{FL} - E_{FR})(1 - 1 + T) = -\frac{2eMT}{h}(E_{FL} - E_{FR})$$

$$(11.201)$$

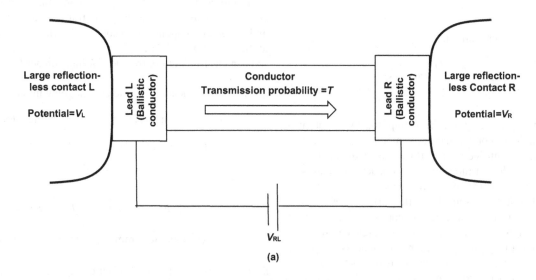

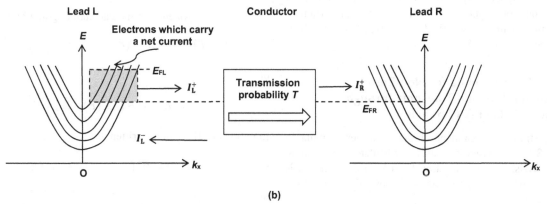

FIGURE 11.7 Conductor with transmission probability T: (a) connected to large reflectionless contacts through ballistic conductor leads, and (b) current components and dispersion relation plots for the lead-conductor-lead structure.

Also, we note that

$$I = I_R^+ = -\frac{2eMT}{h}(E_{FL} - E_{FR}) \qquad (11.202)$$

Let the electrical potentials of the contacts be V_L and V_R. The electrical potentials and quasi-Fermi levels are related as

$$E_{FL} = -eV_L \qquad (11.203)$$

$$E_{FR} = -eV_R \qquad (11.204)$$

Keeping in view eqs. (11.201), (11.202), (11.203), and (11.204), we may write

$$I = I_L^+ - I_L^- = I_R^+ = -\frac{2eMT}{h}(E_{FL} - E_{FR})$$

$$= -\frac{2eMT}{h}\{(-eV_L) - (-eV_R)\}$$

$$= -\frac{2eMT}{h}(eV_R - eV_L)$$

$$= -\frac{2e^2MT}{h}(V_R - V_L) = -\frac{2e^2MT}{h}V_{RL} \quad (11.205)$$

Hence, the conductance is

$$G = \frac{I}{V_{RL}} = \frac{\dfrac{2e^2MT}{h}V_{RL}}{V_{RL}} = \frac{2e^2MT}{h} \qquad (11.206)$$

As noted previously, the limit imposed by quantum mechanics on conductance tantamounts to a contact resistance. With increase in thickness of the conductor, the number of modes in the conductor increases, thereby reducing the contact resistance.

Further, it was assumed that the carriers move only from left to right. So the electrons entering the quantum wire from the left end propagate to the right end where they come to equilibrium with the Fermi energy of this side, which means that the power is dissipated at the right side. When carriers move both ways, the electrons traveling from right to left will come to equilibrium with the Fermi energy at the left end so that power will also be dissipated on the left side.

11.8.2 Landauer Formula for Non-Zero Temperature and Transport through Multiple Energy Channels

The overly simplified treatment of the Landauer formula in Section 11.8.1 was based on several postulates. In a more rigorous analysis, electron injection from both the contacts is considered. Further, transport occurs via several energy channels in the range

$$E_{FL} + \text{a few } k_B T < \text{Energy} < E_{FR} - \text{a few } k_B T \qquad (11.207)$$

with each channel having a different transmission probability.

Let us look at Figure 11.8. The current due to electron influx per unit energy from lead L into the wire is

$$i_L^+(E) = -\frac{2e}{h}M(E)f_L(E) \qquad (11.208)$$

where $f_L(E)$ is the Fermi function of lead L.

The current caused by electron influx per unit energy from lead R into the wire is

$$i_R^-(E) = -\frac{2e}{h}M'(E)f_R(E) \qquad (11.209)$$

where $f_R(E)$ is the Fermi function of lead R and M' is the number of modes in lead R.

Let T be the transmission probability of electron influx from lead L to go into contact R. Then $(1 - T)$ is the transmission probability of electron influx from lead L to go back into contact L. The electron influx $i_L^+(E)$ has two components, one which goes into the contact R with transmission probability T and the other which goes into contact L with transmission probability $(1 - T)$.

Similarly, let T' be the transmission probability of electron influx from lead R to go into contact L. Then $(1 - T')$ is the transmission probability of electron influx from lead R to go back into contact R. Out of the two components of $i_R^-(E)$, one component goes into contact L with transmission probability T' and the other component into contact R with transmission probability $(1 - T')$.

Then electron outflux from lead R into contact R is

$i_R^+(E) = $ Component of left lead current going into contact R

$\qquad$ + Component of right lead current going into contact R

$$= Ti_L^+(E) + (1 - T')i_R^-(E) \qquad (11.210)$$

or

$$i_R^+(E) = Ti_L^+(E) + (1 - T')i_R^-(E) \qquad (11.211)$$

The total electron outflux from lead L into contact L is

$i_L^-(E) = $ Component of left lead current going into contact L

$\qquad$ + Component of right lead current going into contact L

$$= (1 - T)i_L^+(E) + T'i_R^-(E) \qquad (11.212)$$

or

$$i_L^-(E) = (1 - T)i_L^+(E) + T'i_R^-(E) \qquad (11.213)$$

Equation (11.211) can be written as

$$i_R^+(E) = Ti_L^+(E) + i_R^-(E) - T'i_R^-(E) \qquad (11.214)$$

or

$$i_R^+(E) - i_R^-(E) = Ti_L^+(E) - T'i_R^-(E) \qquad (11.215)$$

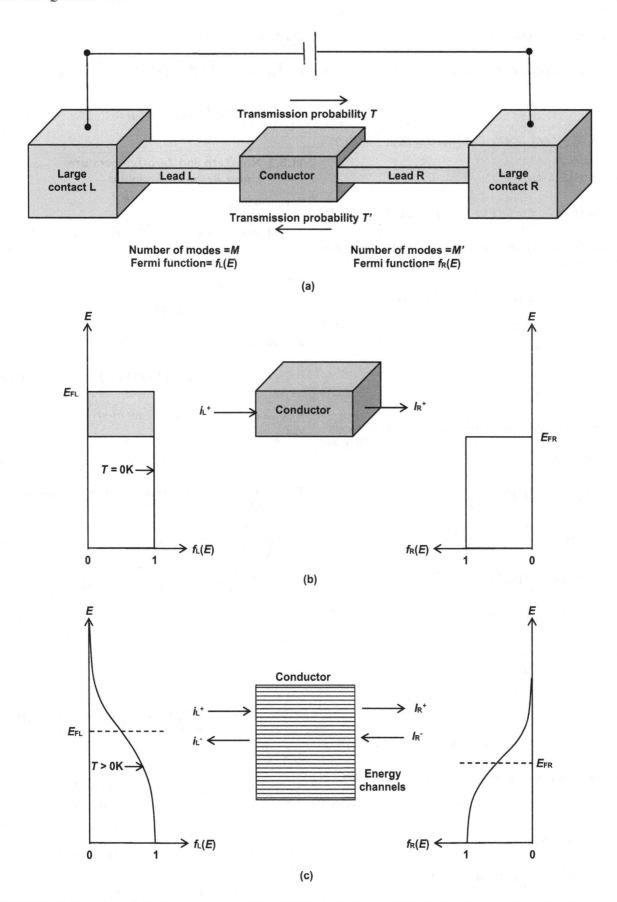

FIGURE 11.8 Conductor connected to large contacts through leads L, R: (a) electrical circuit, (b) distributions of energy of the incident electrons in the leads at zero temperature and (c) electron energy distributions in the leads at non-zero temperature.

Also, eq. (11.213) can be written as

$$i_L^-(E) = i_L^+(E) - Ti_L^+(E) + T'i_R^-(E) \quad (11.216)$$

or

$$i_L^+(E) - i_L^-(E) = Ti_L^+(E) - T'i_R^-(E) \quad (11.217)$$

From eqs. (11.215) and (11.217),

$$i_L^+(E) - i_L^-(E) = i_R^+(E) - i_R^-(E) = Ti_L^+(E) - T'i_R^-(E) \quad (11.218)$$

Thus the total current $i(E)$ at any point through the device may be expressed as

$$i(E) = i_L^+(E) - i_L^-(E) = i_R^+(E) - i_R^-(E)$$
$$= T(E)i_L^+(E) - T'(E)i_R^-(E) \quad (11.219)$$

or

$$i(E) = T(E)i_L^+(E) - T'(E)i_R^-(E) \quad (11.220)$$

showing the dependence of transmission probability on energy.

Substituting for $i_L^+(E)$ and $i_R^-(E)$ from eqs. (11.208) and (11.209) in eq. (11.220) we have

$$i(E) = T(E)\left\{-\frac{2e}{h}M(E)f_L(E)\right\} - T'(E)\left\{-\frac{2e}{h}M'(E)f_R(E)\right\}$$

$$= -T(E)\left\{\frac{2e}{h}M(E)f_L(E)\right\} + T'(E)\left\{\frac{2e}{h}M'(E)f_R(E)\right\}$$

$$= -\frac{2e}{h}\left\{M(E)T(E)f_L(E) - M'(E)T'(E)f_R(E)\right\}$$

$$= -\frac{2e}{h}\left\{\bar{T}(E)f_L(E) - \bar{T}'^{(E)}f_R(E)\right\} \quad (11.221)$$

where

$$\bar{T}(E) = M(E)T(E) \quad (11.222)$$

and

$$\bar{T}'^{(E)} = M'(E)T'(E) \quad (11.223)$$

If there is no inelastic scattering from one energy to another,

$$\bar{T}(E) = \bar{T}'^{(E)} \quad (11.224)$$

Then eq. (11.221) reduces to

$$i(E) = -\frac{2e}{h}\left\{\bar{T}(E)f_L(E) - \bar{T}(E)f_R(E)\right\}$$

$$= -\frac{2e}{h}\bar{T}(E)\left\{f_L(E) - f_R(E)\right\} \quad (11.225)$$

and the total current is

$$I = \int i(E)dE = -\int\left(\frac{2e}{h}\right)\bar{T}(E)\left\{f_L(E) - f_R(E)\right\}dE$$

$$= -\left(\frac{2e}{h}\right)\int\bar{T}(E)\left\{f_L(E) - f_R(E)\right\}dE \quad (11.226)$$

11.8.3 Non-Zero and Zero-Temperature Linear Response Formulae

At equilibrium,

Potential of contact L (V_L) = Potential of contact R (V_R)

$$(11.227)$$

Hence,

$$f_L(E) = f_R(E) \quad (11.228)$$

and consequently

$$I = -\left(\frac{2e}{h}\right)\int\bar{T}(E)(0)dE = 0 \quad (11.229)$$

Suppose the equilibrium state is disturbed by a small amount so that

$$V_L \neq V_R \quad (11.230)$$

The resulting small deviation in current, being proportional to the applied bias $(V_L - V_R)$, can be written as

$$\delta I = -\left(\frac{2e}{h}\right)\int\left[\{\bar{T}(E)\}_{Eq}\delta(f_L - f_R) + (f_L - f_R)_{Eq}\delta\{\bar{T}(E)\}\right]dE$$

$$= -\left(\frac{2e}{h}\right)\int\left[\{\bar{T}(E)\}_{Eq}\delta(f_L - f_R) + 0\times\delta\{\bar{T}(E)\}\right]dE$$

$$(11.231)$$

since

$$(f_L - f_R)_{Eq} = 0 \quad (11.232)$$

$$\therefore \delta I = -\left(\frac{2e}{h}\right)\int\left[\{\bar{T}(E)\}_{Eq}\delta(f_L - f_R)\right]dE \quad (11.233)$$

By a Taylor series expansion

$$\delta(f_L - f_R) = \left(\frac{\partial f}{\partial V}\right)_{Eq}(V_L - V_R) = \left\{-\frac{\partial f_0(E)}{\partial E}\right\}_{Eq}\{e(V_L - V_R)\}$$

$$(11.234)$$

where $f_0(E)$ is the Fermi function at equilibrium given by

$$f_0(E) = \frac{1}{1 + \exp\left(\dfrac{E - E_F}{k_B T}\right)} \quad (11.235)$$

with E_F denoting the Fermi energy. Substituting for $\delta(f_L - f_R)$ from eq. (11.234) in eq. (11.233)

$$\delta I = -\left(\frac{2e}{h}\right)\int\left[\{\bar{T}(E)\}_{Eq}\left\{-\frac{\partial f_0(E)}{\partial E}\right\}_{Eq}\{e(V_L - V_R)\}\right]dE \tag{11.236}$$

or

$$\frac{\delta I}{V_L - V_R} = G = \left(\frac{2e^2}{h}\right)\int\int\left[\{\bar{T}(E)\}_{Eq}\left\{\frac{\partial f_0(E)}{\partial E}\right\}_{Eq}\right]dE \tag{11.237}$$

or

$$G = \left(\frac{2e^2}{h}\right)\int\int\left[\bar{T}(E)\left\{\frac{\partial f_0(E)}{\partial E}\right\}\right]dE \tag{11.238}$$

by dropping the subscript 'Eq' because linear response is a property related to the equilibrium condition.

Equation (11.238) is the linear response formula for conductance at non-zero temperature. At low temperatures,

$$f_0(E) = \theta(E_F - E) \tag{11.239}$$

and

$$-\frac{\partial f_0(E)}{\partial E} = \delta(E_F - E) \tag{11.240}$$

Hence the zero-temperature linear response formula is

$$G = \left(\frac{2e^2}{h}\right)\int\int\left[\bar{T}(E_F)\times 1\right]dE = \left(\frac{2e^2}{h}\right)\sum\bar{T}(E_F) \tag{11.241}$$

11.9 Multi-Terminal Conductors

11.9.1 Büttiker Formula from Landauer's Formula Assuming the Current Carried by Single Energy Level and Zero Temperature

The two-terminal Landauer formula given in eq. (11.205) can be expressed as

$$I = -\frac{2eMT}{h}(E_{FL} - E_{FR}) = -\frac{2e\bar{T}}{h}(E_{FL} - E_{FR}) \tag{11.242}$$

by substituting

$$\bar{T} = MT \tag{11.243}$$

Büttiker asserted in connection with four-terminal measurements that all the probes must be treated on equality basis (Büttiker 1988). Markus Büttiker was a Swiss theoretical physicist. Consider a conductor with multiple terminals. According to Büttiker, the equality of all the probes means that the summation can be extended over all the terminals denoted by subscripts p and q as

$$I_p = -\frac{2e}{h}\sum_q\left(\bar{T}_{q\to p}\ E_{Fp} - \bar{T}_{p\to q}\ E_{Fq}\right) \tag{11.244}$$

Multiplying the numerator and denominator by e

$$I_p = -\frac{2e^2}{h}\sum_q\left(\bar{T}_{q\to p}\ \frac{E_{Fp}}{e} - \bar{T}_{p\to q}\ \frac{E_{Fq}}{e}\right) \tag{11.245}$$

Putting

$$\frac{E_{Fp}}{e} = -V_p \text{ and } \frac{E_{Fq}}{e} = -V_q \tag{11.246}$$

we get

$$\begin{aligned}
I_p &= -\frac{2e^2}{h}\sum_q\left\{-\bar{T}_{q\to p}\ V_p - \left(-\bar{T}_{p\to q}V_q\right)\right\} \\
&= -\frac{2e^2}{h}\sum_q\left(-\bar{T}_{q\to p}\ V_p + \bar{T}_{p\to q}V_q\right) \\
&= \sum_q\left\{\left(-\frac{2e^2}{h}\times-\bar{T}_{q\to p}\ V_p\right) + \left(-\frac{2e^2}{h}\times\bar{T}_{p\to q}V_q\right)\right\} \\
&= \sum_q\left(\frac{2e^2\bar{T}_{q\to p}\ V_p}{h} - \frac{2e^2\bar{T}_{p\to q}V_q}{h}\right) = \sum_q\left(G_{qp}\ V_p - G_{pq}V_q\right)
\end{aligned} \tag{11.247}$$

where

$$G_{qp} = \left(\frac{2e^2}{h}\right)\bar{T}_{q\to p} \tag{11.248}$$

$$G_{pq} = \left(\frac{2e^2}{h}\right)\bar{T}_{p\to q} \tag{11.249}$$

When all the potentials are equal, the current must be zero. This is ensured if

$$G_{qp} = G_{pq} \tag{11.250}$$

Then eq. (11.247) reduces to

$$I_p = \sum_q\left(G_{pq}\ V_p - G_{pq}V_q\right) = \sum_q G_{pq}\left(V_p - V_q\right) \tag{11.251}$$

11.9.2 Multi-Terminal Conductor Formula from Landauer's Formula for the Current Carried by Multiple Energy Levels

For such conductors, eq. (11.244) for the current at terminal p is rewritten as

$$\begin{aligned}
I_p &= \int i_p(E)dE \\
&= -\left(\frac{2e}{h}\right)\int\sum_q\{\bar{T}_{qp}(E)f_p(E) - \bar{T}_{pq}(E)f_q(E)\}dE
\end{aligned} \tag{11.252}$$

where the subscripts p and q are used for different terminals. Also, $\bar{T}_{pq}(E)$ is the total transmission from q to p at energy E and $f_p(E)$ is the Fermi function for terminal p, and $\bar{T}_{qp}(E)$ is the total transmission from p to q at energy E and $f_q(E)$ is the Fermi function for terminal q.

At equilibrium

$$f_p(E) = f_q(E) \tag{11.253}$$

$$\therefore I_p = -\left(\frac{2e}{h}\right)\int \sum_q \left\{\bar{T}_{qp}(E)f_p(E) - \bar{T}_{pq}(E)f_q(E)\right\}dE = 0 \tag{11.254}$$

or

$$\sum_q \left\{\bar{T}_{qp}(E)f_p(E) - \bar{T}_{pq}(E)f_q(E)\right\} = 0 \tag{11.255}$$

or

$$\sum_q \left\{\bar{T}_{qp}(E)f_p(E) - \bar{T}_{pq}(E)f_p(E)\right\} = 0 \tag{11.256}$$

or

$$f_p(E)\sum_q \left\{\bar{T}_{qp}(E) - \bar{T}_{pq}(E)\right\} = 0 \tag{11.257}$$

Since

$$f_p(E) \neq 0 \tag{11.258}$$

$$\therefore \sum_q \left\{\bar{T}_{qp}(E) - \bar{T}_{pq}(E)\right\} = 0 \tag{11.259}$$

or

$$\sum_q \bar{T}_{pq}(E) = \sum_q \bar{T}_{qp}(E) \tag{11.260}$$

which holds away from equilibrium only if inelastic scattering does not take place inside the device.

Thus

$$I_p = -\left(\frac{2e}{h}\right)\sum_q \left\{\bar{T}_{pq}(E)f_p(E) - \bar{T}_{pq}(E)f_q(E)\right\}$$

$$= -\left(\frac{2e}{h}\right)\sum_q \left[\bar{T}_{pq}(E)\left\{f_p(E) - f_q(E)\right\}\right] \tag{11.261}$$

assuming the sum rule to be valid.

11.9.3 Linearization of Multi-Terminal Conductor Response

Applying the procedure followed for two-terminal conductors to the multi-terminal case, we get as in eq. (11.251)

$$I_p = \sum_q G_{pq}\left(V_p - V_q\right) \tag{11.262}$$

where V_p and V_q are the potentials at terminals p and q, and

$$G_{pq} = -\left(\frac{2e^2}{h}\right)\int \bar{T}_{pq}(E)\left\{-\frac{\partial f_0(E)}{\partial E}\right\}dE$$

$$= \left(\frac{2e^2}{h}\right)\int \bar{T}_{pq}(E)\left\{\frac{\partial f_0(E)}{\partial E}\right\}dE \tag{11.263}$$

At low temperatures

$$G_{pq} = \left(\frac{2e^2}{h}\right)\sum_q \bar{T}_{pq}(E_F) \tag{11.264}$$

11.10 Discussion and Conclusions

The object of study in this chapter is a two-terminal quantum wire with macroscopic contacts. The Schrodinger equation is solved by decoupling the electron motion into two components: one along the X-direction under a potential $V(x) = 0$ and another along the Y, Z directions under a finite confinement potential $W(y, z)$. The X-direction equation gives the free particle solution with continuous energy values. The (Y, Z) direction equation is further decomposed into Y-direction and Z-direction parts. Both these parts are treated as particle in infinite potential well problems. The wave function is the product of a function describing the translation of electrons in the X-direction with another function describing their distribution in the (Y, Z) directions. The energy levels are the sum of two terms, one giving continuous energy values and other giving discrete values. Under an applied bias, the distribution functions of the left side and right side contacts described by Fermi-Dirac statistics are shifted relative to each other in proportion to the bias. After determining the transmission coefficient for electrons across the contact/quantum wire interface, the equation for the dependence of current in the quantum wire on electron concentration and temperature is derived. The equilibrium is disturbed by applying a small potential to the wire so that the current is proportional to the voltage. In the zero temperature limit, the Fermi-Dirac distribution function becomes the Heaviside step function. The conductance is the product of a constant ($2e^2/h$) with a summation over transmission coefficients of the electrons which are functions of Fermi level E_F in the contacts and transverse quantum numbers n, m. This conductance formula is called the Landauer formula. The absence of length L of the conductor in this formula indicates the independence of conductance from length. Landauer's formula tells us about the quantization of conductance and prescribes the limiting value beyond which conductance cannot be increased. Besides this deviation from Ohm's law, another discrepancy is the discretization of conductance variations with width of the conductor instead of continuous changes. A cogent interpretation of contact resistance is also provided. In the context of validity domain of Ohm's law, four characteristic lengths, viz. the de Broglie wavelength of electrons, their mean free path, their phase coherence length, and the Debye screening length are decisive parameters.

Another way to arrive at the Landauer formula takes a route from simple to complex situations, commencing from current carried by a single energy level in a quantum wire at zero temperature, and then advancing toward multiple energy levels and non-zero temperature.

Finally, the Büttiker formula is derived from the Landauer formula for current carried by a single energy level at zero temperature and then by multiple energy levels. Thus the Landauer formula is a formula relating the conductance of a 1D conductor to its scattering properties and expressing it as the quantum of conductance multiplied by the sum of the transmission probabilities of all the transport channels in the conductor. The Büttiker formula is the generalization of the Landauer formula to multiple probes.

Illustrative Exercises

11.1 What is the minimum conductance of a ballistic conductor? What is its maximum resistance?

$$77.46 \ \mu S, 12.91 \ k\Omega.$$

11.2 (a) Find the intrinsic Debye screening length in silicon (intrinsic carrier concentration $n_i = 1.5 \times 10^{10} cm^{-3}$). Temperature $T = 300$ K. In which semiconductor device does it play a vital role?

(b) Calculate the extrinsic Debye length in N-doped silicon with donor concentration $N_D = 10^{20} cm^{-3}$. The dielectric constant of silicon $\varepsilon_s = 11.7$. What is its physical significance?

(a) The intrinsic Debye length is given by

$$\left(L_D\right)_{Intrinsic} = \sqrt{\frac{\varepsilon_0 \varepsilon_s k_B T}{2 e^2 N_D}} \tag{11.265}$$

Here

$$N_D = n_i = 1.5 \times 10^{10} \ cm^{-3}$$

$$\therefore L_D = \sqrt{\frac{8.854 \times 10^{-12} \ F \, m^{-1} \times 11.7 \times 1.380649 \times 10^{-23} \ J K^{-1} \times 300}{2 \times \left(1.602 \times 10^{-19} C\right)^2 \times 1.5 \times 10^{10} \times 10^6 \, m^{-3}}}$$

$$= \sqrt{\frac{8.854 \times 11.7 \times 1.380649 \times 300}{2 \times 1.5 \times \left(1.602\right)^2} \times \frac{10^{-12} \times 10^{-23}}{10^{-38} \times 10^{10} \times 10^6 \, m^{-3}}}$$

$$= \sqrt{\frac{42907.1745}{7.6992} \times 10^{-13}} = \sqrt{5.572939 \times 10^3 \times 10^{-13}}$$

$$= 2.3607 \times 10^{-5} \, m = 2.3607 \times 10^{-5} \times 10^6 \, \mu m = 23.61 \mu m \tag{11.266}$$

The intrinsic Debye length plays a meaningful role in the operation of nuclear particle detectors such as α- and β-particle detectors. However, even the intrinsic region in P-I-N diodes has a higher doping

concentration than intrinsic carrier concentration in silicon. So, it becomes inapplicable there.

(b) The extrinsic Debye length is given by

$$\left(L_D\right)_{Extrinsic} = \sqrt{\frac{\varepsilon_0 \varepsilon_s k_B T}{e^2 N_D}} \tag{11.267}$$

Here

$$N_D = 10^{20} \ cm^{-3}$$

$$\therefore L_D = \sqrt{\frac{8.854 \times 10^{-12} \ F \, m^{-1} \times 11.7 \times 1.380649 \times 10^{-23} \ J K^{-1} \times 300}{\left(1.602 \times 10^{-19} C\right)^2 \times 10^{20} \times 10^6 \, m^{-3}}}$$

$$= \sqrt{\frac{8.854 \times 11.7 \times 1.380649 \times 300}{\left(1.602\right)^2} \times \frac{10^{-12} \times 10^{-23}}{10^{-38} \times 10^{20} \times 10^6 \, m^{-3}}}$$

$$= \sqrt{\frac{42907.1745}{2.5664} \times 10^{-23}} = \sqrt{16.7188 \times 10^3 \times 10^{-23}}$$

$$= 4.0889 \times 10^{-10} \, m = \frac{4.0889 \times 10^{-10}}{1 \times 10^{-9}} nm = 0.4089 nm \tag{11.268}$$

The extrinsic Debye length appears more frequently in semiconductor device physics as compared to the intrinsic Debye length because it expresses the distance of decay of the number of electrons $n(x)$ in an electron pulse measured from their number $n(0)$ at the source of the pulse according to the formula

$$n(x) = n(0) \exp\left\{-\frac{x}{\left(L_D\right)_{Extrinsic}}\right\} \tag{11.269}$$

11.3 Calculate the Debye length in copper taking $\varepsilon_{Copper} = 1$ and electron concentration $= 8.49 \times 10^{22} cm^{-3}$.

$$L_D = \sqrt{\frac{\varepsilon_0 \varepsilon_{Copper} k_B T}{e^2 N_D}}$$

$$= \sqrt{\frac{8.854 \times 10^{-12} \ F \, m^{-1} \times 1 \times 1.380649 \times 10^{-23} \ J K^{-1} \times 300}{\left(1.602 \times 10^{-19} C\right)^2 \times 8.49 \times 10^{22} \times 10^6 \, m^{-3}}}$$

$$= \sqrt{\frac{8.854 \times 1 \times 1.380649 \times 300}{8.49 \times \left(1.602\right)^2} \times \frac{10^{-12} \times 10^{-23}}{10^{-38} \times 10^{22} \times 10^6 \, m^{-3}}}$$

$$= \sqrt{\frac{3667.2799}{21.7888} \times 10^{-25}} = \sqrt{16.831 \times 10^{-24}}$$

$$= 4.103 \times 10^{-12} \, m = 4.1 pm \tag{11.270}$$

REFERENCES

Büttiker, M. 1988. Symmetry of electrical conduction. *IBM Journal of Research and Development* 32(3): 317–334.

Datta, S. 1999. *Electronic Transport in Mesoscopic Systems.* Cambridge University Press: Cambridge, UK. pp. 48–116.

Landauer, R. 1957. Spatial variation of currents and fields due to localized scatterers in metallic conduction. *IBM Journal of Research and Development* 1(3): 223–231.

Landauer, R. 1992. Conductance from transmission: Common sense points. *Physica Scripta* T42: 110.

12

Electrons in Quantum Wells

12.1 Sandwich Quantum Well Structures

A quantum well is a potential well that imposes restrictions on the motion of particles in one dimension (e.g. Z-direction). The particles have freedom of motion in the other two directions (X- and Y-directions). As the free motion of carriers is limited to two spatial dimensions, the particles are constrained to occupy a 2D planar region (Nag 2002).

The quantum wells are technological marvels enabled by the progress in semiconductor growth methods whereby it is possible to make layered semiconductor structures with films of precise thickness and controlled quality (Einspruch and Frensley 1994). The transition regions between contiguous films are one or two atomic monolayers thick.

These wells are realized by making sandwiches consisting of an ultra-thin layer of small bandgap material (~40 atomic layers) called the well layer (e.g. GaAs) packed between two barrier layers of large bandgap material (e.g. AlGaAs). In these sandwiches (AlGaAs-GaAs-AlGaAs), there are two heterojunctions (GaAs/AlGaAs) of dissimilar bandgap materials with one heterojunction on each side of the small bandgap material (GaAs). Carrier confinement of electrons and holes in the thin slice of small bandgap material (GaAs) results in restriction and hence the quantization of electron motion in a direction perpendicular to the plane of heterojunctions (GaAs/AlGaAs) and freedom of electron motion in the plane parallel to the heterojunctions (GaAs/AlGaAs). Electronic behavior in this 2D parallel plane is treated as a 'particle-in-a box' problem.

12.2 Band Offsets at Abrupt Heterojunctions

To understand quantum well structures, we resort to the energy band model and enquire how is the energy band diagram modified at the interface between two semiconductor materials. To make this comparison, we must look for some parameter, which is common to the two materials, and in fact, all materials. With respect to this parameter, the energy band diagrams of different materials can be drawn to visualize the overall picture of the structure, notably the changes at the interface. The parameter chosen is the vacuum level of energy E_0, which is the energy of an electron outside a neutral semiconductor material. The energy difference between the bottom of conduction band of a material and the vacuum energy level is known as electron affinity of the material. It is denoted by the symbol χ. Hence, electron affinity of a material is the energy required to remove an electron at the bottom of the conduction band of the material and take it to the vacuum energy level, i.e. to make it a free electron. Knowledge of electron affinities χ_A and χ_B of two materials A, B tells us how the bottoms of the conduction bands E_{CA} and E_{CB} of these materials

are located relative to each other. The discontinuity $\Delta E_{C(A,B)}$ of conduction bands at an abrupt heterojunction of materials A, B is expressed in terms of their electron affinities χ_A and χ_B as

$$\Delta E_{C(A,B)} = E_{C,B} - E_{C,A} = \chi_B - \chi_A \qquad (12.1)$$

The simple procedure of calculating the band offset, as embodied in eq. (12.1), is known as the electron affinity rule.

The bandgap discontinuity is

$$\Delta E_{G(A,B)} = E_{G,A} - E_{G,B} = \Delta E_{C(A,B)} + \Delta E_{V(A,B)} \qquad (12.2)$$

where $E_{G,A}$ and $E_{G,B}$ are the bandgaps of the two materials.

Hence, the discontinuity $\Delta E_{V(A,B)}$ of valence bands is

$$\Delta E_{V(A,B)} = E_{V,B} - E_{V,A} = \Delta E_{G(A,B)} - \Delta E_{C(A,B)} = \Delta E_{G(A,B)}$$

$$- \left(\chi_B - \chi_A \right) = \Delta E_{G(A,B)} - \chi_B + \chi_A = \chi_A - \chi_B + \Delta E_{G(A,B)} \quad (12.3)$$

by applying eq. (12.1).

Looking at Figure 12.1, we notice that the sign of conduction band offset of materials A, B is opposite to that of valence band offset of these materials because the conduction band bottom of material B lies below the conduction band bottom of material A whereas the valence band top of material B lies above the valence band top of material A. Consequently, the lowest conduction band states of the structure appear in material B and the highest valence band states also appear in material B. In other words, the lowest conduction band states and the highest valence band states occur in the same part of the structure, which is here material B. A conspicuous feature of the material composition examined is that the conduction band bottom and valence band top of material B and hence the bandgap of material B (i.e. $E_{G,B}$) are completely nestled within the conduction band bottom and valence band top of material A and hence the bandgap of material A (i.e. $E_{G,A}$). The heterojunction formed by lining up of materials in this manner is called a type I heterojunction. A familiar example of a type I heterojunction is the GaAs/Al$_x$Ga$_{1-x}$As heterojunction for $x < 0.4$. Missous (1993) found that for this heterojunction $\Delta E_C = (0.80 \pm 0.03)x$ eV for $0 \leq x < 0.45$ while $\Delta E_V = (0.51 \pm 0.04)x$ eV for $0 \leq x \leq 1$.

In the material combination shown in Figure 12.2, the lowest conduction band bottom occurs in material B while the highest valence band top is placed in material A. Further, the energy separation between the two is smaller than the lower of the bandgaps $E_{G,A}$ and $E_{G,B}$. This case is referred to as a type II heterojunction. An example of type II heterojunction is AlAs/Al$_x$Ga$_{1-x}$As for $x > 0.4$.

In a type I heterojunction, the electrons and holes in the quantum well are enclosed in one single material. Localization of

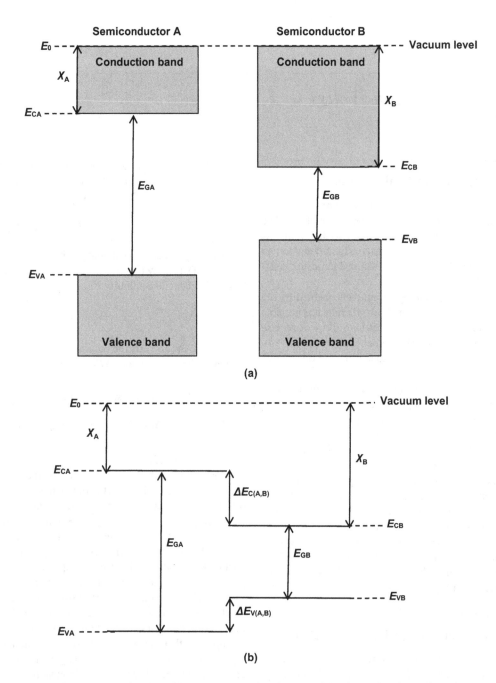

FIGURE 12.1   Band diagrams for type I heterojunction with: (a) separate semiconductors and (b) with semiconductors in contact.

both types of charge carriers in the same region of space leads to faster recombination. In a type II heterojunction, the electrons and holes are confined in different materials. Consequently, the recombination times of electrons and holes are longer.

A third type of heterojunction known as type III is one in which bottom of the conduction band in material B is located below the top of the valence band in material A with an intervening gap between the conduction band bottom and valence band top (Figure 12.3). A line-up with a broken gap is found in the structure InAs/GaSb.

Cautionary remark: It is very important to note here that the band diagrams in Figures 12.1–12.3 have been drawn without taking into consideration any transference of charges at the

interface between the two participating materials. Therefore, these diagrams are simplified representations in the form of step functions without taking any charge transfers across the interface between the two materials. These charge transfers invariably cause bending of bands at the interface. Such diagrams for type I heterojunctions will be presented in Figures 12.4–12.6. These energy band diagrams will provide more rigorous representations of the interfacial phenomena. To a first approximation, the energy band diagrams may be drawn as step functions and subsequently modified by knowledge of charge transfers across the interface for realistic depiction. So, whenever step-function type diagrams are seen, the reader may treat them as first approximation sketches.

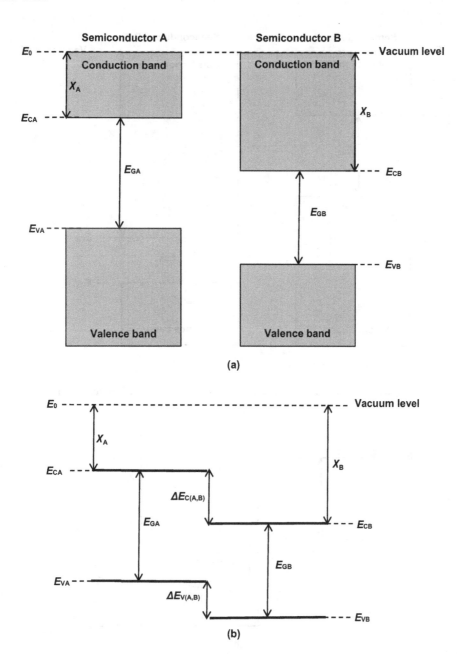

FIGURE 12.2 Band diagrams for type II heterojunction with: (a) separate semiconductors and (b) semiconductors in contact.

12.3 Analysis of a Single Heterojunction of Dissimilar Bandgap Materials

Thus far our attention has remained focused on the band offsets between two materials. Closer scrutiny of the interface reveals that the situation at the interface is not that simple, and therefore the phenomena taking place there must be examined in more detail. Consider a type I heterojunction formed between a wide bandgap semiconductor material and a narrow bandgap semiconductor material for the three cases in which:

i. both the wide bandgap and low bandgap materials are N-doped to form a N-n heterojunction,

ii. the wide bandgap material is N-doped and narrow bandgap material is P-doped (N-p heterojunction), and

iii. the wide bandgap material is P-doped and narrow bandgap material is N-doped (P-n heterojunction). Note that the dopant type in the wide bandgap material is denoted by an uppercase letter and that in the narrow bandgap material by a lowercase letter.

The energy band diagrams are constructed in the same way as for homojunctions (Figures 12.4–12.6). First isolated materials are considered and their energy band diagrams are drawn showing the Fermi levels according to the doping. Then the materials are brought into contact. The Fermi level tends to be the same all throughout the structure. During this process, the charge carriers are transferred from the material in which the Fermi level is higher to the one in which it is located at a lower position. The side from which the charge carriers are supplied is depleted while an accumulation layer is formed on the side, which receives the

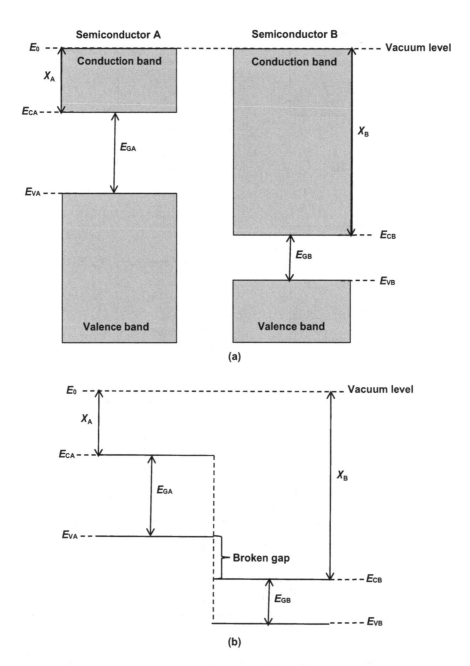

FIGURE 12.3 Band diagrams for type III heterojunction with: (a) separate semiconductors and (b) semiconductors in contact.

charge carriers. Consequent to the transference of charge carriers, an electric field and potential are established between the two sides. This built-in potential V_{bi} equals the difference in work functions ϕ_A, ϕ_B of the two isolated bulk materials:

$$V_{bi} = \phi_B - \phi_A \qquad (12.4)$$

The work function ϕ of a material is the energy difference between the Fermi level E_F and the vacuum energy level E_0 for the material:

$$\phi = E_F - E_0 \qquad (12.5)$$

The built-in field balances the transfer of charge carriers. Eventually, a state of thermodynamic equilibrium is set up.

For the case of Figure 12.4, in the equilibrium position, the conduction band edge moves up on the depletion side because of the emptying of electron states and moves down on the accumulation side due to filling of electron states. As a result, the bands are bent at the interface. Similar band bending occurs in Figure 12.5 but in a more aggravated form. In Figure 12.6, the valence band edge moves downward on the P-side and upward on the n-side.

In all cases, the transference of charge carriers across the interface causes the spatial separation of charge carriers from the respective ionized dopants, be they positively charged donor ions for electrons or negatively charged acceptor ions for holes. This spatial separation reduces the scattering of charge carriers from the dopant ions and thereby enhances the carrier mobility. Modulation doping is a technique of doping the semiconductor in such a way that the free charge carriers are physically separated from the

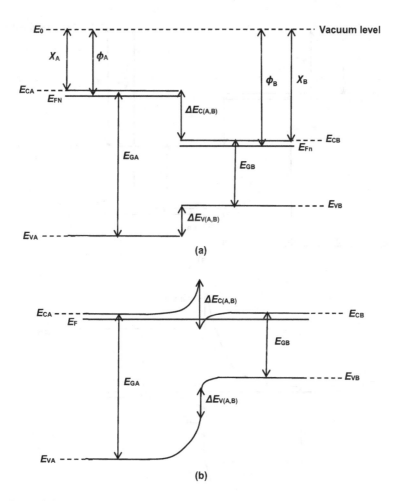

FIGURE 12.4 N-n heterojunction: (a) before contact and (b) after contact.

parent impurity ions so that the mobility reduction induced by ionized impurity scattering is eliminated, an effect which appreciably decreases the mobility in heavily doped semiconductors.

Another noticeable effect is that although both the materials were insulators at low temperatures, a conducting channel is now available at the interface down to $T = 0$ K. The carriers are confined in this channel by the bending of the bands. The channel extends over a length of several nanometers in the Z-direction. In this channel, the carriers can move freely in the X- and Y-directions but their motion in the Z-direction is quantized. The situation is described as the formation of a 2D electron gas in an approximately triangular quantum well. Two main phenomena contribute to the well formation, namely the discontinuity of bandgap and the setting up of electrostatic potential at the interface.

12.4 Heterojunction Equations

It is expedient to write down the basic equations describing the physics of the electrons at the heterojunction.

12.4.1 Poisson's Equation for the Heterojunction

Poisson's equation is a partial differential equation of central importance in physics. It is applied to relate the electrostatic potential $\Phi(z)$ in the Z-direction perpendicular to the interface to the total electrical charge density $\rho_{\text{Total}}(z)$ producing the potential:

$$\frac{d^2\Phi(z)}{dx^2} = -\frac{\rho_{\text{Total}}(z)}{\varepsilon_0 \varepsilon_r} \qquad (12.6)$$

where ε_0 is the permittivity of free space and ε_r is the relative permittivity of the medium.

Total electrical charge density in the Z-direction $\rho_{\text{Total}}(z)$

= Electrical charge density of the

impurity ions in the Z-direction $\rho_{\text{Impurities}}(z)$

+ Electrical charge density of electrons in the

Z-direction $\rho_{\text{Electrons}}(z)$

= Charge density of positively charged donor ions $\rho_{\text{Donors}}(z)$

+ Charge density of negatively charged

acceptor ions $\rho_{\text{Acceptors}}(z) + \rho_{\text{Electrons}}(z)$

= $+qN_D(z) - qN_A(z) + \rho_{\text{Electrons}}(z)$ \qquad (12.7)

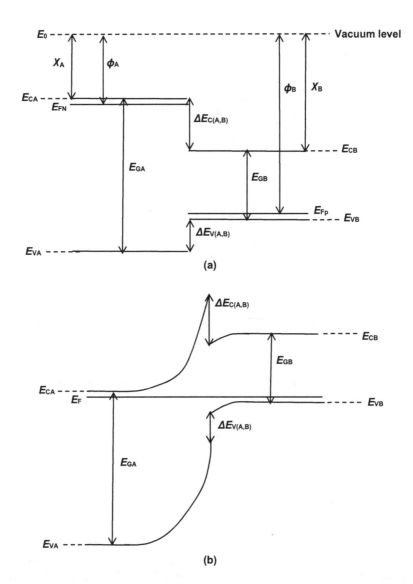

FIGURE 12.5 N-p heterojunction: (a) before contact and (b) after contact.

where q is the elementary charge, $N_D(z)$ is the concentration of donor impurities in the Z-direction, and $N_A(z)$ is the concentration of acceptor impurities in the Z-direction. Note that in this chapter we have used the symbol 'q' for electronic charge. Both the symbols 'e' and 'q' are commonly used in the literature, and so it is useful to be accustomed to either. But whatever symbol is used must be clearly stated.

Let the wave function of the electron be $\Psi_\nu(r)$ where ν denotes the set of quantum numbers of the electron. Then

Probability of finding the electron of the

state represented by ν at the point $\mathbf{r}$ is $= \Psi_\nu(\mathbf{r})$ (12.8)

If the energy-dependent electron distribution function is $f(E_\nu)$

Probability of occupation of the

energy level E_ν by the electron is $= f(E_\nu)$ (12.9)

Then the density of negatively charged electrons is

$\rho_{\text{Electrons}}(z)$

(Elementary charge × Probability of finding

the electron of the state represented by ν at

$= -\sum_\nu$ the point $\mathbf{r}$ × Probability of occupation of

the energy level E_ν by the electron)

$= -\sum_\nu q\Psi_\nu(\mathbf{r})f(E_\nu)$ (12.10)

and from eqs. (12.7) and (12.10),

$\rho_{\text{Total}}(z) = +qN_D(z) - qN_A(z) - \sum_\nu q\Psi_\nu(\mathbf{r})f(E_\nu)$ (12.11)

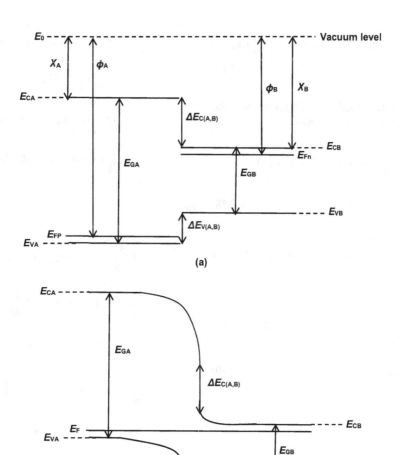

FIGURE 12.6   P-n heterojunction: (a) before contact and (b) after contact.

Substituting for $\rho_{\text{Total}}(z)$ in eq. (12.6) we get

$$\frac{d^2\Phi(z)}{dx^2} = -\frac{\left\{+qN_D(z)-qN_A(z)-\sum_v q\Psi_v(\mathbf{r})f(E_v)\right\}}{\varepsilon_0\varepsilon_r}$$

$$= \frac{\sum_v q\Psi_v(\mathbf{r})f(E_v)-qN_D(z)+qN_A(z)}{\varepsilon_0\varepsilon_r} \qquad (12.12)$$

12.4.2 Schrodinger's Equation for the Heterojunction

The 3D time-independent Schrodinger equation to find the energy levels E_n of the quantum-mechanical system is

$$\left\{-\frac{\hbar^2}{2m}\nabla^2 + V(\mathbf{r})\right\}\Psi(\mathbf{r}) = E\Psi(\mathbf{r}) \qquad (12.13)$$

In the layered structure of the heterojunction, the potential energy $V(\mathbf{r})$ depends only on the Z-coordinate. Hence, it is appropriate to write

$$V(\mathbf{r}) = V(z) \qquad (12.14)$$

so that Schrodinger's equation reduces to

$$\left\{-\frac{\hbar^2}{2m}\left(\frac{\partial^2}{\partial x^2}+\frac{\partial^2}{\partial y^2}+\frac{\partial^2}{\partial z^2}\right)+V(z)\right\}\Psi(x,\,y,\,z) = E\Psi(x,\,y,\,z)$$

$$(12.15)$$

or

$$\left\{-\frac{\hbar^2}{2m}\left(\frac{\partial^2}{\partial x^2}+\frac{\partial^2}{\partial y^2}+\frac{\partial^2}{\partial z^2}\right)+V(z)\right\}\Psi_v(x,\,y,\,z) = E_v\Psi_v(x,\,y,\,z)$$

$$(12.16)$$

We note that the electrons are left free to move in the X- and Y-directions by the potential energy. In the absence of potential, the wave functions will be plane waves. This suggests that plane waves can be tried for motion along X- and Y-directions (eq. (5.14)). The electrostatic potential $\Phi(z)$ in the Z-direction is independent of x- and y-coordinates. So, the wave function $\Psi_v(x,\,y,\,z)$ can be factorized as

$$\Psi_v(x,\ y,\ z) = \frac{1}{\sqrt{S}}\exp(ik_x x)\exp(ik_y y)u_j(z) \text{ where } v \equiv \{j, k_x, k_y\} \tag{12.17}$$

where S is the area of the heterojunction. The $u_j(z)$ factor is the wave function factor along the Z-direction. Also,

$$\Psi_v(x,\ y,\ z)\Psi_v^*(x,\ y,\ z) = \frac{1}{\sqrt{S}}\exp(ik_x x)\exp(ik_y y)u_j(z)$$

$$\times \frac{1}{\sqrt{S}}\exp(-ik_x x)\exp(-ik_y y)u_j(z) = \frac{1}{S}|u_j(z)|^2 \tag{12.18}$$

Substituting the wave function $\Psi_v(x,\ y,\ z)$ from eq. (12.17) in eq. (12.16), we confirm

$$\left\{-\frac{\hbar^2}{2m}\left(\frac{\partial^2}{\partial x^2} + \frac{\partial^2}{\partial y^2} + \frac{\partial^2}{\partial z^2}\right) + V(z)\right\}\frac{1}{\sqrt{S}}\exp(ik_x x)\exp(ik_y y)u_j(z)$$

$$= E_v\frac{1}{\sqrt{S}}\exp(ik_x x)\exp(ik_y y)u_j(z) \tag{12.19}$$

or

$$-\frac{\hbar^2}{2m}\left(\frac{\partial^2}{\partial x^2} + \frac{\partial^2}{\partial y^2} + \frac{\partial^2}{\partial z^2}\right)\frac{1}{\sqrt{S}}\exp(ik_x x)\exp(ik_y y)u_j(z)$$

$$+ V(z)\frac{1}{\sqrt{S}}\exp(ik_x x)\exp(ik_y y)u_j(z)$$

$$= E_v\frac{1}{\sqrt{S}}\exp(ik_x x)\exp(ik_y y)u_j(z) \tag{12.20}$$

or

$$-\frac{\hbar^2}{2m}\frac{\partial^2}{\partial x^2}\left\{\frac{1}{\sqrt{S}}\exp(ik_x x)\exp(ik_y y)u_j(z)\right\}$$

$$-\frac{\hbar^2}{2m}\frac{\partial^2}{\partial y^2}\left\{\frac{1}{\sqrt{S}}\exp(ik_x x)\exp(ik_y y)u_j(z)\right\}$$

$$-\frac{\hbar^2}{2m}\frac{\partial^2}{\partial z^2}\left\{\frac{1}{\sqrt{S}}\exp(ik_x x)\exp(ik_y y)u_j(z)\right\}$$

$$+ V(z)\frac{1}{\sqrt{S}}\exp(ik_x x)\exp(ik_y y)u_j(z)$$

$$= E_v\frac{1}{\sqrt{S}}\exp(ik_x x)\exp(ik_y y)u_j(z) \tag{12.21}$$

or

$$\frac{\hbar^2}{2m}\times\frac{1}{\sqrt{S}}\times\exp(ik_x x)\exp(ik_y y)u_j(z)\times k_x^2 + \frac{\hbar^2}{2m}\times\frac{1}{\sqrt{S}}$$

$$\times\exp(ik_x x)\exp(ik_y y)u_j(z)\times k_y^2$$

$$-\frac{\hbar^2}{2m}\frac{\partial^2}{\partial z^2}\left\{\frac{1}{\sqrt{S}}\exp(ik_x x)\exp(ik_y y)u_j(z)\right\}$$

$$+ V(z)\frac{1}{\sqrt{S}}\exp(ik_x x)\exp(ik_y y)u_j(z)$$

$$= E_v\frac{1}{\sqrt{S}}\exp(ik_x x)\exp(ik_y y)u_j(z) \tag{12.22}$$

since

$$-\frac{\hbar^2}{2m}\frac{\partial^2}{\partial x^2}\left\{\frac{1}{\sqrt{S}}\exp(ik_x x)\exp(ik_y y)u_j(z)\right\}$$

$$= -\frac{\hbar^2}{2m}\times\frac{1}{\sqrt{S}}\times\exp(ik_y y)u_j(z)\times\frac{\partial^2}{\partial x^2}\{\exp(ik_x x)\}$$

$$= -\frac{\hbar^2}{2m}\times\frac{1}{\sqrt{S}}\times\exp(ik_y y)u_j(z)\times\frac{\partial}{\partial x}\left[\frac{\partial}{\partial x}\{\exp(ik_x x)\}\right]$$

$$= -\frac{\hbar^2}{2m}\times\frac{1}{\sqrt{S}}\times\exp(ik_y y)u_j(z)\times\frac{\partial}{\partial x}\left[\{\exp(ik_x x)\}\times(ik_x)\right]$$

$$= -\frac{\hbar^2}{2m}\times\frac{1}{\sqrt{S}}\times\exp(ik_y y)u_j(z)\times(ik_x)\times\frac{\partial}{\partial x}\left[\{\exp(ik_x x)\}\right]$$

$$= -\frac{\hbar^2}{2m}\times\frac{1}{\sqrt{S}}\times\exp(ik_y y)u_j(z)\times(ik_x)\times\left[\{\exp(ik_x x)\}\times(ik_x)\right]$$

$$= -\frac{\hbar^2}{2m}\times\frac{1}{\sqrt{S}}\times\exp(ik_x x)\exp(ik_y y)u_j(z)\times(ik_x)\times(ik_x)$$

$$= -\frac{\hbar^2}{2m}\times\frac{1}{\sqrt{S}}\times\exp(ik_x x)\exp(ik_y y)u_j(z)\times(ik_x)^2$$

$$= -\frac{\hbar^2}{2m}\times\frac{1}{\sqrt{S}}\times\exp(ik_x x)\exp(ik_y y)u_j(z)\times\left(\sqrt{-1}k_x\right)^2$$

$$= -\frac{\hbar^2}{2m}\times\frac{1}{\sqrt{S}}\times\exp(ik_x x)\exp(ik_y y)u_j(z)\times\left(-1\times k_x^2\right)$$

$$= \frac{\hbar^2}{2m}\times\frac{1}{\sqrt{S}}\times\exp(ik_x x)\exp(ik_y y)u_j(z)\times k_x^2 \tag{12.23}$$

and similarly

$$-\frac{\hbar^2}{2m}\frac{\partial^2}{\partial y^2}\left\{\frac{1}{\sqrt{S}}\exp(ik_x x)\exp(ik_y y)u_j(z)\right\}$$

$$= \frac{\hbar^2}{2m}\times\frac{1}{\sqrt{S}}\times\exp(ik_x x)\exp(ik_y y)u_j(z)\times k_y^2 \tag{12.24}$$

Dividing both sides of eq. (12.22) by $\frac{1}{\sqrt{S}}\exp(ik_x x)\exp(ik_y y)$, we have

$$\frac{\hbar^2}{2m} \times u_j(z) \times k_x^2 + \frac{\hbar^2}{2m} \times u_j(z) \times k_y^2 - \frac{\hbar^2}{2m} \frac{\partial^2}{\partial z^2} u_j(z)$$
$$+ V(z) u_j(z) = E_v u_j(z) \tag{12.25}$$

or

$$\frac{\hbar^2 k_x^2}{2m} u_j(z) + \frac{\hbar^2 k_y^2}{2m} u_j(z) - \frac{\hbar^2}{2m} \frac{\partial^2}{\partial z^2} u_j(z) + V(z) u_j(z) = E_v u_j(z) \tag{12.26}$$

or

$$\left\{ \frac{\hbar^2 k_x^2}{2m} + \frac{\hbar^2 k_y^2}{2m} - \frac{\hbar^2}{2m} \frac{\partial^2}{\partial z^2} + V(z) \right\} u_j(z) = E_v u_j(z) \tag{12.27}$$

Cancellation of exponential functions from both sides confirms the correctness of the guessed wave function in eq. (12.17). Only functions of z are left behind. Transposing the k_x and k_y terms of energy of plane waves to the right-hand side allows collection of energy terms together. Thus

$$\left\{ -\frac{\hbar^2}{2m} \frac{\partial^2}{\partial z^2} + V(z) \right\} u_j(z) = E_v u_j(z) - \left(\frac{\hbar^2 k_x^2}{2m} + \frac{\hbar^2 k_y^2}{2m} \right) u_j(z) \tag{12.28}$$

or

$$\left\{ -\frac{\hbar^2}{2m} \frac{\partial^2}{\partial z^2} + V(z) \right\} u_j(z) = \left\{ E_v - \left(\frac{\hbar^2 k_x^2}{2m} + \frac{\hbar^2 k_y^2}{2m} \right) \right\} u_j(z) \tag{12.29}$$

Substituting ε_j for the energy

$$\varepsilon_j = E_v - \left(\frac{\hbar^2 k_x^2}{2m} + \frac{\hbar^2 k_y^2}{2m} \right) \tag{12.30}$$

we obtain from eq. (12.29)

$$\left\{ -\frac{\hbar^2}{2m} \frac{\partial^2}{\partial z^2} + V(z) \right\} u_j(z) = \varepsilon_j u_j(z) \tag{12.31}$$

It must be emphasized that this Schrodinger equation is a 1D equation in the variable z (Davies 1998). The X- and Y-dimensions have been eliminated.

Now,

Total energy $V(z)$ = Electrostatic potential energy $\{-q\Phi(z)\}$

+ Pseudopotential energy due to band

discontinuity at the heterojunction

$$\{\Delta E_{C(A,B)}\} \tag{12.32}$$

or

$$V(z) = -q\Phi(z) + \Delta E_{C(A,B)} \tag{12.33}$$

ΔE_C may be expressed as

$$\Delta E_{C(A,B)} = -V_{bi}\theta(z) \tag{12.34}$$

where V_{bi} is the built-in potential of the heterojunction and $\theta(z)$ is the Heaviside step function defined as

$$\theta(z) = 0 \text{ for } z < 0 \tag{12.35}$$

and

$$\theta(z) = u(z) \text{ for } z \geq 0 \tag{12.36}$$

Substitution for $V(z)$ from eq. (12.33) in eq. (12.31) begets

$$\left\{ -\frac{\hbar^2}{2m} \frac{\partial^2}{\partial z^2} - q\Phi(z) + \Delta E_{C(A,B)} \right\} u_j(z) = \varepsilon_j u_j(z) \tag{12.37}$$

12.4.3 Electron Concentration Equation

From eq. (12.30), the total energy of the electron is written as

$$E_v = \varepsilon_j + \left(\frac{\hbar^2 k_x^2}{2m} + \frac{\hbar^2 k_y^2}{2m} \right) \tag{12.38}$$

To calculate the electron concentration, we write

$$\sum_v [\Psi_v(\mathbf{r})]^2 f(E_v) = \sum_v \frac{1}{S}[u_j(z)]^2 f(E_v) \tag{12.39}$$

by making use of eq. (12.18).

Applying eq. (12.38), the right-hand side of eq. (12.39) becomes

$$\sum_v \frac{1}{S}[u_j(z)]^2 f(E_v) = \sum_{S,j,k_x,k_y} \frac{1}{S}[u_j(z)]^2 f\left\{ \varepsilon_j + \left(\frac{\hbar^2 k_x^2}{2m} + \frac{\hbar^2 k_y^2}{2m} \right) \right\}$$
$$= \frac{1}{S} \sum_{S,j,k_x,k_y} [u_j(z)]^2 f\left\{ \varepsilon_j + \left(\frac{\hbar^2 k_x^2}{2m} + \frac{\hbar^2 k_s^2}{2m} \right) \right\} = \sum_j [u_j(z)]^2 n_{S,j} \tag{12.40}$$

for the level j where the sheet density of electrons is

$$n_{S,j}(E_F) \equiv \frac{1}{S} \sum_{S,k_x,k_y} f\left\{ \varepsilon_j + \left(\frac{\hbar^2 k_x^2}{2m} + \frac{\hbar^2 k_y^2}{2m} \right) \right\} \tag{12.41}$$

12.4.4 The 2D Density of States and Sheet Density of Electrons

The 2D density of states function gives the number of states that are available in the quantum well per unit area in an interval of energy lying between E and $E + dE$. Consider an infinite rectangular well potential. This well has sides of length L_x, L_y. Suppose that electrons of mass m are confined in the well. Then the Schrodinger equation for the electrons is laid down as

$$\left(-\frac{\hbar^2}{2m} \nabla^2 \right) \Psi(x, y) = E\Psi(x, y) \quad (12.42)$$

Writing the Laplacian operator in terms of partial derivatives in x, y as

$$-\frac{\hbar^2}{2m} \left(\frac{\partial^2}{\partial x^2} + \frac{\partial^2}{\partial y^2} \right) \Psi(x, y) = E\Psi(x, y) \quad (12.43)$$

which can be put in the form

$$-\frac{\hbar^2}{2m} \left(\frac{\partial^2 \Psi(x, y)}{\partial x^2} + \frac{\partial^2 \Psi(x, y)}{\partial y^2} \right) = E\Psi(x, y) \quad (12.44)$$

or

$$\frac{\partial^2 \Psi(x, y)}{\partial x^2} + \frac{\partial^2 \Psi(x, y)}{\partial y^2} = -\frac{2mE\Psi(x, y)}{\hbar^2} \quad (12.45)$$

or

$$\frac{\partial^2 \Psi(x, y)}{\partial x^2} + \frac{\partial^2 \Psi(x, y)}{\partial y^2} + \frac{2mE\Psi(x, y)}{\hbar^2} = 0 \quad (12.46)$$

or

$$\frac{\partial^2 \Psi(x, y)}{\partial x^2} + \frac{\partial^2 \Psi(x, y)}{\partial y^2} + \left(\sqrt{\frac{2mE}{\hbar^2}} \right)^2 \Psi(x, y) = 0 \quad (12.47)$$

or

$$\frac{\partial^2 \Psi(x, y)}{\partial x^2} + \frac{\partial^2 \Psi(x, y)}{\partial y^2} + k^2 \Psi(x, y) = 0 \quad (12.48)$$

where the constant k is given by

$$k = \sqrt{\frac{2mE}{\hbar^2}} \quad (12.49)$$

By separation of variables, the wave function becomes

$$\Psi(x, y) = \Psi_x(x)\Psi_y(y) \quad (12.50)$$

Substituting for $\Psi(x, y)$ from eq. (12.50) in eq. (12.48) we get

$$\frac{\partial^2 \{\Psi_x(x)\Psi_y(y)\}}{\partial x^2} + \frac{\partial^2 \{\Psi_x(x)\Psi_y(y)\}}{\partial y^2} + k^2 \Psi_x(x)\Psi_y(y) = 0 \quad (12.51)$$

or

$$\Psi_y(y)\frac{\partial^2 \Psi_x(x)}{\partial x^2} + \Psi_x(x)\frac{\partial^2 \Psi_y(y)}{\partial y^2} + k^2 \Psi_x(x)\Psi_y(y) = 0 \quad (12.52)$$

Dividing both sides of eq. (12.52) by $\Psi_x(x)\Psi_y(y)$,

$$\frac{1}{\Psi_x(x)} \frac{\partial^2 \Psi_x(x)}{\partial x^2} + \frac{1}{\Psi_y(y)} \frac{\partial^2 \Psi_y(y)}{\partial y^2} + k^2 = 0 \quad (12.53)$$

Since k is a constant, the validity of eq. (12.53) for all possible x and y terms is preserved as long as each individual term including $\Psi_x(x)$ and $\Psi_y(y)$ equals a constant. Hence,

$$\frac{1}{\Psi_x(x)} \frac{\partial^2 \Psi_x(x)}{\partial x^2} = -k_x^2 \quad (12.54)$$

and

$$\frac{1}{\Psi_y(y)} \frac{\partial^2 \Psi_y(y)}{\partial y^2} = -k_y^2 \quad (12.55)$$

where

$$k_x^2 + k_y^2 = k^2 \quad (12.56)$$

The solution to the wave functions consists of sine and cosine functions (eqs. (5.58) and (5.69))

$$\Psi_x(x) = A\sin(k_x x) + B\cos(k_x x) \quad (12.57)$$

and

$$\Psi_y(y) = C\sin(k_y y) + D\cos(k_y y) \quad (12.58)$$

where A, B, C, and D are constants

From the arguments given in Section 5.2.2, only sine terms are permissible (eq. (5.75)) so that

$$\Psi_x(x) = A\sin(k_x x) \quad (12.59)$$

and

$$\Psi_y(y) = C\sin(k_y y) \quad (12.60)$$

When the boundary conditions are enforced

At $x = 0$, $\Psi_x(x) = A\sin(k_x \times 0) = 0 \quad (12.61)$

and

$$\text{At } y = 0, \ \Psi_y(y) = C \sin(k_y y) = 0 \qquad (12.62)$$

$$\text{At } x = L_x, \ \Psi_x(L_x) = A \sin(k_x L_x) = 0 \qquad (12.63)$$

and

$$\text{At } y = L_y, \Psi_y(L_y) = C \sin(k_y L_y) = 0 \qquad (12.64)$$

Only those wave vectors are allowed which satisfy (eq. (5.84))

$$k_x L_x = \pi n_x \text{ where } n = \pm 1, 2, 3,... \qquad (12.65)$$

and

$$k_y L_y = \pi n_y \text{ where } n = \pm 1, 2, 3,... \qquad (12.66)$$

or

$$k_x = \frac{\pi n_x}{L_x} \qquad (12.67)$$

and

$$k_y = \frac{\pi n_y}{L_y} \qquad (12.68)$$

The permitted states are plotted as a grid of points in k-space which is a 2D visualization of directions of wave vectors (Figure 12.7). Separation between the states $= \pi/L_x$ in the X-direction and $= \pi/L_y$ in the Y-direction. Then

$$\text{Area of single state in } k\text{-space is} = \left(\frac{\pi}{L_x}\right)\left(\frac{\pi}{L_y}\right) = \frac{\pi^2}{L_x L_y} \qquad (12.69)$$

and

$$\text{Area of circular disc in } k\text{-space is} = \pi k^2 \qquad (12.70)$$

Hence,

Number of filled states in the circular disc

$$= N = \frac{\text{Area of circular disc}}{\text{Area of single state}} \times 2 \times \left(\frac{1}{2} \times \frac{1}{2}\right)$$

$$= \left(\frac{1}{2}\right)\frac{\text{Area of circular disc}}{\text{Area of single state}} \qquad (12.71)$$

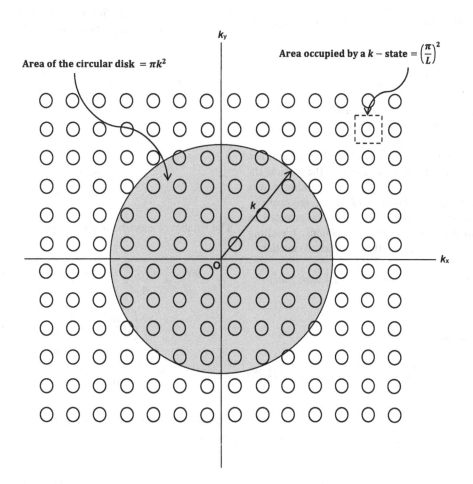

Area of the circular disk $= \pi k^2$

Area occupied by a k – state $= \left(\frac{\pi}{L}\right)^2$

FIGURE 12.7 Construction of k-space in two dimensions for calculation of density of states of free electrons. Rigid boundary conditions are used. The allowed states are represented as a grid of points separated by a distance of π/L.

where factor 2 takes care of two possible spins of the electron, the first 1/2 factor corrects for redundancy in counting identical states $\pm n_x$ and the second 1/2 factor does the same for $\pm n_y$.

From eqs. (12.69), (12.70), and (12.71),

$$N = \left(\frac{1}{2}\right)\underbrace{\frac{\pi k^2}{\pi^2}}_{L_x L_y} = \left(\frac{1}{2}\right) \times \pi k^2 \times \frac{L_x L_y}{\pi^2} = \frac{k^2 L_x L_y}{2\pi} \quad (12.72)$$

Substituting for k from eq. (12.49) in eq. (12.72),

$$N = \frac{k^2 L_x L_y}{2\pi} = \left(\sqrt{\frac{2mE}{\hbar^2}}\right)^2 \times \frac{L_x L_y}{2\pi} = \frac{mEL_x L_y}{\hbar^2 \pi} \quad (12.73)$$

The density of states per unit energy is

$$\frac{dN}{dE} = \frac{d}{dE}\left(\frac{mEL_x L_y}{\hbar^2 \pi}\right) = \frac{mL_x L_y}{\hbar^2 \pi} \quad (12.74)$$

Therefore, the 2D density of states per unit area per unit energy is

$$\{g(E)\}_{2D} = \frac{\dfrac{mL_x L_y}{\hbar^2 \pi}}{L_x L_y} = \frac{m}{\pi\hbar^2} \quad (12.75)$$

Writing m^* in place of m for the electron rest mass, the equation for 2D density of states is

$$\{g(E)\}_{2D} = \frac{m^*}{\pi\hbar^2} \quad (12.76)$$

Interestingly, the 2D density of states is energy-independent. A significant number of states become available immediately on reaching the top of energy bandgap.

We can see that with the chosen fixed boundary conditions, the wave functions are those obtained earlier for the particle-in-a box problem. These wave functions represent standing waves and do not carry any current. Permitted k values are positive. For periodic boundary conditions, the k must be selected in a different way. In place of sine/cosine functions, travelling waves can be chosen conforming to the exponential function

$$\exp\{i(k_x + k_y)L\} = \exp\{i(k_x + k_y)0\} = 1 = \exp(2\pi n i) \quad (12.77)$$

for which the admissible k values are

$$k_x = \frac{2\pi n}{L}, k_y = \frac{2\pi m}{L} \quad (12.78)$$

where n can be zero, and positive or negative integers

$$n = 0, \pm 1, \pm 2, \pm 3, \dots \quad (12.79)$$

and also

$$m = 0, \pm 1, \pm 2, \pm 3, \dots \quad (12.80)$$

These values are located at double the separation for values obtained for fixed boundary conditions but positive as well as negative signs are allowed. The density of states is found by placing the values of k along a line at spacings of $2\pi/L$. Since each wave function can be correlated with two values of spin, the total number of available states for electrons becomes twice as much. Then

$$\text{Area occupied by a } k\text{-state} = \frac{2\pi}{L} \times \frac{2\pi}{L} = \left(\frac{2\pi}{L}\right)^2 \quad (12.81)$$

The number of k-states lying inside an annulus enclosed by radii k and $k + dk$ is determined (Figure 12.8).

$$\text{Area of the annulus} = \pi(k+dk)^2 - \pi k^2 = \pi k^2 + 2\pi kdk$$

$$+ (dk)^2 - \pi k^2 = 2\pi kdk + (dk)^2 \approx 2\pi kdk \quad (12.82)$$

from which the number of states between k and $k + dk$ at a given energy is given by

$$dN_{2D} = 2 \times \frac{\text{Area of the annulus}}{\text{Area occupied by a state}} \quad (12.83)$$

where factor 2 is for electron spin, as noted earlier. Putting the two areas in this equation yields

$$dN_{2D} = 2 \times \frac{2\pi kdk}{\left(\dfrac{2\pi}{L}\right)^2} = \frac{4\pi kdk \times L^2}{4\pi^2} = \frac{L^2}{\pi} kdk \quad (12.84)$$

or

$$\frac{dN_{2D}}{dk} = \frac{L^2}{\pi} k \quad (12.85)$$

The number of states between k and $k + dk$ per unit energy is

$$\frac{dN_{2D}}{dE} = \frac{dN_{2D}}{dk} \times \frac{dk}{dE} = \frac{L^2}{\pi} k \times \frac{dk}{dE} = \frac{L^2}{\pi}\sqrt{\frac{2mE}{\hbar^2}}$$

$$\times \frac{d}{dE}\left(\sqrt{\frac{2mE}{\hbar^2}}\right) = \frac{L^2}{\pi}\sqrt{\frac{2mE}{\hbar^2}} \times \frac{1}{2}\left(\frac{2mE}{\hbar^2}\right)^{1/2-1}$$

$$\times \frac{2m}{\hbar^2} = \frac{L^2 m}{\pi\hbar^2} \quad (12.86)$$

The number of states between k and $k + dk$ per unit area per unit energy is

$$\{g(E)\}_{2D} = \frac{\dfrac{L^2 m}{\pi\hbar^2}}{L^2} = \frac{m}{\pi\hbar^2} \quad (12.87)$$

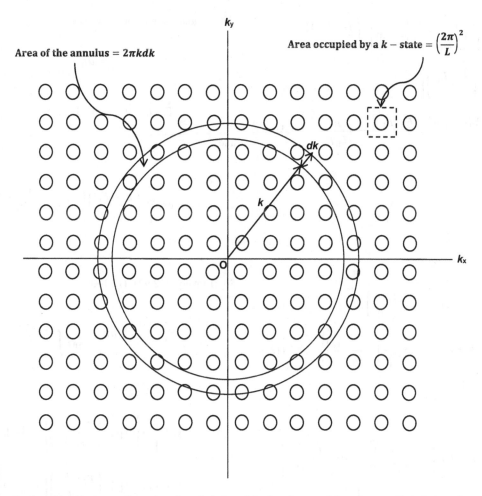

FIGURE 12.8 Visualization of k-space in two dimensions for calculation of density of states of free electrons. Periodic boundary conditions are used. The allowed states are represented as a grid of points separated by a distance of $2\pi/L$. The annular rings of radii k, $k + dk$ correspond to energies E, $E + dE$.

which is same as eq. (12.75).

According to Fermi-Dirac statistics, the sheet density of electrons in a 2D band starting at ε_j is

$$n_{S,j}(E_F) = \int_0^\infty \{g(E)\}_{2D} f(E, E_F, T) dE = \int_{\varepsilon_j}^\infty \left(\frac{m^*}{\pi\hbar^2}\right) \left\{\frac{1}{1 + \exp\left(\dfrac{E - E_F}{k_B T}\right)}\right\} dE = \left(\frac{m^*}{\pi\hbar^2}\right) \int_{\varepsilon_j}^\infty \left\{\frac{1}{1 + \exp\left(\dfrac{E - E_F}{k_B T}\right)}\right\} dE \quad (12.88)$$

Let us consider

$$\int_{\varepsilon_j}^\infty \left\{\frac{1}{1 + \exp\left(\dfrac{E - E_F}{k_B T}\right)}\right\} dE = \int_{\varepsilon_j}^\infty \frac{1}{\exp\left(\dfrac{E - E_F}{k_B T}\right)\left[\exp\left\{-\left(\dfrac{E - E_F}{k_B T}\right)\right\} + 1\right]} dE = \int_{\varepsilon_j}^\infty \frac{\exp\left\{-\left(\dfrac{E - E_F}{k_B T}\right)\right\}}{\left[\exp\left\{-\left(\dfrac{E - E_F}{k_B T}\right)\right\} + 1\right]} dE \quad (12.89)$$

Putting

$$\eta = \exp\left\{-\left(\frac{E - E_F}{k_B T}\right)\right\} \quad (12.90)$$

$$\frac{d\eta}{dE} = \frac{d}{dE}\left[\exp\left\{-\left(\frac{E-E_F}{k_B T}\right)\right\}\right] = \exp\left\{-\left(\frac{E-E_F}{k_B T}\right)\right\}$$

$$\times \frac{d}{dE}\left\{-\left(\frac{E-E_F}{k_B T}\right)\right\}$$

$$= \eta \times \frac{-1\times k_B T - \{0\times-(E-E_F)\}}{(k_B T)^2} = -\frac{\eta}{k_B T} \tag{12.91}$$

$$\therefore d\eta = -\frac{\eta}{k_B T}dE \tag{12.92}$$

or

$$dE = -k_B T\left(\frac{d\eta}{\eta}\right) \tag{12.93}$$

$$\text{For } E = \varepsilon_j, \quad \eta = \exp\left\{-\left(\frac{\varepsilon_j - E_F}{k_B T}\right)\right\} \tag{12.94}$$

and

$$\text{For } E = \infty, \quad \eta = \exp\left\{-\left(\frac{\infty - E_F}{k_B T}\right)\right\} = \exp\left\{-\left(\frac{\infty}{k_B T}\right)\right\}$$

$$= \frac{1}{\exp\left\{+\left(\frac{\infty}{k_B T}\right)\right\}} = \frac{1}{\infty} = 0 \tag{12.95}$$

Hence, from eqs. (12.90) and (12.93)–(12.95)

$$\int_{\varepsilon_j}^{\infty} \frac{\exp\left\{-\left(\frac{E-E_F}{k_B T}\right)\right\}}{\left[\exp\left\{-\left(\frac{E-E_F}{k_B T}\right)\right\}+1\right]} dE$$

$$= \int_{\exp\left\{-\left(\frac{\varepsilon_j-E_F}{k_B T}\right)\right\}}^{0}\left(\frac{\eta}{\eta+1}\right)\times\left\{-k_B T\left(\frac{d\eta}{\eta}\right)\right\}$$

$$= -k_B T\int_{\exp\left\{-\left(\frac{\varepsilon_j-E_F}{k_B T}\right)\right\}}^{0}\frac{d\eta}{\eta+1}$$

$$= k_B T\int_{0}^{\exp\left\{-\left(\frac{\varepsilon_j-E_F}{k_B T}\right)\right\}}\frac{d\eta}{\eta+1} \tag{12.96}$$

Let

$$\xi = \eta + 1 \tag{12.97}$$

Differentiating both sides, we get

$$d\xi = d\eta \tag{12.98}$$

$$\text{For } \eta = 0, \ \xi = 0+1 = 1 \tag{12.99}$$

and

$$\text{For } \eta = \exp\left\{-\left(\frac{\varepsilon_j - E_F}{k_B T}\right)\right\}, \ \xi = \exp\left\{-\left(\frac{\varepsilon_j - E_F}{k_B T}\right)\right\}+1 \tag{12.100}$$

So, from eqs. (12.97)–(12.100)

$$k_B T\int_{0}^{\exp\left\{-\left(\frac{\varepsilon_j-E_F}{k_B T}\right)\right\}}\frac{d\eta}{\eta+1} = k_B T\int_{1}^{\exp\left\{-\left(\frac{\varepsilon_j-E_F}{k_B T}\right)\right\}+1}\frac{d\xi}{\xi}$$

$$= (k_B T)\left|\ln(\xi)\right|_{1}^{\exp\left\{-\left(\frac{\varepsilon_j-E_F}{k_B T}\right)\right\}+1}$$

$$= (k_B T)\left[\ln\left\{\exp\left\{-\left(\frac{\varepsilon_j - E_F}{k_B T}\right)\right\}+1\right\}-\ln(1)\right]$$

$$= (k_B T)\left[\ln\left\{\exp\left\{-\left(\frac{\varepsilon_j - E_F}{k_B T}\right)\right\}+1\right\}-0\right]$$

$$= k_B T\ln\left\{\exp\left\{-\left(\frac{\varepsilon_j - E_F}{k_B T}\right)\right\}+1\right\} \tag{12.101}$$

Applying eq. (12.101), eq. (12.88) reduces to

$$n_{S,j}(E_F) = \left(\frac{m^*}{\pi\hbar^2}\right)\times k_B T\ln\left\{\exp\left\{-\left(\frac{\varepsilon_j - E_F}{k_B T}\right)\right\}+1\right\}$$

$$= \left(\frac{m^* k_B T}{\pi\hbar^2}\right)\ln\left\{1+\exp\left(\frac{E_F - \varepsilon_j}{k_B T}\right)\right\} \tag{12.102}$$

At absolute zero, the Fermi-Dirac distribution function transforms to a step function represented by the Heaviside unit step function. In the zero temperature limit

$$f(E, E_F, T = 0) = \theta\left(E_F^0 - E\right) \tag{12.103}$$

where E_F^0 is the Fermi level at zero temperature. Then

$$n_{S,j}\left(E_F^0\right) = \int_{\varepsilon_j}^{E_F^0}\left(\frac{m^*}{\pi\hbar^2}\right)\theta\left(E_F^0 - E\right)dE = \left(\frac{m^*}{\pi\hbar^2}\right)\theta\left(E_F^0 - E\right)\int_{\varepsilon_j}^{E_F^0}dE$$

$$= \left(\frac{m^*}{\pi\hbar^2}\right)\theta\left(E_F^0 - E\right)|E|_{\varepsilon_j}^{E_F^0}$$

$$= \left(\frac{m^*}{\pi\hbar^2}\right)\left\{\theta\left(E_F^0 - E\right)\right\}\times\left(E_F^0 - E\right)$$

$$= \left(\frac{m^*}{\pi\hbar^2}\right)\left(E_F^0 - E\right)\theta\left(E_F^0 - E\right) \qquad (12.104)$$

for $\varepsilon_j = E$.

12.4.5 Boundary Conditions

For localized electron states, the boundary conditions are

$$\varepsilon_j(z) \to 0 \text{ as } z \to \pm\infty \qquad (12.105)$$

For electrostatic potential,

$$\frac{d\Phi}{dz} \to 0 \text{ as } z \to \pm\infty \qquad (12.106)$$

Equations (12.12), (12.37), (12.41), (12.102), and (12.104) combined with the boundary conditions lay down the mathematical framework of the problem describing the formation of electron gas in the quantum well as well as the quantization of electron energies. It is evident that the physical description of the heterojunctions is done using Poisson's and Schrodinger's equations. These equations are too complex to allow analytical solutions. So, numerical solutions are called for.

12.4.6 Self-Consistent Solution of Schrodinger's and Poisson's Equations

Self-consistency implies obtaining results through a bidirectional coupling between the electrostatics and quantum mechanical aspects until they are agreeable to both sides. Integration of Poisson's equation gives the electrical potential for a given charge density. This potential contributes to the potential energy term in the Schrodinger equation. Integration of the Schrodinger equation gives the wave function of the electron and hence the charge density for a given potential. This charge density contributes to Poisson's equation. The two-way coupled Poisson and Schrodinger equations are solved repeatedly until the difference between a fresh result and preceding result falls below a predecided error margin. The procedure of self-consistent solution typically involves the following steps:

i. Initially, the charge density is obtained from the best estimate using physical arguments. This charge density is used to find the electrical potential by solving Poisson's equation.

ii. The electrical potential is an input to the potential energy term in the Schrodinger equation. The Schrodinger equation is solved to determine the eigen energies and normalized wave functions.

iii. From the wave functions, the particle density is calculated using a statistically weighted sum of probability densities. Feeding the particle density in the charge density equation, the charge density is re-calculated.

iv. Poisson's equation is re-solved to get the new potential value. This potential value is compared with the previous value. If the difference is small, this is the final result.

v. If the difference is large, the earlier calculations are done again. The iterative procedure is carried out until satisfactory agreement is achieved.

Several authors have pursued the self-consistent approach to investigate the electrical properties of quantum wells (Ando 1982, Tan et al 1990, Ram-Mohan 2004, Dubrovskiy and Zubkov 2017). Self-consistent Poisson-Schrodinger analysis reveals the following general features of heterojunctions:

i. Conducting channels are formed with electron concentration in the range $n_s = 10^{11}$–10^{12}cm^{-2} at any temperature inclusive of 0 K.

ii. The electron energies are quantized. The height of the lowest energy level ε_1 with energy of typically few tens of meV increases with the electron concentration.

iii. The electron channel is 6–10 nm wide. The spatial scale of electron confinement decreases with increase in confining potential and the electron concentration.

iv. The electrons are spatially separated from the parent donor impurities in the wide bandgap barrier material. The probability of finding electrons in the barrier is very small but increases with increasing electron confinement.

12.5 Discussion and Conclusions

We start with the basic question: What are quantum wells and how are they made? Then we take the help of the energy band model to understand the band offsets at the interfaces between semiconductors of different bandgaps. Electron affinity is a material property, which makes the comparison of band diagrams of two materials relatively easy. The electron affinity rule is applied for this purpose. There are principally three types of heterojunctions: type I, type II, and type III. It is noted that the carrier recombination rate is faster in type I than in type II heterojunctions because of carrier confinement in the same material in type I and in different materials in type II. A closer inspection of the interfacial region is done by drawing the band diagrams of type I heterojunction under three different cases of formation of N-n, N-p, and P-n heterojunctions with the uppercase letter denoting the wider bandgap semiconductor and vice versa. Charge carriers are transferred across the interface resulting in charge depletion on one side and charge accumulation on the opposite side. In all situations, the charge carriers are separated from their parent dopant ions leading to a doping technique in which a region has a large number of free carriers without any associated dopant impurities and hence the free carriers are unperturbed by ionized impurity scattering. This technique known as modulation doping provides a channel containing higher mobility than achievable by conventional doping. Another advantage is that the channel remains conducting down to 0 K despite the fact that the semiconductor materials themselves become insulating at this temperature.

While writing down the governing equations for the heterojunction, the first equation which comes to mind is Poisson's equation. In this equation, the charge density is expressed in terms of the electron wave function. The second basic equation is the 3D time-independent Schrodinger equation (TISE) in which the potential energy is z-coordinate dependent. Plane wave solutions are attempted in X- and Y-directions and verified by factorizing the wave function and substituting in the TISE. The resulting equation becomes 1D containing Z-direction only. The potential energy term contains contributions from the electrostatic potential and the discontinuity at the heterojunction. The third equation is the equation for electron concentration. It contains the sheet density of electrons for which the 2D density-of-states equation is derived. Then the boundary conditions are laid out.

Because Poisson's and Schrodinger's equations mentioned earlier cannot be solved analytically, the self-consistent solution strategy is pursued in which Poisson's equation is integrated to obtain the potential for a specified charge density. This potential is an input to the potential energy term of the Schrodinger equation. When the Schrodinger equation is integrated, the wave function is found. This gives the charge density for a particular potential. The charge density is again an input to Poisson's equation. The procedure is repeated until the difference between two consecutive results is minimized. By self-consistent solution of Poisson's and Schrodinger's equations, it is deciphered that conducting channels of width 6–10 nm are formed. In these channels, the electron concentration is 10^{11}–10^{12} cm^{-2}. The quantization of electron energies is also discerned.

Illustrative Exercises

12.1 (a) Electrons are trapped in a GaAs quantum well with AlGaAs barriers on either side. If electrons have an effective mass $m* = 0.067 \times$ free electron mass m_0, find the density of states for electrons in the quantum well.

(b) If the quantum well is filled up to an energy of 10 meV, how many states are filled at 0 K?

(c) Does the 2D density of states depend on energy E?

(d) Write the formulae for 1D and 3D density of states and show their energy dependence.

(a) The 2D density of states is given by eq. (12.76)

$$\left[\left\{g(E)\right\}_{2D}\right]_{Electrons} = \frac{m^*_{e,DOS}}{\pi\hbar^2} = \frac{m^*_{e,DOS}}{\pi\left(\dfrac{h}{2\pi}\right)^2} = \frac{m^*_{e,DOS}}{\dfrac{h^2}{4\pi}} = \frac{4\pi m^*_{e,DOS}}{h^2}$$

$$= \frac{4 \times 3.14159 \times 0.067 \times 9.109 \times 10^{-31} \text{ kg}}{\left(6.626 \times 10^{-34}\right)^2 \text{ Js} \times \text{kg m}^2 \text{ s}^{-2} \times \text{s}}$$

$$= \frac{7.6692872 \times 10^{-31}}{43.903876 \times 10^{-68} \text{ m}^2 \text{ J}}$$

$$= 1.746836 \times 10^{36} \times \frac{1}{\text{m}^2 \text{ J}} = 1.746836 \times 10^{36}$$

$$\times \frac{1}{\left(100 \text{ cm}\right)^2 \left(\dfrac{1}{1.602 \times 10^{-19} \text{ eV}}\right)}$$

$$= 1.746836 \times 10^{36} \times \frac{1}{\left(100 \text{ cm}\right)^2 \left(\dfrac{1}{1.602 \times 10^{-19}} \text{ eV}\right)}$$

$$= 1.746836 \times 10^{36} \times \frac{10^{-4} \times 1.602 \times 10^{-19}}{\text{cm}^2 \text{ (eV)}} = \frac{2.798 \times 10^{13}}{\text{cm}^2 \left(10^3 \text{ meV}\right)}$$

$$= 2.798 \times 10^{10} \text{ cm}^{-2} \text{ meV}^{-1} \approx 2.8 \times 10^{10} \text{ cm}^{-2} \text{ meV}^{-1}$$

(12.107)

(b)

Number of electron states filled up to 10 meV at 0 K is

$= $ Density of electron states cm^{-2} meV^{-1} × 10 meV

$= 2.8 \times 10^{10}$ cm^{-2} meV^{-1} × 10 meV $= 2.8 \times 10^{11}$ cm^{-2} (12.108)

(c) No.

(d)

$$\left\{g(E)\right\}_{1D} = \frac{1}{\pi\hbar}\sqrt{\frac{2m^*_{e,DOS}}{E - E_C}}; \left\{g(E)\right\}_{1D} \propto \frac{1}{\sqrt{E - E_C}}$$ (12.109)

and

$$\left\{g(E)\right\}_{3D} = \frac{m^*_{e,DOS}}{\pi^2\hbar^3}\sqrt{2m^*_{e,DOS}\left(E - E_C\right)}; \left\{g(E)\right\}_{3D} \propto \sqrt{E - E_C}$$ (12.110)

where $E - E_C$ is the energy relative to the bottom of conduction band.

12.2 Determine the number of holes cm^{-2} in a GaAs quantum well with AlGaAs barriers on both sides if holes have an effective mass $m* = 0.47 \times$ free electron mass m_0. The quantum well is filled up to an energy of 7 eV and the temperature is 0 K.

$$\left[\left\{g(E)\right\}_{2D}\right]_{Holes} = \frac{m^*_{h,DOS}}{\pi\hbar^2} = \frac{m^*_{h,DOS}}{\pi\left(\dfrac{h}{2\pi}\right)^2}$$

$$= \frac{m^*_{h,DOS}}{\dfrac{h^2}{4\pi}} = \frac{4\pi m^*_{h,DOS}}{h^2}$$

$$= \frac{4\times3.14159\times0.47\times9.109\times10^{-31}\ \text{kg}}{\left(6.626\times10^{-34}\right)^2\ \text{Js}\times\text{kg}\,\text{m}^2\,\text{s}^{-2}\times\text{s}}$$

$$= \frac{53.799477\times10^{-31}}{43.903876\times10^{-68}\ \text{m}^2\,\text{J}}\times\frac{1}{\text{m}^2\,\text{J}}$$

$$= 1.22539\times10^{37}\times\frac{1}{\left(100\ \text{cm}\right)^2\left(\dfrac{1}{1.602\times10^{-19}\ \text{eV}}\right)}$$

$$= 1.22539\times10^{37}\times\frac{1}{\left(100\ \text{cm}\right)^2\left(\dfrac{1}{1.602\times10^{-19}}\ \text{eV}\right)}$$

$$= 1.22539\times10^{37}\times\frac{10^{-4}\times1.602\times10^{-19}}{\text{cm}^2\,(\text{eV})}$$

$$= \frac{1.963\times10^{14}}{\text{cm}^2\left(10^3\ \text{meV}\right)} = 1.963\times10^{11}\ \text{cm}^{-2}\,\text{meV}^{-1}$$

$$\approx 1.96\times10^{11}\ \text{cm}^{-2}\,\text{meV}^{-1} \qquad (12.111)$$

Number of holes filled in the states up to 7 eV at 0 K is

$$= \text{Density of hole states cm}^{-2}\ \text{meV}^{-1}\times 7\ \text{meV}$$

$$= 1.96\times10^{11}\ \text{cm}^{-2}\,\text{meV}^{-1}\times 7\ \text{eV} = 1.372\times10^{12}\ \text{cm}^{-2} \qquad (12.112)$$

REFERENCES

Ando, T. 1982. Self-consistent results for a GaAs/Al$_x$Ga$_{1-x}$As heterojunction. I. Subband structure and light-scattering spectra. *Journal of the Physical Society of Japan* 51(12): 3893–3899.

Davies, J. H. 1998. *The Physics of Low-Dimensional Semiconductors: An Introduction.* Cambridge University Press: Cambridge, UK, pp. 130–135.

Dubrovskiy, S. V. and V. I. Zubkov. 2017. Self-Consistent solution of Schrodinger and Poisson equations by means of numerical methods in the LabVIEW development environment. In *2017 IEEE Conference of Russian Young Researchers in Electrical and Electronic Engineering (EIConRus)*, 1–3 February 2017, St. Petersburg, Russia, pp. 1388–1390.

Einspruch, N. G. and W. R. Frensley (Eds.) 1994. *Heterostructures and Quantum Devices.* Academic Press, Inc.: San Diego, CACalifornia, 452 pages.

Missous, M. 1993. Conduction and valence band offsets at the GaAs/AlGaAs heterostructure interface In: Adachi S. (ed.), *Properties of Aluminum Gallium Arsenide.* INSPEC, IEE: London, UK, pp. 73–76.

Nag, B. R. 2002. *Physics of Quantum Well Devices.* Springer Science+Business Media B. V.: Springer Netherlands, 297 pages.

Ram-Mohan, L. R. 2004. The Schrödinger–Poisson self-consistency in layered quantum semiconductor structures. *Journal of Applied Physics* 95(6): 3081–3092.

Tan, I-H., G. L. Snider, L. D. Chang and E. L. Hu. 1990. A self-consistent solution of Schrödinger–Poisson equations using a nonuniform mesh. *Journal of Applied Physics* 68: 4071–4076.

Part IV

Green's Function Method for Nanoelectronic Device Modeling

13

Dirac Delta and Green's Function Preliminaries

An essential prerequisite to understanding the Green's function is the knowledge of the Dirac delta function. George Green was a British miller. He was a mathematical physicist (1793–1841). Paul Adrien Maurice Dirac was a renowned English theoretical physicist (1902–1984). He received the Nobel Prize in physics in 1933 for his work on atomic theory.

13.1 Dirac Delta Function

13.1.1 Describing Variations in Space

When dealing with variations in space, the delta function, also called the Dirac delta function or impulse function, denoted by the symbol $\delta(x - s)$ is a function located at a distance $x = s$ from the origin and obtained from (Barton 1999, Royston 2008)

$$\delta(x - s) = \text{Limit of a sequence of functions } f_n(x) \text{ for}$$

$$n = 1 \text{ to infinity} = f_n(x)|_{n=1}^{\infty} = \lim_{n \to \infty} f_n(x) \tag{13.1}$$

The delta function is not exactly a function but falls under a different category of mathematical objects called a generalized function or distribution, which maps a function to a number.

The function $\delta(x - s)$ has the following properties:

$$\delta(x - s) = 0 \text{ for } x \neq s \tag{13.2}$$

and

$$\delta(x - s) = \infty \text{ for } x = s \tag{13.3}$$

$$\int_{-\infty}^{\infty} \delta(x - s) dx = 1 \tag{13.4}$$

and

$$\int_{-\infty}^{\infty} f(x) \delta(x - s) dx = f(s) \tag{13.5}$$

If the delta function is located at the origin $s = 0$, eqs. (13.2)–(13.5) reduce to the form

$$\delta(x) = 0 \text{ for } x \neq 0 \tag{13.6}$$

and

$$\delta(x) = \infty \text{ for } x = 0 \tag{13.7}$$

$$\int_{-\infty}^{\infty} \delta(x) dx = 1 \tag{13.8}$$

and

$$\int_{-\infty}^{\infty} f(x) \delta(x) dx = f(0) \tag{13.9}$$

The Dirac delta functions $\delta(x - s)$, $\delta(0)$, and $\delta(x + s)$ are represented by arrows as shown in Figure 13.1.

For easy conception, the delta function is imagined as a spike or harpoon of infinite height, infinitesimal width, and area of unity. Visualize it as the limiting value of a series of rectangular functions of height h and width $1/h$; hence area $= h \times 1/h = 1$. As $h \to \infty$, $1/h \to 0$. The function becomes taller and thinner but its area always remains unity (Figure 13.2).

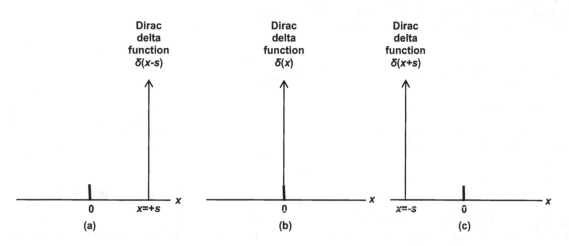

FIGURE 13.1 Representation of the Dirac delta functions for space: (a) $\delta(x - s)$, (b) $\delta(0)$, and (c) $\delta(x + s)$. At $x = +s$, $\delta(x - s) = \delta(s - s) = \delta(0) = \infty$; at $x = 0$, $\delta(x) = \delta(0) = \infty$; and at $x = -s$, $\delta(x + s) = \delta(-s + s) = \delta(0) = \infty$.

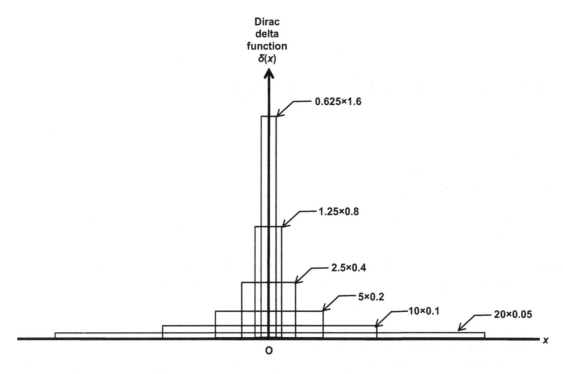

FIGURE 13.2   Visualization of the Dirac delta function as the limiting value of rectangular boxcar functions of decreasing width and increasing height.

A similar plot for the Gaussian function

$$F_\alpha(x) = \frac{1}{2\sqrt{\pi\alpha}} \exp\left(-\frac{x^2}{4\alpha}\right) \qquad (13.10)$$

is shown in Figure 13.3 where the function attains a greater height, reaching a peak around $x = 0$, and also becomes narrower when smaller values of the constant α are taken. The family of Gaussian functions leads to a delta function:

$$\lim_{\alpha \to \infty} \{F_\alpha(x)\} = \delta(x) \qquad (13.11)$$

Consider eq. (13.5). We note that $\delta(x-s) = 0$ everywhere except at $x = s$; therefore the value of the function $f(x) = f(s)$ only matters. We need not bother about the other values of $f(x)$ because they will be multiplied by the zero value of $\delta(x-s)$ and the product will be zero. From this reasoning, we replace $f(x)$ by $f(s)$ to get

$$\int_{-\infty}^{\infty} f(x)\delta(x-s)dx = \int_{-\infty}^{\infty} f(s)\delta(x-s)dx \qquad (13.12)$$

Taking out $f(s)$ as a constant

$$\int_{-\infty}^{\infty} f(x)\delta(x-s)dx = f(s)\int_{-\infty}^{\infty}\delta(x-s)dx = f(s)\times 1 = f(s) \qquad (13.13)$$

by applying eq. (13.4).

Proceeding on similar lines, the value of the function $f(x) = f(0)$ is only important for us because $\delta(x)$ vanishes everywhere except at $x = 0$. For $x \neq 0$, $\delta(x) = 0$. So multiplication of $f(x)$ value for $x \neq 0$ with zero value of $\delta(x)$ will give zero. The constant value $f(0)$ may be taken out of the integral sign

$$\int_{-\infty}^{\infty} f(x)\delta(x)dx = \int_{-\infty}^{\infty} f(0)\delta(x)dx$$

$$= f(0)\int_{-\infty}^{\infty}\delta(x)dx = f(0)\times 1 = f(0) \qquad (13.14)$$

with the help of eq. (13.8).

13.1.2 Describing Variations in Time

When we are describing variations in time, a delta function $\delta(t-\tau)$ displaced from the origin by a time $t = \tau$ is (Figure 13.4)

$$\delta(t-\tau) = 0 \text{ for } t \neq \tau \qquad (13.15)$$

and

$$\delta(t-\tau) = \infty \text{ for } t = \tau \qquad (13.16)$$

Moreover,

$$\int_{-\infty}^{\infty}\delta(t-\tau)dt = 1 \qquad (13.17)$$

and

$$\int_{-\infty}^{\infty} f(t)\delta(t-\tau)dt = f(\tau) \qquad (13.18)$$

For a delta function located at the origin $t = 0$, the equations are

$$\delta(t) = 0 \text{ for } t \neq 0 \qquad (13.19)$$

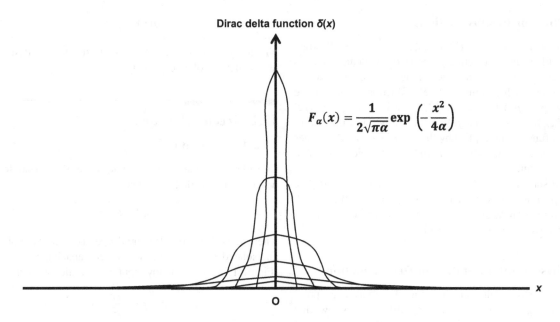

FIGURE 13.3 Imagining the Dirac delta function as the limiting value of Gaussian functions. The function tapers and the peak rises with decreasing values of α.

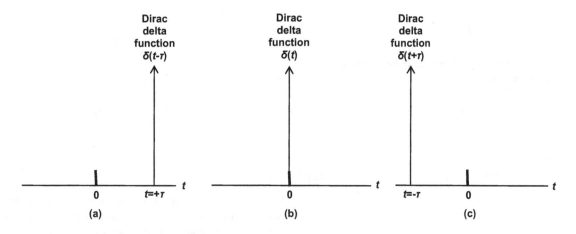

FIGURE 13.4 Representation of the Dirac delta functions for time: (a) $\delta(t - \tau)$, (b) $\delta(0)$, and (c) $\delta(t + \tau)$. At $t = +\tau$, $\delta(t - \tau) = \delta(\tau - \tau) = \delta(0) = \infty$; at $t = 0$, $\delta(t) = \delta(0) = \infty$; and at $t = -\tau$, $\delta(t + \tau) = \delta(-\tau + \tau) = \delta(0) = \infty$.

and

$$\delta(t) = \infty \text{ for } t = 0 \tag{13.20}$$

$$\int_{-\infty}^{\infty} \delta(t)\,dt = 1 \tag{13.21}$$

and

$$\int_{-\infty}^{\infty} f(t)\delta(t)\,dt = f(0) \tag{13.22}$$

Consider eq. (13.18). $\delta(t - \tau) = 0$ everywhere except at $t = \tau$; therefore the value of the function $f(t) = f(\tau)$ only has significance. Whatever be the other values of $f(t)$, they will be multiplied by

the zero value of $\delta(t - \tau)$ and the product will be zero. With these arguments, we replace $f(t)$ by $f(\tau)$ to obtain

$$\int_{-\infty}^{\infty} f(t)\delta(t - \tau)\,dt = \int_{-\infty}^{\infty} f(\tau)\delta(t - \tau)\,dt \tag{13.23}$$

Now $f(\tau)$ being a constant is taken out of the integral sign giving

$$\int_{-\infty}^{\infty} f(\tau)\delta(t - \tau)\,dt = f(\tau)\int_{-\infty}^{\infty} \delta(t - \tau)\,dt = f(\tau) \times 1 = f(\tau) \tag{13.24}$$

where eq. (13.17) has been applied.

Identically, we can write

$$\int_{-\infty}^{\infty} f(t)\delta(t)\,dt = f(0)\int_{-\infty}^{\infty} \delta(t)\,dt = f(0) \times 1 = f(0) \tag{13.25}$$

13.1.3 Noteworthy Observations

The definitions in eqs. (13.2)–(13.5) and (13.15)–(13.18) illustrate the sifting, filtering, or selecting property of the delta functions $\delta(x - s)$, $\delta(x)$, $\delta(t - \tau)$, $\delta(t)$, when they are multiplied with another function $f(x)$ or $f(t)$. It is found that delta functions sieve out and give the values of $f(x)$ at $x = s$, i.e. $f(s)$; $f(x)$ at $x = 0$, i.e. $f(0)$; $f(t)$ at $t = \tau$, i.e. $f(\tau)$; $f(t)$ at $t = 0$, i.e. $f(0)$, respectively. Thus upon integration of a function multiplied by the delta function, the values of the function at $x = s$, $x = 0$, $t = \tau$, and $t = 0$ are picked out and given as an output.

Note that spikes of the delta functions $\delta(x - s)$, $\delta(x)$, $\delta(t - \tau)$, and $\delta(t)$ appear, respectively, at $x - s = 0$ or $x = s$; $x = 0$; $t - \tau = 0$ or $t = \tau$; and $t = 0$ These are the locations at which the arguments of the respective delta functions become zero.

13.1.4 Physical Interpretation of Delta Function

In physics, the Dirac delta function is used to represent a point source of charge or mass (Greenberg 2015). We know that the charge density $\rho(\mathbf{x})$ describes a continuous distribution of charges in 3D space. The total charge Q of the charge distribution is obtained by integrating $\rho(\mathbf{x})$ over space as

$$Q = \int \rho(\mathbf{x}) d^3 x \qquad (13.26)$$

Let us find the charge density of a single point charge q located at the point s. For this point charge

$$\rho(\mathbf{x}) = \infty \text{ at } x = s \qquad (13.27)$$

and

$$\rho(\mathbf{x}) = 0 \text{ for } x \neq s \qquad (13.28)$$

and

$$q = \int \rho(\mathbf{x}) d^3 x \qquad (13.29)$$

because the total charge is q.

All the earlier conditions are satisfied by writing

$$\rho(\mathbf{x}) = q\delta(x - s) \qquad (13.30)$$

which illustrates the use of the delta function in describing the charge density of a point charge.

13.2 Green's Function

13.2.1 Definition

Consider an isolated system without any external input, and described by the differential equation

$$\mathcal{L}\{x(t)\} = 0 \qquad (13.31)$$

where $\mathcal{L}$ is a linear differential operator. An example of such an isolated system is the resistance-capacitance (R-C) circuit (Figure 13.5). Without any input, the circuit equation is

$$RI + \frac{Q}{C} = 0 \qquad (13.32)$$

where I is the current and Q is the charge. In presence of input, there will a term V_{Battery} for battery voltage on the right-hand side of eq. (13.32).

In differential form,

$$R\frac{dQ}{dt} + \frac{Q}{C} = 0 \qquad (13.33)$$

or

$$\left(R\frac{d}{dt} + \frac{1}{C}\right)Q(t) = 0 \qquad (13.34)$$

or

$$\mathcal{L}\{Q(t)\} = 0 \qquad (13.35)$$

Suppose an input is applied to the system. Let this input be a delta function $\delta(t)$, i.e., $V_{\text{Battery}} = \delta(t)$. If the resulting response function to $\delta(t)$ input is $g(t)$, we can write

$$\mathcal{L}\{g(t)\} = \delta(t) \qquad (13.36)$$

The response function $g(t)$ representing the solution for the delta function as input is known as the Green's function for this input

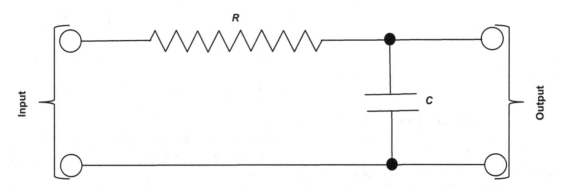

FIGURE 13.5   The resistance-capacitance (RC) circuit.

with the given boundary conditions. Thus the Green's function is the solution of a differential equation under the given boundary conditions with the Dirac delta function as the input.

For an arbitrary input $f(t)$, the response is different from $g(t)$; suppose it is $x(t)$:

$$\mathcal{L}\{x(t)\} = f(t) \tag{13.37}$$

This $x(t)$ representing the general solution to an arbitrary input $f(t)$ is the convolution of $f(t)$ with $g(t)$ given by

$$x(t) = (f * g)(t) = \int_0^t f(\tau)g(t-\tau)d\tau \tag{13.38}$$

The convolution of function $f(t)$ with another function $g(t)$ to produce a third function $x(t)$ may be understood as a linear superposition of the response function $g(t-\tau)$, each of which is multiplied with impulse $f(\tau)\,d\tau$, and is visualized as a weighted sum of input values.

The Green's function $g(t-\tau)$ is denoted by the symbol $G(t, \tau)$:

$$g(t-\tau) \equiv G(t,\tau) \tag{13.39}$$

Then eq. (13.38) becomes

$$x(t) = \int_0^t f(\tau)G(t,\tau)d\tau \tag{13.40}$$

or

$$x(t) = \int_0^t G(t,\tau)f(\tau)d\tau \tag{13.41}$$

which is the general solution for an arbitrary input $f(t)$ in terms of the Green's function $G(t, \tau)$. The Green's function is the kernel or core of this integral. Hence, the Green's function of a given differential operator is the kernel of the integral operator which is the inverse of the given differential operator. It is a tool used for solving ordinary and nonhomogeneous differential equations with initial or boundary conditions. It is also called the response function.

13.2.2 How Green's Function Works?

Consider the series of discontinuous strips of impulse $f(\tau_n)$. For each strip, the shifted impulse response $G(t, \tau)$ is to be applied to produce $x(t)$:

$$x(t) = \sum_n G(t,\tau_n)f(\tau_n)\Delta\tau \tag{13.42}$$

The Green's function takes care of the shifting. It takes the input function $f(\tau)$ and works on each infinitesimal segment by applying the impulse response of the system at the appropriate time shift $(t - \tau)$, e.g. with time shift $\tau = 0$, the equation is

$$G(t,\tau) = g(t-\tau) \Rightarrow G(t,0) = g(t-0) \Rightarrow G(t,0) = g(t) \tag{13.43}$$

The time shift is inserted after each input action. In this way, the Green's function operates on all the input segments and provides the behavior of the system toward an arbitrary input function.

13.2.3 Characterization of the Response of a System by Green's Function

The laws of physics due to Newton, Laplace, Poisson, Maxwell, Schrodinger, Einstein are all written in the language of differential equations, e.g. Maxwell's equations are differential equations describing the changes of electric and magnetic fields with space and time variables. Generally a differential operator $\mathcal{L}$ acts on some function $u(x)$ of interest to us with the given source term $f(x)$ written on the right-hand side of the equation, which is force in mechanics or electrical charge and current densities in electrodynamics:

$$\mathcal{L}\{u(x)\} = f(x) \tag{13.44}$$

If the source term

$$f(x) = \text{Dirac delta function} \tag{13.45}$$

then the function of interest

$$u(x) = \text{Green's function of the operator } \mathcal{L} \tag{13.46}$$

Thus the Green's function characterizes the response of the system to a point source. For the sake of illustration, let us consider an example from electrostatics. According to Gauss's law (Figure 13.6), the divergence of electric field $\mathbf{E}$ at any point in free space equals the charge density $\rho(\mathbf{x})$ divided by permittivity ε_0 of free space:

$$\nabla \cdot \mathbf{E} = \frac{\rho(\mathbf{x})}{\varepsilon_0} \tag{13.47}$$

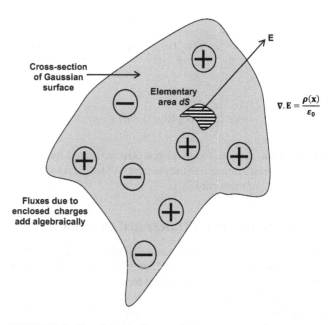

FIGURE 13.6 Gauss's law in electrostatics.

Since electric field $\mathbf{E}$ = Negative gradient of potential $V(x)$,

$$\mathbf{E} = -\nabla V(x) \qquad (13.48)$$

Equation (13.47) reduces to

$$\nabla \cdot \left\{ -\nabla V(x) \right\} = \frac{\rho(\mathbf{x})}{\varepsilon_0} \qquad (13.49)$$

or

$$-\nabla^2 V(x) = \frac{\rho(\mathbf{x})}{\varepsilon_0} \qquad (13.50)$$

Here the source is

$$f(x) = \rho(\mathbf{x}) \qquad (13.51)$$

the differential operator is

$$\mathcal{L} = -\nabla^2 \qquad (13.52)$$

and the quantity of interest is

$$u(x) = V(x) \qquad (13.53)$$

Let the source be a point charge q located at position s. Then from previous arguments,

$$\rho(\mathbf{x}) = q\delta(x - s) \qquad (13.54)$$

from equation (13.30). The potential produced by a point charge q at position s, denoted by $V_{pc}(x, s)$, satisfies the equation

$$-\nabla^2 V_{pc}(x,s) = \frac{\rho(\mathbf{x})}{\varepsilon_0} = \frac{q\delta(x-s)}{\varepsilon_0} \qquad (13.55)$$

or

$$\frac{\varepsilon_0}{q} \times -\nabla^2 V_{pc}(x,s) = \delta(x - s) \qquad (13.56)$$

or

$$-\nabla^2 \left\{ \left(\frac{\varepsilon_0}{q} \right) V_{pc}(x,s) \right\} = \delta(x - s) \qquad (13.57)$$

which means that the electrical potential generated by a point charge = Green's function of the operator $-\nabla^2$ taking into account the constants of proportionality.

13.2.4 Linearity of the Differential Operator

The effect of any distribution representing a source, e.g. a distribution of charges, is found by summation or integration of the effect of a point source, provided the distribution is continuous. Hence the acquisition of the knowledge of behavior of the system toward a point source helps in predicting the response of the system to a distribution of source. This is possible because

we are permitted to add together the individual contributions. The very fact that addition is allowed implies linearity of the differential operator. This is an essential condition for meaningful analysis.

13.2.5 Verifying the Solution

Suppose we are given the Green's function $G(x, s)$ for the differential equation

$$\mathcal{L}\left\{ u(x) \right\} = f(x) \qquad (13.58)$$

The boundary conditions are also provided.
The solution to eq. (13.58) can be expressed as

$$u(x) = \int f(s)G(x,s)ds \qquad (13.59)$$

To authenticate that $u(x)$ is a solution, we substitute for $u(x)$ in the differential equation. The aim is to check whether the equation is satisfied by this value of solution; then clearly

$$\mathcal{L}\left\{ u(x) \right\} = \mathcal{L}\left\{ \int f(s)G(x,s)ds \right\} \qquad (13.60)$$

Since integration is carried out over s, obviously the result will be a function of x. Remember that the differential operator $\mathcal{L}$ is an operator with respect to the variable x. It is not connected with variable s. So, it does not matter if it is moved past the integral over s. Moreover, only the Green's function is x-dependent because the source function is determined at s. Hence,

$$\mathcal{L}\left\{ u(x) \right\} = \int f(s)\left\{ \mathcal{L}G(x,s) \right\}ds \qquad (13.61)$$

Let us now apply the property of the Green's function that

$$\mathcal{L}\left\{ G(x,s) \right\} = \delta(x - s) \qquad (13.62)$$

where $G(x, s)$ is the response function to the $\delta(x - s)$ input.
This gives

$$\mathcal{L}\left\{ u(x) \right\} = \int f(s)\delta(x - s)ds \qquad (13.63)$$

It is evident that $\delta(x - s) = 0$ for all s except $s = x$; therefore only the value $f(s) = f(x)$ of the function has significance. The remaining values of $f(s)$ will be multiplied by the zero value of $\delta(x - s)$. They will all become zero. So, they lose their importance and it becomes reasonable to replace $f(s)$ by $f(x)$ and write

$$\mathcal{L}\left\{ u(x) \right\} = \int f(x)\delta(x - s)ds \qquad (13.64)$$

Pulling $f(x)$ out of the integration sign leads us to

$$\mathcal{L}\left\{ u(x) \right\} = f(x)\int \delta(x - s)ds \qquad (13.65)$$

We note that

$$\int f(s)\delta(x-s)ds = \int f(s)\delta(s-x)ds \qquad (13.66)$$

because

$$\delta(x-s) = \infty \text{ when } x = s \text{ and } \delta(x-s) = 0 \text{ when } x \neq s \qquad (13.67)$$

and

$$\delta(s-x) = \infty \text{ when } s = x \text{ and } \delta(s-x) = 0 \text{ when } s \neq x \qquad (13.68)$$

$$\therefore \delta(x-s) = \delta(s-x) \qquad (13.69)$$

so that from equations (13.63), (13.66), (13.65), (13.69) and (13.4), we have

$$\mathcal{L}\{u(x)\} = \int f(s)\delta(s-x)ds = f(x)\int \delta(x-s)ds$$

$$= f(x)\int \delta(s-x)ds = f(x)\times 1 \qquad (13.70)$$

showing that the solution is given by

$$u(x) = \int f(s)G(x,s)ds \qquad (13.71)$$

Once the Green's function is known, the solution can be found for the given source term by means of integration, offering a much easier method than by direct solution of the differential equation.

13.3 Retarded and Advanced Green's Functions

13.3.1 Retarded Green's Function

The Green's function provides the response at any point inside or outside a conductor caused by excitation at any other point (Datta 1999). If Ψ is the wave function and S the excitation due to a wave incident from one of the leads of the conductor,

$$[E - H_{op}]\Psi = S \qquad (13.72)$$

where E is the energy and H_{op} is the Hamiltonian operator. The Green's function is written as

$$G = [E - H_{op}]^{-1} \qquad (13.73)$$

i.e.,

$$[E - H_{op}]G(x,x') = \delta(x-x') \qquad (13.74)$$

The Green's function $G(x, x')$ is the wave function at x obtained from the application of unit excitation at x'.

If U_0 is the potential energy,

$$[H_{op}] = -\frac{\hbar^2}{2m}\frac{\partial^2}{\partial x^2} + U_0 \qquad (13.75)$$

by substitution for $[H_{op}]$ from eq. (13.75) in eq. (13.74), we have

$$\left(E - U_0 + \frac{\hbar^2}{2m}\frac{\partial^2}{\partial x^2}\right)G(x,x') = \delta(x-x') = 0 \text{ for } x \neq x' \qquad (13.76)$$

For $x > x'$, the solution is

$$G(x,x') = A^+ \exp(+ik|x-x'|) \qquad (13.77)$$

For $x < x'$, the solution is

$$G(x,x') = A^- \exp(-ik|x-x'|) \qquad (13.78)$$

where A^+ and $A-$ are the amplitudes of the waves and

$$k = \sqrt{\frac{2m}{\hbar^2}(E - U_0)} \qquad (13.79)$$

following procedures in Chapter 5.

Irrespective of the values of A^+, A^-, the solution satisfies eq. (13.76) at all points except at $x = x'$. In order to satisfy eq. (13.76) at $x = x'$, the Green's function must be continuous, i.e.,

$$[G(x,x')]_{x=x'^+} = [G(x,x')]_{x=x'^-} \qquad (13.80)$$

or

$$[A^+ \exp(+ik|x-x'|)]_{x=x'^+} = [A^- \exp(-ik|x-x'|)]_{x=x'^-} \qquad (13.81)$$

or

$$A^+ \exp(+0) = A^- \exp(-0) \qquad (13.82)$$

$$\therefore A^+ = A^- = A \qquad (13.83)$$

Further, the derivative of $G(x, x')$ must be discontinuous by $2m/\hbar^2$, i.e.,

$$\left[\frac{\partial G(x,x')}{\partial x}\right]_{x=x'^+} - \left[\frac{\partial G(x,x')}{\partial x}\right]_{x=x'^-} = \frac{2m}{\hbar^2} \qquad (13.84)$$

or

$$\left[\frac{\partial\{A^+ \exp(+ik|x-x'|)\}}{\partial x}\right]_{x=x'^+}$$

$$- \left[\frac{\partial\{A^- \exp(-ik|x-x'|)\}}{\partial x}\right]_{x=x'^-} = \frac{2m}{\hbar^2} \qquad (13.85)$$

or

$$[A^+(ik)\exp(+ik|x-x'|)]_{x=x'^+}$$

$$- [A^-(-ik)\exp(-ik|x-x'|)]_{x=x'^-} = \frac{2m}{\hbar^2} \qquad (13.86)$$

or

$$A^+(ik)\exp(+ik \times 0) - A^-(-ik)\exp(-ik \times 0) = \frac{2m}{\hbar^2} \qquad (13.87)$$

or

$$ik(A^+ + A^-) = \frac{2m}{\hbar^2} \qquad (13.88)$$

or

$$2Aik = \frac{2m}{\hbar^2} \qquad (13.89)$$

from eq. (13.83). Hence,

$$A = \frac{2m}{\hbar^2} \times \frac{1}{2ik} = \frac{m}{i\hbar^2 k} = \frac{i}{i^2} \frac{m}{\hbar \times \hbar k} = \frac{i}{-1} \frac{1}{\hbar \times \frac{\hbar k}{m}} = -\frac{i}{\hbar v} \qquad (13.90)$$

where

$$v = \frac{\hbar k}{m} \qquad (13.91)$$

from eq. (2.23) using $p = mv$.

Hence from eq. (13.77), the Green's function is given by

$$G(x,x') = -\frac{i}{\hbar v}\exp(+ik|x-x'|) \qquad (13.92)$$

To check whether this value of Green's function satisfies eq. (13.76), we rewrite this equation as

$$\left\{\frac{2m}{\hbar^2}(E - U_0) + \frac{\partial^2}{\partial x^2}\right\}G(x,x') = 0 \qquad (13.93)$$

or

$$\left\{k^2 + \frac{\partial^2}{\partial x^2}\right\}G(x,x') = 0 \qquad (13.94)$$

using eq. (13.79). Hence,

$$\frac{\partial^2 G(x,x')}{\partial x^2} + k^2 G(x,x') = 0 \qquad (13.95)$$

Putting the value of $G(x,x')$ from eq. (13.92) in eq. (13.95),

$$\text{Left-hand side} = \frac{\partial^2 G(x,x')}{\partial x^2} + k^2 G(x,x')$$

$$= \frac{\partial^2}{\partial x^2}\left\{-\frac{i}{\hbar v}\exp(+ik|x-x'|)\right\} + k^2 G(x,x')$$

$$= \frac{\partial}{\partial x}\left[\frac{\partial}{\partial x}\left\{-\frac{i}{\hbar v}\exp(+ik|x-x'|)\right\}\right] + k^2 G(x,x')$$

$$= \frac{\partial}{\partial x}\left\{-\frac{i}{\hbar v} \times ik\exp(+ik|x-x'|)\right\} + k^2 G(x,x')$$

$$= -\frac{i}{\hbar v} \times ik \times ik\exp(+ik|x-x'|) + k^2 G(x,x')$$

$$= -\frac{i^3}{\hbar v}k^2\exp(+ik|x-x'|) + k^2 G(x,x')$$

$$= -\frac{i \times i^2}{\hbar v}k^2\exp(+ik|x-x'|) + k^2 G(x,x')$$

$$= -\frac{i \times -1}{\hbar v}k^2\exp(+ik|x-x'|) + k^2 G(x,x')$$

$$= \frac{i}{\hbar v}k^2\exp(+ik|x-x'|) + k^2 G(x,x')$$

$$= \frac{i}{\hbar v}k^2\exp(+ik|x-x'|) - \frac{i}{\hbar v}k^2\exp(+ik|x-x'|) = 0$$

$$(13.96)$$

showing that eq. (13.95) is satisfied by the Green's function given in eq. (13.92). In the last step of eq. (13.96), the value of $G(x,x')$ has been put in the second term of eq. (13.95) from eq. (13.92) while previously we differentiated twice by putting this value in the first term of eq. (13.95) from eq. (13.92). Judging from the positive sign of the exponential factor, this solution represents outgoing waves that originate from the point of excitation (Figure 13.7). It is referred to as the retarded Green's function, denoted by G^R. Thus

$$G^R(x,x') = -\frac{i}{\hbar v}\exp(+ik|x-x'|) \qquad (13.97)$$

13.3.2 Advanced Green's Function

Besides the earlier solution, there is another Green's function which satisfies eq. (13.95). This solution is

$$G(x,x') = +\frac{i}{\hbar v}\exp(-ik|x-x'|) \qquad (13.98)$$

Let us confirm if this is true by putting its value in eq. (13.95). As before, we shall start by putting the value of $G(x,x')$ from eq. (13.98) in the first term of eq. (13.95) and in the last step we shall

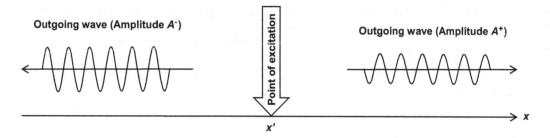

FIGURE 13.7   Retarded Green's function for an infinite one-dimensional wire.

put the value of $G(x, x')$ from eq. (13.98) in the second term of eq. (13.95). Then the left-hand side of eq. (13.95) becomes

$$\text{Left-hand side} = \frac{\partial^2 G(x,x')}{\partial x^2} + k^2 G(x,x')$$

$$= \frac{\partial^2}{\partial x^2} \left\{ +\frac{i}{\hbar v} \exp\left(-ik|x-x'|\right) \right\} + k^2 G(x,x')$$

$$= \frac{\partial}{\partial x} \left[\frac{\partial}{\partial x} \left\{ +\frac{i}{\hbar v} \exp\left(-ik|x-x'|\right) \right\} \right] + k^2 G(x,x')$$

$$= \frac{\partial}{\partial x} \left\{ +\frac{i}{\hbar v} \times -ik \exp\left(-ik|x-x'|\right) \right\} + k^2 G(x,x')$$

$$= +\frac{i}{\hbar v} \times -ik \times -ik \exp\left(-ik|x-x'|\right) + k^2 G(x,x')$$

$$= +\frac{i^3}{\hbar v} k^2 \exp\left(-ik|x-x'|\right) + k^2 G(x,x')$$

$$= +\frac{i \times i^2}{\hbar v} k^2 \exp\left(-ik|x-x'|\right) + k^2 G(x,x')$$

$$= +\frac{i \times -1}{\hbar v} k^2 \exp\left(-ik|x-x'|\right) + k^2 G(x,x')$$

$$= -\frac{i}{\hbar v} k^2 \exp\left(-ik|x-x'|\right) + k^2 G(x,x')$$

$$= -\frac{i}{\hbar v} k^2 \exp\left(-ik|x-x'|\right) + \frac{i}{\hbar v} k^2 \exp\left(-ik|x-x'|\right) = 0$$

$$(13.99)$$

After proving that the earlier solution satisfies eq. (13.95), we note from the negative sign of the exponential factor that it represents incoming waves disappearing at the point of excitation as opposed to the outgoing waves from the excitation point in the foregoing case (Figure 13.8). This value of Green's function is known as the advanced Green's function, denoted by G^A. Thus

$$G^A(x,x') = +\frac{i}{\hbar v} \exp\left(-ik|x-x'|\right) \qquad (13.100)$$

Briefly stated, eq. (13.95), which is the modified form of eq. (13.76), has two solutions. These solutions correspond to different boundary conditions. The two solutions defined for the two boundary conditions are named as the retarded and advanced Green's functions. They are denoted by the symbols G^R and G^A. The retarded function G^R refers to outgoing waves while the advanced function G^A pertains to incoming waves.

13.3.3 Inclusion of Boundary Conditions in Green's Function Equations

By adding an infinitesimal energy part $i\eta$ to the energy ($\eta > 0$), the boundary conditions can be included in the equation itself (Foster and Hofer 2006). The retarded Green's function $G^R(x, x')$ is the only acceptable solution of the modified form of equation (13.74)

$$\left[E - H_{\text{op}} + i\eta \right] G^R(x,x') = \delta(x-x') \qquad (13.101)$$

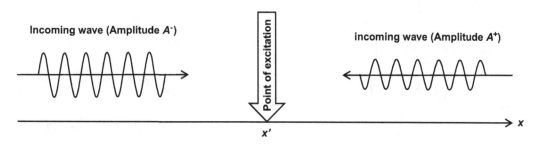

FIGURE 13.8   Advanced Green's function for an infinite one-dimensional wire.

The small imaginary part of energy adds a positive imaginary component to the wave number k. The modified wave number is

$$k' = \sqrt{\frac{2m(E - U_\text{o} + i\eta)}{\hbar^2}} = \sqrt{\frac{2m(E - U_\text{o})}{\hbar^2} + \frac{2mi\eta}{\hbar^2}}$$

$$= \sqrt{\frac{1}{\hbar^2}\{2m(E - U_\text{o}) + 2mi\eta\}}$$

$$= \sqrt{\frac{2m(E - U_\text{o})}{\hbar^2}\left(1 + \frac{i\eta}{E - U_\text{o}}\right)} = \sqrt{\frac{2m(E - U_\text{o})}{\hbar^2}}\left(1 + \frac{i\eta}{E - U_\text{o}}\right)^{1/2}$$

$$= k\left(1 + \frac{1}{2}\frac{i\eta}{E - U_\text{o}} + \dots\right) \cong k\left(1 + \frac{1}{2}\frac{i\eta}{E - U_\text{o}}\right) = k\left(1 + \frac{i\delta}{2}\right)$$

$$\tag{13.102}$$

where

$$\delta = \frac{\eta}{E - U_\text{o}} \tag{13.103}$$

Putting the new value of k, i.e. k' from eq. (13.102) in eq. (13.100) for advanced Green's function

$$G^\text{A}(x, x') = +\frac{i}{\hbar v}\exp\left(-ik'|x - x'|\right)$$

$$= +\frac{i}{\hbar v}\exp\left\{-ik\left(1 + \frac{i\delta}{2}\right)|x - x'|\right\}$$

$$= +\frac{i}{\hbar v}\exp\left\{\left(-ik - \frac{i^2 k\delta}{2}\right)|x - x'|\right\}$$

$$= +\frac{i}{\hbar v}\exp\left[\left\{-ik - \frac{(-1)k\delta}{2}\right\}|x - x'|\right]$$

$$= +\frac{i}{\hbar v}\exp\left\{\left(-ik + \frac{k\delta}{2}\right)|x - x'|\right\} \tag{13.104}$$

Thus the imaginary $i\eta$ component makes the advanced Green's function grow indefinitely as one moves away from the excitation point by virtue of the positive term $+k\delta/2$ in the exponential factor. Since a proper solution must be bounded, the retarded Green's function is left as the only appropriate solution:

$$G^\text{R}(x, x') = -\frac{i}{\hbar v}\exp\left(+ik'|x - x'|\right)$$

$$= -\frac{i}{\hbar v}\exp\left\{+ik\left(1 + \frac{i\delta}{2}\right)|x - x'|\right\}$$

$$= -\frac{i}{\hbar v}\exp\left\{\left(+ik + \frac{i^2 k\delta}{2}\right)|x - x'|\right\}$$

$$= -\frac{i}{\hbar v}\exp\left[\left\{+ik + \frac{(-1)k\delta}{2}\right\}|x - x'|\right]$$

$$= -\frac{i}{\hbar v}\exp\left\{\left(+ik - \frac{k\delta}{2}\right)|x - x'|\right\} \tag{13.105}$$

It does not grow indefinitely because of the negative term $-k\delta/2$ in the exponential factor.

Similarly, consider the equation

$$\left[E - H_\text{op} - i\eta\right]G^\text{A}(x, x') = \delta(x - x') \tag{13.106}$$

Here

$$k' = \sqrt{\frac{2m(E - U_\text{o} - i\eta)}{\hbar^2}} = k\left(1 - \frac{1}{2}\frac{i\eta}{E - U_\text{o}}\right) = k\left(1 - \frac{i\delta}{2}\right) \tag{13.107}$$

Hence from eq. (13.97),

$$G^\text{R}(x, x') = -\frac{i}{\hbar v}\exp\left(+ik'|x - x'|\right)$$

$$= -\frac{i}{\hbar v}\exp\left\{+ik\left(1 - \frac{i\delta}{2}\right)|x - x'|\right\}$$

$$= -\frac{i}{\hbar v}\exp\left\{\left(ik - \frac{ik \times i\delta}{2}\right)|x - x'|\right\}$$

$$= -\frac{i}{\hbar v}\exp\left\{\left(ik - \frac{i^2 k\delta}{2}\right)|x - x'|\right\}$$

$$= -\frac{i}{\hbar v}\exp\left[\left\{ik - \frac{(-1)k\delta}{2}\right\}|x - x'|\right]$$

$$= -\frac{i}{\hbar v}\exp\left\{\left(ik + \frac{k\delta}{2}\right)|x - x'|\right\} \tag{13.108}$$

so that, as earlier, the indefinite growth of this function due to $+k\delta/2$ term precludes its acceptability.

The advanced Green's function $G^\text{A}(x, x')$ is the only acceptable solution of the equation

$$G^\text{A}(x, x') = +\frac{i}{\hbar v}\exp\left(-ik'|x - x'|\right)$$

$$= +\frac{i}{\hbar v}\exp\left\{-ik\left(1 - \frac{i\delta}{2}\right)|x - x'|\right\}$$

$$= +\frac{i}{\hbar v}\exp\left\{\left(-ik + \frac{i^2 k\delta}{2}\right)|x - x'|\right\}$$

$$= +\frac{i}{\hbar v}\exp\left[\left\{-ik + \frac{(-1)k\delta}{2}\right\}|x - x'|\right]$$

$$= +\frac{i}{\hbar v}\exp\left\{\left(-ik - \frac{k\delta}{2}\right)|x - x'|\right\} \tag{13.109}$$

because it remains finite due to the negative term $-k\delta/2$.

In consequence, the retarded and advanced Green's function may be defined as

$$G^R(x,x') = \left[E - H_{op} + i\eta\right]^{-1} \text{ with } \eta \to 0^+ \quad (13.110)$$

$$G^A(x,x') = \left[E - H_{op} - i\eta\right]^{-1} \text{ with } \eta \to 0^+ \quad (13.111)$$

13.4 Discussion and Conclusions

The Dirac delta function $\delta(x) = 0$ is a function with continuous independent variable x which is zero everywhere except at $x = 0$. Its integral over all real numbers is 1. It can be shifted to $x = s$ by writing $\delta(x - s)$. It can be thought of as a function on the real line which is zero everywhere except at the origin where it has a very large value. So, it has a non-zero value only at one position. It can also be looked upon as an approximation to a rectangular pulse whose pulse width approaches zero. It is not a function in the orthodox sense because no function defined for real numbers has the properties enshrined in the Dirac delta function. On a rigorous level, it can be thought of as a distribution and construed as a mathematical object to represent idealized abstractions such as a point mass, a point charge, or an impulsive force exerted by the bat on a cricket ball. Hence, it is called an impulse function. The discrete counterpart of the Dirac delta function is the unit impulse function or Kronecker delta, which is a sequence of numbers $\delta[m]$ having value $\delta[m] = 1$ for $m = 0$ and $\delta[m] = 0$ for $m \neq 0$.

Green's function is a technical tool that helps to find a particular solution of linear inhomogeneous differential equations by finding the solution to a forcing function such as an impulse function or Dirac delta function. The solution of the differential equation for the delta function as input is designated as Green's function. Thus the differential equation is

Differential operator (Response function)

= Inhomogeneous source or forcing function   (13.112)

When source or forcing function = Dirac delta function, response function = Green's function; hence

Differential operator (Green's function)

= Dirac delta function as inhomogeneous source

or forcing function   (13.113)

The particular solution for an arbitrary source or forcing function is

Response for arbitrary inhomogeneous source or forcing function

= Convolution of Green's function with source or

forcing function   (13.114)

Thus Green's functions are the building blocks for finding a particular solution because they are solutions to the delta function inputs. The solution is then found by satisfying boundary conditions.

An example of a one-dimensional conductor is considered taking constant potential energy and zero vector potential. When the response is found at any point inside or outside the conductor due to excitation at any other point, we find that there are two solutions satisfying the Green's function equation. One Green's function solution represents outgoing waves from the excitation point. It is named the retarded Green's function. The other Green's function solution represents incoming waves moving toward and disappearing at the excitation point. It is called the advanced Green's function.

The boundary conditions are elegantly included in the Green's function equations by employing the artifice of addition of a small imaginary component to the energy in the wave number. This component is $+i\eta$ for the retarded Green's function and $-i\eta$ for the advanced Green's function. With inclusion of $+i\eta$, the retarded Green's function remains finite and valid on moving away from the excitation point while the advanced Green's function diverges and becomes invalid. Inclusion of $-i\eta$ causes the indefinite growth of the retarded Green's function so that the advanced Green's function is left behind as the only valid solution.

The ideas of retarded and advanced Green's functions need more elaboration. Retarded means 'slow or delayed'. Retarded Green's function $G^R(\mathbf{r}, t; \mathbf{r}', t')$ works from cause to effect; hence it is said to be causal. It describes the outgoing waves in the form of scalar potentials $\Phi(\mathbf{r}, t)$ and vector potentials $\mathbf{A}(\mathbf{r}, t)$ produced at space-time point $(\mathbf{r}, t)$ by charge density $\rho(\mathbf{r}', t')$ and current density $\mathbf{J}(\mathbf{r}', t')$ sources at space-time point $(\mathbf{r}', t')$ such that $(\mathbf{r}, t) > (\mathbf{r}', t')$, i.e. $\Phi(\mathbf{r}, t)$ and $\mathbf{A}(\mathbf{r}, t)$ are generated at space-time point $(\mathbf{r}, t)$ by disturbance in the form of Dirac δ-function at space-time point $(\mathbf{r}', t')$. So, the cause at space-time point $(\mathbf{r}', t')$ precedes the effect at space-time point $(\mathbf{r}, t)$. The potentials are zero for $(\mathbf{r}, t) < (\mathbf{r}', t')$. Thus the retarded Green's function propagates into the future.

The advanced Green's function $G^A(\mathbf{r}, t; \mathbf{r}', t')$ works from effect to cause; hence it is anti-causal. It describes incoming waves in the form of scalar potentials $\Phi(\mathbf{r}, t)$ and vector potentials $\mathbf{A}(\mathbf{r}, t)$ at space-time point $(\mathbf{r}, t)$ converging to disturbance in the form of Dirac δ-function at space-time point $(\mathbf{r}', t')$, i.e. $(\mathbf{r}, t) < (\mathbf{r}', t')$. The potentials are zero for $(\mathbf{r}, t) > (\mathbf{r}', t')$. Thus the advanced Green's function propagates into the past.

An analogy with stretched string further clarifies the difference between the retarded and advanced Green's functions. Initially, the string is at rest and a Dirac δ-function type disturbance is applied causing vibrations in the string. The retarded Green's function $G^R(x, t; x', t')$ gives the vibratory response of the string at space-time point (x, t) to Dirac δ-function type disturbance at space-time point (x', t'), i.e. $(x, t) > (x', t')$. The advanced Green's function $G^A(x, t; x', t')$ gives the receding vibrations of the string toward its initial rest configuration at space-time point (x, t) when Dirac δ-function type disturbance is applied at a space-time point (x', t'), i.e. $(x, t) < (x', t')$.

Generally, the causal retarded Green's function is relevant but the anti-causal advanced Green's function is also of theoretical interest.

Illustrative Exercises

13.1 What are the values of the Dirac delta function $\delta(x)$ of a one-dimensional variable x for $x = 5, 9, 12, 35, 100, 0$?

Values of $\delta(x)$ for $x = 5, 9, 12, 35, 100$ are all zero. But the value of $\delta(x)$ for $x = 0$ is ∞.

13.2 How do you write the delta function to convey the idea that a particle is located at the position $x = 9$?

$$\delta(x-9) = \infty \text{ because } \delta(x-9) = \delta(9-9)$$

$$= \delta(0) = \infty \text{ and } \delta(x) \text{ for } x = 0 \text{ is } \infty \quad (13.115)$$

13.3 (a) Show graphically that the derivative of the Heaviside step function defined as

$$H(x) = 0 \text{ when } x \langle 0, \ H(x) = 1 \text{ when } x \rangle 0 \quad (13.116)$$

is the Dirac delta function.

(b) Prove that the Heaviside step function is obtained when integration is performed on the Dirac delta function.

The derivative of the Heaviside step function is given by

$$\frac{dH(x)}{dx} = \lim_{x \to 0} \frac{\Delta H}{\Delta x} \quad (13.117)$$

It represents the slope or gradient of the curve at a point. Let us examine its values for $x < 0$, $x > 0$, and $x = 0$.

For values of $x > 0$ or $x < 0$, the curve is a flat line with zero slope. Hence, the derivative of the Heaviside step function is zero. Also, in eq. (13.117), H remains constant. So, $\Delta H = 0$; hence

$$\frac{dH(x)}{dx} = 0 \quad (13.118)$$

Let us now consider two points: x_- at a distance $-L/2$ from $x = 0$ and x_+ at a distance $+L/2$ from $x = 0$. Then the distance

$$\Delta x = x_+ - x_- = +\frac{L}{2} - \left(-\frac{L}{2}\right) = \frac{L}{2} + \frac{L}{2} = L \quad (13.119)$$

Since by eq. (13.116), $\Delta H = 1$, from eq. (13.117), the derivative of the Heaviside step function is

$$\frac{dH(x)}{dx} = \lim_{x \to 0} \frac{\Delta H}{\Delta x} = \lim_{L \to 0} \frac{1}{L} = \infty \quad (13.120)$$

Thus

$$\frac{dH}{dx} = 0 \text{ when } x \langle 0, \ \frac{dH}{dx} = 0 \text{ when } x \rangle 0 \text{ and}$$

$$\frac{dH}{dx} = \infty \text{ when } x = 0 \quad (13.121)$$

The function dH/dx has exactly the same properties as the Dirac delta function

$$\delta(x) = 0 \text{ for } x \neq 0 \text{ and } \delta(x) = \infty \text{ for } x = 0 \quad (13.122)$$

So we can assert that the derivative of the Heaviside step function (the unit step function) is the Dirac delta function (the impulse function).

(c) Let us evaluate the integral of the delta function to any point x_+ for which $x > 0$,

$$\int_{-\infty}^{x_+} \delta(x)dx = \int_{-\infty}^{x_+} \frac{dH}{dx} dx$$

$$= \int_{-\infty}^{x_+} dH = |H|_{-\infty}^{x_+} = H(x_+) - H_{-\infty}$$

$$= H(x > 0) - H_{-\infty} = 1 - 0 = 1 \quad (13.123)$$

Thus integration of the delta function to a point $x > 0$ yields a value of unity, which is the value for the Heaviside step function.

REFERENCES

Barton, G. 1999. *Elements of Green's Functions and Propagation: Potentials, Diffusion and Waves.* Oxford University Press: Oxford, pp. 7–40.

Datta, S. 1999. *Electronic Transport in Mesoscopic Systems.* Cambridge University Press: Cambridge, UK, pp. 117–174, 293–344.

Foster, A. and W. A. Hofer. 2006. *Scanning Probe Microscopy: Atomic Scale Engineering by Forces and Currents.* Springer Science +Business Media, LLC: New York, pp. 66–67.

Greenberg, M. D. *Applications of Green's Functions in Science and Engineering.* Dover Publications: Mineola, NY, pp. 11–12.

Royston, A. 2008. Notes on the Dirac Delta and Green Functions, pp. 1–8, http://theory.uchicago.edu/~sethi/Teaching/P221-F2008/DiracandGreenNotes%2808%29.pdf Accessed 8th September 2017.

14

Method of Finite Differences and Self-Energy of the Leads

Discretization is the process of converting continuous variables, functions, and equations into distinctly disconnected numerical data that can be fed into digital computers as numbers. Thus it is the first step toward making the equations amenable to digital processing for implementation on computers.

14.1 Discretization Methods

Calculation of the Green's function for an arbitrarily shaped conductor is done by solving the differential equation for the Green's function for arbitrary potential energy $U(\mathbf{r})$ and vector potential $\mathbf{A}(\mathbf{r})$ (Datta 1999):

$$\left[E - H_{op} + i\eta\right]G^R(\mathbf{r},\mathbf{r}') = \delta(\mathbf{r}-\mathbf{r}') \tag{14.1}$$

where

$$H_{op}(\mathbf{r}) = \frac{(i\hbar\nabla + e\mathbf{A})^2}{2m} + U(\mathbf{r}) \tag{14.2}$$

A useful technique to solve this equation consists in discretizing the spatial coordinate. Then the Green's function becomes a matrix as

$$G^R(\mathbf{r},\mathbf{r}') \Rightarrow G^R(i,j) \tag{14.3}$$

where the indices i, j represent points on a discrete lattice.

The differential equation for the Green's function becomes a matrix equation

$$\left[(E+i\eta)I - H\right]G^R = I \tag{14.4}$$

where $[I]$ is the identity matrix and $[H]$ is the matrix representation of H_{op}. The Green's function G^R is evaluated by numerical inversion of the matrix as

$$G^R = \left[(E+i\eta)I - H\right]^{-1} \tag{14.5}$$

14.2 One-Dimensional Matrix Representation of the Hamiltonian Operator

Let us consider a discrete lattice in which the points are located at $x = ja$ where a is the spacing between points and j is an integer (Figure 14.1). The Hamiltonian operator for a function $F(x)$ of x is written as

$$\left(H_{op}F\right)_{x=ja} = \left(-\frac{\hbar^2}{2m}\frac{\partial^2 F}{\partial x^2}\right)_{x=ja} + U_{x=ja}F_{x=ja} \tag{14.6}$$

The derivative operators are approximated according to the finite difference method (FDM), which is a numerical method for solving differential equations by representing derivatives with differences. By such discretization, the differential equations are transformed into algebraic equations solvable by matrix algebra.

The first derivative of $F(x)$ with respect to x is

$$\left(\frac{\partial F}{\partial x}\right)_{x=\left(j+\frac{1}{2}\right)a} = \frac{1}{a}\left\{F_{x=(j+1)a} - F_{x=ja}\right\} \tag{14.7}$$

The second derivative of $F(x)$ with respect to x is

$$\begin{aligned}
\left(\frac{\partial^2 F}{\partial x^2}\right)_{x=ja} &= \frac{1}{a}\left\{\left(\frac{\partial F}{\partial x}\right)_{x=\left(j+\frac{1}{2}\right)a} - \left(\frac{\partial F}{\partial x}\right)_{x=\left(j-\frac{1}{2}\right)a}\right\} \\
&= \frac{1}{a}\left[\frac{1}{a}\left\{F_{x=(j+1)a} - F_{x=ja}\right\} - \frac{1}{a}\left\{F_{x=ja} - F_{x=(j-1)a}\right\}\right] \\
&= \frac{1}{a^2}\left[\left\{F_{x=(j+1)a} - F_{x=ja}\right\} - \left\{F_{x=ja} - F_{x=(j-1)a}\right\}\right] \\
&= \frac{1}{a^2}\left\{F_{x=(j+1)a} - 2F_{x=ja} + F_{x=(j-1)a}\right\}
\end{aligned} \tag{14.8}$$

With the help of eq. (14.8), eq. (14.6) for the Hamiltonian operator can be written as

$$\begin{aligned}
\left(H_{op}F\right)_{x=ja} &= -\frac{\hbar^2}{2m}\left(\frac{\partial^2 F}{\partial x^2}\right)_{x=ja} + U_{x=ja}F_{x=ja} \\
&= -\frac{\hbar^2}{2m} \times \frac{1}{a^2}\left\{F_{x=(j+1)a} - 2F_{x=ja} + F_{x=(j-1)a}\right\} + U_{x=ja}F_{x=ja} \\
&= -\frac{\hbar^2}{2ma^2}F_{x=(j+1)a} + \frac{2\hbar^2}{2ma^2}F_{x=ja} - \frac{\hbar^2}{2ma^2}F_{x=(j-1)a} \\
&\quad + U_{x=ja}F_{x=ja} \\
&= -tF_{x=(j+1)a} + 2tF_{x=ja} - tF_{x=(j-1)a} + U_{x=ja}F_{x=ja} \\
&= U_{x=ja}F_{x=ja} + 2tF_{x=ja} - tF_{x=(j-1)a} - tF_{x=(j+1)a} \\
&= \left(U_{x=ja} + 2t\right)F_{x=ja} - tF_{x=(j-1)a} - tF_{x=(j+1)a}
\end{aligned} \tag{14.9}$$

where

$$t = \frac{\hbar^2}{2ma^2} \tag{14.10}$$

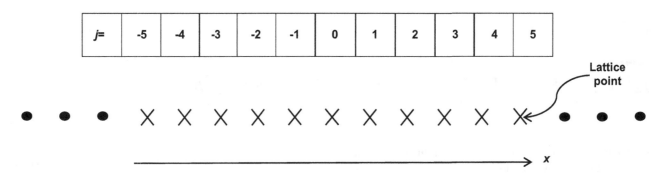

FIGURE 14.1 Discretization of an infinite linear chain into a lattice with points represented by crosses located at $x = ja$ where j acquires integral values only.

To write the matrix for the Hamiltonian operator, eq. (14.9) is written in the form

$$\left(H_{op}F\right)_{x=ja} = \sum_i H_{ji}F_i \tag{14.11}$$

where

$$H_{ji} = U_i + 2t \text{ when } i = j \tag{14.12}$$

$$H_{ji} = -t \text{ when } i \text{ and } j \text{ are nearest neighbors} \tag{14.13}$$

and

$$H_{ji} = 0 \text{ when } i \neq j \text{ nor } i \text{ and } j \text{ are nearest neighbors} \tag{14.14}$$

On these guidelines of placing (potential energy $+2t$) as diagonal elements and linking of nearest neighbors through $-t$, the Hamiltonian operator is written in matrix form as

$$\left[H_{op}\right] = \begin{bmatrix} \cdots & -t & 0 & 0 & 0 \\ -t & U_{-1}+2t & -t & 0 & 0 \\ 0 & -t & U_0+2t & -t & 0 \\ 0 & 0 & -t & U_1+2t & -t \\ 0 & 0 & 0 & -t & \cdots \end{bmatrix}$$

$$\tag{14.15}$$

14.3 Dispersion Relation and Velocity for a Discrete Lattice

The Schrodinger equation for a discrete wire can be expressed as

$$\left(U_0+2t\right)\Psi_{x=ja} - t\Psi_{x=(j-1)a} - t\Psi_{x=(j+1)a} = E\Psi_{x=ja} \tag{14.16}$$

by replacing F with Ψ in eq. (14.9).

Equation (14.16) is satisfied by the solution

$$\Psi_{x=ja} = \exp\left(ikx\right)_{x=ja} \tag{14.17}$$

when

$$\left(U_0+2t\right)\exp(ikx)_{x=ja} - t\left\{\exp(ikx)_{x=(j-1)a}\right\} - t\left\{\exp(ikx)_{x=(j+1)a}\right\}$$

$$= E\exp(ikx)_{x=ja} \tag{14.18}$$

or,

$$\left(U_0+2t\right)\exp(ikja) - t\exp\left\{ik(j-1)a\right\} - t\exp\left\{ik(j+1)a\right\}$$

$$= E\exp(ikja) \tag{14.19}$$

Dividing both sides of eq. (14.19) by $\exp(ikja)$,

$$\frac{\left(U_0+2t\right)\exp(ikja)}{\exp(ikja)} - \frac{t\exp\left\{ik(j-1)a\right\} + t\exp\left\{ik(j+1)a\right\}}{\exp(ikja)}$$

$$= \frac{E\exp(ikja)}{\exp(ikja)} \tag{14.20}$$

or

$$\left(U_0+2t\right) - \frac{t\exp\left\{ik(j-1)a\right\}}{\exp(ikja)} - \frac{t\exp\left\{ik(j+1)a\right\}}{\exp(ikja)} = E$$

$$\tag{14.21}$$

or

$$\left(U_0+2t\right) - t\exp\left\{ik(j-1)a - ikja\right\}$$

$$- t\exp\left\{ik(j+1)a - ikja\right\} = E \tag{14.22}$$

or

$$\left(U_0+2t\right) - t\exp(-ika) - t\exp(+ika) = E \tag{14.23}$$

or

$$(U_0 + 2t) - t\{\exp(-ika) + \exp(+ika)\} = E \qquad (14.24)$$

or

$$(U_0 + 2t) - 2t\left\{\frac{\exp(-ika) + \exp(+ika)}{2}\right\} = E \qquad (14.25)$$

or

$$(U_0 + 2t) - 2t\cos(ka) = E \qquad (14.26)$$

or

$$U_0 + 2t\{1 - \cos(ka)\} = E \qquad (14.27)$$

which is the dispersion relation for the discrete lattice. From eqs. (9.138) and (14.27), the electron velocity is

$$v = \frac{1}{\hbar}\frac{\partial E}{\partial k} = \frac{1}{\hbar}\frac{\partial[U_0 + 2t\{1 - \cos(ka)\}]}{\partial k} = \frac{2t}{\hbar}\frac{\partial}{\partial k}\{1 - \cos(ka)\}$$

$$= \frac{2t}{\hbar}\left[0 - \{-\sin(ka)\} \times a\right] = \frac{2at}{\hbar}\sin(ka) \qquad (14.28)$$

Hence we can write

$$\hbar v_m = 2at\sin(k_m a) \qquad (14.29)$$

14.4 Two-Dimensional Matrix Representation of the Hamiltonian Operator

By extension of eq. (14.11) to two dimensions, the elements of matrix $[H]$ are given by the equations

$$[H]_{ij} = U(\mathbf{r_i}) + zt \quad \text{if} \quad i = j \qquad (14.30)$$

$$[H]_{ij} = -\bar{t}_{ij} \quad \text{if} \quad i, j \text{ are closest neighbors} \qquad (14.31)$$

and

$$[H]_{ij} = 0 \text{ otherwise} \qquad (14.32)$$

where

$$\bar{t}_{ij} = t\exp\left\{\frac{ie\mathbf{A}(\mathbf{r_i} - \mathbf{r_j})}{\hbar}\right\} \qquad (14.33)$$

Here z is the number of closest neighbors; $\mathbf{r_i}, \mathbf{r_j}$ are the position vectors for lattice sites i, j and

$$\hbar = \frac{h}{2\pi} \qquad (14.34)$$

Equation (14.33) is explained by the fact that in a magnetic field, the amplitude of a particle's motion along a particular route = its amplitude along the same path in absence of the field $\times$ exp (the line integral of $\frac{ie}{\hbar} \times$ vector potential $\mathbf{A}$). So,

$$\langle \mathbf{r_i} \mid \mathbf{r_j} \rangle_{A \neq 0} = \langle \mathbf{r_i} \mid \mathbf{r_j} \rangle_{A=0} \exp\left(\int_{\mathbf{r_j}}^{\mathbf{r_i}} \left(\frac{ie}{\hbar}\right)\mathbf{A}.d\mathbf{l}\right)$$

$$= \langle \mathbf{r_i} \mid \mathbf{r_j} \rangle_{A=0} \exp\left\{\left(\frac{ie}{\hbar}\right)\mathbf{A}(\mathbf{r_i} - \mathbf{r_j})\right\} \qquad (14.35)$$

The vector $\mathbf{A}$ is evaluated at the point

$$\mathbf{r} = \frac{\mathbf{r_i} + \mathbf{r_j}}{2} \qquad (14.36)$$

If $\mathbf{A} = 0$,

$$\bar{t}_{ij} = t\exp\left\{\frac{ie \times 0 \times (\mathbf{r_i} - \mathbf{r_j})}{\hbar}\right\} = t\exp(0) = t \qquad (14.37)$$

as for one-dimensional matrix representation.

14.5 Matrix Truncation

Because the system under study is an open system with non-reflecting boundaries connected to leads extending up to infinity, the

$$\text{Matrix} = \left[(E + i\eta)I - H\right] \qquad (14.38)$$

to be inverted as

$$G^{\mathrm{R}} = \left[(E + i\eta)I - H\right]^{-1} \qquad (14.39)$$

for evaluating the retarded Green's function which is infinite-dimensional. For a closed system with fully reflecting boundaries, the matrix must be truncated at some point. Let us consider a conductor described by a Hamiltonian H_C (Figure 14.2). The lead p connected to this conductor is described by H_p. The point p_i is a point in the lead p adjacent to the point i inside the conductor. The coupling is done through the matrix τ_p.

Representing the isolated conductor matrix (G_C) by

$$[EI - H_C] \qquad (14.40)$$

with the isolated lead matrix (G_p) by

$$\left[(E + i\eta)I - H_p\right] \qquad (14.41)$$

and noting that the coupling matrix (G_{pC} or G_{Cp}) is τ_p, partitioning of the Green's function is performed into sub-matrices as

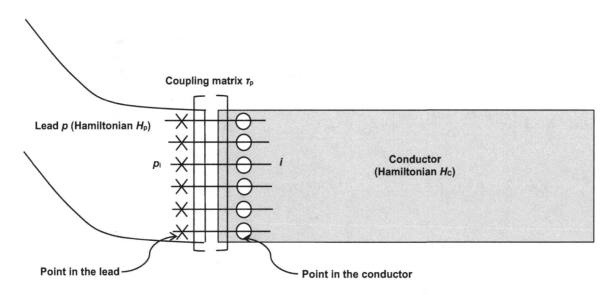

FIGURE 14.2 Connection of a conductor (Hamiltonian H_C) to a lead p (Hamiltonian H_p) through a coupling matrix τ_p with the point in lead adjoining the point i in the conductor labeled as p_i.

$$\begin{bmatrix} G_p & G_{pC} \\ G_{Cp} & G_C \end{bmatrix} = \begin{bmatrix} (E+i\eta)I - H_p & \tau_p \\ \tau_p^+ & EI - H_C \end{bmatrix}^{-1} = \frac{1}{[(E+i\eta)I - H_p][EI - H_C] - \tau_p^+\tau_p} \begin{bmatrix} EI - H_C & -\tau_p \\ -\tau_p^+ & (E+i\eta)I - H_p \end{bmatrix}$$

$$= \begin{bmatrix} \dfrac{EI - H_C}{[(E+i\eta)I - H_p][EI - H_C] - \tau_p^+\tau_p} & \dfrac{-\tau_p}{[(E+i\eta)I - H_p][EI - H_C] - \tau_p^+\tau_p} \\[4mm] \dfrac{-\tau_p^+}{[(E+i\eta)I - H_p][EI - H_C] - \tau_p^+\tau_p} & \dfrac{(E+i\eta)I - H_p}{[(E+i\eta)I - H_p][EI - H_C] - \tau_p^+\tau_p} \end{bmatrix} \quad (14.42)$$

Equation (14.42) is obtained by applying the rule for inversion of a 2×2 matrix (Stover et al. 1999–2019):

$$\begin{bmatrix} a & b \\ c & d \end{bmatrix}^{-1} = \frac{1}{ad - bc} \begin{bmatrix} d & -b \\ -c & a \end{bmatrix} \quad (14.43)$$

by swapping the positions of a and d, affixing minus sign before b and c, and dividing by the determinant ($ad - bc$). Also the rule for scalar multiplication of matrix is followed. Every entry or element in the matrix is multiplied by the given scalar.

There are non-zero values of the coupling matrix only for adjacent points i, p_i.

$$\tau_p(p_i, i) = t \quad (14.44)$$

By comparing the elements of the matrices on the left- and right-hand sides of eq. (14.42), one can write

$$G_{pC} = \frac{-\tau_p}{[(E+i\eta)I - H_p][EI - H_C] - \tau_p^+\tau_p} \quad (14.45)$$

and

$$G_C = \frac{(E+i\eta)I - H_p}{[(E+i\eta)I - H_p][EI - H_C] - \tau_p^+\tau_p} \quad (14.46)$$

Hence,

$$[(E+i\eta)I - H_p]G_{pC} + [\tau_p]G_C = [(E+i\eta)I - H_p]$$

$$\times \frac{-\tau_p}{[(E+i\eta)I - H_p][EI - H_C] - \tau_p^+\tau_p} + \tau_p$$

$$\times \frac{(E+i\eta)I - H_p}{[(E+i\eta)I - H_p][EI - H_C] - \tau_p^+\tau_p} = 0 \quad (14.47)$$

i.e.,

$$[(E+i\eta)I - H_p]G_{pC} + [\tau_p]G_C = 0 \quad (14.48)$$

This equation is obtained simply by putting the values of G_{pC} and G_C from eqs. (14.45) and (14.46) into the expression on the left-hand side of eq. (14.47).

Also, by inputting the values of G_{pC} and G_C from eqs. (14.45) and (14.46), we move through the steps below:

$$[EI - H_C]G_C + [\tau_p^+]G_{pC} = [EI - H_C]$$

$$\times \frac{\left[(E+i\eta)I - H_p\right]}{\left[(E+i\eta)I - H_p\right]\left[EI - H_C\right] - \tau_p^+\tau_p} + \tau_p^+$$

$$\times \frac{(-\tau_p)}{\left[(E+i\eta)I - H_p\right]\left[EI - H_C\right] - \tau_p^+\tau_p}$$

$$= \frac{\left[EI - H_C\right]\left[(E+i\eta)I - H_p\right]}{\left[(E+i\eta)I - H_p\right]\left[EI - H_C\right] - \tau_p^+\tau_p}$$

$$- \frac{\tau_p^+ \times \tau_p}{\left[(E+i\eta)I - H_p\right]\left[EI - H_C\right] - \tau_p^+\tau_p}$$

$$= \frac{\left[EI - H_C\right]\left[(E+i\eta)I - H_p\right] - \tau_p^+ \times \tau_p}{\left[(E+i\eta)I - H_p\right]\left[EI - H_C\right] - \tau_p^+\tau_p} = [1]$$

$$= \begin{bmatrix} 1 & 0 \\ 0 & 1 \end{bmatrix} = I \tag{14.49}$$

i.e.,

$$(EI - H_C)G_C + [\tau_p^+]G_{pC} = I \tag{14.50}$$

Equation (14.48) helps to find G_{pC}

$$\left[(E+i\eta)I - H_p\right]G_{pC} = -\tau_p G_C \tag{14.51}$$

or

$$G_{pC} = -\tau_p G_C \left[(E+i\eta)I - H_p\right]^{-1} \tag{14.52}$$

This equation may be written as

$$G_{pC} = -\tau_p G_C \, g_p^R \tag{14.53}$$

by putting

$$\left[(E+i\eta)I - H_p\right]^{-1} = g_p^R$$

= Green's function for the isolated semi-infinite lead

$$\tag{14.54}$$

where g_p^R is determined analytically.

Substituting for G_{pC} from eq. (14.53) into eq. (14.50)

$$(EI - H_C)G_C + \tau_p^+ \times -\tau_p G_C \, g_p^R = I \tag{14.55}$$

or

$$(EI - H_C)G_C - \tau_p^+ g_p^R \tau_p G_C = I \tag{14.56}$$

or

$$\left[EI - H_C - \tau_p^+ g_p^R \tau_p\right]G_C = I \tag{14.57}$$

$$\therefore G_C = \left[EI - H_C - \tau_p^+ g_p^R \tau_p\right]^{-1} \tag{14.58}$$

The matrices in this equation are finite matrices of dimension $(C \times C)$. The symbol C stands for the number of points inside the conductor. Recalling that the coupling matrix τ_p is zero everywhere inside the conductor with the exception of points p_i, p_j in the lead nearby the points i, j inside the conductor, from eq. (14.44) we have

$$\left[\tau_p^+ g_p^R \tau_p\right]_{i,j} = (t) g_p^R (t) = t^2 g_p^R (p_i, p_j) \tag{14.59}$$

So, from eq. (14.58)

$$G_C = \left[EI - H_C - t^2 g_p^R (p_i, p_j)\right]^{-1} \tag{14.60}$$

If the leads are taken as independent, their effects can be added. Then

$$G_C = \left[EI - H_C - \sum_p t^2 g_p^R (p_i, p_j)\right]^{-1} = \left[EI - H_C - \sum_p \Sigma_p^R (i, j)\right]^{-1}$$

$$= \left[EI - H_C - \Sigma^R\right]^{-1} \tag{14.61}$$

where

$$\Sigma_p^R (i, j) = t^2 g_p^R (p_i, p_j) \tag{14.62}$$

and

$$\Sigma^R = \sum_p \Sigma_p^R (i, j) = \sum_p t^2 g_p^R (p_i, p_j) \tag{14.63}$$

Replacing the symbol G_C by G^R, we get

$$G^R = \left[EI - H_C - \Sigma^R\right]^{-1} \tag{14.64}$$

This Green's function is a statement of electron propagation between two points inside a conductor by taking the effects of leads into consideration. The term accounting for the leads is Σ^R. The term can be looked upon as an effective Hamiltonian which is born through the interaction of the conductor with the leads. An identical term is used for description of electron-phonon and electron-electron interactions. This term is called self-energy. By analogy, Σ^R is the self-energy arising from the leads. Since the leads are simple inert objects, a precise description of self-energy from the leads is provided by Σ^R. Similar remarks are not applicable to self-energy arising from interactions of electrons with phonons or electrons. That self-energy represents only an approximate description.

The question may be asked: What is the self-energy for electrons moving through a material? It is the potential acting on an electron owing to its interactions with the surrounding medium. To give an example, an electron repels other proximate electrons during the course of its motion. This, in turn, alters the potential felt by the moving electron itself. Such kind of effect that an electron produces in its environment is clustered together under the broad term 'self energy'. Thus self-energy of an electron is the contribution of the energy of the electron that originates from the interaction of the electron with the system, of which it is a part.

14.6 Self-Energy due to the Leads

To find the self-energy Σ^R from eq. (14.63), we must determine the Green's function g_p^R for an isolated lead of length L, i.e. due to a wire that terminates on one side at $x = 0$ in an infinite potential at which the wave function $\Psi(x, y) = 0$. A confining potential $U(y)$ is applied in the Y-direction. The Schrodinger equation is

$$\left\{ -\frac{\hbar^2}{2m}\frac{\partial^2}{\partial x^2} - \frac{\hbar^2}{2m}\frac{\partial^2}{\partial y^2} + U(y) \right\} \Psi(x, y) = 0 \quad (14.65)$$

Since the Green's function is obtained by forming a linear combination of wave functions, we find the wave function for the wire.

14.6.1 Wave Function of a Wire Terminating on One Side

The wave function is written as

$$\Psi(x, y) = \Psi(x)\Psi(y) \quad (14.66)$$

Then from eq. (14.65)

$$-\frac{\hbar^2}{2m}\frac{\partial^2\{\Psi(x)\Psi(y)\}}{\partial x^2} - \frac{\hbar^2}{2m}\frac{\partial^2\{\Psi(x)\Psi(y)\}}{\partial y^2}$$
$$+ U(y)\Psi(x)\Psi(y) = 0 \quad (14.67)$$

or

$$-\frac{\hbar^2}{2m}\Psi(y)\frac{\partial^2\{\Psi(x)\}}{\partial x^2} - \frac{\hbar^2}{2m}\Psi(x)\frac{\partial^2\{\Psi(y)\}}{\partial y^2}$$
$$+ U(y)\Psi(x)\Psi(y) = 0 \quad (14.68)$$

Dividing both sides of eq. (14.68) by $\Psi(x)\Psi(y)$, we get

$$-\frac{\hbar^2}{2m}\frac{1}{\Psi(x)}\frac{\partial^2\{\Psi(x)\}}{\partial x^2} - \frac{\hbar^2}{2m}\frac{1}{\Psi(y)}\frac{\partial^2\{\Psi(y)\}}{\partial y^2} + U(y) = 0$$
$$(14.69)$$

or

$$-\frac{\hbar^2}{2m}\frac{1}{\Psi(x)}\frac{\partial^2\{\Psi(x)\}}{\partial x^2} = -\left[-\frac{\hbar^2}{2m}\frac{1}{\Psi(y)}\frac{\partial^2\{\Psi(y)\}}{\partial y^2} + U(y) \right]$$
$$(14.70)$$

or

Function of x only = Function of y only

$$= \text{Constant} = \beta^2 \text{ (suppose)} \quad (14.71)$$

Hence,

$$-\frac{\hbar^2}{2m}\frac{1}{\Psi(x)}\frac{\partial^2\{\Psi(x)\}}{\partial x^2} = \beta^2 \quad (14.72)$$

From Section 5.2 for particle-in-a box problem (eqs. (5.76), (5.97)),

$$\Psi(x) = \sqrt{\frac{2}{L}}\sin(\beta x) \quad (14.73)$$

From eqs. (14.73) and (14.66),

$$\Psi(x, y) = \sqrt{\frac{2}{L}}\Psi(y)\sin(\beta x) \quad (14.74)$$

where $\Psi(y)$ is the solution of

$$-\left[-\frac{\hbar^2}{2m}\frac{1}{\Psi(y)}\frac{\partial^2\{\Psi(y)\}}{\partial y^2} + U(y) \right] = \beta^2 \quad (14.75)$$

or

$$-\frac{\hbar^2}{2m}\frac{\partial^2\{\Psi(y)\}}{\partial y^2} + U(y)\Psi(y) = -\beta^2\Psi(y) = \varepsilon_{m,0}\Psi(y)$$
$$(14.76)$$

enabling us to write the wavefunction as

$$\Psi_{m,\beta}(x) = \Psi(x)\Psi(y) = \sqrt{\frac{2}{L}}\Psi(y)\sin(\beta x) \quad (14.77)$$

and

$$\varepsilon_{m,\beta} = \varepsilon_{m,0} + \frac{\hbar^2\beta^2}{2m} \quad (14.78)$$

14.6.2 Wave Function Expansion

Wave functions with different eigenvalues are said to be orthogonal. Orthogonality may be stated as

$$\int_{-\infty}^{+\infty} \Psi_\beta^*(\mathbf{r})\Psi_\alpha(\mathbf{r}) = 0 \text{ for } \beta \neq \alpha \quad (14.79)$$

meaning that two members of a set of wave functions complying with the earlier integral constraint constitute an orthogonal set of wave functions. In addition, if each member of the set of wave functions is normalized, the wave functions constitute an orthonormal set. For normalization,

$$\int_{-\infty}^{+\infty} \Psi_\beta^*(\mathbf{r})\Psi_\alpha(\mathbf{r}) = 1 \text{ for } \beta = \alpha \quad (14.80)$$

Both the conditions given in eqs. (14.79) and (14.80) can be compactly written as

$$\int_{-\infty}^{+\infty} \Psi_\beta^*(\mathbf{r})\Psi_\alpha(\mathbf{r}) = \delta_{\beta\alpha} \quad (14.81)$$

where the Kronecker delta function

$$\delta_{\beta\alpha} = 0 \text{ if } \beta \neq \alpha \text{ and } \delta_{\beta\alpha} = 1 \text{ if } \beta = \alpha \quad (14.82)$$

Thus orthornormality embodies both orthogonality and normalization. By the principle of superposition, the Green's function can be expanded as a linear combination of wave functions

$$G^R\left(\mathbf{r},\mathbf{r}'\right) = \sum_\alpha C_\alpha\left(\mathbf{r}'\right)\Psi_\alpha\left(\mathbf{r}\right) \tag{14.83}$$

where the coefficients $C_\alpha\left(\mathbf{r}'\right)$ are to be determined.

The Green's function equation is

$$\left(E - H_{op} + i\eta\right)G^R\left(\mathbf{r},\mathbf{r}'\right) = \delta\left(\mathbf{r}-\mathbf{r}'\right) \tag{14.84}$$

where $\delta(\mathbf{r}-\mathbf{r}')$ is the Dirac delta function (Hoskings 2009).

Substituting the expanded form of $G^R(\mathbf{r},\mathbf{r}')$ in eq. (14.83) into the Green's function in eq. (14.84) we get

$$\left(E - H_{op} + i\eta\right)\sum_\alpha C_\alpha\left(\mathbf{r}'\right)\Psi_\alpha\left(\mathbf{r}\right) = \delta\left(\mathbf{r}-\mathbf{r}'\right) \tag{14.85}$$

or

$$\sum_\alpha\left(E - H_{op} + i\eta\right)\{C_\alpha\left(\mathbf{r}'\right)\Psi_\alpha\left(\mathbf{r}\right)\} = \delta\left(\mathbf{r}-\mathbf{r}'\right) \tag{14.86}$$

or

$$\sum_\alpha\left(E\Psi_\alpha\left(\mathbf{r}\right) - H_{op}\Psi_\alpha\left(\mathbf{r}\right) + i\eta\Psi_\alpha\left(\mathbf{r}\right)\right)C_\alpha\left(\mathbf{r}'\right) = \delta\left(\mathbf{r}-\mathbf{r}'\right) \tag{14.87}$$

or

$$\sum_\alpha\left[E\Psi_\alpha\left(\mathbf{r}\right) - \varepsilon_\alpha\Psi_\alpha\left(\mathbf{r}\right) + i\eta\Psi_\alpha\left(\mathbf{r}\right)\right]C_\alpha\left(\mathbf{r}'\right) = \delta\left(\mathbf{r}-\mathbf{r}'\right) \tag{14.88}$$

because

$$H_{op}\Psi_\alpha\left(\mathbf{r}\right) = \varepsilon_\alpha\Psi_\alpha\left(\mathbf{r}\right) \tag{14.89}$$

and H_{op} acts only on $\mathbf{r}$, not on $\mathbf{r}'$; ε_α are the eigenvalues. Eq. (14.88) may be written as

$$\sum_\alpha\left(E - \varepsilon_\alpha + i\eta\right)C_\alpha\left(\mathbf{r}'\right)\Psi_\alpha\left(\mathbf{r}\right) = \delta\left(\mathbf{r}-\mathbf{r}'\right) \tag{14.90}$$

Multiplying both sides of eq. (14.90) by $\Psi_\alpha^*\left(\mathbf{r}\right)$, we have

$$\sum_\alpha\left(E - \varepsilon_\alpha + i\eta\right)C_\alpha\left(\mathbf{r}'\right)\Psi_\alpha\left(\mathbf{r}\right)\Psi_\alpha^*\left(\mathbf{r}\right) = \delta\left(\mathbf{r}-\mathbf{r}'\right)\Psi_\alpha^*\left(\mathbf{r}\right) \tag{14.91}$$

Integrating over $\mathbf{r}$

$$\int_{-\infty}^{+\infty}\sum_\alpha\left(E - \varepsilon_\alpha + i\eta\right)C_\alpha\left(\mathbf{r}'\right)\Psi_\alpha\left(\mathbf{r}\right)\Psi_\alpha^*\left(\mathbf{r}\right)d\mathbf{r}$$

$$= \int_{-\infty}^{+\infty}\Psi_\alpha^*\left(\mathbf{r}\right)\delta\left(\mathbf{r}-\mathbf{r}'\right)d\mathbf{r} \tag{14.92}$$

or

$$\sum_\alpha\left(E - \varepsilon_\alpha + i\eta\right)C_\alpha\left(\mathbf{r}'\right)\int_{-\infty}^{+\infty}\Psi_\alpha\left(\mathbf{r}\right)\Psi_\alpha^*\left(\mathbf{r}\right)d\mathbf{r}$$

$$= \int_{-\infty}^{+\infty}\Psi_\alpha^*\left(\mathbf{r}\right)\delta\left(\mathbf{r}-\mathbf{r}'\right)d\mathbf{r} \tag{14.93}$$

or

$$\sum_\alpha\left(E - \varepsilon_\alpha + i\eta\right)C_\alpha\left(\mathbf{r}'\right)\times 1 = \Psi_\alpha^*\left(\mathbf{r}'\right) \tag{14.94}$$

by applying eqs. (14.80) and (13.5). From eq. (14.94), pulling out the constants from summation sign, we can write

$$C_\alpha\left(\mathbf{r}'\right) = \frac{\Psi_\alpha^*\left(\mathbf{r}'\right)}{E - \varepsilon_\alpha + i\eta} \tag{14.95}$$

14.6.3 Green's Function of the Wire

Substituting for the coefficient $C_\alpha(\mathbf{r}')$ from eq. (14.95) in eq. (14.83), the Green's function is expressed as the summation

$$G^R\left(\mathbf{r},\mathbf{r}'\right) = \sum_\alpha\frac{\Psi_\alpha\left(\mathbf{r}\right)\Psi_\alpha^*\left(\mathbf{r}'\right)}{E - \varepsilon_\alpha + i\eta} \tag{14.96}$$

The Green's function $G^R(x, y; x, y')$ between two points having the same x-coordinate is required. So we set $x = x'$ to get from eq. (14.74)

$$\Psi_\alpha\left(x, y\right) = \sqrt{\frac{2}{L}}\Psi_m\left(y\right)\sin\left(\beta x\right) \tag{14.97}$$

and

$$\Psi_\alpha\left(x, y'\right) = \sqrt{\frac{2}{L}}\Psi_m\left(y'\right)\sin\left(\beta x\right) = \Psi_\alpha^*\left(x, y'\right) \tag{14.98}$$

because complex conjugate of a real number is the number itself. From eq. (14.78)

$$\varepsilon_\alpha = \varepsilon_{m,0} + \frac{\hbar^2\beta^2}{2m} \tag{14.99}$$

Combining eqs. (14.96), (14.97), (14.98), and (14.99)

$$\begin{aligned}G^R\left(x, y; x', y'\right) &= \sum_\alpha\frac{\sqrt{\frac{2}{L}}\Psi_m\left(y\right)\sin\left(\beta x\right)\sqrt{\frac{2}{L}}\Psi_m\left(y'\right)\sin\left(\beta x\right)}{E - \varepsilon_{m,0} - \frac{\hbar^2\beta^2}{2m} + i\eta}\\ &= \sqrt{\frac{2}{L}}\times\sqrt{\frac{2}{L}}\sum_m\sum_{\beta>0}\frac{\Psi_m\left(y\right)\sin\left(\beta x\right)\Psi_m\left(y'\right)\sin\left(\beta x\right)}{E - \varepsilon_{m,0} - \frac{\hbar^2\beta^2}{2m} + i\eta}\\ &= \frac{2}{L}\sum_m\sum_{\beta>0}\frac{\Psi_m\left(y\right)\Psi_m\left(y'\right)\sin^2\left(\beta x\right)}{E - \varepsilon_{m,0} - \frac{\hbar^2\beta^2}{2m} + i\eta}\end{aligned} \tag{14.100}$$

Since the number of allowed states in one dimension in the range β to $\beta+d\beta$ is

$$N_{1D}d\beta = 2\left(\frac{L}{2\pi}\right)d\beta \tag{14.101}$$

with factor 2 accounting for the spin, the total number of filled states is

$$N = \int_{-\infty}^{+\infty} N_{1D} \times \text{Distribution function} \times d\beta$$

$$= \int_{-\infty}^{+\infty} 2\left(\frac{L}{2\pi}\right) \times \text{Distribution function} \times d\beta$$

$$= \int_{-\infty}^{+\infty} \left(\frac{L}{\pi}\right) \times \text{Distribution function} \times d\beta \quad (14.102)$$

Hence, the summation over β in eq. (4.100) can be replaced with integration to give

$$G^R(x,y;x',y') = \frac{2}{L} \times \frac{L}{\pi} \sum_m \int_{-\infty}^{\infty} \frac{\Psi_m(y)\Psi_m(y')\sin^2(\beta x)}{E - \varepsilon_{m,0} - \dfrac{\hbar^2\beta^2}{2m} + i\eta} d\beta$$

$$= \frac{2}{\pi} \sum_m \int_{-\infty}^{\infty} \frac{\Psi_m(y)\Psi_m(y')\sin^2(\beta x)}{E - \varepsilon_{m,0} - \dfrac{\hbar^2\beta^2}{2m} + i\eta} d\beta$$

$$= \frac{2}{\pi} \sum_m \Psi_m(y)\Psi_m(y') \int_{-\infty}^{\infty} \frac{\sin^2(\beta x)}{E - \varepsilon_{m,0} - \dfrac{\hbar^2\beta^2}{2m} + i\eta} d\beta$$

$$(14.103)$$

But

$$\sin^2(\beta x) = \left\{ \frac{\exp(i\beta x) - \exp(-i\beta x)}{2i} \right\}^2$$

$$= \frac{\exp(2i\beta x) + \exp(-2i\beta x) - 2}{4i^2} = \frac{\exp(2i\beta x) + \exp(-2i\beta x) - 2}{4 \times -1}$$

$$= -\frac{\exp(2i\beta x) + \exp(-2i\beta x) - 2}{4} = \frac{2 - \exp(2i\beta x) - \exp(-2i\beta x)}{4}$$

$$(14.104)$$

For the retarded Green's function (equation (13.97)), for outgoing waves we take only the positive exponential function in equation (14.104) and decay the negative exponential function for incoming waves to unity (exp (-0)), obtaining

$$\sin^2(\beta x) = \frac{2 - \exp(2i\beta x) - 1}{4} = \frac{1 - \exp(2i\beta x)}{4} \quad (14.105)$$

Substituting for $\sin^2(\beta x)$ from eq. (14.105) in eq. (14.103)

$$G^R(x,y;x',y') = \frac{2}{\pi} \times \frac{1}{4} \sum_m \Psi_m(y)\Psi_m(y') \int_{-\infty}^{+\infty} \frac{\{1 - \exp(2i\beta x)\}}{E - \varepsilon_{m,0} - \dfrac{\hbar^2\beta^2}{2m} + i\eta} d\beta$$

$$= \frac{1}{2\pi} \sum_m \Psi_m(y)\Psi_m(y') \int_{-\infty}^{+\infty} \frac{\{1 - \exp(2i\beta x)\}}{E - \varepsilon_{m,0} - \dfrac{\hbar^2\beta^2}{2m} + i\eta} d\beta$$

$$(14.106)$$

14.6.4 Contour Integration

The denominator of the integral in eq. (14.106) is

$$E - \varepsilon_{m,0} - \frac{\hbar^2\beta^2}{2m} + i\eta = -\frac{\hbar^2\beta^2}{2m} + E - \varepsilon_{m,0} + i\eta$$

$$= -\frac{\hbar^2}{2m}\left\{ \beta^2 - \frac{2m(E - \varepsilon_{m,0})}{\hbar^2} - \frac{2m}{\hbar^2}i\eta \right\}$$

$$= -\frac{\hbar^2}{2m}\left\{ \beta^2 - \frac{2m(E - \varepsilon_{m,0})}{\hbar^2} - \frac{2m(E - \varepsilon_{m,0})}{\hbar^2}\frac{i\eta}{E - \varepsilon_{m,0}} \right\}$$

$$= -\frac{\hbar^2}{2m}\left\{ \beta^2 - \frac{2m(E - \varepsilon_{m,0})}{\hbar^2}\left(1 + \frac{i\eta}{E - \varepsilon_{m,0}}\right) \right\}$$

$$= -\frac{\hbar^2}{2m}\left\{ \beta^2 - k_m^2\left(1 + \frac{i\eta}{E - \varepsilon_{m,0}}\right) \right\} \quad (14.107)$$

where

$$k_m^2 = \frac{2m(E - \varepsilon_{m,0})}{\hbar^2} \quad (14.108)$$

Applying eq. (14.107), eq. (106) is changed to

$$G^R(x,y;x',y')$$

$$= \frac{1}{2\pi} \sum_m \Psi_m(y)\Psi_m(y') \int_{-\infty}^{+\infty} \frac{\{1 - \exp(2i\beta x)\}}{-\dfrac{\hbar^2}{2m}\left\{ \beta^2 - k_m^2\left(1 + \dfrac{i\eta}{E - \varepsilon_{m,0}}\right) \right\}} d\beta$$

$$= -\frac{1}{2\pi} \times \frac{2m}{\hbar^2} \sum_m \Psi_m(y)\Psi_m(y') \int_{-\infty}^{+\infty} \frac{\{1 - \exp(2i\beta x)\}}{\left\{ \beta^2 - k_m^2\left(1 + \dfrac{i\eta}{E - \varepsilon_{m,0}}\right) \right\}} d\beta$$

$$= -\frac{m}{\pi\hbar^2} \sum_m \Psi_m(y)\Psi_m(y') \int_{-\infty}^{+\infty} \frac{\{1 - \exp(2i\beta x)\}}{\left\{ \beta^2 - k_m^2\left(1 + \dfrac{i\eta}{E - \varepsilon_{m,0}}\right) \right\}} d\beta$$

$$(14.109)$$

The integrand has two poles. These poles are located at

$$\beta = \pm k_m \sqrt{1 + \frac{i\eta}{E - \varepsilon_{m,0}}} = \pm k_m\left(1 + \frac{i\eta}{E - \varepsilon_{m,0}}\right)^{\frac{1}{2}}$$

$$= \pm k_m\left\{ 1 + \frac{1}{2}\left(\frac{i\eta}{E - \varepsilon_{m,0}}\right) + \dots \right\}$$

$$\cong \pm k_m\left\{ 1 + \frac{1}{2}\left(\frac{i\eta}{E - \varepsilon_{m,0}}\right) \right\} = \pm k_m\left(1 + \frac{1}{2}i\delta\right) \quad (14.110)$$

The pole

$$\beta = +k_m \left\{ 1 + \frac{1}{2}\left(\frac{i\eta}{E - \varepsilon_{m,0}}\right) \right\} \tag{14.111}$$

is in the upper half plane with

$$\sum \text{Residue} = \left[\beta - k_m \left\{ 1 + \frac{1}{2}\left(\frac{i\eta}{E - \varepsilon_{m,0}}\right) \right\} \right]$$

$$\times \left[-\frac{m}{\pi\hbar^2} \sum_m \Psi_m(y)\Psi_m(y') \frac{\{1 - \exp(2i\beta x)\}}{\left[\beta^2 - k_m^2 \left\{ 1 + \frac{1}{2}\left(\frac{i\eta}{E - \varepsilon_{m,0}}\right) \right\}^2 \right]} \right]_{\beta = k_m \left\{1 + \frac{1}{2}\left(\frac{i\eta}{E - \varepsilon_{m,0}}\right)\right\}}$$

$$= \left[-\frac{m}{\pi\hbar^2} \sum_m \Psi_m(y)\Psi_m(y') \frac{\{1 - \exp(2i\beta x)\}}{\beta + k_m \left\{ 1 + \frac{1}{2}\left(\frac{i\eta}{E - \varepsilon_{m,0}}\right) \right\}} \right]_{\beta = k_m \left\{1 + \frac{1}{2}\left(\frac{i\eta}{E - \varepsilon_{m,0}}\right)\right\}}$$

$$= -\frac{m}{\pi\hbar^2} \sum_m \Psi_m(y)\Psi_m(y') \frac{\{1 - \exp(2i\beta x)\}}{+k_m \left\{ 1 + \frac{1}{2}\left(\frac{i\eta}{E - \varepsilon_{m,0}}\right) \right\} + k_m \left\{ 1 + \frac{1}{2}\left(\frac{i\eta}{E - \varepsilon_{m,0}}\right) \right\}}$$

$$= -\frac{m}{\pi\hbar^2} \sum_m \Psi_m(y)\Psi_m(y') \frac{\{1 - \exp(2i\beta x)\}}{2k_m} = -\frac{m}{2\pi\hbar^2} \sum_m \Psi_m(y)\Psi_m(y') \frac{\{1 - \exp(2i\beta x)\}}{k_m} \tag{14.112}$$

in the limit

$$\frac{i\eta}{E - \varepsilon_{m,0}} \to 0 \tag{14.113}$$

In eq. (14.112)

$$1 - \exp(2i\beta x) = \frac{\exp(i\beta x)\exp(-i\beta x) - \exp(i\beta x)\exp(i\beta x)}{2i} \times 2i$$

$$= \frac{\exp(-i\beta x) - \exp(i\beta x)}{2i} \times \exp(i\beta x) \times 2i$$

$$= -\frac{\exp(i\beta x) - \exp(-i\beta x)}{2i} \times \exp(i\beta x) \times 2i$$

$$= -\sin(\beta x) \times \exp(i\beta x) \times 2i = -2i\sin(\beta x)\exp(i\beta x) \tag{14.114}$$

From eqs. (14.112) and (14.114)

$$\sum \text{Residue} = -\frac{m}{2\pi\hbar^2} \sum_m \Psi_m(y)\Psi_m(y') \times \frac{-2i\sin(\beta x)\exp(i\beta x)}{k_m}$$

$$= \left(-\frac{m}{2\pi\hbar^2} \times -2i \right) \sum_m \Psi_m(y)\Psi_m(y') \frac{\sin(\beta x)\exp(i\beta x)}{k_m}$$

$$= \left(\frac{mi}{\pi\hbar^2} \right) \sum_m \Psi_m(y)\Psi_m(y') \frac{\sin(\beta x)\exp(i\beta x)}{k_m} \tag{14.115}$$

Applying Cauchy's residue theorem (Watson 2012, Weisstein 1999–2019), eqs. (14.112) and (14.114) give

$$G^R(x, y; x', y') = 2\pi i \sum \text{Residue} = 2\pi i \times \left(\frac{mi}{\pi\hbar^2} \right)$$

$$\times \sum_m \Psi_m(y)\Psi_m(y') \frac{\sin(\beta x)\exp(i\beta x)}{k_m}$$

$$= 2i^2 \times \left(\frac{m}{\hbar^2} \right) \sum_m \Psi_m(y)\Psi_m(y') \frac{\sin(\beta x)\exp(i\beta x)}{k_m}$$

$$= -2 \times \left(\frac{m}{\hbar^2} \right) \sum_m \Psi_m(y)\Psi_m(y') \frac{\sin(\beta x)\exp(i\beta x)}{k_m} \tag{14.116}$$

Putting $\beta = k_m$ and rearranging

$$G^R(x, y; x', y') = -2 \times \left(\frac{m}{\hbar^2}\right) \sum_m \Psi_m(y)\Psi_m(y') \frac{\sin(k_m x)\exp(ik_m x)}{k_m}$$

$$= -\sum_m \frac{2\sin(k_m x)}{\frac{\hbar^2 k_m}{m}} \Psi_m(y)\exp(ik_m x)\Psi_m(y')$$

$$= -\sum_m \frac{2\sin(k_m x)}{\hbar v_m} \Psi_m(y)\exp(ik_m x)\Psi_m(y') \qquad (14.117)$$

where from eq. (2.23) and $p = mv_m$

$$v_m = \frac{\hbar k_m}{m} \qquad (14.118)$$

In the earlier mathematical treatment, the x- and y-coordinates are assumed to be continuous. The Green's function is required on a discrete lattice between any two points along an edge of the lead. In this lattice, the distance between the consecutive points is a. The Green's function for a discrete lattice g^R is obtained from eq. (14.117) by setting x, y, and y' to the selected values, i.e.,

$$x = a; y = p_i \text{ and } y' = p_j \qquad (14.119)$$

Hence we get

$$g_p^R(p_i, p_j) = -\sum_m \frac{2\sin(k_m a)}{\hbar v_m} \Psi_m(p_i)\exp(ik_m a)\Psi_m(p_j)$$

$$\qquad (14.120)$$

Substituting for $\hbar v_m$ from eq. (14.29) in eq. (14.120), we get

$$g_p^R(p_i, p_j) = -\sum_m \frac{2\sin(k_m a)}{2at\sin(k_m a)} \Psi_m(p_i)\exp(ik_m a)\Psi_m(p_j)$$

$$= -\sum_m \frac{1}{at} \Psi_m(p_i)\exp(ik_m a)\Psi_m(p_j) \qquad (14.121)$$

Now, substituting for $g_p^R(p_i, p_j)$ from eq. (14.121) in eq. (14.62), we write the self-energy equation as

$$\sum_p^R(i, j) = t^2 g_p^R(p_i, p_j) = t^2 \times -\frac{1}{at} \sum_m \Psi_m(p_i)\exp(ik_m a)\Psi_m(p_j)$$

$$= -\frac{t}{a} \sum_m \Psi_m(p_i)\exp(ik_m a)\Psi_m(p_j) \qquad (14.122)$$

14.7 Discussion and Conclusions

The Green's function for an irregularly shaped conductor is calculated by solving its differential equation for given potential energy and vector potential by discretization of the spatial coordinate. As a result, the Green's function becomes a matrix and its differential equation changes to a matrix equation. Then the Green's function is obtained by inversion of a matrix. This matrix contains the Hamiltonian operator which must be expressed in matrix form. Taking a linear chain of points at fixed locations and spacing, the second derivative in the Hamiltonian operator is expressed as an algebraic equation and the Hamiltonian operator is stated as a summation from which the matrix is written with defined procedure. In this matrix, each site is linked to its closest neighbor through a coupling or hopping parameter $(-t)$ between adjacent lattice points while each diagonal element is (potential energy $+2t$). This kind of matrix is referred to as a tight-binding matrix in the atomic orbital representation of the Hamiltonian. The reason for calling it tight-binding is that it contains only the diagonal and the coupling parameters to the nearest neighbors. The Schrodinger equation for a wire represented as a discrete lattice is satisfied by a plane wave solution provided an energy equation is obeyed. This equation is the dispersion relation for the discrete wire. The electron velocity is determined from the energy. The matrix representation of the Hamiltonian operator is extended to two dimensions by modification of nearest neighbor coupling for non-zero vector potential. Once the Hamiltonian matrix is derived, the Green's function simply involves a matrix inversion operation over the matrix equation containing the Hamiltonian operator. Here a problem arises because inversion of the infinite-dimensional matrix obtained for the open system with non-reflecting boundaries is not feasible. So the matrix must be truncated to describe a closed system with fully reflecting boundaries. For truncation, the overall matrix of the system is broken into sub-matrices for the isolated conductor and the isolated lead along with coupling matrices. Equations for the coupling and isolated lead matrices straightway follow. The sub-matrix for the isolated conductor now contains finite matrices having dimensions dependent on the number of points inside the conductor. The equation for the Green's function of the conductor is written down noting that the leads are independent and their effects can be added. In this equation, the effect of leads is taken into account through the term Σ^R, which is an effective Hamiltonian-like entity produced when the conductor interacts with the leads. It is interpreted as the self-energy due to the leads, and has to be assessed only for points at the interface of the conductor with lead.

The equation for the Green's function can be used only when we know the self-energy of the leads. But to evaluate the self-energy of the leads, it is required to determine the Green's function for an isolated lead. For this calculation, the Schrodinger equation is solved for a wire terminating on one side $x = 0$ in an infinite potential with a confining potential $U(y)$ in the Y-direction. The Green's function is obtained by linearly combining the wave functions obtained by solving the Schrodinger equation. For superposition, the Green's function is expressed as a summation over the product of coefficients C_α with the wave functions. The coefficients C_α are determined by plugging in the Green's function in summation form into the equation for the Green's function of the conductor with the leads. The summation over the constant β is replaced with integration by taking into consideration the density of states in one dimension. Contour integration is performed employing Cauchy's residue theorem.

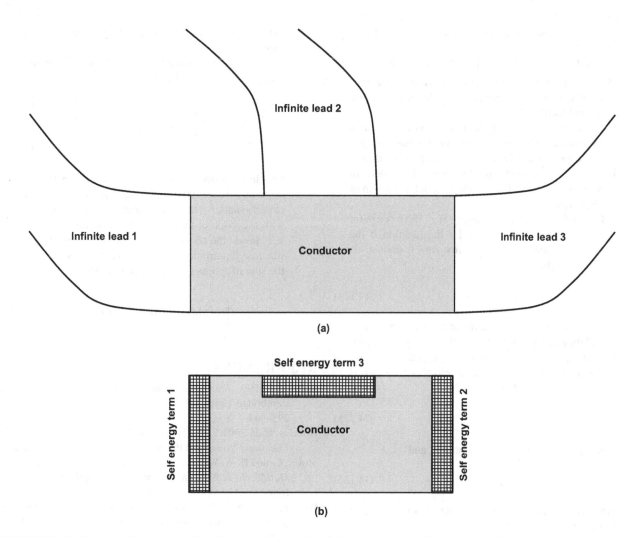

FIGURE 14.3 Replacement of a conductor with infinite leads by equivalent finite conductor with self-energy terms: (a) conductor with infinite leads and (b) substituent finite conductor.

Finally, transformation from the continuous coordinates to discrete lattice gives the required Green's function of the wire.

Thus retarded Green's function calculation was performed using a discrete representation of points in space. Proceeding from a conductor with infinite leads, we carried out matrix truncation to get the Green's function for a finite-size conductor. This Green's function contains a Hamiltonian for the finite-size conductor and a term representing the self-energy function of the leads. The self-energy function is non-zero only for points of the conductor adjoining the leads. Thus the effects of the leads are incorporated at the boundaries of the conductor itself (Figure 14.3).

Illustrative Exercises

14.1 What is an identity matrix?

Also called a unit matrix or elementary matrix, it is a square matrix of order $n \times n$ in which all the diagonal elements are unity and all the remaining elements are zeroes.

14.2 Give an example of an identity matrix of order 6×6.

A matrix having m rows and n columns is said to be a matrix of order $m \times n$.

$$[I] = \begin{vmatrix} 1 & 0 & 0 & 0 & 0 & 0 \\ 0 & 1 & 0 & 0 & 0 & 0 \\ 0 & 0 & 1 & 0 & 0 & 0 \\ 0 & 0 & 0 & 1 & 0 & 0 \\ 0 & 0 & 0 & 0 & 1 & 0 \\ 0 & 0 & 0 & 0 & 0 & 1 \end{vmatrix} \qquad (14.123)$$

14.3 What is the importance of identity matrix?

Multiplication of a matrix [A] by the identity matrix [I] yields the matrix [A] itself. So, the identity matrix occupies the same place in the set of matrices as the integer "1" in the set of integers. Multiplication of a matrix by the identity matrix does not produce any change in exactly the same way as multiplication of a number by integer "1" does not cause any changes. So, identity matrix is for matrix multiplication, the same entity as the integer "1" is for numerical multiplication.

14.4 What is the result of multiplication of a matrix by its inverse?

It is the identity matrix [I], i.e. $[A][A]^{-1} = [I]$.

14.5 What is an analytic function and what are its zeroes and poles? How are the zeroes and poles of an analytic function identified?

A function $f(z)$ is analytic in a region R if it has a derivative at each point of the region R and also $f(z)$ is single valued. Equivalently, the function $f(z)$ is analytic if its Taylor series about a point z_0 in its domain converges to the function for every z_0 in the domain of the function.

For a function which is analytic at z_0, there is a point z_0 called the zero of order m for the function if the function f is zero at z_0 and its first $(m-1)$ derivatives are zero at z_0 but

$$f^m(z_0) \neq 0 \qquad (14.124)$$

For a function which is analytic at z_0, there is a point z_0 called the pole of order m for the function if its reciprocal, i.e. $1/f$ has a zero of order m at the point z_0.

An analytic function $f(z)$ has a zero of order m at z_0 if $f(z)$ can be expressed as

$$f(z) = g(z)(z - z_0)^m \qquad (14.125)$$

where $g(z)$ is an analytic function at z_0 and

$$g(z_0) \neq 0 \qquad (14.126)$$

An analytic function $f(z)$ has a pole of order m at z_0 if $f(z)$ can be expressed as

$$f(z) = \frac{g(z)}{(z - z_0)^m} \qquad (14.127)$$

where $g(z)$ is an analytic function at z_0, and

$$g(z_0) \neq 0 \qquad (14.128)$$

14.6 What is residue of a function? State Cauchy's residue theorem.

The residue of a function $f(z)$ which has a pole of order m at z_0 is given by

$$\text{Res}(f, z_0) = \lim_{z \to z_0} \left[\frac{1}{(m-1)!} \frac{d^{m-1}}{dz^{m-1}} \left\{ (z - z_0)^m f(z) \right\} \right] \qquad (14.129)$$

Cauchy's residue theorem states that: For a function $f(z)$ which is analytic on a simple positively oriented closed contour γ and everywhere inside this contour, with exception of the finite number of points z_1, z_2, z_3, ...z_n inside the above contour, the line integral of the function $f(z)$ around the closed contour γ is $2\pi i$ times the sum of residues of $f(z)$ at the points:

$$\left| \oint f(z)dz \right|_\gamma = 2\pi i \sum_{k=1}^{n} \text{Res}(f, z_k) \qquad (14.130)$$

REFERENCES

Datta, S. 1999. *Electronic Transport in Mesoscopic Systems.* Cambridge University Press: Cambridge, UK, pp. 117–174, 293–344.

Hoskins, R. F. 2009. *Delta Functions: Introduction to Generalized Functions.* Woodhead Publishing: Oxford, UK, 280 pages.

Stover, C. and E. W. Weisstein. 1999–2019. "Matrix Inverse." From *MathWorld*-A Wolfram Web Resource, http://mathworld.wolfram.com/MatrixInverse.html.

Watson, G. N. 2012. *Complex Integration and Cauchy's Theorem, Dover Books on Mathematics.* Dover Publications: New York, 96 pages.

Weisstein, E. W. 1999–2019. "Residue Theorem." From *MathWorld*-A Wolfram Web Resource, http://mathworld.wolfram.com/ResidueTheorem.html.

15

Non-Equilibrium Green's Function (NEGF) Formalism

The non-equilibrium Green's function (NEGF) formalism, also called Keldysh theory, is a generalized microscopic theory of quantum transport unifying quantum dynamics with statistical elucidation of dissipative interactions (Keldysh 1965). Its strength lies in providing a framework for description of quantum transport inclusive of interactions. It is a useful formalism for simulation of nanoscale electronic devices. Notwithstanding, for non-interacting transport, the conceptual sophistication of NEGF approach is superfluous because it becomes equivalent to the Landauer-Büttiker formalism. The perspective of Datta (1999, 2000, 2002, 2015) will be followed all throughout.

15.1 Density Matrix and Correlation Function

In a semi-classical portrayal, a distribution function $f(\mathbf{k})$ is used to specify the number of electrons occupying a particular state $\mathbf{k}$. This idea becomes insufficient in quantum-mechanical picture because of the necessity of specification of the phase relationship among the various states. Therefore, a density matrix $\rho(\mathbf{k}, \mathbf{k}')$ is used for including information about phase correlations. The density matrix is expressed as

$$\rho(\mathbf{k},\mathbf{k}') = \psi_\mathbf{k}\psi_\mathbf{k}^* \quad (15.1)$$

for an electron having a wave function $\sum_\mathbf{k} \Psi_\mathbf{k} \mid \mathbf{k} >$.

This matrix can be transformed to a real space representation and a time-varying density matrix $\rho(\mathbf{k}, \mathbf{k}'; t)$ can be defined. However, this concept lacks generality. Therefore, it must be generalized into a two-time correlation function G^n which correlates the amplitude in state $\mathbf{k}$ at time t with that in state $\mathbf{k}'$ at time t':

$$\rho(\mathbf{k},\mathbf{k}';t) \rightarrow G^n(\mathbf{k},\mathbf{k}';t,t') \quad (15.2)$$

Under steady-state conditions, the correlation function depends primarily on the time difference $t - t' = \tau$. So,

$$G^n(\mathbf{k},\mathbf{k}';t,t') \rightarrow G^n(\mathbf{k},\mathbf{k}';\tau) \quad (15.3)$$

The forward Fourier transform of the correlation function is

$$G^n(\mathbf{k},\mathbf{k}';\omega) = \int_{-\infty}^{+\infty} G^n(\mathbf{k},\mathbf{k}';\tau)\exp(-i\omega\tau)d\tau$$

$$= \int_{-\infty}^{+\infty} G^n(\mathbf{k},\mathbf{k}';\tau)\exp\left(-\frac{iE}{\hbar}\tau\right)d\tau \quad (15.4)$$

or

$$G^n\left(\mathbf{k},\mathbf{k}';\frac{E}{\hbar}\right) = \int_{-\infty}^{+\infty} G^n(\mathbf{k},\mathbf{k}';\tau)\exp\left(-\frac{iE}{\hbar}\tau\right)d\tau \quad (15.5)$$

where

$$\omega = \text{Angular frequency} = 2\pi \times \text{Regular frequency}\ (\nu) = 2\pi\nu \quad (15.6)$$

and since energy E of a particle is proportional to frequency

$$E = h\nu \quad (15.7)$$

with Planck's constant h as the constant of proportionality, we can write the frequency ν as

$$\nu = \frac{E}{h} \quad (15.8)$$

and

$$\therefore \omega = 2\pi\nu = \frac{2\pi E}{h} = \frac{E}{\dfrac{h}{2\pi}} = \frac{E}{\hbar} \quad (15.9)$$

According to scaling property

$$G^n\left(\mathbf{k},\mathbf{k}';\hbar \times \frac{E}{\hbar}\right) = \frac{1}{\hbar}\int_{-\infty}^{+\infty} G^n(\mathbf{k},\mathbf{k}';\tau)\exp\left(-\frac{iE}{\hbar}\tau\right)d\tau \quad (15.10)$$

Using eq. (15.10), an energy-dependent correlation function $G^n(\mathbf{k},\mathbf{k}';E)$ is introduced.

$$G^n(\mathbf{k},\mathbf{k}';E) = \int_{-\infty}^{+\infty} G^n(\mathbf{k},\mathbf{k}';\tau)\exp(-iE\tau)d\tau$$

$$= \frac{1}{\hbar}\int_{-\infty}^{+\infty} G^n(\mathbf{k},\mathbf{k}';\tau)\exp\left(-\frac{iE}{\hbar}\tau\right)d\tau \quad (15.11)$$

If we set $t = t'$, the density matrix $\rho(\mathbf{k}, \mathbf{k}, t)$ in eq. (15.2) reduces to the correlation function:

$$\rho(\mathbf{k},\mathbf{k}';t) = \left| G^n(\mathbf{k},\mathbf{k}';t,t')\right|_{t=t'} \quad (15.12)$$

For $t = t'$, the inverse Fourier transform of the correlation function given in eq. (15.11) is

$$\left| G^n\left(\mathbf{k}, \mathbf{k}'; t, t'\right) \right|_{t=t'} = \frac{1}{2\pi} \int_{-\infty}^{+\infty} G^n\left(\mathbf{k}, \mathbf{k}'; E\right) \exp(+iE\tau) dE$$

$$= \frac{1}{2\pi} \int_{-\infty}^{+\infty} G^n\left(\mathbf{k}, \mathbf{k}'; E\right) \exp(0) dE$$

$$= \frac{1}{2\pi} \int_{-\infty}^{+\infty} G^n\left(\mathbf{k}, \mathbf{k}'; E\right) \times 1 \times dE$$

$$= \frac{1}{2\pi} \int_{-\infty}^{+\infty} G^n\left(\mathbf{k}, \mathbf{k}'; E\right) dE \qquad (15.13)$$

because $\tau = t - t' = 0$. It may be pointed out that the earlier mentioned correlation function is only applicable to steady-state transport but the full correlation function is used for discussing time-varying transport.

Equation (15.13) allows us to write

$$\left| G^n\left(\mathbf{k}, \mathbf{k}'; t, t'\right) \right|_{\mathbf{k}=\mathbf{k}',\ t=t'} = \frac{1}{2\pi} \int_{-\infty}^{+\infty} G^n\left(\mathbf{k}, \mathbf{k}; E\right) dE \quad (15.14)$$

In real space, the equation for electron density $n(\mathbf{r})$ is written as

$$n(\mathbf{r}) = 2\left(\text{for } 2 \text{ spin components}\right) \times \frac{1}{2\pi} \int_{-\infty}^{+\infty} G^n\left(\mathbf{r}, \mathbf{r}; E\right) dE$$

$$= 2 \times \int_{-\infty}^{+\infty} n(r; E) dE \qquad (15.15)$$

where

$$n(r; E) = \text{Electron density per unit energy} = \frac{1}{2\pi} G^n\left(\mathbf{r}, \mathbf{r}; E\right)$$

$$(15.16)$$

$$\therefore\ 2\pi n(r; E) = G^n\left(\mathbf{r}, \mathbf{r}; E\right) \qquad (15.17)$$

The inflow of electrons may be looked upon as an outflow of holes to describe that a hole correlation function G^p can be defined like the electron correlation function G^n. The difference between the electron correlation function G^n and the hole correlation function G^p can be stated with respect to creation operator $a_{\mathbf{k}}$ and annihilation operator $a_{\mathbf{k}}^+$. But in a one-particle description, the two operators reduce to the single particle wave function and its complex conjugate. From commutation of wave functions, G^n and G^p reduce to the product $\Psi_{\mathbf{k}'}^*(t') \Psi_{\mathbf{k}}(t)$.

Remember that the electron and hole correlation functions are the analogs of the distribution functions $f(E)$ and $1 - f(E)$ in classical phraseology, respectively.

15.2 Scattering Functions

The function $S^{out}(\mathbf{k}, t)$ in semi-classical theory, expressing the rate of scattering of electrons out of an initially full state $\mathbf{k}$, is generalized to include phase correlations by forming an outscattering function dependent on time coordinates:

$$S^{out}\left(\mathbf{k}, t\right) \rightarrow \Sigma^{out}\left(\mathbf{k}, \mathbf{k}'; t, t'\right) \qquad (15.18)$$

which is Fourier-transformed to obtain an energy-dependent outscattering function $\Sigma^{out}(\mathbf{k}, \mathbf{k}'; E)$ applicable to steady-state conditions. This is possible because in this situation, the scattering function depends only on the difference τ of two instants of time t and t'.

$$\Sigma^{out}\left(\mathbf{k}, \mathbf{k}'; t, t'\right) \rightarrow \Sigma^{out}\left(\mathbf{k}, \mathbf{k}'; \tau\right) \rightarrow \Sigma^{out}\left(\mathbf{k}, \mathbf{k}'; E\right) \quad (15.19)$$

Similar to $S^{out}(\mathbf{k}, t)$, an inscattering function is defined as

$$S^{in}\left(\mathbf{k}, t\right) \rightarrow \Sigma^{in}\left(\mathbf{k}, \mathbf{k}'; t, t'\right) \qquad (15.20)$$

and a Fourier-transformed energy-dependent inscattering function $\Sigma^{in}(\mathbf{k}, \mathbf{k}'; E)$ is formed for dealing with steady-state problems

$$\Sigma^{in}\left(\mathbf{k}, \mathbf{k}'; t, t'\right) \rightarrow \Sigma^{in}\left(\mathbf{k}, \mathbf{k}'; \tau\right) \rightarrow \Sigma^{in}\left(\mathbf{k}, \mathbf{k}'; E\right) \quad (15.21)$$

Note that an electron inscattering function behaves as a hole outscattering function.

The four matrices G^n, G^p, Σ^{out}, Σ^{in} greatly facilitate the description of charge transport including phase correlations. They are Hermitian or self-adjoint matrices meaning that they are complex square matrices equal to their own conjugate transposes. Their diagonal elements are purely real. For a device represented by a set of N nodes, each of these matrices is an $N \times N$ matrix at a particular energy E.

15.3 Green's Function and Self-Energy

In addition to the four aforementioned functions, another set of four functions are necessary for the language of electron motion in the conductor. The retarded and advanced Green's function G^R, G^A and self-energies Σ^R, Σ^A aid in describing the dynamics of electrons inside the conductor. The retarded Green's function G^R tells us how the electron coherently evolves from the instant of its injection till its coherence is lost. This loss of coherence takes place either by disappearance of electrons into a lead or by their scattering into a different state via electron-electron or electron-phonon interactions. Then the electron starts moving on a different route. The self-energy Σ^R represents the influence of the leads as well as those of the various interactions undergone by the electrons on their dynamics. In matrix form,

$$\left[EI - H_C - \sum{}^R \right] G^R = I \qquad (15.22)$$

which is the same as

$$G^R = \left[EI - H_C - \sum{}^R \right]^{-1} \qquad (15.23)$$

where $[I]$ is the identity matrix and $[H_C]$ is the matrix representation of the Hamilton operator H_{op} for the conductor. The symbol Σ^R denotes an effective Hamiltonian originating from the interaction of the conductor with the leads. Another component of Σ^R arises from the interaction of electrons with phonons and other electrons inside the conductor. So Σ^R is referred to as the self-energy due to the lead-conductor coupling and the various interactions taking place in the conductor.

The advanced Green's function G^A = Hermitian adjoint of the corresponding retarded function G^R

or

$$G^A = \left[G^R \right]^+ \qquad (15.24)$$

The advanced self-energy Σ^A = Hermitian adjoint of the corresponding retarded function Σ^R

or

$$\Sigma^A = \left[\Sigma^R \right]^+ \qquad (15.25)$$

The spectral function

$$A = i\left[G^R - G^A \right] \qquad (15.26)$$

is a generalized density of states = the sum of electron and hole densities. Hence, it is evident that A = sum of electron and hole correlation functions, i.e.,

$$A = G^n + G^p \qquad (15.27)$$

The function A describes the nature of permitted electron states, irrespective of whether they are occupied or not whereas G^n and G^p tell us regarding the number of states filled and those vacant. Thus A gives the total number of states available while their occupancy or vacancy numbers are stated by G^n and G^p.

The function Γ representing the rate of loss of carriers by scattering is expressed as

$$\Gamma = i\left[\sum{}^R - \sum{}^A \right] = \sum{}^{in} + \sum{}^{out} \qquad (15.28)$$

where Σ^{in}, Σ^{out} are the inscattering and outscattering functions. Physically, it means that the inverse lifetime of the electron = rate of its scattering plus the rate of its blocking electrons from other states making effort for its displacement.

A useful relationship between A, Γ, G^R, G^A is

$$A = G^R \Gamma G^A = G^A \Gamma G^R \qquad (15.29)$$

To prove this identity, we proceed from eq. (15.23),

$$G^R = \left[EI - H_C - \sum{}^R \right]^{-1} \qquad (15.30)$$

or

$$\left[G^R \right]^{-1} = EI - H_C - \sum{}^R \qquad (15.31)$$

The Hermitian adjoint of G^R is

$$G^A = \left[EI - H_C - \sum{}^A \right]^{-1} \qquad (15.32)$$

or

$$\left[G^A \right]^{-1} = EI - H_C - \sum{}^A \qquad (15.33)$$

From eqs. (15.31) and (15.33), we have

$$\left[G^R \right]^{-1} - \left[G^A \right]^{-1} = EI - H_C - \sum{}^R - EI + H_C + \sum{}^A$$
$$= \sum{}^A - \sum{}^R = i\Gamma \qquad (15.34)$$

because from eq. (15.28)

$$i\Gamma = i^2 \left[\sum{}^R - \sum{}^A \right] = -1\left[\sum{}^R - \sum{}^A \right] = \sum{}^A - \sum{}^R \qquad (15.35)$$

We multiply eq. (15.34) by G^R from the left and by G^A from the right to obtain

$$G^R \left[G^R \right]^{-1} G^A - G^R \left[G^A \right]^{-1} G^A = G^R i\Gamma G^A = iG^R \Gamma G^A \qquad (15.36)$$

or

$$G^A - G^R = iG^R \Gamma G^A \qquad (15.37)$$

or

$$i\left[G^A - G^R \right] = i^2 G^R \Gamma G^A \qquad (15.38)$$

or

$$i\left[G^A - G^R \right] = -1 \times G^R \Gamma G^A \qquad (15.39)$$

or

$$-i\left[G^A - G^R \right] = G^R \Gamma G^A \qquad (15.40)$$

or

$$i\left[G^R - G^A \right] = G^R \Gamma G^A \qquad (15.41)$$

or

$$A = G^{R}\Gamma G^{A} \qquad (15.42)$$

from eq. (15.26).

Multiplying eq. (15.34) by G^{A} from the left and G^{R} from the right, we get

$$G^{A}\left[G^{R}\right]^{-1}G^{R} - G^{A}\left[G^{A}\right]^{-1}G^{R} = G^{A}i\Gamma G^{R} \qquad (15.43)$$

or

$$G^{A} - G^{R} = iG^{A}\Gamma G^{R} \qquad (15.44)$$

or

$$i\left[G^{A} - G^{R}\right] = i^{2}G^{A}\Gamma G^{R} \qquad (15.45)$$

or

$$i\left[G^{A} - G^{R}\right] = -1 \times G^{A}\Gamma G^{R} \qquad (15.46)$$

or

$$i\left[G^{R} - G^{A}\right] = G^{A}\Gamma G^{R} \qquad (15.47)$$

or

$$A = G^{A}\Gamma G^{R} \qquad (15.48)$$

from eq. (15.26).

We combine eqs. (15.42) and (15.48) to write

$$A = G^{R}\Gamma G^{A} = G^{A}\Gamma G^{R} \qquad (15.49)$$

which is eq. (15.29).

15.4 NEGF Kinetic Equations

The equations of paramount importance in NEGF formalism are kinetic equations. These equations connect the correlation functions G^{n}, G^{p} with the scattering functions Σ^{in}, Σ^{out}.

For electrons,

Electron correlation function $G^{n} \propto$ Inscattering function $\sum^{in}$

$$\qquad (15.50)$$

because the inscattering function gives the rate of electron entry. Hence,

$$G^{n} = G^{R}\sum^{in}G^{A} \qquad (15.51)$$

For holes,

Hole correlation function $G^{p} \propto$ Outscattering function $\sum^{out}$

$$\qquad (15.52)$$

or

$$G^{p} = G^{R}\sum^{out}G^{A} \qquad (15.53)$$

To derive eq. (15.51), let us write down the matrix equations (15.22)–(15.23) in position representation as

$$\left[E(\mathbf{r}) - H_{C}(\mathbf{r})\right]G^{R}(\mathbf{r},\mathbf{r}') - \int \Sigma^{R}(\mathbf{r},\mathbf{r}_{1})G^{R}(\mathbf{r}_{1},\mathbf{r}')d\mathbf{r}_{1} = \delta(\mathbf{r}-\mathbf{r}')$$

$$\qquad (15.54)$$

where the Green's function is the wave function at the point $\mathbf{r}$ produced by a unit excitation at the point $\mathbf{r}'$ (Figure 15.1).

Instead of the delta function on the right-hand side, let us consider a source term at $\mathbf{r}'$ denoted by $S(\mathbf{r}')$. Since the Green's function

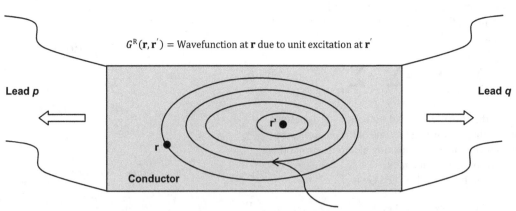

$G^{R}(\mathbf{r},\mathbf{r}') =$ Wavefunction at $\mathbf{r}$ due to unit excitation at $\mathbf{r}'$

Lead p

Lead q

Conductor

Ripples created by the excitation at r'

FIGURE 15.1 Meaning of the function $G^{R}(\mathbf{r},\mathbf{r}')$: In a conductor with leads p and q, an excitation at the point $\mathbf{r}'$ sets up ripples propagating outward, and the wave function at the point $\mathbf{r}$ arising from unit excitation at $\mathbf{r}'$ represents the Green's function at $\mathbf{r}$ caused by this excitation at $\mathbf{r}'$.

G^R = Wavefunction Ψ due to a delta function source (15.55)

Equation (15.54) is converted into

$$[E(\mathbf{r}) - H_C(\mathbf{r})]\Psi(\mathbf{r}, \mathbf{r}') - \int \Sigma^R(\mathbf{r}, \mathbf{r}_1)\Psi(\mathbf{r}_1, \mathbf{r}')d\mathbf{r}_1 = S(\mathbf{r}')$$ (15.56)

by writing wave functions in the places of Green's functions. Here

$$\Psi(\mathbf{r}, \mathbf{r}') = \int G^R(\mathbf{r}, \mathbf{r}')S(\mathbf{r}')d\mathbf{r}_1$$ (15.57)

and

$$\Psi(\mathbf{r}_1, \mathbf{r}') = \int G^R(\mathbf{r}_1, \mathbf{r}')S(\mathbf{r}')d\mathbf{r}_1$$ (15.58)

On these lines, we can write the wave function

$\Psi(\mathbf{r}, \mathbf{r}_1)$ = Wave function at $\mathbf{r}$ due to source term $S(\mathbf{r}_1)$ at $\mathbf{r}_1$

$$= \int G^R(\mathbf{r}, \mathbf{r}_1)S(\mathbf{r}_1)d\mathbf{r}_1$$ (15.59)

The complex conjugate can be written as

$\Psi(\mathbf{r}', \mathbf{r}_1')^*$ = Complex conjugate wave function at $\mathbf{r}'$

due to conjugate source term $S(\mathbf{r}_1')^*$ at $\mathbf{r}_1'$

$$= \int G^R(\mathbf{r}', \mathbf{r}_1')^* S(\mathbf{r}_1')^* d\mathbf{r}_1'$$ (15.60)

We form the product

$$\Psi(\mathbf{r}, \mathbf{r}_1)\Psi(\mathbf{r}', \mathbf{r}_1')^* = \iint G^R(\mathbf{r}, \mathbf{r}_1)S(\mathbf{r}_1)d\mathbf{r}_1 G^R(\mathbf{r}', \mathbf{r}_1')^* S(\mathbf{r}_1')^* d\mathbf{r}_1'$$

$$= \iint G^R(\mathbf{r}, \mathbf{r}_1)G^R(\mathbf{r}', \mathbf{r}_1')^* S(\mathbf{r}_1)S(\mathbf{r}_1')^* d\mathbf{r}_1 d\mathbf{r}_1'$$ (15.61)

Because the interrelationship between wave functions is expressed by the Green's function while the association between inscattering functions is represented by the source terms, we may state that

$$\Psi(\mathbf{r}, \mathbf{r}_1)\Psi(\mathbf{r}', \mathbf{r}_1')^* = G^n(\mathbf{r}, \mathbf{r}')$$ (15.62)

and

$$S(\mathbf{r}_1)S(\mathbf{r}_1')^* = \Sigma^{in}(\mathbf{r}_1, \mathbf{r}_1')$$ (15.63)

Combining eqs. (15.61), (15.62), and (15.63),

$$G^n(\mathbf{r}, \mathbf{r}') = \iint G^R(\mathbf{r}, \mathbf{r}_1)G^R(\mathbf{r}', \mathbf{r}_1')^* \Sigma^{in}(\mathbf{r}_1, \mathbf{r}_1')d\mathbf{r}_1 d\mathbf{r}_1'$$

$$= \iint G^R(\mathbf{r}, \mathbf{r}_1)\Sigma^{in}(\mathbf{r}_1, \mathbf{r}_1')G^R(\mathbf{r}', \mathbf{r}_1')^* d\mathbf{r}_1 d\mathbf{r}_1'$$

$$= \iint G^R(\mathbf{r}, \mathbf{r}_1)\Sigma^{in}(\mathbf{r}_1, \mathbf{r}_1')G^A(\mathbf{r}_1', \mathbf{r}')d\mathbf{r}_1 d\mathbf{r}_1'$$ (15.64)

since

$$G^R(\mathbf{r}', \mathbf{r}_1')^* = G^A(\mathbf{r}_1', \mathbf{r}')$$ (15.65)

In matrix form, eq. (15.64) is written as

$$G^n = G^R \Sigma^{in} G^A$$ (15.66)

Equation (15.53) is an identical equation for holes.

15.5 The Evolution of NEGF Equations from Schrodinger's Equation

In quantum transport, the starting point is Schrodinger's equation

$$E\Psi = H\Psi$$ (15.67)

where the eigenvalues of the matrix H are the energy levels in the channel. The channel has contacts called the source and drain terminals and a voltage is applied across the channel (Figure 15.2). To describe this electrical transport, it is inadequate to use quantum mechanics alone. One must include the entropy-driven processes. Semi-classical transport treatment employs Newtonian mechanics together with scattering processes which lead to the famous Boltzmann H theorem. On similar lines, for quantum transport, additional terms are included with H for entropy-driven processes. These terms are collectively placed under self-energy functions Σ. The self-energy functions consist of the effects of contacts as well as interactions of electrons with surroundings. Thus the Schrodinger equation is modified as (Datta 2015)

$$E\Psi = H\Psi + \Sigma\Psi + S$$ (15.68)

The matrix Σ differs from the matrix H. The matrix H is Hermitian to conserve probability whereas the matrix Σ is non-Hermitian. So, it represents a loss of electrons. Hence, $\Sigma\Psi$ represents the outflow of electrons and will drain the system of all electrons.

Σ = Self − energy for contact 1 + Self − energy for contact 2

+ Self − energy for interaction of electrons with their

surroundings = $\Sigma_1 + \Sigma_2 + \Sigma_0$ (15.69)

This electron loss is taken care of by inclusion of a source or supply term S in eq. (15.68) which represents the inflow of electrons from the contacts. Thus electrons from the S term replenish the losses incurred.

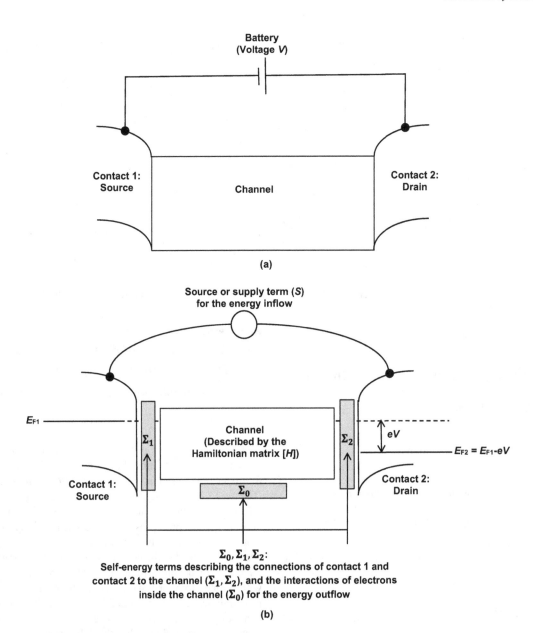

FIGURE 15.2 Non-equilibrium Green's function-based quantum transport model for current flow: (a) basic circuit and (b) its equivalent representation showing the various terms in the model.

The NEGF equations can be written from this modified Schrodinger equation. Bringing all the Ψ's in eq. (15.68) to the left-hand side

$$E\Psi - H\Psi - \Sigma\Psi = S \qquad (15.70)$$

The equation for the wave function Ψ in terms of the source term S is

$$\{\Psi\} = [EI - H - \Sigma]^{-1}\{S\} = G^{R}\{S\} \qquad (15.71)$$

from eq. (15.23).

Then, an equation for probability density $\Psi\Psi^*$ is written down. This is necessary because if there are multiple sources, then their strengths cannot be added coherently. Therefore, the effects of these sources cannot be combined by summation. Incoherence among sources will cause mutual interference.

$$\Psi\Psi^* = G^{R} SS^* G^{A} \qquad (15.72)$$

Here,

$$\Psi\Psi^* = G^{n} = \text{Electron density} \qquad (15.73)$$

and

$$SS^* = \Sigma^{in}$$

= Strength of the source term representing current inflow
$$\qquad (15.74)$$

Thus the non-equilibrium equation is

$$G^{n} = G^{R}\Sigma^{in}G^{A} \qquad (15.75)$$

where

$$\Sigma^{\text{in}} = \text{Current inflow from contact 1}$$

$$+ \text{Current inflow from contact 2}$$

$$+ \text{Current inflow from surroundings}$$

$$= \Sigma_1^{\text{in}} + \Sigma_2^{\text{in}} + \Sigma_0^{\text{in}} \tag{15.76}$$

The equation for holes is derived on similar reasoning. Note that eq. (15.75) is the same as eq. (15.66).

15.6 Equilibrium Solution

Noting that the spectral function A resembles the density of states, the correlation function G^n, G^p are the equivalents of electron and hole densities, and at equilibrium all the states are filled in accordance with a single Fermi function $f_0(E)$, we can write for electrons

$$G^n(E) = f_0(E) A(E) \tag{15.77}$$

For holes,

$$G^p(E) = \{1 - f_0(E)\} A(E) \tag{15.78}$$

From eqs. (15.51) and (15.77),

$$G^R \sum{}^{\text{in}}(E) G^A = f_0(E) A(E) \tag{15.79}$$

Substituting for $A(E)$ from eq. (15.29) into eq. (15.79),

$$G^R \sum{}^{\text{in}}(E) G^A = f_0(E) G^R \Gamma G^A \tag{15.80}$$

$$\therefore \sum{}^{\text{in}}(E) = f_0(E) \Gamma(E) \tag{15.81}$$

From eqs. (15.53) and (15.78),

$$G^R \sum{}^{\text{out}}(E) G^A = \{1 - f_0(E)\} A(E) \tag{15.82}$$

Substituting for $A(E)$ from eq. (15.29) into eq. (15.82),

$$G^R \sum{}^{\text{out}}(E) G^A = \{1 - f_0(E)\} G^R \Gamma G^A \tag{15.83}$$

$$\therefore \sum{}^{\text{out}}(E) = \{1 - f_0(E)\} \Gamma(E) \tag{15.84}$$

15.7 Self-Energy and Scattering Functions due to Interactions within the Conductor

For non-equilibrium cases, simultaneous solution of the kinetic equations (15.51) and (15.53) is done with eqs. (15.22)–(15.25)

for the Green's functions. Let us now turn our attention to the electron-electron and electron-phonon interactions taking place inside the conductor. These interactions influence both the self-energy and the scattering functions. They have two components, one originating from interactions with the leads, as already discussed, and the other arising from phase-breaking interactions taking place inside the conductor. If the interactions with the leads and those inside the conductor are assumed to be independent of each other,

Total retarded self energy

= Self-energy due to interaction with lead p

+ Self-energy from phase-breaking interactions

in the conductor (15.85)

or

$$\Sigma^R = \sum_p \sum_p^R + \Sigma_\phi^R \tag{15.86}$$

The function with subscript p originates from interactions with lead p. That with subscript ϕ arises from phase-breaking interactions inside the conductor.

Similarly,

Total inscattering function

= Inscattering function due to interaction with lead p

+ Inscattering function from phase-breaking interactions

in the conductor (15.87)

or

$$\Sigma^{\text{In}} = \sum_p \sum_p^{\text{In}} + \Sigma_\phi^{\text{In}} \tag{15.88}$$

Total outscattering function

= Outscattering function due to interaction with lead p

+ Outscattering function from phase-breaking interactions

in the conductor (15.89)

or

$$\Sigma^{\text{Out}} = \sum_p \sum_p^{\text{Out}} + \Sigma_\phi^{\text{Out}} \tag{15.90}$$

Equations for self-energy and scattering functions are derived from perturbation theory and become more intricate with decreasing approximation. Electron-electron interactions are discussed in the Hartree-Fock approximation whereas electron-phonon interactions are discussed in the self-consistent Born approximation. Here we state the main results of this analysis without entering into complex details.

15.7.1 Electron-Electron Interactions

The Hartree-Fock method is one of the simplest theories of many-particle systems. In the Hartree-Fock approximation, the self-energy equation contains two terms, one due to the Hartree potential and the other arising from exchange potential. The Hartree potential term is obtained by considering the motion of each electron independent of other electrons under the influence of the average electric field of other electrons together with that due to the atoms. The exchange potential term is obtained from the repulsion between electrons with parallel spins because no electrons with parallel spins can reside in the same quantum state according to Pauli's exclusion principle. Thus the self-energy function is written as

$$\Sigma_\phi^R(\mathbf{r},\mathbf{r}';E) = U_H(\mathbf{r})\delta(\mathbf{r}-\mathbf{r}') + \Sigma_F(\mathbf{r},\mathbf{r}')$$

$$= \frac{1}{2\pi}\iint G^n(\mathbf{r},\mathbf{r}';E)\frac{e^2}{4\pi\varepsilon_0\varepsilon_r|\mathbf{r}-\mathbf{r}'|}d\mathbf{r}'\,dE$$

$$- \int G^{ns}(\mathbf{r},\mathbf{r}';E)\frac{e^2}{4\pi\varepsilon_0\varepsilon_r|\mathbf{r}-\mathbf{r}'|}dE \quad (15.91)$$

where the first term represents the Hartree potential and the second term the exchange potential. The letter 's' in the 'G^{ns}' indicates that the exchange potential is felt by an electron only in the presence of an electron of the same spin.

The inscattering and outscattering functions are not affected by the electron-electron interaction in the Hartree-Fock approximation.

15.7.2 Electron-Phonon Interactions

The self-energy function due to phase-breaking interactions is written as

$$\Sigma_\phi^R(\mathbf{r},\mathbf{r}';E) = -\Gamma_\phi^H(\mathbf{r},\mathbf{r}';E) + \frac{i}{2}\Gamma_\phi(\mathbf{r},\mathbf{r}';E) \quad (15.92)$$

$\Gamma_\phi^H(\mathbf{r},\mathbf{r}';E) = $ Hilbert transform of $\Gamma_\phi(\mathbf{r},\mathbf{r}';E)$

$$= \text{Cauchy principal value}\int\frac{\Gamma_\phi(\mathbf{r},\mathbf{r}';E')}{E-E'}\,dE' \quad (15.93)$$

and

$$\Gamma_\phi(\mathbf{r},\mathbf{r}';E) = \Sigma_\phi^{In}(\mathbf{r},\mathbf{r}';E) + \Sigma_\phi^{Out}(\mathbf{r},\mathbf{r}';E) \quad (15.94)$$

The Cauchy principal value of a function $f(x)$ having a single pole at $x = a$ is

$$P.V.\int_{-\infty}^{+\infty}f(x)dx = \lim_{\varepsilon\to 0}\left\{\int_{-\infty}^{a-\varepsilon}f(x)dx + \int_{a+\varepsilon}^{+\infty}f(x)dx\right\} \quad (15.95)$$

In the lowest order perturbation theory, the inscattering function due to phase-breaking interactions is given by

$$\Sigma_\phi^{In}(\mathbf{r},\mathbf{r}';E) = \int D(\mathbf{r},\mathbf{r}';\hbar\omega)G^n(\mathbf{r},\mathbf{r}';E-\hbar\omega)d(\hbar\omega) \quad (15.96)$$

and the outscattering function due to phase-breaking interactions is

$$\Sigma_\phi^{Out}(\mathbf{r},\mathbf{r}';E) = \int D(\mathbf{r},\mathbf{r}';\hbar\omega)G^p(\mathbf{r},\mathbf{r}';E+\hbar\omega)d(\hbar\omega) \quad (15.97)$$

where the positive value of $\hbar\omega$ applies to absorption and the negative value to emission. The function $D(\mathbf{r},\mathbf{r}';\hbar\omega)$ representing the spatial correlation and energy spectrum of the phase-breaking scatterers is expressed as

$$D(\mathbf{r},\mathbf{r}';\hbar\omega) = \sum_q|V_q|^2\left[\exp\{-iq(\mathbf{r}-\mathbf{r}')\}N_q\delta(\omega-\omega_q)\right.$$
$$\left. + \exp\{+iq(\mathbf{r}-\mathbf{r}')\}(N_q+1)\delta(\omega+\omega_q)\right] \quad (15.98)$$

where V_q denotes the potential felt by an electron due to a phonon with wave vector q and N_q is the number of phonons having wave vector q and frequency ω_q. Note that the symbol 'q' here denotes the wave vector, not the electronic charge.

For a single phonon with wave vector q and frequency ω_q,

$$|D(\mathbf{r},\mathbf{r}';\hbar\omega)|_{\text{Single phonon}} = |V_q|^2\left[\exp\{-iq(\mathbf{r}-\mathbf{r}')\}N_q\delta(\omega-\omega_q)\right.$$
$$\left. + \exp\{+iq(\mathbf{r}-\mathbf{r}')\}(N_q+1)\delta(\omega+\omega_q)\right] \quad (15.99)$$

and

$$|\Sigma_\phi^{In}(\mathbf{r},\mathbf{r}';E)|_{\text{Single phonon}}$$

$$= |V_q|^2\left[\exp\{-iq(\mathbf{r}-\mathbf{r}')\}N_qG^n(\mathbf{r},\mathbf{r}';E-\hbar\omega_q)\right.$$
$$\left. + \exp\{+iq(\mathbf{r}-\mathbf{r}')\}(N_q+1)G^n(\mathbf{r},\mathbf{r}';E+\hbar\omega_q)\right] \quad (15.100)$$

The inscattering function at an energy E consists of two parts, one due to phonon absorption by electrons with energy $(E - \hbar\omega_q)$ and the other due to phonon emission by electrons with energy $(E + \hbar\omega_q)$. The phonon absorption process is proportional to N_q and the emission process to $N_q + 1$.

Equation (15.96) represents a convolution integral. By Fourier transformation, it may be changed from frequency domain described by ω to time domain described by τ. According to the convolution theorem, the Fourier transform of a convolution of two functions is the product of their Fourier transforms. For eq. (15.96), this product may be written as

Fourier transform of $\Sigma_\phi^{\text{In}}(\mathbf{r}, \mathbf{r}'; E)$

$\quad$ = Fourier transform of $D(\mathbf{r}, \mathbf{r}'; \hbar\omega)$

$\quad\quad \times$ Fourier transform of $G^n(\mathbf{r}, \mathbf{r}'; E - \hbar\omega)$ $\quad$ (15.101)

or

$$\Sigma_\phi^{\text{In}}(\mathbf{r}, \mathbf{r}'; \tau) = \bar{D}(\mathbf{r}, \mathbf{r}'; \tau) G^n(\mathbf{r}, \mathbf{r}'; \tau) \quad (15.102)$$

where $\bar{D}$ is the Fourier-transformed function.

15.8 Terminal Current

The probability density $P(r, t)$ in quantum mechanics is related to the wave function $\Psi(r, t)$ as

$$P(r, t) = \Psi^*(r,t)\Psi(r,t) \quad (15.103)$$

The time derivative of the probability density is

$$\frac{\partial P(r, t)}{\partial t} = \frac{\partial}{\partial t}\left\{\Psi^*(r,t)\Psi(r,t)\right\}$$

$$= \Psi^*(r,t)\frac{\partial \Psi(r,t)}{\partial t} + \Psi(r,t)\frac{\partial \Psi^*(r,t)}{\partial t} \quad (15.104)$$

By modifying eq. (3.7), the time-dependent Schrodinger equation for a particle of charge e moving in an electromagnetic field (electric potential V, vector potential $\mathbf{A}$) is

$$-\left(\frac{\hbar}{i}\right)\frac{\partial \Psi(r,t)}{\partial t} = \frac{1}{2m}\left\{\left(\frac{\hbar}{i}\right)\nabla - e\mathbf{A}\right\}$$

$$\cdot\left\{\left(\frac{\hbar}{i}\right)\nabla - e\mathbf{A}\right\}\Psi(r,t) + eV\Psi(r,t) \quad (15.105)$$

or

$$\frac{\partial \Psi(r,t)}{\partial t} = -\left(\frac{i}{\hbar}\right)\frac{1}{2m}\left\{\left(\frac{\hbar}{i}\right)\nabla - e\mathbf{A}\right\}$$

$$\cdot\left\{\left(\frac{\hbar}{i}\right)\nabla - e\mathbf{A}\right\}\Psi(r,t) - \left(\frac{i}{\hbar}\right)eV\Psi(r,t) \quad (15.106)$$

$$\therefore \Psi^*(r,t)\frac{\partial \Psi(r,t)}{\partial t} = -\left(\frac{i}{\hbar}\right)\Psi^*(r,t)\frac{1}{2m}\left\{\left(\frac{\hbar}{i}\right)\nabla - e\mathbf{A}\right\}$$

$$\cdot\left\{\left(\frac{\hbar}{i}\right)\nabla - e\mathbf{A}\right\}\Psi(r,t)$$

$$-\left(\frac{i}{\hbar}\right)eV\Psi^*(r,t)\Psi(r,t) \quad (15.107)$$

To get $\dfrac{\partial \Psi^*(r,t)}{\partial t}$, we take the complex conjugate of eq. (15.105) by changing the sign of imaginary part and writing Ψ^* in place of Ψ to get

$$\left(\frac{\hbar}{i}\right)\frac{\partial \Psi^*(r,t)}{\partial t} = \frac{1}{2m}\left\{-\left(\frac{\hbar}{i}\right)\nabla - e\mathbf{A}\right\}$$

$$\cdot\left\{-\left(\frac{\hbar}{i}\right)\nabla - e\mathbf{A}\right\}\Psi^*(r,t) + eV\Psi^*(r,t) \quad (15.108)$$

or

$$\frac{\partial \Psi^*(r,t)}{\partial t} = \left(\frac{i}{\hbar}\right)\frac{1}{2m}\left[-\left\{\left(\frac{\hbar}{i}\right)\nabla + e\mathbf{A}\right\}\right]$$

$$\cdot\left[-\left\{\left(\frac{\hbar}{i}\right)\nabla + e\mathbf{A}\right\}\right]\Psi^*(r,t) + \left(\frac{i}{\hbar}\right)eV\Psi^*(r,t)$$

$$= \left(\frac{i}{\hbar}\right)\frac{1}{2m}\left\{\left(\frac{\hbar}{i}\right)\nabla + e\mathbf{A}\right\}$$

$$\cdot\left\{\left(\frac{\hbar}{i}\right)\nabla + e\mathbf{A}\right\}\Psi^*(r,t) + \left(\frac{i}{\hbar}\right)eV\Psi^*(r,t) \quad (15.109)$$

$$\therefore \Psi(r,t)\frac{\partial \Psi^*(r,t)}{\partial t} = \left(\frac{i}{\hbar}\right)\frac{1}{2m}\Psi(r,t)\left\{\left(\frac{\hbar}{i}\right)\nabla + e\mathbf{A}\right\}$$

$$\cdot\left\{\left(\frac{\hbar}{i}\right)\nabla + e\mathbf{A}\right\}\Psi^*(r,t)$$

$$+ \left(\frac{i}{\hbar}\right)eV\Psi(r,t)\Psi^*(r,t) \quad (15.110)$$

From eqs. (15.104), (15.107), and (15.110), we have

$$\frac{\partial P(r,t)}{\partial t} = -\left(\frac{i}{\hbar}\right)\Psi^*(r,t)\frac{1}{2m}\left\{\left(\frac{\hbar}{i}\right)\nabla - e\mathbf{A}\right\}\cdot\left\{\left(\frac{\hbar}{i}\right)\nabla - e\mathbf{A}\right\}\Psi(r,t) - \left(\frac{i}{\hbar}\right)eV\Psi^*(r,t)\Psi(r,t)$$

$$+\left(\frac{i}{\hbar}\right)\Psi(r,t)\frac{1}{2m}\left\{\left(\frac{\hbar}{i}\right)\nabla + e\mathbf{A}\right\}\left\{\left(\frac{\hbar}{i}\right)\nabla + e\mathbf{A}\right\}\Psi^*(r,t) + \left(\frac{i}{\hbar}\right)eV\Psi(r,t)\Psi^*(r,t)$$

$$= -\left(\frac{i}{\hbar}\right)\Psi^*(r,t)\frac{1}{2m}\left\{\left(\frac{\hbar}{i}\right)\nabla - e\mathbf{A}\right\}\cdot\left\{\left(\frac{\hbar}{i}\right)\nabla\Psi(r,t) - e\mathbf{A}\Psi(r,t)\right\} - \left(\frac{i}{\hbar}\right)eV\Psi^*(r,t)\Psi(r,t)$$

$$+\left(\frac{i}{\hbar}\right)\Psi(r,t)\frac{1}{2m}\left\{\left(\frac{\hbar}{i}\right)\nabla + e\mathbf{A}\right\}\cdot\left\{\left(\frac{\hbar}{i}\right)\nabla\Psi^*(r,t) + e\mathbf{A}\Psi^*(r,t)\right\} + \left(\frac{i}{\hbar}\right)eV\Psi(r,t)\Psi^*(r,t)$$

$$= -\left(\frac{i}{\hbar}\right)\Psi^*(r,t)\frac{1}{2m}\cdot\left\{\left(\frac{\hbar^2}{i^2}\right)\nabla^2\Psi(r,t) - \left(\frac{\hbar}{i}\right)e\nabla\mathbf{A}\Psi(r,t) - \left(\frac{\hbar}{i}\right)e\mathbf{A}\nabla\Psi(r,t) - e\mathbf{A}\left(\frac{\hbar}{i}\right)\nabla\Psi(r,t) + e^2\mathbf{A}^2\Psi(r,t)\right\}$$

$$-\left(\frac{i}{\hbar}\right)eV\Psi^*(r,t)\Psi(r,t)$$

$$+\left(\frac{i}{\hbar}\right)\Psi(r,t)\frac{1}{2m}\left\{\left(\frac{\hbar^2}{i^2}\right)\nabla^2\Psi^*(r,t) + \left(\frac{\hbar}{i}\right)e\nabla\mathbf{A}\Psi^*(r,t) + \left(\frac{\hbar}{i}\right)e\mathbf{A}\nabla\Psi^*(r,t) + e\mathbf{A}\left(\frac{\hbar}{i}\right)\nabla\Psi^*(r,t) + e^2\mathbf{A}^2\Psi^*(r,t)\right\}$$

$$+\left(\frac{i}{\hbar}\right)eV\Psi(r,t)\Psi^*(r,t)$$

$$= \left\{-\left(\frac{i}{\hbar}\right)\Psi^*(r,t)\frac{1}{2m}\times\left(\frac{\hbar^2}{i^2}\right)\nabla^2\Psi(r,t)\right\} + \left\{\left(\frac{i}{\hbar}\right)\Psi^*(r,t)\frac{1}{2m}\times\left(\frac{\hbar}{i}\right)e\nabla\mathbf{A}\Psi(r,t)\right\} + \left\{\left(\frac{i}{\hbar}\right)\Psi^*(r,t)\frac{1}{2m}\times\left(\frac{\hbar}{i}\right)e\mathbf{A}\nabla\Psi(r,t)\right\}$$

$$+\left\{\left(\frac{i}{\hbar}\right)\Psi^*(r,t)\frac{1}{2m}\times e\mathbf{A}\left(\frac{\hbar}{i}\right)\nabla\Psi(r,t)\right\} + \left\{-\left(\frac{i}{\hbar}\right)\Psi^*(r,t)\frac{1}{2m}\times e^2\mathbf{A}^2\Psi(r,t)\right\} - \left(\frac{i}{\hbar}\right)eV\Psi^*(r,t)\Psi(r,t)$$

$$+\left\{\left(\frac{i}{\hbar}\right)\Psi(r,t)\frac{1}{2m}\times\left(\frac{\hbar^2}{i^2}\right)\nabla^2\Psi^*(r,t)\right\} + \left\{\left(\frac{i}{\hbar}\right)\Psi(r,t)\frac{1}{2m}\times\left(\frac{\hbar}{i}\right)e\nabla\mathbf{A}\Psi^*(r,t)\right\} + \left\{\left(\frac{i}{\hbar}\right)\Psi(r,t)\frac{1}{2m}\times\left(\frac{\hbar}{i}\right)e\mathbf{A}\nabla\Psi^*(r,t)\right\}$$

$$+\left\{\left(\frac{i}{\hbar}\right)\Psi(r,t)\frac{1}{2m}\times e\mathbf{A}\left(\frac{\hbar}{i}\right)\nabla\Psi^*(r,t)\right\} + \left\{\left(\frac{i}{\hbar}\right)\Psi(r,t)\frac{1}{2m}\times e^2\mathbf{A}^2\Psi^*(r,t)\right\} + \left(\frac{i}{\hbar}\right)eV\Psi(r,t)\Psi^*(r,t)$$

$$= -\left(\frac{\hbar}{i}\right)\Psi^*(r,t)\frac{1}{2m}\nabla^2\Psi(r,t) + \Psi^*(r,t)\frac{1}{2m}e\nabla\mathbf{A}\Psi(r,t) + \Psi^*(r,t)\frac{1}{2m}e\mathbf{A}\nabla\Psi(r,t) + \Psi^*(r,t)\frac{1}{2m}e\mathbf{A}\nabla\Psi(r,t)$$

$$-\left(\frac{i}{\hbar}\right)\Psi^*(r,t)\frac{1}{2m}e^2\mathbf{A}^2\Psi(r,t) - \left(\frac{i}{\hbar}\right)eV\Psi^*(r,t)\Psi(r,t) + \left(\frac{\hbar}{i}\right)\Psi(r,t)\frac{1}{2m}\nabla^2\Psi^*(r,t) + \Psi(r,t)\frac{1}{2m}e\nabla\mathbf{A}\Psi^*(r,t)$$

$$+\Psi(r,t)\frac{1}{2m}e\nabla\mathbf{A}\Psi^*(r,t) + \Psi(r,t)\frac{1}{2m}e\mathbf{A}\nabla\Psi^*(r,t) + \Psi(r,t)\frac{1}{2m}e\mathbf{A}\nabla\Psi^*(r,t)$$

$$+\left(\frac{i}{\hbar}\right)\Psi(r,t)\frac{1}{2m}e^2\mathbf{A}^2\Psi^*(r,t) + \left(\frac{i}{\hbar}\right)eV\Psi(r,t)\Psi^*(r,t)$$

$$= -\left(\frac{\hbar}{i}\right)\Psi^*(r,t)\frac{1}{2m}\nabla^2\Psi(r,t) + \left(\frac{\hbar}{i}\right)\Psi(r,t)\frac{1}{2m}\nabla^2\Psi^*(r,t) + \Psi(r,t)\frac{1}{2m}e\nabla\mathbf{A}\Psi^*(r,t) + \Psi(r,t)\frac{1}{2m}e\nabla\mathbf{A}\Psi^*(r,t)$$

$$+\Psi^*(r,t)\frac{1}{2m}e\mathbf{A}\nabla\Psi(r,t) + \Psi^*(r,t)\frac{1}{2m}e\mathbf{A}\nabla\Psi(r,t) + \Psi(r,t)\frac{1}{2m}e\mathbf{A}\nabla\Psi^*(r,t) + \Psi(r,t)\frac{1}{2m}e\mathbf{A}\nabla\Psi^*(r,t)$$

$$= -\left(\frac{\hbar}{i}\right)\Psi^*(r,t)\frac{1}{2m}\nabla^2\Psi(r,t) + \left(\frac{\hbar}{i}\right)\Psi(r,t)\frac{1}{2m}\nabla^2\Psi^*(r,t) + \Psi(r,t)\frac{2}{2m}e\nabla\mathbf{A}\Psi^*(r,t)$$

$$+\Psi^*(r,t)\frac{2}{2m}e\mathbf{A}\nabla\Psi(r,t) + \Psi(r,t)\frac{2}{2m}e\mathbf{A}\nabla\Psi^*(r,t) \tag{15.111}$$

Now

$$-\nabla.\left[\frac{1}{2m}\Psi^*(r,t)\left\{\left(\frac{\hbar}{i}\right)\nabla-e\mathbf{A}\right\}\Psi(r,t)+\frac{1}{2m}\Psi(r,t)\left\{\left(-\frac{\hbar}{i}\right)\nabla-e\mathbf{A}\right\}\Psi^*(r,t)\right]$$

$$=-\nabla.\left[\frac{1}{2m}\Psi^*(r,t)\left\{\left(\frac{\hbar}{i}\right)\nabla-e\mathbf{A}\right\}\Psi(r,t)-\frac{1}{2m}\Psi(r,t)\left\{\left(\frac{\hbar}{i}\right)\nabla+e\mathbf{A}\right\}\Psi^*(r,t)\right]$$

$$=-\nabla.\left[\frac{1}{2m}\Psi^*(r,t)\left(\frac{\hbar}{i}\right)\nabla\Psi(r,t)-\frac{1}{2m}\Psi^*(r,t)e\mathbf{A}\Psi(r,t)-\frac{1}{2m}\Psi(r,t)\left(\frac{\hbar}{i}\right)\nabla\Psi^*(r,t)-\frac{1}{2m}\Psi(r,t)e\mathbf{A}\Psi^*(r,t)\right]$$

$$=-\frac{1}{2m}\nabla\Psi^*(r,t)\left(\frac{\hbar}{i}\right)\nabla\Psi(r,t)-\frac{1}{2m}\Psi^*(r,t)\left(\frac{\hbar}{i}\right)\nabla^2\Psi(r,t)$$

$$+\frac{1}{2m}\nabla\Psi^*(r,t)e\mathbf{A}\Psi(r,t)+\frac{1}{2m}\Psi^*(r,t)e\nabla\mathbf{A}\Psi(r,t)+\frac{1}{2m}\Psi^*(r,t)e\mathbf{A}\nabla\Psi(r,t)$$

$$+\frac{1}{2m}\nabla\Psi(r,t)\left(\frac{\hbar}{i}\right)\nabla\Psi^*(r,t)+\frac{1}{2m}\Psi(r,t)\left(\frac{\hbar}{i}\right)\nabla^2\Psi^*(r,t)+\frac{1}{2m}\nabla\Psi(r,t)e\mathbf{A}\Psi^*(r,t)$$

$$+\frac{1}{2m}\Psi(r,t)e\nabla\mathbf{A}\Psi^*(r,t)+\frac{1}{2m}\Psi(r,t)e\mathbf{A}\nabla\Psi^*(r,t)$$

$$=-\frac{1}{2m}\Psi^*(r,t)\left(\frac{\hbar}{i}\right)\nabla^2\Psi(r,t)+\frac{1}{2m}\Psi(r,t)\left(\frac{\hbar}{i}\right)\nabla^2\Psi^*(r,t)+\frac{1}{2m}\Psi^*(r,t)e\nabla\mathbf{A}\Psi(r,t)$$

$$+\frac{1}{2m}\Psi(r,t)e\nabla\mathbf{A}\Psi^*(r,t)+\frac{1}{2m}\nabla\Psi^*(r,t)e\mathbf{A}\Psi(r,t)$$

$$+\frac{1}{2m}\Psi(r,t)e\mathbf{A}\nabla\Psi^*(r,t)+\frac{1}{2m}\Psi^*(r,t)e\mathbf{A}\nabla\Psi(r,t)+\frac{1}{2m}\nabla\Psi(r,t)e\mathbf{A}\Psi^*(r,t)$$

$$=-\frac{1}{2m}\Psi^*(r,t)\left(\frac{\hbar}{i}\right)\nabla^2\Psi(r,t)+\frac{1}{2m}\Psi(r,t)\left(\frac{\hbar}{i}\right)\nabla^2\Psi^*(r,t)$$

$$+\frac{2}{2m}\Psi^*(r,t)e\nabla\mathbf{A}\Psi(r,t)+\frac{2}{2m}\nabla\Psi^*(r,t)e\mathbf{A}\Psi(r,t)+\frac{2}{2m}\Psi^*(r,t)e\mathbf{A}\nabla\Psi(r,t) \qquad (15.112)$$

Comparison of eqs. (15.111) and (15.112) shows that

$$\frac{\partial P(r,t)}{\partial t}=-\nabla.\left[\frac{1}{2m}\Psi^*(r,t)\left\{\left(\frac{\hbar}{i}\right)\nabla-e\mathbf{A}\right\}\Psi(r,t)+\frac{1}{2m}\Psi(r,t)\left\{\left(-\frac{\hbar}{i}\right)\nabla-e\mathbf{A}\right\}\Psi^*(r,t)\right]$$

$$=-\nabla.\left[\frac{1}{2m}\Psi^*(r,t)\left\{\left(\frac{\hbar}{i}\right)\nabla-e\mathbf{A}\right\}\Psi(r,t)+\frac{1}{2m}\Psi(r,t)\left(\left\{\left(+\frac{\hbar}{i}\right)\nabla-e\mathbf{A}\right\}\Psi(r,t)\right)^*\right]$$

$$=-\nabla.\left[\frac{1}{2m}\Psi^*(r,t)(p-e\mathbf{A})\Psi(r,t)+\frac{1}{2m}\Psi(r,t)\left\{(p-e\mathbf{A})\Psi(r,t)\right\}^*\right]$$

$$=-\nabla.\left[\frac{\Psi^*(r,t)(p-e\mathbf{A})\Psi(r,t)}{2m}+\frac{\Psi(r,t)\left\{(p-e\mathbf{A})\Psi(r,t)\right\}^*}{2m}\right] \qquad (15.113)$$

or

$$\frac{\partial P(r,t)}{\partial t}+\nabla.\left[\frac{\Psi^*(r,t)(p-e\mathbf{A})\Psi(r,t)}{2m}+\frac{\Psi(r,t)\left\{(p-e\mathbf{A})\Psi(r,t)\right\}^*}{2m}\right]=0 \qquad (15.114)$$

where

$$p = \left(\frac{\hbar}{i}\right)\nabla = \left(\frac{i\hbar}{i^2}\right)\nabla = \left(\frac{i\hbar}{-1}\right)\nabla = -i\hbar\nabla \qquad (15.115)$$

The continuity equation is

$$\frac{\partial P(r,t)}{\partial t} + \nabla \cdot \mathbf{J}_P = 0 \qquad (15.116)$$

where $\mathbf{J}_P$ is the probability current density. From eqs. (15.114) and (15.116), the probability current density is

$$\mathbf{J}_P = \frac{\Psi^*(r,t)(p-e\mathbf{A})\Psi(r,t)}{2m} + \frac{\Psi(r,t)\{(p-e\mathbf{A})\Psi(r,t)\}^*}{2m} \qquad (15.117)$$

Multiplying throughout by electronic charge e, the electric current density is

$$\mathbf{J} = e\left[\frac{\Psi^*(r,t)(\mathbf{p}-e\mathbf{A})\Psi(r,t)}{2m} + \frac{\Psi(r,t)\{(\mathbf{p}-e\mathbf{A})\Psi(r,t)\}^*}{2m}\right]$$
$$= \frac{e}{2m}\left[\Psi^*(r,t)(\mathbf{p}-e\mathbf{A})\Psi(r,t) + \Psi(r,t)\{(\mathbf{p}-e\mathbf{A})\Psi(r,t)\}^*\right] \qquad (15.118)$$

which may be written as

$$\mathbf{J}(\mathbf{r},E) = \frac{e}{2m}\left[\Psi^*(\mathbf{r})\{\mathbf{p}-e\mathbf{A}(\mathbf{r})\}\Psi(\mathbf{r}) + \Psi(\mathbf{r})\left(\{\mathbf{p}-e\mathbf{A}(\mathbf{r})\}\Psi(\mathbf{r})\right)^*\right]$$
$$= \frac{e}{2m}\left[\Psi^*(\mathbf{r})\mathbf{p}\Psi(\mathbf{r}) - \Psi^*(\mathbf{r})e\mathbf{A}(\mathbf{r})\Psi(\mathbf{r}) + \Psi(\mathbf{r})\left(\{-\mathbf{p}-e\mathbf{A}(\mathbf{r})\}\Psi^*(\mathbf{r})\right)\right] \qquad (15.119)$$

where the sign of imaginary component $p = -i\hbar\nabla$ is changed. Hence,

$$\mathbf{J}(\mathbf{r},E) = \frac{e}{2m}\left[\Psi^*(\mathbf{r})\mathbf{p}\Psi(\mathbf{r}) - \Psi^*(\mathbf{r})e\mathbf{A}(\mathbf{r})\Psi(\mathbf{r}) - \Psi(\mathbf{r})\mathbf{p}\Psi^*(\mathbf{r}) - \Psi(\mathbf{r})e\mathbf{A}(\mathbf{r})\Psi^*(\mathbf{r})\right]$$
$$= \frac{e}{2m}\left[\Psi^*(\mathbf{r})\mathbf{p}\Psi(\mathbf{r}) - \Psi(\mathbf{r})\mathbf{p}\Psi^*(\mathbf{r}) - \Psi^*(\mathbf{r})e\mathbf{A}(\mathbf{r})\Psi(\mathbf{r}) - \Psi(\mathbf{r})e\mathbf{A}(\mathbf{r})\Psi^*(\mathbf{r})\right] \qquad (15.120)$$

which is written as

$$\mathbf{J}(\mathbf{r},E) = \frac{e}{2m}\left[\Psi^*(\mathbf{r}')\mathbf{p}\Psi(\mathbf{r}) - \Psi(\mathbf{r})\mathbf{p}'\Psi^*(\mathbf{r}') - \Psi^*(\mathbf{r}')e\mathbf{A}(\mathbf{r})\Psi(\mathbf{r}) - \Psi(\mathbf{r})e\mathbf{A}(\mathbf{r})\Psi^*(\mathbf{r}')\right]_{\mathbf{r}'=\mathbf{r}} \qquad (15.121)$$

where

$$\mathbf{p} = -i\hbar\nabla \text{ and } \mathbf{p}' = -i\hbar\nabla' \qquad (15.122)$$

the gradient operator ∇ operates on $\mathbf{r}$ and ∇' on $\mathbf{r}'$. Equation (15.121) can be recast as

$$\mathbf{J}(\mathbf{r},E) = \frac{e}{2m}\left[\{(\mathbf{p}-\mathbf{p}')\Psi(\mathbf{r})\Psi^*(\mathbf{r}')\}_{\mathbf{r}'=\mathbf{r}} - 2e\mathbf{A}(\mathbf{r})\{\Psi(\mathbf{r})\Psi^*(\mathbf{r}')\}_{\mathbf{r}'=\mathbf{r}}\right]$$
$$= \frac{e}{2m}\{(\mathbf{p}-\mathbf{p}')\Psi(\mathbf{r})\Psi^*(\mathbf{r}')\}_{\mathbf{r}'=\mathbf{r}} - \frac{e}{2m} \times 2e\mathbf{A}(\mathbf{r})\{\Psi(\mathbf{r})\Psi^*(\mathbf{r}')\}_{\mathbf{r}'=\mathbf{r}} \qquad (15.123)$$
$$= \frac{e}{2m}\{(\mathbf{p}-\mathbf{p}')\Psi(\mathbf{r})\Psi^*(\mathbf{r}')\}_{\mathbf{r}'=\mathbf{r}} - \frac{e^2}{m}\mathbf{A}(\mathbf{r})\{\Psi(\mathbf{r})\Psi^*(\mathbf{r}')\}_{\mathbf{r}'=\mathbf{r}}$$

Now

Wave function $\Psi(\mathbf{r})\Psi^*(\mathbf{r})$ = Electron density per unit energy $n(\mathbf{r},E) = \dfrac{\text{Electron correlation function } G^n(\mathbf{r},\mathbf{r}';E)}{2\pi} \qquad (15.124)$

In eq. (15.123), replacing the wave function with the correlation function according to eq. (15.124)

$$
\begin{aligned}
\mathbf{J}(\mathbf{r},E) &= \frac{e}{2m}\left\{(\mathbf{p}-\mathbf{p}')\frac{G^{n}(\mathbf{r},\mathbf{r}';E)}{2\pi}\right\}_{\mathbf{r}'=\mathbf{r}} - \frac{e^{2}}{m}\mathbf{A}(\mathbf{r})\left\{\frac{G^{n}(\mathbf{r},\mathbf{r}';E)}{2\pi}\right\}_{\mathbf{r}'=\mathbf{r}} \\
&= \frac{e}{2m}\left[\{-i\hbar\nabla-(-i\hbar\nabla')\}\frac{G^{n}(\mathbf{r},\mathbf{r}';E)}{2\pi}\right]_{\mathbf{r}'=\mathbf{r}} - \frac{e^{2}}{m}\mathbf{A}(\mathbf{r})\left\{\frac{G^{n}(\mathbf{r},\mathbf{r}';E)}{2\pi}\right\}_{\mathbf{r}'=\mathbf{r}} \\
&= \frac{e}{2m}\left[\{-i\hbar\nabla+(i\hbar\nabla')\}\frac{G^{n}(\mathbf{r},\mathbf{r}';E)}{2\pi}\right]_{\mathbf{r}'=\mathbf{r}} - \frac{e^{2}}{m}\mathbf{A}(\mathbf{r})\left\{\frac{G^{n}(\mathbf{r},\mathbf{r}';E)}{2\pi}\right\}_{\mathbf{r}'=\mathbf{r}} \\
&= -\frac{i\hbar e}{2m}\left[\{\nabla-\nabla'\}\frac{G^{n}(\mathbf{r},\mathbf{r}';E)}{2\pi}\right]_{\mathbf{r}'=\mathbf{r}} - \frac{e^{2}}{m}\mathbf{A}(\mathbf{r})\left\{\frac{G^{n}(\mathbf{r},\mathbf{r}';E)}{2\pi}\right\}_{\mathbf{r}'=\mathbf{r}} \\
&= \left[-\frac{i\hbar e}{2m}\left\{\nabla-\nabla'\right\}\frac{G^{n}(\mathbf{r},\mathbf{r}';E)}{2\pi}\right\} - \frac{e^{2}}{m}\mathbf{A}(\mathbf{r})\left\{\frac{G^{n}(\mathbf{r},\mathbf{r}';E)}{2\pi}\right\}\right]_{\mathbf{r}'=\mathbf{r}}
\end{aligned} \tag{15.125}
$$

Integration of $\mathbf{J}(\mathbf{r}, E)$ over the cross section S_{p} of the contact separating the conductor from the lead p yields the terminal current per unit energy as

$$
i_{p}(E) = \int \mathbf{J}(\mathbf{r},E)dS_{\mathrm{p}} \tag{15.126}
$$

15.9 Direct Determination of Terminal Currents without Current Density Calculation

From eq. (15.121)

$$
\begin{aligned}
\nabla\cdot\mathbf{J}(\mathbf{r},E) &= \frac{e}{2m}\nabla\cdot\left[\Psi^{*}(\mathbf{r})\mathbf{p}\Psi(\mathbf{r})-\Psi(\mathbf{r})\mathbf{p}\Psi^{*}(\mathbf{r})-\Psi^{*}(\mathbf{r})e\mathbf{A}(\mathbf{r})\Psi(\mathbf{r})-\Psi(\mathbf{r})e\mathbf{A}(\mathbf{r})\Psi^{*}(\mathbf{r})\right] \\
&= \frac{e}{2m}\nabla\cdot\left[\Psi^{*}(\mathbf{r})(-i\hbar\nabla)\Psi(\mathbf{r})-\Psi(\mathbf{r})(-i\hbar\nabla)\Psi^{*}(\mathbf{r})-\Psi^{*}(\mathbf{r})e\mathbf{A}(\mathbf{r})\Psi(\mathbf{r})-\Psi(\mathbf{r})e\mathbf{A}(\mathbf{r})\Psi^{*}(\mathbf{r})\right] \\
&= \frac{e}{2m}\nabla\cdot\left[-i\hbar\Psi^{*}(\mathbf{r})\nabla\Psi(\mathbf{r})+i\hbar\Psi(\mathbf{r})\nabla\Psi^{*}(\mathbf{r})-\Psi^{*}(\mathbf{r})e\mathbf{A}(\mathbf{r})\Psi(\mathbf{r})-\Psi(\mathbf{r})e\mathbf{A}(\mathbf{r})\Psi^{*}(\mathbf{r})\right] \\
&= \frac{e}{2m}\left[\begin{array}{l}-i\hbar\nabla\Psi^{*}(\mathbf{r})\nabla\Psi(\mathbf{r})-i\hbar\Psi^{*}(\mathbf{r})\nabla^{2}\Psi(\mathbf{r})+i\hbar\nabla\Psi(\mathbf{r})\nabla\Psi^{*}(\mathbf{r})+i\hbar\Psi(\mathbf{r})\nabla^{2}\Psi^{*}(\mathbf{r})-\nabla\Psi^{*}(\mathbf{r})e\mathbf{A}(\mathbf{r})\Psi(\mathbf{r})-\Psi^{*}(\mathbf{r})e\mathbf{A}(\mathbf{r})\nabla\Psi(\mathbf{r}) \\ -e\Psi^{*}(\mathbf{r})\Psi(\mathbf{r})\nabla\mathbf{A}(\mathbf{r})-\nabla\Psi(\mathbf{r})e\mathbf{A}(\mathbf{r})\Psi^{*}(\mathbf{r})-\Psi(\mathbf{r})e\mathbf{A}(\mathbf{r})\nabla\Psi^{*}(\mathbf{r})-e\Psi(\mathbf{r})\Psi^{*}(\mathbf{r})\nabla\mathbf{A}(\mathbf{r})\end{array}\right] \\
&= \frac{e}{2m}\left[i\hbar\{\Psi(\mathbf{r})\nabla^{2}\Psi^{*}(\mathbf{r})-\Psi^{*}(\mathbf{r})\nabla^{2}\Psi(\mathbf{r})\}-2\Psi(\mathbf{r})e\mathbf{A}(\mathbf{r})\nabla\Psi^{*}(\mathbf{r})-2\Psi^{*}(\mathbf{r})e\mathbf{A}(\mathbf{r})\nabla\Psi(\mathbf{r})-2e\Psi(\mathbf{r})\Psi^{*}(\mathbf{r})\nabla\mathbf{A}(\mathbf{r})\right] \\
&= \frac{ei\hbar}{2m}\{\Psi(\mathbf{r})\nabla^{2}\Psi^{*}(\mathbf{r})-\Psi^{*}(\mathbf{r})\nabla^{2}\Psi(\mathbf{r})\}-\frac{e^{2}}{m}\{\Psi(\mathbf{r})\nabla\Psi^{*}(\mathbf{r})+\Psi^{*}(\mathbf{r})\nabla\Psi(\mathbf{r})\}-\frac{e^{2}}{m}\Psi(\mathbf{r})\Psi^{*}(\mathbf{r})\nabla\mathbf{A}(\mathbf{r}) \\
&= \frac{ei\hbar\{\Psi(\mathbf{r})\nabla^{2}\Psi^{*}(\mathbf{r})-\Psi^{*}(\mathbf{r})\nabla^{2}\Psi(\mathbf{r})\}-2e^{2}\{\Psi(\mathbf{r})\nabla\Psi^{*}(\mathbf{r})+\Psi^{*}(\mathbf{r})\nabla\Psi(\mathbf{r})\}-2e^{2}\Psi(\mathbf{r})\Psi^{*}(\mathbf{r})\nabla\mathbf{A}(\mathbf{r})}{2m}
\end{aligned} \tag{15.127}
$$

Multiplying both sides of eq. (15.127) by $i\hbar/e$ we get

$$
\begin{aligned}
\frac{i\hbar}{e}\nabla\cdot\mathbf{J}(\mathbf{r},E) &= \frac{i^{2}\hbar^{2}\{\Psi(\mathbf{r})\nabla^{2}\Psi^{*}(\mathbf{r})-\Psi^{*}(\mathbf{r})\nabla^{2}\Psi(\mathbf{r})\}-2i\hbar e\{\Psi(\mathbf{r})\nabla\Psi^{*}(\mathbf{r})+\Psi^{*}(\mathbf{r})\nabla\Psi(\mathbf{r})\}-2i\hbar e\Psi(\mathbf{r})\Psi^{*}(\mathbf{r})\nabla\mathbf{A}(\mathbf{r})}{2m} \\
&= \frac{-\hbar^{2}\{\Psi(\mathbf{r})\nabla^{2}\Psi^{*}(\mathbf{r})-\Psi^{*}(\mathbf{r})\nabla^{2}\Psi(\mathbf{r})\}-2i\hbar e\{\Psi(\mathbf{r})\nabla\Psi^{*}(\mathbf{r})+\Psi^{*}(\mathbf{r})\nabla\Psi(\mathbf{r})\}-2i\hbar e\Psi(\mathbf{r})\Psi^{*}(\mathbf{r})\nabla\mathbf{A}(\mathbf{r})}{2m}
\end{aligned} \tag{15.128}
$$

But

$$\left\{\mathbf{p}-e\mathbf{A}(\mathbf{r})\right\}^2\Psi(\mathbf{r})=\left\{\mathbf{p}-e\mathbf{A}(\mathbf{r})\right\}\cdot\left\{\mathbf{p}-e\mathbf{A}(\mathbf{r})\right\}\Psi(\mathbf{r})=\left\{-i\hbar\nabla-e\mathbf{A}(\mathbf{r})\right\}\cdot\left\{-i\hbar\nabla-e\mathbf{A}(\mathbf{r})\right\}\Psi(\mathbf{r})$$

$$=\left\{-i\hbar\nabla-e\mathbf{A}(\mathbf{r})\right\}\cdot\left\{-i\hbar\nabla\Psi(\mathbf{r})-e\mathbf{A}(\mathbf{r})\Psi(\mathbf{r})\right\}$$

$$=i^2\hbar^2\nabla^2\Psi(\mathbf{r})+i\hbar e\nabla\mathbf{A}(\mathbf{r})\Psi(\mathbf{r})+i\hbar e\mathbf{A}(\mathbf{r})\nabla\Psi(\mathbf{r})+i\hbar e\mathbf{A}(\mathbf{r})\nabla\Psi(\mathbf{r})+e^2\mathbf{A}^2(\mathbf{r})\Psi(\mathbf{r})$$

$$=-\hbar^2\nabla^2\Psi(\mathbf{r})+2i\hbar e\mathbf{A}(\mathbf{r})\nabla\Psi(\mathbf{r})+i\hbar e\nabla\mathbf{A}(\mathbf{r})\Psi(\mathbf{r})+e^2\mathbf{A}^2(\mathbf{r})\Psi(\mathbf{r})$$

$$\therefore\Psi^*(\mathbf{r})\left\{\mathbf{p}-e\mathbf{A}(\mathbf{r})\right\}^2\Psi(\mathbf{r})=-\hbar^2\Psi^*(\mathbf{r})\nabla^2\Psi(\mathbf{r})+2i\hbar e\mathbf{A}(\mathbf{r})\Psi^*(\mathbf{r})\nabla\Psi(\mathbf{r})+i\hbar e\Psi^*(\mathbf{r})\Psi(\mathbf{r})\nabla\mathbf{A}(\mathbf{r})+e^2\mathbf{A}^2(\mathbf{r})\Psi^*(\mathbf{r})\Psi(\mathbf{r})\qquad(15.129)$$

Similarly,

$$\left[\left\{\mathbf{p}-e\mathbf{A}(\mathbf{r})\right\}^2\Psi(\mathbf{r})\right]^*=\left[\left\{\mathbf{p}-e\mathbf{A}(\mathbf{r})\right\}^2\right]^*\Psi^*(\mathbf{r})=\left\{+i\hbar\nabla-e\mathbf{A}(\mathbf{r})\right\}\left\{+i\hbar\nabla-e\mathbf{A}(\mathbf{r})\right\}\Psi^*(\mathbf{r})$$

$$=i^2\hbar^2\nabla^2\Psi^*(\mathbf{r})-i\hbar e\mathbf{A}(\mathbf{r})\nabla\Psi^*(\mathbf{r})-i\hbar e\Psi^*(\mathbf{r})\nabla\mathbf{A}(\mathbf{r})-i\hbar e\mathbf{A}(\mathbf{r})\nabla\Psi^*(\mathbf{r})+e^2\mathbf{A}^2(\mathbf{r})\Psi^*(\mathbf{r})\qquad(15.130)$$

$$\therefore\Psi(\mathbf{r})\left[\left\{\mathbf{p}-e\mathbf{A}(\mathbf{r})\right\}^2\Psi(\mathbf{r})\right]^*=-\hbar^2\Psi(\mathbf{r})\nabla^2\Psi^*(\mathbf{r})-i\hbar e\mathbf{A}(\mathbf{r})\Psi(\mathbf{r})\nabla\Psi^*(\mathbf{r})$$

$$-i\hbar e\Psi(\mathbf{r})\Psi^*(\mathbf{r})\nabla\mathbf{A}(\mathbf{r})-i\hbar e\mathbf{A}(\mathbf{r})\Psi(\mathbf{r})\nabla\Psi^*(\mathbf{r})+e^2\mathbf{A}^2(\mathbf{r})\Psi(\mathbf{r})\Psi^*(\mathbf{r})$$

$$=-\hbar^2\Psi(\mathbf{r})\nabla^2\Psi^*(\mathbf{r})-2i\hbar e\mathbf{A}(\mathbf{r})\Psi(\mathbf{r})\nabla\Psi^*(\mathbf{r})-i\hbar e\Psi(\mathbf{r})\Psi^*(\mathbf{r})\nabla\mathbf{A}(\mathbf{r})+e^2\mathbf{A}^2(\mathbf{r})\Psi(\mathbf{r})\Psi^*(\mathbf{r})$$

$$(15.131)$$

Subtracting eq. (15.129) from eq. (15.131)

$$\Psi(\mathbf{r})\left[\left\{\mathbf{p}-e\mathbf{A}(\mathbf{r})\right\}^2\Psi(\mathbf{r})\right]^*-\Psi^*(\mathbf{r})\left\{\mathbf{p}-e\mathbf{A}(\mathbf{r})\right\}^2\Psi(\mathbf{r})$$

$$=-\hbar^2\Psi(\mathbf{r})\nabla^2\Psi^*(\mathbf{r})-2i\hbar e\mathbf{A}(\mathbf{r})\Psi(\mathbf{r})\nabla\Psi^*(\mathbf{r})-i\hbar e\Psi(\mathbf{r})\Psi^*(\mathbf{r})\nabla\mathbf{A}(\mathbf{r})+e^2\mathbf{A}^2(\mathbf{r})\Psi(\mathbf{r})\Psi^*(\mathbf{r})$$

$$-\left\{-\hbar^2\Psi^*(\mathbf{r})\nabla^2\Psi(\mathbf{r})+2i\hbar e\mathbf{A}(\mathbf{r})\Psi^*(\mathbf{r})\nabla\Psi(\mathbf{r})+i\hbar e\Psi^*(\mathbf{r})\Psi(\mathbf{r})\nabla\mathbf{A}(\mathbf{r})+e^2\mathbf{A}^2(\mathbf{r})\Psi^*(\mathbf{r})\Psi(\mathbf{r})\right\}$$

$$=-\hbar^2\Psi(\mathbf{r})\nabla^2\Psi^*(\mathbf{r})+\hbar^2\Psi^*(\mathbf{r})\nabla^2\Psi(\mathbf{r})-2i\hbar e\mathbf{A}(\mathbf{r})\left\{\Psi(\mathbf{r})\nabla\Psi^*(\mathbf{r})+\Psi^*(\mathbf{r})\nabla\Psi(\mathbf{r})\right\}$$

$$-2i\hbar e\Psi(\mathbf{r})\Psi^*(\mathbf{r})\nabla\mathbf{A}(\mathbf{r})=-\hbar^2\left\{\Psi(\mathbf{r})\nabla^2\Psi^*(\mathbf{r})-\Psi^*(\mathbf{r})\nabla^2\Psi(\mathbf{r})\right\}$$

$$-2i\hbar e\mathbf{A}(\mathbf{r})\left\{\Psi(\mathbf{r})\nabla\Psi^*(\mathbf{r})+\Psi^*(\mathbf{r})\nabla\Psi(\mathbf{r})\right\}-2i\hbar e\Psi(\mathbf{r})\Psi^*(\mathbf{r})\nabla\mathbf{A}(\mathbf{r})\qquad(15.132)$$

Dividing both sides of eq. (15.132) by $2m$,

$$\frac{\Psi(\mathbf{r})\left[\left\{\mathbf{p}-e\mathbf{A}(\mathbf{r})\right\}^2\Psi(\mathbf{r})\right]^*-\Psi^*(\mathbf{r})\left\{\mathbf{p}-e\mathbf{A}(\mathbf{r})\right\}^2\Psi(\mathbf{r})}{2m}$$

$$=\frac{-\hbar^2\left\{\Psi(\mathbf{r})\nabla^2\Psi^*(\mathbf{r})-\Psi^*(\mathbf{r})\nabla^2\Psi(\mathbf{r})\right\}-2i\hbar e\mathbf{A}(\mathbf{r})\left\{\Psi(\mathbf{r})\nabla\Psi^*(\mathbf{r})+\Psi^*(\mathbf{r})\nabla\Psi(\mathbf{r})\right\}-2i\hbar e\Psi(\mathbf{r})\Psi^*(\mathbf{r})\nabla\mathbf{A}(\mathbf{r})}{2m}\qquad(15.133)$$

Comparison of eqs. (15.128) and (15.133) reveals that

$$\frac{i\hbar}{e}\nabla\cdot\mathbf{J}(\mathbf{r},E)$$

$$=\frac{\Psi(\mathbf{r})\Big[\{\mathbf{p}-e\mathbf{A}(\mathbf{r})\}^2\Psi(\mathbf{r})\Big]^*-\Psi^*(\mathbf{r})\{\mathbf{p}-e\mathbf{A}(\mathbf{r})\}^2\Psi(\mathbf{r})}{2m}$$

$$=\frac{\Psi(\mathbf{r})\Big[\{\mathbf{p}-e\mathbf{A}(\mathbf{r})\}^2\Psi(\mathbf{r})\Big]^*}{2m}-\frac{\Psi^*(\mathbf{r})\{\mathbf{p}-e\mathbf{A}(\mathbf{r})\}^2\Psi(\mathbf{r})}{2m}$$

$$(15.134)$$

The Hamiltonian for the conductor contains the vector potential $\mathbf{A}$ as a representative of any magnetic field and the potential energy U for indicating any boundaries, impurities, or applied potential:

$$H_{\mathrm{C}}=\frac{\{\mathbf{p}-e\mathbf{A}(\mathbf{r})\}^2}{2m}+U\qquad(15.135)$$

or

$$\frac{\{\mathbf{p}-e\mathbf{A}(\mathbf{r})\}^2}{2m}=H_{\mathrm{C}}-U\qquad(15.136)$$

Applying eq. (15.136), eq. (15.134) becomes

$$\frac{i\hbar}{e}\nabla\cdot\mathbf{J}(\mathbf{r},E)=\Psi(\mathbf{r})\Big[(H_{\mathrm{C}}-U)\Psi(\mathbf{r})\Big]^*$$

$$-\Psi^*(\mathbf{r})(H_{\mathrm{C}}-U)\Psi(\mathbf{r})$$

$$=\Psi(\mathbf{r})H_{\mathrm{C}}^*\Psi^*(\mathbf{r})-\Psi(\mathbf{r})U\Psi^*(\mathbf{r})$$

$$-\Psi^*(\mathbf{r})H_{\mathrm{C}}\Psi(\mathbf{r})+\Psi^*(\mathbf{r})U\Psi(\mathbf{r})$$

$$=\Psi(\mathbf{r})H_{\mathrm{C}}^*\Psi^*(\mathbf{r})-\Psi^*(\mathbf{r})H_{\mathrm{C}}\Psi(\mathbf{r})\quad(15.137)$$

where U remains unchanged because it does not contain any imaginary component. Equation (15.137) is written as

$$\frac{i\hbar}{e}\nabla\cdot\mathbf{J}(\mathbf{r},E)=\big\{\Psi(\mathbf{r})H_{\mathrm{C}}^*(\mathbf{r}')\Psi^*(\mathbf{r}')-\Psi^*(\mathbf{r}')H_{\mathrm{C}}(\mathbf{r})\Psi(\mathbf{r})\big\}_{\mathbf{r}'=\mathbf{r}}$$

$$=\Big[\{H_{\mathrm{C}}^*(\mathbf{r}')-H_{\mathrm{C}}(\mathbf{r})\}\Psi(\mathbf{r})\Psi^*(\mathbf{r}')\Big]_{\mathbf{r}'=\mathbf{r}}$$

$$=-\Big[\{H_{\mathrm{C}}(\mathbf{r})-H_{\mathrm{C}}^*(\mathbf{r}')\}\Psi(\mathbf{r})\Psi^*(\mathbf{r}')\Big]_{\mathbf{r}'=\mathbf{r}}\quad(15.138)$$

or

$$i\hbar\nabla\cdot\mathbf{J}(\mathbf{r},E)=-e\Big[\{H_{\mathrm{C}}(\mathbf{r})-H_{\mathrm{C}}^*(\mathbf{r}')\}\Psi(\mathbf{r})\Psi^*(\mathbf{r}')\Big]_{\mathbf{r}'=\mathbf{r}}\quad(15.139)$$

As in the preceding, the wave function is replaced with the correlation function using the eq. (15.124). Hence,

$$i\hbar\nabla\cdot\mathbf{J}(\mathbf{r},E)=-e\Big[\{H_{\mathrm{C}}(\mathbf{r})-H_{\mathrm{C}}^*(\mathbf{r}')\}\frac{G^{\mathrm{n}}(\mathbf{r},\mathbf{r}';E)}{2\pi}\Big]_{\mathbf{r}'=\mathbf{r}}\quad(15.140)$$

or

$$i2\pi\hbar\nabla\cdot\mathbf{J}(\mathbf{r},E)=-e\Big[\{H_{\mathrm{C}}(\mathbf{r})-H_{\mathrm{C}}^*(\mathbf{r}')\}G^{\mathrm{n}}(\mathbf{r},\mathbf{r}';E)\Big]_{\mathbf{r}'=\mathbf{r}}\quad(15.141)$$

or

$$i^2 2\pi\hbar\nabla\cdot\mathbf{J}(\mathbf{r},E)=-ie\Big[\{H_{\mathrm{C}}(\mathbf{r})-H_{\mathrm{C}}^*(\mathbf{r}')\}G^{\mathrm{n}}(\mathbf{r},\mathbf{r}';E)\Big]_{\mathbf{r}'=\mathbf{r}}$$

$$(15.142)$$

or

$$-1\times2\pi\hbar\nabla\cdot\mathbf{J}(\mathbf{r},E)=-ie\Big[\{H_{\mathrm{C}}(\mathbf{r})-H_{\mathrm{C}}^*(\mathbf{r}')\}G^{\mathrm{n}}(\mathbf{r},\mathbf{r}';E)\Big]_{\mathbf{r}'=\mathbf{r}}$$

$$(15.143)$$

or

$$\nabla\cdot\mathbf{J}(\mathbf{r},E)=\frac{ie}{2\pi\hbar}\Big[\{H_{\mathrm{C}}(\mathbf{r})G^{\mathrm{n}}(\mathbf{r},\mathbf{r}';E)-H_{\mathrm{C}}^*(\mathbf{r}')G^{\mathrm{n}}(\mathbf{r},\mathbf{r}';E)\}\Big]_{\mathbf{r}'=\mathbf{r}}$$

$$(15.144)$$

or

$$\nabla\cdot\mathbf{J}(\mathbf{r},E)$$

$$=\frac{ie}{2\pi\frac{h}{2\pi}}\Big[\{H_{\mathrm{C}}(\mathbf{r})G^{\mathrm{n}}(\mathbf{r},\mathbf{r}';E)-H_{\mathrm{C}}^*(\mathbf{r}')G^{\mathrm{n}}(\mathbf{r},\mathbf{r}';E)\}\Big]_{\mathbf{r}'=\mathbf{r}}$$

$$=\frac{ie}{h}\Big[\{H_{\mathrm{C}}(\mathbf{r})G^{\mathrm{n}}(\mathbf{r},\mathbf{r}';E)-H_{\mathrm{C}}^*(\mathbf{r}')G^{\mathrm{n}}(\mathbf{r},\mathbf{r}';E)\}\Big]_{\mathbf{r}'=\mathbf{r}}$$

$$(15.145)$$

For convenience, a current operator I_{op} is defined such that its diagonal elements represent the divergence of the current density. Hence.

$$I_{\mathrm{op}}(E)\equiv\frac{ie}{h}\Big[H_{\mathrm{C}}G^{\mathrm{n}}-G^{\mathrm{n}}H_{\mathrm{C}}\Big]=\nabla\cdot\mathbf{J}(\mathbf{r},E)\quad(15.146)$$

Here,

$$H_{\mathrm{C}}G^{\mathrm{n}}-G^{\mathrm{n}}H_{\mathrm{C}}=H_{\mathrm{C}}G^{\mathrm{R}}\Sigma^{\mathrm{in}}G^{\mathrm{A}}-G^{\mathrm{R}}\Sigma^{\mathrm{in}}G^{\mathrm{A}}H_{\mathrm{C}}\quad(15.147)$$

since from eq. (15.51)

$$G^{\mathrm{n}}=G^{\mathrm{R}}\Sigma^{\mathrm{in}}G^{\mathrm{A}}\qquad(15.148)$$

Moreover, from eq. (15.22)

$$\big[EI-H_{\mathrm{C}}-\Sigma^{\mathrm{R}}\big]G^{\mathrm{R}}=I\qquad(15.149)$$

or

$$EG^{\mathrm{R}}-H_{\mathrm{C}}G^{\mathrm{R}}-\Sigma^{\mathrm{R}}G^{\mathrm{R}}=I\qquad(15.150)$$

or

$$H_{\mathrm{C}}G^{\mathrm{R}}=EG^{\mathrm{R}}-\Sigma^{\mathrm{R}}G^{\mathrm{R}}-I\qquad(15.151)$$

The Hermitian conjugate of $H_C G^R$ is the matrix $G^A H_C$ given by

$$G^A H_C = EG^A - G^A \Sigma^A - I \qquad (15.152)$$

Substituting for $H_C G^R$ and $G^A H_C$ from eqs. (15.151) and (15.152) in eq. (15.147)

$$H_C G^n - G^n H_C = \left(EG^R - \Sigma^R G^R - I \right) \Sigma^{in} G^A$$

$$- G^R \Sigma^{in} \left(EG^A - G^A \Sigma^A - I \right)$$

$$= EG^R \Sigma^{in} G^A - \Sigma^R G^R \Sigma^{in} G^A$$

$$- \Sigma^{in} G^A - G^R \Sigma^{in} EG^A + G^R \Sigma^{in} G^A \Sigma^A + G^R \Sigma^{in}$$

$$= -\Sigma^R G^R \Sigma^{in} G^A - \Sigma^{in} G^A + G^R \Sigma^{in} G^A \Sigma^A + G^R \Sigma^{in}$$

$$= -\Sigma^R G^n - \Sigma^{in} G^A + G^n \Sigma^A + G^R \Sigma^{in}$$

$$= G^R \Sigma^{in} - \Sigma^{in} G^A - \Sigma^R G^n + G^n \Sigma^A \qquad (15.153)$$

by applying eq. (15.148). From eqs. (15.146) and (15.153), the modified equation for the current operator is

$$I_{op}(E) = \frac{ie}{h} \left[G^R \Sigma^{in} - \Sigma^{in} G^A - \Sigma^R G^n + G^n \Sigma^A \right]$$

$$= \frac{ie}{h} \left[\Sigma^{in} \left(G^R - G^A \right) - \left(\Sigma^R - \Sigma^A \right) G^n \right] \qquad (15.154)$$

The trace of the current operator representing the sum of all entries in the main diagonal, i.e. the sum of its diagonal elements is

$$Tr \left[I_{op}(E) \right] = \text{Net current outflow per unit energy across}$$

$$\text{an imaginary surface enclosing the conductor}$$

$$= \int \nabla \cdot \mathbf{J}(\mathbf{r}, E) d\mathbf{r}$$

$$= \frac{ie}{h} Tr \left[\Sigma^{in} \left(G^R - G^A \right) - \left(\Sigma^R - \Sigma^A \right) G^n \right] \qquad (15.155)$$

Since from eq. (15.26)

$$i \left[G^R - G^A \right] = \text{Spectral function } (A) \qquad (15.156)$$

and from eq. (15.35)

$$\Sigma^A - \Sigma^R = i\Gamma \text{ or, } i \left(\Sigma^A - \Sigma^R \right) = i^2 \Gamma \text{ or,}$$

$$i \left(\Sigma^A - \Sigma^R \right) = -1 \times \Gamma \text{ or, } i \left(\Sigma^R - \Sigma^A \right) = \Gamma \qquad (15.157)$$

we have

$$Tr \left[I_{op}(E) \right] = \frac{e}{h} Tr \left[\Sigma^{in} A - \Gamma G^n \right]$$

$$= \frac{e}{h} Tr \left[\text{Inscattering function} \right.$$

$$\times \text{ Density of vacant states}$$

$$- \text{ Carrier loss by scattering } \times \text{ Electron density} \right]$$

$$= \frac{e}{h} Tr \left[\text{Rate of inscattering} - \text{Rate of outscattering} \right]$$

$$= \frac{e}{h} Tr \left[\Sigma^{In} G^p - \Sigma^{Out} G^n \right] \qquad (15.158)$$

But we know that the inscattering and outscattering functions contain the contributions from the leads and the interactions in the conductor. So, the trace of the current operator is decomposed into separate components corresponding to the leads and the interactions as:

$$Tr \left[I_{op}(E) \right] = \sum_p i_p(E) + i_\phi(E)$$

$$= \sum_p \frac{e}{h} Tr \left[\Sigma^{In}_p G^p - \Sigma^{Out}_p G^n \right]$$

$$+ \frac{e}{h} Tr \left[\Sigma^{In}_\phi G^p - \Sigma^{Out}_\phi G^n \right] \qquad (15.159)$$

where the term

$$i_p(E) = \frac{e}{h} Tr \left[\Sigma^{In}_p G^p - \Sigma^{Out}_p G^n \right] = \text{Influx from lead } p \qquad (15.160)$$

Integration over energy yields the terminal current.

15.10 Procedure of Solution

The method is based on the tight-binding model. In this model, a discrete lattice of N points is selected in real space covering the conductor with lattice sites labeled by indices i and j (Figure 15.3). Then each relevant quantity, e.g. G^n, G^R, Σ^n, Σ^R is a matrix of $(N \times N)$ dimensions for each energy. For computations, we climb up the NEGF ladder shown in Figure 15.4.

   i. The self-energy function due to the leads $\Sigma^R p(i, j; E)$ is calculated. The inscattering function $\Sigma^{in} p$ and the outscattering function $\Sigma^{out} p$ are obtained individually by assuming that each lead p is in local equilibrium with a Fermi distribution $f_p(E)$. Then inscattering and outscattering functions can be expressed with an equilibrium solution.

   ii. The Green's functions G^R, G^A are calculated using the matrix representation of the Hamiltonian operator H_C.

   iii. The correlation functions G^n, G^p are calculated. The scattering functions are found by adding together the

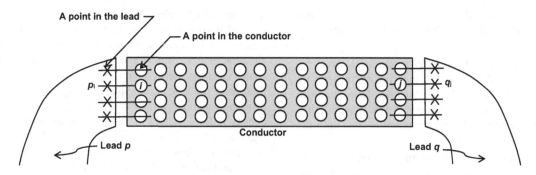

FIGURE 15.3 Discrete lattice in real space spanning the conductor connected to leads p and q.

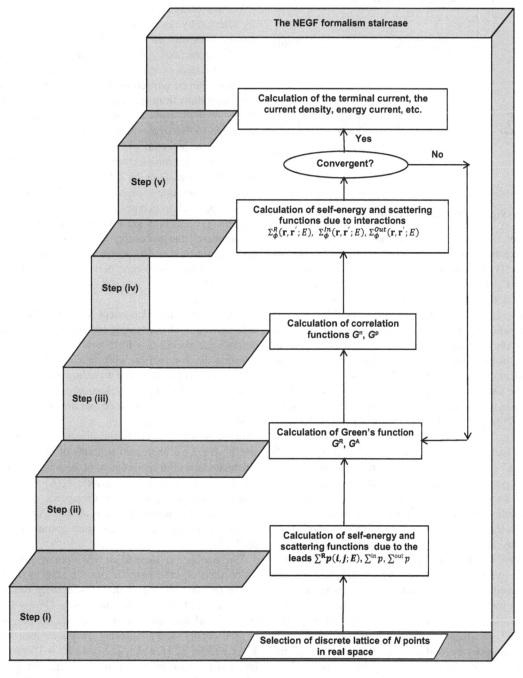

FIGURE 15.4 Iterative procedure for the implementation of non-equilibrium Green's function formalism.

contributions from the interactions and the leads. After each iteration, the contribution from interactions is updated.

iv. The self-energy and scattering functions $\Sigma_\phi^R(\mathbf{r},\mathbf{r}';E)$, $\Sigma_\phi^{In}(\mathbf{r},\mathbf{r}';E)$, $\Sigma_\phi^{Out}(\mathbf{r},\mathbf{r}';E)$ arising from interactions are computed. Then it is checked whether these quantities have changed due to interactions. If yes, steps (ii), (iii), and (iv) are repeated. This repetition is done until the functions converge.

v. Following the convergence of the solution for correlation function, the said function is used to calculate the different parameters such as the terminal current, the current density, energy current, exchange of energy with the reservoir. The current at lead p is given by

$$I_p = 2\int i_p(E)dE \tag{15.161}$$

with

$$i_p = \left(\frac{e}{h}\right)Tr\left[\sum^{in}G^p - \sum^{out}G^n\right] = \left(\frac{e}{h}\right)Tr\left[\sum_p^{In}A - \Gamma_p G^n\right] \tag{15.162}$$

15.11 Discussion and Conclusions

The distribution function $f(\mathbf{k})$ in the semi-classical picture becomes inadequate to describe quantum transport because of the need to specify phase relationships among states. Therefore, a density matrix $\rho(\mathbf{k},\mathbf{k}')$ is defined which is generalized into a time-varying density matrix $\rho(\mathbf{k},\mathbf{k}';t)$. A still more general concept is a two-time correlation function $G^n(\mathbf{k},\mathbf{k}';t,t')$. For steady-state conditions, a correlation function $G^n(\mathbf{k},\mathbf{k}';\tau)$ is used where τ is the time difference $t - t'$. The Fourier transform of the correlation function yields the energy spectrum. So, an energy-dependent correlation function $G^n(\mathbf{r},\mathbf{r};E)$ is used which is related to the electron density per unit energy $n(\mathbf{r};E)$. It is applicable to steady-state transport. Electron and hole correlation functions G^n and G^p are the equivalents of $f(E)$ and $1 - f(E)$, respectively.

The semi-classical outscattering function $S^{out}(\mathbf{k},t)$ is generalized to $\Sigma^{out}(\mathbf{k},\mathbf{k}';t,t')$ to include phase correlations and Fourier transformed to get an energy-dependent outscattering function $\Sigma^{out}(\mathbf{k},\mathbf{k}';E)$ for steady state conditions. Likewise, there are inscattering functions $\Sigma^{in}(\mathbf{k},\mathbf{k}';t,t')$ and $\Sigma^{in}(\mathbf{k},\mathbf{k}';E)$.

Together with the four matrices G^n, G^p, Σ^{out}, Σ^{in}, another set of four functions are used, namely, the retarded and advanced Green's functions G^R, G^A and self-energies Σ^R, Σ^A. The self-energy is taken in a broader perspective to include components from the lead-conductor coupling and from the conductor itself. Besides the spectral function A gives the aggregate states while the function Γ represents the loss of carriers by scattering. The functions A, Γ, G^R, G^A are interrelated.

The kinetic equations for NEGF analysis connect the correlation functions G^n, G^p with the scattering functions Σ^{in}, Σ^{out}. The Schrodinger equation $E\Psi = H\Psi$ in its original form is insufficient to describe electrical transport through a conducting channel having contacts at the two ends called the source and drain terminals. Two additional terms must be inserted in the equation, one for inflow and the other for outflow. The inflow term S is for the energy source and the outflow term $\Sigma\Psi$ is for the dissipative processes taking place during current flow. From the resulting equation for wave function, the equation for probability density follows leading to the equation for electron correlation function and a similar equation for hole correlation function. The equilibrium solutions are expressed in terms of spectral function A and the carrier loss function Γ. Only for non-equilibrium situations, the kinetic equations for electron and hole correlation functions need to be solved simultaneously with the equations for the retarded and advanced Green's functions.

The self-energies Σ^R, Σ^A and the scattering functions Σ^{in}, Σ^{out} are affected by the electron-electron and electron-phonon interactions, both with the leads and inside the conductor. Assuming that the interactions with the leads and those inside the conductor are independent of each other, the self-energies and scattering functions can be written as sum of two separate terms describing interactions from the respective causes. In the Hartree-Fock theory of many-particle systems, the self-energy contains one term due to Hartree potential and another term arising from the exchange potential whereas the scattering functions are not affected by electron-electron interactions. For electron-phonon interactions, equations for self-energy, inscattering and outscattering functions due to phase-breaking interactions are stated.

One approach to derive the equation for terminal current per unit energy $i_p(E)$ is to get it through current density $\mathbf{J}(\mathbf{r}, E)$. In this approach, the probability density is written as the product of the wave function and its complex conjugate, and its time derivative is calculated. The time derivative is the sum of two terms $\Psi^*(r,t)\frac{\partial\Psi(r,t)}{\partial t}$ and $\Psi(r,t)\frac{\partial\Psi^*(r,t)}{\partial t}$. These terms are found by applying the time-dependent Schrodinger equation for a particle moving in an electromagnetic field, which leads to the time derivative equation in a form similar to the current continuity equation. By comparison with the continuity equation, the current density is extracted. The wave function in the current density equation is replaced by the electron correlation function. Then integrating over the cross-sectional area of the contact, the terminal current is found.

In the second approach, the divergence of the current density $\nabla \cdot \mathbf{J}(\mathbf{r}, E)$ is expressed in terms of Hamiltonian H_C of the conductor, and after replacement of the wave function by the correlation function, a current operator $I_{op}(E)$ is introduced whose diagonal elements contain the divergence of the current density. Then the terminal current is obtained by integration of the component of the trace of the current operator corresponding to the leads.

The sequence of steps in an NEGF analysis follows the path: Calculation of self-energy $\Sigma^R p(i,j;E)$ due to the leads, inscattering function $\Sigma^{in}p$ and the outscattering function $\Sigma^{out}p \rightarrow$ Calculation of Green's functions G^R, $G^A \rightarrow$ Calculation of electron and hole correlation functions G^n, $G^p \rightarrow$ Calculation of self-energy and scattering functions due to scattering: $\Sigma_\phi^R(\mathbf{r},\mathbf{r}';E)s$, $\Sigma_\phi^{In}(\mathbf{r},\mathbf{r}';E)$, $\Sigma_\phi^{Out}(\mathbf{r},\mathbf{r}';E) \rightarrow$ If convergent, calculation of terminal current; if non-convergent retrace to the step of Green's function calculations and repeat unless convergent.

Illustrative Exercises

15.1 What is the simplest approach to approximation of electron-electron interactions.

It is by Hartree approximation in which we put

N – electron wavefunction Ψ_{Hartree}

= Product of single electron orbitals $\Psi_i\left(\mathbf{r}_i s_i\right)$ (15.163)

or

$$\Psi_{\text{Hartree}} = \Psi\left(\mathbf{r}_1 s_1, \mathbf{r}_2 s_2, \mathbf{r}_3 s_3, \ldots, \mathbf{r_n} s_n\right)$$

$$= \frac{1}{\sqrt{N}}\,\Psi_1\left(\mathbf{r}_1 s_1\right)\Psi_2\left(\mathbf{r}_2 s_2\right)\Psi_3\left(\mathbf{r}_3 s_3\right)\ldots,\Psi_n\left(\mathbf{r_n} s_n\right)$$

(15.164)

where

$$\Psi_i\left(\mathbf{r}_i s_i\right) = \text{Spatial function} \times \text{Spin function} = \phi_i\left(\mathbf{r}_i\right)\sigma_i\left(s_i\right)$$

(15.165)

and $\sigma = \alpha$ represents clockwise spin while $\sigma = \beta$ represents anticlockwise spin.

15.2 Does Hartree approximation take exchange interactions into account.

No. The exchange interactions are accounted in the Hartree-Fock approximation by expressing the wave function as a product of antisymmetric orbitals:

$$\Psi_{\text{Hartree–Fock}}$$

$$= \frac{1}{\sqrt{N}}\left\{\Psi_1\left(\mathbf{r}_1 s_1\right)\Psi_2\left(\mathbf{r}_2 s_2\right)\ldots,-\Psi_1\left(\mathbf{r}_2 s_2\right)\Psi_2\left(\mathbf{r}_1 s_1\right)\ldots+\ldots\right\}$$

(15.166)

15.3 What is Hartree? Express it in Joules and eV.

It is an atomic unit of energy defined as double the binding energy of an electron in the ground state of a hydrogen atom. It is used in molecular orbital theory.

$$1\ \text{Hartree} = \frac{m_e e^4}{\left(4\pi\varepsilon_0\right)^2 \hbar^2}$$

$$= \frac{9.109\times10^{-31}\ \text{kg}\times\left(1.602\times10^{-19}\ \text{C}\right)^4}{\left(4\times3.14159\times8.854\times10^{-12}\ \text{Fm}^{-1}\right)^2\left(\dfrac{6.626\times10^{-34}\ \text{Js}}{2\times3.14159}\right)^2}$$

$$= \frac{9.109\times\left(1.602\right)^4\times10^{-107}\times\left(2\times3.14159\right)^2}{\left(4\times3.14159\times8.854\times6.626\right)^2\times10^{-92}}\ \text{J}$$

$$= \frac{9.109\times6.586\times39.47835\times10^{-15}}{\left(4\times3.14159\times8.854\times6.626\right)^2}\ \text{J} = \frac{2368.38\times10^{-15}}{543501.6824}\ \text{J}$$

$$= 0.0043576\times10^{-15} = 4.3576\times10^{-18}\ \text{J}$$

$$= \frac{4.3576\times10^{-18}}{1.602\times10^{-19}}\ \text{eV} = 27.2\ \text{eV}$$

(15.167)

REFERENCES

Datta, S. 1999. *Electronic Transport in Mesoscopic Systems*. Cambridge University Press, Cambridge, UK, pp. 117–174, 293–344.

Datta, S. 2000. Nanoscale device modeling: The Green's function method. *Superlattices and Microstructures* 28(4): 253–278.

Datta, S. 2002. The non-equilibrium Green's function (NEGF) formalism: An elementary introduction. *Technical Digest. IEEE International Electron Devices Meeting (IEDM)*, San Francisco, CA, USA, 8–11 December 2002, pp. 703–706.

Datta, S. 2015. Non-equilibrium Green's function (NEGF) method: A different perspective 2015. *International Workshop on Computational Electronics (IWCE)*, West Lafayette, IN, USA, pp. 1–6

Keldysh, L. V. 1965. Diagram technique for nonequilibrium processes. *Soviet Physics JETP* 20(4): 1018–1026.

Part V

Fabrication and Characterization of Nanostructures

16

Fabrication Tools

Nanoelectronics has evolved from microelectronics as a consequence of downscaling of components. It is therefore expedient to review the microelectronics fabrication technology with emphasis on nanolithographic techniques that have paved the road from micro- to nanoelectronics. Primarily, silicon and gallium arsenide have been the elemental and compound semiconductor materials, respectively, from which the magnificent edifices of microelectronics, microelectromechanical systems, nanoelectronics and nanoelectromechanical systems have been built. These materials have been supported by several other materials like germanium, silicon carbide, diamond, and III–V semiconductors, e.g. GaP, InP, GaN. Important nanomaterials include carbon-based nanomaterials such as graphene and CNTs. Of great utility are the transition metal dichalcogenides MX_2 where M is a transition metal and X = S, Se, Te.

In this chapter, the key equipment and processes used in nanofabrication will be described keeping silicon in focus as the mainstream nanoelectronics material. Occasionally, we shall have opportunities to talk about processes relating to remaining materials too.

16.1 Silicon Single-Crystal Growth

Silicon is the second most abundant element in earth's crust (27.7% by mass), oxygen being the first one (46.6% by mass). Despite its abundance, silicon is rarely found in pure form. It occurs mostly as silicates, e.g. in quartz sand.

(i) 98% pure metallurgical grade silicon (MGS or MG-Si: impurity content ~$5 \times 10^{16} cm^{-3}$) is obtained by heating silicon dioxide with coke (carbon) in an electric arc furnace at >1900°C:

$$SiO_2(l) + 2C(s) \rightarrow Si(s) + 2CO(g) \qquad (16.1)$$

(ii) To make electronic grade silicon (EGS or EG-Si: impurity level $\leq 5 \times 10^{13} cm^{-3}$), also called semiconductor grade silicon (SGS), metallurgical grade silicon is reacted with anhydrous HCl gas at 300°C to form liquid trichlorosilane (boiling point 31.5°C):

$$Si(s) + 3HCl(g) \rightarrow HCl_3Si(l) + H_2(g)$$
$$+ By\text{-}products\left(SiH_4, SiCl_4, Si_2Cl_6, H_2SiCl_2\right)$$
$$+ Chlorides\, of\, impurities, e.g., FeCl_3 \qquad (16.2)$$

By fractional distillation of mixture of trichlorosilane and impurity chlorides, pure trichlorosilane is separated from the impurity chlorides. In the Siemens process, the pure trichlorosilane is reacted with hydrogen at 1100°C to form EGS:

$$HCl_3Si(l) + H_2(g) \rightarrow Si(s) + 3HCl(g) \qquad (16.3)$$

which is reverse of the trichlorosilane formation reaction. This EGS is polycrystalline silicon (poly-Si).

(iiia) Poly-Si is converted into monocrystalline silicon using the Czochralski or C-Z method. In this method (Figure 16.1), a seed crystal placed in a seed holder is lowered into the molten poly-Si (melting point 1412°C)+required dopant, contained in a fused silica crucible with a graphite susceptor. The seed is slowly withdrawn using a crystal pulling and simultaneous rotation mechanism when the molten poly-Si coming in contact with seed crystallizes to form single-crystal silicon. The crystal growth is carried out in an inert gas environment. The temperature field is stabilized to form homogeneous single-crystal silicon ingots. The single-crystal silicon ingot is mechanically processed to produce silicon wafers having thickness of several hundred microns. C-Z silicon contains oxygen ($5–10 \times 10^{17} cm^{-3}$) and carbon ($5–10 \times 10^{15} cm^{-3}$) impurities coming from the dissolution of quartz crucible and from the formation of carbon monoxide through reaction of silicon monoxide evaporating from the melt with a graphite susceptor.

(iiib) Poly-Si is transformed into a single crystal form using a float-zone process in which the vertically held EGS rod is passed through an RF heating coil (Figure 16.2). A localized zone of molten silicon is formed from which the single-crystal silicon is grown by contacting with seed crystal from one end. The growth process is carried out in vacuum or under constant purging with an inert gas. As the float zone silicon is not in contact with any crucible, it is free of any impurities from the crucible. The oxygen and carbon content in it is below $5 \times 10^{15} cm^{-3}$. Silicon wafers >200 mm diameter cannot be grown using this method due to difficulties with the floating zone.

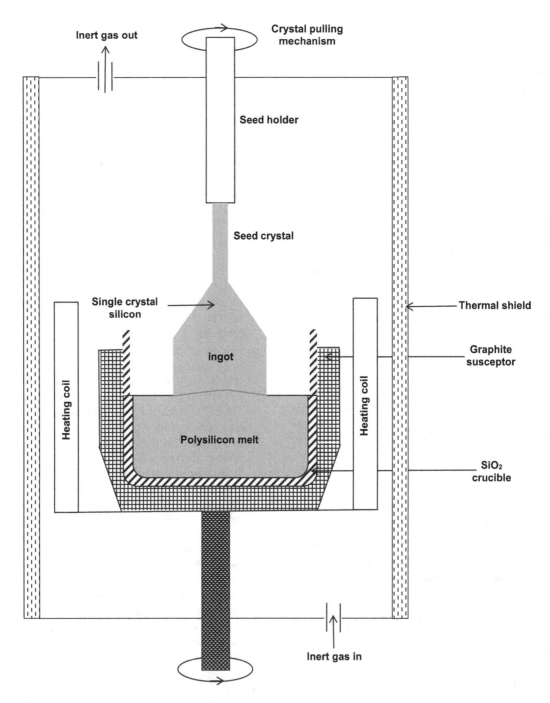

FIGURE 16.1 The Czochralski process of single-crystal silicon growth.

16.2 Thermal Oxidation of Silicon

It is a chemical process carried out in a furnace at 800°C–1200°C either using molecular oxygen (dry oxidation) or steam (wet oxidation) (Figure 16.3):

$$Si + O_2 \rightarrow SiO_2 \tag{16.4}$$

$$Si + 2H_2O \rightarrow SiO_2 + 2H_2 \tag{16.5}$$

Silicon oxidizes in atmospheric oxygen but the oxidation does not extend deep into silicon. Silicon oxidation ceases at a self-limiting native oxide thickness ~2 nm. The oxidizing species (oxygen or OH) diffuses through the pre-existing silicon dioxide to the silicon dioxide-silicon interface. Here it reacts with silicon to form amorphous silicon dioxide. Thus the oxide thickens with growth of more oxide at the silicon/silicon dioxide interface. For every 1 μm of Si oxidized, 2.17 μm of SiO_2 appears. Dry oxide is good quality, high-density oxide. But it grows at a very slow rate. Wet oxide is low-density oxide of inferior quality to dry oxide. But it has a high growth rate, hence used to grow thicker films.

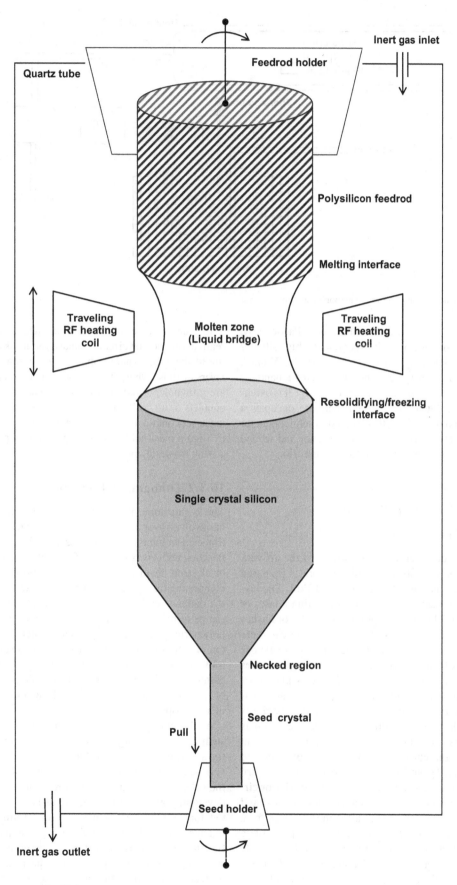

FIGURE 16.2 Float-zone crystal pulling.

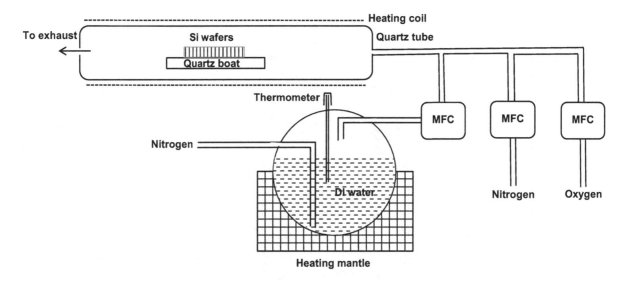

FIGURE 16.3 Arrangement for dry and wet oxidation of silicon.

Silicon dioxide has many attractive properties. Particularly, the DC resistivity of silicon dioxide is exceptionally high ~10^{17} Ω cm. In addition, SiO_2 has a high dielectric strength ~10^7 V cm^{-1}. Even further, it is a large energy gap insulator having a bandgap of 8.9 eV, as opposed to 5 eV for silicon nitride (Si_3N_4). It is extensively used in Si integrated circuits, and therefore occupies a unique niche in silicon processing. It is utilized in semiconductor technology mainly in two forms, first as an insulator and second as a masking material for selective diffusion of impurities.

16.3 Mask Making and Lithography

16.3.1 Mask Making

The device or circuit designer makes a drawing of the artwork to be implemented on the substrate. This artwork is prepared on a computer using a layout software such as LEdit. The layout is fed to a pattern generator machine containing a laser or electron beam writer. The machine produces a series of flashes corresponding to the bright areas in the layout whereas there are no flashes for the dark areas. The flashes are used to expose a resist-coated mask blank in accordance with the geometries on the layout. The mask blank consists of a glass plate covered with 100 nm thick chrome by sputtering. The mask writing is performed by raster scanning in which the substrate is mechanically translated in one direction while the beam is swept in the traverse direction. After developing the latent image in the resist, the chrome film is selectively etched to produce the designed layout on the mask blank. The pattern engraved in the chrome film is used as a mask for device or circuit fabrication. As chrome is a hard material, it can withstand any abrasion during processing work, allowing its usage for a long duration. The mask is often protected from particulate contamination by stretching a thin membrane called pellicle across a frame on one side of the mask and pasted to the mask on the sides. As the pellicle is not in contact with the chrome film, any particles collected on it remain out of focus and do not vitiate the exposure.

In practice, a set of mutually aligned masks is necessary to realize a device or circuit because the masks are made according to the areas where doping is to be done to create diffused layers or those where it is avoided, areas where insulating layers are grown/deposited, areas where metal films are deposited for contacts or any other requirement to be met. All the masks in a mask set contain an alignment mark which is used as a reference to align a mask with respect to the pattern produced in the preceding masking operation.

16.3.2 Lithography Principles

The cornerstone of nanofabrication is nanolithography, which has evolved over decades from micro- to nanoscale through multifaceted technological advances. It works on the basic premise of flooding with energy in some form, either electromagnetic waves or electron beams, a template or blueprint containing a pattern comprising transparent and opaque regions (Figure 16.4). Below the template called the mask is placed the substrate over which the pattern is to be transferred. The substrate is coated with a polymeric material called the resist which is sensitive to the photon or electron energy. Either the resist is softened on exposure (positive resist) or hardened (negative resist). Accordingly, the pattern formed in the resist is the direct replica or inverse replica of the pattern on the mask when the exposed resist is immersed in a developing solution.

A photolithographic operation consists of some key steps such as spin coating the resist on the thin film deposited on the substrate to form a uniform layer of resist on the thin film, soft baking the resist typically at 90°C for ½ h to evaporate the solvents, exposing the photoresist to energy through mask, developing and hard baking the resist, typically at 120°C for ½ h for positive resist or at 140°C for 40 min for negative resist, to adhere it firmly enough. Strong adhesion of photoresist with the thin film on the substrate is essential to withstand the subsequent etching step in which the resist serves as a layer protecting the film from etching in resist-covered areas. The etchant removes the portions of film that are not covered with photoresist. Subsequent to the completion of etching, the resist

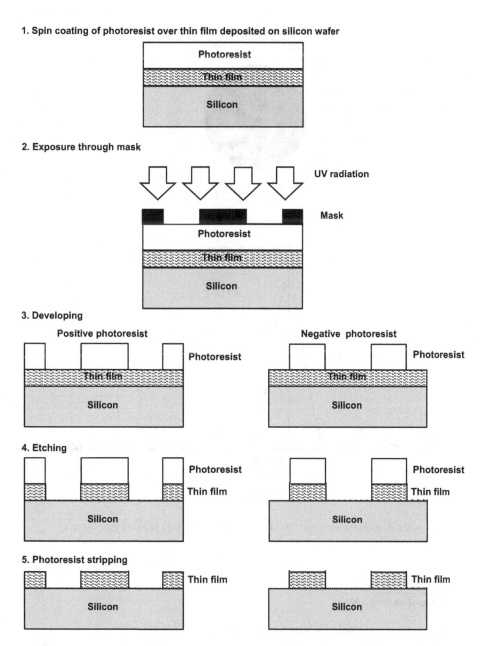

FIGURE 16.4 The photolithographic process using positive and negative photoresists.

is no longer necessary. It is got rid of by stripping it off in a solvent. The pattern formed on the thin film on the substrate is left behind.

16.3.3 Optical Resolution

The progression from microscale to nanoscale resolution has been achieved by understanding the limitations imposed by physics (Totzeck et al 2007). A light wave bends when passing by a corner or through a small aperture. This bending of light around sharp corners is especially conspicuous when the size of the aperture is nearly the same or smaller than the wavelength of light. When light is passed through a small circular aperture, one expects to obtain a bright clear circular spot but instead a fuzzy spot surrounded by circular dark and bright fringes in the form of concentric circular rings is seen (Figure 16.5). It is called the diffraction pattern. The bright circle spot at the center of the fuzzy spot is known as the Airy disk and the concentric rings constitute the Airy pattern.

As a consequence of diffraction, two closely placed point sources of light produce circular images with blurred boundaries and even start overlapping if close enough so that they are no longer separately recognizable. Thus two images merge together and appear as though it was a single image. These effects arise from the small diameter of the light beam, not due to any interaction of the light beam with the aperture.

The minimum lateral separation r between two sources of light for resolution into distinguishable objects is laid down in the Rayleigh criterion (Figure 16.6) which states that two images are just resolvable when the center of diffraction maximum of one, i.e. its Airy disk, overlies the first minimum or dark ring of the diffraction pattern of the other.

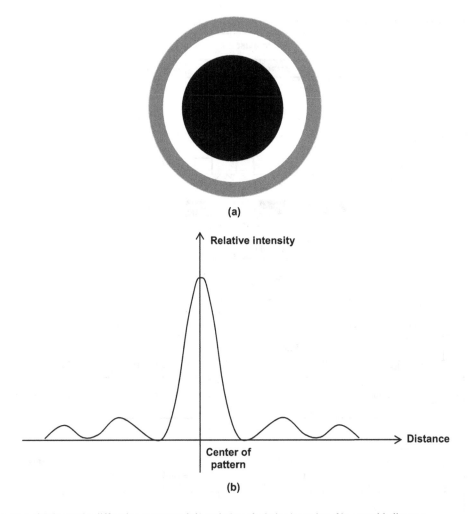

(a)

(b)

FIGURE 16.5 Diffraction of light: (a) the diffraction pattern and (b) variation of relative intensity of image with distance.

$$r = \frac{0.61 \times \text{Wavelength of light}\,(\lambda)}{\text{Numerical aperture of the lens } (\text{NA})}$$

$$= \frac{0.61 \times \text{Wavelength of light}\,(\lambda)}{\text{Refractive index of the surrounding medium}\,(n) \times \left\{ \sin(1/2)\,\text{angle of light beam entering the lens}\,(\theta) \right\}} \tag{16.6}$$

or

$$r = \frac{0.61\lambda}{\text{NA}} = \frac{0.61\lambda}{n\sin\theta} \tag{16.7}$$

The lateral separation r can be decreased either by harnessing smaller wavelength for illumination or by increasing the numerical aperture of the lens, which depends on the refractive index of the surrounding medium.

16.3.4 The 248 and 193 nm Excimer Laser Lithographies

Krypton fluoride (KrF) excimer laser is used to produce deep ultraviolet (DUV) radiation of wavelength 248 nm. For DUV of wavelength 193 nm, the argon fluoride (ArF) excimer laser is used. By using ArF excimer laser immersion lithography, up to 38 nm half-pitch lines are attained. Half-pitch is ½ the minimum center-to-center spacing between interconnect lines.

16.3.5 Immersion Lithography

Immersion lithography is a modification of the customary dry lithography machine using a liquid medium such as ultra-pure deionized water (refractive index = 1.44) or another liquid to fill the air gap between the final lens and the surface of the wafer (Figure 16.7). The separation r is thereby diminished by a factor = the refractive index of the liquid.

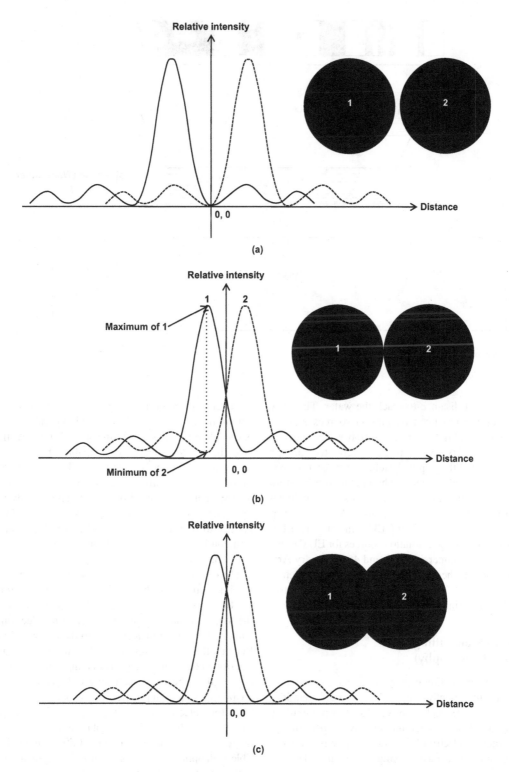

FIGURE 16.6 Explaining Rayleigh's criterion: images of two objects when they are: (a) far apart, (b) just at the Rayleigh's limit, and (c) closer than the Rayleigh limit.

16.3.6 Extreme UV Lithography

From the 193 ArF excimer laser source, the extreme UV (EUV) lithography pushes down the wavelength by a factor of ~14.6 to soft X-ray wavelength of 13.2 nm (Sreenivasan 2017). But the adoption of EUV takes us to a region of physics riddled with practical difficulties because of the different properties of EUV. The EUV is generated by irradiating 30 μm tin micro-droplets with ~25 kW CO_2 laser to create ionized gas plasma providing a 250 W EUV source. As most materials absorb EUV photons, they cannot penetrate any medium. So, the process must be carried out in a high-vacuum chamber in which the EUV photons

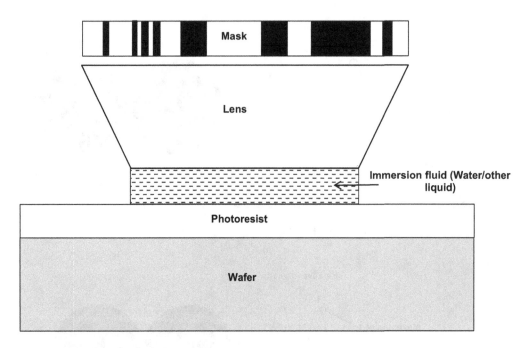

FIGURE 16.7 Immersion lithography.

can traverse a sufficient distance to reach the wafer. The EUV photons themselves are produced by plasma sources and are guided by specialized ultra-reflective multilayer, flat mirrors made of 40 Mo/Si bilayers instead of being refracted through usual lenses. The EUV lithography system is a projection system using reflective optical elements. The photomasks used for EUV lithography work in reflective mode and special pellicles are designed for protection from adhering particles. As the mirrors of the EUV system too absorb 30% EUV, the original EUV source needs to be very strong. Radiation sources for EUV target 250 W against 90 W for immersion ArF and 45 W for dry ArF. Creation of a EUV source therefore remains a gigantic challenge. Ablative action of the high-intensity beam on the mirrors introduces additional engineering hurdles.

16.3.7 Electron-Beam Lithography (E-Beam Lithography)

Habituated to think about lithography as a mask-based process, electron-beam lithography (Figure 16.8) is a different idea because it is maskless lithography (Groves 2013). A pre-existing mask is not necessary. Conventional lithography is a pattern replicator process whereas electron-beam lithography is a pattern generator process, much like mask making. Essentially, electron-beam lithography is a technique used to make masks for use by other types of lithography.

In conventional lithography, the pattern is embodied in the mask. In electron-beam lithography, it is prepared as a computer data file. In the traditional lithography, the pattern is defined by exposure through a mask while in electron-beam lithography, it is directly written on the electron-sensitive resist with an electronic pen. So, the former is an exposure-through-mask technique while the latter is a direct-write technique to draw custom shapes on a surface.

In conventional lithography, different parts of the pattern are simultaneously produced so that it is a parallel operation, which is completed in a smaller time span. Naturally, the throughput is high, making the process capable of large-scale manufacturing of semiconductor devices and circuits. In electron-beam lithography, one part of the pattern is produced and then the next so that it is a serial operation. The serially generated pattern takes an inordinately long time, resulting in low throughput useful only for applications like mask making, small-scale semiconductor device/circuit manufacturing, and research and development.

The operating principle of an electron-beam lithography system is strikingly similar to that of a scanning electron microscope. So, the resolution achievable with electron-beam lithography is the same as that possible in a scanning electron microscope, namely 0.06–0.15 nm according to the energy of electron beam. But this theoretical limit is hardly attained in practice because electron optics is not the solo deciding factor. Interaction of electrons with the resist by scattering and secondary processes plays a vital role so that in practice, the resolution is a few nanometers. Therefore, electron-beam lithography is used for creating nanostructures with sub-10 nm resolution as structures in this size range cannot be made with optical lithography. Resolution possible with optical lithography is inferior by an order of magnitude to that by electron-beam lithography.

The electron-beam lithography can be looked upon as the progeny of the electron microscope. But there are some basic differences between the two. The number of pixels in a high-resolution electron microscope picture is typically 10^6 but the same in a high-accuracy electron-beam pattern can have $\sim 10^{12}$ pixels, which is responsible for its relatively exorbitant price. For the research work, an electron microscope is often converted into an electron-beam lithography system using less expensive accessories.

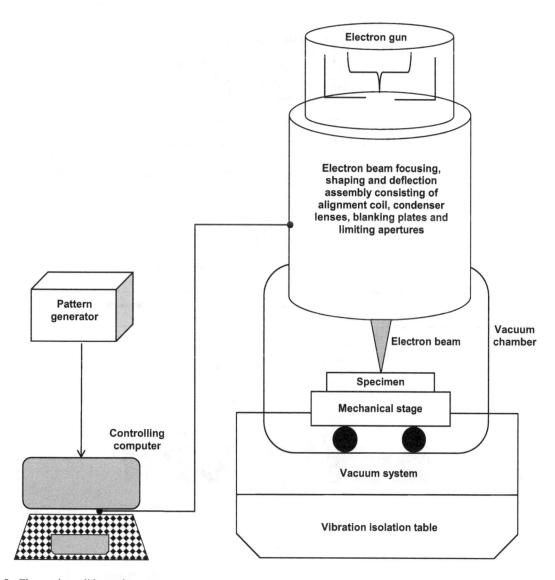

FIGURE 16.8 Electron-beam lithography system.

An electron-beam lithography system is a high-vacuum system with facility to insert the substrate into the vacuum or withdraw it. The substrate is mounted on a platform and can be displaced in X, Y directions accurately.

To produce the electron beam, lower resolution systems use lanthanum fluoride thermionic emission sources while the high-resolution systems use field electron emission sources, e.g. heated W/ZrO_2. The finely focused electron beam serving as the writing pen is produced by an electron optical column consisting of electric and magnetic lenses. Electric lenses are prone to greater aberration. Focusing, blanking or unblanking, and scanning operations are done by an analog electronic circuit while a digital electronics circuit is used to store and transfer the data of the pattern. All the operations are programmed using a high-speed computer equipped with extensive software.

Larger size patterns require the precise movement of the platform to tile the writing fields against each other; this tiling is called stitching.

16.3.8 Ion Beam Lithography

There are three types of ion beam lithography (Watt et al 2005). They are capable of producing sub-100 nm features. Two of these types: focused ion beam (FIB) and proton-beam writing (p-beam writing) are direct-write techniques. Being slow in operation, they are unsuited to large volume fabrication. The third one, ion projection lithography (IPL), mimics traditional lithography and is therefore applicable to bulk production. In FIB (Figure 16.9), a heavy ion beam with energy ~30 keV, usually consisting of Ga^+ ions from a liquid source, is focused down a column similar to the electron microscope and made to fall on the substrate. The pattern is written on the substrate either by its surface modification or sputtering. It is the most mature of the three types of ion beam lithography. Its uniqueness lies in the fact that it is applicable to practically any material because it machines the material by corroding its surface. The p-beam writing uses fast protons produced in a proton accelerator with energies in the MeV range.

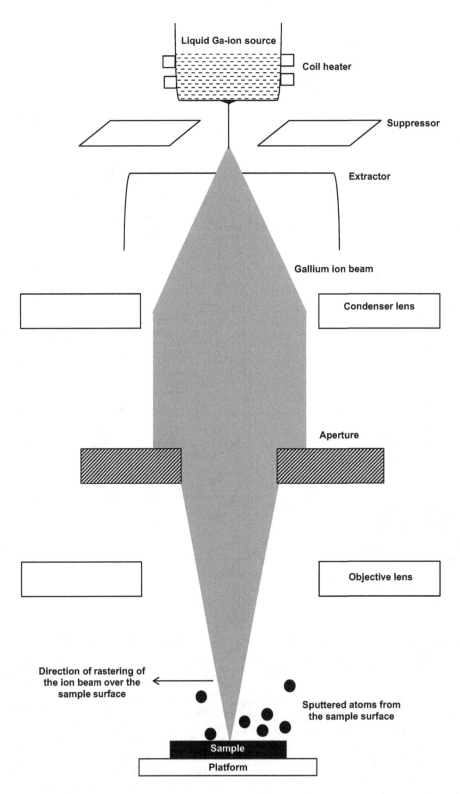

FIGURE 16.9   Focused ion beam technique.

The beam is focused by magnetic quadrupole lenses and scanned by magnetic or electrostatic deflection. The high-energy proton beam penetrates into the resist and can write 3D nanostructures in poly(methyl methacrylate) (PMMA). These nanostructures have a high aspect ratio. The IPL technique illuminates a large-area stencil mask with medium energy ions (50–150 keV) such as protons, He$^+$, and Ar$^+$. The transmitted ion beam is projected through electrostatic lenses on the substrate. The image is reduced by two to three orders of magnitude.

16.3.9 X-Ray Lithography

It is a next-generation lithography technique. Its versions include 1 nm X-ray lithography (XRL) and 0.1 nm deep X-ray lithography (DXRL). The X-rays produced by a synchrotron radiation source are collimated by mirrors or diffractive lenses. The mask for XRL consists of a film of gold on beryllium. The X-ray resists are PMMA, polyvinylidene fluoride (PVDF), acrylic glass, polycarbonate (Jain and Chaubey 2017).

16.3.10 Nanoimprint Lithography (NIL)

Resolution, throughput, and cost are the three driving factors that lead to widespread adoption of a lithography process for bulk volume production. All these three requirements can be met with nanoimprint lithography (NIL) which works by mechanical deformation of a resist using a stamp or mold containing the surface topographical features to be printed (Lan and Ding 2010). The mask for this lithography is made using electron-beam lithography. Two main versions of NIL with sub-10nm resolution are mentioned below (Figure 16.10):

i. Thermoplastic NIL or thermal NIL (T-NIL) or hot embossing lithography: After coating the substrate with resist, the mold is contacted with the substrate. They are pressed together and jointly heated up to the glass transition temperature of the resist. At this temperature, the resist becomes a viscous liquid and can flow to acquire the shape of the mold. Then the mold and the substrate are cooled down and separated when

1. Substrate preparation: spin coating resist over thin film on substrate

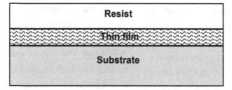

2. Mold preparation

3. Pressing together the mold and the substrate, and heating/UV exposure

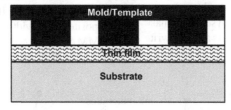

4. Removing the mold from the substrate

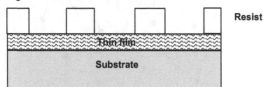

5. Reactive ion etching

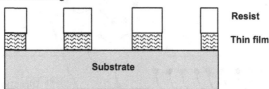

6. Photoresist removal

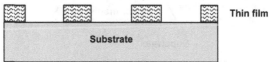

FIGURE 16.10 Sequence of steps in a thermal/UV nanoimprint lithography process.

the pattern is seen in the resist on the substrate. Etching is done to selectively remove the material on the substrate from uncovered areas.

ii. Photo NIL (P-NIL) or UV-NIL: A UV-curable polymer resist is spin-coated on the substrate. A transparent mold made of fused silica or polydimethylsiloxane (PDMS) is pressed against the resist-coated substrate when the resist fills the cavities in the mold. The mold is exposed to UV for hardening the resist. After exposure, the mold is removed leaving behind the substrate with the pattern transferred from the mold. Reactive ion etching (RIE) of the substrate with patterned resist is performed to define the pattern in the substrate.

Both versions involve shaping of the resist by mechanical pressure. Subsequently, the resist is cross-linked by thermal curing or UV exposure. But since T-NIL allows the use of non-transparent molds, it is less expensive than UV-NIL.

Besides planar nanostructures, NIL has been utilized for direct patterning of three-dimensional nanostructures such as sub-40 nm T-gates (Li et al 2001). The mold for the T-gate is prepared by electron-beam lithography. The mold is then used to make an opening in PMMA. The leftover PMMA at the bottom of the pit is etched in oxygen plasma followed by deposition of Ti/Au gate metals and lift-off in acetone.

16.3.11 Dip Pen Nanolithography (DPN)

It is like NIL in the respect that it uses a nanoprobe, viz. the atomic force microscopy (AFM) probe tip, instead of a stamp (Figure 16.11). Non-requirement of the exposure system and mask makes it different from traditional lithography. It is a direct-write constructive lithography involving the transference of a solution of weakly adherent molecules (the molecular ink) from the tip of an ink-coated AFM probe to a solid surface (having stronger affinity to the ink) by capillary action mediated through a water film (the water or solvent meniscus), which naturally forms between the probe and the substrate under atmospheric humidity (Piner et al 1999). Soft and hard materials can be transferred to a variety of substrates using this technique. Alkanethiols can be printed on a thin gold film by dip pen nanolithography (DPN). It has capability of sub-50 nm resolution. Multiple inks can be deposited, either sequentially or parallely by DPN. The size of water meniscus determined by the ambient humidity and the probe dwell time on the substrate together determine the printed feature size. The probe dwell time ranges from milliseconds to seconds. Success of DPN printing is highly dependent on the choice of the suitable ink and substrate combination with regard to chemical affinity because then only the ink will chemisorb and bind on the substrate surface. Proper binding of ink to the substrate initiates a self-regulating mechanism for diffusion of ink to the substrate.

Another version of DPN, the thermal DPN (tDPN), works like a soldering iron. It uses a meltable ink to be used with a heatable AFM tip. The ink is solid at room temperature. The tip is heated to the desired temperature when the ink melts and deposits on the substrate. The speed of the probe tip is a vital parameter. Temperature provides an additional control parameter because the deposition rate of the ink can be varied and the ink can be switched on and off.

Temperature-controlled DPN is well suited to deposition of polymer nanocomposites. Fe_3O_4 magnetic nanoparticles suspended in the PMMA matrix were coated using this method (Lee et al 2009). By changing the speed of the probe and temperature, the deposited line width was adjusted between 78 and 400 nm. The PMMA matrix was etched in oxygen plasma revealing the 10 nm wide row of nanoparticles.

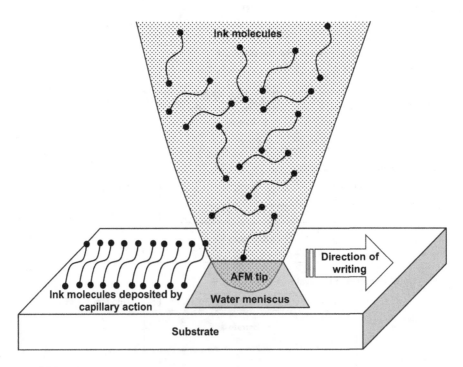

FIGURE 16.11 Dip pen nanolithography.

16.3.12 Block Copolymer Lithography

Optical lithography, electron-beam lithography, and NIL are all top-down approaches for transferring patterns using masks or molded stamps. These constitute only one aspect of lithography. Bottom-up approaches too have caught the attention of researchers. One such strategy uses block copolymers (BCPs), which are macromolecules made of two or more chemically distinct polymer chains A and B called blocks joined together by a covalent bond (Segalman 2005). In a melt, BCPs are thermodynamically driven to segregate into a variety of ordered structures due to immiscibility of the polymer blocks. The ordered structures contain domains of segments A and B with dimensions of 5–50 nm. The parameters deciding the size and shape of the domains are the molecular weight and composition of the BCP. The common shapes of the domains are spheres, cylinders, and lamellae. This property of spontaneous self-assembly of BCPs into nano-sized domains forms the basis of BCP lithography.

As BCPs exhibit many properties similar to the photoresists used in a lithography laboratory, the lithography facilities can be extended to processing with BCPs. Their solubility in various solvents allows them to be spin-coated on different substrate materials, e.g. semiconductors, metals, ceramics, polymers, much like the usual practice with photoresists. Using BCP lithography Shin et al (2010) fabricated hexagonal-shaped nanoparticle arrays of noble metals on 5″ transparent glass substrates (Figure 16.12). The BCP used is polystyrene-block-poly(methyl methacrylate) (PS-b-PMMA).

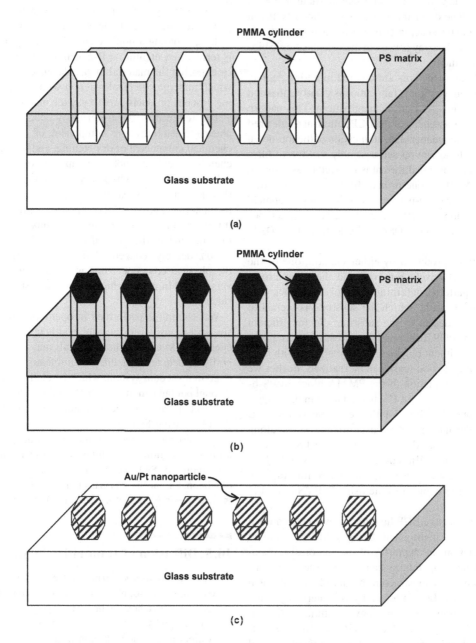

FIGURE 16.12 Fabrication of hexagonal-shaped nanoparticle array on the glass substrate by block copolymer lithography: (a) PS-b-PMMA film formation by spin coating and self-assembly of the polymer, (b) template formation after exposure to UV and rinsing with acetic acid, and (c) nanoparticle array obtained after noble metal film deposition and lifting off the PS matrix.

After chemical cleaning, the surface of the glass substrate is modified by brushing with a random copolymer P(S-r-MMA) to provide equal surface tension to the two polymer blocks in PS-b-PMMA. A 75 nm thick film of Ps-b-PMMA made by blending two components with different number-average molecular weights of PS and PMMA is formed on the substrate by spin coating. Thermal annealing is done at 200°C. The PS-b-PMMA BCP self-assembles into a film containing vertically aligned hexagonal PMMA cylinders surrounded by the PS matrix. The BCP is exposed to UV radiation, which weakens the PMMA and cross-links PS. After rinsing with acetic acid and water, a nanoporous film of PS containing hexagonal pores is obtained. This is the template. A noble metal such as Au and Pt is deposited on this template. Subsequently, the PS matrix is lifted off leaving a metallic nanoparticle array on the glass substrate. The diameter of the nanoparticles is 40 nm and the center-to-center spacing between the nanoparticles is 80 nm. The nanoparticle array is used to make a plasmonic biosensor for prostate cancer.

While the BCP nanostructures are ordered to a high degree, their usefulness in nanoelectronic fabrication is greatly enhanced when the substrate contains a pre-defined pattern. This pattern serves to guide the self-assembly of the BCP pattern at the desired positions in the desired manner. This unification of the top-down nanolithography with bottom-up self-assembly seeking to form the BCP pattern in a controlled format is referred to as directed self-assembly (DSA). It is done principally in two ways, either by delineating a chemical pre-pattern or carving a topographical pre-pattern such as a mold (Galatsis et al 2010). The two methods are known as chemical epitaxy DSA and grapho-epitaxy DSA, respectively.

In the DSA pattern produced by either method, one of the polymer blocks is selectively removed and the other cross-linked. The DSA pattern containing one polymer block acts as a mask or template using which the substrate is etched to transfer the required design to the substrate. So for efficient pattern transfer in BCP lithography, an essential requirement is that the constituent blocks must be etchable with high selectivity. Then only a good mask can be realized. Earlier we have seen how the degradation of the PMMA block in PS-b-PMMA BCP by irradiation with UV helped us to make a mask made of nanoporous PS film. Another example is PS-b-PB BCP. In ozone plasma RIE, the unsaturated bonds along the PB backbone are cleaved. At the same time PS is cross-linked. The ozone-treated film contains only cross-linked PS domains. These are used as a sacrificial mask for transferring pattern to an underlying substrate using RIE (Nunns et al 2013).

A significant advantage of BCP lithography over conventional lithography needs to be highlighted. Being a diffusion-limited process, small variations in chemistry of the photoresist, exposure time, or baking temperatures can cause large deviations in the dimensions of the nanostructure produced by traditional lithography. In contrast, BCP lithography is a thermodynamic process. It progresses until the morphology has attained minimal free energy. Therefore, it is capable of self-correcting or self-healing any dimensional variations in the pattern (Galatsis et al 2010).

16.4 Wet and Dry Etching

It is the removal of a layer or layers of material, either selectively or completely. One way to do this is to immerse the sample in a liquid. This liquid is a chemical solution, which reacts with the material to form a soluble product, thereby consuming it. As the process is carried out in liquid phase, it is called wet etching. The rate of removal of the material is called the etch rate and the chemical solution is known as the etchant. This kind of etching is isotropic with the same etch rate in all directions except for crystalline materials where it is determined by the orientation of the exposed crystal plane (orientation-dependent etching). The etchant may seep laterally below the photoresist to remove the material. Hence, wet etching causes undercutting.

Alternatively, etching may be done without using any liquid phase etchant. Instead, the etchant is in the gaseous state. Such etching avoiding a liquid etchant and using a gas phase etchant is termed dry etching or RIE (Figure 16.13). It is accomplished by exposing the material to gas plasma containing reactive ions, which undergo chemical reaction with the material. The resulting gaseous products carry away the material.

The removal of the material by chemical reaction with gas plasma constitutes one component of dry etching. It is the chemical component of dry etching, and is isotropic like the wet chemical etching. Another component of dry etching is physical in nature. The high-energy ions in the plasma strike against the material with great force to physically tear away pieces of the material. The physical component acting through momentum transfer is anisotropic because by bombarding the material with ions moving in the vertical direction, the horizontal etch rate is considerably reduced relative to the vertical etch rate. Thus the physical component brings anisotropy to etching, which is very useful in making high aspect ratio structures with vertical sidewalls.

Wet etching is performed in open atmosphere. It is inexpensive and is easily implementable although it involves wastage of chemicals. Moreover, the chemicals may be hazardous. It is generally inadequate to define sub-µm feature sizes. Dry etching is done in a vacuum system, which is relatively costly. It gives low throughput but can make sub-100 nm size features.

A wet etchant for silicon is tetramethyl ammonium hydroxide (TMAH) using SiO_2 or Si_3N_4 as mask. Controllable etching of silicon dioxide is done in a buffered oxide etch (BOE), which is a mixture of ammonium fluoride (buffering agent) and HF. For fast etching of SiO_2, HF (49% in water) is used. For dry etching of silicon, CF_4 or SF_6 gas is used. For SiO_2, $CF_4 + H_2$ or $CHF_3 + O_2$ is used. Resist/polymer is etched in O_2 plasma.

16.5 Diffusion of Impurities in Silicon

Three-zone open-tube furnaces are used with solid, liquid, or gaseous impurity sources. In a solid-source system, the source is placed in a source boat in the furnace tube. A carrier gas, nitrogen or oxygen, picks up the source vapors and transports them to the wafers placed downstream in a quartz boat. Sources available in wafer form are stacked alternately with silicon wafers.

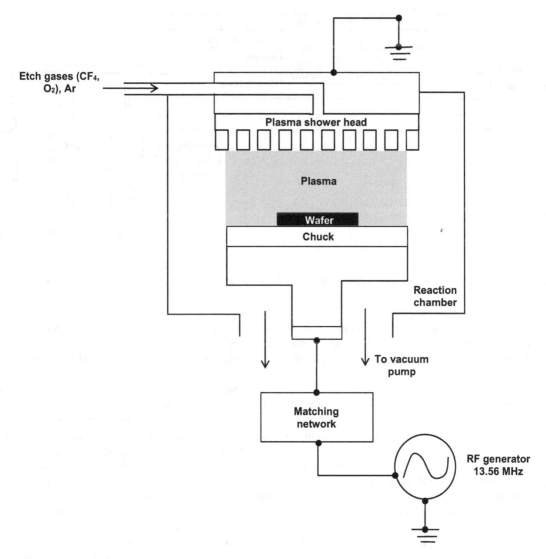

Etch gases (CF₄, O₂), Ar

Plasma shower head

Plasma

Wafer

Chuck

Reaction chamber

To vacuum pump

Matching network

RF generator 13.56 MHz

FIGURE 16.13 Reactive ion etching system.

Liquid sources are filled in bubblers. A carrier gas bubbling through the liquid source carries its vapors to the wafers in the furnace tube. Gaseous sources directly supply the dopant impurity to the wafers. Owing to their high toxicity, special precautions are mandatory for gas exhaust and scrubbing. Three N-type dopants, namely phosphorous, arsenic, and antimony, are used while only one P-type dopant, boron, is used. Gallium and aluminum having a high diffusion coefficient in silicon dioxide are not masked by it; hence they are not used.

For the solid phosphorous pentoxide source, the reaction taking place at the silicon surface is:

$$2P_2O_5 + 5Si \leftrightarrow 4P + 5SiO_2 \qquad (16.8)$$

Phosphorous diffuses into silicon from this layer.

For the liquid phosphorous oxychloride (POCl₃) source, POCl₃ vapors are carried into the furnace tube by bubbling nitrogen carrier gas through POCl₃ (Figure 16.14). Oxygen is also flowed into the furnace tube. Phosphorous pentoxide is deposited on the surface of the wafer by reaction between POCl₃ and oxygen.

Phosphorous diffusion takes place from P_2O_5 into silicon, as in previous case.

$$4POCl_3 + 3O_2 \rightarrow 2P_2O_5 + 6Cl_2 \qquad (16.9)$$

The chlorine gas released in the reaction serves to getter away any contaminants present by forming volatile chlorides.

The gaseous source phosphine being highly toxic and explosive is taken in diluted form with 99.9% Ar or N₂. Inside the furnace tube, it is mixed with oxygen to form P_2O_5, which acts as earlier. Unreacted PH₃ is cleaned from the exhausted gas.

$$2PH_3 + 4O_2 \rightarrow P_2O_5 + 3H_2O \qquad (16.10)$$

The silicon surface reactions for arsenic and antimony solid sources are:

$$2As_2O_3 + 3Si \leftrightarrow 3SiO_2 + 4As \qquad (16.11)$$

$$2Sb_2O_3 + 3Si \leftrightarrow 3SiO_2 + 4Sb \qquad (16.12)$$

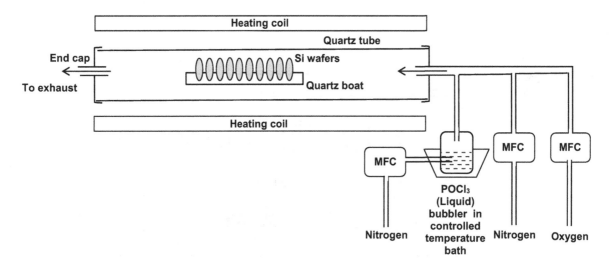

FIGURE 16.14   Phosphorous diffusion furnace arrangement using a liquid dopant source.

A solid source for boron is trimethyl borate. Placed outside the furnace tube, it is cooled below room temperature. Boron diffuses into silicon from boron trioxide formed on the wafer surface.

$$2(CH_3O)_3 B + 9O_2 \rightarrow B_2O_3 + 6CO_2 + 9H_2O (900°C) \quad (16.13)$$

$$2B_2O_3 + 3Si \leftrightarrow 3SiO_2 + 4B \quad (16.14)$$

Another boron solid source is boron nitride which is taken in wafer form. For diffusion, activated source wafers are placed with silicon wafers in the same quartz boat.

Liquid source for boron is boron tribromide.

$$4BBr_3 + 3O_2 \rightarrow 2B_2O_3 + 6Br_2, \ 2B_2O_3 + 3Si \leftrightarrow 3SiO_2 + 4B \quad (16.15)$$

Gaseous source for boron is diborane:

$$B_2H_6 + 3O_2 \rightarrow B_2O_3 + 3H_2O(300°C), 2B_2O_3 + 3Si \leftrightarrow 3SiO_2 + 4B \quad (16.16)$$

or

$$B_2H_6 + 6CO_2 \rightarrow B_2O_3 + 6CO + 3H_2O, 2B_2O_3 + 3Si \leftrightarrow 3SiO_2 + 4B \quad (16.17)$$

16.6 Ion Implantation

Diffusion is an isotropic process which cannot control junction depth and dopant concentration independently. Ion implantation provides independent control of junction depth through ion energy and dopant concentration = ion current × implantation time. An anisotropic doping profile can be produced by ion implantation. Diffusion is a high-temperature process which requires a hard mask such as silicon dioxide. Ion implantation is a room-temperature process which can be done using photoresist as a mask.

Ion implantation is a process to modify the surface of a material by bombarding it with ions of the desired material having sufficient energy (typically 10–500 keV and up to MeV) to penetrate significantly into targeted surface and become embedded, thereby altering its physical and chemical properties. The equipment (Figure 16.15) consists of an ion source (e.g. thermionically emitted electrons from a tungsten filament colliding with dopant gas (PH_3, AsH_3, BF_3) molecules to cause ionization), an ion extraction electrode (which pulls out the ions from the source using a high electric field), an ion selection chamber (where the required ions are separated by a magnetic field mass analyzer by gyro radii of ions so that ions of correct charge-to-mass ratio can pass through a slit), an ion accelerating column (where the ions are electrostatically speeded to final kinetic energy), and a doping chamber (where the ion beam impinges upon and scans the wafer surface).

During post-implantation annealing, the dopant atoms are lodged into substitutional sites in a single crystal structure. Hence they become activated as donor or acceptor impurities. Rapid thermal annealing (RTA) minimizes diffusion of the dopant and gives better thermal budget than furnace annealing.

16.7 Physical Vapor Deposition (PVD)

16.7.1 Vacuum Evaporation

It is a method widely used for deposition of metal films for forming contact electrodes and interconnections in devices. The metal to be deposited, e.g. Al, Al-Si, Au, Ni, Cr, is placed on a crucible in a vacuum chamber evacuated to 10^{-6} to 10^{-7} torr and heated either resistively or with an electron beam (Figure 16.16). After melting, the metal evaporates and condenses on the wafers placed in a sample holder above. Vacuum condition is essential to provide a large mean free path to the metal vapors and to avoid contamination of the deposited film. A turbomolecular pump backed by a scroll pump is preferred due to lower

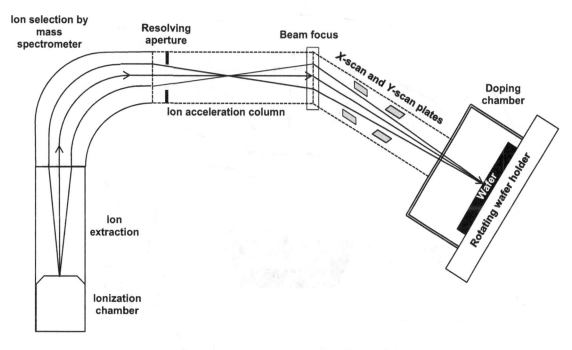

FIGURE 16.15 Stages of ion trajectory in an ion implantation machine.

maintenance needs. Thickness uniformity is ensured if both the source and the substrate lie on the surface of a sphere because the angular dependence of thickness is eliminated by this arrangement. Film thickness is measured with a crystal thickness monitor. Batch processing of multiple wafers with planetary motion is done using a substrate planetary fixture consisting of several dome-shaped substrate holders rotating around their own axis and about the vertical axis.

16.7.2 Molecular Beam Epitaxy (MBE)

The crystalline substrate is heated to a low temperature ~500°C–600°C in high vacuum (10^{-8} torr) or ultra-high vacuum (10^{-12} torr) (Figure 16.17). No carrier gases are used. Generally elemental sources in Knudsen cells or electron-beam evaporators are used to propel the material on the substrate. For growing layered structures, shutters are used. Due to high-vacuum growth conditions, highly directional elemental beams with long mean free paths are obtained. Flux densities in the beams are in the range of 10^{14}–10^{16} atoms $cm^{-2} s^{-1}$. The evaporated atoms do not collide or interact with each until they reach the substrate. Avoidance of carrier gases and use of high vacuum conditions eliminate the chances of contamination by impurities. So, very high purity films are formed. The growth rates are slow, around one monolayer per second. The molecules impinging on the surface condense, adsorb, desorb, dissociate, migrate, and react to form crystals systematically with the same or related arrangement as the substrate, thereby allowing epitaxial growth of films. Thus MBE is an atomic layer-by-layer epitaxial growth of the crystal structure by interaction of atomic or molecular beams with a hot crystalline substrate in an ultra-clean vacuum environment.

MBE differs from the usual evaporation processes because of the precise control of growth conditions. Due to the low substrate temperatures used, material interdiffusion is minimal. Hence, thin-film heterostructures can be realized with sub-nm control in thickness apart from control of composition, doping, and crystal quality. Such tailoring of thin-film structures is known as bandgap engineering. For growing AlGaAs on GaAs substrate, the MBE uses beams of elements Al, Ga, and As along with beam of dopant, e.g. Si for N-type or Be for P-type doping.

The main drawback of MBE is that it cannot be scaled up for industrial manufacturing. Every time the source material is to be filled, the vacuum needs to be broken. Moreover, compound semiconductors containing phosphorous or arsenic are difficult to grow by MBE.

16.7.3 Sputtering

It is a room-temperature process used to deposit thin films of various materials, elements, e.g. Au, Cu or compounds, e.g. SiN, Ta_2O_5, AlN, ZnO, conducting or insulating, on a substrate in a vacuum chamber by producing a DC/RF gaseous plasma in an inert gas from which the positive inert gas ions are accelerated toward the cathode made of the material to be deposited (called the sputtering target) causing ejection of microscopic target particles by momentum transfer, which fall and deposit on the anode on which the substrate is placed (Figure 16.18). Films of conductive materials such as metals are deposited using DC sputtering. High melting point refractory metals can be easily deposited by sputtering.

But for sputter deposition of insulating materials, DC sputtering cannot be used because of the building up of a layer of positive inert gas ions near the target which produces arcing and even stoppage of sputtering. Then RF sputtering comes to rescue. The potentials of the electrodes are alternated at radio frequencies (13.56 MHz) to avoid the positive charge

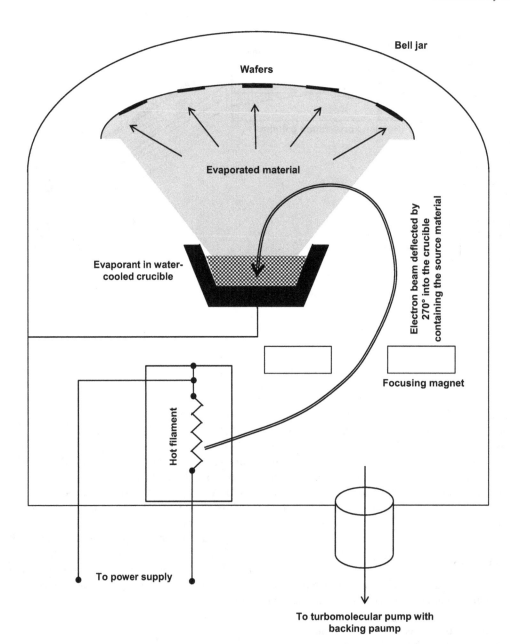

FIGURE 16.16 Electron-beam evaporation system.

buildup at the target. By applying an RF potential to the target, the positive charge accumulation on the target is neutralized because both the positive ions and electrons will now impinge on the target, electrons when the target is positive and positive ions when the target is negative. But electrons have a higher mobility than positive ions. So, a greater number of electrons will reach the target creating a negative bias which will provide continuous sputtering. Films of chemical compounds are obtained either by sputtering from a composite target or by reactive sputtering in which a gas, e.g. oxygen or nitrogen, is introduced. RF sputtering is also used for depositing metal films.

Sputtering being a mechanical process, not a chemical one, offers a broad scope and choices in material deposition, applicable to a diversity of materials.

16.7.4 Laser Ablation Deposition (LAD) or Photoablation Deposition

It is a process in which a laser beam from a high fluence, short wavelength laser source (e.g. ArF, KrF, or XeCl excimer laser giving pulses of few $100\,mJ\,cm^{-2}$), operating either in a pulse or continuous mode, is focused on the surface of the material, metallic or dielectric, to be deposited when the material evaporates or sublimes to form a thin film on a neighboring substrate (Figure 16.19). The process is done in high vacuum to avoid contamination from the environment.

The efficiency of laser ablation process is influenced by the wavelength of the laser beam and the duration and repetition rate of the laser pulse. An important parameter is beam quality governed by its brightness or intensity, focusing, and homogeneity.

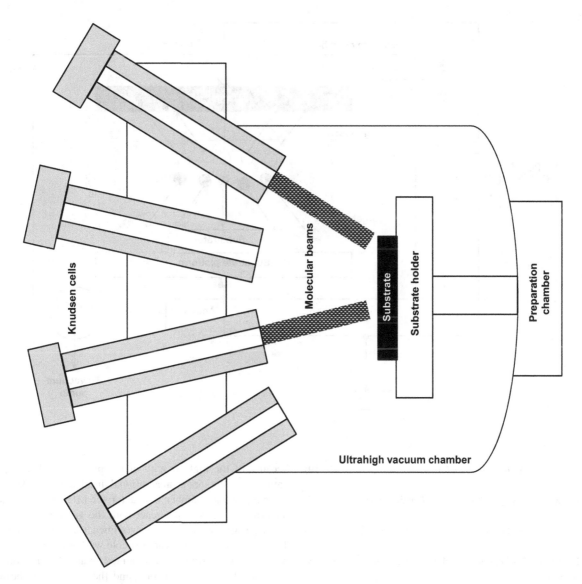

FIGURE 16.17 Molecular beam epitaxy setup.

The size of the beam decides the area to be ablated. The quantity of the material removed is a function of the properties of the material, the wavelength of the laser used, and the pulse duration. As the pulse duration can be varied from femtoseconds to milliseconds, material deposition can be accurately controlled.

16.8 Chemical Vapor Deposition (CVD)

16.8.1 Generic CVD Process

Chemical vapor deposition (CVD) is a synthesis process in which a material is deposited, usually as a thin film, in a solid phase on a heated substrate as a result of chemical reactions between volatile gaseous precursors on/in the vicinity of the substrate (Pierson 1999) (Figure 16.20). CVD reaction chambers are of two types: (i) hot-wall reactor in which the chamber is heated and the substrate placed inside automatically gets heated; (ii) cold-wall reactor in which the substrate is directly

heated either resistively or by induction heating while the chamber remains at a lower temperature. CVD may be carried out in open atmosphere. It is called atmospheric pressure chemical vapor deposition (APCVD). CVD performed at sub-atmospheric pressure is known as low-pressure chemical vapor deposition (LPCVD). CVD done in ultra-high vacuum (10^{-8} torr) is termed ultra-high vacuum chemical vapor deposition (UHVCVD). In a variant of CVD called plasma-enhanced chemical vapor deposition (PECVD), the CVD is carried out in a gaseous plasma (Figure 16.21). PECVD derives the benefit of plasma for lowering the deposition temperature so that films that are deposited at a high temperature in LPCVD can be formed at relatively lower temperatures by PECVD on substrates incapable of withstanding high temperatures.

A few CVD processes are as follows:

Polysilicon LPCVD (25–150 Pa) using silane or trichlorosilane at 600°C–650°C:

$$SiH_4 \rightarrow Si + 2H_2 \tag{16.18}$$

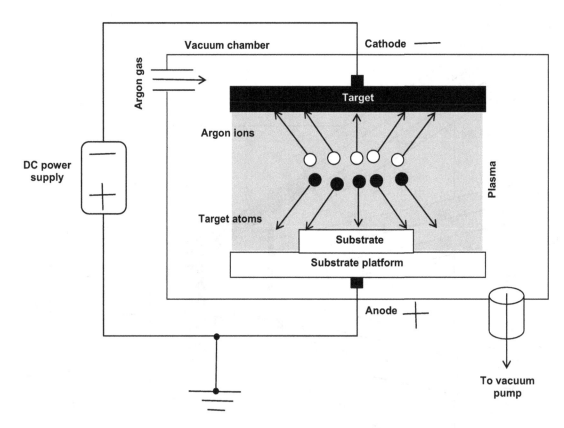

FIGURE 16.18 Principle of DC sputtering.

$$SiHCl_3 \rightarrow Si + Cl_2 + HCl \qquad (16.19)$$

Silicon dioxide APCVD at 300°C–500°C using SiH_4:

$$SiH_4 + O_2 \rightarrow SiO_2 + 2H_2 \qquad (16.20)$$

Silicon dioxide APCVD at 300°C–450°C using tetraethyl orthosilicate (TEOS):

$$Si(C_2H_5O)_4 \rightarrow SiO_2 + \text{by-products} \qquad (16.21)$$

Silicon dioxide PECVD at 350°C using silane and nitrous oxide:

$$SiH_4 + 2N_2O \rightarrow SiO_2 + 2N_2 + 2H_2 \qquad (16.22)$$

16.8.2 Metal-Organic Chemical Vapor Deposition (MOCVD)

The metal-organic chemical vapor deposition (MOCVD) differs from the MBE in several respects. First, the pressure regime in MOCVD is much higher, 15–750 torr as opposed to UHV in MBE. Second, gas phase precursors are used in MOCVD in place of elemental sources in MBE. Third, the sources are transported over the substrate with a carrier gas in MOCVD instead of evaporated beams in MBE. Fourth, the sources can be easily replenished in MOCVD by exchanging a pressurized gas bottle or bubbler as there is no UHV requirement, as in MBE. All these differences make the system scalable for industrial production. So, it is widely used for growing compound semiconductor films. But a stringent safety procedure must be followed for MOCVD

because of the high toxicity of the precursors. The expensive nature of precursors is an additional disadvantage of MOCVD.

The reaction chamber or reactor for MOCVD is made of a material which does not react with the gases used. Generally, quartz, stainless steel, or quartz-lined stainless steel is used. There are two types of reactors: cold wall or hot wall. In a cold-wall reactor, the substrate is mounted on an RF induction heated SiC-coated graphite susceptor and the susceptor is heated. In a hot-wall reactor, the quartz tube reactor itself is heated. It is useful when some gases need to be pre-cracked before reaching the substrate surface. The growth temperature range is typically 500°C–800°C but can be higher, e.g. 1000°C for GaN because of difficulty in thermal cracking of ammonia and requirement of higher energy for gallium-nitrogen bond formation (Lansheng et al 2016). The gas delivery system comprises a clean network of stainless steel tubing with automatic valves and mass flow controllers.

There are two feed lines in the MOCVD reactor (Figure 16.22):

i. One feed line is the metal organic (MO) line, and
ii. The other feed line is the hydride line.

In both the feed lines, hydrogen or nitrogen is used as the carrier gas. Before entering the reactor, the two lines merge together to blend the gases.

Common MO group III precursors are trimethyl alkyls such as trimethylgallium (TMGa) (l), $Ga(CH_3)_3$; trimethylaluminum (TMAl) (l), $Al(CH_3)_3$; trimethylindium (TMIn), $In(CH_3)_3$ (s). Triethyl alkyls like triethylgallium (TEGa) (l), $Ga(C_2H_5)_3$; triethylaluminum (TEAl) (l), $Al(C_2H_5)_3$; triethylindium (TEIn),

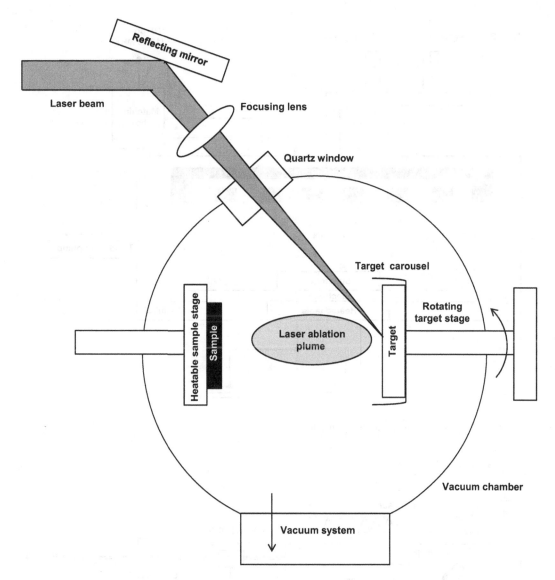

FIGURE 16.19 Laser ablation deposition.

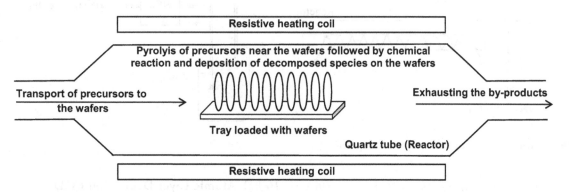

FIGURE 16.20 Main steps in a chemical vapor deposition process.

In($C_2H_5)_3$ (l) are also used. Greater stability and higher vapor pressure makes trimethyl precursors preferable over triethyl precursors. Precursors are kept at 4°C–22°C. Common hydrides are arsine (AsH_3) (g), phosphine (PH_3) (g), stibine (SbH_3) (g), and ammonia (NH_3) (g). Here l = liquid, g = gas, and s = solid.

A III–V compound semiconductor GaAs is grown by pyrolysis of volatile metal organic compound with gaseous hydride (Coleman 1997), e.g.

$$Ga(CH_3)_3 + AsH_3 \rightarrow GaAs + 3CH_4 \qquad (16.23)$$

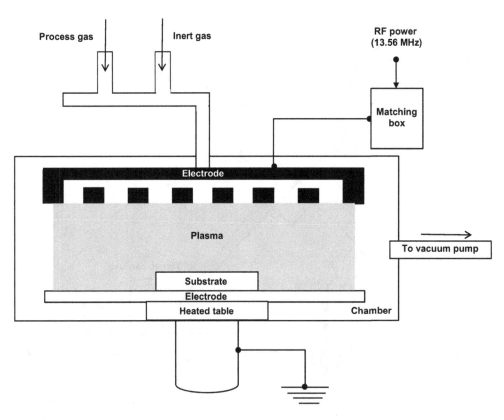

FIGURE 16.21   Apparatus of RF plasma-enhanced chemical vapor deposition.

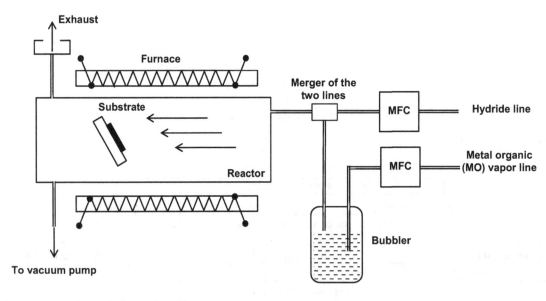

FIGURE 16.22   Metal-organic chemical vapor deposition system.

$$Ga(CH_3)_3 + NH_3 \rightarrow GaN + 3CH_4 \qquad (16.24)$$

Semiconductor alloys are grown by mixing the vapors of alloy constituents to form the desired composition, e.g. the ternary alloy InGaAs is grown as

$$xIn(CH_3)_3 + (1-x)Ga(CH_3)_3 + AsH_3 \rightarrow In_xGa_{1-x}As + 3CH_4$$
$$(16.25)$$

16.8.3 Atomic Layer Deposition (ALD)

It is a vapor phase thin-film deposition technique (Figure 16.23) from gaseous chemical precursors consisting of a sequence of alternating pulses of precursors 1 and 2 called half-reactions of 1 and 2 (Johnson et al 2014). During one half-reaction, precursor 1 is admitted into the reaction chamber under vacuum <1 torr for a specified time duration during which it reacts with the surface of the substrate through a self-limiting reaction which automatically

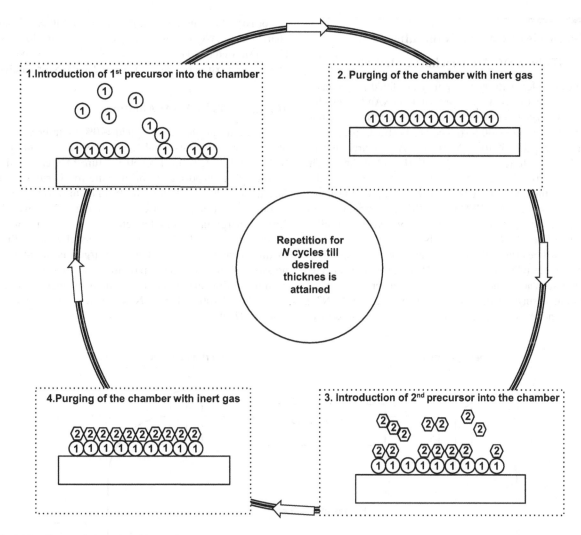

FIGURE 16.23 The atomic layer deposition cycle.

ceases after all the reaction sites on the substrate have been consumed leaving not more than one monolayer of precursor 1 on the surface. Then the residual precursor 1 along with any reaction by-products is fully removed from the chamber by purging it with an inert gas. During the next half-reaction, precursor 2 is introduced into the chamber, again for a limited time duration to allow reaction with precursor 1 already on the substrate to form a layer of the desired material {1, 2} comprising 1 and 2 on the substrate. As earlier, the chamber is purged with an inert gas which cleans the chamber of any remaining precursor 2 or any by-products of the reaction. Thus one half-reaction of precursor 1 and one half-reaction of precursor 2 are completed. In this way, several cycles of precursor 1 half-reaction and precursor 2 half-reaction are carried out until the desired thickness of material {1, 2} made of 1 and 2 has been achieved. Precursors 1 and 2 are never simultaneously present together in the chamber, i.e. they were never overlapped.

Generally, the deposition temperatures in atomic layer deposition (ALD) are <350°C. The temperature of saturated growth depends on the particular ALD reaction. The range of temperatures within which the growth is saturated constitutes the ALD temperature window. Outside the ALD temperature window, the growth rate falls. Also, the process yields non-ALD type growth.

So the ALD must be performed within the ALD window for a specific reaction.

The cyclic, self-saturating nature of ALD half-reactions bestows on this process many desirable features, notably the precision of control in film thickness and composition apart from conformality of coatings. These features act as the principal deciding factors favoring ALD when choosing ALD over competing processes of CVD or sputtering. But ALD also suffers from the drawback of slow speed because of the time consumption in the innumerable pulsing and purging operations, and the large number of sequential cycles necessary to attain a given film thickness. About 100–300 nm thick film is generally deposited in an hour. Nonetheless, ALD offers versatility in the respect that a variety of metallic, semiconducting or insulating films can be deposited. The deposited films can be crystalline or amorphous. Metal oxides such as ZrO_2 and Y_2O_3 and noble metals like platinum can be deposited. The ALD process is used to grow a uniform, pin-hole free film of high-k dielectric for the gate stack of a MOSFET. The gate stack is a multilayer comprising a thin film of silicon oxynitride with $k \sim 4$ followed by a HfO_2 high-k dielectric layer with $k \sim 20$ and capped at the top with a layer that matches with the work function of the gate metal (Kuhn 2012).

16.9 Synthesis of Carbon Nanotubes

16.9.1 CNTs by a DC Arc-Discharge Method

Graphite electrodes with 99.99% purity are fitted in a vacuum chamber (Figure 16.24) at a distance of 1 mm (Ando and Zhao 2006). After evacuation of the chamber by a diffusion pump, helium is filled at 500–700 torr. The arc is struck using a DC power supply capable of 10–50 V, 50–100 A. The arc temperature can be as high as 3000°C–4000°C (Liu et al 2003). The arc current is ~70–80 A. During arc-discharge synthesis lasting for ~1–2 min, the anode is consumed while 10 μm long multi-walled carbon nanotubes (MWCNTs) with diameter 5–30 nm are deposited on the cathode at a yield of 30% by weight, and fullerenes collect on the walls of the chamber. The chamber and the electrodes must be efficiently cooled to prevent over-sintering of carbon nanotubes (CNTs). The MWCNTs obtained are collected after de-pressurization and cooling of the chamber.

To produce single-walled carbon nanotubes (SWCNTs), a graphite anode containing a small percentage of catalyst (Fe or Ni) ~4% is used with a pure graphite cathode for arc discharge. The SWCNTs are deposited in the soot. Journet et al. (1997) showed that large quantities of SWCNTs can be obtained by using a mixture of 1 at. % Y and 4.2 at. % Ni as catalyst.

16.9.2 CNTs by Laser Ablation

SWCNTs are produced in a yield >70% by laser vaporization of graphite rods (Thess et al 1996). These rods contain 98.8 at. % C and 1.2 at. % of a 50:50 mixture of nickel and cobalt powders. The vaporization of graphite rod is carried out in a furnace at 1200°C (Figure 16.25). For uniformity of vaporization, two laser pulses are used: first at 532 nm wavelength, 250 mJ energy in a 5 mm diameter spot and second 50 ns later at 1064 nm, 300 mJ in a 7-mm spot, coaxial with the first spot. In this furnace, a constant flow of argon is maintained at a pressure of 500 torr. Thermal treatment at 1000°C in vacuum is used to sublime out fullerenes. Nanotubes organize into ropes consisting of 100–500 SWCNTs. The diameter of SWCNTs is 1.38 ± 0.02 nm.

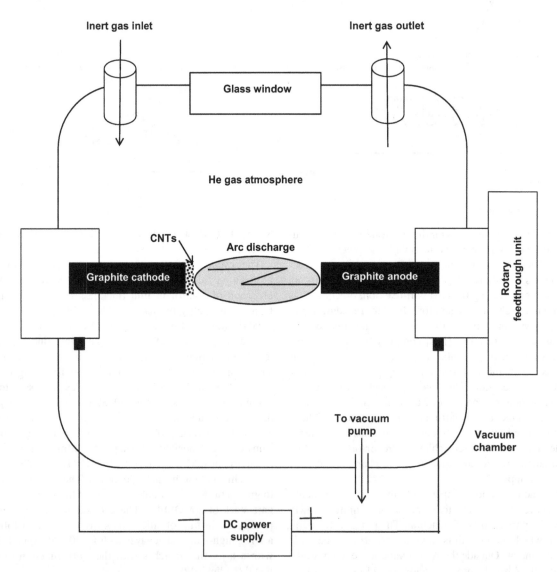

FIGURE 16.24 Electric arc-discharge method of carbon nanotube synthesis.

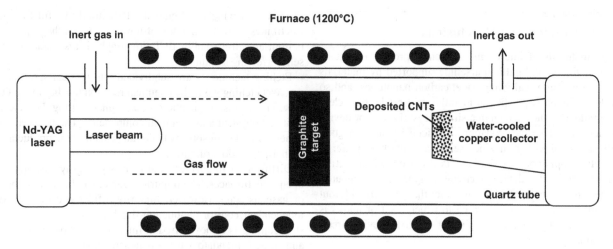

FIGURE 16.25 Carbon nanotube growth by laser ablation.

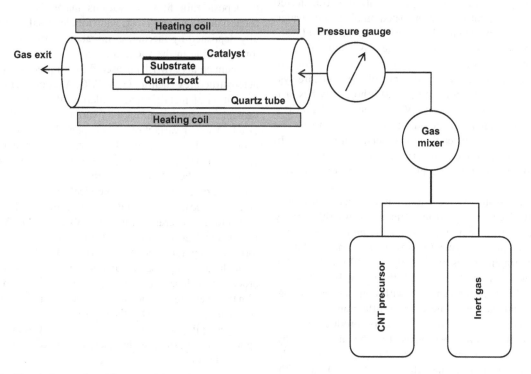

FIGURE 16.26 Chemical vapor deposition reactor for carbon nanotube synthesis.

16.9.3 CVD Growth of CNTs

CVD growth is done at lower temperature than arc-discharge and laser methods. It is also done at ambient pressure. Laser and arc-discharge-grown CNTs are superior to CVD-grown CNTs in crystallinity but CVD offers better yield and purity. Besides CVD is a versatile process allowing the use of various hydrocarbons and substrates besides capability to grow CNTs at defined locations on a patterned substrate (Kumar and Ando 2010). Examples of CNT precursors are methane, ethylene, acetylene, benzene, and carbon monoxide. Catalytic agents include iron, cobalt, and nickel. Mo and Fe/Mo alloys are also used (Yahyazadeh and Khoshandam 2017). In the low-temperature range (600°C–900°C), MWCNT growth is favored while at higher temperatures ~900°C–1200°C, SWCNTs are

formed. At higher temperatures, carbon monoxide and methane are used as the precursors due to their higher stability against self-decomposition. The CNTs are grown by passing the CNT precursor for 1/4–1 h through a quartz tube furnace in which the catalyst-coated substrate is maintained at the chosen temperature (Figure 16.26). In case of a liquid precursor such as benzene, an inert gas is passed through the precursor heated in a flask to transport the vapors to the substrate. The precursor is the feedstock and the catalyst particles are the seeds for CNT growth. The CNTs are collected after cooling down the furnace to room temperature.

The by-products accompanying CNT growth such as amorphous carbon and fullerenes are removed by refluxing the CNTs in nitric acid to oxidize these by-products. The residual metal catalyst particles are also removed by this cleaning.

16.10 Discussion and Conclusions

Favorite materials of micro/nanoelectronics process engineers have been silicon and gallium arsenide supported by materials such as germanium, gallium nitride, carbon nanotubes, and so forth. Raw silicon is purified to metallurgical grade, next electronic grade, and then grown in a single crystal form for device manufacturing. Thermal oxidation of silicon produces a good quality silicon dioxide film which serves as a masking material for selective doping of impurities and is also used as a dielectric.

The semiconductor device or circuit designer uses computer-aided design tools to create the artwork for the geometrical layout to be patterned on silicon according to which a set of masks is made by a pattern generator machine for the given device/circuit specifications.

The critical process step determining the feature sizes of the geometry is lithography in which the pattern is transferred from the mask to a photoresist film coated on the silicon wafer. According to the Rayleigh resolution criterion, the optical resolution can be increased by decreasing the wavelength of radiation for exposure of photoresist and by increasing the refractive index of the medium adjacent to the lens of the exposure system. From this understanding, the optical lithography has progressed from 248 to 193 nm deep UV and then to 13.2 nm EUV lithography. Also water has been used as the medium near the lens in place of the air gap leading to immersion lithography. In EUV lithography, complications arise in the creation of EUV source and the use of reflective elements instead of refractive lenses.

Moving apart from UV radiation, a new concept in lithography is introduced by electron-beam maskless lithography in which an electron-beam raster scans the photoresist on the wafer. Theoretically capable to provide sub-nm resolution, e-beam lithography gives practically sub-10 nm resolution because of electron interactions with the resist. In opposition to the parallel processing achieved with conventional lithography, the serial nature of the e-beam lithography makes it a slow process which is not useful for mass production. Like electron-beam lithography, two types of ion beam lithography, the FIB and proton beam types, are slow processes but the third type, IPL, is faster. Then there are 1 and 0.1 nm lithographies using X-rays, which are being developed as next-generation tools.

Breaking away from the practices of using UV, electron or proton beams, or X-rays for exposing the photoresist through a mask for delineating the pattern, the technology has moved to the principle of mechanically deforming the resist with a stamp or mold. In this NIL, two types, viz. thermal-NIL and UV-NIL, are prominent. Using an AFM probe tip in place of a stamp, dip pen nanolithography has emerged. Ink transference from the AFM tip to the substrate takes place by capillary action via a water meniscus. In the thermal dip pen, a meltable ink is used.

In a revolutionary change from top-down to bottom-up approach, BCP lithography exploits the self-assembling propensity of BCPs such as PS-b-PMMA BCP into a film containing vertically aligned PMMA cylinders embedded in the PS matrix. After UV exposure, the PMMA is removed leaving a nanoporous PS matrix serving as a template. Particularly useful for nanoelectronics is the application of top-down nanolithography with bottom-up self-assembly. BCP lithography is also endowed with self-correcting advantage.

Etching techniques are subdivided into two classes, wet and dry. Wet etching works by chemical reaction in a liquid medium. It is generally isotropic and produces undercutting. Dry etching contains physical and chemical components. The physical component provides anisotropy to etching enabling the production of high aspect ratio structures.

Diffusion of impurities in silicon is generally carried out in open-tube furnaces. The impurity sources can be in solid, liquid, or gaseous state. However, impurity diffusion cannot provide independent control of junction depth and concentration. This shortcoming is removed by ion implantation. Post-implantation annealing is mandatory for dopant activation.

Many physical vapor deposition (PVD) processes are used to deposit thin films of various materials on the substrate. Interconnections and metal pads for electrical contacts are usually deposited by thermal evaporation in vacuum ~10^{-6} to 10^{-7} torr. Working in the vacuum range ~10^{-8} to 10^{-12} torr, molecular beam epitaxy (MBE) is used for high-precision deposition of heterostructures such as GaAs/AlGaAs. For room-temperature deposition of metallic and insulating films, sputtering offers a valuable alternative through its various forms: DC, pulsed DC, or RF. Another variant of PVD is pulsed laser deposition which is done by shinning a laser beam on a target material in a high vacuum chamber when the target material evaporates to form a film on the substrate placed nearby.

CVD processes involve a chemical reaction between volatile precursors near the substrate resulting in film deposition upon it. There are several variants of CVD, viz. APCVD, LPCVD, UHVCVD, PECVD. Instead of transportation of sources in the form of evaporated beams in MBE, MOCVD does the same through gaseous phase precursors and carrier gases, making the process upscalable for industrial exploitation. When precision of film thickness, its composition, and conformability are crucial to device properties, ALD is the answer.

To get a glimpse of non-silicon technology, we briefly touched upon CNT deposition techniques, viz. arc discharge, laser ablation, and CVD.

Illustrative Exercises

16.1   Find the junction depth for infinite-source boron diffusion in an N-type silicon wafer of background concentration 5×10^{16} cm^{-3}, carried out at a temperature of 1050°C for 90 min, given that at this temperature, the diffusion coefficient of boron in silicon is 4.64×10^{-14} cm^2 s^{-1}, solid solubility of boron in silicon is 2.4×10^{20} cm^{-3} and erf^{-1}(0.9998) = 2.63.

For an infinite impurity source, the diffusion profile obeys the complementary error function (erfc) according to the equation

$$N(x,t) = N_s \, \text{erfc}\left(\frac{x}{2\sqrt{Dt}}\right) \qquad (16.26)$$

where $N(x, t)$ is the impurity concentration at a depth x at time t, N_s is the impurity concentration at the surface of the wafer as determined by the solid solubility of the dopant in silicon at diffusion temperature, D is the diffusion coefficient of impurity in silicon at the diffusion temperature, and t is the time for which the diffusion is performed. As we are calculating the junction depth, $N(x, t) = 5 \times 10^{16}\,\text{cm}^{-3}$, $N_s = 2.4 \times 10^{20}\,\text{cm}^{-3}$, $D = 4.64 \times 10^{-14}\,\text{cm}^2\text{s}^{-1}$, and $t = 90\,\text{min} = 90 \times 60 = 5400\,\text{s}$.

Eq. (16.26) is rewritten as

$$\frac{N(x,t)}{N_s} = \text{erfc}\left(\frac{x}{2\sqrt{Dt}}\right) = 1 - \text{erf}\left(\frac{x}{2\sqrt{Dt}}\right) \qquad (16.27)$$

or

$$\text{erf}\left(\frac{x}{2\sqrt{Dt}}\right) = 1 - \frac{N(x,t)}{N_s} \qquad (16.28)$$

or

$$\frac{x}{2\sqrt{Dt}} = \text{erf}^{-1}\left\{1 - \frac{N(x,t)}{N_s}\right\} \qquad (16.29)$$

or

$$x = 2\sqrt{Dt}\,\text{erf}^{-1}\left\{1 - \frac{N(x,t)}{N_s}\right\}$$

$$= 2\sqrt{4.64 \times 10^{-14} \times 5400}\,\text{erf}^{-1}\left\{1 - \frac{5 \times 10^{16}}{2.4 \times 10^{20}}\right\}$$

$$= 3.1658 \times 10^{-5}\,\text{erf}^{-1}\left\{1 - 2.083 \times 10^{-4}\right\}$$

$$= 3.1658 \times 10^{-5}\,\text{erf}^{-1}\left\{0.999792\right\}$$

$$\approx 3.1658 \times 10^{-5}\,\text{erf}^{-1}\left\{0.9998\right\}$$

$$= 3.1658 \times 10^{-5} \times 2.63 = 8.33 \times 10^{-5}\,\text{cm}$$

$$= 8.33 \times 10^{-5} \times 10^4\,\mu\text{m} = 0.83\,\mu\text{m} \qquad (16.30)$$

16.2   A two-step phosphorous diffusion is performed into a P-type silicon wafer of concentration $1 \times 10^{16}\,\text{cm}^{-3}$. Pre-deposition is done at 900°C for ½ h; diffusion coefficient of phosphorous in silicon at 900°C is $7.3 \times 10^{-16}\,\text{cm}^2\text{s}^{-1}$ and its solid solubility at this temperature is $7 \times 10^{20}\,\text{cm}^{-3}$. Drive-in is done at 1100°C for 1h; diffusion coefficient of phosphorous in silicon at 1100°C is $1.5 \times 10^{-13}\,\text{cm}^2\text{s}^{-1}$. What are the resulting surface concentration of phosphorous and junction depth?

The impurity dose Q introduced into silicon during pre-deposition is

$$Q = 2N_s\sqrt{\frac{D_1 t_1}{\pi}} \qquad (16.31)$$

where N_s is the surface concentration of impurity as determined by its solid solubility limit at the pre-deposition temperature, D_1 is the diffusion coefficient of the impurity in silicon at the pre-deposition temperature, and t_1 is the time for which pre-deposition is carried out. Here, $N_s = 7 \times 10^{20}\,\text{cm}^{-3}$, $D_1 = 7.3 \times 10^{-16}\,\text{cm}^2\text{s}^{-1}$, and $t_1 = \frac{1}{2}\,\text{h} = 30\,\text{min} = 30 \times 60 = 1800\,\text{s}$. Hence, the dose introduced during the pre-deposition is

$$Q = 2 \times 7 \times 10^{20}\sqrt{\frac{7.3 \times 10^{-16} \times 1800}{3.14}}$$

$$= 1.4 \times 10^{21} \times 64.69 \times 10^{-8} = 9.06 \times 10^{14}\,\text{cm}^{-2} \qquad (16.32)$$

After pre-deposition, the surface concentration changes. The impurity diffusion profile obeys the Gaussian function

$$N = N_0 \exp\left(-\frac{x^2}{4D_2 t_2}\right) \qquad (16.33)$$

where $N(x, t)$ is the concentration of impurity at a depth x below the surface of silicon, and N_0 is the surface concentration after drive-in given by

$$N_0 = \frac{Q}{\sqrt{\pi D_2 t_2}} \qquad (16.34)$$

D_2 is the diffusion coefficient of the impurity in silicon at the temperature of drive-in and t_2 is the time for which drive-in is performed. Here $D_2 = 1.5 \times 10^{-13}\,\text{cm}^2\text{s}^{-1}$ and $t_2 = 1\,\text{h} = 60\,\text{min} = 60 \times 60 = 3600\,\text{s}$. When $x = $ junction depth x_j, $N = $ background doping of silicon wafer, here $1 \times 10^{16}\,\text{cm}^{-3}$. Then eq. (16.33) becomes

$$N = N_0 \exp\left(-\frac{x_j^2}{4D_2 t_2}\right) \qquad (16.35)$$

or

$$\frac{N}{N_0} = \exp\left(-\frac{x_j^2}{4D_2 t_2}\right) \qquad (16.36)$$

Taking natural logarithm of both sides

$$\ln\left(\frac{N}{N_0}\right) = -\frac{x_j^2}{4D_2 t_2} \qquad (16.37)$$

or

$$x_j^2 = -4D_2 t_2 \ln\left(\frac{N}{N_0}\right) \qquad (16.38)$$

Equation (16.34) will be used for calculating the surface concentration after drive-in. Putting the value of surface concentration obtained, eq. (16.38) will be used for determining the junction depth. From eq. (16.34),

Putting the value of Q from eq. (16.32)

$$N_0 = \frac{Q}{\sqrt{\pi D_2 t_2}} = \frac{9.06 \times 10^{14}}{\sqrt{3.14 \times 1.5 \times 10^{-13} \times 3600}}$$

$$= \frac{9.06 \times 10^{14}}{\sqrt{16.956 \times 10^{-10}}} = \frac{9.06 \times 10^{14}}{4.12 \times 10^{-5}} = 2.2 \times 10^{19} \text{ cm}^{-3} \qquad (16.39)$$

Applying eq. (16.38) and eq. (16.39)

$$x_j^2 = -4 D_2 t_2 \ln\left(\frac{N}{N_0}\right)$$

$$= -4 \times 1.5 \times 10^{-13} \times 3600 \times \ln\left(\frac{1 \times 10^{16}}{2.2 \times 10^{19}}\right) = -21600 \times 10^{-13}$$

$$\times \ln\left(4.55 \times 10^{-4}\right) = -21600 \times 10^{-13} \times -7.695 = 1.66212 \times 10^{-8}$$

$$(16.40)$$

$$\therefore x_j = \sqrt{1.66212 \times 10^{-8}} = 1.289 \times 10^{-4} \text{ cm}$$

$$= 1.289 \times 10^{-4} \times 10^4 \text{ μm} = 1.29 \text{ μm} \qquad (16.41)$$

16.3 (a) A deep UV lithography equipment uses a 157 nm F2 laser source. The numerical aperture of the lens is 0.6. What is the Rayleigh resolution r? Define depth of field DOF and obtain its value? (b) Do the practical values of Rayleigh resolution and depth of resolution match with the calculations?

a. The Rayleigh resolution is expressed in terms of the wavelength λ of light and numerical aperture NA of the lens by eq. (16.7) as

$$r = \frac{0.61\lambda}{\text{NA}} = \frac{0.61 \times 157 \text{ nm}}{0.6} = 159.62 \text{ nm} \qquad (16.42)$$

The depth of field is a distance in the vertical direction across the examined sample that yields satisfactory image quality. It is measured from above and below the focal plane, and is given by

$$\text{DOF} = \pm\frac{\lambda}{2(\text{NA})^2} = \pm\frac{157 \text{ nm}}{2 \times (0.6)^2} = \pm 218.056 \text{ nm} \quad (16.43)$$

b. Practically, Rayleigh resolution may differ from calculated values and is expressed as

$$r_{\text{Pracical}} = \frac{k_1 \lambda}{\text{NA}} \text{ where } 0.6 < k_1 < 0.8 \qquad (16.44)$$

where k_1 is an experimental parameter determined by the equipment and resist used. Similarly

$$\text{DOF}_{\text{Practical}} = \pm\frac{k_2 \lambda}{(\text{NA})^2} \qquad (16.45)$$

where k_2 is an experimental parameter <1.

REFERENCES

Ando, Y. and X. Zhao. 2006. Synthesis of carbon nanotubes by arc-discharge method. *New Diamond and Frontier Carbon Technology* 16(3): 123–137.

Coleman, J.J. 1997. Metal organic chemical vapor deposition for optoelectronic devices. *Proceedings of the IEEE* 85(11): 1715–1729.

Galatsis, K., K.L. Wang, M. Ozkan, C.S. Ozkan, Y. Huang, J.P. Chang, H.G. Monbouquette, Y. Chen, P. Nealey and Y. Botros. 2010. Patterning and templating for nanoelectronics. *Advanced Materials* 22: 769–778.

Groves, T.R. 2013. Chapter 3: Electron-beam lithography. In: Feldman M. (Ed.), *Nanolithography: The Art of Fabricating Nanoelectronic and Nanophotonic Devices*. Woodhead Publishing: Sawston, pp. 80–115.

Jain, N.K. and S.K. Chaubey. 2017. Review of miniature gear manufacturing. *Comprehensive Materials Finishing* 1: 504–538.

Johnson, R.W., A. Hultqvist and S.F. Bent. 2014. A brief review of atomic layer deposition: From fundamentals to applications. *Materials Today* 17(5): 236–246.

Journet, C., W.K. Maser, P. Bernier, A. Loiseau, M. Lamy de la Chapelle, S. Lefrant, P. Deniard, R. Leek and J.E. Fischer. 1997. Large-scale production of single-walled carbon nanotubes by the electric-arc technique. *Nature* 388: 756–758.

Kuhn, K.J. 2012. Considerations for ultimate CMOS scaling. *IEEE Transactions on Electron Devices* 59(7): 1813–1828.

Kumar, M. and Y. Ando. 2010. Chemical vapor deposition of carbon nanotubes: A review on growth mechanism and mass production. *Journal of Nanoscienece and Nanotechnology* 10: 3739–3758.

Lan, H. and Y. Ding. 2010. Chapter 23: Nanoimprint lithography. In: Wang M. (Ed.), *Lithography*. InTech: Croatia, pp. 457–494.

Lansheng, F., G. Runqiu and Z. Jincheng. 2016. The effect of substrate temperature on chemical reactions during GaN growth in a vertical MOCVD. *Proceedings of the 2016 4th International Conference on Machinery, Materials and Information Technology Applications (ICMMITA 2016)*, December 10–11, 2016, Xi'an, China, Advances in Computer Science Research, vol. 71, pp. 1621–1624.

Lee, W.K., Z. Dai, W.P. King and P.E. Sheehan. 2009. Maskless nanoscale writing of nanoparticle-polymer composites and nanoparticle assemblies using thermal nanoprobes. *Nano Letters* 10(1): 129–133.

Li, M., L. Chen and S.Y. Chou. 2001. Direct three-dimensional patterning using nanoimprint lithography. *Applied Physics Letters* 78(21): 3322–3324.

Liu, X., C. Lee, S. Han, C. Li and C. Zhou. 2003. Chapter 1: Carbon nanotubes: Synthesis, devices and integrated systems. In: Reed M.A. and T. Lee (Eds.), *Molecular Nanoelectronics*. American Scientific Publishers: Stevenson Ranch, CA, pp. 1–20.

Nunns, A., J. Gwyther and I. Manners. 2013. Inorganic block copolymer lithography. *Polymer* 54: 1269–1284.

Pierson, H.O. 1999. *Handbook of Chemical Vapor Deposition (CVD): Principles, Technology and Applications*. Noyes Publications: Norwich, NJ, p. 36.

Piner, R.D., J. Zhu, F. Xu, S. Hong and C.A. Mirkin. 1999. "Dippen" nanolithography. *Science* 283(5402): 661–663.

Segalman, R.A. 2005. Patterning with block copolymer thin films. *Materials Science and Engineering R* 48: 191–226.

Shin, D.O., J.-R. Jeong, T.H. Han, C.M. Koo, H.-J. Park, Y.T. Lim and S.O. Kim. 2010. A plasmonic biosensor array by block copolymer lithography. *Journal of Materials Chemistry* 20: 7241–7247.

Sreenivasan, S.V. 2017. Nanoimprint lithography steppers for volume fabrication of leading-edge semiconductor integrated circuits. *Microsystems and Nanoengineering* 3(17075): 1–19.

Thess, A., R. Lee, P. Nikolaev, H. Dai, P. Petit, J. Robert, C. Xu, Y.H. Lee, S.G. Kim, A.G. Rinzler, D.T. Colbert, G.E. Scuseria, D. Tomanek, J.E. Fischer and R.E. Smalley. 1996. Crystalline ropes of metallic carbon nanotubes. *Science* 273: 483–487.

Totzeck, M., W. Ulrich, A. Göhnermeier and W. Kaiser. 2007. Pushing deep ultraviolet lithography to its limits. *Nature Photonics* 1: 629–631.

Watt, F., A.A. Bettiol, J.A. Van Kan, E.J. Teo and M.B.H. Breese. 2005. Ion beam lithography and nanofabrication: A review. *International Journal of Nanoscience* 4(3): 269–286.

Yahyazadeh, A. and B. Khoshandam. 2017. Carbon nanotube synthesis via the catalytic chemical vapor deposition of methane in the presence of iron, molybdenum, and iron–molybdenum alloy thin layer catalysts. *Results in Physics* 7: 3826–3837.

17

Characterization Facilities

Various process-monitoring tools used in a semiconductor fabrication laboratory will be presented.

17.1 Four-Point Probe for Sheet Resistance Measurements

The four-point probe consists of four equispaced metal probes, generally made of tungsten and supported by springs (Figure 17.1). The interprobe spacing s is typically 1 mm. The semiconductor sample is placed on a mechanical stage below the probes and the probes are brought in contact with the surface of the sample. The spring action ensures a pressure contact,

and the pressure is adjusted to avoid any damage to the sample. The current I is supplied through the outer two probes from a high impedance source. The voltage drop V across the inner two probes is measured.

17.1.1 Bulk Sample

For the probe setup shown in Figure 17.2, it is assumed that metal tip is infinitesimally small. Regarding the sample, it is postulated to extend to semi-infinite dimensions laterally. About the current, it is assumed that it emanates in the form of a spherical protrusion of radius r from the tips of the outer probes. We apply the well-known formulae (see eq. (7.37)):

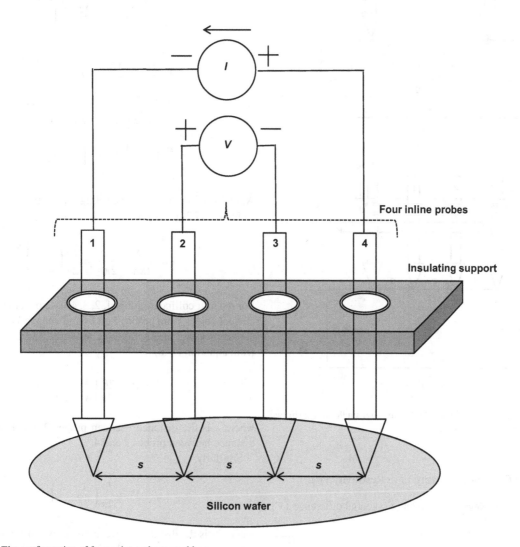

FIGURE 17.1 The configuration of four-point probe assembly.

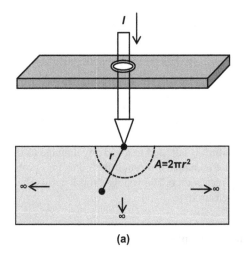

(a)

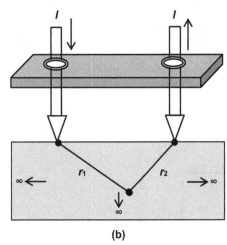

(b)

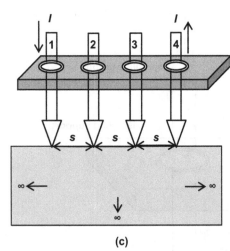

(c)

FIGURE 17.2 Calculation of voltages due to: (a) a single probe, (b) two probes, and (c) four collinear probes.

Electric field (E) = Current density (J) × Resistivity (ρ)

$\quad$ = Negative gradient of potential difference (V)

$$= -\frac{dV}{dr} \qquad (17.1)$$

Current density $(J) = \dfrac{\text{Current } (I)}{\text{Cross-sectional area } (A)} \qquad (17.2)$

To calculate the voltage due to a single probe, as shown in Figure 17.2 (a), we note that the cross-sectional area under consideration is the surface area of the semi-sphere given by

$$A = \frac{4\pi r^2}{2} = 2\pi r^2 \qquad (17.3)$$

So, from eq. (17.2)

$$J = \frac{I}{2\pi r^2} \qquad (17.4)$$

and from eqs. (17.1) and (17.4)

$$-\frac{dV}{dr} = J\rho = \frac{I\rho}{2\pi r^2} \qquad (17.5)$$

or

$$dV = -\frac{I\rho}{2\pi}\left(\frac{dr}{r^2}\right) \qquad (17.6)$$

$$\therefore V = \int_0^V dV = -\int_\infty^r \frac{I\rho}{2\pi}\left(\frac{dr}{r^2}\right) = -\frac{I\rho}{2\pi}\int_\infty^r r^{-2}\, dr$$

$$= -\frac{I\rho}{2\pi}\left|\frac{r^{-2+1}}{-2+1}\right|_\infty^r = -\frac{I\rho}{2\pi \times -1}\left|r^{-1}\right|_\infty^r$$

$$= \frac{I\rho}{2\pi}\left(\frac{1}{r} - \frac{1}{\infty}\right) = \frac{I\rho}{2\pi}\left(\frac{1}{r} - 0\right) = \frac{I\rho}{2\pi r} \qquad (17.7)$$

Applying equation (17.7), the voltage due to two probes, as shown in Figure 17.2 (b), is given by

$$V = \frac{I\rho}{2\pi r_1} - \frac{I\rho}{2\pi r_2} = \frac{I\rho}{2\pi}\left(\frac{1}{r_1} - \frac{1}{r_2}\right) \qquad (17.8)$$

For the four collinear probes 1, 2, 3, 4, as shown in Figure 17.2(c), current comes from probes 1 and 4 and voltage V_{23} is determined across probes 2 and 3. Looking at Figure 17.2(c), we can write from equation (17.8)

$$V_2 = \frac{I\rho}{2\pi}\left(\frac{1}{s} - \frac{1}{2s}\right) \qquad (17.9)$$

where s is the distance between probes 2 and 1, and $2s$ is the distance between probes 2 and 4.

$\quad$ Similarly,

$$V_3 = \frac{I\rho}{2\pi}\left(\frac{1}{2s} - \frac{1}{s}\right) \qquad (17.10)$$

where $2s$ is the distance between probes 3 and 1, and s is the distance between probes 3 and 4. Now from eqs. (17.9) and (17.10),

$$V_{23} = V_2 - V_3 = \frac{I\rho}{2\pi}\left(\frac{1}{s} - \frac{1}{2s}\right) - \frac{I\rho}{2\pi}\left(\frac{1}{2s} - \frac{1}{s}\right)$$

$$= \frac{I\rho}{2\pi}\left(\frac{1}{s} - \frac{1}{2s} - \frac{1}{2s} + \frac{1}{s}\right) = \frac{I\rho}{2\pi}\left(\frac{2-1-1+2}{2s}\right)$$

$$= \frac{I\rho}{2\pi}\left(\frac{2}{2s}\right) = \frac{I\rho}{2\pi s} \qquad (17.11)$$

Putting

$$V_{23} = V \qquad (17.12)$$

we get

$$V = \frac{I\rho}{2\pi s} \qquad (17.13)$$

from which

$$\rho = 2\pi s \frac{V}{I}\, \Omega\,\text{cm} \qquad (17.14)$$

Since s is known from instrument specifications, knowledge of V and I gives ρ.

Alternatively, from the relation

$$\text{Resistance }(R) = \frac{\text{Resistivity }(\rho) \times \text{Length }(l)}{\text{Cross-sectional area }(A)} \qquad (17.15)$$

differential resistance is

$$dR = \rho\frac{dr}{A} = \rho\frac{dr}{2\pi r^2} \qquad (17.16)$$

so that

$$R = \int_0^R dR = \int_{r_1}^{r_2} \rho\frac{dr}{2\pi r^2} = \frac{\rho}{2\pi}\int_s^{2s}\frac{dr}{r^2} = \frac{\rho}{2\pi}\left|\frac{r^{-2+1}}{-2+1}\right|_s^{2s}$$

$$= \frac{\rho}{2\pi \times -1}\left(\frac{1}{2s} - \frac{1}{s}\right)$$

$$= \frac{\rho}{2\pi \times -1}\left(\frac{1-2}{2s}\right) = \frac{\rho}{4\pi s} \qquad (17.17)$$

Keeping in view the superposition of current at the outer two probes, we may write

$$R = \frac{V}{2I} \qquad (17.18)$$

Hence, from equations (17.17) and (17.18)

$$\frac{V}{2I} = \frac{\rho}{4\pi s} \qquad (17.19)$$

or

$$\rho = \frac{4\pi s V}{2I} = 2\pi s\left(\frac{V}{I}\right) \qquad (17.20)$$

17.1.2 Thin Sheet

The difference from the preceding case of the bulk sample is that the current flow occurs across circular rings instead of spherical protrusions. For a thin sheet of thickness $t \ll s$, the concerned cross-sectional area is

$$A = 2\pi r t \qquad (17.21)$$

The differential resistance can be written as

$$dR = \rho\frac{dr}{A} = \rho\frac{dr}{2\pi r t} \qquad (17.22)$$

and

$$R = \int_0^R dR = \int_{r_1}^{r_2} \rho\frac{dr}{2\pi r t} = \frac{\rho}{2\pi t}\int_s^{2s}\frac{dr}{r}$$

$$= \frac{\rho}{2\pi t}\left|\ln(r)\right|_s^{2s} = \frac{\rho}{2\pi t}\{\ln(2s) - \ln(s)\}$$

$$= \frac{\rho}{2\pi t}\ln\left(\frac{2s}{s}\right) = \frac{\rho}{2\pi t}\ln 2 \qquad (17.23)$$

As for the bulk sample, from superposition of current, applying equations (17.18) and (17.23)

$$R = \frac{V}{2I} = \frac{\rho}{2\pi t}\ln 2 \qquad (17.24)$$

or

$$\rho = \frac{\pi t}{\ln 2}\left(\frac{V}{I}\right) \qquad (17.25)$$

Here values of t, V, and I are needed to measure ρ.

17.2 X-Ray Diffraction (XRD) Crystallography

It is an analytical technique to determine the precise position/arrangement of atoms in the lattice of a crystalline material in 3D space by providing information regarding unit cell dimensions. It is also used for identification of chemical compounds in a sample from the uniqueness of the diffraction pattern. The diffraction pattern serves as a chemical finger print which reveals the identity of the material by comparison with a database of reference diffraction patterns. Thus the technique provides information on both the crystallographic structure and chemical composition of a sample.

When a beam of X-rays interacts with the atoms of a target material, the X-rays undergo scattering. The diffraction of X-rays follows Bragg's law. The Bragg's equation is (eq. (2.17))

$$2d\sin\theta = n\lambda \qquad (17.26)$$

where d denotes the distance between interatomic planes of the crystal, θ is the angle subtended by the X-ray beam with the direction of atomic planes in the crystal, n is an integer, and λ is

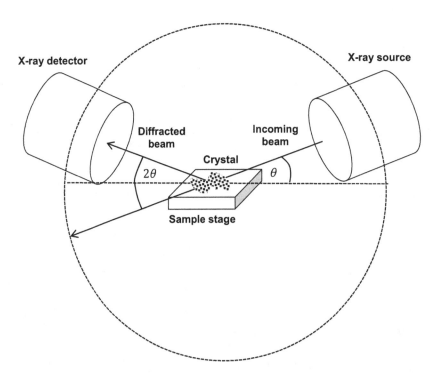

FIGURE 17.3 Measurement setup of the X-ray diffraction study.

the wavelength of X-rays. Depending on the path difference, the X-rays interfere constructively or destructively. The intensities of the diffracted waves are governed by the shape and size of the unit cell of the material because the crystalline material behaves as a diffraction grating.

An X-ray diffractometer (XRD) (Figure 17.3) has three principal components: a source of X-rays, a holder for mounting the sample, and a detector of X-rays. The X-rays generated by a cathode ray tube are filtered for monochromaticity. The collimated and concentrated beam is directed toward the sample at an angle θ. The X-ray detector makes a recording of the intensity of diffracted X-rays at an angle 2θ away from the trajectory of incident X-rays. The sample is subjected to scanning through a range of angles so that X-ray intensities from all possible directions are collected. The detector angle always remains at an angle 2θ with respect to the X-ray source path. In order that different directions are included, the material is usually taken in a powdered, homogenized form to enable speedier implementation.

17.3 Scanning Electron Microscope (SEM)

A scanning electron microscope (SEM) is a microscope capable of <1 nm resolution which operates by raster scanning a specimen with a focused beam of high-energy electrons (5–100 keV) in vacuum (Zhou et al. 2006).

17.3.1 Secondary and Backscattered Electrons

When the electron beam impinges on the specimen surface, various types of signals are produced (Figure 17.4). Among these signals, the signals from two types of electrons, the secondary

electrons (SEs) and reflected or backscattered electrons (BSEs), are used for imaging while element-specific X-ray emission is used for the chemical characterization of the specimen.

The SEs are produced by inelastic collisions of the incident electron beam with electrons in surface atoms when they knock off these electrons from their atomic shells. As they originate from near-surface regions of the specimen, they provide information about the surface of the specimen.

The BSEs are those primary electrons reflected back from deeper regions below the surface due to elastic collisions between the incident beam and the specimen atoms. They therefore supply information about these deeper regions. As BSEs are born from electron-atom collisions, atoms of a heavier element (higher atomic number) will give more scattered electrons. So, the intensity of BSE beam varies with the atomic number of the element in the specimen.

17.3.2 SEM Micrography

For SEM micrography, the position of the incident beam is combined with detected SE/BSE information to construct an image of the specimen in which edges are bright and recessed regions look dark. Magnification is the ratio of the area of the displayed image on the screen to the scanned area of the specimen. The SE image reveals the surface topography of the specimen. The BSE image is sensitive to the atomic numbers of the elements in the specimen. A heavier element (e.g. 26 Fe) therefore appears brighter in the BSE image than a lighter element (e.g. 12 Mg). So the BSE image contains two types of information, first about the surface topography and second on the composition of the specimen. The BSE image tells about the distribution of elements in the specimen, and not about their identity. As the region of

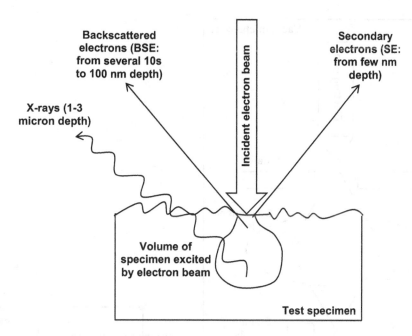

FIGURE 17.4 Signals produced by the interaction between the incident electron beam and the sample, at the surface and below up to 3 μm depth.

production of BSEs is several tens of nm deep inside the surface, they provide poorer spatial resolution than SEs. Bearing in mind that the incident electron beam can be focused to a very small size, it is ostensive that much higher spatial resolution <1 nm is achievable with SEs than with BSEs.

17.3.3 Energy Dispersive X-Ray (EDX) Analysis

Let us now turn our attention to the X-ray emission during SEM. Consequent upon the emission of each SE from a specimen atom, an electron vacancy or hole is created. If this vacancy is in an inner shell, the atom is in an unstable state. For stabilization, the electron vacancy is filled by an electron jumping from a higher energy shell. The difference of energies between the two shells is released as an X-ray photon. The energies of emitted X-rays from a specimen are characteristic of the elements present in it. They therefore serve as the unique signatures or finger prints of the elements, helping in their identification. Thus qualitative and quantitative data about the chemical composition of the specimen are recorded. The compositional data form the basis of the energy dispersive X-ray (EDX) analysis.

17.3.4 SEM Components

The main components of an SEM are (Figure 17.5):

i. The electron source: It is usually a thermionic emission gun made of V-shaped 100 μm long resistively heated tungsten filament or a solid-state cerium hexaboride or lanthanum hexaboride crystal. A field-emission gum is made of a tungsten wire with a fine tip ~100 nm.

ii. Condenser lenses to focus the electron beam.

iii. Scanning coils to deflect the beam along the *X*- and *Y*-axes for scanning the specimen.

iv. SE and BSE detectors: SE detection is done using an Everhart-Thornley (E-T) detector while the BSE are detected using a P-N junction diode. The E-T detector consists of a scintillator enclosed in a Faraday cage placed inside the chamber of the microscope containing the specimen. A small positive voltage (300 V) is applied to the Faraday cage to attract the low energy <50 eV SEs. The large positive bias (10 kV) on the scintillator accelerates the electrons and converts them to photons, which are carried down a light pipe from the scintillator inside the evacuated SEM specimen chamber to the photomultiplier tube outside the chamber for amplification.

In the reverse-biased P-N junction diode detector, the BSEs generate electron-hole pairs. These are carried to oppositely biased electrodes across which a current is produced according to the amount of absorbed BSEs.

v. X-ray detector: It is a solid-state semiconductor detector in which electron-hole pairs are produced. The higher the energy of the incident X-rays, the larger the number of electron-hole pairs generated. On cooling with liquid N_2, the energy resolution is <165 eV. A Peltier-cooled P-I-N diode is sometimes used for X-ray detection.

vi. Sample chamber: It is a chamber on a vibration-free platform containing a stage on which the specimen is kept. It is equipped with translation, rotation, and tilt facilities.

vii. Vacuum chamber: It houses the different components of SEM as the SEM works in vacuum. This vacuum = 10^{-3} to 10^{-4} Pa.

viii. Image display unit: Usually a cathode ray tube or liquid-crystal display is used.

ix. Computer and software.

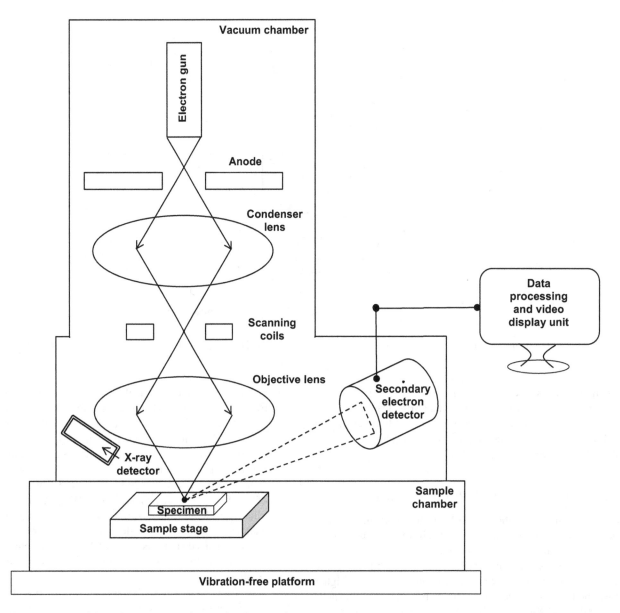

FIGURE 17.5 Layout of the main components of a scanning electron microscope.

17.4 Transmission Electron Microscope (TEM)

A transmission electron microscope (TEM) has similar components as that of the SEM, viz. the electron source, electromagnetic and electrostatic lenses, housed in a vacuum chamber (Figure 17.6) but there are vital differences (Pennycook and Nellist 2011):

i. The SEM works in a reflection mode and the imaging signals are SEs and BSEs, giving a 3D image of the surface of the sample. Samples of any arbitrary thickness are allowed.

The TEM operates in a transmission mode. The electron beam passes through an ultrathin sample of thickness <150 nm generating a two-dimensional (2D) projection image based on its transmission characteristics, thereby revealing its inner structure, crystalline, and stress state.

ii. In the SEM, the acceleration voltage of the scanning electron beam is 1–30 kV. In the TEM, the accelerating voltage is much higher ~60–300 kV because the beam has to penetrate the sample.

iii. The optimal spatial resolution of the SEM is ~0.5 nm while the same for TEM is <50 pm. The resolution of the SEM is inadequate to image individual atoms, as is possible with the TEM.

iv. The maximum magnification given by the SEM is 1–2×10^6X while the TEM provides $>5 \times 10^7$X.

v. The SEM produces an image on the PC screen by capturing and counting electrons with detectors. The TEM generates an image directly on a phosphor-coated screen or PC screen with a CCD camera.

vi. Sample preparation is fairly complex in the TEM but is relatively easier in the SEM.

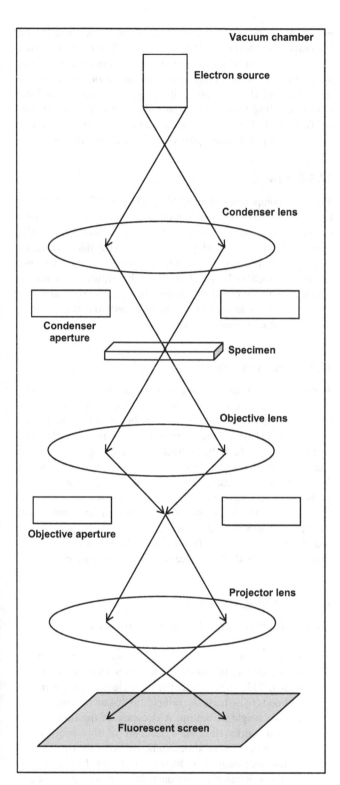

FIGURE 17.6 Transmission electron microscope.

17.5 Scanning Tunneling Microscope (STM)

As evident from its name, a scanning tunneling microscope (STM) is a microscope in which a sharp metal tip made of W, Pt-Ir, or Au is scanned over the surface of a conducting sample

at a sufficiently close distance. This scanning is usually done in ultra-high vacuum. During scanning, the tunneling current between the tip (one electrode) and the sample (second electrode) is measured by applying a small voltage across the two electrodes (Voigtländer 2015).

17.5.1 Dependence of the Tunneling Current on Sample-to-Tip Spacing

For tunneling, the vacuum gap represents the potential barrier. The height of this barrier is the work function Φ of the two surfaces. When the applied bias is smaller than the work function Φ, the barrier is approximated to be rectangular. This rectangle has a width z = the distance between the tip and the sample and height = Φ. Then the tunneling current is given by

$$I_t(z) = I_0 \exp(-2kz) \tag{17.27}$$

where

$$I_0 = \text{Function}\left(\text{Applied voltage} \times \text{Density of states}\right.$$
$$\left. \text{in the sample and the tip}\right) \tag{17.28}$$

and

$$k = \sqrt{\frac{2m\Phi}{\hbar^2}} \tag{17.29}$$

m is the mass of the electron and $\hbar$ is the reduced Planck's constant. The tunneling current depends on the applied voltage, and the width and the height of the potential barrier. Taking a typical value of $\Phi = 4\,\text{eV}$ for a metal, and applying equation (17.29)

$$k = \sqrt{\frac{2 \times 9.109 \times 10^{-31}\,\text{kg} \times 4\,\text{V} \times 1.602 \times 10^{-19}\,\text{C}}{\left(1.05457 \times 10^{-34}\,\text{Js}\right)^2}}$$

$$= \frac{10.80467 \times 10^{-25}}{1.05457 \times 10^{-34}} \frac{\sqrt{\text{kg} \times \text{V} \times \text{C}}}{\text{Joule} \times \text{s}}$$

$$= 10.246 \times 10^9 \frac{\sqrt{\text{kg} \times \dfrac{\text{Joule}}{\text{C}} \times \text{C}}}{\text{Joule} \times \text{s}}$$

$$= 1.025 \times 10^{10} \sqrt{\frac{\text{kg}}{\text{Joule}}} \times \frac{1}{\text{s}} = 1.025 \times 10^{10} \sqrt{\frac{\text{kg}}{\dfrac{\text{kgm}^2}{\text{s}^2}}} \times \frac{1}{\text{s}}$$

$$= 1.025 \times 10^{10} \sqrt{\text{kg} \times \frac{\text{s}^2}{\text{kgm}^2}} \times \frac{1}{\text{s}}$$

$$= 1.025 \times 10^{10} \frac{1}{\text{m}} = 1.025 \times 10^{10}\,\text{m}^{-1} \tag{17.30}$$

From equation (17.27), the tunneling current for $z = 0.1$ nm $= 0.1 \times 10^{-9}\,\text{m}$ is

$$\left|I_t(z)\right|_{0.1\,nm} = I_0 \exp\left(-2 \times 1.025 \times 10^{10} \times 0.1 \times 10^{-9}\right)$$

$$= I_0 \exp(-2.05) \qquad (17.31)$$

Similarly, the tunneling current for $z = 0.2\,nm = 0.2 \times 10^{-9}\,m$ is

$$\left|I_t(z)\right|_{0.2\,nm} = I_0 \exp\left(-2 \times 1.025 \times 10^{10} \times 0.2 \times 10^{-9}\right)$$

$$= I_0 \exp(-4.1) \qquad (17.32)$$

Hence, from equations (17.32) and (17.31), the ratio of the tunneling currents is

$$\frac{\left|I_t(z)\right|_{0.2\,nm}}{\left|I_t(z)\right|_{0.1\,nm}} = \frac{I_0 \exp(-4.1)}{I_0 \exp(-2.05)} = \exp(-2.05)$$

$$= 0.1287 = 0.13 \qquad (17.33)$$

Taking $\Phi = 5\,eV$ (Au) and repeating the calculations with equation (17.29), we get

$$k = \sqrt{\frac{2 \times 9.109 \times 10^{-31}\,kg \times 5V \times 1.602 \times 10^{-19}\,C}{\left(1.05457 \times 10^{-34}\,Js\right)^2}}$$

$$= \frac{12.08 \times 10^{-25}}{1.05457 \times 10^{-34}} = 11.455 \times 10^9\,m^{-1}$$

$$= 1.15 \times 10^{10}\,m^{-1} \qquad (17.34)$$

Then by equation (17.27)

$$\left|I_t(z)\right|_{0.1\,nm} = I_0 \exp\left(-2 \times 1.15 \times 10^{10} \times 0.1 \times 10^{-9}\right)$$

$$= I_0 \exp(-2.3) \qquad (17.35)$$

and

$$\left|I_t(z)\right|_{0.2\,nm} = I_0 \exp\left(-2 \times 1.15 \times 10^{10} \times 0.2 \times 10^{-9}\right)$$

$$= I_0 \exp(-4.6) \qquad (17.36)$$

$$\therefore \frac{\left|I_t(z)\right|_{0.2\,nm}}{\left|I_t(z)\right|_{0.1\,nm}} = \frac{I_0 \exp(-4.6)}{I_0 \exp(-2.3)} = \exp(-2.3)$$

$$= 0.1003 \approx 0.1 \qquad (17.37)$$

These calculations show that for a metal having work function = 4 eV, the tunneling current decreases by a ratio of 0.13 when the sample-tip distance is increased from 0.1 to 0.2 nm. For higher metal work function = 5 eV, this ratio becomes 0.1 for the same change in sample-to-tip spacing. Further for the 5 eV work function case, the ratio for every 0.1 nm increment of the distance being 0.1, over the range of atomic diameter 0.3 nm, the ratio

will be $0.1 \times 0.1 \times 0.1 = 1/1000$, considering the three 0.1 nm intervals, viz., 0–0.1, 0.1–0.2, and 0.2–0.3. The examples reveal the extreme sensitivity of the tunneling current to sample-to-tip spacing in the sub-nm range, which is the reason for the usefulness of the STM for nanoscale imaging. The strong dependence of the tunneling current on distance shows that the current flowing from the last atom in the tip to the first atom in the specimen or vice versa is dominant, providing single atom imaging.

17.5.2 STM Components

The main components of an STM are (Figure 17.7): the metal tip for scanning the sample; the sample stage on which the surface to be examined is placed, and X- and Y-motions can be imparted; a piezoelectric tube containing a piezoelement capable of deformation by applying a voltage and production of voltage by deformation using which the height of the tip can be adjusted; a system for coarse adjustment of the tip-sample distance; a current amplifier (nA); a feedback loop; a vibration isolation/protection arrangement and a controlling computer.

17.5.3 Imaging Procedure

To begin the measurement, the coarse adjustment system is used to set the sample-tip distance at a predecided value. This distance z is generally 0.4–0.7 nm. It lies between the two ranges: $0.3 < z < 1$ nm at which the force of interaction is attractive and $0 < z < 0.3$ nm at which the interactive force changes to repulsive nature. Once this setting has been done, the tunneling current flow is started by biasing the sample at a fixed voltage. As the tip is raster scanned in the X, Y-directions over the surface of the specimen, the tunneling current varies in accordance with the ups and downs, curvatures and undulations, depressions and elevations on the surface because the tip-sample distance changes in step with the surface topographical features.

17.5.4 Imaging Modes

The instrument is operated in one of the two modes:

i. Constant current, variable height mode: When the tunneling current increases, the feedback loop produces a signal. This signal feeds the piezoelectric tube which responds to increase the sample-tip distance by increasing the height of the tip. A decrease of the tunneling current works through a reversal of this sequence through the feedback mechanism to bring the tip closer to the specimen by reduction of the tip height. The feedback electronics continuously alters the tip height z to maintain the tunneling current constant by inputting the required voltage to the piezoelectric tube at each (x, y) position of the tip. The changes in the z values are combined with various (x, y) locations to generate a picture of the surface details of the specimen. The piezoelectricity-induced movements for height changes are sluggish so that this mode is slow.

ii. Constant height, variable current mode: This is a method to avoid the slow piezoelectric movements by

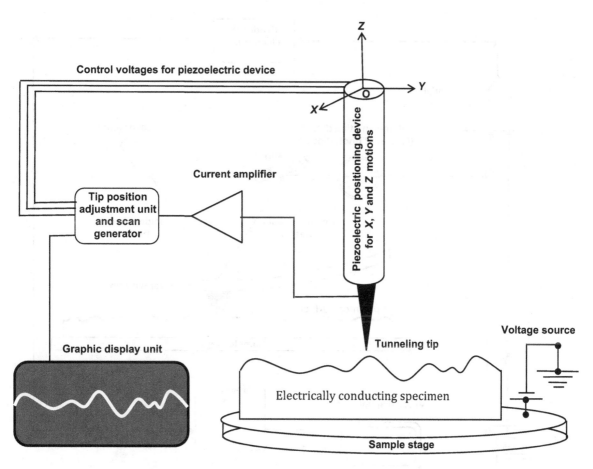

FIGURE 17.7 Scanning tunneling microscope components and their functions.

adjusting the tip at a convenient height over the sample. Keeping the tip height fixed, the tip is moved in a horizontal plane for scanning the specimen. The tunneling current varies as the tip scans the sample in the *X*-, *Y*-directions in step with the tip-sample distance variations. The image of the scanned surface is generated from the changes in the tunneling current at different (*x*, *y*) positions.

For highly precise measurements of irregular surfaces, the constant current mode is preferred although it takes a longer time. On the opposite side, for relatively smooth or nearly flat surfaces, the constant height mode is chosen for speedier measurements. So, the experimenter judiciously selects the proper mode as demanded by the circumstances.

Nonetheless, the indispensable requirement that the sample and the tip must be electrically conducting or semiconducting poses a serious limitation on the utilization of the STM. Another limitation is the necessity of a vacuum environment for measurements, representing an obvious encumbrance.

17.6 Atomic Force Microscope (AFM)

The AFM (Figure 17.8) serves to remove the hurdles of the vacuum pre-condition and the constraint that the sample must be electrically conducting (Eaton and West 2010, Voigtländer 2015).

It is a microscope that can provide measurements in diverse environments in open air, vacuum, high temperatures, humid ambients, liquids, and for materials that are conductors, semiconductors, insulators, or soft biological samples in their natural habitat (Vahabi et al. 2013).

17.6.1 Competition with Other Microscopes

To compare with competing microscopes, the optical microscope provides magnification of ~10^3X, an electron microscope ~10^5X, and an AFM ~10^6X. But the AFM provides measurements in three dimensions, along the horizontal *X*-*Y* plane and the vertical *Z*-direction while an optical or electron microscope can generate only 2D images; they cannot measure in the *Z*-direction. Laterally, the resolution of the AFM is low ~30nm but vertically, it is up to 0.1nm.

17.6.2 Imaging Principle

The AFM works on the principle of feeling or touching. Even if we close our eyes, we can get an idea of a surrounding surface by gently moving our fingers over it. An AFM does exactly that. It is a surface profilometer with a fine sharp tip that is run over a surface to get sensation of its contours and thereby build a 3D map of the surface. The tip attached to the free end of a 50–400 μm long flexible microcantilever beam made of silicon or silicon nitride is

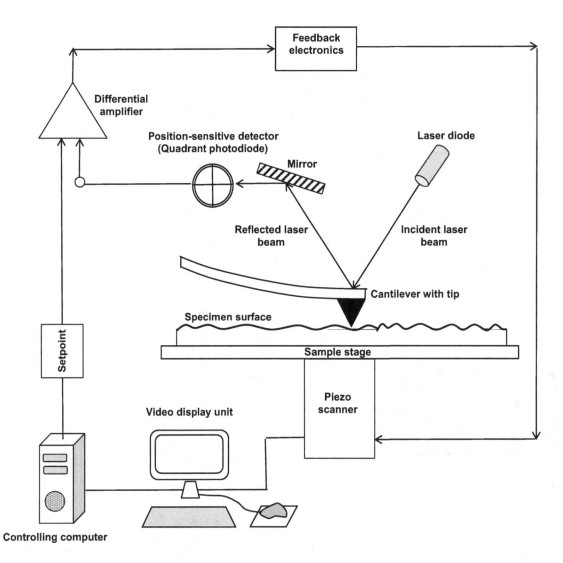

FIGURE 17.8   Layout and working of an atomic force microscope.

a 10 μm tall pyramid with end radius 5–20 nm. The force F acting on the tip is given by Hooke's law of elasticity:

$$F = -kx \qquad (17.38)$$

The symbol k stands for the spring constant of the cantilever beam. It is also called stiffness. The displacement or deformation of the tip of the cantilever beam is denoted by x. The spring constant is therefore defined as the ratio of force acting on the cantilever to its displacement.

The spring constant k is expressed in terms of the Young's modulus Y of the cantilever material and dimensions of the rectangular cantilever. If w is the width, t is the thickness, and L is the length of the cantilever beam, k can be calculated from the equation

$$k = \frac{Ywt^3}{4L^3} \qquad (17.39)$$

The AFM is called by this name because it measures the force between the tip and the specimen. The cantilever is deflected by

this force. The atomic-scale interaction between the tip and the sample is transformed into a macroscale phenomenon of deflection of the cantilever by the interactive force.

At a large distance, the intermolecular force due to van der Waals interaction is of attractive nature because of:

i. dipole-dipole interaction between molecules with permanent dipole moments,

ii. permanent dipole-induced dipole interaction between polar and non-polar molecules, and

iii. induced dipole-induced dipole interaction called London dispersion.

At a very small distance, the intermolecular force changes from attractive to repulsive nature due to repulsion of electron shells of the atoms with overlapping of their electron density distributions and repulsion of their atomic nuclei at a very short distance. The intermolecular interaction is approximated by an empirical potential called the Lennard-Jones potential:

$$U(r) = 4\varepsilon \left\{ \left(\frac{\sigma}{r}\right)^{12} - \left(\frac{\sigma}{r}\right)^{6} \right\} = \varepsilon \left\{ \left(\frac{r_{min}}{r}\right)^{12} - 2\left(\frac{r_{min}}{r}\right)^{6} \right\}$$

$$(17.40)$$

In this equation, ε is the depth of the potential well and σ is the distance at which the intermolecular potential becomes zero. The symbols r and r_{min} are defined as follows: r is the distance between the centers of the molecules and r_{min} is the distance at which the potential attains the minimum value. This minimum value corresponds to the equilibrium separation of the molecules:

$$r_{min} = \sqrt[6]{2}\sigma \qquad (17.41)$$

At $r > r_{min}$, attractive force dominates, $(\sigma/r)^{6}$ term. For $r < r_{min}$, repulsive force takes over, $(\sigma/r)^{12}$ term.

17.6.3 Measurement of Deflection of the Cantilever

The deflection produced by the force acting on the cantilever is measured with great accuracy using an optical technique. A laser beam is focused on the free end of the cantilever. From the free end, it suffers reflection and is incident on a photodiode acting as a position-sensitive photodetector (PSPD).

A piezoelectric element made of quartz, barium titanate, or lead zirconate titanate (PZT) is fitted to the cantilever. It can be used to vibrate the cantilever at a desired oscillation frequency. The sample is kept on a stage which has an X, Y, Z drive to move the sample in these directions.

17.6.4 Usage and Operational Modes

The AFM is used in two ways:

  i. Constant force, variable height: The feedback system is applied to change the tip-sample spacing for keeping the force constant.
 ii. Constant height, variable force: The feedback system is not applied. It is only used when surface roughness is low to achieve a high scan speed.

There are mainly two operational modes for AFM imaging:

  i. Static mode or contact mode and
 ii. Dynamic mode, also called AC mode or vibration mode, with two subclasses:
   a. Non-contact mode and
   b. Tapping mode or intermittent contact mode.

In the contact mode, the tip is dragged over the specimen surface. The tip-sample distance is <0.5 nm and the tip-sample force is in the repulsive regime with force ~10^{-9}N. The force is adjusted by pushing the tip against the specimen surface with the help of a piezoelectric positioning element. The deflection of the tip is measured. It is compared against a set deflection in a DC feedback amplifier. If it is same, no voltage is applied by the feedback mechanism to the piezoelement. If it is different, the required voltage is applied to the piezoelement so that the set deflection value is obtained, either by raising or lowering the tip.

This voltage is a measure of the hills and valleys, and therefore provides Z-direction measurements at different (x, y) positions on the sample surface. It is used to construct the topographical image of the surface. The disadvantage of the contact mode is that the frictional force between the tip and the sample can damage the sample or alter its morphology. It is therefore not recommended for fragile samples.

The non-contact mode offers an opportunity to overcome the difficulty faced in the contact mode. The tip is kept at a height of 0.5–2 nm above the surface during scanning. The tip-sample force ~10^{-12}N is now in the attractive regime. Disappointingly, the attractive force is feebler than the repulsive force. So, the DC measurements will not give accurate results and AC methods must be resorted to. The tip is given a small oscillation at 10^5 Hz using the piezoelectric element with the amplitude ~10 nm. The van der Waals force is maximum in the distance range of 1–10 nm.

The resonant frequency of a cantilever vibrating near the sample surface is affected by the interatomic forces and can be expressed in terms of an effective spring constant which takes into account the effect of the force gradient

$$f = \frac{1}{2\pi} \sqrt{\frac{\text{Effective spring constant}}{\text{Mass}}}$$

$$= \frac{1}{2\pi} \sqrt{\frac{\text{Mechanical spring constant} - \text{Spatial force gradient}}{\text{Mass}}}$$

$$(17.42)$$

Near the sample surface, the force gradient acting on the cantilever increases due to which its resonance frequency decreases. Because the amplitude is maximum at the resonant frequency, the decrease in the resonant frequency affects the amplitude at a given frequency. By operating the cantilever at a frequency proximate to but higher than its free-space resonant frequency and bringing it closer to the surface, the lowered resonant frequency being around the resonant frequency, a large change in amplitude will occur. Noting that this change in amplitude is caused by the force gradient which, in turn, is the result of the tip-sample spacing, and employing a feedback loop to maintain the amplitude constant by varying the tip-sample spacing and thereby the force gradient, a topographical map of the surface is created from the amount of scanner z movement necessary to keep amplitude constant.

The tapping mode combines the advantages of both contact and non-contact modes. Contacting provides a higher resolution and non-contacting prevents damage to the sample from frictional and adhesion forces. The tip alternately touches the surface for better image resolution and lifts off to avoid dragging. When the tip passes over an elevated region on the surface, the amplitude decreases and when it passes over a depressed region, the amplitude increases. The feedback loop adjusts the tip-sample separation to maintain a constant amplitude and the image is constructed, as in the non-contact mode.

17.7 Discussion and Conclusions

A ubiquitous simple equipment in all semiconductor laboratories is the four-point probe (4PP) instrument. X-ray crystallography is the most favored technique for determination of atomic and molecular structures of crystals by deciphering the arrangements of atoms in 3D space.

The SEM is capable of sub-nm resolution. The SEM mainly uses signals produced from SEs and BSEs for image production. As SEs come from shallower regions of the specimen, they reveal near-surface topography. BSEs coming from deep inside the specimen give information about the deeper region but at a lower spatial resolution. However, additional information on composition of the sample is available from BSEs. The production of X-rays during SEM imaging is yet another feather in its cap in the form of providing EDX analytical tool. The SEM works in a vacuum environment using arrangements for electron focusing and deflection along with SE, BSE, and X-ray detectors. Using identical components with some variations, the TEM works in the transmission mode as opposed to the reflection mode in the SEM. The TEM provides resolution <50 pm and much higher magnification than the SEM but suffers from difficulty in sample preparation.

As the name suggests, the STM works by scanning the metallic tip over the sample. While scanning, it measures the tunneling current between the tip and the sample to construct its topographical image. The extremely high sensitivity of the tip-to-sample tunneling current to the tip-to-sample spacing bestows on the STM the capability for single atom imaging. It is operated in either of the two modes: constant height, variable current and constant current, variable height. The requirements of performing measurements in vacuum and restriction to conducting specimens impose serious limitations on the STM.

Removing both obstacles confronting the STM, the AFM also provides 3D imaging with a typical lateral resolution of 30 nm but with a low vertical resolution of ~0.1 nm. The AFM derives its name from the fact that it measures the force between its tip and the sample. The force is attractive at large tip-sample distances, changing to repulsive nature at shorter distances. The cantilever deflection is generally measured optically. AFM is used in contact, non-contact, and tapping modes. The contact mode provides more sensitivity because the repulsive force is higher in magnitude but can damage the specimen. The non-contact mode is less sensitive because the attractive force is smaller than the repulsive force. The tapping mode combines the merits of both contact and non-contact modes. It utilizes the change in resonant frequency of the cantilever and the consequent change in its amplitude of vibration to build the image.

Illustrative Exercises

17.1. In a four-point probe measurement setup, the spacing (s) between the probes is 1 mm. During measurement on a silicon wafer of thickness (t) = 500 μm, the ratio of voltage and current readings yields a resistance value $R = 50$ Ω. Find the resistivity (ρ) of the wafer.

The approximate formula in eq. (17.25) holds for $t \ll s$, which is untrue here. An accurate formula for resistivity obtained by applying correction for sample thickness is

$$\rho = R\frac{\pi t}{\ln\left\{\dfrac{\sinh\left(\dfrac{t}{s}\right)}{\sinh\left(\dfrac{t}{2s}\right)}\right\}} = 50\,\Omega \times \frac{3.14 \times \left(\dfrac{500}{10^4}\right)\text{cm}}{\ln\left\{\dfrac{\sinh\left(\dfrac{500\,\mu\text{m}}{1000\,\mu\text{m}}\right)}{\sinh\left(\dfrac{500\,\mu\text{m}}{2\times1000\,\mu\text{m}}\right)}\right\}}$$

$$= 50 \times \frac{3.14 \times 0.05}{\ln\left\{\dfrac{\sinh(0.5)}{\sinh(0.25)}\right\}}\,\Omega\text{cm}$$

$$= \frac{7.85}{\ln\left(\dfrac{0.5211}{0.2526}\right)}\,\Omega\text{cm} = \frac{7.85}{0.724}\,\Omega\text{cm} = 10.84\,\Omega\text{cm} \qquad (17.43)$$

Note that in the limit $t \ll s$,

$$\sinh\left(\frac{t}{s}\right) = \frac{t}{s} \qquad (17.44)$$

So eq. (17.43) reduces to

$$\rho = R\frac{\pi t}{\ln\left\{\dfrac{\dfrac{t}{s}}{\dfrac{t}{2s}}\right\}} = R\frac{\pi t}{\ln\left(\dfrac{t}{s}\times\dfrac{2s}{t}\right)} = R\frac{\pi t}{\ln(2)} = \frac{\pi t}{\ln(2)}\left(\frac{V}{I}\right) \qquad (17.45)$$

which is eq. (17.25). From this approximate equation

$$\rho = R\frac{\pi t}{\ln(2)} = 50\,\Omega \times \frac{3.14 \times \left(\dfrac{500}{10^4}\right)\text{cm}}{0.69315} = 11.325\ \Omega\text{cm} \qquad (17.46)$$

17.2. (a) An electron microscope is operated at an energy of 100 keV. Determine the velocity and wavelength of electrons, and the resolution of the microscope if the numerical aperture is 0.8?

(b) Compare the velocity of electrons with the velocity of light. How are the wavelength and resolution values affected by consideration of relativistic effects?

(a) By de Broglie's formula eq. (2.16), the wavelength λ of the electrons is

$$\lambda = \frac{h}{p} = \frac{h}{mv} \qquad (17.47)$$

where p is the momentum of the electron, m its mass, and v its velocity.

The kinetic energy of the electrons is

$$E = \frac{1}{2}mv^2 \qquad (17.48)$$

The energy of electrons under an accelerating potential V is

$$E = eV \qquad (17.49)$$

where e is the electronic charge.

Combining eqs. (17.48) and (17.49), we get

$$\frac{1}{2}mv^2 = eV \qquad (17.50)$$

from which

$$v^2 = \frac{2eV}{m} \qquad (17.51)$$

or

$$v = \sqrt{\frac{2eV}{m}} = \sqrt{\frac{2 \times 1.602 \times 10^{-19} \times 100 \times 1000}{9.109 \times 10^{-31}}}$$

$$= \sqrt{3.5174 \times 10^{16}} = 1.8755 \times 10^8 \, \text{ms}^{-1} \qquad (17.52)$$

Substituting for v from eq. (17.52) in eq. (17.47), we have

$$\lambda = \frac{h}{m \times \sqrt{\dfrac{2eV}{m}}} = \frac{h}{\sqrt{2meV}}$$

$$= \frac{6.626 \times 10^{-34}}{\sqrt{2 \times 9.109 \times 10^{-31} \times 1.602 \times 10^{-19} \times 100 \times 1000}}$$

$$= \frac{6.626 \times 10^{-34}}{\sqrt{2 \times 9.109 \times 1.602 \times 10^{-45}}} = \frac{6.626 \times 10^{-34}}{\sqrt{2.9185 \times 10^{-44}}}$$

$$= \frac{6.626 \times 10^{-34}}{1.70836 \times 10^{-22}} = 3.8786 \times 10^{-12} \, \text{m} \approx 3.88 \, \text{pm}$$

$$(17.53)$$

If NA is the numerical aperture of the microscope lens, the resolution of the microscope is given by the well-known formula (eq. (16.7))

$$r = \frac{0.61\lambda}{\text{NA}} = \frac{0.61 \times 3.88 \, \text{pm}}{0.8} = 2.96 \ \text{pm} \qquad (17.54)$$

(b) From eq. (17.52),

$$\frac{\text{Velocity of electrons } (v)}{\text{Velocity of light } (c)} = \frac{1.8755 \times 10^8 \, \text{ms}^{-1}}{3 \times 10^8 \, \text{ms}^{-1}}$$

$$\times 100\% = 62.52\% \qquad (17.55)$$

Relativistic effects are accounted for in the amended formula for wavelength of electrons

$$\lambda = \frac{h}{\sqrt{2meV}} \times \left(1 + \frac{eV}{2mc^2}\right)^{-0.5}$$

$$= 3.88 \, \text{pm} \times \left(1 + \frac{1.602 \times 10^{-19} \times 100 \times 1000}{2 \times 9.109 \times 10^{-31} \times \left(3 \times 10^8\right)^2}\right)^{-0.5}$$

$$= 3.88 \, \text{pm} \times \left(1 + \frac{1.602 \times 10^{-14}}{2 \times 9.109 \times 9 \times 10^{-15}}\right)^{-0.5}$$

$$= 3.88 \, \text{pm} \times \left(1 + 0.0977\right)^{-0.5} = 3.70 \, \text{pm} \qquad (17.56)$$

So, the corrected resolution is

$$r = \frac{0.61\lambda}{\text{NA}} = \frac{0.61 \times 3.70 \, \text{pm}}{0.8} = 2.82 \, \text{pm} \qquad (17.57)$$

Thus by inclusion of relativistic effects, the wavelength of electrons is decreased and the resolution of microscope is improved.

REFERENCES

Eaton, P. and P. West. 2010. *Atomic Force Microscopy.* Oxford University Press: Oxford, pp. 1–81.

Pennycook, S.J. and P.D. Nellist (Eds.). 2011. *Scanning Transmission Electron Microscopy: Imaging and Analysis.* Springer Science+Business Media, LLC: New York, 764 pages.

Vahabi, S., B. Nazemi Salman and A. Javanmard 2013. Atomic force microscopy application in biological research: A review study. *Iranian Journal of Medical Sciences* 38(2): 76–83.

Voigtländer, B. 2015. *Scanning Probe Microscopy: Atomic Force Microscopy and Scanning Tunneling Microscopy.* Springer-Verlag: Berlin, pp. 145–308.

Zhou, W., R.P. Apkarian, Z.L. Wang and D. Joy. 2006. Fundamentals of scanning electron microscopy (SEM), In: Zhou, W. and Z.L. Wang (Eds.), *Scanning Microscopy for Nanotechnology: Techniques and Applications.* Springer Science+Business Media, LLC: New York, pp. 1–40.

Part VI

Exemplar Nanoelectronic Devices

18

Resonant Tunneling Diodes

The resonant tunneling diode (RTD) is a valuable device for THz electronic communication, indeed the fastest electronic device for continuous-wave terahertz frequency generation at room temperature (Zawawi 2015). An oscillation frequency of 1.55 THz has been obtained in RTD oscillators integrated with a 16 µm long slot antenna (Maekawa et al 2014).

18.1 The Constituent Layers

18.1.1 Structure and Juxtapositioning of Layers

The main part of these diodes is a quantum well interposed between potential barrier layers on its two sides (Figure 18.1). In the GaAs-AlGaAs diodes, the well is made of undoped GaAs and the barrier layers are formed from undoped AlGaAs. In this material system, GaAs has a smaller bandgap = 1.424 eV at 300 K. AlGaAs is a larger bandgap material.

The energy gap of $Al_xGa_{1-x}As$ alloy varies with aluminum content x. For $x < 0.45$, it acts as a direct bandgap semiconductor with energy gap E_{Direct} given by (see reference: NSM Archive)

$$E_{Direct} = 1.424 + 1.247x \quad \text{for } x < 0.45 \quad (18.1)$$

For $x > 0.45$, it becomes an indirect bandgap material whose energy gap $E_{Indirect}$ is expressed as (NSM Archive)

$$E_{Indirect} = 1.9 + 0.125x + 0.143\ x^2 \quad \text{for } x > 0.45 \quad (18.2)$$

At $x = 0$, $E_{Direct} = 1.424$ eV by eq. (18.1) while at $x = 1$, $E_{Indirect} = 2.168$ eV by eq. (18.2).

Typical structural parameters of an RTD are: well thickness = 5 nm, thickness of barrier layers = 1.5–5 nm. In a symmetric construction, the two barrier layers are of equal thickness but these thicknesses need not be identical and asymmetric diodes are also made.

The barrier layer-quantum well-barrier layer structure is sandwiched between heavily doped layers of the same material as the well. On both sides, undoped layers of the well material separate the barrier layer-quantum well-barrier layer structure and the heavily doped layers. The heavily doped layers serve as the contacts of the RTD. They are known as the emitter and collector regions. The additional undoped separation layers are called spacer layers. They are intentionally introduced to play a vital role. They prevent the scattering of electrons from the ionized impurities that would otherwise take place if the contact layers directly touched the barrier layers.

18.1.2 Stuffing a 2D System between Two 3D Systems

Effectively, an RTD consists of a quantum well (made of a narrow bandgap semiconductor material) between two potential barriers (made of wide bandgap semiconductor material) with heavily doped contacts (made of narrow bandgap material) lying beyond the potential barriers and serving as emitter and collector terminals; extra spacer layers (made of narrow bandgap material) intervening the potential barrier/emitter and potential barrier/collector layers are placed to avoid carrier scattering by doped contacts. In the quantum well, the electrons can move to and fro remaining confined in one plane but not perpendicular to that plane. The in-plane free electron motion constitutes a two-dimensional (2D) system. In contrast, in the contacts, the electrons can move freely in all the three dimensions, which is a 3D system. Therefore, a 2D system is crammed between two 3D systems. Moreover, there is a difference between the type of energy levels admissible in the well and in the contacts. In the well, only discrete energy levels are permitted whereas in the contacts, they are continuous.

The quantized energy levels in the well are ascribed a special name. They are referred by the name 'quasi-bound states'; 'quasi' has the meaning, 'apparently but not really, i.e., seemingly'. The underlying reason for being called 'quasi' is that a finite probability exists for electron tunneling out of the well. Let the energy of the lowest quasi-bound state be denoted by ε_1.

Due to the different materials used, the fabrication of this kind of diode involves heteroepitaxy, the process of epitaxial deposition of a crystal of a material on a substrate of different material. Technological advancements enabling the realization of these diodes include molecular beam epitaxy (MBE), metal-organic vapor phase epitaxy (MOVPE), and metal-organic MBE (MOMBE), also called chemical beam epitaxy (CBE) combining the advantages of MBE and MOVPE.

18.2 Operational Modes

The modes of device operation can be understood with reference to its energy-band diagrams (Figure 18.2).

18.2.1 Without External Bias: Equilibrium Condition

During the device design phase, care is taken to make sure that the quasi-bound energy level ε_1 in the quantum well is positioned above the Fermi energy level E_F in the emitter. In this circumstance, the probability of tunneling of electrons through the double barriers is extremely low. Further, the device design ensures that the temperature-induced transport of carriers over the twin

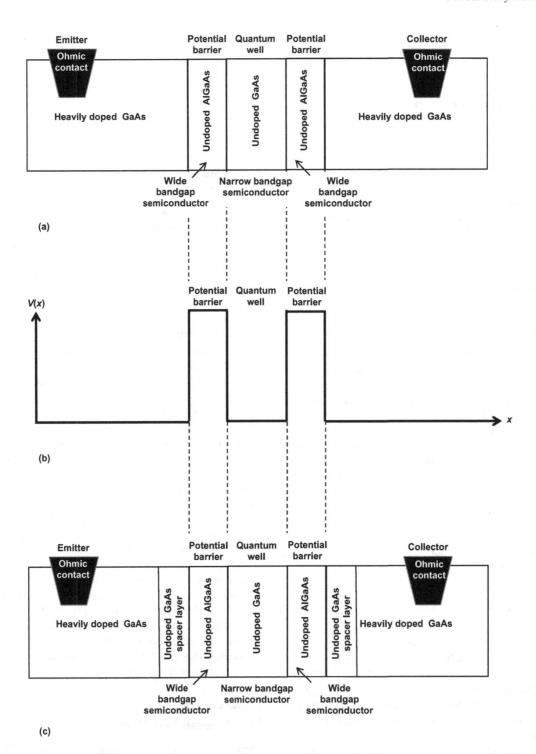

FIGURE 18.1   Resonant tunneling diode: (a) and (c) without and with GaAs spacer layers respectively, and (b) electrical potential $V(x)$-distance (x) diagram.

potential barriers is minimal. Consequently, the output current is low.

18.2.2 Low External Bias: In-Resonance Condition

On applying a bias, the energy level ε_1 in the quantum well undergoes a downward shift. When the applied bias reaches

a certain value, the energy level ε_1 comes in alignment with the Fermi energy level E_F in the emitter. This is the beginning of the condition favoring the tunneling of electrons across the barriers. The condition lasts as long as the energy level ε_1 remains below the Fermi energy E_F in the emitter contact but above the bottom of the conduction band in the emitter contact during the increase of external bias. There is a finite

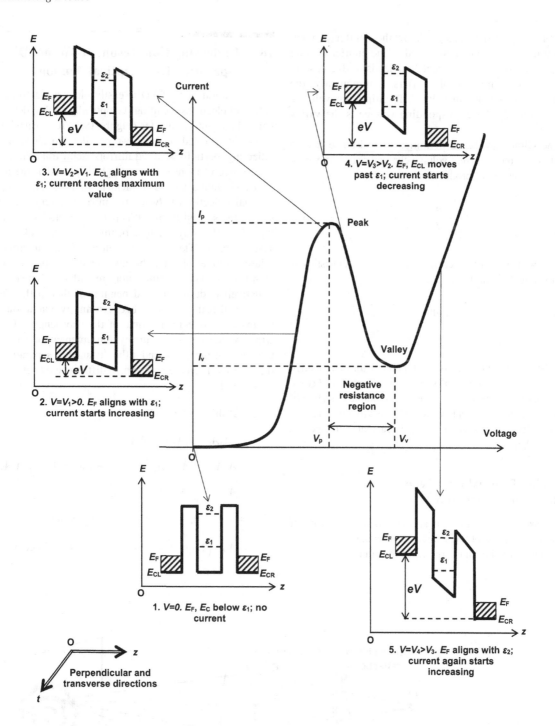

FIGURE 18.2 Current-voltage characteristics of a resonant tunneling diode and energy-band diagrams corresponding to different portions of the characteristics.

probability of tunneling of carriers whose kinetic energy E_z of perpendicular motion is

$$E_z = \frac{\hbar^2 k_z^2}{2m^*} = \varepsilon_1 \qquad (18.3)$$

where $\hbar$ is the reduced Planck's constant, k_z is the component of the wave vector of the carrier in the Z-direction, and m^* is the effective mass of the carrier. The current increases reaching a peak value I_p as the voltage is raised; the associated voltage is the

peak voltage V_p. The stage of resonant transmission of carriers through the barriers is attained.

18.2.3 High External Bias: Off-Resonance Condition

As the external bias is continuously raised, the energy level ε_1 moves below the bottom of the conduction band in the emitter contact although it is still below the Fermi energy E_F in the emitter contact. The upsetting of the alignment among the energy level ε_1, the Fermi energy E_F in the emitter contact, and the bottom of the conduction band in the emitter contact from its favored position of current flow leads to decrease in the probability of tunneling of carriers. Consequently, the current falls with increase in voltage. Its minimum value is the valley current I_v; the corresponding voltage is the valley voltage V_v.

Recall the Ohm's law-based definition of resistance R of a conductor having linear current-voltage characteristics, in which application of a voltage Φ_0 produces a current I:

$$R = \frac{\Phi_0}{I} \tag{18.4}$$

A conductor with non-linear current-voltage characteristics is described by a differential resistance

$$R_d = \frac{d\Phi_0}{dI} \tag{18.5}$$

Normally, the differential resistance of a conductor is a positive quantity because with increase of voltage Φ_0, the current I rises. But an RTD in the off-resonance conduction exhibits a decline of current with voltage increment, which is a negative differential resistance (NDR) behavior. As a result, its current-voltage characteristic is N-shaped.

18.2.4 Higher External Bias: Second Resonance Condition

Further increase of external bias is accompanied by increase of current due to alignment of the second quasi-bound energy level ε_2 in the well with the Fermi energy E_F in the emitter.

18.3 Understanding Resonant Tunnel Diode Operation from Optical Analogy

The operation of an RTD is easily understood from the wave nature of electrons recalling the optically analog device called a Fabry-Perot interferometer (Figure 18.3). It consists of a reflective cavity formed between two parallel, closely spaced, highly reflecting, partially silvered mirrors (semi-transparent mirrors). The source of monochromatic light is placed at the focal plane of a collimating lens. The incident light suffers from multiple internal reflections between the mirrors. Every time that light strikes the second mirror, it is partly reflected and partly transmitted. In this way, a single beam of light is broken down into several beams. Thus in the region behind the mirrors, we have a large number of offset beams which have traversed different path lengths. These beams interfere with each other, either constructively or destructively depending on their path difference. If the optical path difference between any two adjacent emerging beams is an integral multiple of the wavelength of light, constructive interference occurs by reinforcement of waves. Let θ be the angle of incidence and d the distance of separation between the mirrors. Then the optical path difference Δl between the rays E_1 and E_2 is

$\Delta l = $ Path of ray $E_2 - $ Path of ray E_1

$= (A_0A_1 + A_1A_2 + A_2A_3) - (A_0A_1 + A_1C)$

$= A_0A_1 + A_1A_2 + A_2A_3 - A_0A_1 - A_1C = A_1A_2 + A_2A_3 - A_1C$

$= A_1A_2 + A_2A_3 - BA_3 = A_1A_2 + A_2B$

$= A_1A_2 + A_1A_2 \cos 2\theta = A_1A_2(1 + \cos 2\theta)$

$= A_1A_2(2\cos^2\theta) = 2(A_1A_2\cos\theta)\cos\theta = 2d\cos\theta \tag{18.6}$

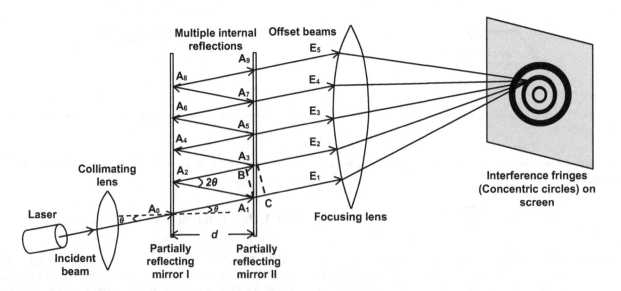

FIGURE 18.3   Fabry-Perot interferometer.

If n is the refractive index of the medium between the mirrors,

$$\Delta l = 2nd \cos \theta \qquad (18.7)$$

Hence the condition for maximum light intensity is

$$\Delta l = 2nd\cos \theta = m\lambda \qquad (18.8)$$

where m is an integer and λ is the wavelength of light. For air, $n = 1$ so that

$$2d\cos \theta = m\lambda \qquad (18.9)$$

or,

$$d\cos \theta = m\left(\frac{\lambda}{2}\right) \qquad (18.10)$$

When collected by a focusing lens, these interfering beams produce a pattern of alternately bright and dark bands in the form of concentric circles or rings on the screen placed behind the lens. These are known as interference fringes.

Similar to the air cavity in the Fabry-Perot interferometer is the quantum well in the RTD. Like the semi-transparent mirrors of the Fabry-Perot interferometer are the potential barriers of the resonant tunnel diode. Instead of the light waves in the interferometer, we have electron waves in the diode.

When multiples of half the electron wavelength match the distance $d\cos\theta$ for separation d between the barriers, the transparency of the double barrier to electron waves is maximum. This is the so-called resonance condition of the diode and is attained by applying a suitable bias between the contacts, thereby aligning the Fermi energy level E_F on the emitter side with the energy level ε_1 on the well side. Then a high intensity current flows through the diode. In this particular situation, the two potential barriers become almost invisible to electrons so that the transmission of electron wave is unity and its reflection is zero. A striking quantum-mechanical phenomenon! But subsequent increase in bias disturbs the alignment lowering the current. However, the current rises again rapidly due to thermionic emission over the potential barrier, and subsequently on reaching the matching condition of E_F of the emitter with the energy level ε_2 of the well.

18.4. Parameters of the Resonant Tunneling Diode

18.4.1 Peak-to-Valley Ratio

In order that the NDR characteristic of an RTD is clearly defined and distinctly visible, it is necessary that its peak current I_p should be as high as possible and the valley current I_v sufficiently low. The parameter quantifying this desirable feature is the peak-to-valley ratio (PVR) of the diode, which is theoretically possible to be as high as 1000 but is much lower in fabricated diodes. The reason is the criticality of thickness of the potential barriers because the diode operation essentially relies on tunneling. It is difficult to deposit very thin layers ~2–4 nm for potential barriers with required precision.

18.4.2 Operational Speed

It depends first on the charging time of the diode and second on the tunneling time. The latter is governed by the lifetime of the quasi-stable state in the quantum well and is very small so that the speed of operation is essentially governed by the charging time of the diode.

18.5 Two Types of Double Barrier Resonant Tunneling

It needs to be emphasized that resonant tunneling differs from the simple tunneling process that we are accustomed to. The difference lies in the fact that resonant tunneling is associated with the presence of quasi-bound or meta-stable states in the classically forbidden region, which is here the quantum well. Each quasi-stationary state is characterized by an energy ε_n and lifetime τ_n. In the ideal case, the scattering processes are assumed to be absent but practically, it is difficult to get rid of them so that the condition for spatial quantization may be written in accordance with the uncertainty principle as

$$\varepsilon_{n+1} - \varepsilon_n = \frac{\hbar}{\tau_s} \qquad (18.11)$$

where ε_n, ε_{n+1} are the energies of the nth and $(n+1)$th quasi-energy levels, respectively, and τ_s is the scattering time of the electron in the 2-D subband related to the nth quasi-energy level.

The comparison between the magnitudes of the scattering time τ_s and the lifetime τ_n determines the nature of resonant tunneling taking place. Two cases arise and consequently, there are two types of resonant tunneling:

i. Coherent tunneling: This occurs when $\tau_n \ll \tau_s$ because the electron does not suffer any scattering and the phase of the electron wave function at any point in space shows continuity in time.

ii. Incoherent or sequential tunneling: It happens when $\tau_n \gg \tau_s$ since the phase of the electron wave function becomes haphazard or chaotic by its scattering, and hence unpredictable.

18.5.1 Coherent Tunneling

A single wave function $\psi(\mathbf{r}, z)$ is used for the electron throughout the structure. An electron in a well occupying a quasi-stationary state moves within the well for a finite period of time = the lifetime τ_n of the state. The energy of the quasi-stationary state is not discrete but broadened. By the uncertainty principle, the relation between the full width of the energy Γ_n of quasi-stationary state at half maximum and the lifetime τ_n is

$$\Gamma_n = \frac{\hbar}{\tau_n} \qquad (18.12)$$

Description of the interaction of the electron with the double barrier structure is formulated in terms of the transmission coefficient $T(E_z)$ representing the probability of tunneling of the electron through the structure:

$$T(E_z) = \frac{\text{Probability flux of the transmitted wave}}{\text{Probability flux of the incident wave}}$$

$$= \frac{\text{Outgoing probability current density}}{\text{Incoming probability current density}} \quad (18.13)$$

The transmission coefficient of a double barrier resonant tunneling structure has the same form as that of a Fabry-Perot interferometer. When the transmission coefficient of the interferometer is plotted as a function of phase difference, the resulting transmission spectrum consists of a series of discrete peaks corresponding to the spacing between the mirrors for which an integral number of half wavelengths of light fit within the Fabry-Perot cavity.

Like the Fabry-Perot interferometer, the transmission coefficient of the double barrier structure shows a sharp peak at the energy of each quasi-bound level in the quantum well. Physical interpretation of the peak-shaped contour is that the coherent wave function of the electron interferes with itself inside the quantum well consequent upon its multiple reflections between the two potential barriers. For an energy value of electron matching with a quasi-energy level in the quantum well, the interference is constructive. This interference produces an enhanced transmitted wave. $T(E_z)$ has a unity value. At the same time, the reflected wave outside the structure is annulled. But for an energy value of electron different from the quasi energy level in the quantum well, destructive interference takes place. As a result, $T(E_z) = 0$ and no wave is transmitted through the structure. The transmission coefficient $T(E_z)$ can be expressed as a Breit-Wigner distribution, a continuous probability distribution, also called Lorentz distribution, used to model resonances in high-energy particle physics:

$$T_{\text{Coherent}}(E_z) = \frac{(\Gamma_n/2)^2}{(E_z - \varepsilon_n)^2 + (\Gamma_n/2)^2} \quad (18.14)$$

where E_z is the energy of the electron and ε_n is the energy of the quasi-bound state.

When $E_z = \varepsilon_n$

$$T_{\text{Coherent}}(E_z) = \frac{(\Gamma_n/2)^2}{(0)^2 + (\Gamma_n/2)^2} = \frac{(\Gamma_n/2)^2}{(\Gamma_n/2)^2} = 1 \quad (18.15)$$

Let us calculate the current flowing in an RTD under an applied bias V. Application of the bias V separates the Fermi energies on the left and right sides of the tunneling barrier by eV. This is evident from the band diagram of the tunneling barrier under applied bias shown in Figure 18.2. It is assumed that the Hamiltonian on either side of the barrier can be resolved into two components, viz. the perpendicular (longitudinal) and transverse components. The former is named as the z-component and the latter as the t-component. Choice of the zero reference of potential energy in the system is to be made. This zero is selected as the conduction band minimum on the left, written as

$$E_{C,L} = 0 \quad (18.16)$$

where $E_{C,L}$ is the conduction band minimum on the left.

With the earlier assumption and supposition, the energy of the electron before tunneling is

$$E_{z,L;t,L} = E_{z,L} + E_{t,L} = \frac{\hbar^2 k_{z,L}^2}{2m^*} + \frac{\hbar^2 k_{t,L}^2}{2m^*} \quad (18.17)$$

when the electron is on the left side of the barrier. Here $k_{z,L}$ and $k_{t,L}$ are the components of the wave vector along the Z-direction and transverse direction, respectively, on the left side.

Similarly, the energy of the electron after tunneling becomes

$$E_{z,R;t,R} = E_{z,R} + E_{t,R} = \frac{\hbar^2 k_{z,R}^2}{2m^*} + \frac{\hbar^2 k_{t,R}^2}{2m^*} + E_{C,R} \quad (18.18)$$

when the electron has moved to the right side of the barrier. Here $k_{z,R}$ and $k_{t,R}$ are the components of the wave vector along the Z-direction and transverse direction, respectively, on the right side. Also, $E_{C,R}$ is the conduction band minimum on the right.

Assumption of conservation of transverse momentum during tunneling from left to right implies that

$$k_{t,L} = k_{t,R} \quad (18.19)$$

Hence the transverse energies on the two sides of the barrier are related as

$$E_{t,L} = E_{t,R} \quad (18.20)$$

So, the energy components in the Z-direction on the left and right sides are

$$E_{z,L;z,R} = \frac{\hbar^2 k_{z,L}^2}{2m^*} = \frac{\hbar^2 k_{z,R}^2}{2m^*} + E_{C,R} \quad (18.21)$$

Absorption of electrons by the contacts is postulated to be perfect. Coherent tunneling coerces the condition of maintenance of phase of the electron throughout its propagation across the structure. An electron setting out on its journey from the left contact at a given energy E impinges on the barrier and has a certain transmission coefficient $T(E)$ of crossing the barrier and moving to the other side with the same energy E and transverse momentum. It reaches the right contact where it loses its phase coherence and excess energy via inelastic collisions with the Fermi sea of electrons in this contact. Eventually, it is absorbed in the right contact. Thus

Aggregate current flowing through the double barrier structure

= Number of electrons transmitted across

   the barriers per unit time from left to right

− the Number of electrons transmitted across

   the barriers per unit time from right to left   (18.22)

For the energy E with z-component E_z, the incident current density $j_{\text{Incident, Left}}$ perpendicular to the barrier from the left from electrons in an infinitesimally small volume $dk_{z,L}d^2k_{t,L}$ of momentum space is given by

$$j_{\text{Incident, Left}} = -\text{Electronic charge } (e) \times \text{Density of states (DOS)}$$

$$\text{in } \mathbf{k}\text{-space}\left[\rho\left(k_{z,L}, k_{t,L}\right)\right]$$

$$\times \text{Electron distribution function on}$$

$$\text{the left side of barriers}\left[f_L\left(k_{z,L}, k_{t,L}\right)\right]$$

$$\times \text{Electron velocity in } Z\text{-direction from}$$

$$\text{the left side }\left[v_z\left(k_{z,L}\right)\right] \times dk_{z,L} \times d^2k_{t,L} \quad (18.23)$$

or,

$$j_{\text{Incident, Left}} = -e\rho\left(k_{z,L}, k_{t,L}\right)f_L\left(k_{z,L}, k_{t,L}\right)v_z\left(k_{z,L}\right)dk_{z,L}d^2k_{t,L} \quad (18.24)$$

where

$$\rho\left(k_{z,L}, k_{t,L}\right) = \frac{2}{(2\pi)^3} \quad (18.25)$$

and

$$v_z\left(k_{z,L}\right) = \left(\frac{1}{\hbar}\right)\frac{\partial E\left(k_{z,L}\right)}{\partial k_{z,L}} = \left(\frac{1}{\hbar}\right)\frac{\partial\left(\frac{\hbar^2 k_{z,L}^2}{2m^*}\right)}{\partial k_{z,L}}$$

$$= \left(\frac{1}{\hbar}\right)\times\frac{\hbar^2}{2m^*}\times\frac{\partial\left(k_{z,L}^2\right)}{\partial k_{z,L}} = \left(\frac{1}{\hbar}\right)\times\frac{\hbar^2}{2m^*}\times 2k_{z,L} = \frac{\hbar k_{z,L}}{m^*} \quad (18.26)$$

Substituting for $\rho(k_{z,L}, k_{t,L})$ from eq. (18.25) and $v_z(k_{z,L})$ from eq. (18.26) into eq. (18.24) we get

$$j_{\text{Incident, Left}} = -e\times\frac{2}{(2\pi)^3}f_L\left(k_{z,L}, k_{t,L}\right)\times\frac{\hbar k_{z,L}}{m^*}\times dk_{z,L}d^2k_{t,L}$$

$$= -\frac{2e\hbar}{(2\pi)^3 m^*}f_L\left(k_{z,L}, k_{t,L}\right)k_{z,L}dk_{z,L}d^2k_{t,L} \quad (18.27)$$

The transmitted current density from the left side of the barriers is simply

$$j_{\text{Transmitted, Left}} = j_{\text{Incident, Left}} \times T\left(k_{z,L}\right) \quad (18.28)$$

where $T(k_{z,L})$ is the transmission coefficient.

From eqs. (18.27) and (18.28), the transmitted current density from the left contact to the right contact is

$$j_{\text{Transmitted, Left}} = -\frac{2e\hbar}{(2\pi)^3 m^*}f_L\left(k_{z,L}, k_{t,L}\right)k_{z,L}dk_{z,L}d^2k_{t,L} \times T\left(k_{z,L}\right)$$

$$= -\frac{2e\hbar}{(2\pi)^3 m^*}T\left(k_{z,L}\right)f_L\left(k_{z,L}, k_{t,L}\right)k_{z,L}dk_{z,L}d^2k_{t,L} \quad (18.29)$$

Likewise, the transmitted current density from the right contact to the left contact may be expressed as

$$j_{\text{Transmitted, Right}} = -\frac{2e\hbar}{(2\pi)^3 m^*}T\left(k_{z,R}\right)f_R\left(k_{z,R}, k_{t,R}\right)k_{z,R}dk_{z,R}d^2k_{t,R} \quad (18.30)$$

The symmetry of the transmission coefficient enables us to write

$$T\left(k_{z,L}\right) = T\left(k_{z,R}\right) = T\left(k_z\right) \quad (18.31)$$

At a given energy E_z, this equality takes the form

$$T\left(E_{z,L}\right) = T\left(E_{z,R}\right) = T\left(E_z\right) \quad (18.32)$$

in view of the inter-relationship between E_z and k_z.

From eq. (18.21)

$$E_{z,L;z,R} = \frac{\hbar^2 k_{z,L}^2}{2m^*} \quad (18.33)$$

Differentiating both sides of eq. (18.33) with respect to $k_{z,L}$,

$$\frac{dE_{z,L;z,R}}{dk_{z,L}} = \frac{d}{dk_{z,L}}\left(\frac{\hbar^2 k_{z,L}^2}{2m^*}\right) \quad (18.34)$$

or,

$$\frac{dE_{z,L;z,R}}{dk_{z,L}} = \frac{\hbar^2}{2m^*}\times 2k_{z,L} = \left(\frac{\hbar^2}{m^*}\right)k_{z,L} \quad (18.35)$$

or,

$$k_{z,L}dk_{z,L} = \left(\frac{m^*}{\hbar^2}\right)dE_{z,L;z,R} \quad (18.36)$$

Again from eq. (18.21),

$$E_{z,L;z,R} = \frac{\hbar^2 k_{z,R}^2}{2m^*} + E_{C,R} \quad (18.37)$$

Differentiating both sides with respect to $k_{z,R}$,

$$\frac{dE_{z,L;z,R}}{dk_{z,R}} = \frac{\hbar^2}{2m^*}\times 2k_{z,R} + 0 = \left(\frac{\hbar^2}{m^*}\right)k_{z,R} \quad (18.38)$$

or,

$$k_{z,R}dk_{z,R} = \left(\frac{m^*}{\hbar^2}\right)dE_{z,L;z,R} \quad (18.39)$$

Comparing eqs. (18.36) and (18.39),

$$k_{z,L}dk_{z,L} = k_{z,R}dk_{z,R} = \left(\frac{m^*}{\hbar^2}\right)dE_{z,L;z,R} = \left(\frac{m^*}{\hbar^2}\right)dE_z \quad (18.40)$$

We note that that the cross sections of the volume element are same for left and right sides. Because

$$dk_{t,L} = dk_{t,R} = dk_t \qquad (18.41)$$

$$\therefore d^2k_{t,L} = d^2k_{t,R} = d^2k_t \qquad (18.42)$$

Applying eqs. (18.31), (18.40), and (18.42), eq. (18.29) for leftward transmitted current density is rewritten as

$$j_{\text{Transmitted,Left}} = -\frac{2e\hbar}{(2\pi)^3 m^*} T(E_z) f_L(k_{z,L}, k_{t,L}) \left(\frac{m^*}{\hbar^2}\right) dE_z d^2k_t$$

$$= -\frac{2e}{(2\pi)^3 \hbar} T(E_z) f_L(k_{z,L}, k_{t,L}) dE_z d^2k_t \qquad (18.43)$$

In the same manner, applying eqs. (18.31), (18.40), and (18.42), eq. (18.30) for rightward transmitted current density is rewritten as

$$j_{\text{Transmitted,Right}} = -\frac{2e\hbar}{(2\pi)^3 m^*} T(E_z) f_R(k_{z,R}, k_{t,R}) \left(\frac{m^*}{\hbar^2}\right) dE_z d^2k_t$$

$$= -\frac{2e}{(2\pi)^3 \hbar} T(E_z) f_R(k_{z,R}, k_{t,R}) dE_z d^2k_t \qquad (18.44)$$

For the elementary volume considered, the total current density j_{Total} in the direction of voltage drop is

$$j_{\text{Total}} = j_{\text{Transmitted,Right}} - j_{\text{Transmitted,Left}}$$

$$= -\frac{2e}{(2\pi)^3 \hbar} T(E_z) f_R(k_{z,R}, k_{t,R}) dE_z d^2k_t$$

$$-\left\{-\frac{2e}{(2\pi)^3 \hbar} T(E_z) f_L(k_{z,L}, k_{t,L}) dE_z d^2k_t\right\}$$

$$= \frac{2e}{(2\pi)^3 \hbar} T(E_z) dE_z d^2k_t \left\{f_L(k_{z,L}, k_{t,L}) - f_R(k_{z,R}, k_{t,R})\right\}$$

$$(18.45)$$

In cylindrical coordinates, the areal element in the plane perpendicular to the direction of electron transport has an area

$$d^2k_t = dxdy = k_t dk_t d\theta \qquad (18.46)$$

Substituting for d^2k_t from eq. (18.46) in eq. (18.45)

$$j_{\text{Total}} = \frac{2e}{(2\pi)^3 \hbar} T(E_z) dE_z k_t dk_t d\theta \left\{f_L(k_{z,L}, k_{t,L}) - f_R(k_{z,R}, k_{t,R})\right\}$$

$$(18.47)$$

To calculate the total current density over the whole volume, integration is to be performed with respect to the variable E_z in the Z-direction and the variable k_t in the transverse direction. The variable k_t may be transformed into energy E_t:

Differentiating both sides of the following equation with respect to k_t

$$E_t = \frac{\hbar^2 k_t^2}{2m^*} \qquad (18.48)$$

we get

$$\frac{dE_t}{dk_t} = \frac{d}{dk_t}\left(\frac{\hbar^2 k_t^2}{2m^*}\right) = \left(\frac{\hbar^2}{2m^*}\right) \times 2k_t = \frac{\hbar^2 k_t}{m^*} \qquad (18.49)$$

or,

$$dk_t = \frac{m^*}{\hbar^2 k_t} dE_t \qquad (18.50)$$

The distribution functions are also written in terms of energies E_z, E_t in place of wave vectors k_z, k_t. Hence, from equations (18.47) and (18.50)

$$j_{\text{Total}} = \frac{2e}{(2\pi)^3 \hbar} T(E_z) dE_z k_t \times \frac{m^*}{\hbar^2 k_t} dE_t$$

$$\times d\theta \left\{f_L(E_{z,L}, E_{t,L}) - f_R(E_{z,R}, E_{t,R})\right\}$$

$$= \frac{2em^*}{(2\pi)^3 \hbar^3} T(E_z) dE_z dE_t d\theta \left\{f_L(E_z, E_t) - f_R(E_z, E_t)\right\}$$

$$(18.51)$$

For full volume calculation, the limits of integration are imposed: E_z varies from 0 to ∞, tunneling below $E_z = 0$ being forbidden. E_t also varies from 0 to ∞. The limits for θ are 0 and 2π. The current density is now denoted by the capital letter 'J'. Thus

$$J_{\text{Total}} = \frac{2em^*}{(2\pi)^3 \hbar^3} \int_0^\infty T(E_z) dE_z \int_0^{2\pi} d\theta$$

$$\times \int_0^\infty \left\{f_L(E_z, E_t) - f_R(E_z, E_t)\right\} dE_t \qquad (18.52)$$

But

$$\int_0^{2\pi} d\theta = [\theta]_0^{2\pi} = 2\pi - 0 = 2\pi \qquad (18.53)$$

Further, assuming that the distribution functions are represented by the Fermi-Dirac statistics determined by the Fermi levels $E_{F,L}$ and $E_{F,R}$ in the bulk on the left and right sides of the barriers, we may write

$$f_L(E_z, E_t) = \frac{1}{1 + \exp\left(\dfrac{E_z + E_t - E_{F,L}}{k_B T}\right)} \qquad (18.54)$$

and

$$f_R\left(E_z,E_t\right)=\cfrac{1}{1+\exp\left(\cfrac{E_z+E_t-E_{F,R}}{k_BT}\right)}$$

$$=\cfrac{1}{1+\exp\left\{\cfrac{E_z+E_t-\left(E_{F,L}-eV\right)}{k_BT}\right\}}$$

$$=\cfrac{1}{1+\exp\left(\cfrac{E_z+E_t-E_{F,L}+eV}{k_BT}\right)}\qquad(18.55)$$

because

$$f_L\left(E_z,E_t\right)=f_R\left(E_z,E_t\right)+eV\qquad(18.56)$$

where V is the voltage applied to the diode.

Combining eqs. (18.52), (18.53), (18.54), and (18.55), we have

$$J_{Total}=\frac{2em^*\times2\pi}{(2\pi)^3\hbar^3}\int_0^\infty T\left(E_z\right)dE_z\int_0^\infty\left\{f_L\left(E_z,E_t\right)-f_R\left(E_z,E_t\right)\right\}dE_t$$

$$=\frac{4\pi em^*}{(2\pi)^3\hbar^3}\int_0^\infty T\left(E_z\right)dE_z\int_0^\infty\left\{f_L\left(E_z,E_t\right)-f_R\left(E_z,E_t\right)\right\}dE_t$$

$$=\frac{4\pi em^*}{(2\pi)^3\hbar^3}\int_0^\infty T\left(E_z\right)dE_z\times$$

$$\int_0^\infty\left\{\cfrac{1}{1+\exp\left(\cfrac{E_z+E_t-E_{F,L}}{k_BT}\right)}-\cfrac{1}{1+\exp\left(\cfrac{E_z+E_t-E_{F,L}+eV}{k_BT}\right)}\right\}dE_t\qquad(18.57)$$

Let us consider the integral

$$\int_0^\infty\cfrac{dE_t}{1+\exp\left(\cfrac{E_z+E_t-E_{F,L}}{k_BT}\right)}$$

which is the first term in equation (18.57).

Here,

$$\cfrac{dE_t}{1+\exp\left(\cfrac{E_z+E_t-E_{F,L}}{k_BT}\right)}=\cfrac{dE_t}{\exp\left(\cfrac{E_z+E_t-E_{F,L}}{k_BT}\right)\left[\exp\left\{-\left(\cfrac{E_z+E_t-E_{F,L}}{k_BT}\right)\right\}+1\right]}$$

$$=\cfrac{\left[\exp\left\{-\left(\cfrac{E_z+E_t-E_{F,L}}{k_BT}\right)\right\}\right]dE_t}{\left[\exp\left\{-\left(\cfrac{E_z+E_t-E_{F,L}}{k_BT}\right)\right\}+1\right]}\qquad(18.58)$$

But

$$\frac{d}{dE_t}\left(\ln\left[\exp\left\{-\left(\cfrac{E_z+E_t-E_{F,L}}{k_BT}\right)\right\}+1\right]\right)=\cfrac{1}{\exp\left\{-\left(\cfrac{E_z+E_t-E_{F,L}}{k_BT}\right)\right\}+1}\times\exp\left\{-\left(\cfrac{E_z+E_t-E_{F,L}}{k_BT}\right)\right\}$$

$$\times\frac{-1\times k_BT-\left\{-\left(E_z+E_t-E_{F,L}\right)\right\}\times0}{\left(k_BT\right)^2}$$

$$=-\cfrac{\exp\left\{-\left(\cfrac{E_z+E_t-E_{F,L}}{k_BT}\right)\right\}}{\exp\left\{-\left(\cfrac{E_z+E_t-E_{F,L}}{k_BT}\right)\right\}+1}\times\frac{1}{k_BT}\qquad(18.59)$$

$$\therefore\cfrac{\left[\exp\left\{-\left(\cfrac{E_z+E_t-E_{F,L}}{k_BT}\right)\right\}\right]dE_t}{\left[\exp\left\{-\left(\cfrac{E_z+E_t-E_{F,L}}{k_BT}\right)\right\}+1\right]}=-\left(k_BT\right)\times d\left(\ln\left[\exp\left\{-\left(\cfrac{E_z+E_t-E_{F,L}}{k_BT}\right)\right\}+1\right]\right)\qquad(18.60)$$

and

$$\int_0^\infty \frac{\left[\exp\left\{-\left(\frac{E_z + E_t - E_{F,L}}{k_B T}\right)\right\}\right] dE_t}{\left[\exp\left\{-\left(\frac{E_z + E_t - E_{F,L}}{k_B T}\right)\right\} + 1\right]} = -(k_B T) \times \int_0^\infty d\left(\ln\left[\exp\left\{-\left(\frac{E_z + E_t - E_{F,L}}{k_B T}\right)\right\} + 1\right]\right)$$

$$= -(k_B T)\left|\ln\left[\exp\left\{-\left(\frac{E_z + E_t - E_{F,L}}{k_B T}\right)\right\} + 1\right]\right|_0^\infty = -(k_B T)$$

$$\times\left[\ln\left(\exp\left\{-\left(\frac{E_z + \infty - E_{F,L}}{k_B T}\right)\right\} + 1\right) - \ln\left(\exp\left\{-\left(\frac{E_z + 0 - E_{F,L}}{k_B T}\right)\right\} + 1\right)\right]$$

$$= -(k_B T) \times\left[\ln(0+1) - \ln\left(\exp\left\{-\left(\frac{E_z + 0 - E_{F,L}}{k_B T}\right)\right\} + 1\right)\right]$$

$$= (k_B T) \times \ln\left[\exp\left\{-\left(\frac{E_z + 0 - E_{F,L}}{k_B T}\right)\right\} + 1\right]$$

$$= (k_B T) \times \ln\left[\exp\left\{-\left(\frac{E_z - E_{F,L}}{k_B T}\right)\right\} + 1\right] = (k_B T)\ln\left\{\exp\left(\frac{E_{F,L} - E_z}{k_B T}\right) + 1\right\} \tag{18.61}$$

Following similar steps we evaluate the integral

$$\int_0^\infty \frac{dE_t}{1 + \exp\left(\frac{E_z + E_t - E_{F,L} + eV}{k_B T}\right)} = (k_B T)$$

$$\times \ln\left[\exp\left\{-\left(\frac{E_z - E_{F,L} + eV}{k_B T}\right)\right\}\right]$$

$$= (k_B T)\ln\left\{\exp\left(\frac{E_{F,L} - E_z - eV}{k_B T}\right) + 1\right\} \tag{18.62}$$

which is the second term in equation (18.57).

Substituting for the integrals from eqs. (18.61) and (18.62) into eq. (18.57), we get

$$J_{\text{Total}} = \left\{\frac{4\pi e m^*}{(2\pi)^3 \hbar^3}\int_0^\infty T(E_z) dE_z\right\} \times (k_B T)$$

$$\times\left[\ln\left\{\exp\left(\frac{E_{F,L} - E_z}{k_B T}\right) + 1\right\} - \ln\left\{\exp\left(\frac{E_{F,L} - E_z - eV}{k_B T}\right) + 1\right\}\right]$$

$$= \left\{\frac{4\pi e m^* k_B T}{(2\pi)^3 \hbar^3}\int_0^\infty T(E_z) dE_z\right\}\ln\frac{\left\{\exp\left(\frac{E_{F,L} - E_z}{k_B T}\right) + 1\right\}}{\left\{\exp\left(\frac{E_{F,L} - E_z - eV}{k_B T}\right) + 1\right\}}$$

$$\tag{18.63}$$

which is the Tsu-Esaki formula (Tsu and Esaki 1973).

18.5.2 Sequential Tunneling

Tunneling current calculations based on coherent tunneling mechanism showed divergence from experimental results on two aspects, namely the resonance peak was sharper than the observed one and the PVR of the RTD was higher by more than 100 times than the measured value. These discrepancies were ascribed to the non-inclusion of incoherent scattering in the analysis (Lüth 1996). In reality, a coherent wave function is non-existent over the whole RTD structure. The phase of the wave function is altered during inelastic scattering resulting in loss of coherence. Incoherent tunneling is taken into account by generalizing the transmission coefficient obtained from Breit-Wigner formalism. By consideration of inelastic scattering, the transmission coefficient given by Stone and Lee 1985 is:

$$T_{\text{Incoherent}}(E_z) = \frac{\frac{1}{4}(\Gamma_n \Gamma_T)}{(E_z - \varepsilon_n)^2 + \left(\frac{\Gamma_T}{2}\right)^2} \tag{18.64}$$

where Γ_n is the full width of the energy at half maximum when the transmission is purely coherent and

$$\Gamma_T = \Gamma_n + \Gamma_s \tag{18.65}$$

Γ_s is the full width of the energy at half maximum for inelastic scattering and Γ_T is the total full width at half maximum. The uncertainty principle relates Γ_s with scattering time τ_s:

$$\Gamma_s = \frac{\hbar}{\tau_s} \tag{18.66}$$

Note that Γ_n in eq. (18.14) is replaced by a larger quantity Γ_T in eq. (18.64). The effect of considering inelastic scattering is that PVR of the RTD diminishes with respect to that calculated with coherent tunneling. The extent of decrease obeys the equation

$$(PVR)_{\text{With scattering}} = (PVR)_{\text{Without scattering}} \times \frac{\Gamma_n}{\Gamma_n + \Gamma_s} \quad (18.67)$$

providing a more closer agreement with the experimental data on current-voltage characteristics of the diode. Notwithstanding the satisfactory agreement, the physical explanation of the scattering mechanisms involved is necessary (Lüth 1996).

18.6 Competition of the Resonant Tunneling Diode with Other Devices

Some interesting facts about RTDs vis-à-vis related devices are:

i. In a MOSFET, the charge carriers (electrons for N-channel device and holes for P-channel device) flow through the channel (inversion layer) by drift mechanism under the electric field applied between source and drain terminals. In a normal tunnel diode, the carrier flow mechanism is tunneling and the flow medium is a depletion region. In a resonant tunnel diode, the carrier transport mechanism is resonant tunneling and the carriers move through quasi-stable states in the quantum well.

ii. A normal tunnel diode suffers from the drawback of a high reverse leakage current. But in a symmetrical RTD structure, the emitter and collector regions are identical. Owing to the similar dopants and doping concentrations used on the emitter and collector sides, the current-voltage characteristics are the same on both sides. Thus the excessively high leakage current problem encountered in a normal tunnel diode is avoided. The RTD has a better rectifying property.

iii. The transport mechanism of the RTDs allows picosecond switching speeds. Comparatively lower switching speeds provided by normal tunnel diodes along with their protracted valley region makes them unfavorable for designing high-performance circuits. The picosecond switching and sharper transitions available with RTDs make them the preferred candidates for such circuits (Mazumder et al 1998). Tunneling is intrinsically a superfast phenomenon. It is not restricted by the transit time of carriers. Therefore, RTDs represent one of the fastest electronic devices ever made.

18.7 Discussion and Conclusions

RTDs have exhibited THz oscillation frequencies. Picosecond switching and sharper transitions achieved with resonant tunnel diodes combined with their superior rectifying property endow them with considerable operational advantages over simple tunnel diodes (Mizuta and Tanoue 1995). A quantum well ~5 nm

thick is made by heteroepitaxy using two semiconductors of different bandgaps. Heavily doped layers act as the emitter and collector contact regions but they do not directly interface with the layers forming the quantum well; instead undoped regions separate them from the contacts. For a small voltage applied between the emitter and the collector, a low current flows from the emitter to the collector. This current is produced mainly from non-resonant tunneling across the barriers. The leakage current between surface states too contributes to this current. When the applied voltage is increased, at a particular voltage the conduction band of the emitter region is aligned with the resonant level in the quantum well. Then the current increases to the peak value. With further increase of applied voltage, this alignment is offset leading to decrease in current with voltage. As a result, a negative resistance effect is observed.

The operation of the RTD is immediately obvious by drawing a comparison with the equivalent optical device called the Fabry-Perot interferometer with the electron waves replacing light waves, the potential barriers (made of wide bandgap semiconductors) serving as the partially silvered mirrors, and the quantum well substituting the air cavity. The barriers become transparent to electron waves whenever the electron wavelength/2 or its multiple equals the distance between the barriers. Then the current is maximum.

Resonant tunneling takes place through quasi-bound states in the quantum well. When the scattering time is much longer than the lifetime of the quasi-bound state, the electron is not scattered and its phase remains unaffected. The tunneling is coherent. But when the scattering time is much smaller than the lifetime of the state, the electron phase is disturbed and the tunneling becomes incoherent.

During coherent tunneling, the electron is described by a single wave function throughout the structure. By the uncertainty principle, the product of the full width of the energy Γ_n of quasi-bound state at half maximum and the lifetime τ_n of the state equals the reduced Planck's constant. At an energy equal to the energy of quasi-bound state, the transmission coefficient of the double barrier structure exhibits a peak resulting from the constructive interference of electron waves. The transmission coefficient versus energy of the electron graph can be modeled as a Breit-Wigner distribution so that the transmission coefficient is unity when the energy of the electron in the Z-direction equals the energy of the quasi-bound state.

We calculate the current flowing through the diode by finding the components of the electron energy in the Z-direction, both on the left and right sides of the barrier. This calculation is accomplished by resolving the Hamiltonian into components in two directions, viz. the Z-direction and the transverse direction. Then the current is expressed as a difference between the number of electrons transmitted across the barriers per unit time from left to right and the number of electrons transmitted across the barriers per unit time from right to left. The incident current densities for both directions are calculated. They are the products of six factors: electronic charge, density of states in k-space, electron distribution function on the left side/right side, velocity of electrons in the Z-direction, length, and cross section of the volume element in k-space. The transmitted current densities are obtained by multiplying the incident current densities with respective transmission coefficients. These coefficients are taken to be equal from symmetry considerations. By differentiation of the equations for energy components of the electron on the left and right sides of the barrier in the Z-direction, $k_{z,L}dk_{z,L}$ and $k_{z,R}dk_{z,R}$ are determined and substituted in the equations for transmitted current densities. The cross sections of the volume element are taken to be equal for left and right sides. This areal element is changed to cylindrical coordinates. The wave vector k_t is converted into energy E_t and the equation is rewritten with energies in place of wave vectors. For calculation over full volume, the angular limits for integration are 0 to 2π and energy from 0 to ∞. Equations for the Fermi-Dirac distribution functions are inserted into the equation for current density. The result of this integration gives the Tsu-Esaki formula for coherent tunneling.

The experimental results deviated from the theoretical formula derived assuming coherent tunneling in two respects, viz. the sharpness of the resonance peak and the large difference in the measured and calculated PVR. Non-consideration of incoherent scattering was the cause of this deviation. To bring harmony between theory and experiment, the transmission coefficient equation was modified by including inelastic scattering. A closer agreement of measured results with theoretical predictions was found.

Illustrative Exercises

18.1 An RTD (Figure 18.4) with a symmetrical structure comprises two AlGaAs barriers of height 0.1083 eV each on the two sides of a GaAs quantum well of width 5 nm. The effective mass of the electron in GaAs = 0.067 × free electron mass. Calculate the ground state energy of the electrons in the quantum well. At what voltage will the first peak current value be observed in the current-voltage characteristic of the diode?

We recall the following equations (eqs. (5.192), (5.149), (5.131)) from Chapter 5 on the finite quantum well:

$$k\tan\left(\frac{1}{2}kL\right) = \alpha \qquad (18.68)$$

$$k = \sqrt{\frac{2m}{\hbar^2}E} \qquad (18.69)$$

and

$$\alpha = \sqrt{\frac{2m}{\hbar^2}\{V_0(x) - E\}} \qquad (18.70)$$

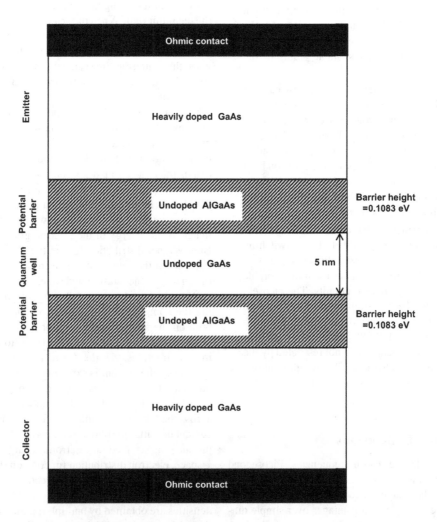

FIGURE 18.4 Resonant tunneling structure for Illustrative Exercise 18.1.

where E, the energy of the particle, is $< V_0$, the height of the barrier. Substituting for k and α from eqs. (18.69) and (18.70) in eq. (18.68) we get

$$\sqrt{\frac{2m}{\hbar^2}E} \times \tan\left(\frac{1}{2}\sqrt{\frac{2m}{\hbar^2}E} \times L\right) = \sqrt{\frac{2m}{\hbar^2}\{V_0(x)-E\}} \tag{18.71}$$

or,

$$\tan\left(\frac{1}{2}\sqrt{\frac{2m}{\hbar^2}E} \times L\right) = \frac{\sqrt{\frac{2m}{\hbar^2}\{V_0(x)-E\}}}{\sqrt{\frac{2m}{\hbar^2}E}} = \sqrt{\frac{\{V_0(x)-E\}}{E}} \tag{18.72}$$

or,

$$\tan\left(\frac{\sqrt{2mE}}{2\hbar}L\right) = \sqrt{\frac{\{V_0(x)-E\}}{E}} \tag{18.73}$$

which we write as

$$\tan\left(\frac{\sqrt{2m^*E}}{2\hbar}L\right) = \sqrt{\frac{\{V_0(x)-E\}}{E}} \tag{18.74}$$

where m^* is the effective mass of the electrons in GaAs $= 0.067\ m_0$. Equation (18.74) is a transcendental equation containing the variable energy E on both sides. This equation is solved graphically as described in Chapter 5. Let us guess a solution $E = 0.055$ eV $= 0.055 \times 1.602 \times 10^{-19}$ J, and investigate whether it is true. Taking the well width $L = 5 \times 10^{-9}$ m, we find for the two sides of eq. (18.74) that

and

$$\text{Right-hand side} = \sqrt{\frac{\{V_0(x)-E\}}{E}} = \sqrt{\frac{0.1083-0.055}{0.055}} = 0.9844 \tag{18.76}$$

So the solution $E = 0.055$ eV is verified. Hence, the electrons in the ground state of the quantum well will possess an energy $= 55$ meV.

Obviously, the energy 0.055 eV is lower than the barrier height (0.1083 eV). The first current peak will commence when sufficient bias has been applied to enable the energy of the ground state of an electron in the wall of the quantum well on the emitter side to be aligned with the minimum energy of the conduction band of the emitter. Because the barrier height in the symmetrical diode structure is 0.1083 eV, this alignment will take place at an electron energy

$$E_{\text{First peak}} = 2 \times 0.055\text{ eV} = 0.110\text{ eV} = 110\text{ meV} \tag{18.77}$$

The corresponding voltage is

$$V_{\text{First peak}} = \frac{E_{\text{First peak}}}{e} = \frac{110\text{ meV}}{e} = 110\text{ mV} \tag{18.78}$$

18.2 In the previous problem, suppose the distance between the two potential barriers of the RTD is 18 nm. Calculate the lifetime of an electron in the intervening space between the barriers.

From eq. (1.9), the probability that an electron with energy 0.055 eV is able to tunnel across the 18 nm thick barrier of height 0.1083 eV is given by eq. (18.79):

$$\text{Left-hand side} = \tan\left(\frac{\sqrt{2 \times 0.067 \times 9.109 \times 10^{-31}\text{ kg} \times 0.055 \times 1.602 \times 10^{-19}\text{ J}}}{2 \times \frac{6.626 \times 10^{-34}\text{ J.s}}{2\pi}} \times 5 \times 10^{-9}\text{ m}\right)$$

$$= \tan\left(\frac{\sqrt{2 \times 0.067 \times 9.109 \times 10^{-31}\text{ kg} \times 0.055 \times 1.602 \times 10^{-19}\text{ J}}}{2 \times 6.626 \times 10^{-34}\text{ J.s}} \times 2\pi \times 5 \times 10^{-9}\text{ m}\right)$$

$$= \tan\left(\frac{\sqrt{2 \times 0.067 \times 9.109 \times 0.055 \times 1.602 \times 10^{-50}}}{6.626 \times 10^{-34}} \times \pi \times 5 \times 10^{-9} \times \frac{\sqrt{\text{kg} \times \text{J}} \times \text{m}}{\text{J.s}}\right)$$

$$= \tan\left(\frac{\sqrt{2 \times 0.067 \times 9.109 \times 0.055 \times 1.602}}{6.626 \times 10^{-34}} \times 10^{-25} \times \pi \times 5 \times 10^{-9} \times \frac{\sqrt{\text{kg}} \times \text{m}}{\sqrt{\text{J}}.\text{s}}\right)$$

$$= \tan\left(\frac{0.32794}{6.626 \times 10^{-34}} \times 10^{-34} \times \pi \times 5 \times \frac{\sqrt{\text{kg}} \times \text{m}}{\sqrt{\text{kg} \times \text{m}^2 \times \text{s}^{-2}}.\text{s}}\right)$$

$$= \tan\left(\frac{0.32794}{6.626} \times \pi \times 5 \frac{\sqrt{\text{kg}} \times \text{m}}{\sqrt{\text{kg}} \times \text{m} \times \text{s}^{-1} \times \text{s}}\right) = \tan\left(\frac{0.32794}{6.626} \times \pi \times 5\right)$$

$$= \tan(0.2475\pi) = \tan(0.2475 \times 180°) = \tan(44.55°) = 0.9844 \tag{18.75}$$

$$\left(P\right)_{r=18\,\text{nm},\ E=0.1083\,\text{eV}} = \exp\left\{\frac{-4r\pi\sqrt{2m(V-E)}}{h}\right\}$$

$$= \exp\left\{\frac{-4\times18\times10^{-9}\times3.14159\sqrt{2\times9.109\times10^{-31}\text{kg}\times0.067\times(0.1083-0.055)\times1.602\times10^{-19}\text{J}}}{6.626\times10^{-34}\text{J}}\right\}$$

$$= \exp\left\{\frac{-226.19\times10^{-9}\sqrt{2\times9.109\times0.067\times0.0533\times1.602\times10^{-50}}}{6.626\times10^{-34}}\right\}$$

$$= \exp\left\{\frac{-226.19\times10^{-9}\sqrt{0.10422\times10^{-50}}}{6.626\times10^{-34}}\right\}$$

$$= \exp\left\{\frac{-226.19\times10^{-9}\times0.3228\times10^{-25}}{6.626\times10^{-34}}\right\} = \exp\left\{\frac{-226.19\times0.3228}{6.626}\right\} = \exp(-11.019)$$

$$= 1.639\times10^{-5} \tag{18.79}$$

The 'attempt frequency' of the electron having energy 0.055 eV is given by

$$v = \frac{E}{h} = \frac{0.055\times1.602\times10^{-19}}{6.626\times10^{-34}}$$

$$= 0.0133\times10^{15}\,\text{s} = 1.33\times10^{13}\,\text{s}^{-1} \tag{18.80}$$

Hence the lifetime of an electron between the two barriers of the RTD is

$$\tau = \frac{1}{\text{Attempt frequency}\times\text{Probability of tunneling}}$$

$$= \frac{1}{1.33\times10^{13}\times1.639\times10^{-5}}$$

$$= 0.4587\times10^{-8} = 4.6\times10^{-9}\ \text{s} \tag{18.81}$$

REFERENCES

Douglas Stone, A. and P.A. Lee. 1985. Effect of inelastic processes on resonant tunneling in one dimension. *Physical Review Letters* 54(11): 1196–1199.

Lüth, H. 1996. Tunneling in semiconductor nanostructures: Physics and devices. *Acta Physica Polonica A* 90(4): 667–679.

Maekawa, T., H. Kanaya, S. Suzuki and M. Asada. 2014. Frequency increase in terahertz oscillation of resonant tunnelling diode up to 1.55 THz by reduced slot-antenna length. *Electronics Letters* 50(17): 1214–1216.

Mazumder, P., S. Kulkarni, M. Bhattacharya, J.P. Sun and G.I. Haddad. 1998. Digital circuit applications of resonant tunneling device. *Proceedings of the IEEE* 86(4): 664–686.

Mizuta, H. and T. Tanoue. 1995. *The Physics and Applications of Resonant Tunneling Diodes.* Cambridge University Press: Cambridge, pp. 1–87.

NSM Archive. Aluminium Gallium Arsenide (AlGaAs). http://www.ioffe.ru/SVA/NSM/Semicond/AlGaAs/bandstr.html

Tsu, R. and L. Esaki. 1973. Tunneling in a finite superlattice. *Applied Physics Letters* 22(11): 562.

Zawawi M.A.B.M. 2015. Advanced $In_{0.8}Ga_{0.2}As/AlAs$ Resonant Tunneling Diodes for Applications in Integrated mm-waves MMIC Oscillators. Ph.D Thesis, University of Manchester, p. 18.

19

Nanoscale MOSFETs and Similar Devices

19.1 Short-Channel Effects in a Conventional Planar Bulk MOSFET

If a MOSFET is progressively miniaturized, a stage comes when the channel length of the MOSFET and the depletion regions at its source and drain junctions become comparable in size (D'Agostino and Quercia 2000, Lundstrom 2017). The device at this stage possessing similar dimensions of the channel and depletion regions is said to be a short-channel MOSFET (Figure 19.1). The dimensional shrinkage of the MOSFET influences the device performance detrimentally in a multiplicity of ways which are collectively known as short-channel effects because they come into play only when the channel is drastically shortened. These effects pose formidable technological challenges which must be carefully obviated for satisfactory behavior of the small-size MOSFETs. The effects accrued with the MOSFET downscaling are briefly presented in the forthcoming subsections.

19.1.1 Drain-Induced Barrier Lowering (DIBL) and Threshold Voltage Roll-Off

A potential barrier exists between the source and the channel. The source-channel barrier is the outcome of the balance of drift and diffusion currents flowing across these two regions. This barrier must be controlled by the voltage applied to the gate to achieve maximum transconductance. On applying the gate voltage, the barrier is surmounted by an electron moving from the source to the channel. In a long-channel MOSFET, the source and hence the source-channel barrier are located far away from the drain. So, the barrier is less susceptible to the influence of the drain and is electrostatically shielded from the drain voltage by the voltage applied between the gate and the substrate. But in a short-channel device, the height of the potential energy barrier between the source and channel regions is decreased when the drain voltage is increased because the source and the drain are in proximity. This leads to injection of more electrons from the source and a resultant upswing in the drain current. The effect is even more

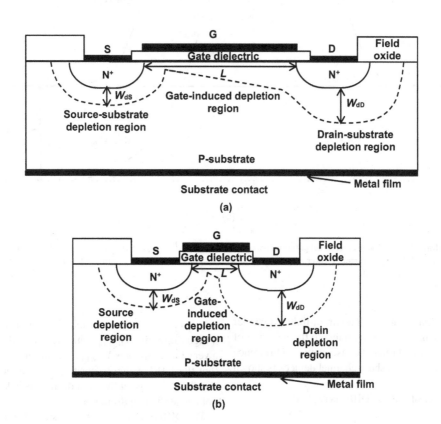

FIGURE 19.1   Channel lengths and depletion regions in the two types of MOSFETs: (a) long-channel type and (b) short-channel type.

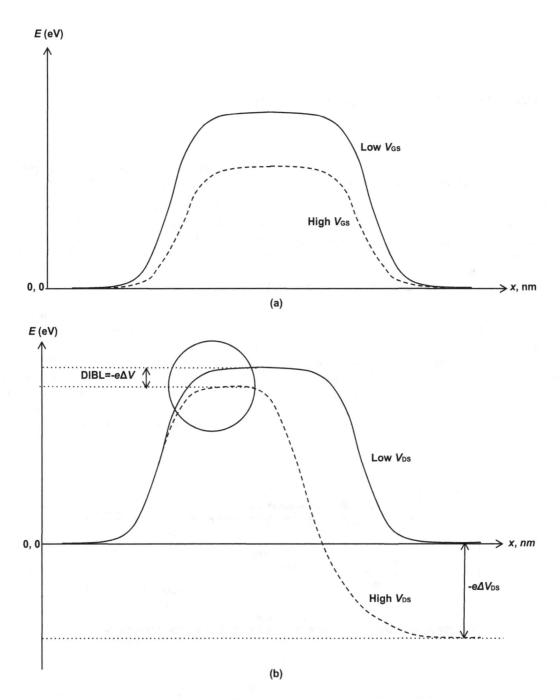

FIGURE 19.2 Gate and drain effects on source-channel potential barrier: (a) V_{DS}-independent gate-controlled lowering of source-channel potential barrier in a long-channel MOSFET, and (b) V_{DS}-dependent lowering of source-channel potential barrier in a short-channel MOSFET. The encircled region in (b) reveals the drain-induced barrier lowering effect.

conspicuous at higher drain voltages. The drain-induced barrier lowering (DIBL), as the name suggests, refers to the lowering of the source-channel barrier and related consequences. Difference in behaviors of long-channel and short-channel devices is shown in Figure 19.2.

The magnitude or intensity of the DIBL is expressed as

$$DIBL = -\frac{V_{Th}^{High\,DD} - V_{Th}^{Low\,DD}}{V_{High\,DD} - V_{Low\,DD}} \qquad (19.1)$$

where $V_{Th}^{High\,DD}$ is the threshold voltage of MOSFET measured at a high drain bias $V_{High\,DD}$ while $V_{Low\,DD}$ is the threshold voltage at a very low drain bias $V_{Low\,DD}$. The negative sign is placed in order that the DIBL value is positive. This is necessary because the threshold voltage at the high drain bias is less than the threshold voltage at the low drain bias.

The DIBL may be easily understood in terms of the mechanism of the depletion region formation during device operation. Let us examine more closely how the depletion region is produced and controlled in a long-channel MOSFET vis-à-vis the short-channel version. In a long-channel MOSFET, the depletion region below the gate oxide is primarily produced and controlled by the application of gate-to-substrate voltage. The potential profile here is one-dimensional (1D). But the small separation between the source and drain terminals does not allow the same to happen in a short-channel device. In a short-channel device, the depletion regions at the source-substrate and drain-substrate junctions enlarge and spread underneath the gate. Encroachment of the depletion regions from the source and drain sides under the gate makes the potential profile two-dimensional (2D). The effects of these potentials get intermingled and the drain–source potential may dominate over the gate–source voltage.

Examining more closely, it is found that the independence of the threshold voltage from the drain voltage is lost because the burden of balancing the charge in the depletion region below the gate is now shared between the gate-substrate and drain–source biases instead of being solely governed by the gate bias. Charge sharing takes place among the source, drain, and gate terminals. Effectively, the depletion region is a collaborative creation of the gate–source voltage and the drain–source voltage. The drain-to-source voltage starts to control the depletion region in association with the gate-to-source voltage. Sometimes, the drain voltage can turn on the transistor irrespective of the gate voltage. As this task must solely be performed by the gate bias, the gate loses control over the depletion region formation. In other words, the gating action of the device is crippled.

The effect is observed as a threshold voltage reduction and increase of output conductance of the field-effect transistor (FET) with drain voltage. Reduction of the source-channel potential barrier permits current flow between the source and the drain at a gate–source voltage less than the threshold voltage. Consequently, the device turns on prematurely. The smooth fall of the threshold voltage with the channel length shortening is referred to as threshold voltage roll-off.

In the subthreshold regime where the carrier concentration is low, the carrier population-gradient dependent diffusion current predominates over the carrier-concentration determined drift current. Hence, the carrier diffusion may continue after turning off the device due to lowering of the source-channel potential barrier by DIBL, making it difficult to switch off the MOSFET. As this diffusion current, which is essentially a leakage current of the MOSFET, depends exponentially on the source-channel potential barrier, this current abnormally rises. Hence, the device shows excessive leakage.

In the extreme case, the depletion regions formed at the source-substrate and drain-substrate junctions merge together causing punchthrough breakdown when current flows from the source to the drain regardless of any gate voltage (Figure 19.3). If the relative permittivity of silicon is denoted by ε_s, the punchthrough voltage V_{PT} for an N-channel MOSFET of channel length L with P-substrate doping concentration N_A is given by

$$V_{PT} = \frac{eN_A L^2}{2\varepsilon_0 \varepsilon_s} \tag{19.2}$$

showing that the punchthrough voltage decreases as the square of the channel length. An increase in N_A is helpful to combat the V_{PT} reduction in a short-channel MOSFET.

Let us now estimate the reduction of threshold voltage in a short-channel MOSFET. We recall that the threshold voltage of a MOSFET consists of three main components:

i. The flat-band voltage V_{FB}, which is the voltage required to maintain the energy bands in the flat condition. In the absence of any fixed charges in the MOSFET gate dielectric (SiO_2) or at the oxide-silicon interface, it is given by

$$V_{FB} = \phi_m - \phi_s \tag{19.3}$$

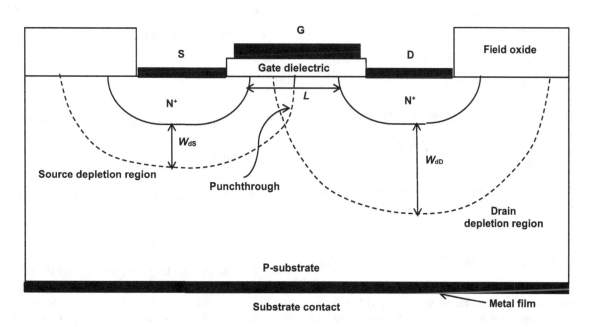

FIGURE 19.3 Punchthrough breakdown by merging together the depletion regions produced by source-substrate and drain-substrate junctions at a high source–drain bias.

where ϕ_m and ϕ_s denote the work functions of the metal and the semiconductor, respectively.

ii. Two times the bulk potential ($2\phi_B$). The bulk potential is the potential change of the semiconductor in conversion from the intrinsic state (carrier concentration $= n_i$) to the doped condition with the required concentration of the dopant (N_A). Considering an N-channel MOSFET, the P-body layer is inverted at the potential ϕ_B marking the onset of inversion but the concentration of electrons in the surface channel becomes equal to the acceptor doping concentration in the bulk P-layer at the potential $2\phi_B$, which is named the strong inversion condition. The bulk potential is expressed as

$$\phi_B = \left(\frac{k_B T}{e}\right) \ln \left(\frac{N_A}{n_i}\right) \qquad (19.4)$$

where k_B is the Boltzmann constant, T is the temperature in Kelvin scale, and e is the elementary charge.

iii. The voltage drop across the silicon dioxide film due to the charge in the depletion region underneath the gate. It is

the gate dielectric undergoes depletion. But in a short-channel device, the triangular regions on the two sides of the rectangle come under the influence of drain–source potential (Figure 19.4). If we remove these triangles from the rectangle, we get a smaller trapezium-shaped region which is under gate control. Further, the trapezium is asymmetric with the larger size of the depletion region near the drain than on the source side because the positive drain voltage reverse biasing the N$^+$-P junction falls as we recede away from the drain toward the source due to the voltage dropped across the semiconductor. As the channel length is decreased, the trapezium decreases in area. As a result, the gate potential has to exert influence on a smaller area of the depletion region, which it may do very easily leading to a lowering of the threshold voltage.

The extent of threshold voltage reduction is found by subtracting the triangular areas from the area of the rectangle. With reference to the threshold voltage V_{Th} of the long-channel MOSFET, the threshold voltage of the short-channel device may be expressed as

$$(V_{Th})_{\text{Short channel}} = V_{Th} - \Delta V_{Th} \qquad (19.10)$$

$$V_{ox} = \frac{\text{Charge}}{\text{Capacitance}} = \frac{\text{Depletion region charge per unit area} \left(Q_{\text{Depletion}}\right)}{\text{Capacitance of the metal-oxide} - \text{semiconductor capacitor per unit area} \left(C_{ox}\right)} = \frac{eN_A W}{C_{ox}} \qquad (19.5)$$

where W is the width of the depletion region which is a function of the applied voltage ($V = 2\phi_B$ for strong inversion). We know that V and W are related as

$$V = \frac{eN_A W^2}{2\varepsilon_0 \varepsilon_s} \qquad (19.6)$$

where ε_0 is the permittivity of vacuum and ε_s is the dielectric constant of silicon. From eq. (19.6),

$$W = \sqrt{\frac{2\varepsilon_0 \varepsilon_s V}{eN_A}} = \sqrt{\frac{2\varepsilon_0 \varepsilon_s (2\phi_B)}{eN_A}} = 2\sqrt{\frac{\varepsilon_0 \varepsilon_s \phi_B}{eN_A}} \qquad (19.7)$$

Substituting for W from eq. (19.7) in eq. (19.5) we get

$$V_{ox} = \frac{eN_A}{C_{ox}} \times 2\sqrt{\frac{\varepsilon_0 \varepsilon_s \phi_B}{eN_A}} = \frac{2\sqrt{eN_A \varepsilon_0 \varepsilon_s \phi_B}}{C_{ox}} \qquad (19.8)$$

Summing up the contributions from the three components of threshold voltage given in eqs. (19.3), (19.4), and (19.8), we can write the threshold voltage as

$$V_{Th} = \phi_m - \phi_s + 2\phi_B + \frac{2\sqrt{eN_A \varepsilon_0 \varepsilon_s \phi_B}}{C_{ox}} \qquad (19.9)$$

We focus our attention on the fourth term in eq. (19.9). As we know, this term represents the contribution from the depletion layer charge. On applying a gate potential to a long-channel MOSFET, the rectangular region in the P-substrate underlying

where ΔV_{Th} is the reduction in threshold voltage caused by the encroachment of the drain–source bias-induced depletion region in the rectangular region directly below the gate. Taking a simplified geometrical representation shown in Figure 19.4(b) we note that

Area of the rectangular region in a long – channel MOSFET

$$= \text{Length} \times \text{Breadth} = LW_{dm} \qquad (19.11)$$

and

Area of the trapezoidal region in a short-channel MOSFET

$$= \frac{1}{2}\left(\text{Sum of parallel sides}\right)$$

$$\times \left(\text{Perpendicular distance between them}\right) = \frac{1}{2}(L + l) \times W_{dm}$$

$$(19.12)$$

where L is the channel length measured from the edge of the source to the drain edge, l is the length of the smaller parallel side of the trapezium away from the oxide-silicon interface, and W_{dm} is the vertical depth of the depletion region from the oxide-silicon surface into the P-substrate. Looking at the above-mentioned diagram, we obtain

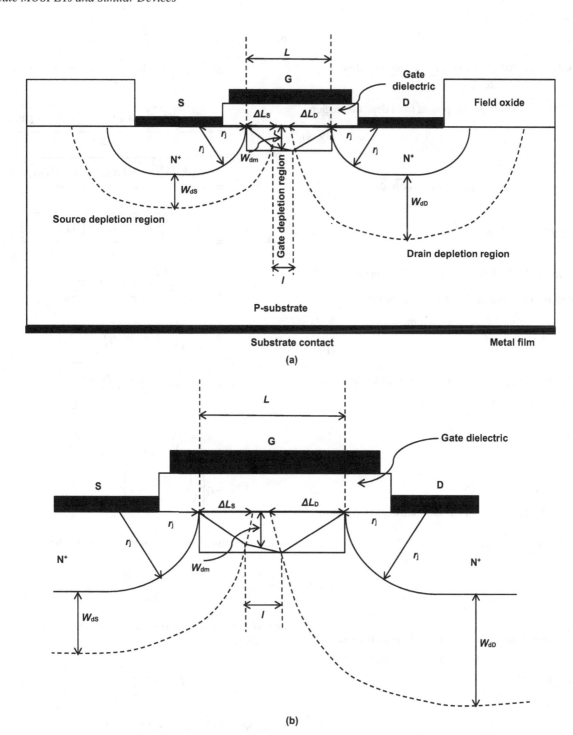

FIGURE 19.4 The entanglement of the source and drain with gate for gaining domination over the depletion region under the gate: (a) view of the full structure showing the rectangular-, triangular-, and trapezium-shaped regions, and (b) enlarged view of the relevant portion for explaining the symbols used.

$$\frac{\text{Area of rectangle} - \text{Area of trapezium}}{\text{Area of rectangle}} = \frac{LW_{dm} - \frac{1}{2}(L + l) \times W_{dm}}{LW_{dm}} = \frac{LW_{dm}}{LW_{dm}} - \frac{1}{2LW_{dm}}(L + l)W_{dm}$$

$$= 1 - \frac{L + l}{2L} = \frac{2L - (L + l)}{2L} = \frac{L - l}{2L}$$

$$= \frac{\Delta L_S + l + \Delta L_D - l}{2L} = \frac{\Delta L_S + \Delta L_D}{2L} \qquad (19.13)$$

Evidently,

$$\Delta V_{Th} = \frac{\text{Area difference between rectangle and trapezium}}{\text{Rectangular area}}$$

$$\times \text{ Threshold voltage component from depletion region}$$

$$(19.14)$$

or,

$$\Delta V_{Th} = \frac{\Delta L_S + \Delta L_D}{2L} \times \frac{2\sqrt{eN_A \varepsilon_0 \varepsilon_s \phi_B}}{C_{ox}} \qquad (19.15)$$

using eqs. (19.13) and (19.8).

Let

$$r_j = \text{Junction depth of the N}^+ \text{source or N}^+ \text{drain diffusion}$$

$$= \text{Radius of the quarter circular arc representing}$$

$$\text{the edge of the source or drain diffusion}$$

$$(19.16)$$

$$W_{dS} = \text{Depth of the N}^+ \text{source} - \text{P substrate depletion region}$$

$$= \sqrt{\frac{2\varepsilon_0 \varepsilon_s (0 + \phi_{bi})}{eN_A}} = \sqrt{\frac{2\varepsilon_0 \varepsilon_s \phi_{bi}}{eN_A}}$$

$$(19.17)$$

and

$$W_{dD} = \text{Depth of the N}^+ \text{drain} - \text{P substrate depletion region}$$

$$= \sqrt{\frac{2\varepsilon_0 \varepsilon_s (V_{DS} + \phi_{bi})}{eN_A}}$$

$$(19.18)$$

where

$$\phi_{bi} = \text{Built-in potential of the N}^+ - \text{P region from the source to}$$

$$\text{the substrate or from the drain to the substrate}$$

$$= \frac{k_B T}{e} \ln\left(\frac{N_A N_D}{n_i^2}\right)$$

$$(19.19)$$

N_A, N_D are the doping concentrations of the P-substrate and source/drain regions, n_i is the intrinsic carrier concentration in silicon, V_{DS} is the drain–source voltage, k_B is the Boltzmann constant, T is the temperature, and e is the electronic charge. Let ΔL_S be the lateral extent of the depletion region near the source and ΔL_D be the lateral extent of depletion region near the drain.

By Pythagoras theorem

$$\left(W_{dS} + r_j\right)^2 = \left(\Delta L_S + r_j\right)^2 + W_{dm}^2 \qquad (19.20)$$

or

$$W_{dS}^2 + 2W_{dS}r_j + r_j^2 = \Delta L_S^2 + 2\Delta L_S r_j + r_j^2 + W_{dm}^2 \qquad (19.21)$$

or

$$\Delta L_S^2 + 2\Delta L_S r_j + W_{dm}^2 - W_{dS}^2 - 2W_{dS}r_j = 0 \qquad (19.22)$$

$$\therefore \Delta L_S = \frac{-2r_j \pm \sqrt{\left(2r_j\right)^2 - 4 \times 1 \times \left(W_{dm}^2 - W_{dS}^2 - 2W_{dS}r_j\right)}}{2}$$

$$= \frac{-2r_j \pm \sqrt{4\left\{\left(r_j\right)^2 - \left(W_{dm}^2 - W_{dS}^2 - 2W_{dS}r_j\right)\right\}}}{2}$$

$$= \frac{-2r_j \pm 2\sqrt{\left\{\left(r_j\right)^2 + 2W_{dS}r_j - \left(W_{dm}^2 - W_{dS}^2\right)\right\}}}{2}$$

$$\cong \frac{-2r_j \pm 2\sqrt{\left\{\left(r_j\right)^2 + 2W_{dS}r_j - 0\right\}}}{2}$$

$$= -r_j \pm \sqrt{\left\{\left(r_j\right)^2 + 2W_{dS}r_j\right\}}$$

$$= -r_j \pm \sqrt{\left(r_j\right)^2 \left\{\frac{\left(r_j\right)^2}{\left(r_j\right)^2} + \frac{2W_{dS}r_j}{\left(r_j\right)^2}\right\}}$$

$$= -r_j \pm r_j \sqrt{1 + \frac{2W_{dS}}{r_j}} = r_j \sqrt{1 + \frac{2W_{dS}}{r_j}} - r_j$$

$$= r_j\left(\sqrt{1 + \frac{2W_{dS}}{r_j}} - 1\right) \qquad (19.23)$$

where the minus sign preceding r_j is rejected because

$$r_j \sqrt{1 + \frac{2W_{dS}}{r_j}} > r_j \qquad (19.24)$$

and ΔL_S cannot be negative. Also, $W_{dm} \approx W_{dS}$.

Similarly,

$$\Delta L_D = r_j\left(\sqrt{1 + \frac{2W_{dD}}{r_j}} - 1\right) \qquad (19.25)$$

From eqs. (19.15), (19.23), and (19.25)

$$\Delta V_{Th} = \frac{r_j\left(\sqrt{1 + \frac{2W_{dS}}{r_j}} - 1\right) + r_j\left(\sqrt{1 + \frac{2W_{dD}}{r_j}} - 1\right)}{2L} \times \frac{2\sqrt{eN_A \varepsilon_0 \varepsilon_s \phi_B}}{C_{ox}}$$

$$= \frac{r_j}{L}\left\{\left(\sqrt{1 + \frac{2W_{dS}}{r_j}} - 1\right) + \left(\sqrt{1 + \frac{2W_{dD}}{r_j}} - 1\right)\right\} \frac{\sqrt{eN_A \varepsilon_0 \varepsilon_s \phi_B}}{C_{ox}}$$

$$= \frac{r_j}{L}\left(\sqrt{1 + \frac{2W_{dS}}{r_j}} + \sqrt{1 + \frac{2W_{dD}}{r_j}} - 2\right) \frac{\sqrt{eN_A \varepsilon_0 \varepsilon_s \phi_B}}{C_{ox}}$$

$$(19.26)$$

The decrease in V_{Th} can be expressed as

$$\Delta V_{Th} = \frac{r_j}{L}\left(\sqrt{1+\frac{2W_{dm}}{r_j}}+\sqrt{1+\frac{2W_{dm}}{r_j}}-2\right)\frac{\sqrt{eN_A\varepsilon_0\varepsilon_s\phi_B}}{C_{ox}}$$

$$(19.27)$$

since

$$W_{dS} \approx W_{dD} \approx W_{dm} \tag{19.28}$$

Hence,

$$\Delta V_{Th} = \frac{r_j}{L}\left(2\sqrt{1+\frac{2W_{dm}}{r_j}}-2\right)\frac{\sqrt{eN_A\varepsilon_0\varepsilon_s\phi_B}}{C_{ox}}$$

$$(19.29)$$

$$= \frac{2r_j}{L}\left(\sqrt{1+\frac{2W_{dm}}{r_j}}-1\right)\frac{\sqrt{eN_A\varepsilon_0\varepsilon_s\phi_B}}{C_{ox}}$$

$$\therefore \Delta V_{Th} \to 0 \text{ as } r_j \to 0, \text{ or } W_{dm} \to 0, \text{ or } C_{ox} \to \infty \quad (19.30)$$

This means that the threshold voltage shift can be minimized by reducing r_j, i.e. using shallower source/drain junctions and by increasing the oxide capacitance by using a thinner oxide film. A thinner gate oxide also makes the gate effect more predominant over the drain and is thereby helpful. Another method involves replacing the silicon dioxide with a high dielectric constant insulator like hafnium oxide (HfO_2) having relative permittivity of 25 against 3.9 for SiO_2.

But in MOSFETs with shallow source/drain junctions, the parasitic resistance associated with source/drain junctions

$$R_{Parasitic} \propto \frac{\text{Sheet resistance of source or drain region}}{r_j} \quad (19.31)$$

increases with lowering of r_j. So a design trade-off must be made. Generally, shallow extensions of source/drain diffusions are used with deeper source/drain regions for effective reduction of r_j without appreciably increasing the parasitic resistances.

Using eq. (19.30), the DIBL effect is also mitigated by using shallow source and drain junctions and by increasing the doping concentration of the substrate in the vicinity of the source and drain junctions whereby the depletion region thicknesses W_{dS}, W_{dD} in the substrate are decreased. Such doping is called halo doping and is done by halo implants. These implants are also known as pocket implants. During these implants, the dopants are placed contiguous to the source and drain regions exactly below the active channel.

19.1.2 Carrier Velocity Saturation

At a critical field of 1×10^6 V m^{-1} for P-type silicon, the drift velocity of carriers reaches the maximum value of 10^5 m s^{-1} due to the effects of scattering (Figure 19.5(a)). The linear relationship between carrier velocity and electric field no longer holds so that the mobility of carriers decreases. As a result, the drain current and hence the transconductance in saturation mode fall.

To avoid velocity saturation, the bias voltages must be lowered as the device dimensions are scaled down.

The carrier drift velocity v_d is expressed as a function of the applied electric field E as

$$v_d = \frac{\mu E}{1+\dfrac{E}{E_C}} \tag{19.32}$$

where μ is the carrier mobility and E_C is the critical electric field at which the velocity becomes constant with variation of field. At low fields $E \ll E_C$

$$\frac{E}{E_C} \ll 1 \tag{19.33}$$

So,

$$v_d = \frac{\mu E}{1+0} = \mu E \tag{19.34}$$

Since μ is a constant, the drift velocity varies linearly with the field.

At high fields $E \gg E_C$

$$v_d = \frac{\mu E}{1+\dfrac{E}{E_C}} = \frac{\mu}{\dfrac{1}{E}+\dfrac{1}{E_C}} \tag{19.35}$$

Since

$$\frac{1}{E} \ll \frac{1}{E_C} \tag{19.36}$$

we have

$$v_d = \frac{\mu}{0+\dfrac{1}{E_C}} = \mu E_C \tag{19.37}$$

or

$$(v_d)_{sat} = \mu E_C \tag{19.38}$$

Since μ is constant and E_C also has a fixed value for a semiconductor, the drift velocity saturates to a constant value.

The earlier relation may also be interpreted as follows: If there is no velocity saturation, there will be no critical field, i.e. the critical field will be infinite.

$$\lim_{E_C\to\infty} v_d = \lim_{E_C\to\infty} \frac{\mu E}{1+\dfrac{E}{E_C}} = \frac{\mu E}{1+\dfrac{E}{\infty}} = \frac{\mu E}{1+0} = \mu E \quad (19.39)$$

On the opposite side, when there is velocity saturation, we write

$$\lim_{E_C\to0} v_d = \lim_{E_C\to0} \frac{\mu E}{1+\dfrac{E}{E_C}} = \frac{\mu E E_C}{E_C+E} = \frac{\mu E E_C}{0+E} = \mu E_C \quad (19.40)$$

To understand the effect of velocity saturation on MOSFET behavior, let us derive the classical current–voltage equation for a long-channel N-MOSFET, viz.

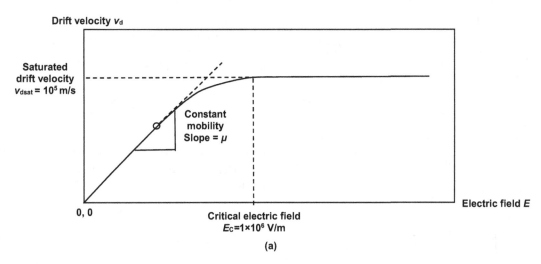

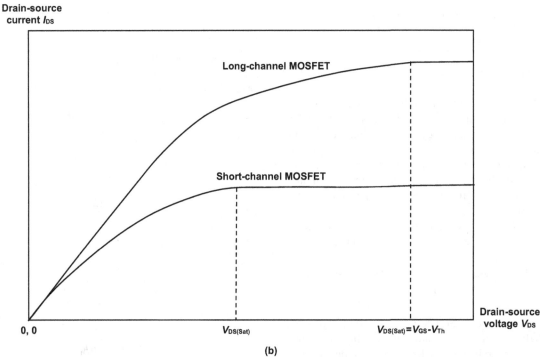

FIGURE 19.5 Velocity saturation and its effect on current–voltage characteristics of a MOSFET: (a) Drift velocity–electric field curve for P-type silicon and (b) I_{DS}–V_{DS} characteristics of long-channel and short-channel MOSFETs.

$$(I_{DS})_{\text{Long channel}} = C_{ox}\left(\frac{W}{L}\right)\mu_n\left\{(V_{GS}-V_{Th})V_{DS}-\frac{V_{DS}^2}{2}\right\} \quad (19.41)$$

by using eq. (19.32) for the dependence of electron drift velocity on electric field. Here C_{ox} is the capacitance of silicon dioxide per unit area, W is the width of the channel, L is the channel length, μ_n is the mobility of electrons in the channel, I_{DS} is the drain–source current, V_{DS} is the drain–source voltage, V_{GS} is the gate–source voltage, and V_{Th} is the threshold voltage.

The long-channel MOSFET equation (19.41) is derived on the assumption that drift velocity v_d of carriers varies linearly with electric field E and the mobility μ_n is the constant of proportionality. This assumption becomes invalid in a short-channel MOSFET at moderate fields because of reduction in the channel length.

In a MOSFET, for $V_{GS}>V_{Th}$ and $V_{GS}\gg V_{DS}$, the oxide capacitance for a slice of length dx and width W (area Wdx) is

$$C = C_{ox}Wdx \quad (19.42)$$

The charge induced in the channel at the voltage $V(x)$ is

$$\text{Charge} = \text{Capacitance} \times \text{Voltage} \quad (19.43)$$

or

$$dQ(x) = -C_{ox}Wdx\left\{V_{GS}-V_{Th}-V(x)\right\} \quad (19.44)$$

During depletion and inversion, the surface charge in a P-type semiconductor is negative.

The electronic current in the channel is

$$I_{DS} = \text{Time derivative of charge} = \frac{dQ(x)}{dt}$$

$$= \frac{d}{dt}\left[-C_{ox}Wdx\{V_{GS} - V_{Th} - V(x)\}\right]$$

$$= -C_{ox}W\frac{dx}{dt}\{V_{GS} - V_{Th} - V(x)\}$$

$$= -C_{ox}Wv_d\{V_{GS} - V_{Th} - V(x)\} \tag{19.45}$$

since

$$\frac{dx}{dt} = v_d \tag{19.46}$$

The drain current in the conventional current direction is

$$I_{DS} = C_{ox}Wv_d\{V_{GS} - V_{Th} - V(x)\} \tag{19.47}$$

Putting the value of v_d from eq. (19.32) in eq. (19.47) with correct sign, we get

$$I_{DS} = C_{ox}W\left(-\frac{\mu_n E}{1+\frac{E}{E_C}}\right)\{V_{GS} - V_{Th} - V(x)\}$$

$$= -\frac{C_{ox}W\mu_n E\{V_{GS} - V_{Th} - V(x)\}}{1+\frac{E}{E_C}}$$

$$= -\frac{C_{ox}W\mu_n\{V_{GS} - V_{Th} - V(x)\}}{1+\frac{1}{E_C}\times-\frac{dV}{dx}}\times-\frac{dV}{dx}$$

$$= \frac{C_{ox}W\mu_n\{V_{GS} - V_{Th} - V(x)\}}{1-\frac{1}{E_C}\times\frac{dV}{dx}}\times\frac{dV}{dx} \tag{19.48}$$

since

$$E = -\frac{dV}{dx} \tag{19.49}$$

Minus sign is used in the earlier equation for v_d because the electron moves opposite to the direction of the field. The mobility of electrons in a silicon MOSFET is $\mu_n = 700\,\text{cm}^2\text{V}^{-1}\text{s}^{-1}$ because the carriers undergo surface scattering at the silicon-silicon dioxide interface and are thereby slowed down.

Equation (19.48) gives

$$I_{DS}\left(1-\frac{1}{E_C}\frac{dV}{dx}\right)dx = C_{ox}W\mu_n\{V_{GS} - V_{Th} - V(x)\}dV \tag{19.50}$$

or

$$I_{DS}dx - \frac{I_{DS}}{E_C}\frac{dV}{dx}dx = C_{ox}W\mu_n\{V_{GS} - V_{Th} - V(x)\}dV \tag{19.51}$$

Integrating from the source to the drain between length 0 to L and voltage 0 to V_{DS}, we get

$$\int_0^L I_{DS}\,dx - \int_0^{V_{DS}}\frac{I_{DS}dV}{E_C} = \int_0^{V_{DS}} C_{ox}W\mu_n\{V_{GS} - V_{Th} - V(x)\}dV$$

$$= C_{ox}W\mu_n\int_0^{V_{DS}}\{V_{GS} - V_{Th} - V(x)\}dV \tag{19.52}$$

or

$$I_{DS}|x|_0^L - \frac{I_{DS}}{E_C}\int_0^{V_{DS}}dV = C_{ox}W\mu_n$$

$$\times\int_0^{V_{DS}}(V_{GS} - V_{Th})dV - C_{ox}W\mu_n\int_0^{V_{DS}}V(x)dV \tag{19.53}$$

or

$$I_{DS}(L-0) - \frac{I_{DS}}{E_C}|V|_0^{V_{DS}} = C_{ox}W\mu_n(V_{GS} - V_{Th})$$

$$\times\int_0^{V_{DS}}dV - C_{ox}W\mu_n\left|\frac{V^2}{2}\right|_0^{V_{DS}} \tag{19.54}$$

or

$$I_{DS}L - \frac{I_{DS}}{E_C}(V_{DS} - 0) = C_{ox}W\mu_n(V_{GS} - V_{Th})|V|_0^{V_{DS}}$$

$$- C_{ox}W\mu_n\left(\frac{V_{DS}^2}{2} - 0\right) \tag{19.55}$$

or

$$I_{DS}L - \frac{I_{DS}V_{DS}}{E_C} = C_{ox}W\mu_n(V_{GS} - V_{Th})(V_{DS} - 0) - C_{ox}W\mu_n\frac{V_{DS}^2}{2} \tag{19.56}$$

or

$$I_{DS}L - \frac{I_{DS}V_{DS}L}{LE_C} = C_{ox}W\mu_n(V_{GS} - V_{Th})(V_{DS}) - C_{ox}W\mu_n\frac{V_{DS}^2}{2} \tag{19.57}$$

or

$$\left(I_{DS} - \frac{I_{DS}V_{DS}}{LE_C}\right)L = C_{ox}W\mu_n(V_{GS} - V_{Th})V_{DS} - C_{ox}W\mu_n\frac{V_{DS}^2}{2} \tag{19.58}$$

or

$$\left(I_{DS} + \frac{I_{DS}E}{E_C}\right)L = C_{ox}W\mu_n\left\{(V_{GS} - V_{Th})V_{DS} - \frac{V_{DS}^2}{2}\right\} \tag{19.59}$$

since

$$-\frac{V_{DS}}{L} = E \tag{19.60}$$

Equation (19.59) is written as

$$I_{DS}\left(1+\frac{E}{E_C}\right)L = C_{ox}W\mu_n\left\{(V_{GS}-V_{Th})V_{DS}-\frac{V_{DS}^2}{2}\right\} \quad (19.61)$$

or

$$I_{DS}\left(1+\frac{E}{E_C}\right) = C_{ox}\left(\frac{W}{L}\right)\mu_n\left\{(V_{GS}-V_{Th})V_{DS}-\frac{V_{DS}^2}{2}\right\} \quad (19.62)$$

Thus we arrive at the current–voltage equation for a short-channel MOSFET

$$(I_{DS})_{\text{Short channel}} = \left(\frac{1}{1+\dfrac{E}{E_C}}\right)C_{ox}\left(\frac{W}{L}\right)\mu_n\left\{(V_{GS}-V_{Th})V_{DS}-\frac{V_{DS}^2}{2}\right\}$$

$$= \left(\frac{1}{1+\dfrac{E}{E_C}}\right)(I_{DS})_{\text{Long channel}}$$

$$(19.63)$$

using eq. (19.41).

For a short-channel MOSFET having $L=100\,$nm, even at a low applied voltage of 0.5 V, the field E is 0.5 V/100×10^{-9} m$=5\times10^6$ V m^{-1}, which is five times larger than the critical field ($E_C=10^6$ V m^{-1}); hence the current decreases relative to the long-channel device.

To determine the saturation voltage for field-dependent velocity, let us look for the situation in which the change in V_{DS} does not alter I_{DS} and denote the corresponding voltage by $V_{DS(Sat)}$, i.e. when

$$\frac{dI_{DS}}{dV_{DS}} = 0 \quad (19.64)$$

$$V_{DS} = V_{DS(Sat)} \quad (19.65)$$

From eq. (19.63),

$$\frac{d}{dV_{DS}}\left[\left(\frac{1}{1+\dfrac{E}{E_C}}\right)C_{ox}\left(\frac{W}{L}\right)\mu_n\left\{(V_{GS}-V_{Th})V_{DS}-\frac{V_{DS}^2}{2}\right\}\right]=0$$

$$(19.66)$$

Putting

$$E = -\frac{V_{DS}}{L} \quad (19.67)$$

and from eq. (19.38) with negative sign prefixed for electron moving opposite to the electric field direction

$$E_C = -\frac{v_{Sat}}{\mu_n} \quad (19.68)$$

we get

$$\frac{d}{dV_{DS}}\left[\left(\frac{1}{1+\dfrac{-\dfrac{V_{DS}}{L}}{-\dfrac{v_{Sat}}{\mu_n}}}\right)C_{ox}\left(\frac{W}{L}\right)\mu_n\left\{(V_{GS}-V_{Th})V_{DS}-\frac{V_{DS}^2}{2}\right\}\right]=0$$

$$(19.69)$$

or

$$\frac{d}{dV_{DS}}\left[\frac{C_{ox}\left(\dfrac{W}{L}\right)\mu_n\left\{(V_{GS}-V_{Th})V_{DS}-\dfrac{V_{DS}^2}{2}\right\}}{1+\dfrac{V_{DS}\mu_n}{Lv_{Sat}}}\right]=0 \quad (19.70)$$

or

$$\frac{C_{ox}\left(\dfrac{W}{L}\right)\mu_n\left\{(V_{GS}-V_{Th})\times1-\dfrac{2V_{DS}}{2}\right\}\left(1+\dfrac{V_{DS}\mu_n}{Lv_{Sat}}\right)-\left(0+\dfrac{1\times\mu_n}{Lv_{Sat}}\right)C_{ox}\left(\dfrac{W}{L}\right)\mu_n\left\{(V_{GS}-V_{Th})V_{DS}-\dfrac{V_{DS}^2}{2}\right\}}{\left(1+\dfrac{V_{DS}\mu_n}{Lv_{Sat}}\right)^2}=0 \quad (19.71)$$

or

$$\frac{C_{ox}\left(\dfrac{W}{L}\right)\mu_n\left[(V_{GS}-V_{Th}-V_{DS})\left(1+\dfrac{V_{DS}\mu_n}{Lv_{Sat}}\right)-\dfrac{\mu_n}{Lv_{Sat}}\left\{(V_{GS}-V_{Th})V_{DS}-\dfrac{V_{DS}^2}{2}\right\}\right]}{\left(1+\dfrac{V_{DS}\mu_n}{Lv_{Sat}}\right)^2}=0 \quad (19.72)$$

or

$$\left(V_{GS} - V_{Th} - V_{DS} \right) \left(1 + \frac{V_{DS}\mu_n}{Lv_{Sat}} \right) - \frac{\mu_n}{Lv_{Sat}}$$

$$\times \left\{ \left(V_{GS} - V_{Th} \right) V_{DS} - \frac{V_{DS}^2}{2} \right\} = 0 \qquad (19.73)$$

or

$$V_{GS} - V_{Th} - V_{DS} + V_{GS}\frac{V_{DS}\mu_n}{Lv_{Sat}} - V_{Th}\frac{V_{DS}\mu_n}{Lv_{Sat}} - \frac{V_{DS}^2\mu_n}{Lv_{Sat}}$$

$$- \frac{\mu_n}{Lv_{Sat}}V_{GS}V_{DS} + V_{Th}V_{DS}\frac{\mu_n}{Lv_{Sat}} + \frac{\mu_n}{Lv_{Sat}}\frac{V_{DS}^2}{2} = 0 \qquad (19.74)$$

or

$$- \frac{\mu_n}{Lv_{Sat}}\frac{V_{DS}^2}{2} - V_{DS} + V_{GS} - V_{Th} = 0 \qquad (19.75)$$

or

$$\frac{\mu_n}{Lv_{Sat}}\frac{V_{DS}^2}{2} + V_{DS} - \left(V_{GS} - V_{Th} \right) = 0 \qquad (19.76)$$

or

$$V_{DS}^2 + 2\frac{Lv_{Sat}}{\mu_n}V_{DS} - \left(V_{GS} - V_{Th} \right)\frac{2Lv_{Sat}}{\mu_n} = 0 \qquad (19.77)$$

which is a quadratic equation in V_{DS} and of the form

$$ax^2 + bx = c = 0 \qquad (19.78)$$

with the solution

$$x = \frac{-b \pm \sqrt{b^2 - 4ac}}{2a} \qquad (19.79)$$

Hence, the solution to eq. (19.77) is

$$V_{DS} = \frac{-\frac{2Lv_{Sat}}{\mu_n} \pm \sqrt{\left(\frac{2Lv_{Sat}}{\mu_n} \right)^2 - 4(1) \times -\left(V_{GS} - V_{Th} \right)\frac{2Lv_{Sat}}{\mu_n}}}{2 \times 1}$$

$$= \frac{-\frac{2Lv_{Sat}}{\mu_n} \pm \sqrt{\left(\frac{2Lv_{Sat}}{\mu_n} \right)^2 + 4\left(V_{GS} - V_{Th} \right)\frac{2Lv_{Sat}}{\mu_n}}}{2 \times 1}$$

$$= \frac{-\frac{2Lv_{Sat}}{\mu_n} \pm \sqrt{\left(\frac{2Lv_{Sat}}{\mu_n} \right)^2 + 4\left(V_{GS} - V_{Th} \right)\frac{\left(\frac{2Lv_{Sat}}{\mu_n} \right)^2}{\frac{2Lv_{Sat}}{\mu_n}}}}{2}$$

$$= \frac{-\frac{2Lv_{Sat}}{\mu_n} \pm \sqrt{\left(\frac{2Lv_{Sat}}{\mu_n} \right)^2 \left\{ 1 + 4\left(V_{GS} - V_{Th} \right)\frac{1}{\frac{2Lv_{Sat}}{\mu_n}} \right\}}}{2}$$

$$= \frac{-\frac{2Lv_{Sat}}{\mu_n} \pm \left(\frac{2Lv_{Sat}}{\mu_n} \right)\sqrt{\left\{ 1 + 4\left(V_{GS} - V_{Th} \right)\frac{\mu_n}{2Lv_{Sat}} \right\}}}{2}$$

$$= -\frac{Lv_{Sat}}{\mu_n} \pm \left(\frac{Lv_{Sat}}{\mu_n} \right)\sqrt{\left\{ 1 + 2\left(V_{GS} - V_{Th} \right)\frac{\mu_n}{Lv_{Sat}} \right\}} \qquad (19.80)$$

Since $V_{DS} > 0$, the negative sign is dropped. Also, $V_{DS(Sat)}$ is put for V_{DS}, as mentioned earlier, giving

$$V_{DS(Sat)} = -\frac{Lv_{Sat}}{\mu_n} + \left(\frac{Lv_{Sat}}{\mu_n} \right)\sqrt{\left\{ 1 + 2\left(V_{GS} - V_{Th} \right)\frac{\mu_n}{Lv_{Sat}} \right\}} = \left(\frac{Lv_{Sat}}{\mu_n} \right)\left[\sqrt{\left\{ 1 + 2\left(V_{GS} - V_{Th} \right)\frac{\mu_n}{Lv_{Sat}} \right\}} - 1 \right]$$

$$= \frac{\left(\frac{Lv_{Sat}}{\mu_n} \right)\left[\sqrt{\left\{ 1 + 2\left(V_{GS} - V_{Th} \right)\frac{\mu_n}{Lv_{Sat}} \right\}} - 1 \right]\left[\sqrt{\left\{ 1 + 2\left(V_{GS} - V_{Th} \right)\frac{\mu_n}{Lv_{Sat}} \right\}} + 1 \right]}{\left[\sqrt{\left\{ 1 + 2\left(V_{GS} - V_{Th} \right)\frac{\mu_n}{Lv_{Sat}} \right\}} + 1 \right]}$$

$$= \frac{\left(\frac{Lv_{Sat}}{\mu_n} \right)\left[\left(\sqrt{\left\{ 1 + 2\left(V_{GS} - V_{Th} \right)\frac{\mu_n}{Lv_{Sat}} \right\}} \right)^2 - (1)^2 \right]}{\left[1 + \sqrt{\left\{ 1 + 2\left(V_{GS} - V_{Th} \right)\frac{\mu_n}{Lv_{Sat}} \right\}} \right]} \qquad (19.81)$$

$$= \frac{\left(\frac{Lv_{Sat}}{\mu_n} \right)\left\{ 1 + 2\left(V_{GS} - V_{Th} \right)\frac{\mu_n}{Lv_{Sat}} - 1 \right\}}{1 + \sqrt{\left\{ 1 + 2\left(V_{GS} - V_{Th} \right)\frac{\mu_n}{Lv_{Sat}} \right\}}} = \frac{\left(\frac{Lv_{Sat}}{\mu_n} \right) \times 2\left(V_{GS} - V_{Th} \right)\frac{\mu_n}{Lv_{Sat}}}{1 + \sqrt{\left\{ 1 + 2\left(V_{GS} - V_{Th} \right)\frac{\mu_n}{Lv_{Sat}} \right\}}} = \frac{2\left(V_{GS} - V_{Th} \right)}{1 + \sqrt{\left\{ 1 + 2\left(V_{GS} - V_{Th} \right)\frac{\mu_n}{Lv_{Sat}} \right\}}}$$

Again, since

$$2(V_{GS} - V_{Th})\frac{\mu_n}{Lv_{Sat}} > 0 \tag{19.82}$$

$$\sqrt{\left\{1 + 2(V_{GS} - V_{Th})\frac{\mu_n}{Lv_{Sat}}\right\}} > 1 \tag{19.83}$$

$$\therefore 1 + \sqrt{\left\{1 + 2(V_{GS} - V_{Th})\frac{\mu_n}{Lv_{Sat}}\right\}} > 2 \tag{19.84}$$

Hence from equation (19.81)

$$V_{DS(Sat)} < (V_{GS} - V_{Th}) \tag{19.85}$$

Note that for a long-channel MOSFET, the following conditions are satisfied during saturation so that in the saturation mode,

$$V_{GS} > V_{Th} \tag{19.86}$$

$$V_{GS} - V_{Th} = V_{DSAT} < V_{DS} \tag{19.87}$$

Clearly, the short-channel MOSFET enters the saturation mode before V_{DS} reaches $V_{GS}-V_{Th}$. It shows an extended saturation region starting from a lower value of V_{DS}. (Figure 19.5(b)).

Further, from the first line of eq. (19.81)

$$V_{DS(Sat)} = -\frac{Lv_{Sat}}{\mu_n} + \left(\frac{Lv_{Sat}}{\mu_n}\right)\left\{1 + 2(V_{GS} - V_{Th})\frac{\mu_n}{Lv_{Sat}}\right\}^{1/2} \tag{19.88}$$

By Taylor's theorem

$$(1+x)^{1/2} = 1 + \frac{x}{2} - \frac{x^2}{8} + \frac{x^3}{16} - \cdots \tag{19.89}$$

$$\therefore V_{DS(Sat)} = -\frac{Lv_{Sat}}{\mu_n} + \left(\frac{Lv_{Sat}}{\mu_n}\right)\left[1 + \frac{1}{2}\times 2(V_{GS} - V_{Th})\frac{\mu_n}{Lv_{Sat}} - \frac{1}{8}\right.$$

$$\left.\times\left\{2(V_{GS} - V_{Th})\frac{\mu_n}{Lv_{Sat}}\right\}^2 + \cdots\right]$$

$$= -\frac{Lv_{Sat}}{\mu_n} + \left(\frac{Lv_{Sat}}{\mu_n}\right) + (V_{GS} - V_{Th})\frac{\mu_n}{Lv_{Sat}}$$

$$\times\left(\frac{Lv_{Sat}}{\mu_n}\right) - \frac{1}{8}\times 4\left\{(V_{GS} - V_{Th})\frac{\mu_n}{Lv_{Sat}}\right\}^2\left(\frac{Lv_{Sat}}{\mu_n}\right) + \cdots$$

$$= V_{GS} - V_{Th} - \frac{(V_{GS} - V_{Th})^2}{2}\left(\frac{\mu_n}{Lv_{Sat}}\right) + \cdots$$

$$= (V_{GS} - V_{Th})\left\{1 - \frac{(V_{GS} - V_{Th})\left(\frac{\mu_n}{Lv_{Sat}}\right)}{2} + \cdots\right\} \tag{19.90}$$

To know how the long-channel MOSFET equation is modified by inserting the $V_{DS(Sat)}$ expression obtained with inclusion of the effect of velocity saturation, we write from equation (19.41)

$$I_{DS} = C_{ox}\left(\frac{W}{L}\right)\mu_n\left\{(V_{GS} - V_{Th})V_{DS} - \frac{V_{DS}^2}{2}\right\}$$

$$= C_{ox}\left(\frac{W}{L}\right)\mu_n\left\{(V_{DS})V_{DS} - \frac{V_{DS}^2}{2}\right\}$$

$$= \left(\frac{1}{2}\right)C_{ox}\left(\frac{W}{L}\right)\mu_n V_{DS}^2$$

$$= \left(\frac{1}{2}\right)C_{ox}\left(\frac{W}{L}\right)\mu_n V_{DS(Sat)}^2 \tag{19.91}$$

putting

$$V_{GS} - V_{Th} = V_{DS} \tag{19.92}$$

and

$$V_{DS} = V_{DS(Sat)} \tag{19.93}$$

Substituting for $V_{DS(Sat)}$ from eq. (19.90) in eq. (19.91), we have

$$I_{DS} = \left(\frac{1}{2}\right)C_{ox}\left(\frac{W}{L}\right)\mu_n\left[(V_{GS} - V_{Th})\left\{1 - \frac{(V_{GS} - V_{Th})\left(\frac{\mu_n}{Lv_{Sat}}\right)}{2} + \cdots\right\}\right]^2$$

$$= \left(\frac{1}{2}\right)C_{ox}\left(\frac{W}{L}\right)\mu_n (V_{GS} - V_{Th})^2\left\{1 - 2\times\frac{(V_{GS} - V_{Th})\left(\frac{\mu_n}{Lv_{Sat}}\right)}{2} + \cdots\right\}$$

$$= \left(\frac{1}{2}\right)C_{ox}\left(\frac{W}{L}\right)\mu_n (V_{GS} - V_{Th})^2\left\{1 - (V_{GS} - V_{Th})\left(\frac{\mu_n}{Lv_{Sat}}\right) + \cdots\right\}$$

$$= \left(\frac{1}{2}\right)C_{ox}\left(\frac{W}{L}\right)\mu_n (V_{GS} - V_{Th})^2\left\{\frac{1}{1 + (V_{GS} - V_{Th})\left(\frac{\mu_n}{Lv_{Sat}}\right)}\right\} \tag{19.94}$$

since

$$1 - x = (1+x)^{-1} = \frac{1}{1+x} \tag{19.95}$$

Thus the well-known equation for long-channel MOSFET operation in the saturation region, viz.,

$$I_{DS} = \left(\frac{1}{2}\right)C_{ox}\left(\frac{W}{L}\right)\mu_n (V_{GS} - V_{Th})^2 \tag{19.96}$$

is modified for velocity saturation by multiplication with the

$$\text{Correction factor} = \frac{1}{1 + (V_{GS} - V_{Th})\left(\dfrac{\mu_n}{Lv_{Sat}}\right)} \tag{19.97}$$

It will be shown that the effect of velocity saturation on an ideal long-channel MOSFET obeying the square law is modeled by including a resistance R_{SX} in series with the source terminal. Let the gate–source voltage of the ideal long-channel MOSFET be V'_{GS}. For this MOSFET, we may write the current–voltage equation as

$$I_{DS} = \left(\frac{1}{2}\right)C_{ox}\left(\frac{W}{L}\right)\mu_n\ (V'_{GS} - V_{Th})^2 \tag{19.98}$$

Let the gate–source voltage of the MOSFET affected by velocity saturation be V_{GS}, and suppose this voltage is written as the sum

$$V_{GS} = V'_{GS} + \text{Voltage drop across a}$$
$$\text{series resistor } R_{SX} = V'_{GS} + I_{DS}R_{SX} \tag{19.99}$$

or

$$V'_{GS} = V_{GS} - I_{DS}R_{SX} \tag{19.100}$$

Then from eqs. (19.98) and (19.100), the operation of velocity-saturated MOSFET is described by the ideal long-channel MOSFET obeying the relation

From eqs. (19.94) and (19.103), it is easy to write

$$1 + (V_{GS} - V_{Th})\left(\frac{\mu_n}{Lv_{Sat}}\right) = 1 + R_{SX}C_{ox}\left(\frac{W}{L}\right)\mu_n\ (V_{GS} - V_{Th}) \tag{19.104}$$

or

$$(V_{GS} - V_{Th})\left(\frac{\mu_n}{Lv_{Sat}}\right) = R_{SX}C_{ox}\left(\frac{W}{L}\right)\mu_n\ (V_{GS} - V_{Th}) \tag{19.105}$$

or

$$\frac{1}{v_{Sat}} = R_{SX}C_{ox}W \tag{19.106}$$

giving the expression for the resistor R_{SX} as

$$R_{SX} = \frac{1}{v_{Sat}\ C_{ox}W} \tag{19.107}$$

Thus velocity saturation affects a MOSFET in the same way as if a resistor R_{SX} is connected in series with the source of an ideal long-channel MOSFET obeying the square law (Figure 19.6). Therefore, parasitic resistance reduction becomes more crucial in short-channel MOSFETs.

In a simplified first-order model for velocity saturation, one starts with the pre-supposition that $V_{DS(Sat)}$ is attained at a value when the electric field equals the critical field so that we can write

$$I_{DS(\text{Velocity-saturated})} = \left(\frac{1}{2}\right)C_{ox}\left(\frac{W}{L}\right)\mu_n\ (V_{GS} - I_{DS}R_{SX} - V_{Th})^2 = \left(\frac{1}{2}\right)C_{ox}\left(\frac{W}{L}\right)\mu_n\ \{(V_{GS} - V_{Th}) - I_{DS}R_{SX}\}^2$$

$$\tag{19.101}$$

$$= \left(\frac{1}{2}\right)C_{ox}\left(\frac{W}{L}\right)\mu_n\ \{(V_{GS} - V_{Th})^2 - 2(V_{GS} - V_{Th})I_{DS}R_{SX} + (I_{DS}R_{SX})^2\}$$

$$= \left(\frac{1}{2}\right)C_{ox}\left(\frac{W}{L}\right)\mu_n\ \{(V_{GS} - V_{Th})^2 - 2(V_{GS} - V_{Th})I_{DS}R_{SX}\}$$

neglecting the term $(I_{DS}R_{SX})^2$. Equation (19.101) is rewritten as

$$I_{DS(\text{Velocity-saturated})} = \left(\frac{1}{2}\right)C_{ox}\left(\frac{W}{L}\right)\mu_n(V_{GS} - V_{Th})^2 - \left(\frac{1}{2}\right)C_{ox}\left(\frac{W}{L}\right)\mu_n\{2(V_{GS} - V_{Th})I_{DS}R_{SX}\}$$

$$\tag{19.102}$$

$$= \left(\frac{1}{2}\right)C_{ox}\left(\frac{W}{L}\right)\mu_n(V_{GS} - V_{Th})^2\left(1 - \frac{2I_{DS}R_{SX}}{V_{GS} - V_{Th}}\right)$$

Substituting for I_{DS} for long-channel MOSFET in saturation mode from eq. (19.96) in eq. (19.102) we get

$$I_{DS(\text{Velocity-saturated})} = \left(\frac{1}{2}\right)C_{ox}\left(\frac{W}{L}\right)\mu_n(V_{GS} - V_{Th})^2\left\{1 - \frac{2R_{SX}}{V_{GS} - V_{Th}} \times \left(\frac{1}{2}\right)C_{ox}\left(\frac{W}{L}\right)\mu_n\ (V_{GS} - V_{Th})^2\right\} = \left(\frac{1}{2}\right)C_{ox}\left(\frac{W}{L}\right)\mu_n(V_{GS} - V_{Th})^2$$

$$\times\left\{1 - R_{SX}C_{ox}\left(\frac{W}{L}\right)\mu_n\ (V_{GS} - V_{Th})\right\} = \frac{\left(\frac{1}{2}\right)C_{ox}\left(\frac{W}{L}\right)\mu_n(V_{GS} - V_{Th})^2}{\left\{1 - R_{SX}C_{ox}\left(\frac{W}{L}\right)\mu_n\ (V_{GS} - V_{Th})\right\}^{-1}} = \frac{\left(\frac{1}{2}\right)C_{ox}\left(\frac{W}{L}\right)\mu_n(V_{GS} - V_{Th})^2}{1 + R_{SX}C_{ox}\left(\frac{W}{L}\right)\mu_n\ (V_{GS} - V_{Th})}$$

$$\tag{19.103}$$

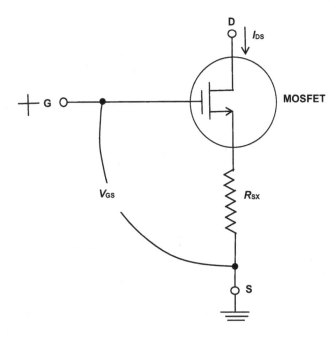

$$k_1 = C_{ox} W \mu_n \left(\frac{v_{Sat}}{\mu_n} \right) \qquad (19.112)$$

and

$$k_2 = \frac{1}{2} V_{DS(Sat)} \qquad (19.113)$$

are constants. Equation (19.111) shows the linear dependence of saturated drain–source current $I_{DS(Sat)}$ on $(V_{GS} - V_{Th})$ of a short-channel MOSFET in opposition to the normal square law dependence without velocity saturation. Although the model is first order only, it reveals the impact of velocity saturation on MOSFET as a change in MOSFET transfer characteristic from the square law to linear variation. Another major difference from the classical MOSFET equation is that the drain–source current is independent of the channel length because of the absence of L in constant k_1 whereas we know that it is normally inversely proportional to the channel length.

FIGURE 19.6 Modeling velocity saturation in a MOSFET by connecting a resistance R_{SX} in series with the source terminal.

$$I_{DS} = I_{DS(Sat)} \qquad (19.108)$$

when

$$V_{DS} = V_{DS(Sat)} = -LE_C = -L\left(-\frac{v_{Sat}}{\mu_n} \right) = \frac{Lv_{Sat}}{\mu_n} \qquad (19.109)$$

since

$$E_C = -\frac{V_{DS(Sat)}}{L} \qquad (19.110)$$

Then using eq. (19.68) and assuming the I_{DS} equation (19.41) for the long-channel MOSFET to be valid, put $V_{DS} = V_{DS(Sat)}$ in eq. (19.41) getting

$$\begin{aligned} I_{DS(Sat)} &= C_{ox} \frac{W}{L} \mu_n \left\{ (V_{GS} - V_{Th}) V_{DS(Sat)} - \frac{V_{DS(Sat)}^2}{2} \right\} \\ &= C_{ox} \frac{W}{L} \mu_n V_{DS(Sat)} \left\{ (V_{GS} - V_{Th}) - \frac{1}{2} V_{DS(Sat)} \right\} \\ &= C_{ox} \frac{W}{L} \mu_n \left(\frac{Lv_{Sat}}{\mu_n} \right) \left\{ (V_{GS} - V_{Th}) - \frac{1}{2} V_{DS(Sat)} \right\} \\ &= C_{ox} W \mu_n \left(\frac{v_{Sat}}{\mu_n} \right) \left\{ (V_{GS} - V_{Th}) - \frac{1}{2} V_{DS(Sat)} \right\} \\ &= k_1 \left\{ (V_{GS} - V_{Th}) - k_2 \right\} \qquad (19.111) \end{aligned}$$

where equation (19.109) for $V_{DS(Sat)}$ is applied,

19.1.3 Hot Carrier Effects

Hot carrier injection is caused by the acceleration of electrons and holes in regions of high electric field in the FET device. These high-velocity carriers originate either via the drain avalanche, through the channel, the substrate, or by secondary mechanisms. Hot carriers injected into the gate oxide, either by direct tunneling through thin oxide or by field-assisted Fowler-Nordheim tunneling, are trapped inside the oxide. The resulting space charge accumulation shifts the threshold voltage of the device.

Formation of hot carriers is inhibited by decreasing the electric field between the channel and the drain. This is achieved by reducing the doping concentration in the drain region proximate to the channel as compared to the main drain area. The resulting FET structure is called the lightly doped drain (LDD) structure (Figure 19.7). To avoid process complexity, the doping concentration in the source region near the channel is also taken less than that in the main source area. At the same time, P+ halo or pocket implants are formed for suppressing punchthrough.

As DIBL effect minimization needs a high substrate doping and hot carrier effect is subdued by using an LDD, both requirements are met by adopting a super-steep retrograde (SSR) channel doping profile. The retrograded channel profile (RCP) is a stepped profile with a lightly doped region below the surface and a heavily doped region buried deeper inside the substrate (De and Osburn 1999).

19.1.4 Carrier Mobility Degradation by Surface Scattering

In a MOSFET, the carrier motion from the source to the drain takes place under the combined simultaneous influence of a horizontal source-to-drain electric field determined by the applied drain–source voltage V_{DS} and a vertical gate-to-source electric field dependent on the applied gate-to-source voltage V_{GS}. On decreasing the channel length, the lateral component of the horizontal source-to-drain field increases. To compensate the effect of this increased lateral component, the vertical gate-to-source

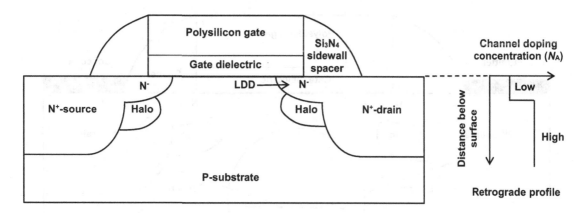

FIGURE 19.7 Lightly doped N⁻-source/drain extensions, P⁺ halo implants, and retrograde doping profile in an N-channel MOSFET.

field must be raised. Consequently, the carriers in the channel impinge with higher velocities at the silicon-silicon dioxide interface. The enhanced carrier collisions with interface lower the carrier mobility in the channel through scattering effects (Figure 19.8). Therefore, the electric biases must be precisely decreased in accordance with the MOSFET dimensional shrinkage to minimize the mobility decline.

19.1.5 Regenerative Feedback between Avalanche Breakdown and Parasitic Bipolar Transistor

The high drain–source electric field in a short-channel MOSFET may trigger avalanche breakdown by carrier multiplication through impact ionization and electron–hole pair generation. This breakdown is supported by the parasitic bipolar transistor formed by the N⁺-source, the P-substrate, and the N⁺-drain in an N-channel MOSFET (Figure 19.9). The source-substrate junction is the emitter-base junction of this transistor. The substrate-drain junction acts as its base-collector junction. The electrons flowing in the MOSFET channel migrate to the drain while the holes move to the substrate. At high drain currents, the source-substrate junction becomes forward biased and the amplified substrate-drain current adds to the avalanche current at the drain by a positive feedback mechanism lowering the drain–source breakdown voltage. Proper scaling therefore ensures that

the electric field is restricted within safe limit by reducing the applied biases proportionately.

19.2 From Bulk Silicon-MOSFET to SOI-MOSFET Technology

Introspection of the short-channel effects reveals that they weaken the controlling ability of the gate over the channel and create undesirable leakage currents that seriously impair the functioning of the MOSFET by making it difficult to deplete the channel for current cessation. Therefore, the elimination of these effects required a technological innovation. A major technological breakthrough involved the use of silicon-on-insulator (SOI) wafers as the starting manufacturing material for MOSFETs in place of the bulk silicon wafers used for fabricating the traditional planar MOSFETs. The SOI wafers have a trilayer structure consisting of a bottom thick silicon handle layer, a buried oxide layer, and a thin top silicon active device layer. The SOI wafer is made by wafer bonding or separation-by-implantation of oxygen (SIMOX) techniques.

The buried oxide reduces the leakage current flow across the source/substrate and drain/substrate junctions. It also decreases the coupling capacitance between the channel and the substrate, thereby providing faster device operation.

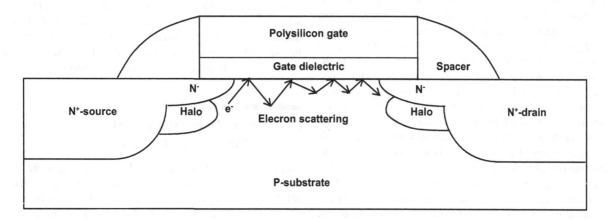

FIGURE 19.8 Electron mobility reduction caused by enhanced scattering of electrons at the Si/SiO₂ interface by the vertical gate-to-source field of a small-geometry MOSFET.

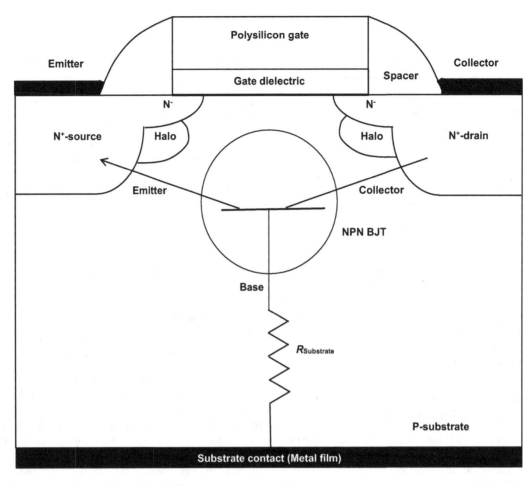

FIGURE 19.9   Parasitic NPN bipolar junction transistor in a MOSFET.

There are two principal variants of SOI-MOSFET: the partially depleted SOI-MOSFET (PD-SOI-MOSFET) and fully depleted SOI-MOSFET (FD-SOI-MOSFET) (Rahou et al. 2013).

19.2.1 PD-SOI-MOSFET

As the name implies, the PD-MOSFET has a thicker silicon body which is not fully depleted during operation (Figure 19.10). The residual undepleted silicon region acts like a floating body (a body with floating potential) because it resides over the insulating buried oxide. When the drain–source voltage is increased, electrons flowing from the source to the drain create electron–hole pairs near the drain. The electrons migrate to the drain while the holes are attracted to the lowest potential region which is the floating body. With further increase in drain–source voltage, the source-floating body potential rises and this diode becomes forward biased injecting more carriers into the drain. Consequently, the threshold voltage decreases. The output drain–source current rises. The transconductance too increases. The phenomenon is visible as a kink in the current–voltage characteristics of the device and is accordingly named as the kink effect. The kink effect is the main shortcoming of the PD-MOSFET. Notwithstanding, the PD-MOSFET allows easy manufacturing by processes compatible with the bulk silicon counterpart, which is a noteworthy advantage of this structure.

19.2.2 FD-SOI-MOSFET

The FD-MOSFET variant (Figure 19.11) is able to get rid of the kink effect problem. In this MOSFET, the silicon channel layer is very thin. By virtue of its ultrathin silicon body, it operates by maintaining the channel fully in the depletion mode. There is no undepleted region left so that there is no floating body. Moreover, its channel is made of undoped silicon. The movement of carriers in undoped silicon is considerably easier than in doped materials. Hence, the carrier mobility is enhanced enabling higher drain–source current and transconductance. Overall, the FD-MOSFET displays the current–voltage characteristics better than the PD-MOSFET. However, the realization of precise silicon thickness during its manufacturing is a critical process step.

19.3 From 2D MOSFET to 3D MOSFET: The FinFET

Further refinement of the MOSFET targeting more aggressive downscaling paved the way toward the development of a 3D architecture resembling the shape of the back fin of a fish to stabilize the motion of the fish during swimming, hence called a FinFET (Figure 19.12). Hitherto, the source, drain, and gate regions of the MOSFET were confined to two dimensions

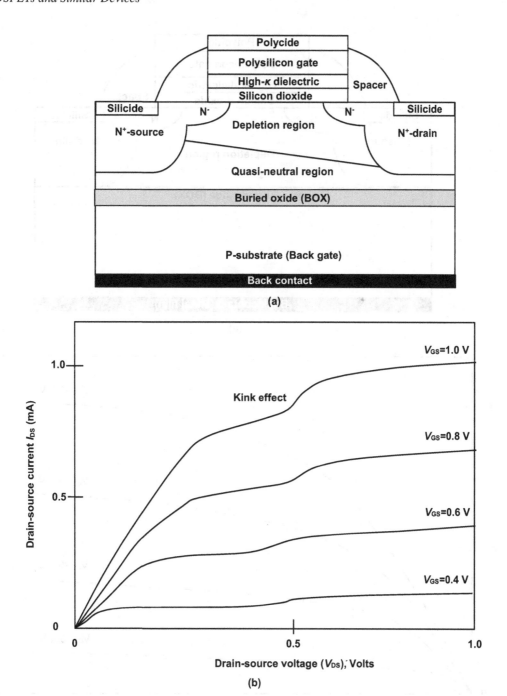

FIGURE 19.10 Partially depleted MOSFET: (a) schematic cross section and (b) output characteristics.

mainly because of process simplicity. In the FinFET structure, the source–drain silicon layer serving as the channel projects outward from the body of the substrate instead of the standard practice of its construction below the surface. The gate dielectric and the electrode are wrapped around the source–drain silicon so that the gate can exercise its electrostatic influence from the left and right sides of the silicon in addition to its upper surface. In the planar MOSFET, only the top surface of the silicon was available for gating action.

Although the FinFET has a single gate comprising the left, right, and top portions, it can be viewed as a multigate device, actually a three-gate device by counting its three constituent

segments (Colinge 2004). Needless to say the FinFET gate working from the three sides of silicon is more effective than the planar MOSFET gate operating only from one side. Therefore, it is a befitting remedy for short-channel effects.

The drain–source current is determined by the channel aspect ratio (channel width/channel length). For the FinFET, the channel width is expressed as

$$\text{Channel width} = 2 \times \text{Fin height} + \text{Fin width} \qquad (19.114)$$

Therefore, the channel width is increased by taking wider and higher fins but there is a limit to increasing either of

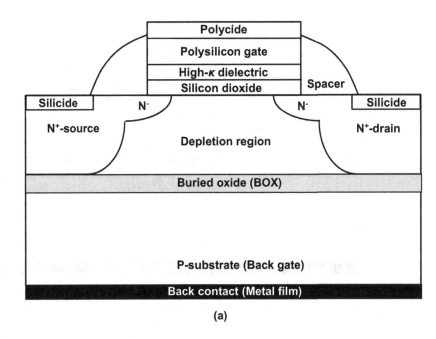

(a)

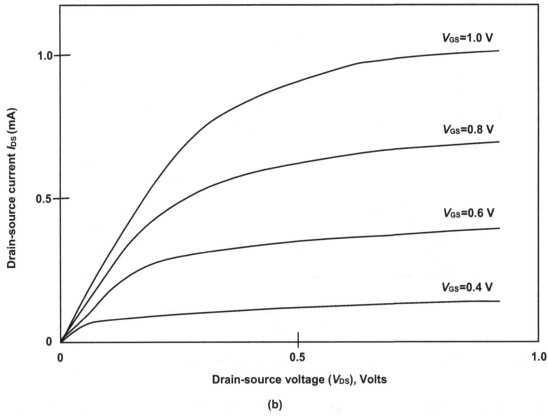

(b)

FIGURE 19.11 Fully depleted MOSFET: (a) schematic cross section and (b) output characteristics.

these parameters. In a broader fin, the gate loses its effective-
ness while higher fins are problematic due to their fragility.
Alternatively, the drain–source current can be augmented by
using several fins connected in parallel. A multiple fin design

with *n* fins can provide *n* times the current available with a
single fin.

In the beginning, the FinFETs were fabricated on SOI wafers
but FinFETs have also been realized on bulk silicon wafers.

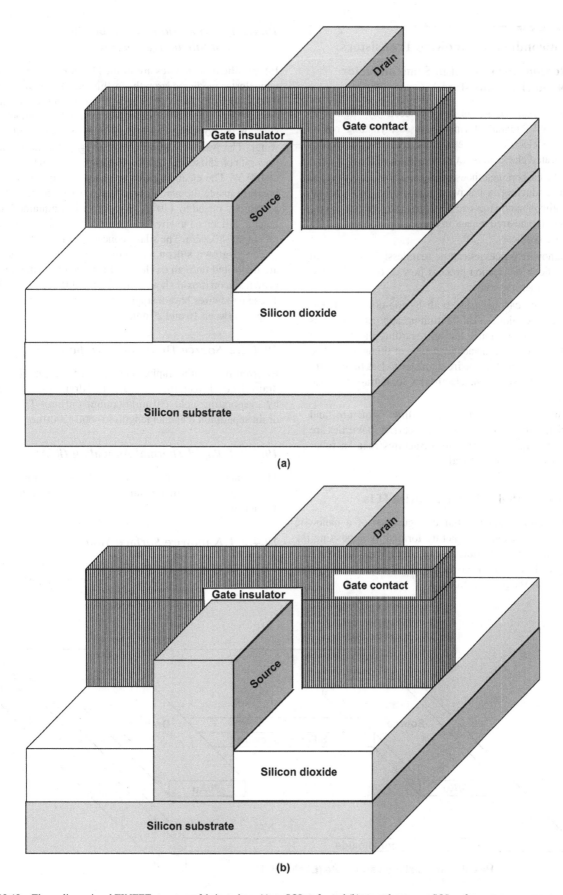

FIGURE 19.12 Three-dimensional FINFET structures fabricated on: (a) an SOI wafer and (b) a regular or non-SOI wafer.

19.4 Semiconductor Nanowire Transistors

19.4.1 Reasons for Interest in Semiconductor Nanowire Transistors

i. Silicon nanowires can be produced by high-yield repeatable, manufacturing processes, yielding reproducible electrical properties for the development of integrated electronics (Lu et al. 2008).

ii. Unlike carbon nanotubes, which may be conducting or semiconducting in behavior, exclusively semiconducting silicon nanowires can be prepared by the incorporation of the desired dopant in the required concentration during synthesis.

iii. Si nanowire processes can be seamlessly introduced in the silicon fabrication process flow as opposed to the use of other materials.

iv. Nanowires can be made both by top-down and bottom-up techniques. The bottom-up approach provides precision dimensional control regarding the nanowire width far beyond the achievements of lithography. The controllability of nanowire diameter to less than 10 nm can enable MOS nanoelectronics for ultra-short gate lengths.

v. Owing to the excellent crystalline structure and smooth surfaces obtained, higher carrier mobilities are expected in nanowire heterostructures than in other nanostructures of identical size.

19.4.2 Back-Gated Silicon Nanowire FETs

The most common and familiar configuration of a nanowire transistor is the back-gated structure formed by depositing the nanowires on a silicon dioxide/degeneratively doped silicon substrate (Figure 19.13). In this geometry, the heavily doped silicon substrate acts as the gate electrode.

19.4.2.1 Preparation and Deposition of Silicon Nanowires

P-type silicon nanowires are formed by boron doping during the nanocluster-mediated growth. In this method (Cui et al. 2001), gold nanoclusters act as the catalytic agent. The vapor-phase reactant is silane gas (SiH_4). Negatively charged Au nanoclusters are stuck on the positively charged poly-L-lysine deposited on Si/SiO_2 wafer. The wafer is cleaned in an oxygen plasma at an oxygen flow rate of 250 sccm. Plasma cleaning is done in 0.7 torr vacuum at 100 W. The cleaned wafer is loaded in a quartz reactor. The pressure inside the reactor is reduced to 100 mtorr and the temperature is raised to 440°C. Argon flow is maintained inside the reactor. Silicon nanowires are grown for 5–10 min at a silane flow rate of 10–80 sccm. The silane concentration is 10% in helium.

The as-grown silicon nanowires from an ethanol suspension are deposited onto an oxidized silicon wafer having a thermally grown silicon dioxide layer of thickness 600 nm (Cui et al. 2003). These nanowires have a single-crystal structure. Their diameter ranges between 10 and 20 nm.

19.4.2.2 Source/Drain Metallization

Electron-beam lithography is employed for creating the source/drain electrode pattern. The source and drain contacts are formed by evaporating gold (50 nm)/titanium (50 nm). The source-to-drain separation (channel length) is ~800–2000 nm.

19.4.2.3 Rapid Thermal Annealing (RTA)

The contacts are annealed at 300°C–600°C for 3 min. The annealing is done in forming gas (10% hydrogen in helium) ambience.

19.4.2.4 Nanowire Surface Modification

It is done with 4-nitrophenyl octadecanoate, $C_{24}H_{39}NO_4$ to form a Si-O-C ester linkage with good stability and non-polar behavior.

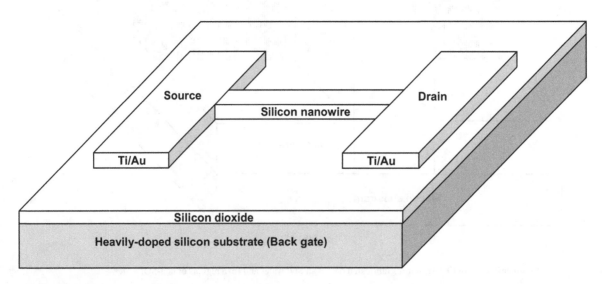

FIGURE 19.13 Back-gated silicon nanowire FET.

19.4.2.5 Nanowire FET Characteristics

Prior to annealing, the two-terminal nanowire resistance shows a wide spread from $< M\Omega$ to $> G\Omega$. The average value of resistance is 160 $M\Omega$. Post annealing, the spread span contracts to 0.1–10 $M\Omega$. The average value is 0.62 $M\Omega$, which is 0.0039 (0.62/160) times less than the value prior to annealing.

Before surface modification, the FET conductance shows a feeble dependence on gate voltage V_G. After surface modification, it acquires sensitivity to V_G. The FET is switched off at $V_G = 2.5$ V. The current on/off ratio is 10^4. The average transconductance and mobility values before chemical modification are 45 nS and 30 cm^{2}V^{-1}s^{-1}, respectively. These values increase to 800 nS and 560 cm^{2}V^{-1}s^{-1} after modification. The corresponding maximum values are 2000 nS and 1350 cm^{2}V^{-1}s^{-1}, respectively (Cui et al. 2003).

19.4.3 Junctionless Multigate Nanowire FETs

Strikingly similar to the FinFET is the gate-all-around (GAA) or surround-gate nanowire FET concept that can be realized using free-standing nanowires. It can be made with semi-cylindrical or fully cylindrical gates. Particularly, the junctionless nanowire FET variant (Figure 19.14) with gate on three sides of the channel region is a strong contender to support the relentless downscaling of the MOSFET.

The junctionless transistor concept (Lee et al. 2009, 2010; Colinge et al. 2010) mimics the operation of the accumulation-mode MOSFET device designed with a thin channel semiconductor channel region. As we know, the accumulation mode MOSFET is a normally on transistor in which there is a high concentration of carriers in the accumulation layer and in the contact areas. Moreover, the polarity of charge carriers in the accumulation layer and the contact areas is same. Therefore, a heavily doped nanowire with comparable carrier concentrations in the nanowire and the end zones is essentially a filamentary resistor with a uniform doping along its length. The middle portion of the nanowire may be covered with an insulating film capped with the metal film for gate contact. Metallized zones can be formed for source/drain contacts near the extremities of the nanowire. The resulting structure is basically a gated resistor structure. It will be conducting in the absence of an applied gate voltage due to the high doping concentration in the nanowire but can be made non-conducting by the application of a gate voltage to deplete the nanowire of carriers. Based on this reasoning, a nanowire FET can be constructed without introducing any P–N junctions. The absence of P–N junctions yields the enormous benefit of process simplification on the one hand and avoids doping-induced defects in the silicon on the other hand, thereby assuring high carrier lifetimes and lower leakage currents in the device.

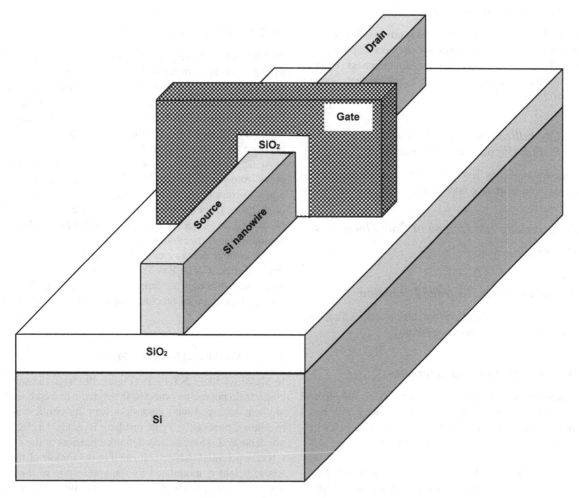

FIGURE 19.14 FINFET-like junctionless multigate silicon nanowire FET.

The process steps undertaken by Colinge et al. 2010 in the fabrication of trigate N-type nanowire transistors are discussed below.

19.4.3.1 Electron-Beam Lithography for Nanowire Delineation

Commencing the process with an SOI wafer, the silicon nanowire is defined by electron-beam lithography. The nanowire in the shape of a nanoribbon is 10 nm thick with width of a few tens of nanometers.

19.4.3.2 Gate Dielectric Formation

Silicon dioxide of thickness 10 nm is grown on the nanowire.

19.4.3.3 Nanowire Doping

For N-type nanowires, arsenic implantation is performed. The dopant concentration is 2×10^{19} atoms cm^{-3}. The high doping concentration in nanowires is indispensable for two reasons. First, a high drain–source current is achieved. Second, a low contact resistance is obtained at the source and drain electrodes.

19.4.3.4 a-Si Deposition on the Gate Oxide and Its Doping

Amorphous silicon is deposited on the SiO_2-covered Si nanowire. This deposition is done in a low-pressure chemical vapor deposition (LPCVD) reactor at 550°C. The a-Si film thickness is 50 nm. This film is heavily P-doped with boron by ion implantation at a dose of 2×10^{14} ions cm^{-2}.

19.4.3.5 Crystallization of a-Si to Make P+-Polysilicon Gate Electrodes

This is done by annealing at 900°C for ½ h in nitrogen atmosphere. By this annealing, the dopants are also activated.

19.4.3.6 Patterning and Etching of Gate Electrodes

Reactive-ion etching is used.

19.4.3.7 Source/Drain Electrode Formation

Following the deposition of a silicon dioxide protective film, contact holes are made. TiW contacts are formed.

19.4.3.8 Nanowire FET Characteristics

A silicon nanowire that is covered on its top surface and left and right surfaces with the gate dielectric and P+ contact layer has a trigate configuration similar to a FinFET. Nanowire FETs outscore classical MOSFETs on three vital parameters:

i. They exhibit near-ideal subthreshold slope of 64 mV/decade which is maintained within a few percent of this value in the temperature range 225–475 K.

The temperature coefficient of the threshold voltage is −1.5 mV/°C. For comparison bulk Si MOSFETs show a slope of ~80 mV/decade while the best SOI trigate devices show a slope of 63 mV/decade.

ii. The leakage current of the junctionless nanowire FET is exceedingly low $< 10^{-15}$ A and the current on/off ratio is 10^6 within $V_G = 0$ V and ± 1 V.

iii. The electron mobility of $100 \, cm^2 V^{-1} s^{-1}$ at 300 K in a nanowire FET is lower than that in a trigate FET but decreases by only 6% against 36% reduction in mobility in a trigate FET at 473 K.

Exactly the same steps are followed in the fabrication of trigate P-type nanowire transistors except that BF_2 is used during ion implantation for P-type doping and a-Si is N+-doped using arsenic ions.

19.5 CNT-FETs

These FETs are of two types (Figure 19.15): Schottky-barrier CNT-FETs (SB-CNT-FETs) and MOSFET-like CNT FETs (Kordrostami and Sheikhi 2010). The two types of FETs differ in the ways contacts are formed at the two ends of the CNT for source and drain electrodes, Schottky contacts or ohmic contacts.

19.5.1 SB-CNT-FETs

In SB-CNT-FETs (Figure 19.15(a)), the source/drain contacts to the CNT are established by metal film deposition so that the source and drain electrodes act as Schottky-barrier diodes. The CNT is covered with the gate dielectric and the gate electrode. When a gate potential is applied, the potential barriers at the source/drain terminals are lowered leading to current flow between the source and the drain.

At negative and low voltages, the tunneling probability of holes from the drain side into the valence band is high so that a hole current flows. At high voltages, the tunneling probability of electrons from the source side to the conduction band is high producing an electronic current. Thus the SB-CNT-FET works by ambipolar carrier transport.

The current delivery capability and hence the transconductance of these CNT-FETs are severely limited by the tunneling mechanism through Schottky barriers. Further, the ambipolar transport mechanism precludes their utilization in CMOS logic circuits.

19.5.2 MOSFET-Like CNT FETs

In MOSFET-like CNT FETs (Figure 19.15(b)), the source/drain contacts do not use any metals. Instead, the two ends of the CNT are heavily doped with potassium forming ohmic contacts. As mentioned previously, a gate insulator layer and an electrode film are deposited. There are no Schottky barriers at the ends of the CNT. A potential barrier is formed in the middle of the channel whose height is modulated by the application of the gate voltage to produce current flow between the two ends of the CNT. This type of CNT FETs operates by the unipolar transport of

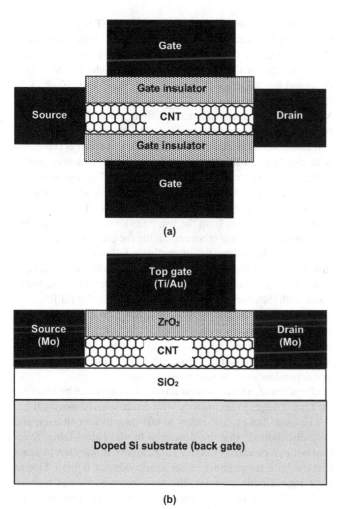

FIGURE 19.15 Types of CNT FET: (a) SB-CNT-FET and (b) MOSFET-like CNT FET with the use of a higher dielectric constant material like ZrO_2 as gate dielectric.

carriers similar to the normal MOSFETs. Due to the absence of SB, the ON-current is high and OFF-leakage current is low (Raychowdhury et al. 2006).

The MOSFET-like CNT-FET fabrication process is as follows: The process (Javey et al. 2002) is based on the technique of patterned growth of a carbon-nanotube bridge suspended across molybdenum electrodes on raised platforms of silicon dioxide (Franklin et al. 2002).

19.5.2.1 Patterned CNT Growth

A doped silicon wafer is taken to serve as the back gate. It is covered with a thermally grown silicon dioxide layer of thickness 500 nm. The oxide layer is the back gate dielectric. A 50 nm thick molybdenum film is deposited by sputtering (Franklin et al. 2002). After defining the Mo pattern by photolithography, molybdenum is removed from the areas uncovered with photoresist by reactive ion etching (RIE). The RIE is done with sulfur hexafluoride (SF_6) and chloropentafluoroethane (C_2ClF_5). Thus two molybdenum electrodes are obtained that will act as the source/drain electrodes. The source-to-drain distance is 3 μm.

Photoresist removal from Mo electrodes is followed by coating with poly(methyl methacrylate) (PMMA) (thickness 1.6 μm). In PMMA, wells are formed by deep UV or electron-beam lithography. An alumina-supported Fe catalyst in methanol is spin coated over developed PMMA film. PMMA is lifted off in acetone leaving behind two Fe catalyst islands on the two Mo electrodes. These Fe catalyst islands will initiate the CNT growth.

The CNT synthesis is performed in a CVD reactor under combined methane (72 mL min^{-1}) and hydrogen (10 mL min^{-1}) flow. The growth temperature is 900°C and the time of CNT growth is 5 min. During the period when the CVD reactor is heated or cooled, pure hydrogen flow is maintained to avoid oxidation of Mo electrodes (Franklin et al. 2002).

19.5.2.2 Zirconium Oxide Top-Gate Formation by Atomic Layer Deposition

For the atomic layer deposition (ALD) process, the precursor used is zirconium (IV) chloride ($ZrCl_4$), the oxidizer is water (H_2O), and the carrier gas is nitrogen (Javey et al. 2002). The base pressure is 10^{-8} torr. The process pressure is 0.5 torr and the temperature is kept at 300°C. During the ALD process, the precursor and oxidizer pulses of duration 1–2 s are alternated. ZrO_2 film of thickness 0.06 nm is deposited in each pulse cycle. The ZrO_2 film thickness is 8 nm.

19.5.2.3 Top-Gate Electrode Deposition

The gate electrode geometry is defined by electron-beam lithography and Ti/Au film of thickness 60 nm is deposited. The photoresist is lifted off to form the top-gate electrode of width 2 μm.

19.5.2.4 MOS-Like CNT-FET Characteristics

The as-formed CNT FETs are P-type devices. The top-gated devices show a subthreshold swing of 70 mV/decade and a transconductance of 12 μS/tube, which is normalized by the CNT width (2 nm) to 8000 S m^{-1}. The hole mobility is 3000 cm^{2}V^{-1}s^{-1}. The back-gated device has a subthreshold swing of 1–2 V/decade.

Heating in molecular hydrogen at 400°C for 1 h transforms the as-formed P-type top-gated CNT FETs into N-type CNT FETs with a subthreshold swing of 70–100 mV/decade. The transconductance is 600 S m^{-1} and the mobility of electrons is 1000 cm^{2}V^{-1}s^{-1} (Javey et al. 2002).

19.6 Discussion and Conclusions

Four types of nanoFETs were covered in the chapter: the Si planar MOSFET, the FinFET, the Si nanowire FET, and the CNT FET. We first considered the silicon MOSFET. A short-channel MOSFET is one in which the channel length has a size comparable with the depletion widths of the source/drain junctions. Short-channel effects are the harmful effects observed in such a MOSFET. A prominent short-channel effect is the lowering of the source-channel potential barrier because this barrier, which was solely controlled by the gate voltage in a long-channel MOSFET, becomes accessible to the drain voltage with the drain terminal coming closer to the source terminal in a short-channel MOSFET. Consequently the threshold voltage of the MOSFET decreases and the leakage current increases. Punchthrough voltage breakdown occurs when the depletion regions of the source and drain junctions touch each other.

The threshold voltage is written as the sum of three components: flat band voltage, twice the bulk potential, and the voltage drop due to charge in the depletion region below the gate oxide. The voltage drop due to charge in the depletion region is affected by DIBL. Instead of a rectangular area to be depleted by the gate, only a trapezoidal area is to be depleted in a short-channel device with the triangular areas cut off from the rectangle falling under the supervision of the source/drain. The threshold voltage reduction ΔV_{Th} is calculated by multiplying the threshold voltage component due to charge in the depletion region with the ratio that is equal to the difference between areas of rectangle and trapezium/the area of rectangle. The equation obtained is recast by finding the expressions for lateral extents of the depletion regions associated with the source and drain ΔL_S and ΔL_D in terms of the junction depth r_j of source/drain diffusion, the depths W_{dS}, W_{dD} of the source/drain depletion regions, and channel length L. To get the final equation, we put, $W_{dS} = W_{dD} = W_{dm}$, the depth of the depletion region in the middle region below the gate. The analysis shows that the junction depth r_j of source/drain diffusion must be reduced and gate oxide capacitance increased to overcome ΔV_{Th} reduction. The gate insulator capacitance is increased by reduction of gate oxide thickness or by use of a higher dielectric constant material as the gate dielectric. Deep source/drain regions with shallow extensions enable r_j diminution without increasing parasitic resistances. Pocket or halo implants with high doping concentration further help by decreasing the depletion region thicknesses W_{dS}, W_{dD}.

The linear relationship between the carrier drift velocity v_d and the electric field crumbles down at a critical electric field E_C so that the relation $v_d = \mu E$ is modified to $v_d = \mu E/(1 + E/E_C)$. The equation for the MOSFET drain–source current is derived by finding the charge induced in the channel at a voltage $V(x)$, determining the derivative of charge with respect to time to obtain the current, putting the modified field-dependent drift velocity in the current equation and integrating over the channel length L from source to drain voltage, when it is noticed that I_{DS} for a short-channel MOSFET equals I_{DS} for a long-channel MOSFET divided by $(1 + E/E_C)$. For a MOSFET with 100 nm channel length, the electric field becomes five times the critical field at a low drain–source voltage equal to 0.5 V.

The saturation voltage for field-dependent velocity is determined from the equation $dI_{DS}/dV_{DS} = 0$ which leads to a quadratic equation in V_{DS}. The negative sign is rejected since $V_{DS} > 0$. When the $V_{DS(Sat)}$ modified for velocity saturation is substituted in the long-channel MOSFET I_{DS} equation, it is seen that the new equation differs from the earlier equation by a multiplicative correction factor. Further reasoning shows that in a short-channel MOSFET, the saturation mode commences from a lower value of V_{DS} than in a long-channel device, indicating a long drawn-out saturation region.

We express the gate–source voltage of a velocity-saturated short-channel MOSFET (V_{GS}) as the sum of the gate–source voltage of a long-channel MOSFET (V'_{GS}) plus the voltage drop across a series resistor ($I_{DS}R_{SX}$). Then the drain–source current

of the velocity saturated short-channel MOSFET (I_{DS})$_{Velocity-saturated}$ is calculated by substituting the gate–source voltage of the long-channel MOSFET (V'_{GS}) which is the difference between the gate–source voltage of the velocity-saturated short-channel MOSFET (V_{GS}) and the voltage drop across the series resistor ($I_{DS}R_{SX}$). In the equation thus obtained I_{DS} is substituted for a long-channel MOSFET. After algebraic manipulation, we arrive at an equation for drain-source current of a velocity-saturated MOSFET containing R_{SX}. The equation reached in this manner is compared with the previously derived equation for drain-souce current containing a corrective multiplication factor. This equation was obtained by substituting the $V_{DS(Sat)}$ modified for velocity saturation in the I_{DS} equation for long-channel MOSFET. Equating the drain-source currents from the two equations, viz., one equation containing R_{SX} and the other containing the correction factor, the equation for the resistor R_{SX} immediately follows. It is thus found that the effect of velocity saturation on a short-channel device can be accounted for and is the same as if a resistance is connected in series with the source terminal of a long-channel MOSFET.

$I_{DS(Sat)}$ is found by putting $V_{DS=VDS(Sat)}$ at a field $E = E_C$. The saturated drain current in a short-channel MOSFET is found to be a linear function of voltage ($V_{GS} - V_{Th}$). In a long-channel MOSFET under saturation condition, the drain-source current has a square law dependence on the gate-source voltage.

Obviation of velocity saturation effects is achieved by lowering the bias voltages in accordance with dimensional downscaling.

The term 'hot carrier' refers to non-equilibrium electron and hole distributions that are described by the Fermi-Dirac function with an elevated effective temperature. It does not indicate that the bulk temperature of the semiconductor is high. Due to their high kinetic energies, the hot electrons can get trapped in the SiO_2 gate dielectric altering the threshold voltage of the MOSFET. To minimize their effect, the channel-drain electric field is decreased by using an LDD structure. The SSR channel doping profile solves both DIBL and hot carrier issues.

To recompense for the increased horizontal source–drain field in a short-channel MOSFET, the vertical gate-to-source field must be raised. The consequential enhanced scattering of carriers at the silicon-silicon dioxide interface lowers the mobility of carriers. This calls for attention to properly scale down voltages when designing small-area MOSFET devices. Keeping voltages within safe limits is also necessary for preventing positive feedback between the avalanche breakdown and the parasitic bipolar transistor in the MOSFET structure.

SOI MOSFET technology reduced the leakage currents in MOSFETs and the coupling capacitance between the channel and the substrate. In a partially-depleted SOI MOSFET, the undepleted silicon acts like a body with floating potential due to which a kink is produced in the current–voltage characteristics. But the structure offers easy manufacturability. Fully-depleted SOI MOSFETs overcome the kink problem but suffer from difficult manufacturability.

In the FinFET, the source–drain channel layer projects outward from the silicon surface. It is covered on the left, right, and top sides with the gate dielectric and metal layers. This arrangement makes the gating action highly efficient. The fin width and height are increased to achieve larger channel widths.

Silicon nanowires can be produced by both top-down and bottom-up techniques. They can be accurately doped as desired whereas carbon nanotubes can be conducting or semiconducting. To fabricate the commonly used back-gated silicon nanowire transistors, silicon nanowires are deposited from an ethanol suspension onto a heavily doped oxidized silicon substrate. Ti/Au source/drain contacts are annealed by rapid thermal annealing (RTA) and the nanowire surface is chemically modified.

The junctionless Si nanowire FET is a normally on-transistor made from a heavily doped silicon nanowire with the middle portion covered with a dielectric film followed by a metallic film. The device does not contain any P–N junctions. After defining a silicon nanoribbon on an SOI wafer, the gate oxide is grown followed by arsenic implantation. A P⁺ polysilicon gate electrode is formed by depositing amorphous silicon films on the gate oxide, along with doping and crystallization by thermal annealing. Source/drain contacts are made with TiW.

In the Schottky-barrier CNT FET, the source and drain electrodes act as Schottky-barrier diodes while the CNT is covered with a gate dielectric and a metal film thereupon. Current flows between the source and the drain on applying a gate voltage due to lowering of Schottky barriers at the source and drain electrodes. The Schottky-barrier CNT FET works by ambipolar carrier transport. The MOSFET-like unipolar CNT FET is made without forming metal contacts at the source and drain ends. Instead, the two ends are heavily doped with potassium. As usual, there is a gate insulator and a gate electrode in this device. Both back-gating and top-gating provisions can be made.

Illustrative Exercises

19.1 A 65 nm technology node silicon transistor has a channel length of 40 nm. Find the punchthrough voltages for P-substrate doping $= 1 \times 10^{16}$ cm^{-3}, 1×10^{17} cm^{-3}, and 1×10^{18} cm^{-3}. The relative permittivity of silicon is 11.7.

We apply eq. (19.2). For P-substrate doping $= 1 \times 10^{16}$ cm^{-3}

$$V_{PT} = \frac{eN_A L^2}{2\varepsilon_0 \varepsilon_s} = \frac{1.602 \times 10^{-19} \times 1 \times 10^{16} \times \left(40 \times 10^{-9} \times 100\right)^2}{2 \times 8.854 \times 10^{-14} \times 11.7}$$

$$= \frac{1.602 \times 1 \times (40)^2 \times 10^{-19} \times 10^{16} \times 10^{-14}}{2 \times 8.854 \times 11.7 \times 10^{-14}}$$

$$= \frac{1.602 \times 1 \times 1600 \times 10^{-3}}{2 \times 8.854 \times 11.7} = \frac{2563.2 \times 10^{-3}}{207.184}$$

$$= 12.37 \times 10^{-3} = 12.37 \text{ mV} \qquad (19.115)$$

For P-substrate doping $= 1 \times 10^{17}$ cm^{-3}

$$V_{PT} = 123.7 \text{mV} \qquad (19.116)$$

For P-substrate doping $= 1 \times 10^{18}$ cm^{-3}

$$V_{PT} = 1237 \text{mV} = 1.237 \text{ V} \qquad (19.117)$$

19.2 An N-channel transistor has a channel length of 30 nm. When the drain–source voltage is 1 V, by what factor is the drain–source current of this short-channel transistor reduced with respect to a long-channel device? The critical electric field is 10^6 V m^{-1}.

The applied electric field is

$$E = \frac{\text{Voltage}}{\text{Distance}} = \frac{1}{30 \times 10^{-9}} = 3.33 \times 10^7 \text{ V m}^{-1} \qquad (19.118)$$

$$\left(I_{DS}\right)_{\text{Short channel}} = \left(\frac{1}{1 + \dfrac{E}{E_C}}\right)\left(I_{DS}\right)_{\text{Long channel}}$$

$$= \left(\frac{1}{1 + \dfrac{3.33 \times 10^7}{10^6}}\right)\left(I_{DS}\right)_{\text{Long channel}}$$

$$= \frac{1}{1 + 33.3}\left(I_{DS}\right)_{\text{Long channel}}$$

$$= \frac{1}{34.3}\left(I_{DS}\right)_{\text{Long channel}}$$

$$= 0.0292\left(I_{DS}\right)_{\text{Long channel}} \qquad (19.119)$$

19.3 An aluminum-gate MOSFET is fabricated using a 1 nm thick hafnium oxide gate dielectric with dielectric constant 25? What is the equivalent oxide thickness and equivalent oxide capacitance per unit area? Find the electron density in the inversion layer at an applied gate voltage of 1.2 V. Given that P-substrate doping is 1×10^{18} cm^{-3}, work function of aluminum $\phi_m = 4.1$ eV, electron affinity of silicon $\chi = 4.05$ eV, energy gap of silicon $E_G = 1.12$ eV, and intrinsic carrier concentration in silicon $n_i = 1 \times 10^{10}$ cm^{-3}, the dielectric constant of silicon dioxide is 3.9 and that of silicon is 11.7.

The equivalent oxide thickness is

$$\text{EOT} = \frac{\text{Dielectric constant of high } \kappa \text{ material}}{\text{Dielectric constant of silicon dioxide}}$$

$$\times \text{Physical thickness of high } \kappa \text{ material}$$

$$= \frac{\varepsilon_{HfO_2}}{\varepsilon_{SiO_2}} t_{HfO_2} = \frac{25}{3.9} \times 1 \times 10^{-9} \times 100 \text{ cm}$$

$$= 6.41 \times 10^{-7} \text{ cm} \qquad (19.120)$$

The equivalent oxide capacitance is

$$C_{\text{Equivalent}} = \frac{\varepsilon_0 \varepsilon_{SiO_2}}{\text{EOT}} = \frac{8.854 \times 10^{-14} \times 3.9}{6.41 \times 10^{-7}}$$

$$= 5.387 \times 10^{-7} \text{ F cm}^{-2} \qquad (19.121)$$

To calculate the threshold voltage, we find that the bulk potential is (eq. (19.4))

$$\phi_B = \left(\frac{k_B T}{e}\right) \ln\left(\frac{N_A}{n_i}\right) = \left(\frac{1.38065 \times 10^{-23} \times 300}{1.602 \times 10^{-19}}\right) \ln\left(\frac{1 \times 10^{18}}{1 \times 10^{10}}\right)$$

$$= 258.5487 \times 10^{-4} \ln\left(10^8\right) = 258.5487 \times 10^{-4} \ln\left(10^8\right)$$

$$= 258.5487 \times 10^{-4} \times 18.42 = 4762.467 \times 10^{-4} = 0.476 \text{ V}$$

$$(19.122)$$

The flat band voltage is (eq. (19.3))

$$V_{FB} = \phi_m - \phi_s = \phi_m - \left(\chi + \frac{E_G}{2} + \phi_B\right) = 4.1 - \left(4.05 + \frac{1.12}{2} + 0.476\right)$$

$$= 4.1 - (4.05 + 0.56 + 0.476) = 4.1 - 5.086 = -0.986 \text{ V}$$

$$(19.123)$$

The potential drop across the oxide is (eq. (19.8))

$$V_{ox} = \frac{2\sqrt{eN_A \varepsilon_0 \varepsilon_{Si} \phi_B}}{C_{Equivalent}}$$

$$= \frac{2\sqrt{1.602 \times 10^{-19} \times 1 \times 10^{18} \times 8.854 \times 10^{-14} \times 11.7 \times 0.476}}{5.387 \times 10^{-7}}$$

$$= \frac{2\sqrt{1.602 \times 1 \times 8.854 \times 11.7 \times 0.476 \times 10^{-15}}}{5.387 \times 10^{-7}}$$

$$= \frac{2 \times \sqrt{78.994 \times 10^{-15}}}{5.387 \times 10^{-7}} = \frac{2 \times 2.81 \times 10^{-7}}{5.387 \times 10^{-7}} = 1.043 \text{ V} \quad (19.124)$$

$$\therefore V_{Th} = V_{FB} + 2\phi_B + V_{ox} = -0.986 + 2 \times 0.476 + 1.043$$

$$= -0.986 + 0.952 + 1.043 = 1.009 \text{ V}$$

$$(19.125)$$

by eq. (19.9).

The electron density in the inversion layer is

$$n = \frac{Q_{Inversion}}{e} = \frac{C_{Equivalent}}{e}\left(V_{GS} - V_{Th}\right) = \frac{5.387 \times 10^{-7} \times (1.2 - 1.009)}{1.602 \times 10^{-19}}$$

$$= \frac{5.387 \times 0.191 \times 10^{-7} \times 10^{19}}{1.602} = 0.6423 \times 10^{12} = 6.42 \times 10^{11} \text{ cm}^{-2}$$

$$(19.126)$$

REFERENCES

Colinge, J.-P. 2004. Multiple-gate SOI MOSFETs. *Solid-State Electronics* 48: 897–905.

Colinge, J.-P., C.-W. Lee, A. Afzalian, N. D. Akhavan, R. Yan, I. Ferain, P. Razavi, B. O'Neill, A. Blake, M. White, A.-M. Kelleher, B. McCarthy and R. Murphy. 2010. February Nanowire transistors without junctions. *Nature Nanotechnology* 1–5, doi: 10.1038/nnano.2010.15.

Cui, Y., L. J. Lauhon, M. S. Gudiksen, J. Wang and C. M. Lieber. 2001. Diameter-controlled synthesis of single-crystal silicon nanowires. *Applied Physics Letters* 78(15): 2214–2216.

Cui, Y., Z. Zhong, D. Wang, W. U. Wang and C. M. Lieber. 2003. High performance silicon nanowire field-effect transistors. *Nano Letters* 3(2): 149–152.

D'Agostino F., D. Quercia. 2000. Short-Channel Effects in MOSFETs, http://www0.cs.ucl.ac.uk/staff/ucacdxq/projects/vlsi/report.pdf.

De, I. and C. M. Osburn. 1999. Impact of super-steep-retrograde channel doping profiles on the performance of scaled devices. *IEEE Transactions on Electron Devices* 46(8): 1711–1717.

Franklin, N. R., Q. Wang, T. W. Tombler, A. Javey, M. Shim, and H. Dai. 2002. Integration of suspended carbon nanotube arrays into electronic devices and electromechanical systems. *Applied Physics Letters* 81(5): 913–915.

Javey, A., H. Kim, M. Brink, Q. Wang, A. Ural, J. Guo, P. Mcintyre, P. Mceuen, M. Lundstrom and H. Dai. 2002. High-κ dielectrics for advanced carbon- nanotube transistors and logic gates. *Nature Materials* 1: 241–246.

Kordrostami, Z. and M. H. Sheikhi. 2010. Fundamental physical aspects of carbon nanotube transistors. In: Marulanda J. M. (Ed.) *Carbon Nanotubes*. InTech: London, UK, pp. 169–186, http://www.intechopen.com/books/carbon-nanotubes/fundamental-physical-aspects-of-carbon-nanotube-transistors.

Lee, C.-W., A. Afzalian, N. D. Akhavan, R. Yan, I. Ferain, and J.-P. Colinge. 2009. Junctionless multigate field-effect transistor. *Applied Physics Letters* 94: 053511-1–053511-2.

Lee, C.-W., A. Borne, I. Ferain, A. Afzalian, R. Yan, N. D. Akhavan, P. Razavi and J.-P. Colinge. 2010. High-temperature performance of silicon junctionless MOSFETs. *IEEE Transactions on Electron Devices* 57(3):620–625.

Lu, W., P. Xie and C. M. Lieber. 2008. Nanowire transistor performance limits and applications. *IEEE Transactions on Electron Devices*, 55(11): 2859–2876.

Lundstrom M. 2017. *Fundamentals of Nanotransistors*. World Scientific Publishing Co. Pte Ltd.: Singapore, 388 pp.

Rahou, F. Z., A. Guen-Bouazza and M. Rahou. 2013. Electrical characteristics comparison between fully-depleted SOI MOSFET and partially-depleted SOI MOSFET using Silvaco software. *Global Journal of Researches in Engineering: Electrical and Electronics Engineering* 13(1): 7.

Raychowdhury, A., A. Keshavarzi, J. Kurtin, V. De and K. Roy. 2006. Carbon nanotube field-effect transistors for high-performance digital circuits—DC analysis and modeling toward optimum transistor structure. *IEEE Transactions on Electron Devices* 53(11): 2711–2717.

20

High-Electron-Mobility Transistors

The device discussed in this chapter is called by various names and acronyms as high-electron-mobility transistor (HEMT), heterostructure field-effect transistor (HFET), modulation-doped FET (MODFET), two-dimensional electron gas FET (TEGFET) and so forth. Examples of high mobility are: a Hall mobility of 6200 cm^{2}V^{-1}s^{-1} at 300 K and 32500 cm^{2}V^{-1}s^{-1} at 77 K in the quantum well of GaAs/N-Al$_x$Ga$_{1-x}$As HEMT at a carrier concentration of 7×10^{11} cm^{-2} showing a 5.5 times mobility advantage relative to the MESFET at 77 K (Mimura et al. 1980), a field-effect mobility of 49300 cm^{2}V^{-1}s^{-1} at 77 K in the enhancement-mode GaAs/N-AlGaAs HEMT with a sheet electron concentration of 5.5×10^{11} cm^{-2} (Hiyamizu et al 1981a), and room-temperature mobility as high as 8600 cm^{2}V^{-1}s^{-1} in selectively doped GaAs/N-AlGaAs heterostructure (Hiyamizu et al. 1981b). The reason for the high mobility is the spatial separation and hence unimpeded motion of the carriers in the channel from the ionized donors producing them, a situation which is obtained by the technique known as modulation doping (Mimura 2005). Hence, the name modulation-doped FET. We mentioned about modulation doping in Chapter 12, Section 12.3. The channel is formed at the heterojunction between two semiconductors of different bandgaps from which

the name 'heterostructrure FET' arises. Further, the channel is a two-dimensional sheet of carriers formed at the interface between the two semiconductors which gives the name 'two-dimensional electron gas FET' (Liu et al. 2016).

As already discussed, the HEMT channel is a thin highly conducting region formed at the interface between a doped wide bandgap semiconductor and an undoped narrow bandgap semiconductor containing electrons transferred from the former to the latter known as two-dimensional electron gas (2DEG) because of its thin sheet-like appearance (Figure 20.1). An undoped spacer layer is included to further keep the free carriers away from the doped wide bandgap semiconductor and prevent any decrease in their mobility from doping effect. The HEMT has a Schottky barrier gate formed with a metal-semiconductor junction like a MESFET or JFET. It has low-resistance ohmic source and drain contacts.

20.1 MOSFET, MESFET, and HEMT

With the MOSFET device firmly established in silicon technology and the MESFET in GaAs technology, it is natural to enquire about the relevance of HEMTs.

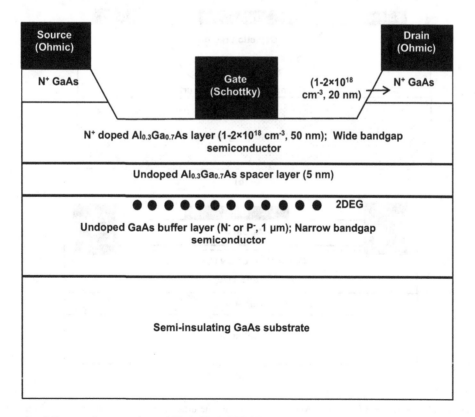

FIGURE 20.1 Cross-sectional diagram of a recessed-gate AlGaAs/GaAs HEMT.

In a MOSFET (Figure 20.2(a)), the voltage applied between the gate electrode and the body modulates the conductivity of an inversion layer or channel formed at the semiconductor-insulator interface. The charge carriers in a MOSFET channel suffer surface scattering whereby the mobility in the channel is reduced to one-half of its bulk value.

In the MESFET (Figure 20.2(b)), the current flow between the source and the drain is controlled by varying the width of the depletion region underneath a reverse-biased metal-semiconductor junction. As the channel is farther from the surface by a distance equal to the width of the depletion region, the carrier mobility in the channel is the same as the bulk value. Higher mobilities are obtained in the MESFET when compared with the MOSFET. But still the carriers are moving in a doped semiconductor layer and their motion is impeded by ionized impurities on the way. So higher mobility is possible if the carriers do not share the same space as the ionized impurities, i.e. carriers move in an undoped region, which is realized in the HEMT.

In the HEMT (Figure 20.2(c)), the carrier motion takes place in an undoped layer. So, the highest mobility of the three architectures is realized.

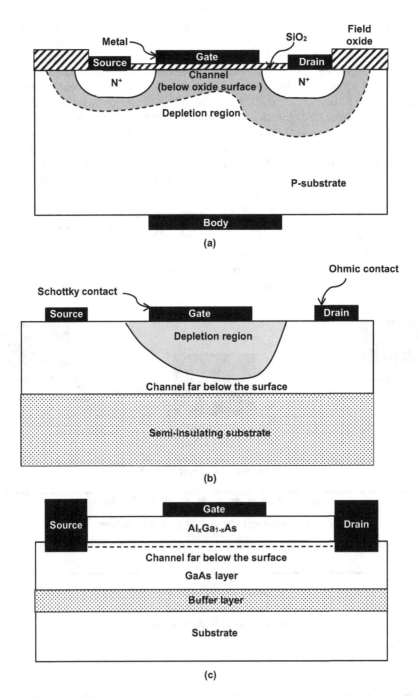

FIGURE 20.2 Drawings of MOSFET, MESFET, and HEMT devices illustrate the differences in the location of the channel in their structures: (a) MOSFET, (b) MESFET, and (c) HEMT.

20.2 HEMT Operation

20.2.1 The 2DEG Channel Formation

Contemplate the situation at the instant when the two materials under consideration, viz. AlGaAs and GaAs, are brought into intimate contact but no charge transfer has taken place across the interface. The AlGaAs layer is N$^+$-doped whereas the GaAs layer is undoped. Practically, it is slightly P-doped. The energy-band diagrams in just contacted condition will be similar to the N-p

heterojunction diagram, which is a redrawn version of Figure 12.1 for N-AlGaAs/p-GaAs. Just when the two materials are brought together, the transition between AlGaAs and GaAs is assumed to be abrupt in the scale of atomic dimensions. Under this assumption, the bands will display sharp discontinuities at the points where the two materials meet (Figure 20.3(a)). But soon after the contact is established, the difference in electron concentrations on the two sides and the difference in electron affinities of the two materials come into play. Electron concentration in heavily doped AlGaAs is much higher than in undoped GaAs.

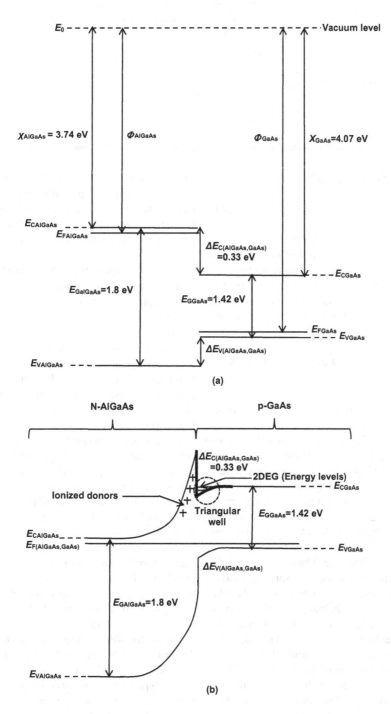

(a)

(b)

FIGURE 20.3 Drawing the energy-band diagram for N$^+$AlGaAs/P-GaAs heterojunction: (a) immediately after contact but before any charge relocation on the two sides of the interface, (b) after charge exchange across the interface.

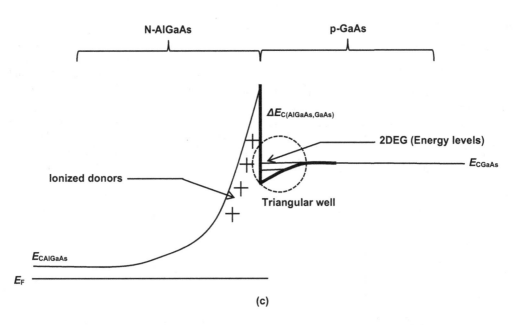

N-AlGaAs   **p-GaAs**

$\Delta E_{C(AlGaAs,GaAs)}$

2DEG (Energy levels)

E_{CGaAs}

Ionized donors

Triangular well

$E_{CAlGaAs}$

E_F

(c)

FIGURE 20.3 (CONTINUED) (c) enlarged view of the triangular well portion. The triangular well region is encircled.

The electron affinity of AlGaAs is given by

$$\eta = 4.07 - 1.1x \text{ eV} \quad \text{for} \quad x < 0.45 \quad (20.1)$$

The AlAs concentration $x = 0.3$ is considered as an optimum value because a lower value of x results in a smaller discontinuity at the heterojunction so that the electron concentration in the potential well is low while a value of x higher than 0.3 causes the formation of undesirable trapping centers in the material that impair device performance. For $x = 0.3$, i.e. for $Al_{0.3}Ga_{0.7}As$, we have

$$\eta = 4.07 - 1.1x = 4.07 - 1.1 \times 0.3 = 4.07 - 0.33 = 3.74 \text{ eV}$$
$$(20.2)$$

The electron affinity of GaAs ($x = 0$) is 4.07 eV, which is higher than for $Al_{0.3}Ga_{0.7}As$ (3.74 eV). On account of both the electron concentration gradient from AlGaAs to GaAs and the higher electron affinity of GaAs than AlGaAs, electron transference takes place from AlGaAs to GaAs. As a consequence of this electron migration to GaAs, the positively charged donor ions are left behind in the AlGaAs layer so that the charge neutrality is disturbed and a positive charge is build up on the AlGaAs side adjoining the AlGaAs/GaAs interface. Similarly, the electrons that have moved toward the GaAs side make a thin negatively charged layer contiguous to the interface. Thus an electric field is set up at the interface of the two semiconductors. A positive potential exists on the AlGaAs side and a negative potential on the GaAs side. Since $E = -eV$, the energy is lowered on the AlGaAs side and raised on the GaAs side. Accordingly the conduction band bends downward on the higher potential AlGaAs side and upward on the lower potential GaAs side (Figure 20.3(b)).

The same can be understood in terms of the Fermi level. Since the Fermi level is the energy level at which the probability of electron occupancy is 50%, the Fermi level moves downward on the AlGaAs side suffering loss of electrons and moves upward on the GaAs side due to gain of electrons. So the energy-band

diagram is drawn to maintain the equality of Fermi levels on both sides, which is possible by band bending. The electron transference phenomenon due to diffusion is opposed by the repulsive action of the electric field and in the state of dynamic equilibrium, the two rates of electron transfer:

i. by diffusion in the concentration gradient, and
ii. by drift in the electric field or the potential gradient, neutralize each other.

In this condition, a thin film of electrons around 10 nm thick is formed on the GaAs side. As can be seen from the band diagram, a quasi-triangular-shaped region is formed at the heterojunction interface on the GaAs. It is the potential well in which the electrons are confined. Recall that the GaAs layer is not doped/slightly P-doped with impurities but it contains a channel of electrons, which can move freely and their motion is not disturbed by collisions with donor impurity ions that are invariably present in any heavily doped N-type semiconductor. So building the heterojunction has provided a means of filling a thin region of the GaAs layer with electrons without introducing any impurities in this layer. These electrons confined in a space comparable to the de Broglie wavelength of the electrons leads to the quantization of electron motion in a direction perpendicular to the heterojunction interface. The restriction of electron motion in one direction leaves them free in the other two directions imparting them a two-dimensional character. So, the thin sheet of electrons confined in the quasi-triangular potential well is called a 2DEG.

The quasi-triangular potential well, shown in a magnified view in Figure 20.3(c), is a dynamic well because the potential profile of the well depends on the charge density inside it, as determined by the Poisson's equation while charge density in the well is obtained from the solution of the Schrodinger equation by inputting the potential profile in the well. In view of this mutual interdependence, numerical techniques must be adopted to determine the electron concentration through a self-consistent solution of Poisson's and Schrodinger equation. The problem can be analytically solved in the limit of triangular well approximation but the best agreement with experiment is found when the self-consistent solution is pursued, as elaborated in Chapter 12, section 12.4.6.

20.2.2 Modulation of the Channel: Enhancement and Depletion Mode HEMTs

From the electrical operation aspect, HEMTs are classified as enhancement and depletion types.

In a HEMT made from materials such as GaAs/AlGaAs that do not carry a net charge induced by interfacial polarization, application of a positive voltage may be necessary for attracting electrons to the channel The thickness of the depletion region formed in the barrier layer under the gate electrode determines whether the device will operate in the normally off or normally on mode. If the thickness of the barrier layer (AlGaAs) in the HEMT structure is taken to be very small, not only the barrier layer is fully depleted, the depletion region extends further beyond and the adjoining channel in the well layer is also depleted. Operation of this HEMT reminds us of the normally off or enhancement-mode MOSFET. So, this HEMT is called an enhancement HEMT or eHEMT. But if the barrier layer (AlGaAs) of the HEMT is sufficiently thick, only part of this layer is depleted and the channel is not affected. So, a channel exists under normal conditions and a negative bias is applied to vary the channel thickness providing a depletion mode operation. Such a HEMT is known as dHEMT.

Although the familiar semiconductor material combination used in HEMTs is: AlGaAs/GaAs, another common pair is GAInN/GaN (Ng and Arulkumaran 2012). Unlike AlGaAs/GaAs, nitrides like GaN and AlGaN have built-in polarization produced by their crystalline structure. When a HEMT is made with GaN as the channel layer and AlGaN as the barrier layer, the difference in polarization between the two layers results in the generation of an uncompensated positive sheet of charge. So the channel is formed at the interface even if no external positive voltage is applied or the AlGaN layer is undoped. Owing to the similarity of this HEMT with a depletion-mode MOSFET, it is referred to as a depletion HEMT or dHEMT.

20.3 Recessed-Gate and Self-Aligned Gate HEMTs

Structurally, the HEMTs are of two types: non-self-aligned recessed-gate and self-aligned T-gate.

20.3.1 Non-Self-Aligned Recessed-Gate HEMT Fabrication

To fabricate a recessed-gate HEMT (Subramanian 1990), a semi-insulating GaAs (SI-GaAs) substrate is taken (Figure 20.4(a), also Figure 20.1). The different layers are grown/deposited in a metal-organic chemical vapor deposition (MOCVD) or a molecular beam epitaxy (MBE) facility. An undoped GaAs buffer layer (~1 μm) is grown over the SI-GaAs substrate. This layer has either N⁻ (in MOCVD) or P⁻ (in MBE due to carbon) polarity

with dopant concentration of 10^{13}–10^{14} cm^{-3}. On the GaAs buffer layer, an undoped $Al_xGa_{1-x}As$ spacer layer (4–10 nm) is deposited. A silicon-doped N⁺-$Al_xGa_{1-x}As$ layer (40–50 nm) with doping concentration of 2×10^{18} cm^{-3} covers the undoped $Al_xGa_{1-x}As$ spacer layer. Finally, a 10 nm thick silicon-doped N⁺-GaAs capping layer is laid on N⁺-$Al_xGa_{1-x}As$ layer. Its doping concentration is 2×10^{18} cm^{-3} like the N⁺-$Al_xGa_{1-x}As$ layer.

After completing the MOCVD or MBE process steps, the individual devices are isolated by mesa etching in cold H_2SO_4:H_2O_2:H_2O = 3:1:1 solution for ½ min up to SI-GaAs. Then the source/drain contact pattern is defined by photolithography; Au:Ge/Ni is evaporated and patterned by lift-off technique. The metal is alloyed by sintering at 480°C for 2 min to form ohmic contacts. The gate area is etched in H_2O:H_3PO_4:H_2O_2 = 50:3:1 for ½ to 1½ min to control the threshold voltage. After exposing and developing the photoresist through a gate metallization mask, Al or Ti/Pd/Au is evaporated and patterned by lift-off. Thus the gate contact is made directly with the N⁺-$Al_xGa_{1-x}As$ layer but the source/drain contacts are deposited over elevated N⁺-doped GaAs islands (Subramanian 1990).

20.3.2 Self-Aligned T-Gate HEMT Fabrication

The salient feature of the self-aligned gate process is that a second ohmic film is deposited over the first ohmic source/drain metallization layer after the Schottky barrier gate metal is deposited (Lee et al., 2004). To facilitate self-alignment, the gate metal is shaped like the letter 'T' with the upper stick of the letter projecting over the surface to provide overhangs (Figure 20.4(b)). These overhangs are used as a shadow mask to deposit the second film. The annealing temperature of the second film should be such that the performance of the Schottky barrier gate film does not deteriorate during annealing.

Device isolation is done to form mesas by chlorine-based dry etching. First Ti/Al/Pd/Au multilayer coating for source/drain contacts is formed to cover the sides of the mesa and reaching up to the GaN layer below the AlGaN layer. Rapid thermal annealing is done at 850°C for ½ min. This is followed by Ni/Au gate metal deposition and patterning into T-shape. The second Ti/Al/Ti/Au film is deposited over the first source/drain contact film with the T-shaped gate metal as a shadow mask. Final annealing is carried out at 750°C for ½ h in nitrogen atmosphere. Thus the source/gate and gate/drain distances are minimized and the alignment is guaranteed by the overhangs of the gate film (Lee et al. 2004).

20.3.3 Pseudomorphic and Metamorphic HEMTs

From growth technology viewpoint, there are two kinds of HEMTs: pseudomorphic and metamorphic. For HEMT fabrication, it is generally difficult to find two semiconductors with precisely matching lattice constants. So a thin film of the second semiconductor is deposited over the first semiconductor. Owing to its small thickness, the film stretches to cover and fit over the slightly unmatched lattice. HEMT fabricated in this way is said to be pseudomorphic HEMT or pHEMT. A material is pseudomorphic if it has the uncharacteristic crystalline form of another material instead of the form usually associated with its composition.

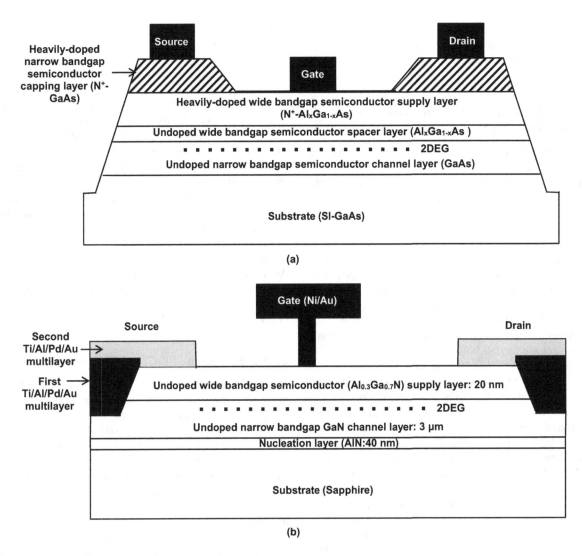

FIGURE 20.4   HEMT structures: (a) non-self-aligned recessed gate and (b) self-aligned T-gate.

The other way is to interpose a buffer layer of graded composition between the two semiconductors so that it can match with the lattice constants of the two semiconductors. HEMT made by this approach is known as metamorphic HEMT or mHEMT. A metamorph is a material whose original structure is changed by pressure or thermal treatment.

20.4 The Sheet Density of Electrons

The sheet electron density trapped in the quantum well of a HEMT plays a vital role in its operation and performance. To evaluate this density, we go through the steps in the subsections below.

20.4.1 Energy Levels of Electrons in a Triangular Well with Infinitely High Potential Wall ΔE_C

Starting from the flat-band condition, the transfer of electrons from AlGaAs (N-type doped wide bandgap semiconductor labeled as N) to GaAs (N-type doped small bandgap semiconductor labeled as n) produces an electric field F in the Z-direction (Figure 20.5) related to electrostatic potential V as

$$F = -\frac{dV}{dz} \tag{20.3}$$

or

$$dV = -Fdz \tag{20.4}$$

or

$$\int dV = -F \int dz \tag{20.5}$$

or

$$V = -Fz \tag{20.6}$$

Hence, the potential energy of the electron is

$$U(z) = -eV = -e \times -Fz = eFz \tag{20.7}$$

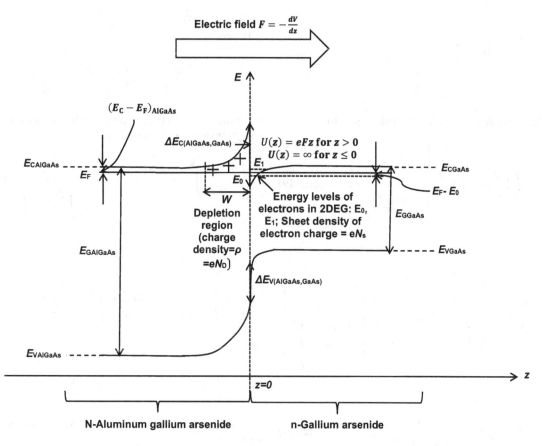

FIGURE 20.5 Energy-band diagram of an N$^+$-doped AlGaAs/N-doped GaAs structure for calculating the sheet density of electrons N_s trapped in the triangular potential well.

The infinite triangular well is defined as

$$U(z) = eFz \quad \text{for} \quad z > 0 \qquad (20.8)$$

and

$$U(z) = \infty \quad \text{for} \quad z \leq 0 \qquad (20.9)$$

For $z > 0$, the time-independent Schrodinger equation (TISE) is

$$-\frac{\hbar^2}{2m_n^*}\frac{\partial^2 \Psi(z)}{\partial z^2} + U(z)\Psi(z) = E\Psi(z) \qquad (20.10)$$

or

$$-\frac{\hbar^2}{2m_n^*}\frac{\partial^2 \Psi(z)}{\partial z^2} + eFz\Psi(z) = E\Psi(z) \qquad (20.11)$$

Here the effective mass of the electron is assumed to be m_n^* across the AlGaAs/GaAs interface, i.e. equal in both semiconductors. The dielectric constant will also be assumed to be same (ε) in both GaAs and AlGaAs.

Dividing both sides of eq. (20.11) by $-\dfrac{\hbar^2}{2m_n^*}$ we get

$$\frac{\partial^2 \Psi(z)}{\partial z^2} - \frac{2m_n^*}{\hbar^2}eFz\Psi(z) = -\frac{2m_n^*}{\hbar^2}E\Psi(z) \qquad (20.12)$$

or

$$\frac{\partial^2 \Psi(z)}{\partial z^2} - \frac{2m_n^*}{\hbar^2}eFz\Psi(z) + \frac{2m_n^*}{\hbar^2}E\Psi(z) = 0 \qquad (20.13)$$

or

$$\frac{\partial^2 \Psi(z)}{\partial z^2} + \frac{2m_n^*}{\hbar^2}(E - eFz)\Psi(z) = 0 \qquad (20.14)$$

or

$$\frac{1}{eF}\frac{\partial^2 \Psi(z)}{\partial z^2} + \frac{2m_n^*}{\hbar^2}\left(\frac{E}{eF} - z\right)\Psi(z) = 0 \qquad (20.15)$$

or

$$\frac{\partial^2 \Psi(z)}{\partial z^2} + \frac{2m_n^* eF}{\hbar^2}\left(\frac{E}{eF} - z\right)\Psi(z) = 0 \qquad (20.16)$$

Put

$$\xi \equiv -\left(\frac{2m_n^* eF}{\hbar^2}\right)^{1/3}\left(\frac{E}{eF} - z\right) \qquad (20.17)$$

or

$$\frac{E}{eF} - z = -\frac{\xi}{\left(\dfrac{2m_n^* eF}{\hbar^2}\right)^{1/3}} \tag{20.18}$$

Differentiating both sides of eq. (20.18) with respect to ξ,

$$\frac{d}{d\xi}\left(\frac{E}{eF} - z\right) = \frac{d}{d\xi}\left\{-\frac{\xi}{\left(\dfrac{2m_n^* eF}{\hbar^2}\right)^{1/3}}\right\} \tag{20.19}$$

or

$$\frac{d}{d\xi}\left(\frac{E}{eF}\right) - \frac{dz}{d\xi} = \frac{d\xi}{d\xi}\left\{-\frac{1}{\left(\dfrac{2m_n^* eF}{\hbar^2}\right)^{1/3}}\right\} + \xi\frac{d}{d\xi}\left\{-\frac{1}{\left(\dfrac{2m_n^* eF}{\hbar^2}\right)^{1/3}}\right\} \tag{20.20}$$

or

$$0 - \frac{dz}{d\xi} = 1 \times -\frac{1}{\left(\dfrac{2m_n^* eF}{\hbar^2}\right)^{1/3}} + \xi \times 0 \tag{20.21}$$

or

$$-\frac{dz}{d\xi} = -\frac{1}{\left(\dfrac{2m_n^* eF}{\hbar^2}\right)^{1/3}} \tag{20.22}$$

or

$$dz = \frac{d\xi}{\left(\dfrac{2m_n^* eF}{\hbar^2}\right)^{1/3}} \tag{20.23}$$

$$\therefore \frac{d\Psi(z)}{dz} = \frac{\dfrac{d\Psi(z)}{d\xi}}{\left(\dfrac{2m_n^* eF}{\hbar^2}\right)^{1/3}} = \left(\frac{2m_n^* eF}{\hbar^2}\right)^{1/3}\frac{d\Psi(z)}{d\xi} \tag{20.24}$$

and

$$\frac{d}{dz}\left(\frac{d\Psi(z)}{dz}\right) = \frac{d}{dz}\left[\left(\frac{2m_n^* eF}{\hbar^2}\right)^{1/3}\frac{d\Psi(z)}{d\xi}\right] = \frac{d}{d\xi}\left[\left(\frac{2m_n^* eF}{\hbar^2}\right)^{1/3}\frac{d\Psi(z)}{d\xi}\right] \times \frac{d\xi}{dz} = \left(\frac{2m_n^* eF}{\hbar^2}\right)^{1/3} \times \frac{d}{d\xi}\left\{\frac{d\Psi(z)}{d\xi}\right\} \times \frac{d\xi}{dz}$$

$$= \left(\frac{2m_n^* eF}{\hbar^2}\right)^{1/3} \times \frac{d^2\Psi(z)}{d\xi^2} \times \frac{d\xi}{dz} = \left(\frac{2m_n^* eF}{\hbar^2}\right)^{1/3} \times \frac{d^2\Psi(z)}{d\xi^2} \times \left(\frac{2m_n^* eF}{\hbar^2}\right)^{1/3} \tag{20.25}$$

from eq. (20.23).

Hence,

$$\frac{d^2\Psi(z)}{dz^2} = \left(\frac{2m_n^* eF}{\hbar^2}\right)^{2/3}\frac{d^2\Psi(z)}{d\xi^2} \tag{20.26}$$

Substituting for $d^2\Psi(z)/dz^2$ from eq. (20.26) and $\{E/(eF) - z\}$ from eq. (20.18) in eq. (20.16) we have

$$\left(\frac{2m_n^* eF}{\hbar^2}\right)^{2/3}\frac{d^2\Psi(z)}{d\xi^2} + \frac{2m_n^* eF}{\hbar^2} \times -\frac{\xi}{\left(\dfrac{2m_n^* eF}{\hbar^2}\right)^{1/3}}\Psi(z) = 0 \tag{20.27}$$

or

$$\left(\frac{2m_n^* eF}{\hbar^2}\right)^{2/3}\frac{d^2\Psi(z)}{d\xi^2} - \left(\frac{2m_n^* eF}{\hbar^2}\right)^{2/3}\xi\Psi(z) = 0 \tag{20.28}$$

or

$$\frac{d^2\Psi(z)}{d\xi^2} - \xi\Psi(z) = 0 \tag{20.29}$$

This equation has the form of the Airy differential equation written as

$$\frac{d^2 y}{dx^2} - xy = 0 \tag{20.30}$$

The general solution of eq. (20.30) can be expressed in terms of the Airy function $A_i(\xi)$ given by

$$A_i(\xi) = \frac{1}{\sqrt{\pi}}\int_0^\infty \cos\left(\frac{1}{3}u^3 + u\xi\right)du \tag{20.31}$$

which is referred to as the Airy integral. Applying eq. (20.17) for ξ

$$A_i(\xi) = A_i\left\{-\left(\frac{2m_n^* eF}{\hbar^2}\right)^{1/3}\left(\frac{E}{eF} - z\right)\right\} \tag{20.32}$$

For an infinite barrier, the boundary condition is

$$\Psi(z) = 0 \quad \text{at} \quad z = 0 \tag{20.33}$$

Hence,

$$A_i \left\{ -\left(\frac{2m_n^* eF}{\hbar^2} \right)^{\frac{1}{3}} \left(\frac{E}{eF} - 0 \right) \right\} = 0 \qquad (20.34)$$

or

$$A_i \left\{ -\left(\frac{2m_n^* eF}{\hbar^2} \right)^{\frac{1}{3}} \left(\frac{E}{eF} \right) \right\} = 0 \qquad (20.35)$$

or

$$A_i \left\{ -\left(\frac{2m_n^*}{\hbar^2} \right)^{\frac{1}{3}} (eF)^{\frac{1}{3}} \left(\frac{E}{eF} \right) \right\} = 0 \qquad (20.36)$$

or

$$A_i \left\{ -\left(\frac{2m_n^*}{\hbar^2} \right)^{\frac{1}{3}} \frac{1}{(eF)^{\frac{2}{3}}} E \right\} = 0 \qquad (20.37)$$

or

$$A_i \left\{ -\left(\frac{2m_n^*}{\hbar^2 e^2 F^2} \right)^{\frac{1}{3}} E \right\} = 0 \qquad (20.38)$$

The first few real roots of this equation are:

$$a_1 = -2.33811, \quad a_2 = -4.08795, \quad a_3 = -5.52086 \qquad (20.39)$$

Asymptotically (line approaching a curve but never touching it),

$$a_n \approx -\left\{ \frac{3\pi}{2} \left(n - \frac{1}{4} \right) \right\}^{\frac{2}{3}} \qquad (20.40)$$

To find the energies of the states in which the electrons are confined, we set

$$-\left(\frac{2m_n^*}{\hbar^2 e^2 F^2} \right)^{\frac{1}{3}} E_n = a_n \qquad (20.41)$$

or

$$E_n = -\frac{a_n}{-\left(\frac{2m_n^*}{\hbar^2 e^2 F^2} \right)^{\frac{1}{3}}} = -\left(\frac{\hbar^2 e^2 F^2}{2m_n^*} \right)^{\frac{1}{3}} a_n \qquad (20.42)$$

which asymptotically takes the form

or

$$\oint F \, dS = \frac{Q_s}{\varepsilon_0 \varepsilon} \qquad (20.45)$$

so that we can write

$$F = \frac{1}{\varepsilon_0 \varepsilon} \frac{Q_s}{S} \qquad (20.46)$$

Let us denote the surface charge per unit area by N_s, i.e.

$$\frac{Q_s}{S} = eN_s \qquad (20.47)$$

where e is the electronic charge and N_s is the carrier concentration in the electron sheet. Then

$$F = \frac{eN_s}{\varepsilon_0 \varepsilon} \qquad (20.48)$$

Substituting for F from eq. (20.48) in eq. (20.43) gives

$$E_n = \left(\frac{\hbar^2}{2m_n^*} \right)^{\frac{1}{3}} \left\{ \frac{3\pi e^2 N_s}{2\varepsilon_0 \varepsilon} \left(n - \frac{1}{4} \right) \right\}^{\frac{2}{3}} \qquad (20.49)$$

20.4.2 Carrier Concentration N_s in 2DEG for One Occupied Confined Energy Level

To find the carrier concentration, let us look into the various energy components which must be added together to obtain ΔE_C. These include:

i. The energy level of electrons: Let us consider the energy level E_1 for which $n = 1$. For this energy level, eq. (20.42) is

$$E_1 = -\left(\frac{\hbar^2 e^2 F^2}{2m_n^*} \right)^{\frac{1}{3}} a_1 = -\left(\frac{\hbar^2 e^2}{2m_n^*} \right)^{\frac{1}{3}} F^{\frac{2}{3}} a_1$$

$$= -\left(\frac{\hbar^2 e^2}{2m_n^*} \right)^{\frac{1}{3}} \left(\frac{eN_s}{\varepsilon_0 \varepsilon} \right)^{\frac{2}{3}} a_1 = -\left(\frac{\hbar^2}{2m_n^*} \right)^{\frac{1}{3}} e^{\frac{2}{3}} \left(\frac{eN_s}{\varepsilon_0 \varepsilon} \right)^{\frac{2}{3}} a_1$$

$$= -\left(\frac{\hbar^2}{2m_n^*} \right)^{\frac{1}{3}} \left(\frac{e^2 N_s}{\varepsilon_0 \varepsilon} \right)^{\frac{2}{3}} a_1 \qquad (20.50)$$

where eq. (20.48) is applied.

$$E_n = -\left(\frac{\hbar^2 e^2 F^2}{2m_n^*} \right)^{\frac{1}{3}} \times -\left\{ \frac{3\pi}{2} \left(n - \frac{1}{4} \right) \right\}^{\frac{2}{3}} = \left(\frac{\hbar^2}{2m_n^*} \right)^{\frac{1}{3}} \times (eF)^{\frac{2}{3}} \times \left\{ \frac{3\pi}{2} \left(n - \frac{1}{4} \right) \right\}^{\frac{2}{3}} = \left(\frac{\hbar^2}{2m_n^*} \right)^{\frac{1}{3}} \left\{ \frac{3\pi eF}{2} \left(n - \frac{1}{4} \right) \right\}^{\frac{2}{3}} \qquad (20.43)$$

using equation (20.40). The electric field created by electron transfer to GaAs, i.e. F at $z = 0$, is found from Gauss law in electrostatics:

$$\text{Surface integral of electric field over a closed surface} = \frac{\text{Net charge enclosed by the surface} (Q_s)}{\text{Permittivity of free space} (\varepsilon_0) \times \text{Relative permittivity of the medium} (\varepsilon)} \qquad (20.44)$$

ii. Energy difference $\left(\Delta E_{N_s} = E_F - E_0\right)$ caused by sheet density of electron charge eN_s formed on the GaAs side

Recall from Section 12.4.4, eq. (12.76) that the two-dimensional density of states is given by

$$g(E)_{2D} = \frac{m_n^*}{\pi \hbar^2} \qquad (20.51)$$

for non-degenerate ground state E_0 and $2g(E)_{2D}$ for states higher than E_1 (spin degeneracy = 2); hence we can write the interface carrier concentration by applying the Fermi-Dirac statistics as

$$N_s = g(E)_{2D} \int_{E_0}^{E_1} \frac{dE}{1 + \exp\left\{\left(\frac{E - E_F}{k_B T}\right)\right\}}$$

$$+ 2g(E)_{2D} \int_{E_1}^{\infty} \frac{dE}{1 + \exp\left\{\left(\frac{E - E_F}{k_B T}\right)\right\}} \qquad (20.52)$$

Consider

$$\int \frac{dE}{1 + \exp\left\{\left(\frac{E - E_F}{k_B T}\right)\right\}}$$

$$= \int \frac{dE}{\exp\left\{\left(\frac{E - E_F}{k_B T}\right)\right\}\left[\exp\left\{-\left(\frac{E - E_F}{k_B T}\right)\right\} + 1\right]}$$

$$= \int \frac{\exp\left\{-\left(\frac{E - E_F}{k_B T}\right)\right\} dE}{\left[\exp\left\{-\left(\frac{E - E_F}{k_B T}\right)\right\} + 1\right]} \qquad (20.53)$$

Let us put

$$\exp\left\{-\left(\frac{E - E_F}{k_B T}\right)\right\} + 1 = u \qquad (20.54)$$

Then

$$\frac{d}{dE}\left[\exp\left\{-\left(\frac{E - E_F}{k_B T}\right)\right\} + 1\right] = \frac{du}{dE} \qquad (20.55)$$

or

$$\exp\left\{-\left(\frac{E - E_F}{k_B T}\right)\right\} \times \frac{(-1+0)(k_B T) + (E - E_F) \times 0}{(k_B T)^2} + 0 = \frac{du}{dE} \qquad (20.56)$$

or

$$\exp\left\{-\left(\frac{E - E_F}{k_B T}\right)\right\} \times \frac{-k_B T}{(k_B T)^2} = \frac{du}{dE} \qquad (20.57)$$

$$\therefore \exp\left\{-\left(\frac{E - E_F}{k_B T}\right)\right\} dE = -\frac{(k_B T)^2}{k_B T} du = -(k_B T) du \qquad (20.58)$$

From eqs. (20.53) and (20.58),

$$\int \frac{dE}{1 + \exp\left\{\left(\frac{E - E_F}{k_B T}\right)\right\}} = (-k_B T) \int \frac{du}{u} = (-k_B T) \ln u$$

$$= (-k_B T) \ln\left[\exp\left\{-\left(\frac{E - E_F}{k_B T}\right)\right\} + 1\right] \qquad (20.59)$$

by putting back the value of u from eq. (20.54).

From eqs. (20.52) and (20.59),

$$N_s = g(E)_{2D} \times \left|(-k_B T) \ln\left[\exp\left\{-\left(\frac{E - E_F}{k_B T}\right)\right\} + 1\right]\right|_{E_0}^{E_1}$$

$$+ 2g(E)_{2D} \times \left|(-k_B T) \ln\left[\exp\left\{-\left(\frac{E - E_F}{k_B T}\right)\right\} + 1\right]\right|_{E_1}^{\infty}$$

$$= g(E)_{2D} \times (-k_B T) \ln\left[\exp\left\{-\left(\frac{E_1 - E_F}{k_B T}\right)\right\} + 1\right] - g(E)_{2D}$$

$$\times (-k_B T) \ln\left[\exp\left\{-\left(\frac{E_0 - E_F}{k_B T}\right)\right\} + 1\right]$$

$$+ 2g(E)_{2D} \times (-k_B T) \ln\left[\exp\left\{-\left(\frac{\infty - E_F}{k_B T}\right)\right\} + 1\right]$$

$$- 2g(E)_{2D} (-k_B T) \ln\left[\exp\left\{-\left(\frac{E_1 - E_F}{k_B T}\right)\right\} + 1\right]$$

$$= -g(E)_{2D} (k_B T) \ln\left[\exp\left\{-\left(\frac{E_1 - E_F}{k_B T}\right)\right\} + 1\right]$$

$$+ g(E)_{2D} (k_B T) \ln\left[\exp\left\{-\left(\frac{E_0 - E_F}{k_B T}\right)\right\} + 1\right]$$

$$+ 2g(E)_{2D} (-k_B T) \ln\left[\exp\left\{-\left(\frac{\infty - E_F}{k_B T}\right)\right\} + 1\right]$$

$$+ 2g(E)_{2D} (k_B T) \ln\left[\exp\left\{-\left(\frac{E_1 - E_F}{k_B T}\right)\right\} + 1\right]$$

$$= g(E)_{2D} (k_B T) \ln\left[\exp\left\{-\left(\frac{E_1 - E_F}{k_B T}\right)\right\} + 1\right]$$

$$+ g(E)_{2D} (k_B T) \ln\left[\exp\left\{-\left(\frac{E_0 - E_F}{k_B T}\right)\right\} + 1\right]$$

$$= g(E)_{2D} (k_B T) \ln \left\{ \exp \left(\frac{E_F - E_1}{k_B T} \right) + 1 \right\}$$

$$+ g(E)_{2D} (k_B T) \ln \left\{ \exp \left(\frac{E_F - E_0}{k_B T} \right) + 1 \right\} \tag{20.60}$$

because

$$2 g(E)_{2D} \times (-k_B T) \ln \left[\exp \left\{ -\left(\frac{\infty - E_F}{k_B T} \right) \right\} + 1 \right]$$

$$= 2 g(E)_{2D} \times (-k_B T) \ln \left[\exp \left(\frac{1}{\frac{\infty - E_F}{k_B T}} \right) + 1 \right]$$

$$= 2 g(E)_{2D} \times (-k_B T) \ln (0 + 1)$$

$$= 2 g(E)_{2D} \times (-k_B T) \ln (1) = 2 g(E)_{2D} \times (-k_B T) \times 0 = 0 \tag{20.61}$$

Thus from eq. (20.60)

$$N_s = g(E)_{2D} (k_B T) \ln \left[\left\{ 1 + \exp \left(\frac{E_F - E_0}{k_B T} \right) \right\} \right.$$

$$\left. \times \left\{ 1 + \exp \left(\frac{E_F - E_1}{k_B T} \right) \right\} \right] \tag{20.62}$$

At low temperatures $T \to 0$,

$$\exp \left(\frac{E_F - E_0}{k_B T} \right) \gg 1 \quad \text{and} \quad \exp \left(\frac{E_F - E_1}{k_B T} \right) \gg 1 \tag{20.63}$$

$$\therefore N_s = g(E)_{2D} (k_B T) \ln \left[\left\{ \exp \left(\frac{E_F - E_0}{k_B T} \right) \right\} \left\{ \exp \left(\frac{E_F - E_1}{k_B T} \right) \right\} \right]$$

$$= g(E)_{2D} (k_B T) \left[\ln \left\{ \exp \left(\frac{E_F - E_0}{k_B T} \right) \right\} \right]$$

$$+ g(E)_{2D} (k_B T) \left[\ln \left\{ \exp \left(\frac{E_F - E_1}{k_B T} \right) \right\} \right]$$

$$= g(E)_{2D} (k_B T) \left(\frac{E_F - E_0}{k_B T} \right) + g(E)_{2D} (k_B T) \left(\frac{E_F - E_1}{k_B T} \right)$$

$$= g(E)_{2D} (k_B T) \left(\frac{E_F - E_0}{k_B T} \right) + 0 = g(E)_{2D} (k_B T) \left(\frac{E_F - E_0}{k_B T} \right)$$

$$= g(E)_{2D} (E_F - E_0) \tag{20.64}$$

for an unoccupied second subband so that the $E_F - E_1$ term in eq. (20.64) term becomes zero. Hence, from eq. (20.51)

$$E_F - E_0 = \frac{N_s}{g(E)_{2D}} = \frac{N_s}{\frac{m_n^*}{\pi \hbar^2}} = \frac{\pi \hbar^2 N_s}{m_n^*} \tag{20.65}$$

iii. Energy difference between the conduction band edge and Fermi level

At equilibrium, the Fermi level is same throughout the structure. Assuming step function, this energy difference may be expressed as

$$\Delta E_{C \to F} = (E_C - E_F)_{AlGaAs} \tag{20.66}$$

iv. Electrical potential energy of the field across the depletion region on the AlGaAs side

An electric charge placed in the field of the depletion region will either move in the direction of the field and perform work, or conversely, work will have to be done to move this charge against the direction of the field. The energy equivalent to this amount of work is stored in the electric field as potential energy. To find the potential energy, let us determine the potential across the depletion region.

The electrostatic potential V across the depletion region is related to the charge density ρ by Poisson's equation

$$\frac{d^2 V}{dz^2} = -\frac{\rho}{\varepsilon_0 \varepsilon} \tag{20.67}$$

or

$$\frac{d}{dz} \left(\frac{dV}{dz} \right) = -\frac{e N_D}{\varepsilon_0 \varepsilon} \tag{20.68}$$

where N_D is the doping concentration in AlGaAs. From eq. (20.68)

$$d \left(\frac{dV}{dz} \right) = -\frac{e N_D}{\varepsilon_0 \varepsilon} dz \tag{20.69}$$

$$\therefore \int d \left(\frac{dV}{dz} \right) = -\frac{e N_D}{\varepsilon_0 \varepsilon} \int dz \tag{20.70}$$

or

$$\frac{dV}{dz} = -\frac{e N_D}{\varepsilon_0 \varepsilon} z + \text{Constant} \tag{20.71}$$

or

$$dV = -\frac{e N_D}{\varepsilon_0 \varepsilon} z dz + \text{Constant} \tag{20.72}$$

$$\therefore \int dV = - \int_0^{W_{AlGaAs}} \frac{e N_D}{\varepsilon_0 \varepsilon} z dz = -\frac{e N_D}{\varepsilon_0 \varepsilon} \left. \frac{z^2}{2} \right|_0^{W_{AlGaAs}}$$

$$= -\frac{e N_D}{\varepsilon_0 \varepsilon} \left\{ \frac{(W_{AlGaAs})^2}{2} - 0 \right\}$$

$$= -\frac{e N_D}{\varepsilon_0 \varepsilon} \times \frac{W_{AlGaAs}^2}{2} = -\frac{e N_D W_{AlGaAs}^2}{2 \varepsilon_0 \varepsilon} \tag{20.73}$$

where W_{AlGaAs} is the width of the depletion region on the AlGaAs side.

Hence, from eq. (20.73) we write

$$V_{Depletion} = -\frac{eN_D W_{AlGaAs}^2}{2\varepsilon_0 \varepsilon} \tag{20.74}$$

from which the electrical potential energy is

$$E_{Depletion} = -eV_{Depletion} = -e \times -\frac{eN_D W_{AlGaAs}^2}{2\varepsilon_0 \varepsilon} \tag{20.75}$$

which may be written as

$$E_{Depletion} = \frac{e^2 N_D^2 W_{AlGaAs}^2}{2\varepsilon_0 \varepsilon N_D} \tag{20.76}$$

For charge neutrality

Electric charge in the depletion region

$$= \text{Electric charge in the 2DEG} \tag{20.77}$$

or from eq. (20.47)

$$eN_D W_{AlGaAs} = eN_s \tag{20.78}$$

from which

$$N_D W_{AlGaAs} = N_s \tag{20.79}$$

or

$$N_D^2 W_{AlGaAs}^2 = N_s^2 \tag{20.80}$$

In view of eq. (20.80), eq. (20.76) becomes

$$E_{Depletion} = \frac{e^2 N_s^2}{2\varepsilon_0 \varepsilon N_D} \tag{20.81}$$

Combining together eqs. (20.50), (20.65), (20.66), and (20.81)

$$\Delta E_C = E_1 + \Delta E_{N_s} + \Delta E_{C \to F} + E_{Depletion}$$

$$= -\left(\frac{\hbar^2}{2m_n^*}\right)^{1/3} \left(\frac{e^2 N_s}{\varepsilon_0 \varepsilon}\right)^{2/3} a_1 + \frac{\pi \hbar^2 N_s}{m_n^*}$$

$$+ \left(E_C - E_F\right)_{AlGaAs} + \frac{e^2 N_s^2}{2\varepsilon_0 \varepsilon N_D} \tag{20.82}$$

Equation (20.82) is an implicit equation in the variable N_s, which is found by solving this equation.

In practice, one may be working with more complicated HEMT structures. For these structures, eq. (20.82) is modified to include the relevant effects. The structural intricacies include:

i. inclusion of undoped AlGaAs spacer layer after doped AlGaAs, i.e. for making the arrangement: doped AlGaAs/undoped AlGaAs/undoped GaAs.

ii. recombination centers in the barrier layer, and

iii. polarization charges in nitride materials, e.g. when considering AlGaN/GaN HEMTs.

20.5 Linear Charge-Control Model of HEMTs

This model is propounded by Delagebeaudeuf and Linh (1982). A HEMT device is considered in which ϕ_M is the work function of the metal used for the Schottky contact (energy difference between the Fermi level and vacuum level) to the large bandgap semiconductor 2 having dielectric constant ε_2 and electron affinity χ_2 (energy difference between the conduction band edge and vacuum level) and total thickness d_2 out of which a small thickness d is left undoped (Figure 20.6). Semiconductor 2 has a doping concentration of N_2. The full thickness d_2 is assumed to be sufficiently small to be fully depleted as a result of blending of depletion regions formed at the Schottky barrier contact and by electron transfer to the small bandgap semiconductor 1. Suppose Φ_B is the Schottky barrier height. Let V_G be the applied reverse bias gate voltage. For a device in which d_2 is not small, V_G is assumed to be sufficiently large to deplete semiconductor 2. So semiconductor 2 is depleted either by virtue of being thin or by application of large reverse bias V_G. Let E_F be the height of the Fermi level from the conduction band edge of the small bandgap semiconductor and let ΔE_C denote the energy difference between the edges of the conduction bands of the two semiconductors.

From the definitions of $e\Phi_M$, $e\Phi_B$, $-eV$, eE_F, $e\chi_2$, and $e\Delta E_C$ in Figure 20.6(c), it is evident that they are inter-related as

$$e\Phi_B + eV + eE_F = e\chi_2 + e\Delta E_C \tag{20.83}$$

or

$$e\Phi_B - eV_G + eE_F = e\chi_2 + e\Delta E_C \tag{20.84}$$

putting for reverse bias

$$V = -V_G \tag{20.85}$$

From eq. (20.84),

$$\Phi_B - V_G + E_F = \chi_2 + \Delta E_C \tag{20.86}$$

or

$$\chi_2 = \Phi_B - V_G - \Delta E_C + E_F \tag{20.87}$$

Assuming fully depleted semiconductor 2 in the charge-control regime, as mentioned earlier, and taking the origin at the heterointerface, the charge concentrations in the two parts of semiconductor (undoped and doped with carrier concentration N_2) are:

$$N_2(x) = 0 \quad \text{for} \quad 0 > x > -d \tag{20.88}$$

and

$$N_2(x) = N_2 \quad \text{for} \quad -d > x > -d_2 \tag{20.89}$$

We arbitrarily choose the potential at $x = 0$

$$V_2(0) = 0 \qquad (20.90)$$

Then by Poisson's equation, the potential at $x = -d_2$ due to carrier concentration $N_2(x)$ is given by

$$\frac{d^2V}{dx^2} = -\frac{\text{Charge density}}{\varepsilon_0\varepsilon_2} = -\frac{eN_2(x)dx}{\varepsilon_0\varepsilon_2} \qquad (20.91)$$

Hence,

$$\int \frac{d^2V}{dx^2} = -\int \frac{eN_2(x)dx}{\varepsilon_0\varepsilon_2} \qquad (20.92)$$

or

$$\int \frac{d}{dx}\left(\frac{dV}{dx}\right) = -\frac{eN_2(x)}{\varepsilon_0\varepsilon_2}\int dx \qquad (20.93)$$

or

$$\frac{dV}{dx} = -\frac{eN_2(x)x}{\varepsilon_0\varepsilon_2} \qquad (20.94)$$

or

$$dV = -\frac{eN_2(x)x}{\varepsilon_0\varepsilon_2}dx \qquad (20.95)$$

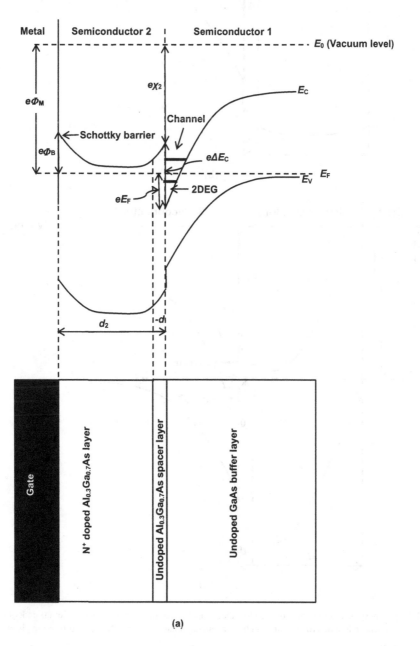

(a)

FIGURE 20.6 Energy-band diagrams of metal-semiconductor 2–semiconductor 1 structure: (a) under equilibrium, (b) under forward bias, and (c) under reverse bias. This is an N-n heterostructure and must be differentiated from the N-p heterostructure given in Figure 20.3.

(Continued)

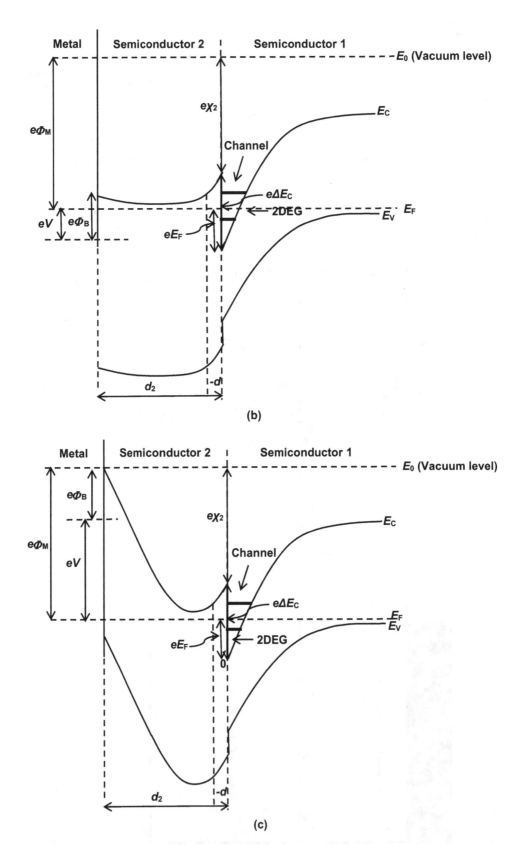

FIGURE 20.6 (CONTINUED)   Energy-band diagrams of metal-semiconductor 2–semiconductor 1 structure: (a) under equilibrium, (b) under forward bias, and (c) under reverse bias. This is an N-n heterostructure and must be differentiated from the N-p heterostructure given in Figure 20.3.

$$\therefore \int_0^U dV = -\frac{eN_2(x)}{\varepsilon_0 \varepsilon_2} \int_0^{-(d_2-d)} x \, dx \qquad (20.96)$$

or

$$|V|_0^U = -\frac{eN_2(x)}{\varepsilon_0 \varepsilon_2} \left.\frac{x^2}{2}\right|_0^{-(d_2-d)} \qquad (20.97)$$

or

$$U - 0 = -\frac{eN_2(x)}{\varepsilon_0 \varepsilon_2} \left[\frac{\{-(d_2-d)\}^2}{2} - 0 \right] \qquad (20.98)$$

or

$$U = -\frac{eN_2(x)(d_2-d)^2}{2\varepsilon_0 \varepsilon_2} \qquad (20.99)$$

The total potential V_2 at $x = -d_2$ is

$$V_2 = U - \text{Potential } V_i \text{ due to interfacial electric field } E_i \qquad (20.100)$$

$$\therefore V_2 = U - V_i = -\frac{eN_2(x)(d_2-d)^2}{2\varepsilon_0 \varepsilon_2} + E_i d_2 \qquad (20.101)$$

since

$$E_i = -\frac{V_i}{d_2} \qquad (20.102)$$

But

$$V_2 = -\chi_2 \qquad (20.103)$$

since first electron affinity is negative, energy being released when an electron is added to a neutral atom. From eqs. (20.101) and (20.103),

$$-\frac{eN_2(x)(d_2-d)^2}{2\varepsilon_0 \varepsilon_2} + E_i d_2 = -\chi_2 \qquad (20.104)$$

or

$$E_i d_2 = \frac{eN_2(x)(d_2-d)^2}{2\varepsilon_0 \varepsilon_2} - \chi_2 \qquad (20.105)$$

or

$$\left(\frac{\varepsilon_2}{d_2}\right) E_i d_2 = \left(\frac{\varepsilon_2}{d_2}\right)\frac{eN_2(x)(d_2-d)^2}{2\varepsilon_0 \varepsilon_2} - \left(\frac{\varepsilon_2}{d_2}\right)\chi_2 \qquad (20.106)$$

or

$$\varepsilon_2 E_i = \left(\frac{\varepsilon_2}{d_2}\right)\zeta - \left(\frac{\varepsilon_2}{d_2}\right)\chi_2 = \left(\frac{\varepsilon_2}{d_2}\right)(\zeta - \chi_2) \qquad (20.107)$$

by putting

$$\zeta = \frac{eN_2(x)(d_2-d)^2}{2\varepsilon_0 \varepsilon_2} \qquad (20.108)$$

Substituting for χ_2 from eq. (20.87) in eq. (20.107)

$$\varepsilon_2 E_i = \left(\frac{\varepsilon_2}{d_2}\right)\{\zeta - (\Phi_B - V_G - \Delta E_C + E_F)\}$$

$$= \left(\frac{\varepsilon_2}{d_2}\right)\{\zeta - \Phi_B + V_G + \Delta E_C - E_F\} \qquad (20.109)$$

From eq. (20.47), we can write the sheet electron charge $\frac{Q_s}{S}$ related to the sheet electron concentration N_s as

$$\frac{Q_s}{S} = eN_s = \varepsilon_2 E_i = \left(\frac{\varepsilon_2}{d_2}\right)\{\zeta - \Phi_B + V_G + \Delta E_C - E_F\} \qquad (20.110)$$

where interface states are ignored. Thus

$$\frac{Q_s}{S} \cong \left(\frac{\varepsilon_2}{d_2}\right)\{V_G - (\Phi_B - \Delta E_C + E_F - \zeta)\} = \left(\frac{\varepsilon_2}{d_2}\right)(V_G - V_{\text{Off}}) \qquad (20.111)$$

where

$$V_{\text{Off}} = \Phi_B - \Delta E_C + E_F - \zeta \qquad (20.112)$$

is the voltage at which the 2DEG is annihilated.

At an applied gate voltage V_G, let $V_{\text{Channel}}(x)$ be the channel voltage at the point x under the gate. Then the effective gate voltage controlling the charge at location x is

$$V_{\text{Effective}}(x) = V_G - V_{\text{Channel}}(x) \qquad (20.113)$$

according to which eq. (20.111) becomes

$$\frac{Q_s}{S} = \left(\frac{\varepsilon_2}{d_2}\right)\{V_G - V_{\text{Channel}}(x) - V_{\text{Off}}\} \qquad (20.114)$$

by replacing V_G in equation (20.111) by $V_{\text{Effective}}$ from equation (20.113). If Z is the width of the gate and $v_d(x)$ is the drift velocity of electrons at position x, the current in the channel is

$$I_{\text{DS}} = \frac{\text{Charge per unit area} \times \text{Gate width} \times \text{Channel length}}{\text{Time to cross the channel length}}$$

$$= \text{Charge per unit area} \times \text{Gate width} \times \text{Electron velocity}$$

$$= \frac{Q_s}{S} Z \, v_d(x) \qquad (20.115)$$

Let us take the electric field dependence of the electron drift velocity from eq. (19.34) as

$$v_d = -\mu_n E \quad \text{for} \quad E < E_C \qquad (20.116)$$

Minus sign is used because the direction of electron motion is opposite to the direction of the electric field.

We write down the current equation for $E < E_C$ by substituting for $\frac{Q_s}{S}$ from eq. (20.114) and for v_d from eq. (20.116) in eq. (20.115)

$$I_{DS} = \left(\frac{\varepsilon_2}{d_2}\right)\{V_G - V_{Channel}(x) - V_{Off}\} Z\, v_d(x)$$

$$= \left(\frac{\varepsilon_2}{d_2}\right)\{V_G - V_{Channel}(x) - V_{Off}\} Z \times (-\mu_n E)$$

$$= -\left(\frac{\varepsilon_2}{d_2}\right)\mu_n Z\{V_G - V_{Channel}(x) - V_{Off}\} \times E$$

$$= -\left(\frac{\varepsilon_2}{d_2}\right)\mu_n Z\{V_G - V_{Channel}(x) - V_{Off}\} \times \left\{-\frac{dV_{Channel}(x)}{dx}\right\}$$

$$= \left(\frac{\varepsilon_2}{d_2}\right)\mu_n Z\{V_G - V_{Channel}(x) - V_{Off}\}\frac{dV_{Channel}(x)}{dx} \tag{20.117}$$

since

$$E = -\frac{dV_{Channel}(x)}{dx} \tag{20.118}$$

Eq. (20.117) is

$$I_{DS} = \left(\frac{\varepsilon_2}{d_2}\right)\mu_n Z\{V_G - V_{Channel}(x) - V_{Off}\}\frac{dV_{Channel}(x)}{dx} \tag{20.119}$$

At a low drain–source voltage in the linear region of operation,

$$V_{Channel}(x) = 0 \tag{20.120}$$

and we can write

$$I_{DS} = \left(\frac{\varepsilon_2}{d_2}\right)\mu_n (V_G - V_{Off}) Z \frac{V_{Channel}(L) - V_{Channel}(0)}{L - 0}$$

$$= \left(\frac{\varepsilon_2}{d_2}\right)\mu_n (V_G - V_{Off}) Z \frac{V_{Channel}(L) - V_{Channel}(0)}{L} \tag{20.121}$$

by putting

$$\frac{dV_{Channel}}{dx} = \frac{(\text{Potential at } x = L) - (\text{Potential at } x = 0)}{(x = L) - (x = 0)}$$

$$= \frac{V_{Channel}(L) - V_{Channel}(0)}{L - 0} \tag{20.122}$$

If R_S is source access resistance and R_D is drain access resistance,

$$V_{Channel}(0) = R_S I_{DS} \tag{20.123}$$

and

$$V_{Channel}(L) = V_{DS} - R_D I_{DS} \tag{20.124}$$

Substituting for $V_{Channel}(L)$ and $V_{Channel}(0)$ from eqs. (20.124) and (20.123) in eq. (20.121), we get

$$I_{DS} = \left(\frac{\varepsilon_2}{d_2}\right)\mu_n (V_G - V_{Off}) Z \frac{V_{DS} - R_D I_{DS} - R_S I_{DS}}{L}$$

$$= \left(\frac{\varepsilon_2}{L d_2}\right)\mu_n (V_G - V_{Off}) Z (V_{DS} - R_D I_{DS} - R_S I_{DS}) \tag{20.125}$$

or

$$I_{DS} L d_2 = \varepsilon_2 \mu_n (V_G - V_{Off}) Z (V_{DS} - R_D I_{DS} - R_S I_{DS}) \tag{20.126}$$

or

$$\frac{I_{DS} L d_2}{I_{DS}} = \frac{\varepsilon_2 \mu_n (V_G - V_{Off}) Z (V_{DS} - R_D I_{DS} - R_S I_{DS})}{I_{DS}} \tag{20.127}$$

or

$$L d_2 = \varepsilon_2 \mu_n (V_G - V_{Off}) Z \left(\frac{V_{DS}}{I_{DS}} - R_D - R_S\right)$$

$$= \varepsilon_2 \mu_n (V_G - V_{Off}) Z \left(\frac{V_{DS}}{I_{DS}}\right) - \varepsilon_2 \mu_n (V_G - V_{Off}) Z (R_D + R_S) \tag{20.128}$$

or

$$\varepsilon_2 \mu_n (V_G - V_{Off}) Z \left(\frac{V_{DS}}{I_{DS}}\right) = L d_2 + \varepsilon_2 \mu_n (V_G - V_{Off}) Z (R_D + R_S) \tag{20.129}$$

$$\therefore \frac{V_{DS}}{I_{DS}} = \frac{L d_2 + \varepsilon_2 \mu_n (V_G - V_{Off}) Z (R_D + R_S)}{\varepsilon_2 \mu_n (V_G - V_{Off}) Z}$$

$$= R_D + R_S + \frac{L d_2}{\varepsilon_2 \mu_n (V_G - V_{Off}) Z} \tag{20.130}$$

If R_{DS} is the drain–source resistance

$$R_{DS} = \frac{V_{DS}}{I_{DS}} = R_D + R_S + \frac{L d_2}{\varepsilon_2 \mu_n (V_G - V_{Off}) Z} \tag{20.131}$$

As the gate voltage V_G increases the drain–source resistance R_{DS} decreases so that the HEMT device acts as a voltage-controlled resistor.

20.6 Discussion and Conclusions

HFET, HEMT, MODFET, and TEGFET are all acronyms for the same device. In a HEMT, the mobility can be as high as $49300\,cm^2\,V^{-1}\,s^{-1}$, and the reason for the high mobility is separation of the free carriers from the parent dopant ions, thus preventing the ions from exerting their degradation effect on mobility through ionized impurity scattering. The gate is the usual Schottky-barrier metal-semiconductor junction. Mobility-wise, HFET comes first followed by MESFET with MOSFET at the end.

On contacting N+-doped AlGaAs layer with undoped GaAs, electron transference takes place due to the concentration gradient from AlGaAs to GaAs and the higher electron affinity of GaAs. Soon, the creation of a positive unbalanced charge buildup on the AlGaAs side and an unbalanced negative charge buildup on the GaAs side sets up an electric field across the interface opposing further electron transference. In the ~10 nm or less thick electron charge sheet on the GaAs side, the electron motion is quantized perpendicular to the interface but free in the other two directions. Hence, it is called a 2D free electron gas. In the energy-band diagram of the AlGaAs-GaAs heterojunction, a quasi-triangular potential well is seen, which is the quantum well.

A HEMT made from a material combination such as AlGaAs/GaAs, which does not have a built-in polarization charge, works as an enhancement HEMT (eHEMT) for a thin AlGaAs layer. A GaInN/GaN HEMT which has a built-in polarization charge works as a depletion or dHEMT.

In the non-aligned gate HEMT process, following layers are sequentially deposited on the SI-GaAs substrate: an undoped GaAs buffer layer, an undoped $Al_xGa_{1-x}As$ spacer layer, an $N^+-Al_xGa_{1-x}As$ layer, and an N^+-GaAs capping layer. Devices are isolated by mesa etching to reach up to SI-GaAs. Source/drain contacts are formed, gate area is opened by etching, and gate metallization is done. As the gate lies sunken within the raised GaAs islands, the structure is of recessed-gate type. In the aligned gate process, first source/drain metallization is done to cover the sides of the mesa up to GaN layer below AlGaN; then gate metallization and patterning are done to obtain a T-shaped gate metal with the top of 'T' projecting over the surface. Second source/drain metallization uses these overhangs of 'T' as a shadow mask.

A HEMT made by depositing a thin film of a semiconductor B over a semiconductor A with the thin film of B adjusting to the lattice structure of A by virtue of being thin is said to be pseudomorphic. A HEMT made by inserting a buffer layer between the two semiconductors to facilitate lattice matching is metamorphic.

To find the electron energy levels in the well, the potential energy of the electron (charge e^-) in the electric field F is expressed as $U = eFz$, and the triangular potential well is defined. The TISE is solved assuming that the electron effective mass is the same in GaAs and AlGaAs, and the dielectric constants of the semiconductors are equal. The equation is found to have the form of the Airy differential equation whose general solution is written in terms of the Airy function. The boundary condition $\Psi(z) = 0$ at $z = 0$ is applied. The energy levels of electrons are found by expressing the roots in an asymptotic form. The equation for energy levels contains the field F, which is expressed in terms of surface charge per unit area eN_s and hence the electron sheet density N_s.

To find N_s, we sum up the energy components that comprise the discontinuity of conduction band edge ΔE_C at the interface of the two semiconductors.

i. We start with energy level E_1 for $n = 1$. So, the first energy component is E_1. Its equation contains N_s.

ii. The second energy component is energy difference $\left(\Delta E_{N_s} = E_F - E_0 \right)$ created by electron charge eN_s transferred to the GaAs side. Here E_0 is the energy of the non-degenerate ground state. The equation for N_s is written from the equations for 2D density of states and the Fermi-Dirac distribution function. Then N_s is obtained by integration. The equation for $E_F - E_0$ is written in the low temperature limit $T \to 0$. Obviously this equation contains N_s.

iii. The third energy component is the energy difference between the conduction band edge and Fermi level $\Delta E_{C \to F}$.

iv. The fourth energy component is the electrical potential energy $E_{depletion}$ of the field across the depletion region on the AlGaAs side. To find the potential energy, the potential V across the depletion region on the AlGaAs side is determined by the integration of Poisson's equation. Then the charge neutrality condition is applied to get the equation for $E_{depletion}$. The electric charge in the depletion region equals the electric charge in the electron sheet. The equation for $E_{depletion}$ also contains N_s.

Finally, the ΔE_C equation is written by adding together contributions from energy components (i)–(iv). We find that in this equation, the terms due to energy components (i), (ii), and (iv) contain N_s. We are unable to express N_s explicitly in terms of other parameters. So, it is an implicit or "mixed-up" equation in N_s. An equation is said to be implicit when the dependent variable is not isolated on one side of the equation in terms of the independent variables or constants. This kind of equation can be solved numerically.

The linear charge-control model of the HEMT was proposed by Delagebeaudeuf and Linh (1982). Consider a HEMT device in which the large bandgap semiconductor 2 has a dielectric constant ε_2, electron affinity χ_2, and thickness d_2. A small portion d of semiconductor 2 is undoped while the remaining $(d_2 - d)$ portion is doped. The metal layer on semiconductor 2 has a work function Φ_M. The Schottky barrier height is Φ_B. Also suppose the discontinuity of the conduction band edge is ΔE_C. The reverse gate voltage V_G is adequate enough to fully deplete semiconductor 2. The distance between the Fermi level and the bottom of the conduction band edge of the small bandgap semiconductor 1 is E_F.

From the energy-band diagram of the HEMT, χ_2 is written in terms of Φ_B, V_G, ΔE_C, and E_F. The total potential V_2 at $x = -d_2$ consists of two components, first from the charge in the doped region $(d_2 - d)$ in semiconductor 2 and second due to the interfacial electric field E_i. The first component is found by the integration of Poisson's equation from 0 to $-(d_2 - d)$ and the second is related to E_i. But $V_2 = -\chi_2$. So we can substitute for χ_2 obtained earlier in the V_2 equation. The resulting equation gives us the sheet electron charge $Q_s = eN_s = \varepsilon_2 E_i$. The charge Q_s contains a factor $\left\{ V_G - \left(\Phi_B - \Delta E_C + E_F - \xi \right) \right\} = (V_G - V_{Off})$. When the portion $\left(\Phi_B - \Delta E_C + E_F - \xi \right)$ in this equation equals V_G, Q_s becomes zero. So the portion $\left(\Phi_B - \Delta E_C + E_F - \xi \right)$ is denoted by V_{off}. Now the charge Q_s can be expressed in terms of the voltage $V_G - V_{Channel}(x) - V_{Off}$. The equation for Q_s enables us to write down the equation for drain–source current I_{DS} for gate width Z and drift velocity $v_d = -\mu_n E$. But $E = -dV_{Channel}(x)/dx$. At a low drain–source voltage, the drain–source resistance R_{DS} can be expressed as the sum of source access resistance R_S, drain access resistance R_D, and a third term containing V_G in the denominator. So R_{DS} decreases as V_G increases, indicating the operation of the HEMT as a voltage-controlled resistor.

Illustrative Exercises

20.1. In a GaAs/AlGaAs HEMT, the electron concentration in the channel formed in GaAs is $5 \times 10^{11} \mathrm{cm}^{-2}$. If the dielectric constant of GaAs is 12.9, what is the electric field setup at the GaAs/AlGaAs interface?

Here $N_s = 5 \times 10^{11} \mathrm{cm}^{-2} = 5 \times 10^{11} \times 10^4 \mathrm{m}^{-2}$.

Eq. (20.48) is applied.

$$\therefore F = \frac{eN_s}{\varepsilon_0 \varepsilon_{GaAs}} = \frac{1.602 \times 10^{-19}\,\mathrm{C} \times 5 \times 10^{11} \times 10^4\,\mathrm{m}^{-2}}{8.854 \times 10^{-12}\,\mathrm{F\,m^{-1}} \times 12.9}$$

$$= 0.07 \times 10^8\,\mathrm{V\,m^{-1}} = 7 \times 10^6\,\mathrm{V\,m^{-1}} \qquad (20.132)$$

20.2. Calculate the energy of the first level E_1 in a GaAs/AlGaAs HEMT if an electric field of $7\,\mathrm{MV\,m^{-1}}$ is created by electron transference from AlGaAs to GaAs. Electron effective mass $= 0.067\,m_0$.

From eq. (20.43),

$$E_{n=1} = \left(\frac{\hbar^2}{2m_n^*}\right)^{1/3} \left\{\frac{3\pi eF}{2}\left(n - \frac{1}{4}\right)\right\}^{2/3}$$

$$= \left\{\frac{\left(1.05457 \times 10^{-34}\,\mathrm{J\,s^{-1}}\right)^2}{2 \times 0.067 \times 9.109 \times 10^{-31}\,\mathrm{kg}}\right\}^{0.33}$$

$$\times \left\{\frac{3 \times 3.14 \times 1.602 \times 10^{-19}\,\mathrm{C} \times 7 \times 10^6\,\mathrm{V\,m^{-1}}}{2}\left(1 - \frac{1}{4}\right)\right\}^{0.67}$$

$$= \left(\frac{1.112 \times 10^{-68}}{1.2206 \times 10^{-31}}\right)^{0.33} \left(3.961 \times 10^{-12}\right)^{0.67}$$

$$= \left(9.11 \times 10^{-38}\right)^{0.33} \left(3.961 \times 10^{-12}\right)^{0.67}$$

$$= 5.96 \times 10^{-13} \times 2.2937 \times 10^{-8} = 1.367 \times 10^{-20}\,\mathrm{J}$$

$$= \frac{1.367 \times 10^{-20}}{1.602 \times 10^{-19}}\,\mathrm{eV} = 0.08533\,\mathrm{eV} \qquad (20.133)$$

REFERENCES

Delagebeaudeuf, D. and N.T. Linh. 1982. Metal-(n) AlGaAs-GaAs two-dimensional electron gas FET. *IEEE Transactions on Electrons Devices ED* 29(6): 955–960.

Hiyamizu, S., T. Mimura, T. Fujii, K. Nanbu and H. Hashimoto. 1981a. Extremely high mobility of two-dimensional electron gas in selectively doped GaAs/N-AlGaAs heterojunction structures grown by MBE. *Japanese Journal of Applied Physics* 20(4): L245–L248.

Hiyamizu, S., K. Nanbu, T. Mimura, T. Fujii and H. Hashimoto. 1981b. Room-temperature mobility of two-dimensional electron gas in selectively doped GaAs/N-AlGaAs heterojunction structures. *Japanese Journal of Applied Physics* 20(5): L378–L380.

Lee, J., D. Liu, H. Kim, M. Schuette, J.S. Flynn, G.R. Brandes and W. Lu. 2004. Self-aligned AlGaN/GaN high electron mobility transistors. *Electronics Letters* 40(19): 2.

Liu Z., T. Huang, Q. Li, X. Lu and X. Zou. 2016. Compound semiconductor materials and devices. *Synthesis Lectures on Emerging Engineering Technologies* 2(3): 1–73.

Mimura, T. 2005. Development of high electron mobility transistor. *Japanese Journal of Applied Physics* 44(12): 8263–8268.

Mimura, T., S. Hiyamizu, T. Fujii and K. Nanbu. 1980. A new field-effect transistor with selectively doped GaAs/n-Al$_x$Ga$_{1-x}$As heterojunctions. *Japanese Journal of Applied Physics* 19(5): L225–L227.

Ng, G.I. and S. Arulkumaran. 2012. GaN HEMTs technology and applications. In: Iniewski, K. (Ed.), *Nano Semiconductors: Devices and Technology*. CRC Press, Taylor & Francis Group, LLC: Boca Raton, FL, pp. 377–414.

Subramanian, S. 1990. High electron mobility transistor. *Bulletin of Materials Science* 13(1–2): 121–133.

21

Single-Electron Transistors

Calculations of charging energy of a conductive island followed by understanding of Coulomb blockade effect and working of tunnel junctions lead us to the single electron transistor.

21.1 Energy Used in Charging a Conductive Island

Let us consider an island of conducting/semiconducting material being charged by a battery. The potential V of the island due to charge q already present on the island is

$$V = \frac{q}{C} \tag{21.1}$$

where C is the capacitance of the island.

The work done to transfer a small charge dq to this island against the electric field of charge q already present on the island is

$$dW = Vdq = \frac{q}{C}dq \tag{21.2}$$

So the total work done to transfer a charge Q is

$$\int_0^W dW = \int_0^Q \frac{q}{C}dq = \frac{1}{C}\int_0^Q q\,dq = \frac{1}{C}\left.\frac{q^2}{2}\right|_0^Q \tag{21.3}$$

or

$$W = \frac{1}{C}\left(\frac{Q^2}{2} - 0\right) = \frac{Q^2}{2C} \tag{21.4}$$

The energy consumed during the charging of the island is stored in it as electrostatic potential energy. Because it is the charging energy, it will be denoted by W_C; hence

$$W_C = \frac{Q^2}{2C} \tag{21.5}$$

The earlier equation may also be derived by taking an uncharged conducting island at zero potential and bringing electrons on it one by one. The first electron experiences a potential $V = 0$. The second electron has to withstand the repulsive force from a potential $0+dV$. The third electron faces a potential $0+2dV$. So work has to be done against the repulsive forces of increasing magnitudes as the potential of the island increases from 0 to V. The charging operation may be performed against an average potential $(0+V)/2 = V/2$ so that the work done is

$$W_C = \frac{QV}{2} \tag{21.6}$$

With the capacitance of the island being C, we have

$$C = \frac{Q}{V} \tag{21.7}$$

or

$$V = \frac{Q}{C} \tag{21.8}$$

so that eq. (21.6) becomes

$$W_C = \frac{Q}{2} \times \frac{Q}{C} = \frac{Q^2}{2C} \tag{21.9}$$

which is the same as eq. (21.5).

21.2 Other Energy Components, Work Done by Voltage Sources, and the Helmholtz Free Energy

Besides the electrostatic charging energy W_C, other characteristic energies associated with the island include the Fermi energy change ΔE_F dependent on the carrier concentration (typically $10^{22}/cm^3$ for metals and increasing from $10^{10}/cm^3$ for intrinsic silicon to $10^{20}/cm^3$ for degenerately doped silicon) and quantum confinement energy W_N so that the total energy is

$$W_\Sigma = W_C + \Delta E_F + W_N \tag{21.10}$$

The work done by voltage sources in the circuit is the summation of the time integral of power supplied to it by the sources and can be expressed in terms of voltage V, current I, and time t as

$$W_V = \sum_{\text{Voltage sources}} \int V(t)I(t)\,dt \tag{21.11}$$

Sequel to any tunneling event, charge flow occurs. It takes place from/to the contacts. The aim of charge flow is to produce equilibrium. The transferred charge is the sum of tunneling and polarization charges, that is, the electron charge tunneling into/out of the island and the polarization charge building up due to change in potential of the island.

The Helmholtz free energy is a thermodynamic potential. A thermodynamic potential is a state function. Together with the equations of state, it describes the equilibrium state of a thermodynamic system. It determines the amount of useful work that can be obtained from a closed system at a constant temperature and volume. Here the free energy is defined as

F = Total energy stored − Work done by voltage sources

$$= W_\Sigma - W_V = W_C + \Delta E_F + W_N - \sum_{\text{Voltage sources}} \int V(t)I(t)dt \tag{21.12}$$

using equations (21.10) and (21.11).

21.3 Capacitance of a Spherical Conductive Island

The electric field E at a point on the surface of a charged conducting sphere carrying a charge Q and having a radius r is obtained from Gauss' law (Figure 21.1). Considering a spherical Gaussian surface of radius $R > r$, the electric flux across the Gaussian surface is

$$\Phi = \text{Electric field} \times \text{Area of the spherical surface}$$

$$= E\left(4\pi R^2\right) = \frac{Q}{\varepsilon_0} \tag{21.13}$$

from which the electric field at a distance $R > r$ is

$$E = \frac{Q}{\varepsilon_0\left(4\pi R^2\right)} = \frac{Q}{4\pi\varepsilon_0 R^2} \tag{21.14}$$

For a point on the surface of the sphere, $R = r$; hence

$$E = \frac{Q}{4\pi\varepsilon_0 r^2} \tag{21.15}$$

The magnitude of the electric field is the same at every point on the surface and its direction points outward. The magnitude is the same as if a point charge Q is placed at the center of the spherical conductor.

The potential at a point on the surface of the spherical island is found by integrating the electric field along the radial direction

$$\int dV = -\int_\infty^r E\,dr = -\int_\infty^r \frac{Q}{4\pi\varepsilon_0 r^2}\,dr \tag{21.16}$$

using eq. (21.15) and

$$E = -\frac{dV}{dr} \tag{21.17}$$

or

$$dV = -E\,dr \tag{21.18}$$

Hence, from equation (21.16)

$$\int dV = -\frac{Q}{4\pi\varepsilon_0}\int_\infty^r r^{-2}\,dr = -\frac{Q}{4\pi\varepsilon_0}\left.\frac{r^{-2+1}}{-2+1}\right|_\infty^r = -\frac{Q}{4\pi\varepsilon_0}\left.\frac{r^{-1}}{-1}\right|_\infty^r$$

$$= \frac{Q}{4\pi\varepsilon_0}\left(r^{-1} - \infty^{-1}\right)$$

$$= \frac{Q}{4\pi\varepsilon_0}\left(r^{-1} - \frac{1}{\infty}\right) = \frac{Q}{4\pi\varepsilon_0}\left(r^{-1} - 0\right) = \frac{Q}{4\pi\varepsilon_0 r} \tag{21.19}$$

or

$$V = \frac{Q}{4\pi\varepsilon_0 r} \tag{21.20}$$

Therefore, the capacitance of the spherical conductive island is

$$C = \frac{Q}{V} = \frac{Q}{\dfrac{Q}{4\pi\varepsilon_0 r}} = Q \times \frac{4\pi\varepsilon_0 r}{Q} = 4\pi\varepsilon_0 r \tag{21.21}$$

or

$$C = 4\pi\varepsilon_0 r \tag{21.22}$$

21.4 Dependence of Charging Energy on the Island Size and Coulomb Blockade Effect

For a spherical conductive island of radius 5 nm (diameter 10 nm),

$$C = 4 \times 3.14159 \times 8.854 \times 10^{-12}\,\text{F m}^{-1} \times 5 \times 10^{-9}\,\text{m}$$

$$= 5.563 \times 10^{-19}\,\text{F} \tag{21.23}$$

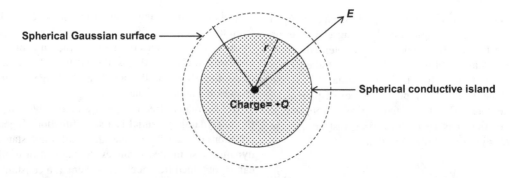

FIGURE 21.1 Electric field due to a charged conducting sphere.

Hence from equation (21.5), the energy stored in the island on receiving a single electron charge $= 1.6 \times 10^{-19}$ C is

$$W = \frac{Q^2}{2C} = \frac{\left(1.6 \times 10^{-19}\right)^2}{2 \times 5.563 \times 10^{-19}} = \frac{2.56 \times 10^{-38}}{1.1126 \times 10^{-18}} = 2.301 \times 10^{-20} \text{ J}$$

$$(21.24)$$

The thermal energy at room temperature ($T = 300$ K) is

$$H = k_B T \qquad (21.25)$$

where k_B is the Boltzmann constant $= 1.3806 \times 10^{-23}$ J K^{-1}.
So

$$H = 1.3806 \times 10^{-23} \times 300 = 4.142 \times 10^{-21} \text{ J} \qquad (21.26)$$

The earlier calculation shows that for a conductive island of ultra-small dimensions in the nanometer range, the capacitance becomes exceedingly small so that the energy change occurring on transferring a single electron to the island may reach an extremely high value and even exceed the thermal energy at room temperature, and by an appreciable margin at extremely low temperatures. So a further incoming electron will encounter a large opposition in moving toward the island. This inhibition of tunneling of an electron to a charged conductor can be overcome by an applied bias. The inhibitory effect is called the Coulomb blockade effect and is observed when the conducting or semi-conducting object is so small in size that addition or removal of an electron produces a sufficiently large change in electrostatic energy comparable to thermal energy, either at room temperature or by cooling down to low temperatures, and so is able to control the transport of a single electron or few electrons into or out of the object, unperturbed by temperature effects.

Obviously, room temperature operation is achievable for islands of radius 5 nm. What is the maximum radius of the spherical island for which charging energy becomes comparable to the thermal energy at room temperature? To find this radius, we equate the charging energy in eq. (21.24) with thermal energy in eq. (21.25) to find C and using this C value determine r for the island:

$$\frac{Q^2}{2C} = k_B T \qquad (21.27)$$

or

$$C = \frac{Q^2}{2k_B T} = \frac{\left(1.6 \times 10^{-19}\right)^2}{2 \times 1.3806 \times 10^{-23} \times 300} = \frac{2.56 \times 10^{-38}}{8.2836 \times 10^{-21}}$$

$$(21.28)$$

$$= 0.309 \times 10^{-17} = 3.09 \times 10^{-18} \text{F}$$

From eq. (21.22),

$$r = \frac{C}{4\pi\varepsilon_0} = \frac{3.09 \times 10^{-18}}{4 \times 3.14159 \times 8.854 \times 10^{-12}} = \frac{3.09 \times 10^{-18}}{1.1126 \times 10^{-10}}$$

$$= 2.777 \times 10^{-8} \text{ m} = 27.77 \times 10^{-9} \text{ m}$$

$$\approx 28 \text{ nm} \qquad (21.29)$$

Nitrogen is a liquid in the zone $-210°C <$ Temperature $< -196°C$. How large island will be required for working at liquid nitrogen temperature (77 K or $-196°C$)? As earlier,

$$C = \frac{Q^2}{2k_B T} = \frac{\left(1.6 \times 10^{-19}\right)^2}{2 \times 1.3806 \times 10^{-23} \times 77}$$

$$= \frac{2.56 \times 10^{-38}}{2.126 \times 10^{-21}} = 1.2 \times 10^{-17} \text{F} \qquad (21.30)$$

giving

$$r = \frac{C}{4\pi\varepsilon_0} = \frac{1.2 \times 10^{-17}}{4 \times 3.14159 \times 8.854 \times 10^{-12}} = \frac{1.2 \times 10^{-17}}{1.1126 \times 10^{-10}}$$

$$= 1.078 \times 10^{-7} \text{ m} = 107.8 \times 10^{-9} \text{ m}$$

$$= 108 \text{ nm} \qquad (21.31)$$

The island radius can be increased from 28 nm for room temperature operation to 108 nm for working at liquid nitrogen temperature. So, the constraint on size is relaxed by a factor of $108/28 = 3.86$.

21.5 Orthodox Theory of Single-Electron Tunneling

The orthodox theory ignores the shape and size of the islands. Tunneling of an electron to an island or from it is considered to be instantaneous. Practically, tunneling takes place in a time interval $\sim 10^{-14}$ s. Following tunneling, the charge redistribution in the island is also postulated to be instantaneous. The energy spectrum of the contacts and leads is assumed to be continuous.

21.5.1 Tunnel Junction

For constructing a single-electron transistor (SET), tunnel junctions are used. A tunnel junction consists of a thin insulating film separating two conductive islands (Figure 21.2). Although electrons are classically forbidden from crossing over from one island to the other, they can do so quantum mechanically because the electrons have a non-zero amplitude in the barrier and hence a finite probability of passing through the barrier. When a tunnel junction is subjected to an applied bias, current flows between the conductive islands by quantum mechanical tunneling, and the tunnel junction behaves as a resistor whose resistance R varies exponentially with the thickness of the insulating film. Since an insulating layer sandwiched between two conductors also exhibits capacitance C, the tunnel junction shows capacitive property in addition to resistive property, and is represented by a parallel combination of resistor R and capacitor C. Tunneling of an electron from one conductive island of the tunnel junction to the other islands builds up a voltage $V = e/C$ on the recipient island which the succeeding electron has to surpass for tunneling. This increase in differential resistance at low bias is an outcome and manifestation of Coulomb blockade.

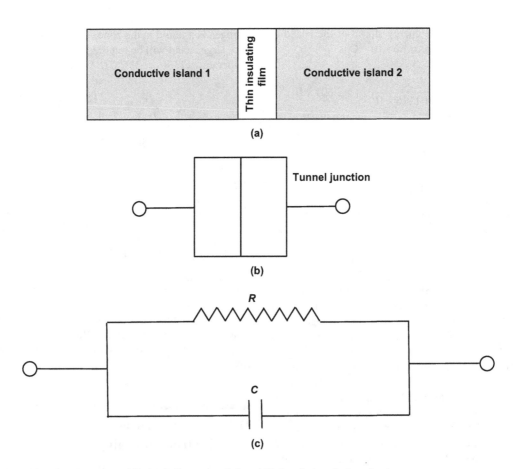

FIGURE 21.2 Tunnel junction: (a) structure, (b) circuit diagram symbol, and (c) electrical equivalent circuit.

21.5.2 Minimum Tunnel Resistance

The Heisenberg uncertainty principle imposes a fundamental limit on the precision of simultaneous measurement of energy and time of a particle. The uncertainties in the energy ΔE and time Δt of an electron obey the principle

$$\Delta E \Delta t \geq h \tag{21.32}$$

where h is Planck's constant. Note that eq. (21.32) is only an approximate form of eq. (2.21) for a first-order estimate. Since the time constant for charging the capacitance C of the tunnel junction through the tunnel resistor R is

$$\tau = RC \tag{21.33}$$

the time for charge fluctuations is from eq. (21.33)

$$\Delta t = RC \tag{21.34}$$

The potential produced on the tunnel junction by charging with one electron is from eq. (21.1)

$$V_e = -\frac{e}{C} \tag{21.35}$$

The corresponding energy is from eq. (21.2)

$$E_e = -eV_e = -e \times -\frac{e}{C} = \frac{e^2}{C} \tag{21.36}$$

This energy will prevent the next electron from tunneling. So, the energy gap for a single electron is from eq. (21.36)

$$\Delta E = \frac{e^2}{C} \tag{21.37}$$

Equations (21.32), (21.34), and (21.37) lead to

$$\left(\frac{e^2}{C}\right) \times (RC) \geq h \tag{21.38}$$

or

$$R \geq \frac{hC}{e^2 C} \tag{21.39}$$

or

$$R \geq \frac{h}{e^2} \tag{21.40}$$

giving

$$R \geq \frac{6.62607 \times 10^{-34} \text{ Js}}{\left(1.602 \times 10^{-19} \text{ C}\right)^2} = \frac{6.62607 \times 10^{-34}}{2.5664 \times 10^{-38}} = 2.58185 \times 10^4$$

$$= 25818.5 \ \Omega \approx 25819 \ \Omega = 25.8 \text{ k}\Omega \tag{21.41}$$

Note that from eq. (1.8) or eq. (11.186), $2e^2/h$ is the conductance quantum, or $h/(2e^2)$ is the resistance quantum.

21.5.3 Conductive Island with Two Tunnel Junctions

Let us look upon a conductive island connected with an ideal voltage source V_s through tunnel junctions J_1 and J_2 (Wasshuber 1997). Figure 21.3 shows this circuit arrangement. If the number of electrons entering the island through junction J_1 is n_1 and the number of electrons leaving the island through junction J_2 is n_2, the number of electrons left behind on the island is

$$n = n_1 - n_2 \qquad (21.42)$$

Since n_1 electrons enter the island through junction J_1, the charge on junction J_1 is

$$q_1 = C_1 V_1 \qquad (21.43)$$

where C_1 is the capacitance of junction J_1 and V_1 is the voltage across it.

Due to n_2 electrons leaving the island through junction J_2, the charge on junction J_2 is

$$q_2 = C_2 V_2 \qquad (21.44)$$

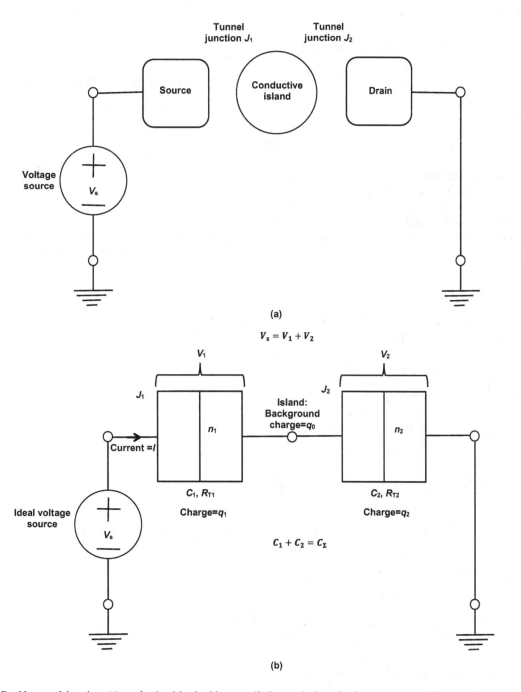

FIGURE 21.3 Double tunnel junction: (a) conductive island with source/drain terminals and voltage source and (b) equivalent circuit diagram and explanation of symbols used.

where C_2 is the capacitance of junction J_2 and V_2 is the voltage across it.

In addition to charges q_1 and q_2, the circuit contains a background charge q_0 arising from stray capacitances. So the total charge on the whole island is

q = Background charge on the island

    $-$Negative charge created by n_1 electrons entering the island

    $+$Positive charge created by n_2 electrons leaving the island

$$= q_0 - q_1 + q_2$$

$$= q_0 - n_1 e + n_2 e = q_0 - (n_1 - n_2)e = q_0 - ne \qquad (21.45)$$

The voltage drops V_1 and V_2 together add up to give the supply voltage V_s as

$$V_s = V_1 + V_2 \qquad (21.46)$$

Also, let us denote

$$C_1 + C_2 = C_\Sigma \qquad (21.47)$$

When one electron tunnels from junction J_1 through the island, junction J_1 acquires a positive potential. The potential of island remains unchanged as it first gains and subsequently loses one electron while the potential of junction J_2 becomes negative due to reception of one electron. So, the tunneling of one electron is described by the equation

Potential of junction J_1 + Potential of island

    $-$ Potential of junction $J_2 = 0 \qquad (21.48)$

or

Potential of junction J_1 = Potential of junction J_2

    $-$ Potential of island $\qquad (21.49)$

or

$$V_1 = \frac{C_2 V_s}{C_1 + C_2} - \frac{q_0 - ne}{C_1 + C_2} = \frac{C_2 V_s - q_0 + ne}{C_1 + C_2} = \frac{C_2 V_s + ne - q_0}{C_\Sigma}$$

$$(21.50)$$

Also,

Potential of junction J_2 = Potential of junction J_1

    + Potential of island $\qquad (21.51)$

or

$$V_2 = \frac{C_1 V_s}{C_1 + C_2} + \frac{q_0 - ne}{C_1 + C_2} = \frac{C_1 V_s + q_0 - ne}{C_1 + C_2} = \frac{C_1 V_s - ne + q_0}{C_\Sigma}$$

$$(21.52)$$

Applying equatons (21.43), (21.44), (21.50) and (21.52), the aggregate energy stored in the two junctions is from eq. (21.5)

$$W_C = \frac{q_1^2}{2C_1} + \frac{q_2^2}{2C_2} = \frac{C_1^2 V_1^2 C_2 + C_2^2 V_2^2 C_1}{2C_1 C_2} = \frac{C_1 C_2 \left(C_1 V_1^2 + C_2 V_2^2\right)}{2C_1 C_2} = \frac{C_1 V_1^2 + C_2 V_2^2}{2}$$

$$= \frac{C_1 \left\{\dfrac{C_2 V_s + (ne - q_0)}{C_\Sigma}\right\}^2 + C_2 \left\{\dfrac{C_1 V_s - (ne - q_0)}{C_\Sigma}\right\}^2}{2}$$

$$= \frac{C_1 \left\{C_2 V_s + (ne - q_0)\right\}^2 + C_2 \left\{C_1 V_s - (ne - q_0)\right\}^2}{2C_\Sigma^2}$$

$$= \frac{C_1 \left\{C_2^2 V_s^2 + 2C_2 V_s (ne - q_0) + (ne - q_0)^2\right\} + C_2 \left\{C_1^2 V_s^2 - 2C_1 V_s (ne - q_0) + (ne - q_0)^2\right\}}{2C_\Sigma^2}$$

$$= \frac{C_1 C_2^2 V_s^2 + 2C_1 C_2 V_s (ne - q_0) + C_1 (ne - q_0)^2 + C_1^2 C_2 V_s^2 - 2C_1 C_2 V_s (ne - q_0) + C_2 (ne - q_0)^2}{2C_\Sigma^2}$$

$$= \frac{C_1 C_2^2 V_s^2 + C_2 (ne - q_0)^2 + C_1^2 C_2 V_s^2 + C_1 (ne - q_0)^2}{2C_\Sigma^2}$$

$$= \frac{C_2 \left\{C_1 C_2 V_s^2 + (ne - q_0)^2\right\} + C_1 \left\{C_1 C_2 V_s^2 + (ne - q_0)^2\right\}}{2(C_1 + C_2)^2} = \frac{(C_2 + C_1)\left\{C_1 C_2 V_s^2 + (ne - q_0)^2\right\}}{2(C_1 + C_2)^2}$$

$$= \frac{C_1 C_2 V_s^2 + (ne - q_0)^2}{2(C_1 + C_2)} = \frac{C_1 C_2 V_s^2 + (ne - q_0)^2}{2C_\Sigma} \qquad (21.53)$$

What happens when one electron $-e$ tunnels through the first tunnel junction? No sooner this occurs, the potential V_1 of the first tunnel junction is altered by

$$\Delta V_1 = -\frac{e}{C_\Sigma} \tag{21.54}$$

The consequent polarization charge is given by

$$q_P = -\frac{eC_1}{C_\Sigma} \tag{21.55}$$

The charge q_1 on the first tunnel junction decreases as the voltage source receives the polarization charge q_P. So the voltage source must replace the electron $-e$ from the first tunnel junction plus the change in polarization charge caused by electron tunneling. The total replacement charge coming through the second tunnel junction is

$$q_R = -\frac{eC_2}{C_\Sigma} \tag{21.56}$$

The work done by the voltage source for one electron tunneling across junction J_1 is

$$w_1 = \text{Charge} \times \text{Voltage} = -\frac{eC_2V_s}{C_\Sigma} \tag{21.57}$$

The work done by the voltage source for n_1 electrons tunneling across junction J_1 is from eq. (21.57)

$$W_1 = \text{Charge} \times \text{Voltage} = -\frac{n_1eC_2V_s}{C_\Sigma} \tag{21.58}$$

Similarly, the work done by the voltage source for n_2 electrons tunneling across junction J_2 is

$$W_2 = \text{Charge} \times \text{Voltage} = -\frac{n_2eC_1V_s}{C_\Sigma} \tag{21.59}$$

The total work done by the voltage source is

$$W = W_1 + W_2 = -\frac{n_1eC_2V_s}{C_\Sigma} - \frac{n_2eC_1V_s}{C_\Sigma} = -\frac{eV_s}{C_\Sigma}(n_1C_2 + n_2C_1) \tag{21.60}$$

The free energy of the circuit is

$$F = W_C - W \tag{21.61}$$

From eqs. (21.53) and (21.60), as applied to eq. (21.61), we get

$$F = \frac{C_1C_2V_s^2 + (ne - q_0)^2}{2C_\Sigma} - \left\{-\frac{eV_s}{C_\Sigma}(n_1C_2 + n_2C_1)\right\}$$

$$= \frac{1}{C_\Sigma}\left\{\frac{C_1C_2V_s^2 + (ne - q_0)^2}{2} + eV_s(n_1C_2 + n_2C_1)\right\} \tag{21.62}$$

Applying eq. (21.62), the free energy changes for an electron tunneling across junction J_1 (inward or outward) are:

i.

$$\Delta F_1^+ = F(n_1 + 1, n_2) - F(n_1, n_2)$$

$$= \frac{1}{C_\Sigma}\left[\frac{C_1C_2V_s^2 + \{(n_1 + 1 - n_2)e - q_0\}^2}{2} + eV_s\{(n_1 + 1)C_2 + n_2C_1\}\right.$$

$$\left. - \frac{C_1C_2V_s^2 + \{(n_1 - n_2)e - q_0\}^2}{2} - eV_s(n_1C_2 + n_2C_1)\right] \tag{21.63}$$

Using equation (21.42) and noting that

$$\{(n_1 + 1 - n_2)e - q_0\}^2 - \{(n_1 - n_2)e - q_0\}^2 = \{(n+1)e - q_0\}^2 - (ne - q_0)^2$$

$$= (n+1)^2e^2 - 2(n+1)eq_0 + q_0^2 - (n^2e^2 - 2neq_0 + q_0^2)$$

$$= n^2e^2 + 2ne^2 + e^2 - 2neq_0 - 2eq_0 + q_0^2 - n^2e^2 + 2neq_0 - q_0^2$$

$$= 2ne^2 + e^2 - 2eq_0 \tag{21.64}$$

and

$$eV_s\{(n_1 + 1)C_2 + n_2C_1\} - eV_s(n_1C_2 + n_2C_1) = eV_s\{(n_1C_2 + C_2 + n_2C_1) - (n_1C_2 + n_2C_1)\}$$

$$= eV_s(n_1C_2 + C_2 + n_2C_1 - n_1C_2 - n_2C_1) = eV_sC_2 \tag{21.65}$$

we have from eq. (21.63)

$$\Delta F_1^+ = \frac{1}{C_\Sigma}\left(\frac{2ne^2 + e^2 - 2eq_0}{2} + eV_sC_2\right) = \frac{1}{C_\Sigma}\left(\frac{e^2}{2} + \frac{2ne^2 - 2eq_0}{2} + eV_sC_2\right)$$

$$= \frac{1}{C_\Sigma}\left(\frac{e^2}{2} + ne^2 - eq_0 + eV_sC_2\right) = \frac{e}{C_\Sigma}\left(\frac{e}{2} + ne - q_0 + V_sC_2\right) = \frac{e}{C_\Sigma}\left(\frac{e}{2} + V_sC_2 + ne - q_0\right)$$

$$= \frac{e}{C_\Sigma}\left\{\frac{e}{2} + (V_sC_2 + ne - q_0)\right\} \tag{21.66}$$

ii.

$$\Delta F_1^- = F(n_1 - 1, n_2) - F(n_1, n_2)$$

$$= \frac{1}{C_\Sigma}\left[\frac{C_1C_2V_s^2 + \{(n_1 - 1 - n_2)e - q_0\}^2}{2} + eV_s\{(n_1 - 1)C_2 + n_2C_1\} - \frac{C_1C_2V_s^2 + \{(n_1 - n_2)e - q_0\}^2}{2} - eV_s(n_1C_2 + n_2C_1)\right] \quad (21.67)$$

Since by equation (21.42),

$$\{(n_1 - 1 - n_2)e - q_0\}^2 - \{(n_1 - n_2)e - q_0\}^2 = \{(n-1)e - q_0\}^2 - (ne - q_0)^2$$

$$= (n-1)^2 e^2 - 2(n-1)eq_0 + q_0^2 - (n^2e^2 - 2neq_0 + q_0^2)$$

$$= n^2e^2 - 2ne^2 + e^2 - 2neq_0 + 2eq_0 + q_0^2 - n^2e^2 + 2neq_0 - q_0^2$$

$$= -2ne^2 + e^2 + 2eq_0 \quad (21.68)$$

and

$$eV_s\{(n_1 - 1)C_2 + n_2C_1\} - eV_s(n_1C_2 + n_2C_1) = eV_s\{(n_1C_2 - C_2 + n_2C_1) - (n_1C_2 + n_2C_1)\}$$

$$= eV_s(n_1C_2 - C_2 + n_2C_1 - n_1C_2 - n_2C_1) = -eV_sC_2 \quad (21.69)$$

we get from eq. (21.67)

$$\Delta F_1^- = \frac{1}{C_\Sigma}\left(\frac{-2ne^2 + e^2 + 2eq_0}{2} - eV_sC_2\right) = \frac{1}{C_\Sigma}\left(\frac{e^2}{2} - \frac{2ne^2 - 2eq_0}{2} - eV_sC_2\right)$$

$$= \frac{1}{C_\Sigma}\left(\frac{e^2}{2} - ne^2 + eq_0 - eV_sC_2\right) = \frac{e}{C_\Sigma}\left(\frac{e}{2} - ne + q_0 - V_sC_2\right) = \frac{e}{C_\Sigma}\left(\frac{e}{2} - V_sC_2 - ne + q_0\right)$$

$$= \frac{e}{C_\Sigma}\left\{\frac{e}{2} - (V_sC_2 + ne - q_0)\right\} \quad (21.70)$$

Similarly, the free energy changes for an electron tunneling across junction J_2 (inward or outward) are:

iii.

$$\Delta F_2^+ = F(n_1, n_2 + 1) - F(n_1, n_2)$$

$$= \frac{1}{C_\Sigma}\left[\frac{C_1C_2V_s^2 + \{(n_1 - n_2 - 1)e - q_0\}^2}{2} + eV_s\{n_1C_2 + (n_2 + 1)C_1\} - \frac{C_1C_2V_s^2 + \{(n_1 - n_2)e - q_0\}^2}{2} - eV_s(n_1C_2 + n_2C_1)\right]$$

$$(21.71)$$

Because

$$\{(n_1 - 1 - n_2)e - q_0\}^2 - \{(n_1 - n_2)e - q_0\}^2 = -2ne^2 + e^2 + 2eq_0 \quad (21.72)$$

and

$$eV_s\{n_1C_2 + (n_2 + 1)C_1\} - eV_s(n_1C_2 + n_2C_1) = eV_s\{(n_1C_2 + n_2C_1 + C_1) - (n_1C_2 + n_2C_1)\} = eV_sC_1 \quad (21.73)$$

we get from eq. (21.71)

$$\Delta F_2^+ = \frac{1}{C_\Sigma} \left(\frac{-2ne^2 + e^2 + 2eq_0}{2} + eV_sC_1 \right) = \frac{1}{C_\Sigma} \left(\frac{e^2}{2} - \frac{2ne^2 - 2eq_0}{2} + eV_sC_1 \right)$$

$$= \frac{1}{C_\Sigma} \left(\frac{e^2}{2} - ne^2 + eq_0 + eV_sC_1 \right) = \frac{1}{C_\Sigma} \left(\frac{e^2}{2} + eV_sC_1 - ne^2 + eq_0 \right)$$

$$= \frac{e}{C_\Sigma} \left(\frac{e}{2} + V_sC_1 - ne + q_0 \right) = \frac{e}{C_\Sigma} \left\{ \frac{e}{2} + \left(V_sC_1 - ne + q_0 \right) \right\} \tag{21.74}$$

iv.

$$\Delta F_2^- = F(n_1, n_2 - 1) - F(n_1, n_2)$$

$$= \frac{1}{C_\Sigma} \left[\frac{C_1C_2V_s^2 + \{(n_1 - n_2 + 1)e - q_0\}^2}{2} + eV_s\{n_1C_2 + (n_2 - 1)C_1\} - \frac{C_1C_2V_s^2 + \{(n_1 - n_2)e - q_0\}^2}{2} - eV_s(n_1C_2 + n_2C_1) \right] \tag{21.75}$$

As

$$\{(n_1 - n_2 + 1)e - q_0\}^2 - \{(n_1 - n_2)e - q_0\}^2 = 2ne^2 + e^2 - 2eq_0 \tag{21.76}$$

and

$$eV_s\{n_1C_2 + (n_2 - 1)C_1\} - eV_s(n_1C_2 + n_2C_1) = eV_s(n_1C_2 + n_2C_1 - C_1 - n_1C_2 - n_2C_1) = -eV_sC_1 \tag{21.77}$$

we have from eq. (21.75)

$$\Delta F_2^- = \frac{1}{C_\Sigma} \left(\frac{2ne^2 + e^2 - 2eq_0}{2} - eV_sC_1 \right)$$

$$= \frac{1}{C_\Sigma} \left(\frac{e^2}{2} + ne^2 - eq_0 - eV_sC_1 \right) = \frac{1}{C_\Sigma} \left(\frac{e^2}{2} - eV_sC_1 + ne^2 - eq_0 \right)$$

$$= \frac{e}{C_\Sigma} \left(\frac{e}{2} - V_sC_1 + ne - q_0 \right) = \frac{e}{C_\Sigma} \left\{ \frac{e}{2} - \left(V_sC_1 - ne + q_0 \right) \right\} \tag{21.78}$$

Combining eqs. (21.66) and (21.70) for (i) and (ii) as well as eqs. (21.74) and (21.78) for (iii) and (iv)

$$\Delta F_1^\pm = \frac{e}{C_\Sigma} \left\{ \frac{e}{2} \pm \left(V_sC_2 + ne - q_0 \right) \right\} \tag{21.79}$$

and

$$\Delta F_2^\pm = \frac{e}{C_\Sigma} \left\{ \frac{e}{2} \pm \left(V_sC_1 - ne + q_0 \right) \right\} \tag{21.80}$$

These equations have a far-reaching significance. The free energies ΔF can acquire positive or negative values. A positive value of free energy implies a transition from a lower energy state to a higher energy state. This is a highly improbable event. On the contrary, a negative value of free energy signifying a transition from a higher to a lower energy state is a highly likely event.

So at room temperature, the permissible transitions are those for which

$$\Delta F_1 < 0, \Delta F_2 < 0, \tag{21.81}$$

The equations contain a term dependent on the supply voltage V_s which can be adjusted to obtain a negative value of ΔF. They tell us that ΔF will remain positive until the supply voltage exceeds a threshold level. As the supply voltage is multiplied by the corresponding junction capacitance in all the equations, the product of the supply voltage and capacitance will determine the condition when free energy will become negative for a junction. Depending on the capacitance of a tunnel junction, negative free energy is achieved at a lower or higher supply voltage.

In all cases where the free energy is positive, there is no tunneling at a low value of the supply voltage. This suppression of tunneling at a low bias is a consequence of and is therefore ascribed to the Coulomb blockade effect.

The Coulomb blockade effect is easily understood by referring to the energy-band diagram of a double tunnel junction, first when the supply voltage is zero and afterward by applying a small supply voltage. As can be seen from Figure 21.4, in the absence of the supply voltage, an energy gap appears. This energy gap owes its origin to the charging energy of the island. Half of this energy gap lies above the original Fermi level and half below it. The energy gap forbids tunneling of electrons from either the left electrode or the right electrode

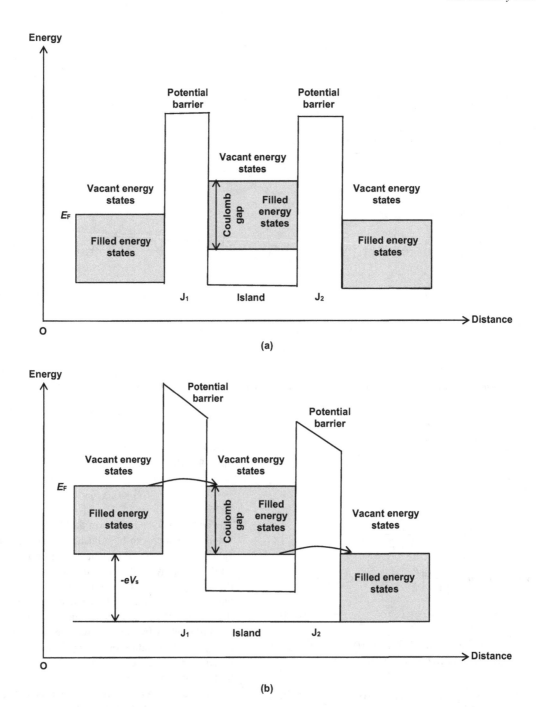

FIGURE 21.4   Energy-band diagram of a double tunnel junction: (a) in thermal equilibrium and (b) after application of bias V_s.

into the island. Electron tunneling out of the island is also prohibited. When a supply voltage exceeding the threshold for Coulomb blockade is applied, the energy gap is surmounted. Then electron tunneling is allowed leading to the flow of current through the double tunnel junction.

Let us see how the Coulomb blockade effect is revealed in the current-voltage characteristic of the double tunnel junction (Figure 21.5). At a low value of the supply voltage, the current across a tunnel junction is zero. As soon as the threshold bias level is exceeded, the current flows. But remember that there are two

tunnel junctions and they may differ in transparency. A junction having a low tunnel resistance is said to be more transparent whereas a high resistance tunnel junction is less transparent.

Suppose junction J_1 has a low tunnel resistance R_{T1} and junction J_2 has a high tunnel resistance R_{T2}, i.e. $R_{T1} < R_{T2}$. When $V_s = 1V$ is reached, an electron tunnels through J_1 but it cannot tunnel through J_2 at this voltage. It can only tunnel through J_2 at a higher $V_s = 2V$ because of the lower transparency of J_2. When $V_s = 2V$, the electron tunnels through J_2. Immediately another electron tunnels through J_1. Such a sequence of

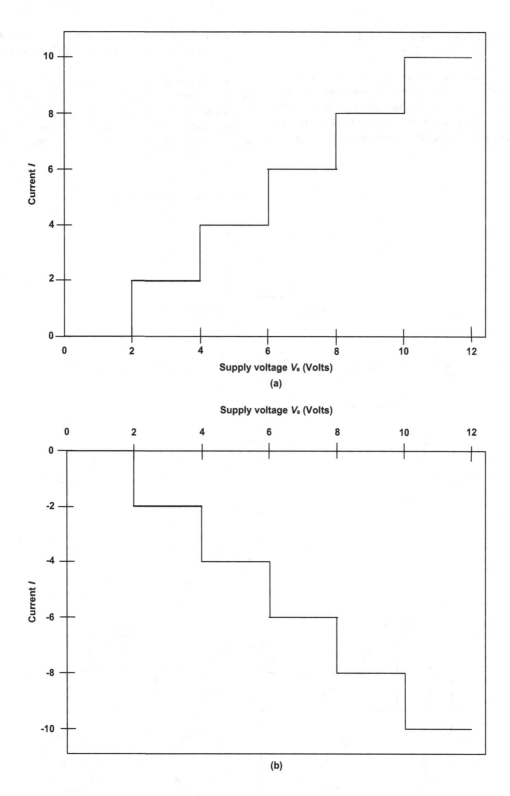

FIGURE 21.5   Current-voltage characteristics of a double tunnel junction when: (a) $R_{T1} < R_{T2}$ and (b) $R_{T1} > R_{T2}$

tunneling events in which the electron tunnels through J_1 at a lower voltage and through J_2 at a higher voltage produces an ascending staircase-like current-voltage characteristic (Figure. 21.5(a)).

Let us think over the situation in which $R_{T1} > R_{T2}$. Then any electron tunneling through J_2 into the island will wait in the island till a higher supply voltage is applied. Then it will tunnel through J_1. Essentially, the current flows in the negative direction but the staircase shape of the current-voltage characteristic is maintained. So, the *I–V* characteristic becomes a descending staircase (Figure 21.5(b)).

21.5.4 Coupling a Gate Electrode to the Double Tunnel Junction

In a double tunnel junction, the current is increased from one step to the next when the supply voltage across the system is increased. What is to be done if a higher current is needed while keeping the supply voltage constant. Then we need a third controlling electrode, the gate electrode with applied voltage V_G.

This electrode is capacitively coupled to the island and the resulting three-terminal device is called a single electron transistor (SET).

Consider the schematic diagram of a SET shown in Figure 21.6 (see reference: Coulomb blockade in single electron transistors). It consists of two tunnel junctions J_1 and J_2 biased by two independent voltage sources V_1 and V_2 so that any one source can be varied without affecting the other source. In addition, it has

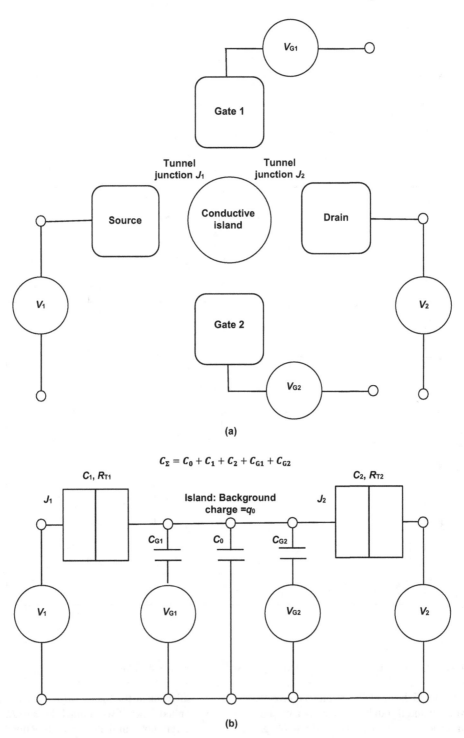

FIGURE 21.6   Single-electron transistor: (a) schematic layout and (b) equivalent circuit representation.

two gate electrodes that too are actuated by independent voltage sources V_{G1} and V_{G2}. One gate terminal is used for changing the background charge on the island and the other during device operation for varying the current flow from junction J_1 through the island to junction J_2.

Then

Charge on the island = q_{Island}

= Charge on capacitor C_1 + Charge on capacitor C_{G1}

+ Charge on capacitor C_{G2} + Charge on capacitor C_2

+ Offset charge due to stray capacitance

$$= C_1V_1 + C_{G1}V_{G1} + C_{G2}V_{G2} + C_2V_2 + (q_0 - ne) \qquad (21.82)$$

Denoting the voltage of the island by $V(n)$ and the stray capacitance in the circuit referred to ground by C_0, we can write

Potential of the island = $V(n) = \dfrac{\text{Total charge}}{\text{Total capacitance}}$

$$= \frac{q_{Island}}{C_1 + C_{G1} + C_{G2} + C_2 + C_0}$$

$$= \frac{C_1V_1 + C_{G1}V_{G1} + C_{G2}V_{G2} + C_2V_2 + (q_0 - ne)}{C_\Sigma}$$

$$= \frac{q_0 - ne + C_1V_1 + C_2V_2 + C_{G1}V_{G1} + C_{G2}V_{G2}}{C_\Sigma} \qquad (21.83)$$

where

$$C_\Sigma = C_0 + C_1 + C_2 + C_{G1} + C_{G2} \qquad (21.84)$$

Suppose a charge $+q$ is added to the island. Then the potential of the island becomes

$$V_q = V(n) + \frac{q}{C_\Sigma} \qquad (21.85)$$

We shall apply equation (21.85) to find the energies required for charge transferences. The electrostatic energy required to

transfer an electron of charge $-e$ from ground potential to the island is from eq. (21.85)

$$W_{-e} = \int_0^{-e} \left(V(n) + \frac{q}{C_\Sigma} \right) dq = \int_0^{-e} V(n)dq +$$

$$\int_0^{-e} \left(\frac{q}{C_\Sigma} \right) dq = V(n)\int_0^{-e} dq + \frac{1}{C_\Sigma}\int_0^{-e} q\,dq$$

$$= V(n)|q|_0^{-e} + \frac{1}{C_\Sigma}\frac{|q^2|}{2}\bigg|_0^{-e} = V(n)(-e - 0) + \frac{1}{C_\Sigma}\left(\frac{e^2}{2} - 0 \right)$$

$$= -eV(n) + \frac{e^2}{2C_\Sigma} \qquad (21.86)$$

The electrostatic energy required to remove an electron of charge $-e$ from the island to ground potential is the same as the energy necessary to transfer a hole of charge $+e$ from ground potential to the island. So, this energy is from eq. (21.85)

$$W_{+e} = \int_0^{+e} \left(V(n) + \frac{q}{C_\Sigma} \right) dq = \int_0^{+e} V(n)dq + \int_0^{+e} \left(\frac{q}{C_\Sigma} \right) dq$$

$$= V(n)\int_0^{+e} dq + \frac{1}{C_\Sigma}\int_0^{+e} q\,dq$$

$$= V(n)|q|_0^{+e} + \frac{1}{C_\Sigma}\frac{|q^2|}{2}\bigg|_0^{+e} = V(n)(e - 0) + \frac{1}{C_\Sigma}\left(\frac{e^2}{2} - 0 \right)$$

$$= eV(n) + \frac{e^2}{2C_\Sigma} \qquad (21.87)$$

Now we calculate the energy changes ΔW during tunneling. Four possible events of electron tunneling can be visualized for the junctions: two events for junction J_1 and two for junction J_2. The envisioned tunneling events for junction J_1 are: an electron from the left lead tunneling rightward from J_1 and an electron from the right lead tunneling leftward from J_1. Similarly, the envisaged tunneling events for junction J_2 include an electron from the left lead tunneling to the right from J_2 and an electron from the right lead tunneling to the left from junction J_2.

$\Delta W_{1R}(n)$ = Energy change during an electron from the left lead tunneling rightward from J_1

= Energy required to bring an electron from the left lead to ground potential $(+eV_1)$

+Energy needed to bring the electron from ground potential to the island $\left(-eV(n) + \dfrac{e^2}{2C_\Sigma} \right)$

$$= +eV_1 - eV(n) + \frac{e^2}{2C_\Sigma} = +eV_1 - e\left(\frac{q_0 - ne + C_1V_1 + C_2V_2 + C_{G1}V_{G1} + C_{G2}V_{G2}}{C_\Sigma} \right) + \frac{e^2}{2C_\Sigma} \qquad (21.88)$$

where equations (21.86) and (21.83) are applied.

Hence,

$$\Delta W_{1R}(n) = +eV_1 - eV(n) + \frac{e^2}{2C_\Sigma}$$

$$= +eV_1 - e\left(\frac{q_0 - ne + C_1V_1 + C_2V_2 + C_{G1}V_{G1} + C_{G2}V_{G2}}{C_\Sigma}\right) + \frac{e^2}{2C_\Sigma}$$

$$(21.89)$$

Arguing on similar lines, the following equations are written down

$$\Delta W_{1L}(n) = -eV_1 + eV(n) + \frac{e^2}{2C_\Sigma}$$

$$= -eV_1 + e\left(\frac{q_0 - ne + C_1V_1 + C_2V_2 + C_{G1}V_{G1} + C_{G2}V_{G2}}{C_\Sigma}\right) + \frac{e^2}{2C_\Sigma}$$

$$(21.90)$$

by applying equations (21.87) and (21.83).

$$\Delta W_{2R}(n) = -eV_2 + eV(n) + \frac{e^2}{2C_\Sigma}$$

$$= -eV_2 + e\left(\frac{q_0 - ne + C_1V_1 + C_2V_2 + C_{G1}V_{G1} + C_{G2}V_{G2}}{C_\Sigma}\right) + \frac{e^2}{2C_\Sigma}$$

$$(21.91)$$

where equations (21.87) and (21.83) are applied.

$$\Delta W_{2L}(n) = +eV_2 - eV(n) + \frac{e^2}{2C_\Sigma}$$

$$= +eV_2 - e\left(\frac{q_0 - ne + C_1V_1 + C_2V_2 + C_{G1}V_{G1} + C_{G2}V_{G2}}{C_\Sigma}\right) + \frac{e^2}{2C_\Sigma}$$

$$(21.92)$$

by applying equations (21.86) and (21.83).

As discussed previously for a double tunnel junction structure, a negative energy change is an essential prerequisite to allow a tunneling event at zero temperature whereas a positive energy change disallows the event. It transpires that certain combinations of supply voltages to the tunnel junctions and gate biases to the island give positive values of energy changes. For these combinations, tunneling in either direction is denied at low temperatures. This denial to tunneling is a clear manifestation of Coulomb blockade. However, tunneling can be enabled by expending energy through variations in gate bias, still keeping supply voltages for the junctions at low values. The intent of a gate bias change is to make the energy change negative. In this way, the SET works by controlling the tunneling across the junctions through gate bias while the supply voltages to the junctions are held at values that do not allow tunneling and also keeping the temperature low. This gated control of the flow of single electrons leads to single electronics (Wasshuber 2001, Goser et al. 2004).

Four equations define the boundaries of the Coulomb blockade region. If we set $\Delta W = 0$ in any of these equations, we obtain a line in the $V_s - V_{G1}$ plane. V_{G1} is the controlling gate for device operation whereas V_{G2} allows the choice of background charge in the island. The four lines obtained from the four equations fully enclose the Coulomb blockade region. Two kinds of supply voltages are used for the tunnel junctions: symmetric and antisymmetric. The symmetric supply voltages are defined as

$$V_1 = -V_2 = V_s \qquad (21.93)$$

while in the case of antisymmetric supplies,

$$V_1 = V_s \text{ and } V_2 = 0 \qquad (21.94)$$

Let us determine the four lines for the antisymmetric supplies taking the charged state $n = 0$ for the island. Applying equation (21.94), equation (21.89) yields

$$+eV_1 - e\left(\frac{q_0 - ne + C_1V_1 + C_2V_2 + C_{G1}V_{G1} + C_{G2}V_{G2}}{C_\Sigma}\right) + \frac{e^2}{2C_\Sigma} = 0$$

$$(21.95)$$

or

$$+eV_s - e\left(\frac{q_0 + C_1V_s + C_{G1}V_{G1} + C_{G2}V_{G2}}{C_\Sigma}\right) + \frac{e^2}{2C_\Sigma} = 0 \qquad (21.96)$$

or

$$+eV_s - \frac{eC_1V_s}{C_\Sigma} = -\frac{e^2}{2C_\Sigma} + e\left(\frac{q_0 + C_{G1}V_{G1} + C_{G2}V_{G2}}{C_\Sigma}\right) \qquad (21.97)$$

or

$$\frac{C_\Sigma V_s - C_1V_s}{C_\Sigma} = -\frac{e}{2C_\Sigma} + \left(\frac{q_0 + C_{G1}V_{G1} + C_{G2}V_{G2}}{C_\Sigma}\right) \qquad (21.98)$$

or

$$V_s(C_\Sigma - C_1) = -\frac{e}{2} + q_0 + C_{G1}V_{G1} + C_{G2}V_{G2} \qquad (21.99)$$

$$\therefore V_s = \frac{q_0 + C_{G1}V_{G1} + C_{G2}V_{G2} - \frac{e}{2}}{C_\Sigma - C_1} \qquad (21.100)$$

From equations (21.90) and (21.94),

$$-eV_s + e\left(\frac{q_0 + C_1V_s + C_{G1}V_{G1} + C_{G2}V_{G2}}{C_\Sigma}\right) + \frac{e^2}{2C_\Sigma} = 0 \qquad (21.101)$$

or

$$-eV_s + e\frac{C_1V_s}{C_\Sigma} + e\left(\frac{q_0 + C_{G1}V_{G1} + C_{G2}V_{G2}}{C_\Sigma}\right) + \frac{e^2}{2C_\Sigma} = 0 \qquad (21.102)$$

or

$$-eV_s + e\frac{C_1 V_s}{C_\Sigma} = -\frac{e^2}{2C_\Sigma} - e\left(\frac{q_0 + C_{G1}V_{G1} + C_{G2}V_{G2}}{C_\Sigma}\right) \quad (21.103)$$

or

$$\frac{-eV_s C_\Sigma}{C_\Sigma} + \frac{eC_1 V_s}{C_\Sigma} = -\frac{e^2}{2C_\Sigma} - e\left(\frac{q_0 + C_{G1}V_{G1} + C_{G2}V_{G2}}{C_\Sigma}\right) \quad (21.104)$$

or

$$-V_s C_\Sigma + C_1 V_s = -\frac{e}{2} - q_0 - C_{G1}V_{G1} - C_{G2}V_{G2} \quad (21.105)$$

or

$$V_s C_\Sigma - C_1 V_s = \frac{e}{2} + q_0 + C_{G1}V_{G1} + C_{G2}V_{G2} \quad (21.106)$$

or

$$V_s(C_\Sigma - C_1) = q_0 + C_{G1}V_{G1} + C_{G2}V_{G2} + \frac{e}{2} \quad (21.107)$$

or

$$V_s = \frac{q_0 + C_{G1}V_{G1} + C_{G2}V_{G2} + \dfrac{e}{2}}{C_\Sigma - C_1} \quad (21.108)$$

Equations (21.91) and (21.94) lead- to

$$e\left(\frac{q_0 + C_1 V_1 + C_{G1}V_{G1} + C_{G2}V_{G2}}{C_\Sigma}\right) + \frac{e^2}{2C_\Sigma} = 0 \quad (21.109)$$

or

$$e\left(q_0 + C_1 V_s + C_{G1}V_{G1} + C_{G2}V_{G2} + \frac{e}{2}\right) = 0 \quad (21.110)$$

or

$$q_0 + C_1 V_s + C_{G1}V_{G1} + C_{G2}V_{G2} + \frac{e}{2} = 0 \quad (21.111)$$

or

$$C_1 V_s = -q_0 - C_{G1}V_{G1} - C_{G2}V_{G2} - \frac{e}{2} \quad (21.112)$$

or

$$V_s = -\frac{q_0 + C_{G1}V_{G1} + C_{G2}V_{G2} + \dfrac{e}{2}}{C_1} \quad (21.113)$$

Equations (21.92) and (21.94) give

$$-e\left(\frac{q_0 + C_1 V_s + C_{G1}V_{G1} + C_{G2}V_{G2}}{C_\Sigma}\right) + \frac{e^2}{2C_\Sigma} = 0 \quad (21.114)$$

or

$$-(q_0 + C_1 V_s + C_{G1}V_{G1} + C_{G2}V_{G2}) + \frac{e}{2} = 0 \quad (21.115)$$

or

$$-C_1 V_s = -\frac{e}{2} + q_0 + C_{G1}V_{G1} + C_{G2}V_{G2} \quad (21.116)$$

or

$$V_s = \frac{-(q_0 + C_{G1}V_{G1} + C_{G2}V_{G2}) + \dfrac{e}{2}}{C_1} \quad (21.117)$$

Figure 21.7 shows the rhombus ABCD drawn with the four lines represented by eqs. (21.100), (21.108), (21.113), and (21.117). This rhombus is known as the Coulomb diamond.

At point A, $V_s = 0$. So, from eq. (21.113)

$$V_s = -\frac{q_0 + C_{G1}V_{G1} + C_{G2}V_{G2} + \dfrac{e}{2}}{C_1} = 0 \quad (21.118)$$

$$\therefore q_0 + C_{G1}V_{G1} + C_{G2}V_{G2} + \frac{e}{2} = 0 \quad (21.119)$$

or

$$C_{G1}V_{G1} = -\left(q_0 + C_{G2}V_{G2} + \frac{e}{2}\right) \quad (21.120)$$

or

$$V_{G1} = -\frac{q_0 + C_{G2}V_{G2} + \dfrac{e}{2}}{C_{G1}} \quad (21.121)$$

The coordinates of point A are $\left(-\dfrac{q_0 + C_{G2}V_{G2} + \dfrac{e}{2}}{C_{G1}}, 0\right)$.

At point C, $V_s = 0$. So, from eq. (21.117)

$$V_s = -\frac{q_0 + C_{G1}V_{G1} + C_{G2}V_{G2} - \dfrac{e}{2}}{C_1} = 0 \quad (21.122)$$

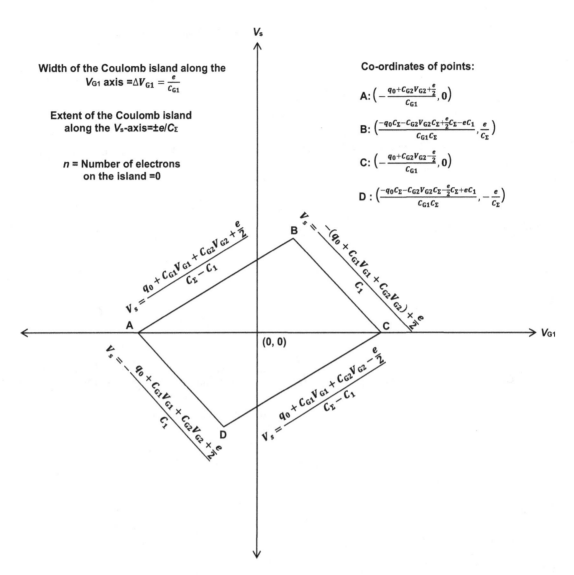

FIGURE 21.7 Defining the borders of the Coulomb blockade region for an asymmetrically-biased SET for $n = 0$. Inside the rhombus, the free energy change is positive so that tunneling is prohibited.

$$\therefore -q_0 - C_{G1}V_{G1} - C_{G2}V_{G2} + \frac{e}{2} = 0 \qquad (21.123)$$

or

$$C_{G1}V_{G1} = -q_0 - C_{G2}V_{G2} + \frac{e}{2} \qquad (21.124)$$

or

$$V_{G1} = \frac{-q_0 - C_{G2}V_{G2} + \frac{e}{2}}{C_{G1}} \qquad (21.125)$$

The coordinates of point C are $\left(-\dfrac{q_0 + C_{G2}V_{G2} - \dfrac{e}{2}}{C_{G1}}, 0 \right)$.

Distance between points A and C is

$$\Delta V_{G1} = \sqrt{\left[\left\{ -\frac{q_0 + C_{G2}V_{G2} + \frac{e}{2}}{C_{G1}} \right\} - \left\{ -\frac{q_0 + C_{G2}V_{G2} - \frac{e}{2}}{C_{G1}} \right\} \right]^2 + (0-0)^2}$$

$$= \sqrt{\left(\frac{-q_0 - C_{G2}V_{G2} - \frac{e}{2}}{C_{G1}} + \frac{q_0 + C_{G2}V_{G2} - \frac{e}{2}}{C_{G1}} \right)^2} = \sqrt{\left(\frac{-2\frac{e}{2}}{C_{G1}} \right)^2}$$

$$= \sqrt{\frac{e^2}{C_{G1}^2}} = \frac{e}{C_{G1}} \qquad (21.126)$$

This means that the width of the Coulomb diamond along the V_{G1} axis is e/C_{G1}.

Point B is the point of intersection of lines represented by eqs. (21.108) and (21.117). Hence,

$$V_s = \frac{q_0 + C_{G1}V_{G1} + C_{G2}V_{G2} + \dfrac{e}{2}}{C_\Sigma - C_1} = \frac{-(q_0 + C_{G1}V_{G1} + C_{G2}V_{G2}) + \dfrac{e}{2}}{C_1}$$

(21.127)

or

$$(C_\Sigma - C_1)\left\{-(q_0 + C_{G1}V_{G1} + C_{G2}V_{G2}) + \frac{e}{2}\right\}$$
$$= C_1\left(q_0 + C_{G1}V_{G1} + C_{G2}V_{G2} + \frac{e}{2}\right)$$

(21.128)

or

$$(C_\Sigma - C_1)\left(-q_0 - C_{G1}V_{G1} - C_{G2}V_{G2} + \frac{e}{2}\right)$$
$$= C_1\left(q_0 + C_{G1}V_{G1} + C_{G2}V_{G2} + \frac{e}{2}\right)$$

(21.129)

or

$$-q_0 C_\Sigma - C_{G1}V_{G1}C_\Sigma - C_{G2}V_{G2}C_\Sigma + \frac{e}{2}C_\Sigma + q_0 C_1$$
$$+ C_{G1}V_{G1}C_1 + C_{G2}V_{G2}C_1 - \frac{e}{2}C_1$$
$$= q_0 C_1 + C_{G1}V_{G1}C_1 + C_{G2}V_{G2}C_1 + \frac{e}{2}C_1$$

(21.130)

or

$$C_{G1}V_{G1}C_\Sigma = -q_0 C_\Sigma - C_{G2}V_{G2}C_\Sigma + \frac{e}{2}C_\Sigma - \frac{e}{2}C_1 - \frac{e}{2}C_1$$
$$= -q_0 C_\Sigma - C_{G2}V_{G2}C_\Sigma + \frac{e}{2}C_\Sigma - eC_1$$

(21.131)

or

$$V_{G1} = \frac{-q_0 C_\Sigma - C_{G2}V_{G2}C_\Sigma + \dfrac{e}{2}C_\Sigma - eC_1}{C_{G1}C_\Sigma}$$

(21.132)

Putting this value of V_{G1} in eq. (21.117), we have

$$V_s = \frac{-(q_0 + C_{G1}V_{G1} + C_{G2}V_{G2}) + \dfrac{e}{2}}{C_1} = \frac{-q_0 - C_{G1}\left(\dfrac{-q_0 C_\Sigma - C_{G2}V_{G2}C_\Sigma + \dfrac{e}{2}C_\Sigma - eC_1}{C_{G1}C_\Sigma}\right) - C_{G2}V_{G2} + \dfrac{e}{2}}{C_1}$$

$$= \frac{-q_0 + \dfrac{q_0 C_\Sigma + C_{G2}V_{G2}C_\Sigma - \dfrac{e}{2}C_\Sigma + eC_1}{C_\Sigma} - C_{G2}V_{G2} + \dfrac{e}{2}}{C_1}$$

$$= \frac{\dfrac{-2q_0 C_\Sigma + 2q_0 C_\Sigma + 2C_{G2}V_{G2}C_\Sigma - eC_\Sigma + 2eC_1 - 2C_{G2}V_{G2}C_\Sigma + eC_\Sigma}{2C_\Sigma}}{C_1}$$

$$= \frac{-2q_0 C_\Sigma + 2q_0 C_\Sigma + 2C_{G2}V_{G2}C_\Sigma - eC_\Sigma + 2eC_1 - 2C_{G2}V_{G2}C_\Sigma + eC_\Sigma}{2C_\Sigma C_1} = \frac{2eC_1}{2C_\Sigma C_1} = \frac{e}{C_\Sigma}$$

(21.133)

We find that the coordinates of point B are

$$\left(\frac{-q_0 C_\Sigma - C_{G2}V_{G2}C_\Sigma + \dfrac{e}{2}C_\Sigma - eC_1}{C_{G1}C_\Sigma}, \frac{e}{C_\Sigma}\right).$$

Point D being the point of intersection of lines represented by eqs. (21.100) and (21.113), we can write

$$V_s = \frac{q_0 + C_{G1}V_{G1} + C_{G2}V_{G2} - \dfrac{e}{2}}{C_\Sigma - C_1} = \frac{-q_0 - C_{G1}V_{G1} - C_{G2}V_{G2} - \dfrac{e}{2}}{C_1}$$

(21.134)

or

$$(C_\Sigma - C_1)\left(-q_0 - C_{G1}V_{G1} - C_{G2}V_{G2} - \frac{e}{2}\right)$$
$$= C_1\left(q_0 + C_{G1}V_{G1} + C_{G2}V_{G2} - \frac{e}{2}\right)$$

(21.135)

or

$$-q_0 C_\Sigma - C_{G1}V_{G1}C_\Sigma - C_{G2}V_{G2}C_\Sigma - \frac{e}{2}C_\Sigma$$
$$+ q_0 C_1 + C_{G1}V_{G1}C_1 + C_{G2}V_{G2}C_1 + \frac{e}{2}C_1$$
$$= q_0 C_1 + C_{G1}V_{G1}C_1 + C_{G2}V_{G2}C_1 - \frac{e}{2}C_1$$

(21.136)

or

$$C_{G1}V_{G1}C_\Sigma = -q_0 C_\Sigma - C_{G2}V_{G2}C_\Sigma - \frac{e}{2}C_\Sigma + \frac{e}{2}C_1 + \frac{e}{2}C_1$$
$$= -q_0 C_\Sigma - C_{G2}V_{G2}C_\Sigma - \frac{e}{2}C_\Sigma + eC_1$$

(21.137)

or

$$V_{G1} = \frac{-q_0 C_\Sigma - C_{G2}V_{G2}C_\Sigma - \dfrac{e}{2}C_\Sigma + eC_1}{C_{G1}C_\Sigma}$$

(21.138)

From eqs. (21.113) and (21.138),

$$
V_s = \frac{-q_0 - C_{G1}V_{G1} - C_{G2}V_{G2} - \dfrac{e}{2}}{C_1} = \frac{-q_0 - C_{G1}\left(\dfrac{-q_0 C_\Sigma - C_{G2}V_{G2}C_\Sigma - \dfrac{e}{2}C_\Sigma + eC_1}{C_{G1}C_\Sigma}\right) - C_{G2}V_{G2} - \dfrac{e}{2}}{C_1}
$$

$$
= \frac{-q_0 + \dfrac{q_0 C_\Sigma + C_{G2}V_{G2}C_\Sigma + \dfrac{e}{2}C_\Sigma - eC_1}{C_\Sigma} - C_{G2}V_{G2} - \dfrac{e}{2}}{C_1}
$$

$$
= \frac{-q_0 + \dfrac{2q_0 C_\Sigma + 2C_{G2}V_{G2}C_\Sigma + eC_\Sigma - 2eC_1}{2C_\Sigma} - C_{G2}V_{G2} - \dfrac{e}{2}}{C_1}
$$

$$
= \frac{\dfrac{-2q_0 C_\Sigma + 2q_0 C_\Sigma + 2C_{G2}V_{G2}C_\Sigma + eC_\Sigma - 2eC_1 - 2C_{G2}V_{G2}C_\Sigma - eC_\Sigma}{2C_\Sigma}}{C_1}
$$

$$
= \frac{-2q_0 C_\Sigma + 2q_0 C_\Sigma + 2C_{G2}V_{G2}C_\Sigma + eC_\Sigma - 2eC_1 - 2C_{G2}V_{G2}C_\Sigma - eC_\Sigma}{2C_\Sigma C_1} = -\frac{2eC_1}{2C_\Sigma C_1} = -\frac{e}{C_\Sigma} \tag{21.139}
$$

From the earlier calculation, coordinates of point D are

$$
\left(\frac{-q_0 C_\Sigma - C_{G2}V_{G2}C_\Sigma - \dfrac{e}{2}C_\Sigma + eC_1}{C_{G1}C_\Sigma}, -\frac{e}{C_\Sigma} \right).
$$

Thus the Coulomb diamond extends to $\pm e/C_\Sigma$ along the V_s-axis. Equations (21.100), (21.108), (21.113), and (21.117) along with the coordinates of points A, B, C, and D determined in the foregoing analysis completely define the Coulomb island.

As already shown, the Coulomb diamond is e/C_{G1} wide along the V_{G1} axis. Performing similar calculations for $n = -1, -2, -3,\ldots$ and $n = +1, +2, +3, \ldots$, a graph is generated between V_s and V_{G1}. It is called the stability plot of the SET (Figure 21.8). In this graph, the shaded regions are the stable regions. If the voltages V_1 and V_2 applied to the tunnel junctions are such that Coulomb blockade prevails, and the gate bias V_{G1} is raised, current sharply increases at points situated at intervals of e/C_{G1} on the V_{G1} axis producing oscillations called Coulomb oscillations. This alternation of tunneling and non-tunneling events tantamounts to cutting through the stability plot at intervals of e/C_{G1}. Hence, these oscillations have a periodicity of e/C_{G1}.

21.6 Single-Electron Transistor Fabrication by Nanowire-Based Process

Dubuc et al. (2008) reported a process for SET fabrication based on nanowire fabrication. In this process (Figure 21.9), a trench is dug in the SiO_2 layer on a Si substrate. In the center of the trench, a titanium wire is made perpendicular to the surface of the trench. This titanium wire is oxidized to form TiO_2. Consequently, a linear titanium pillar is erected, which is covered with TiO_2 on the left, right, and top sides. Then titanium metal is deposited over the oxidized titanium wire

covering the pillar and the surrounding trench. Thus we get the structure $SiO_2/Ti/TiO_2/Ti$. When the surface of the resulting layered structure is planarized by chemical mechanical polishing to reach the SiO_2 surface, the SET is formed with titanium island and tunnel junctions on either side. In this SET, TiO_2 film is the dielectric with titanium source and drain on the two sides of the central titanium island. The developed process is capable of provision of subattofarad resolution. It is useful for fabricating SETs that can work up to 430 K, which matches the operating temperature limits of traditional field-effect transistors (FETs).

21.7 Discussion and Conclusions

Energy used in charging a conductive island is stored in it as electrostatic potential energy. Besides this energy, the island may also have Fermi energy and quantum confinement energy. In an electrical circuit, the work done by voltage sources is found by adding together the integrals over time of power supplied by the various sources in the circuit. The difference of energy stored minus the work done is the free energy.

A spherical conductive island of radius r has a capacitance $C = 4\pi\varepsilon_0 r$. Energy stored in a spherical island having 5 nm radius is 2.3×10^{-20} J, which is larger than the thermal energy at room temperature $= 4.1 \times 10^{-21}$ J. An extremely small island of a few nm radius has a very low capacitance and therefore can be charged to sufficiently large energy that the energy change during tunneling of an electron to the island is greater than the thermal energy at room temperature and much greater than the thermal energy at a low temperature. Thus the approaching electron experiences a vehement opposition from the charged island. This is the Coulomb blockade effect. If we work at room temperature, the maximum island radius must be 28 nm. But if we are contended

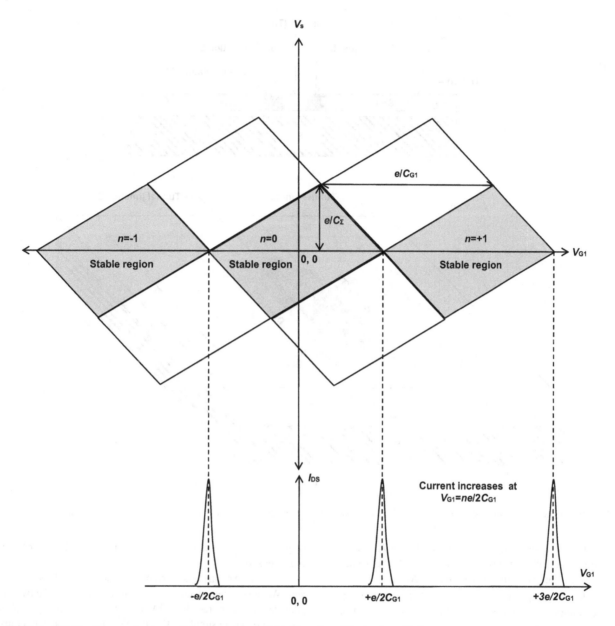

FIGURE 21.8 Stability diagram of a single-electron transistor and the generation of Coulomb oscillations.

in working at liquid nitrogen temperature, we can take islands as large as 108 nm.

In the orthodox theory of SET, the tunneling time and the subsequent distribution of charge on the island are assumed to be instantaneous. The electrons in contacts/leads can acquire continuous energy values.

Two conductive islands separated by a thin insulating film constitute a tunnel junction. The equivalent circuit of a tunnel junction contains a parallely connected resistor R and a capacitor C. In the uncertainty principle, if we put $\Delta E = e^2/C$ and $\Delta t =$ time constant $= RC$, the capacitance C cancels out and we can find the minimum tunnel resistance to be 25819 Ω.

The double tunnel junction consists of two tunnel junctions J_1 and J_2 to a conductive island. J_1 and J_2 are connected in series across an ideal voltage source V_s so that $V_s = V_1 + V_2 =$ sum of voltage drops across J_1 and J_2. Also the total capacitance

$C = C_1 + C_2 =$ sum of capacitances of J_1 and J_2. The total charge q on the island = Background charge + Negative charge due to n_1 electrons entering the island + Positive charge due to n_2 electrons leaving the island = $q_0 - q_1 + q_2$. We want to derive the equation for free energy of the circuit. For this derivation, equations for V_1 and V_2 are written down. The total energy W_C stored in the system and the total work W done by the voltage source are found out. Then free energy of the circuit $F = W_C - W$. Now there are four possibilities:

i. an electron may tunnel into the island from J_1 to make the number of electrons tunneling = $n_1 + 1$; the associated free energy change is denoted by ΔF_1^+.

ii. an electron may tunnel out of the island from J_1 to make the number of electrons tunneling = $n_1 - 1$; the free energy change is denoted by ΔF_1^-.

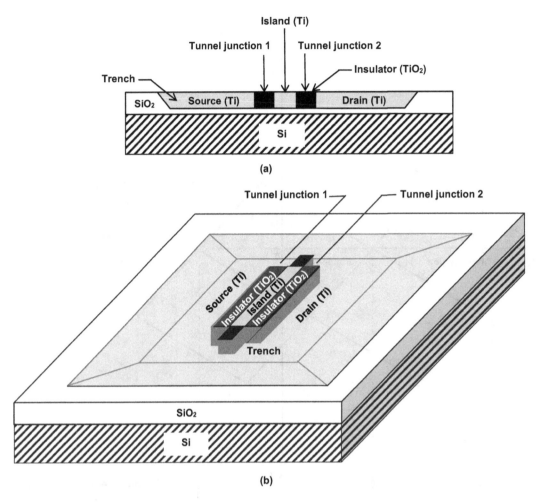

FIGURE 21.9 Single-electron transistor with titanium island and source/drain contacts: (a) cross-sectional diagram and (b) 3D drawing.

iii. an electron may tunnel out of the island from J_2 to make the number of electrons tunneling $= n_2 + 1$; the associated free energy change is denoted by ΔF_2^+.

iv. an electron may tunnel into the island from J_2 to make the number of electrons tunneling $= n_2 - 1$; the free energy change is denoted by ΔF_2^-.

Thus we get four equations for ΔF_1^+, ΔF_1^-, ΔF_2^+, and ΔF_2^-. Tunneling chances are remote if $\Delta F_1 > 0$, $\Delta F_2 > 0$ indicating transitions from lower to higher energy states but tunneling probabilities are high if $\Delta F_1 < 0$, $\Delta F_2 < 0$ indicating transitions from higher to lower energy states. How to get negative free energy? The ΔF equations contain a term containing V_s which can be increased by taking a higher supply voltage value. Then ΔF becomes negative. The subdual of tunneling at low V_s is explained to be due to Coulomb blockade. Tunneling can be initiated by applying a high V_s to triumph over this blockade.

In the energy-band diagram of the double tunnel junction, current cannot flow in the absence of bias due to the energy gap. The current begins to flow when the applied bias overcomes the energy gap. The current-voltage characteristic of the double tunnel junction is an ascending staircase when $R_{T1} < R_{T2}$. It becomes a descending staircase when $R_{T2} < R_{T1}$ due to reversal of the current direction. In both cases, the current

changes from one step to the next when the supply voltage V_s is increased.

How can we control the current without altering the supply voltage? We need a gate electrode. For this analysis, we take two independent voltage sources V_1 and V_2 for tunnel junctions J_1 and J_2. We also take two gate supplies, V_{G1} for feeding the input signal during device operation and V_{G2} for bias setting. We write the equations for charge on the island q_{island} and potential of the island $V(n)$. When a charge $+q$ is added to the island, its potential changes. So we can write equations for energy required to bring an electron from ground potential to the island (W_{-e}) and energy required to remove an electron from the island to ground potential (W_{+e}). We note the four possible tunnel events in the circuit. Each event is associated with an energy change ΔW. With the help of equations obtained previously, we write down the equation for:

i. energy change $\Delta W_{1R}(n)$ for an electron tunneling right through junction J_1;

ii. energy change $\Delta W_{1L}(n)$ for an electron tunneling left through junction J_1;

iii. energy change $\Delta W_{2R}(n)$ for an electron tunneling right through junction J_2; and

iv. energy change $\Delta W_{2L}(n)$ for an electron tunneling left through junction J_2.

For combinations of supply voltages and gate biases, all four energy changes are positive leading to Coulomb blockade. Then the gate voltage can be applied to make the energy changes negative and thereby move an electron either way through the junctions. This is single transistor action.

Antisymmetric supplies are considered for which $V_1 = V_s$, $V_2 = 0$. The case $n = 0$ is taken. Four boundaries of the Coulomb blockade regions are determined by putting $\Delta W = 0$ in these four equations. The resulting four lines define a rhombus ABCD called the Coulomb island. The coordinates of points A, C, B, D are determined. The distance AC is calculated from which it is found that the Coulomb diamond extends up to a width e/C_{G1} along the V_{G1} axis. Also the Coulomb diamond extends to $\pm e/C_\Sigma$ along the supply voltage V_s axis. Identical calculations for $n = -1, 0, +1$ yield the stability plot of the SET in which current increases abruptly at intervals of e/C_{G1}. These oscillations of current are known as Coulomb oscillations.

Details of nanowire-based process for the practical realization of the SET reported by Dubuc et al. (2008) are described.

Illustrative Exercises

21.1 Helium liquefies at −268.9°C. What is the size of the conducting island required for single electronics at this temperature.

$$-268.9°C = -268.9°C + 273.15\,K = 4.25\,K$$

From eq. (21.27)

$$C = \frac{Q^2}{2k_BT} = \frac{\left(1.6 \times 10^{-19}\right)^2}{2 \times 1.3806 \times 10^{-23} \times 4.25}$$

$$= \frac{2.56 \times 10^{-38}}{1.1735 \times 10^{-22}} = 2.18 \times 10^{-16}\,F \qquad (21.140)$$

and from eq. (21.22)

$$r = \frac{C}{4\pi\varepsilon_0} = \frac{2.18 \times 10^{-16}}{4 \times 3.14159 \times 8.854 \times 10^{-12}}$$

$$= \frac{2.18 \times 10^{-16}}{1.1126 \times 10^{-10}} = 1.959 \times 10^{-6}\,m = 1.959\,\mu m \qquad (21.141)$$

21.2 Starting from the argument that the time t for which an electron resides on a conductive island is much greater than the quantum uncertainty Δt, show that the minimum tunnel resistance R_T for single electron charging is $R_T \gg h/e^2$.

Since

$$t \gg \Delta t \qquad (21.142)$$

from the uncertainty principle

$$t \gg \Delta t \geq \frac{h}{\Delta E} \qquad (21.143)$$

If the applied voltage is V, the supplied energy is eV, which is obviously much greater than the energy uncertainty ΔE, so

$$eV \gg \Delta E \qquad (21.144)$$

or

$$\Delta E \ll eV \qquad (21.145)$$

Combining eqs. (21.142), (21.143), and (21.145)

$$t \geq \frac{h}{\Delta E} \gg \frac{h}{eV} \qquad (21.146)$$

For moderate bias and temperature, a maximum of one extra electron can reside on the conductive island at any time. The current due to one electron is

$$I = \frac{\text{Charge}}{\text{Time}} = \frac{e}{t} \qquad (21.147)$$

or

$$t = \frac{e}{I} \qquad (21.148)$$

From eqs. (21.146) and (21.148),

$$\frac{e}{I} \gg \frac{h}{eV} \qquad (21.149)$$

or

$$e^2V \gg hI \qquad (21.150)$$

or

$$\frac{V}{I} \gg \frac{h}{e^2} \qquad (21.151)$$

Hence,

$$R_T = \frac{V}{I} \gg \frac{h}{e^2} \qquad (21.152)$$

21.3 The parasitic or stray capacitance is the undesired, unavoidable capacitance between the two parts of an electronic device or circuit or between the two independent circuits by virtue of their proximity to each other. Such a stray capacitance between two independent circuits can be in the attofarad range as the circuits are far apart, e.g. the capacitance between two conductors of area $1\,mm^2$ placed $10\,m$ apart in air will be

$$C = \frac{8.854 \times 10^{-12} \times 1 \times 10^{-6}}{10} = 8.854 \times 10^{-19}\,F \approx 1\,aF$$

$$(21.153)$$

Will any Coulomb blockade effect be observed across such conductors or circuits due to the small capacitance?

No. Although the capacitance of the tunnel junction formed is small due to the large distance between the two circuits, at the same time the resistance of the tunnel junction is infinite across which tunneling will not take place. Such a situation will not lead to the manifestation of Coulomb blockade which is the repulsive opposition to an incoming tunneling electron to a conductor offered by electrons already present on the conductor. Thus in the absence of tunneling, Coulomb blockade will not be observed.

21.4 As the dimensions of a quantum dot are shrunk, a stage comes when its size becomes comparable to the de Broglie wavelength of electrons and the energy levels of the electrons on the quantum dot will be quantized. Is quantization of energy levels an essential condition for the observation of Coulomb blockade faced by an incoming electron?

No, because the concept of Coulomb blockade does not presuppose any energy quantization. It involves Coulomb electrostatic repulsion and tunneling.

REFERENCES

Coulomb Blockade in Single-Electron Transistors, https://lampx. tugraz.at/~hadley/ss2/set/transistor/coulombblockade.php.

Dubuc, C., J. Beauvais and D. Drouin. 2008. A nanodamascene process for advanced single-electron transistor fabrication. *IEEE Transactions on Nanotechnology* 7: 68–73.

Goser, K., P. Glösekötter and J. Dienstuhl. 2004. Ch.13:Single-Electron Transistor (SET). In: *Nanoelectronics and Nanosystems: From Transistors to Molecular and Quantum Devices.* Springer: Berlin, Heidelberg, pp. 209–223.

Wasshuber, C. 1997. The double tunnel junction. In: Dissertation about Single–Electron Devices and Circuits, http://www.iue. tuwien.ac.at/phd/wasshuber/node22.html

Wasshuber, C. 2001. *Computational Single-Electronics.* Springer-Verlag Wien GmbH: New York, pp. 1–8.

Heterostructure Optoelectronic Devices

The operation of optoelectronic devices is based on the electrical and optical properties of materials (Vasko and Kuznetsov 1999). They work with electrical and optical signals. In this chapter, a series of devices with a P-I-N diode configuration are surveyed. The peculiarity of these devices is that in each of these devices, a single quantum well or a multiple quantum well (MQW) is incorporated in the intrinsic region. The quantum wells (QW) bring forth a multiplicity of advanced features in the device by confining electrons in two dimensions, which can otherwise move freely in three dimensions.

22.1 Heterojunction Laser Diode

In a forward-biased P-I-N diode, electrons are injected from the N-side and holes are injected from the P-side into the central intrinsic (I) region, which is the active region of the device. The pumping electric current is decreased if the thickness of the region with excess carrier concentration is small. Because this thickness determines the laser threshold current, it plays a crucial role in laser diode design.

In homojunction P-I-N structures (Figure 22.1), there is a minimum limit beyond which the thickness of this region cannot be reduced. The diffusion length of carriers in the given semiconductor material imposes this limit. As soon as the carriers enter the intrinsic region, the electrons diffuse to the P-side and the holes to the N-side because of the concentration gradients. In the P-side, the minority carrier electrons recombine with the majority carrier holes. Similarly, in the N-side, the minority-carrier holes recombine with majority carrier electrons. The average life span of a carrier before its recombination is called its minority-carrier

lifetime τ, and is defined as the time taken by the excess carrier population to decrease to $1/e$ times its initial value. The distance traversed by the carrier during its lifetime is the diffusion length of the carrier, and is given by

$$L_D = \sqrt{D\tau} \tag{22.1}$$

where D is the diffusion coefficient of the carrier.

Practically, the diffusion length of carriers in direct bandgap materials is around a few microns. Therefore, the minimum width of the semiconductor region with excess carrier concentration is of this order in the case of homojunction P-I-N diodes. In order to decrease this width to smaller values, we take the help of heterojunctions. These junctions enable the localization of non-equilibrium electrons and holes in smaller widths than the diffusion length.

22.1.1 Double Heterostructure Laser

As the name implies, this laser comprises two heterojunctions (Figure 22.2): one heterojunction is between N-type $Al_{0.3}Ga_{0.7}As$ layer below and undoped P-GaAs layer above it and the other heterojunction is formed between the undoped P-GaAs layer below and the P-type $Al_{0.3}Ga_{0.7}As$ layer above it. Thus the laser is essentially a sandwich made with two $Al_{0.3}Ga_{0.7}As$ cladding layers (each typically $1\,\mu m$ thick) and a P-GaAs middle layer ($0.15\,\mu m$). Electrical contacts are made to both the cladding layers. Note that the P-type $Al_{0.3}Ga_{0.7}As$/undoped P-GaAs/N-type $Al_{0.3}Ga_{0.7}As$ together constitute a P-I-N diode, which by forward biasing will impel holes and electrons from the cladding layers into the central layer.

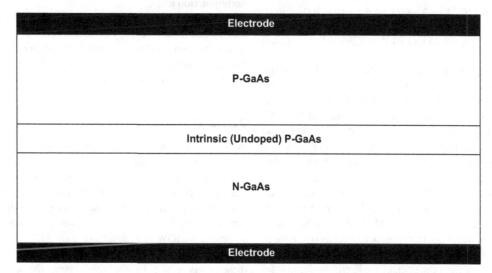

FIGURE 22.1 Homojunction laser diode.

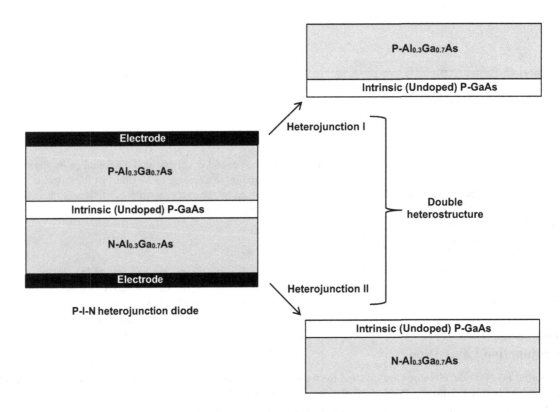

FIGURE 22.2   Double heterostructure laser diode showing the two constituent heterojunctions.

Bandgap is related to refractive index. The larger bandgap semiconductor has a smaller refractive index and vice versa. Let us compare the infrared refractive indices of the two semiconductors. At 300 K, the bandgap of $Al_{0.3}Ga_{0.7}As$ is 1.8 eV while that of GaAs is 1.424 eV. At 300 K, the infrared refractive index of $Al_{0.3}Ga_{0.7}As$ is

$$n = 3.3 - 0.53x + 0.09x^2 = 3.3 - 0.53 \times 0.3 + 0.09 \times 0.3^2$$
$$= 3.3 - 0.159 + 0.0081 = 3.149 \tag{22.2}$$

while that of GaAs is

$$n = 3.255\left(1 + 4.5 \times 10^{-5} T\right) = 3.255\left(1 + 4.5 \times 10^{-5} \times 300\right)$$
$$= 3.255(1 + 0.0135) = 3.255 \times 1.0135 = 3.299 \tag{22.3}$$

Looking at the relative bandgaps and refractive indices of the semiconductor materials in this sandwich, it is evident that the double heterostructure performs two roles: electron/hole confinement and light confinement. The electrons injected from the N-type $Al_{0.3}Ga_{0.7}As$ into the undoped/lightly P-doped GaAs and the holes injected by the P-type $Al_{0.3}Ga_{0.7}As$ into undoped/lightly P-doped GaAs will be confined in the GaAs layer due to the difference in bandgaps of $Al_{0.3}Ga_{0.7}As$ and GaAs. In this way, the double heterostructure provides electrical carrier confinement. Further, the photons formed by electron-hole radiative recombination in the GaAs layer will be confined in this layer because it will act as a waveguide owing to difference in refractive indices of $Al_{0.3}Ga_{0.7}As$ and GaAs. Any photons falling on the cladding layers at angles greater than the critical angle will undergo total internal reflection.

The main idea behind building a double heterostructure is to make a barrier preceding the bottom part of the structure before the N-layer to stop hole diffusion and another barrier prior to the top part of the structure before the P-layer to prevent electrons from diffusing. The central intrinsic part is accessible to both the types of carriers, holes and electrons. By implementing the double heterostructure, the thickness limit on the region containing excess carriers is no longer the diffusion length of carriers, as in the case of homojunctions. Now it is the thickness of the middle GaAs layer and this thickness can be reduced to the extent allowed by technology. Overall, the performance characteristics of the laser are significantly ameliorated than a homojunction laser.

The operation of the double heterostructure laser diode can be understood from its energy-band diagrams shown in Figure 22.3 under thermal equilibrium and forward biased conditions. On forward biasing, the electrons and holes injected into the intrinsic/lightly P-doped GaAs layer recombine to emit photons.

22.1.2 Quantum Well Laser

When the thickness of the active region in the double heterostructure laser is reduced to the order of the de Broglie wavelength, quantum effects become predominant. A laser with a narrow active region ~10 nm or less comparable to the de Broglie wavelength is called a QW laser because the thin active region acts as a QW. When several such wells are included in the laser structure, a MQW laser results (Figure 22.4).

In a bulk laser, the carriers are free for movement in three dimensions while in the QW laser, the carrier motion is two-dimensional.

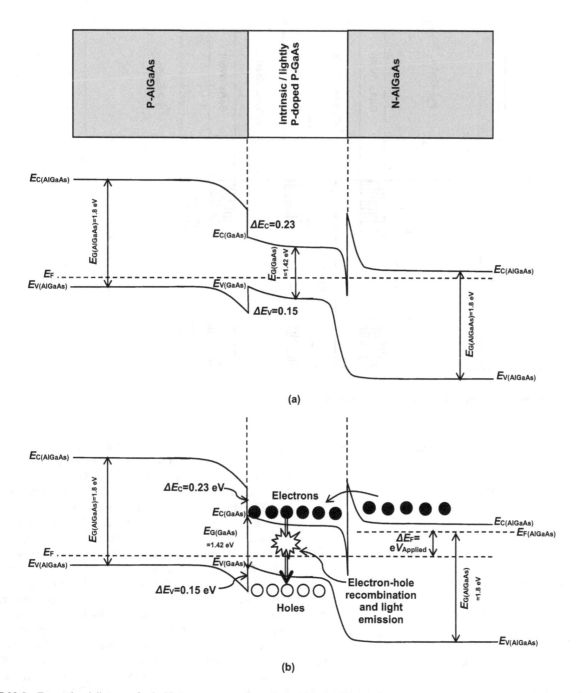

FIGURE 22.3 Energy-band diagram of a double heterostructure laser diode: (a) under thermal equilibrium and (b) under forward bias.

For the bulk laser, the number of electron states at an energy level E per unit volume per unit energy called the density of states (DoS) is given by

$$D(E)_{3D} = \sum_{i=l,h} \frac{m_r^i}{\pi^2 \hbar^3} \sqrt{2m_r^i (E - E_G)} \qquad (22.4)$$

where the effective mass m_r of the transition is obtained from

$$\frac{1}{m_r} = \frac{1}{m_C} + \frac{1}{m_V} \qquad (22.5)$$

m_C, m_V being the conduction band and valence band masses, respectively. E_G is the bandgap of the material, and symbols l, h above m_r denote light and heavy holes, respectively. Also $E > E_G$.

In the case of the QW laser, in the direction normal to the QW plane, only discrete energy values are permitted. Assuming an infinitely deep well, the permissible energy levels are from eq. (5.100)

$$E_n = \frac{n^2 \pi^2 \hbar^2}{2mL_z^2} = \frac{\hbar^2}{2m}\left(\frac{n\pi}{L_z}\right)^2 = \frac{\hbar^2 \mathbf{k}^2}{2m} \qquad (22.6)$$

where n is an integer, L_z is the thickness of the QW, and $\mathbf{k}$ is the wave vector. Taking the energy at the top of the valence band to

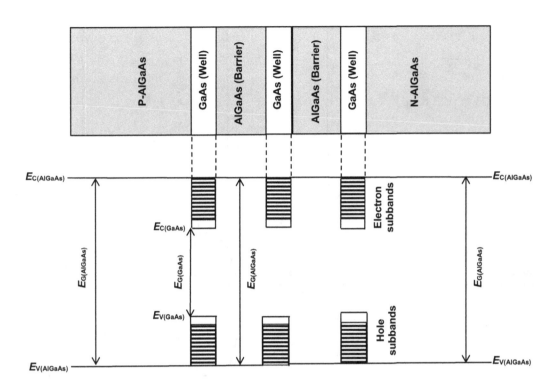

FIGURE 22.4 Simplified energy band diagram of the double heterostructure quantum well laser diode under thermal equilibrium.

be zero, the allowed energies for an electron in the conduction band are

$$E = E_G + E_n^C \qquad (22.7)$$

where

$$E_n^C = E_n \quad \text{for} \quad m = m_C \qquad (22.8)$$

The allowed energies for a hole in the valence band are

$$E = -E_n^V \qquad (22.9)$$

where

$$E_n^V = E_n \quad \text{for} \quad m = m_V \qquad (22.10)$$

So from eqs. (22.6), (22.7) and (22.9), the allowed transition energies are

$$E = E_G + E_n^C - \left(-E_n^V\right) + \frac{\hbar^2 \mathbf{k}^2}{2m} = E_G + E_n^C + E_n^V + \frac{\hbar^2 \mathbf{k}^2}{2m} \quad (22.11)$$

Although the 2D density of states is constant (eq. (12.76)), the DoS for a quantum well is a step function. The steps occur at the energy of each quantized level. The DoS for a quantum well is given by

$$D(E)_{2D} = \sum_{i=l,h} \sum_{n=1}^{\infty} \frac{m_r^i}{\pi \hbar^2 L_z} \theta \left(E - E_G - E_n^C - E_{n,i}^V \right) \quad (22.12)$$

where θ is the Heaviside function.

The plot of the DoS with respect to energy for a bulk laser is continuous and parabolic in shape with DoS proportional to the square root of energy. The same for a laser with a QW is like a staircase consisting of a series of steps (Spencer et al. 1997) (see Figure 22.5). For the bulk laser, the DoS is zero at the band edge and increases with energy. For the QW, the DoS is non-zero at the band edge. It rises abruptly at a particular energy value. Then it becomes constant and remains independent of energy within a certain energy interval. After this energy interval, the DoS again rises steeply and remains flat up to a specific energy value. Essentially, the DoS-energy graph contains a repetition of vertical ascents with flat top regions. The ascents occur after certain energy intervals. Thus the DoS has a sharp upswing and is concentrated at the band edge in a QW laser. In opposition, the DoS vanishes at the band edge and rises smoothly with energy in a bulk laser (Nag 2000). In addition to the difference in DoS, another striking feature is the co-existence of electrons and holes in the small volume of thin QW making it easier for electrons to find holes and recombine with them. Both these differences lead to several advantageous properties of QW lasers that are not found in bulk lasers. These include the low threshold current requirement, less vulnerability of its characteristics to temperature variations, higher efficiency, and the ability to tune the wavelength of emitted light of a QW laser by changing the thickness of the QW in place of the material composition change necessary in a bulk laser.

The volume of the active region in a QW laser is drastically reduced relative to a bulk double heterostructure laser. Although the quantum mechanical effects are harnessed by this laser, the optical confinement is sacrificed because the thin quantum mechanical region has poor waveguiding property. The lack of optical confinement is replenished by inserting a layer, graded in refractive index between the QW and each cladding layer for waveguiding purpose. The resulting laser is

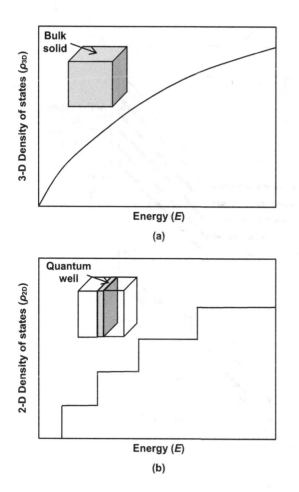

FIGURE 22.5 Density of states versus energy graphs for different lasers: (a) bulk laser and (b) quantum well laser.

known as the graded index separate confinement heterostructure (GRIN-SCH) laser due to assignment of optical responsibility to the graded index fiber.

The decrease in volume must also be compensated if more power is needed. The scheme adopted for this compensation involves the embedding of a plurality of GaAs QWs (8–10 nm thick) in the active region of the laser interleaved with $Al_{0.3}Ga_{0.7}As$ barrier layers (10–20 nm thick to avoid coupling). Thus evolves the concept of MQW lasers. A laser with N QWs of equal thickness has N times the gain of a laser with a single QW but the beneficial effect of high gain is accompanied with disadvantage of higher threshold current density which equals a constant plus a term proportional to the number of wells (Jasim 2009). A single well laser is found to be the best when a low threshold current is essential but in applications demanding a high gain and large output power, MQWs are used, compromising the threshold current density.

22.2 Quantum Well and Barrier Layer Structures in LEDs

In MQW LEDs (Figures 22.6 and 22.7), one must enquire about the effects of the thickness of quantum barriers (QBs) and QWs as well as those of doping the barriers and grading

the InN composition of wells. Equally important is deciding the optimum number of wells. The motions of electrons and holes are restricted in the thin QW in the direction perpendicular to the well.

22.2.1 Effect of Thickness of the Quantum Barrier Layer in Multiple Quantum Well GaN LEDs

Lin et al. (2013) fabricated LEDs, each with three periods of QWs and QBs. In these LEDs, the thickness of the $Ga_{0.85}In_{0.15}N$ QW is 4.2 nm. To study the effects of QB thickness, three thicknesses of the QB are used: 24.5, 9.1, and 3.6 nm. As QB thickness decreases, the electric field in the QB increases but that in the QW decreases. This electric field is the polarization-induced electric field created in III-nitride materials by its wurtzite structure. The diminished electric field in the QW is favorable for mitigation of efficiency droop of LED at high currents, as we shall see below.

Consequent upon the prevalence of the polarization-induced field, charges build up at the QW/QB interfaces. When the field in the well is high, the internal quantum efficiency of the LED is degraded through the quantum-confined Stark effect (QCSE). The QCSE (Figure 22.8) is an optical modulation mechanism concerned with the influence of an external electric field on the absorption/emission spectrum of a QW. Without any externally applied field, the electrons and holes occupy a discrete set energy subbands. But when the field is applied, the electron states move to lower energies and hole states to higher energies whereby the number of permitted absorption/emission frequencies of light decreases. Furthermore, the overlapping of the wave functions of electrons and holes is reduced. The overlap integral diminishes with the transference of electrons and holes to the opposite sides of the QW. Quantum confinement becomes less efficient. As a result, the recombination efficiency worsens.

Moreover, the electronic leakage current increases because the electrons reside for a shorter time above the well. Efficiency droop is observed at high injection levels. The efficiency sag at 200 mA is 18.1% for 24.5 nm thick QB LED, 9.4% for 9.1 nm thick QB LED, and 0.8% for 3.6 nm thick QB LED. Onset currents for efficiency droops are 35.2, 48.2, and 158 mA, respectively, for the three LEDs. Thus the LED operation is seriously impacted with a thick QB causing a high electric field in the well (Lin et al 2013).

22.2.2 Suppression of the QCSE by Silicon Doping of Quantum Barriers

Wang et al. (2017) found that doping of four QWs of eight-period $Ga_{0.8}In_{0.2}N/GaN$ MQW LEDs with silicon subdues the QCSE. In these LEDs, the QW thickness is 2.5 nm and the QB thickness is 8 nm. Silicon doping lowers the barrier height in the four QBs resulting in the creation of a soft confinement potential in the QWs. These LEDs have a lower forward voltage (3.77 V) and breakdown voltage (−6.77 V) than those without silicon doping. Carrier localization in the QWs is therefore enhanced leading to the homogeneity of carrier distribution in the wells. Leakage of carriers and non-radiative Auger recombination processes are minimized. Light output at high injection levels is thus boosted (Wang et al. 2017).

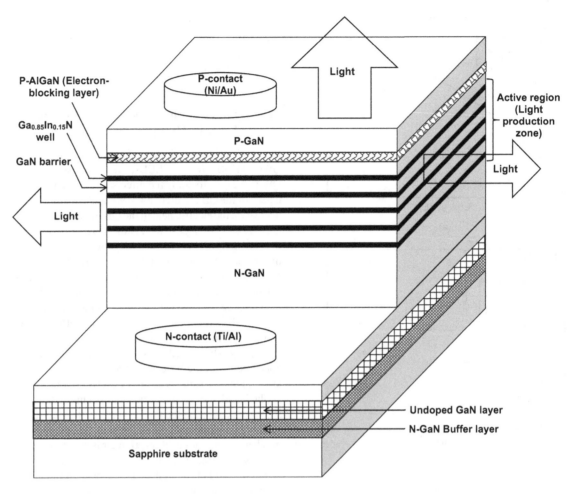

FIGURE 22.6   Different layers in the GaN MQW LED structure.

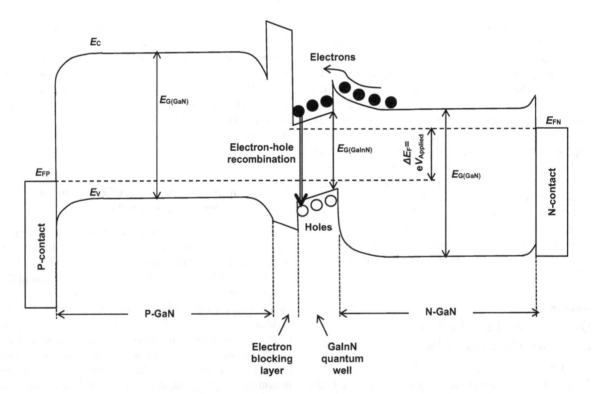

FIGURE 22.7   Energy-band diagram of forward-biased GaN LED.

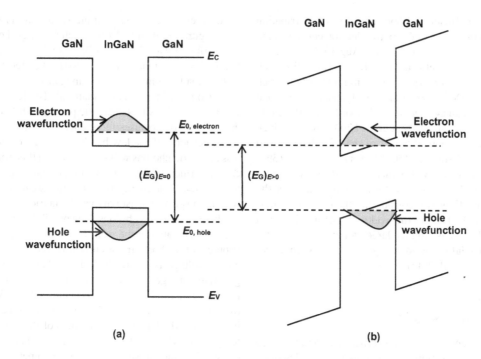

FIGURE 22.8 Quantum-confined Stark effect: (a) without electric field ($E = 0$), and (b) with applied electric field ($E > 0$).

22.2.3 Effect of the Number of Quantum Wells on LED Performance

In this regard, there is a varied opinion but it is unanimous that there is an optimum number of wells beyond which increase in well number lowers the LED performance.

Zainal et al. (2007) performed a simulation study on the influence of the number of QWs on the emission efficiency of LEDs using SILVACO/ATLAS software. The QW material is $In_{0.13}Ga_{0.87}N$ and the QB material is $In_{0.01}Ga_{0.99}N$. During the simulation, the number of wells is increased from 1 to 5. The single QW LED was adjudged to be the best with performance deteriorating as the number of wells increased. The energy-band diagrams of the simulated structures revealed that the barrier height of the MQW LED structure was higher both at P-GaN/active region interface and active region/N-GaN interface and more prominently at the latter interface than for the single well structure, thereby hampering the entry of carriers into the active region. In the MQW LEDs, the carrier distribution is inhomogeneous among the QWs with maximum electron and hole population in the first QW. Therefore the maximum efficiency of light emission is found for the first well. The last three wells were almost ineffective due to lesser number of carriers (Zainal et al 2007).

Meel et al. (2017) optimized a 5-well LED design with 3 nm thick InGaN well and 6 nm thick GaN barrier using SILVACO/ATLAS simulation program. They found that when the number of wells is decreased below 5, the radiative efficiency falls as less number of carriers is captured. But increasing the number of wells above 5 too lowers the efficiency because the holes are unable to reach the well on the N-GaN side owing to their lower mobility (Meel et al. 2017).

Choi et al. (2015) fabricated LEDs with 5, 6, and 7 QWs using $In_{0.05}Ga_{0.95}N$ as well material and $Al_{0.15}Ga_{0.85}N$ as barrier

material. The wells are 3 nm thick and the barriers are 5 nm thick. The 7-well LED displayed superior electrical/optical properties than the 6-well LED and the 6-well LED was better than the 5-well LED. The features compared are current–voltage characteristics, distribution of light pattern, external quantum efficiency, and red shift of the peak wavelength in electroluminescence spectra. Forward voltage drop of the 5-well LED at 350 mA current is 3.94 V which decreases to 3.78 V for the 6-well LED and 3.77 V for the 7-well LED. The relative normalized external quantum efficiency for 5-well LED at 50 A/cm² is 100%, for the 6-well LED it is 122.3%, and for the 7-well LED it is 132.2%. The improvement in performance with increase in the number of QWs is ascribed to the larger effective volume available with greater number of wells (Choi et al. 2015).

22.2.4 Overcoming Auger Recombination-Engendered Efficiency Reduction in MQW LEDs by Linearly Graded InN Composition Profile

Customarily thin QWs are used in MQW LEDs to obviate the QCSE. The thin wells are able to spatially restrict electrons and holes in small volumes. But at the same time they have a derogatory effect on the efficiency. The non-radiative recombination rate increases as the cube of the carrier concentration and therefore a high concentration promotes this type of recombination. Therefore, a technique needs to be evolved that concomitantly decreases the detrimental effects of both factors, viz. the QCSE and the Auger recombination.

Zhang et al. (2014) demonstrated numerically and experimentally that by using a larger well thickness and linearly grading the InN composition along the [0001] growth orientation in the InGaN/GaN 3-period QWs, a higher optical power was obtained together with a lower efficiency droop. The electron distribution is more uniform in the wider well. So, the electron concentration at high currents is less, thereby delimiting Auger recombination and associated detrimental effect. The QB was 12 nm thick but two thicknesses of QWs were tried: 3 and 5 nm. In the 5 nm thick quantum well, the InN composition was linearly diminished from 0.15 to 0.08 with an average of 0.115 in order that the emission wavelength was the same as that of the LED with 3 nm thick QW. It is found that the optical power of the LED having well with graded InN composition at 150 A/cm² is raised by 29.39% relative to the ungraded InN well structure. At the same current density, the droop is 31.83% for the former and 39.23% for the latter LED structure. However, considering the electroluminescence spectra of the LEDs, the full width at half maximum in terms of current density is slightly larger for graded InN QW LEDs, offsetting trivially the advantages gained by compositional grading (Zhang et al. 2014).

22.3 Quantum Well Solar Cell

The solar spectrum encompasses the broad wavelength range from $\lambda_1 = 290$ nm to $\lambda_2 = 2500$ nm corresponding to energies from

$$E_1 = \frac{hc}{\lambda_1} = \frac{6.626 \times 10^{-34} \times 3 \times 10^8}{290 \times 10^{-9} \times 1.602 \times 10^{-19}}$$

$$= \frac{19.878 \times 10^{-26}}{464.58 \times 10^{-28}} = 4.278 \text{ eV} \qquad (22.13)$$

to

$$E_2 = \frac{hc}{\lambda_2} = \frac{6.626 \times 10^{-34} \times 3 \times 10^8}{2500 \times 10^{-9} \times 1.602 \times 10^{-19}}$$

$$= \frac{19.878 \times 10^{-26}}{4005 \times 10^{-28}} = 0.4963 \approx 0.5 \text{ eV} \qquad (22.14)$$

where eq. (2.30) is applied.

The solar spectrum is subdivided into three bands: ultraviolet radiation (290–380 nm and 2%), visible radiation (380–780 nm and 51%), and infrared radiation (780–2500 nm and 47%).

Silicon has a bandgap of 1.11 eV. Part of the solar spectrum having energy >1.11 eV will produce electron-hole pairs in silicon while the part below this limit will be ineffective. So, to tap the solar energy from the part of the solar spectrum with energy <1.1 eV, we have to make the solar cell from a material which has a lower bandgap than silicon. This means that for covering the full solar spectrum, we need solar cells of various semiconductor materials having different values of bandgaps. These solar cells can be connected in tandem to make a combined solar cell utilizing the full solar spectrum. In this combined solar cell, there are solar cells built made from materials of increasing bandgaps. Light is made to fall from the side in which the lowest bandgap solar cell is placed. The remaining unused part of the spectrum reaches the second cell and part of it is used by it depending on its bandgap. The leftover spectrum radiation moves to the third cell and in this way, the full spectrum is used for conversion into electrical energy. Such a cell is considerably more efficient than a cell made from a single bandgap material.

An alternative to the multiple solar cells is the MQW solar cell (Figures 22.9 and 22.10), which achieves the same objective in a different way (Naho 2013). In a QW, the well layer and the barrier layer are made from materials of different bandgaps. The well is made from a narrow bandgap material and the barrier layer from a wider bandgap material. The lower bandgap semiconductor exploits less energetic photons while the higher bandgap semiconductor uses more energetic photons for conversion into electricity. If the low bandgap semiconductor is absent, the less energetic photons will be wasted. On this principle, the wells can be formed from a variety of materials of different bandgaps to avail the light energy from the full solar spectrum and cells with high spectral sensitivity can be fabricated.

There are two possibilities. One possibility is that the electron-hole pairs created in the well recombine. Then they do not contribute to the short circuit current of the cell; instead the open circuit voltage is diminished. The other possibility is that the generated charge carriers escape from the well and take part in photocurrent production. This is achieved by making the barriers thin enough so that the carriers escape from the well by tunneling through the barrier with the help of thermal energy at room temperature (Ekins-Daukes et al. 2009). Thus the cell must be designed in such a way that the two competing and counteracting mechanisms of carrier recombination and thermally assisted tunneling take place at the appropriate rates for the successful operation of the cell. The wells are usually 2–5 nm thick and the barrier thickness is generally 10 nm or less. To prevent carrier loss by recombination, the minority-carrier lifetime in the semiconductor material must be long. It is determined by the quality of the material and absence of defects.

The short circuit current of the cell is determined by the lower bandgap material of the well while the output voltage depends on the higher bandgap material of the barrier, the recombination in the well and well/barrier interfaces. The absorption edge and spectral response of the cell is controlled by the width and depth of the well. The deeper the well, the longer the wavelengths and hence lower the frequencies absorbed. Insertion of a larger number of wells raises the photocurrent provided the recombination of carriers is kept within limits (Abdelkrim et al. 2016).

22.4 Discussion and Conclusions

In a laser diode, the thickness of the region with excess carrier concentration determines the laser threshold current and must be kept small to minimize the threshold current. But in a homojunction, this thickness cannot be decreased below the diffusion length of carriers. This minimization can be achieved with heterojunctions.

A double heterostructure comprising the three layers: P-type $Al_{0.3}Ga_{0.7}As$ layer /undoped GaAs layer/N-type $Al_{0.3}Ga_{0.7}As$ layer provides both carrier and light confinement and considerably improves laser performance. But when the central undoped GaAs layer thickness is reduced to ≤ 10 nm, a QW is formed in the central region and quantum-mechanical effects appear. The DoS for a bulk laser is zero at the band edges and rises smoothly into a continuous parabola while the same for a QW laser is concentrated at the band edges and has a staircase ascent. These differences impart many useful features to the QW laser, principally

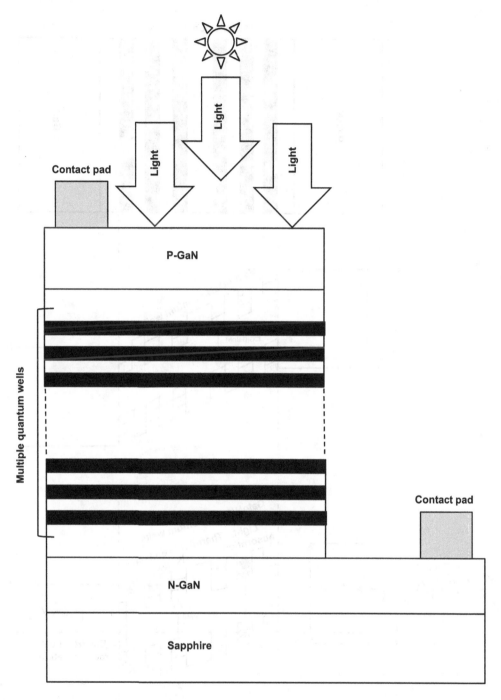

FIGURE 22.9 Multi-quantum well solar cell.

the decreased threshold current and lower thermal sensitivity. However, the optical confinement is degraded because of the smaller volume of the active region. For better optical confinement, a GRIN-SCH is employed. For more power, MQWs are necessary.

QWs also serve as the key elements of LEDs. For a thick barrier layer of the QW, electric field is low in the barrier but high in the well. For a thin barrier, the field is high in the barrier but low in the well. A low field in the well is desirable for decreasing the efficiency droop by the QCSE as well as for increasing the current for onset of droop (Lin et al. 2013). The QCSE effect is also assuaged by doping the barrier layers resulting in a soft confinement potential due to barrier height lowering (Wang et al. 2017).

Although MQW LEDs are required for higher power, simulations show that the single QW LED gives better overall performance, and the performance decreases with increase in the number of wells (Zainal et al. 2007). A 5-well LED design was optimized by Meel et al. (2017). Choi et al. (2015) found that the relative normalized external quantum efficiency increased from 5-well LED to 6-well LED and further for the 7-well LED. Looking at divergence of results, it can be stated that the number of wells must be limited to the extent essentially needed for an application.

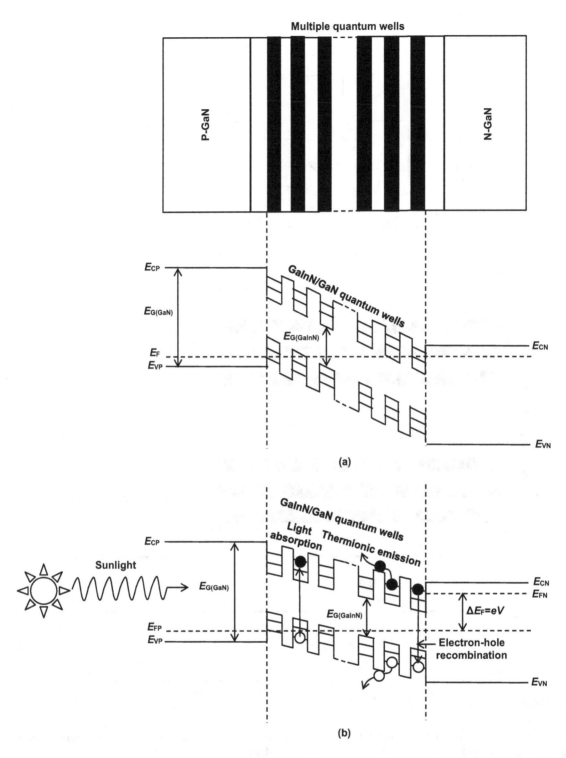

FIGURE 22.10 Energy-band diagrams of multi-quantum well solar cell: (a) under thermal equilibrium and (b) under illumination with solar rays.

A thin well is favored for diminishing the QCSE but is also prone to increased Auger recombination. The first factor increases the efficiency and the second decreases it. So a graded InN composition profile has been suggested for providing higher optical power and lower efficiency droop (Zhang et al. 2014).

The solar spectrum spans over a broad wavelength range which tantamounts to a wide energy range from 0.5 to 4.278 eV.

A solar cell made of silicon utilizes only the part of radiation in the solar spectrum having energies >1.1 eV. For utilization of the remaining part, a cell made of a lower bandgap material must be combined with it. This leads to the idea of connection of multiple solar cells, each made with a different bandgap material. Equivalent performance is realized by making a MQW solar cell. This cell is made using well/barrier layers from materials of different bandgaps. The design of such a MQW solar cell calls for careful attention to choosing barrier thickness for successful operation.

Illustrative Exercises

22.1　(a)　What parameter decides the thickness of the region of excess carrier concentration in a homojunction diode?

(b)　The diffusion coefficient of electrons in gallium arsenide is $200\,cm^2s^{-1}$. What is the thickness of the region of excess carrier concentration in a GaAs homojunction diode if the minority carrier lifetime is: (i) 10^{-9} s and (ii) 10^{-7}s?

Answer: (a) The diffusion length of carriers.

(b)　(i)

$$L_D = \sqrt{D\tau} = \sqrt{200\ cm^2\ s^{-1} \times 1\times 10^{-9}\ s} = 4.47\times 10^{-4}\ cm$$

$$= 4.47\times 10^{-6}\ m = 4.47\ \mu m \qquad (22.15)$$

(ii)

$$L_D = \sqrt{D\tau} = \sqrt{200\ cm^2 s^{-1} \times 1\times 10^{-7}\ s} = 4.47\times 10^{-3}\ cm$$

$$= 4.47\times 10^{-5}\ m = 44.7\ \mu m \qquad (22.16)$$

22.2　Find the power output of an LED of internal quantum efficiency 70% emitting a radiation of 850 nm at a forward current of 10 mA.

Power output of LED

$$= \frac{\text{Number of electrons injected into the depletion region} \times \text{Fraction of electrons converted into photons} \times \text{Energy per photon}}{\text{Time}} \qquad (22.17)$$

or

$$P = \frac{N}{t} \times \eta \times h\nu \qquad (22.18)$$

where N is the number of electrons, t is the time, η is the internal quantum efficiency of the LED, h is Planck's constant, and ν is the frequency of radiation emitted.

If e is the electronic charge, the forward current of LED is

$$I_F = \frac{Ne}{t} \qquad (22.19)$$

or

$$\frac{N}{t} = \frac{I_F}{e} \qquad (22.20)$$

From eqs. (22.18) and (22.20),

$$P = \frac{I_F}{e} \times \eta \times h\nu \qquad (22.21)$$

If c is the velocity of photons emitted and λ is their wavelength

$$\nu = \frac{c}{\lambda} \qquad (22.22)$$

Hence,

$$P = \frac{I_F}{e} \times \eta \times h\frac{c}{\lambda} = \frac{10\times 10^{-3}}{1.602\times 10^{-19}} \times \frac{70}{100} \times \frac{6.626\times 10^{-34} \times 3\times 10^8}{850\times 10^{-9}}$$

$$= \frac{10\times 70\times 6.626\times 3}{1.602\times 100\times 850}\times 10^{-1} = \frac{13914.6}{136170}\times 10^{-1}$$

$$= 0.01022\ W = 10.22\ mW \qquad (22.23)$$

22.3　(a)　What is the maximum wavelength in the solar spectrum which a silicon solar cell can convert into electricity? Bandgap of silicon = 1.1 eV. Will 1500 nm wavelength solar radiation be able to produce electron-hole pairs in silicon?

(b)　Compare with germanium (bandgap = 0.66 eV).

(c)　Will a gallium arsenide (bandgap = 1.424 eV) solar cell be useful for solar radiation of wavelength 2000 nm?

(a)

Maximum wavelength for silicon is from eq. (2.30)

$$\lambda_{Si} = \frac{hc}{E} = \frac{6.626\times 10^{-34} \times 3\times 10^8}{1.1\times 1.602\times 10^{-19}} = 1.128\times 10^{-6}\ m$$

$$= \frac{1.128\times 10^{-6}}{10^{-9}}\ nm = 1128\ nm \qquad (22.24)$$

1500 nm wavelength solar radiation has energy

$$E = \frac{hc}{\lambda} = \frac{4.135667\times 10^{-15}\ eV - s\times 3\times 10^8\ ms^{-1}}{1500\times 10^{-9}\ m}$$

$$= 0.00827\times 10^2\ eV = 0.827\ eV \qquad (22.25)$$

Clearly, 1500 nm wavelength has less energy (0.827 eV) than the bandgap of silicon (1.1 eV). So, it will not be able to break the bonds to produce electron-hole pairs in a Si solar cell.

(b) From eq. (2.30)

$$\lambda_{\text{Ge}} = \frac{hc}{E} = \frac{6.626 \times 10^{-34} \times 3 \times 10^8}{0.66 \times 1.602 \times 10^{-19}} = 1.88 \times 10^{-6}\,\text{m}$$

$$= \frac{1.88 \times 10^{-6}}{10^{-9}}\,\text{nm} = 1880\,\text{nm} \qquad (22.26)$$

As shown in part (a) above, 1500 nm wavelength has an energy of 0.827 eV which is greater than the bandgap of germanium and so will be effective in producing electron-hole pairs in a Ge solar cell. Moreover 1500 nm is less than the maximum wavelength 1880 nm for a Ge solar cell; hence it is effective.

(c) A GaAs solar cell will convert part of the solar spectrum having energy greater than its bandgap of 1.424 eV. Now from eq. (2.30)

$$\lambda_{\text{Ge}} = \frac{hc}{E} = \frac{6.626 \times 10^{-34} \times 3 \times 10^8}{1.424 \times 1.602 \times 10^{-19}} = 8.714 \times 10^{-7}\,\text{m}$$

$$= \frac{8.714 \times 10^{-7}}{10^{-9}}\,\text{nm} = 871.4\,\text{nm} \qquad (22.27)$$

Hence, the maximum wavelength of solar radiation used by a GaAs solar cell is 871.4 nm. So it will not be useful for converting 2000 nm wavelength into energy.

REFERENCES

Abdelkrim, T., B. Kamel, B. Abderrahmane, A.M. Benyoucef. 2016. Numerical simulation of multi-quantum well solar cells GaAs/AlGaAs. In: Kadja, M., A. Zaatri, Z. Nemouchi, R. Bessaih, S. Benissaad and K. Talbi (Eds.), Third International Conference on Energy, Materials, Applied Energetics and Pollution. *ICEMAEP2016*, October 30–31, 2016, Constantine, Algeria, pp. 574–579.

Choi, H.-S., D.-G. Zheng, H. Kim and J.-I. Shim. 2015. Effects of the number of quantum wells on the performance of near-ultraviolet light-emitting diodes. *Journal of the Korean Physical Society* 66(10): 1554–1558.

Ekins-Daukes, N.J., J. Adams, I.M. Ballard, K.W.J. Barnham, B. Browne, J.P. Connolly, T. Tibbits, G. Hill and J.S. Roberts. 2009. Physics of quantum well solar cells. Physics and simulation of optoelectronic devices XVII. In Osinski, M., B. Witzigmann, F. Henneberger, Y. Arakawa (Eds.). *Proceedings of SPIE*, Vol. 7211, pp. 72110L-1–72110L-11.

Jasim, S.S. 2009. Performance optimization of multi-quantum wells laser used in optical communications. *Iraqi Journal of Applied Physics Letters* 2(2): 11–14.

Lin, G.-B., D-Y. Kim, Q. Shan, J. Cho, E.F. Schubert, H. Shim, C. Sone, and J.K. Kim. 2013. Effect of quantum barrier thickness in the multiple-quantum-well active region of GaInN/GaN light-emitting diodes. *IEEE Photonics Journal* 5(4): 1600207 (8 pages).

Meel, K., P. Mahala and S. Singh. 2017. Design and fabrication of multi quantum well based GaN/InGaN blue LED. *3rd International Conference on Communication Systems (ICCS-2017) IOP Conf*erence *Series: Materials Science and Engineering* 331(218): 1–5.

Nag, B.R. 2000. *Physics of Quantum Well Devices*. Kluwer Academic Publishers: Dordrecht, p. 202.

Naho, I. 2013. *Multi-quantum well solar cell and method of manufacturing multi-quantum well solar cell*. European patent application: EP 2 768 029 A1, pp. 1–20.

Spencer, R.M., J. Greenberg, L.F. Eastman, C.-Y. Tsai and S.S. O'Keefe. 1997. High-speed direct modulation of semiconductor lasers. *International Journal of High Speed Electronics and Systems* 8(3): 417–456.

Vasko, F.T., A.V. Kuznetsov. 1999. Heterostructure-based optoelectronic devices. In: Vasko, F.T., A.V. Kuznetsov (Eds.), *Electronic States and Optical Transitions in Semiconductor Heterostructures*. Springer Science +Business Media: New York, pp. 291–320.

Wang, H.-C., M.-C. Chen, Y.-S. Lin, M.-Y. Lu, K.-I. Lin and Y.-C. Cheng. 2017. Optimal silicon doping layers of quantum barriers in the growth sequence forming soft confinement potential of eight-period $In_{0.2}Ga_{0.8}N/GaN$ quantum wells of blue LEDs. *Nano Express* 12: 591 (8 pages).

Zainal, N., Z. Hassan, H.A. Hassan and M.R. Hashim. 2007. Comparative study of single and multiple quantum wells of $In_{0.13}Ga_{0.87}N$ based LED by simulation method. *Optoelectronics and Advanced Materials – Rapid Communications* 1(8): 404–407.

Zhang, Z.-H., W. Liu, Z. Ju, S.T. Tan, Y. Ji, Z. Kyaw, X. Zhang, L. Wang, X. W. Sun and H.V. Demir. 2014. InGaN/GaN multiple-quantum-well light-emitting diodes with a grading InN composition suppressing the Auger recombination. *Applied Physics Letters* 105: 033506-1–033506-5.

Index

Printed in the United States
by Baker & Taylor Publisher Services